UNITS

Mass $1kg = 1000g = 0.001$ metric ton $= 2.20462\ lb_m$
$1lb_m = 16oz = 5E\text{-}4ton = 453.6g = 0.45359kg$

Length $1m = 100cm = 1000mm = 1E6\mu m = 1E9nm = 39.370in = 3.2808ft = 1.0936yd$
$1ft = 12in = 0.30480m = 30.480cm$
$1in = 2.5400cm$

Volume $1m^3 = 1000L = 1E6cm^3 = 1E6ml = 35.315ft^3 = 264.17gal$
$1ft^3 = 1728.0in^3 = 7.4805gal = 0.028317m^3 = 28.317L = 28317ml$

Force $1N = 1kg\cdot m/s^2 = 1E5dynes = 1E5g\cdot cm/s^2 = 0.22481lb_f$
$1lb_f = 32.174lb_m\cdot ft/s^2 = 4.4482N$

Pressure $1atm = 1.01325E5\ N/m^2(Pa) = 1.01325bar = 760mm_{Hg\ at\ 0°C} = 33.9ft_{H2O\ at\ 4°C} = 14.696psia$
$1bar = 0.1MPa = 0.98692atm = 14.504psia = 750.06mm_{Hg\ at\ 0°C} = 10.197m_{H2O\ at\ 4°C}$

Energy $1J = 1N\cdot m = 1E7ergs = 1\ kgm^2/s^2 = 0.23899cal = 0.73756ft\cdot lb_f = 1E7g\cdot cm^2/s^2 = 1\ MPa\cdot cm^3$
$1kJ = 0.94805\ Btu = 2.7778E\text{-}4kW\cdot h$

Power $1W = 1J/s = 0.23899cal/s = 0.73756ft\cdot lb_f/s = 3.4144Btu/h$
$1hp = 550\ ft\cdot lb_f/s = 0.70726Btu/s = 0.74570kW$

Gas Constant, R
     $= 8.3143\ J/(mole\cdot K) = 8.3143\ m^3Pa/(mole\cdot K) = 83.143\ cm^3bar/(mole\cdot K)$
     $= 8,314.3\ cm^3kPa/(mole\cdot K)$
     $= 8.3143\ cm^3MPa/(mole\cdot K) = 82.056\ cm^3atm/(mole\cdot K) = 1.9870\ cal/(mole\cdot K)$
     $= 1.9870\ Btu/(lbmole\cdot R)$
     $= 10.731\ ft^3psia/(lbmole\cdot R)$

Gravitational Constants at sea level
     $g = 9.8066\ m/s^2$ $g/g_c = 9.8066\ N/kg$ $g_c = 1\ (kg\cdot m/s^2)/N$
     $g = 32.174\ ft/s^2$ $g/g_c = 1\ lb_f/lb_m$ $g_c = 32.174\ [(lb_m\cdot ft)/s^2]/lb_f$

Software and supplementary information is available at www.egr.msu.edu/~lira/thermtxt.htm

INTRODUCTORY CHEMICAL ENGINEERING THERMODYNAMICS

INTRODUCTORY CHEMICAL ENGINEERING THERMODYNAMICS

J. Richard Elliott
Carl T. Lira

Prentice Hall
Upper Saddle River, NJ 07458
www.prenhallprofessional.com

Library of Congress Cataloging-in-Publication Data

Elliott, J. Richard

    Introductory chemical engineering thermodynamics / J. Richard Elliott, Jr., Carl T. Lira

        p. cm. _ (Prentice-Hall international series in the physical and chemcial engineering sciences)

    Includes bibliographical references and index.

    ISBN 0-13-011386-7

    1. Thermodynamics. 2. Chemical engineering. I. Lira, Carl T.

II. Title. III. Series.

TP149.E45 1998

  660_ .2969_dc21   98-49402

                                          CIP

Editorial/production supervision: *Vincent J. Janoski*
Acquisitions editor: *Bernard Goodwin*
Marketing manager: *Danny Hoyt*
Manufacturing manager: *Alan Fischer*
Cover design director: *Jerry Votta*

©1999 by J. Richard Elliott & Carl T. Lira

Published by Prentice Hall PTR
Prentice-Hall, Inc.
Upper Saddle River, NJ 07458

Prentice Hall books are widely used by corporations and government agencies
for training, marketing, and resale.

The publisher offers discounts on this book when ordered in bulk quantities.
For more information, contact: Corporate Sales Department, Phone: 800-382-3419;
Fax: 201-236-7141; E-mail: corpsales@prenhall.com; or write: Prentice Hall PTR,
Corp. Sales Dept., One Lake Street, Upper Saddle River, NJ 07458.

TRADEMARKS: Aspen Plus is a trademark of Aspen Technology, Inc.; Microsoft is a registered trademark of
the Microsoft Corporation; Excel is a trademark of the Microsoft Corporation; Mathematica is a registered
trademark of Wolfram Research; Mathcad is a registered trademark of MathSoft, Inc.

All products or services mentioned in this book are the trademarks or service marks of their respective
companies or organizations. Screen shots reprinted by permission from Microsoft Corporation.

ISBN 0-13-011386-7

Text printed in the United States at Hamilton Printing in Castleton, New York

13th printing, January 2008

Prentice-Hall International (UK) Limited, *London*
Prentice-Hall of Australia Pty. Limited, *Sydney*
Prentice-Hall Canada Inc., *Toronto*
Prentice-Hall Hispanoamericana, S.A., *Mexico*
Prentice-Hall of India Private Limited, *New Delhi*
Prentice-Hall Singapore Pte. Ltd., *Singapore*
Prentice-Hall of Japan, Inc., *Tokyo*
Editora Prentice-Hall do Brasil, Ltda., *Rio de Janeiro*

CONTENTS

PREFACE

"No happy phrase of ours is ever quite original with us; there is nothing of our own in it except some slight change born of our temperament, character, environment, teachings and associations."

Mark Twain

Thank you for your interest in our book. We have developed this book to address ongoing evolutions in applied thermodynamics and computer technology. Molecular perspective is becoming more important in the refinement of thermodynamic models for fluid properties and phase behavior. Molecular simulation is increasingly used for exploring and improving fluid models. While many of these techniques are still outside the scope of this text, these new technologies will be important to practicing engineers in the near future, and an introduction to the molecular perspective is important for this reason. We expect our text to continue to evolve with the chemical engineering field.

Computer technology has made process simulators commonplace in most undergraduate curriculums and professional work environments. This increase in computational flexibility has moved many of the process calculations from mainframe computers and thermodynamic property experts to the desktop and practicing engineers and students. This increase in computational ability also increases the responsibility of the individuals developing process simulations to choose meaningful models for the components in the system because most simulators provide even more options for thermodynamic models than we can cover in this text. We have included background and comparison on many of the popular thermodynamic models to address this issue.

Computational advances are also affecting education. Thus we have significant usage of equations of state throughout the text. We find these computational tools remove much of the drudgery of repetitive calculations, which permits more class time to be spent on the development of theories, molecular perspective, and comparisons of alternative models. We have included FORTRAN, Excel spreadsheets, TI85, and HP48 calculator programs to complement the text. The programs are summarized in the appendices.

(a) Solutions to cubic equations of state are no longer tedious with the handheld calculators available today for about $100. We provide programs for calculation of thermodynamic properties via the Peng-Robinson equation, vapor pressure programs, Peng-Robinson K-ratios and bubble pressures of mixtures, and van Laar and UNIFAC activity coefficients as well as several other utility programs. Our choice of the HP48 calculator is due to its being one of the first to provide a computer interface for downloading programs from a PC and provide calculator-to-calculator communication, which facilitates distribution of the programs. If all students in the class have access to these engineering calculators, as practiced at the University of Akron, questions on exams can be designed to apply to these programs directly. This obviates the need for traditional methods of reading charts for departure functions and K-ratios and enables treatment of modern methods like equations of state and UNIFAC.

(b) Spreadsheets have also improved to the point that they are powerful tools for solving engineering problems. We have chosen to develop spreadsheets for Microsoft® Excel because of the widespread availability. Certainly Mathcad®, Mathematica®, and other software could be used, but none has the widespread availability of spreadsheets. We have found the solver within Excel to provide a good tool for solving a wide variety of problems. We provide spreadsheets for thermodynamic properties, phase and reaction equilibria.

(c) High-level programming is still necessary for more advanced topics. For these applications, we provide compiled programs for thermodynamic properties and phase behavior. For an associating system, such as an alcohol, we provide the ESD equation of state. These programs are menu-driven and do not require knowledge of a computer language.

In a limited number of instances, we provide FORTRAN source code. We provide FORTRAN code because of our own abilities to program faster in FORTRAN, although other languages are finding increasing popularity in the engineering community. We have tried to avoid customization of the code for a specific FORTRAN compiler, which improves portability to other operating platforms but also limits the "bells and whistles" that a specific interface could provide. These programs provide a framework for students and practicing engineers to customize for their own applications.

Energy and entropy balances are at the heart of process engineering calculations. We develop these approaches first using the ideal gas law or thermodynamic tables, then revisit the topics after developing equation-of-state techniques for thermodynamic properties. We are well aware of the concern that students often apply the ideal gas law inappropriately. Therefore we clearly mark equations using the ideal gas law or assuming a temperature-independent heat capacity. From a pedagogical standpoint, we are faced with the issues of developing first and second law balances, equations of state (and their departure functions) for fluid properties, and then combining the principles. We have found it best that students quickly develop ability and confidence in application of the balances with simple calculational procedures before introducing the equation of state. The balance concepts are typically more easily grasped and are essential for extension to later courses in the curriculum. Another benefit of this approach is that the later development of the equation of state can be directly followed by departure functions, and the reasons for needing properties such as enthalpy and entropy are well understood from the earlier emphasis on the balances. This enables students to focus on the development of the departure functions without being distracted by not completely understanding how these properties will be used.

Fugacity is another property which is difficult to understand. We have tried to focus on the need for a property which is a natural function of T and P, and also stress how it is related to departure functions. There are many ways to calculate fugacities (which provides many trees to block the view of the forest), and we have tried to provide tables and diagrams to show the inter-relations between fugacity coefficients, activity coefficients, ideal gases, ideal solutions, and real solutions.

A distinct feature of this text is its emphasis on molecular physics at the introductory level. Our perspective is that this background must be made available to students in an integrated manner, but it is up to instructors to decide the level of emphasis for the entire spectrum of their students. We have organized this material such that it may be covered as a supplementary reading assignment or as a homework and test assignment. With the latter emphasis, it is possible to formulate a graduate course based on this text.

Throughout the text, we have used text boxes to highlight important statements and equations. Boxed equations are not always final results of derivations. In some cases, the boxes highlight mathematical definitions of important intermediate results that might be useful for homework problems.

We consider the examples to be an integral part of the text, and we use them to illustrate important points. In some cases, derivations and important equations are within an example because the equations are model-specific (e.g., ideal gas). Examples are often cross-referenced and are therefore listed in the table of contents.

There are many marginal notes throughout the text. Where you find a ❶, it means that an important point is made, or a useful equation has been introduced. Where you find a ⓗ or ⓣ, it means that a calculator program is available to assist in calculations. The calculator programs are sometimes not necessary, but extremely helpful. Where you find a ▤, it means that an Excel spreadsheet or a compiled program is available. In some cases, the program is simply convenient, but typically you will find that these calculations are tedious without the program. For calculator or PC icons, the program names are given by the icons. See the computer appendix or the readme files for specific program instructions.

We periodically update computer software and the computer appendix. The latest software is available from our website http://www.egr.msu.edu/~lira/thermtxt.htm. We hope you find our approaches helpful in your learning and educational endeavors. We welcome your suggestions for further improvements and enhancements. You may contact us easily at the email addresses below. Unfortunately, we will be unable to personally respond to all comments, although we will try.

NOTES TO STUDENTS

Computer programs facilitate the solution to homework problems, but should not be used to replace an understanding of the material. Always understand exactly which formulas are required before turning to the computer. Before using the computer, we recommend that you know how to solve the problem by hand calculations. If you do not understand the formulas in the spreadsheets it is a good indication that you need to do more studying before using the program so that the structure of the spreadsheet will make sense. When you understand the procedures, it should be obvious which spreadsheet cells will help you to the answer, and which cells are intermediate calculations. It is also helpful to rework example problems from the text using the software.

ACKNOWLEDGMENTS

We would like to thank the many people who helped this work find its way to the classroom. We express appreciation to Professors Joan Brennecke, Mike Matthews, Bruce Poling, Ross Taylor, and Mark Thies, who worked with early versions of the text and provided suggestions for improvement. We are also greatly indebted to Dave Hart for proofreading an early version. There are many students who suffered through error-prone preliminary versions, and we thank them all for their patience and vision of the common goal of an error-free book. CTL would like to thank Ryoko Yamasaki for her work in typing many parts of the manuscript and problem solutions. CTL also thanks family members Gail, Nicolas, and Adrienne for their patience while the text was prepared, as many family sacrifices helped make this book possible. JRE thanks family members Guliz, Serra, and Eileen for their similar forbearance. We acknowledge Dan Friend and NIST, Boulder for contributions to the steam tables and thermodynamic charts. Lastly, we acknowledge the influences of the many authors of previous thermodynamics texts. We hope we have done justice to this distinguished tradition, while simultaneously bringing deeper insight to a broader audience.

Carl T. Lira, Michigan State University, lira@egr.msu.edu

J.Richard Elliott, University of Akron, dickelliott@uakron.edu

NOTATION

General Symbols

a — Activity, or dimensional equation of state parameter or energetic parameter, or heat capacity or other constant

A — Intensive Helmholtz energy, or dimensionless constant for equation of state, or Antoine, Margules or other constant

b — Dimensional equation of state parameter or heat capacity or other constant

B — Virial coefficient, or dimensionless constant for other equation of state, or Antoine or other constant

$C, c,...$ — Constants, c is a shape factor for the ESD equation of state

C_P — Intensive constant pressure heat capacity

C_V — Intensive constant volume heat capacity

F — Feed

f — Pure fluid fugacity

$\hat{f}_i$ — Fugacity of component in mixture

G — Intensive Gibbs energy or mass velocity

g — Gravitational constant (9.8066 m/s^2) or radial distribution function

g_c — Gravitational conversion factor (1 kg-m/N-s^2) (32.174[(1b$_m$ · ft)/s^2]/1b$_f$)

H — Intensive enthalpy

K_a — Reaction equilibrium constant

k — Boltzmann's constant = R/N_A

k_{ij} — Binary interaction coefficient (Eqn. 10.10)

K — Distribution coefficient (vapor-liquid, liquid-liquid, etc.)

m — Mass

N — Number of molecules

n — Number of moles

N_A — Avogadro's number = 6.0222 E 23 mol^{-1}

P — Pressure

$\underline{Q}$ — Extensive heat transfer across system boundary

Q	Intensive heat transfer across system boundary
q	Quality (% vapor)
R	Gas constant ($8.3143 \text{ cm}^3\text{-MPa/mole-K}$)
S	Intensive entropy
T	Temperature
U	Intensive internal energy
u	Velocity, or pair potential energy function
V	Intensive volume
$\underline{W}$	Extensive work done at boundary
W	Intensive work done at boundary
Z	Compressibility factor
z	Height

Greek Symbols

α	Isobaric coefficient of thermal expansion, also Peng-Robinson equation of state parameter, and also an ESD equation of state variable
ε	Potential energy parameter
η	Thermal, compressor, pump, turbine or expander efficiency, or reduced density, b/V
γ	C_P/C_V
γ_i	Activity coefficient
κ_T	Isothermal compressibility
κ_S	Isentropic compressibility
κ	Parameter for Peng-Robinson equation of state
λ	Universal free volume fraction
μ_i	Chemical potential for a component in a mixture

ω	Acentric factor
φ	Pure fluid fugacity coefficient
$\hat{\varphi}_i$	Component fugacity coefficient in a mixture
Φ_i	Volume fraction of component i
ρ	Intensive density
σ	Molecular diameter
ξ	Reaction Coordinate

Operators

Δ	Denotes change: (1) for a closed system, denotes (final state – initial state); (2) for an open steady-state system denotes (outlet state – inlet state)
$\ln$	Natural logarithm (base e)
$\log$	Common logarithm (base 10)
∂	Partial differential
Π	Cumulative product operator
Σ	Cumulative summation

Special Notation

ig	Equation applies to ideal gas only
$*$	Equation assumes heat capacity is temperature-independent.
$-$ as in $\overline{H}_i$	Partial molar property
$\_$ as in $\underline{U}$	Extensive property
$\hat{}$	Mixture property
$[\equiv]$	Specifies units for variable
$\equiv$	Equivalence or definition
$\approx$	Approximately equal to

Subscripts

b	Property at normal boiling point temperature

EC	Expansion/contraction work		*f*	Property at final state
f	Denotes property for formation of molecule from atoms in their naturally occurring pure molecular form		*fus*	Fusion (melting) process
			i	Property at initial state
			α,β	Denotes phase of interest
gen	Generated entropy		*ig*	Ideal gas property
i	Component in a mixture		*in*	Property at inlet (open system)
m	Property at melting point		*is*	Ideal solution property
mix	Used with Δ to denote property change on mixing		*L*	Liquid phase
r	Property reduced by critical constant		o	Standard state
			out	Property at outlet (open system)
R	Reference state		*S*	Solid phase
S	Shaft work		*sat*	Saturation property

Superscripts

			V	Vapor phase
∞	Infinite dilution		*vap*	Vaporization process
E	Excess property for a mixture			

FIRST AND SECOND LAWS

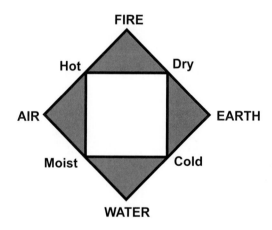

FIRE

Hot   Dry

AIR   EARTH

Moist   Cold

WATER

Aristotle, 384–322 B.C.

The ancient Greeks thought that there were only four elements: earth, air, fire, and water. As a matter of fact, you can explain a large number of natural phenomena with little more. The first and second laws of thermodynamics can be developed and illustrated quite completely with just solid blocks(earth), ideal gases (air), steam property tables (water), and heat (fire). Without significantly more effort, we can include a number of other "elements": methane, carbon dioxide, and several refrigerants. These additional species are quite common, and charts that are functionally equivalent to the steam property tables are readily available.

The first and second laws provide the foundation for all of thermodynamics, and their importance should not be underestimated. Many engineering disciplines typically devote an entire semester to the "earth, air, fire, and water" concepts. This knowledge is so fundamental and so universal that it is essential to any applied scientist. Nevertheless, chemical engineers must quickly lay this foundation and move on to other issues covered in Units II, III, and IV. The important thing for chemical engineers to anticipate as they move through Unit I is that the principles are at the core of the entire text and it will be necessary to integrate information from Unit I in the later units. The key is to follow the methods of applying systematically the first law (energy balance) and the second law (entropy balance). Watch carefully how the general equations are quickly reduced to the specific problem at hand. Especially watch how the systems of equations are developed to match the unknown variables in the problem. Learn to perform similar reductions quickly and accurately for yourself. It takes practice, but thorough knowledge of that much will help immensely when it comes to Unit II.

CHAPTER

1

INTRODUCTION

"Aside from the logical and mathematical sciences, there are three great branches of natural science which stand apart by reason of the variety of far reaching deductions drawn from a small number of primary postulates. They are mechanics, electromagnetics, and thermodynamics.

These sciences are monuments to the power of the human mind; and their intensive study is amply repaid by the aesthetic and intellectual satisfaction derived from a recognition of order and simplicity which have been discovered among the most complex of natural phenomena. . . Yet the greatest development of applied thermodynamics is still to come. It has been predicted that the era into which we are passing will be known as the chemical age; but the fullest employment of chemical science in meeting the various needs of society can be made only through the constant use of the methods of thermodynamics."

Lewis and Randall (1923)

Lewis and Randall eloquently summarized the broad significance of thermodynamics as long ago as 1923. They went on to describe a number of the miraculous scientific developments of the time and the relevant roles of thermodynamics. Historically, thermodynamics has guided the development of steam engines, refrigerators, nuclear power plants, and rocket nozzles, to name just a few. The principles remain important today in the refinement of alternative refrigerants, heat pumps and improved turbines, and also in technological advances including computer chips, superconductors, advanced materials, and bioengineered drugs. These latter day "miracles" on first thought might appear to have little to do with power generation and refrigeration cycles. However, as Lewis and Randall point out, the implications of the postulates of thermodynamics are far reaching and will continue to be important in the development of even newer technologies. Much of modern thermodynamics focuses on characterization of the properties of mixtures, as their constituents partition into stable phases, inhomogeneous domains, and/or react. The capacity of thermodynamics to bring "quantitative precision in place of the old, vague ideas"[1] is as germane today as it was then.

---

1. Lewis, G.N., Randall, M. *Thermodynamics and the Free Energy of Chemical Substances,* McGraw-Hill, NY, 1923.

Before overwhelming you with the details that comprise thermodynamics, we outline a few "primary postulates" as clearly as possible and put them into the context of what we will refer to as classical equilibrium thermodynamics. In their simplest human terms, our primary premises can be expressed as:

1. You can't get something for nothing.

2. Generation of disorder results in lost work.[1]

Occasionally, it may seem that we are discussing principles that are much more sophisticated. But the fact is that all of our discussions can be reduced to these fundamental principles. When you find yourself in the midst of a difficult problem, it may be helpful to remember that. We will see that coupling these two observations with some slightly sophisticated reasoning (mathematics included) leads to many clear and reliable insights about a wide range of subjects from energy crises to high-tech materials, from environmental remediation to biosynthesis. The bad news is that the level of sophistication required is not likely to be instantly assimilated by the average student. The good news is that many students have passed this way before, and the proper course is about as well-marked as one might hope.

There is less than universal agreement on what comprises "thermodynamics." If we simply take the word apart, "thermo" sounds like "thermal" which ought to have something to do with heat, temperature, or energy, and "dynamics" ought to have something to do with movement. And if we could just leave the identification of thermodynamics as the study of "energy movements," it would be sufficient for the purposes of this text. Unfortunately, such a definition would not clarify what makes thermodynamics distinct from, say, transport phenomena or kinetics, so we should spend some time clarifying the definition of thermodynamics in this way before moving on to the definitions of temperature, heat, energy, etc.

The definition of thermodynamics as the study of energy movements has evolved considerably to include classical thermodynamics, quantum thermodynamics, statistical thermodynamics, and irreversible thermodynamics as well as equilibrium and non-equilibrium thermodynamics. Classical thermodynamics has the general connotation of referring to the implications of constraints related to multivariable calculus as developed by J.W. Gibbs. We spend a significant effort applying these insights in developing generalized equations for the thermodynamic properties of pure substances. Statistical thermodynamics focuses on the idea that knowing the precise states of 10^{23} atoms is not practical and prescribes ways of computing the average properties of interest based on very limited measurements. We touch on this principle in our introduction to entropy and in our ultrasimplified kinetic theory of ideal gases, but we will generally refrain from detailed formulation of all the statistical averages. Irreversible thermodynamics and non-equilibrium thermodynamics emphasize the ways atoms and energy move over periods of time. At this point, it becomes clear that such a broad characterization of thermodynamics would overlap with transport phenomena and kinetics in a way that would begin to be confusing at the introductory level. Nevertheless, these fields of study represent legitimate subtopics within the general domain of thermodynamics.

These considerations should give some idea of the narrowness and the breadth that are possible within the general study of thermodynamics. This text will try to find a happy medium. One general unifying principle about the perspective offered by thermodynamics is that there are certain properties that are invariant with respect to time. For example, the process of diffusion may indicate some changes in the system with time, but the diffusion coefficient can be considered to be a property

---

1. The term "lost work" refers to the loss of capability to perform work, and is discussed in more detail in Sections 2.4, 3.3, and 3.4.

which only depends on a temperature, density, and composition profile. A thermodynamicist would consider the diffusion process as something straightforward given the diffusion coefficient, and focus on understanding the diffusion coefficient. A transport specialist would just estimate the diffusion coefficient as best as he could and get on with it. A kineticist would want to know how fast the diffusion was relative to other processes involved. In more down to earth terms, if we were touring about the countryside, the thermodynamicists would want to know where we were going, the transport specialists would want to know how long it takes to get there, and the kineticists would want to know how fast the gasoline was running out.

Through the study of thermodynamics we utilize a few basic concepts: energy, entropy, and equilibrium. The ways in which these are related to each other and to temperature, pressure, and density are best understood in terms of the molecular mechanisms which provide the connections. These connections, in turn, can be summarized by the equation of state, our quantitative description of the substance. Showing how energy and entropy evolve into insight about molecular characteristics and their impacts on process applications is the primary goal of this text. These insights should stick with you long after you have forgotten how to estimate any particular thermodynamic property, a heat capacity or activity coefficient, for example. We will see how assuming an equation of state and applying the rules of thermodynamics leads to accurate and extremely general insights relevant to many applications.

1.1   THE MOLECULAR NATURE OF ENERGY

Having described the basic meaning of thermodynamics through its emphases and how it relates to other fields of study, we must proceed systematically to describe some of the terms that will be applied over the long term.

Energy is a word which applies to many aspects of the system. Its formal definition is in terms of the capability to perform work. We will not quantify the potential for work until the next chapter, but you should have some general understanding of work from your course in introductory physics. Energy may take the form of kinetic energy or potential energy, and it may refer to energy of a macroscopic or a molecular scale.

Energy is the sum total of all capacity for doing work that is associated with matter: kinetic, potential, submolecular (i.e., molecular rearrangements by reaction), or subatomic (i.e., ionization, fission, fusion).

Kinetic energy is the energy associated with motion of a system. Motion can be classified as translational, rotational or vibrational.

Potential energy is the energy associated with a system due to its position in a force field.

In the study of "energy movements," we must continually ask, "How much energy is here now, and how much is there?" In the process, we need to establish a point for beginning our calculations. According to the definition above, we might intuitively represent zero internal energy by a perfect vacuum. But then, knowing the internal energy of a single proton inside the vacuum would require knowing how much energy it takes to make a proton from nothing. Since this is not entirely practical, this intuitive choice is not a good choice. This is essentially the line of reasoning that gives rise to the necessity of defining reference states. It should be clear from this reasoning that there is no absolute value of energy that is always the most convenient; there are only changes in energy from one reference state to another. Thus it is common in the study of thermodynamics to speak of energy changes relative to some reference conditions that are likely to apply throughout any particular process of interest. These reference conditions may change from, say, defining the internal energy of liquid water to be zero at 0.01°C (as in the steam tables) to setting it equal to zero for the

❶ Energy will be tabulated relative to a convenient reference state.

elements of hydrogen and oxygen (as in the heat of reaction), depending on the situation. In studies of activity and chemical reactions, we will need to be somewhat careful in specifying the exact nature of the reference state. Since we will focus on changes in kinetic energy, potential energy, and energies of reaction, we will not need to specify reference states any more fundamental than the elements, thus do not consider subatomic particles.

Potential Energy

Kinetic energy is commonly covered in detail during introductory physics courses so we will assume it is already familiar. Students may be somewhat less comfortable with the concept of potential energy, however, especially as it relates to molecular interactions. Therefore, we spend some time with it here.

Macroscopic Potential Energy

Potential energy is associated with the "work" of moving a system some distance through a force field. On the macroscopic scale, we are well aware of the effect of gravity. This is a very common force field. As an example, the earth and the moon are two spherical bodies which are attracted by a force which varies as r^{-2}. The potential energy represents the work of moving the two bodies closer together or farther apart, which is simply the integral of the force over distance. So the potential function varies as r^{-1}. These kinds of operations should be familiar from a course in introductory physics. They are the same at the microscopic level except that the forces vary with position according to different laws.

Intermolecular Potential Energy

In consideration of potential energy on a molecular scale, it is important to realize that uncharged atoms do exert forces on each other. For a rigorous description, the origin of the intermolecular potential must be traced back to the solution of Schrödinger's quantum mechanics equation for the motions of electrons around nuclei. We cannot afford that level of rigor in this course. In fact, that level of rigor is not currently available for some of the complex molecules to be considered in this course. On the other hand, there are several generalities which arise from this rigorous approach which can be appreciated from a somewhat intuitive perspective. For instance, atoms are comprised of dense nuclei of positive charge with electron densities of negative charge built around the nucleus in shells. The outermost shell is referred to as the valence shell. Technically, we could say that this insight represents a generalization based on solutions of Schrödinger's equation (which was in turn based on compiled experimental data detailing the differences between subatomic particles and the macroscopic particles which Newton's laws accurately describe). This much should be familiar from an elementary course in chemistry or physics. Another generalization which may be less familiar is that electron density often tends to concentrate in lobes in the valence shell for elements like C, N, O, F, S, Cl. These lobes may be occupied by protons that are tetrahedrally coordinated as in CH_4, or they may be occupied by unbonded electron pairs that fill out the tetrahedral coordination as in NH_3 or H_2O. These elements (H, C, N, O, F, S, Cl) and some noble gases like He, Ne, and Ar provide virtually all of the building blocks for the molecules to be considered in this text. The two generalizations discussed here are sufficient to describe what we need to know about intermolecular forces at the introductory level.

By considering the implications of atomic structure and atomic collisions, it is possible to develop subclassifications of intermolecular forces. These are:

1. Electrostatic forces between charged particles (ions) and between permanent dipoles, quadrupoles, and higher multipoles.

❶ Engineering model potentials permit representation of attractive and repulsive forces in a tractable form.

2. Induction forces between a permanent dipole (or quadrupole) and an induced dipole.

3. Forces of attraction (dispersion forces) and repulsion between nonpolar molecules.

4. Specific (chemical) forces leading to association and complex formation, especially evident in the case of hydrogen bonding.

Further, attractive forces will be quantified by negative numerical values, and repulsive forces will be characterized by positive numerical values.

Electrostatic Forces

The force between two point charges described by Coulomb's Law is very similar to the law of gravitation and should be familiar from elementary courses in chemistry and physics,

$$F \propto \frac{q_i q_j}{r^2}$$

where q_i and q_j are the charges, and r is the separation of centers. Upon integration, $u = \int F dr$, the potential energy is proportional to inverse distance,

$$u \propto \frac{q_i q_j}{r}$$

If all molecules were perfectly spherical and rigid, the only way that these electrostatic interactions could come into play is through the presence of ions. But a molecule like NH_3 is not perfectly spherical. NH_3 has three protons on one side and a lobe of electron density in the unbonded valence shell electron pair. This permanent asymmetric distribution of charge density gives rise to a permanent dipole on the NH_3 molecule. This means that ammonia molecules lined up with the electrons facing each other will repel while molecules lined up with the electrons facing the protons will attract. Since electrostatic energy drops off as r^{-1}, one might expect that the impact of these forces would be long range. Fortunately, with the close proximity of the positive charge to the negative charge in a molecule like NH_3, the charges tend to cancel each other as the molecule spins and tumbles about through a fluid. This spinning and tumbling makes it reasonable to consider a spherical average of the intermolecular energy as a function of distance that may be constructed by averaging over all orientational angles between the molecules at each distance. In a large collection of molecules randomly distributed to each other, this averaging approach gives rise to many cancellations, and the net impact is that

$$u_{dipole-dipole} = \frac{-\varepsilon_{dipole}}{r^6 kT} \qquad 1.1$$

This surprisingly simple result is responsible for a large part of the attractive energy between polar molecules. This energy is attractive because the molecules tend to spend somewhat more time lined up attractively than repulsively, and the r^{-6} power arises from the many cancellations that occur in a fluid.

Induction Forces

When a molecule with a permanent dipole approaches a molecule with no dipole, the positive charge of the dipolar molecule will tend to pull electron density away from the nonpolar molecule and "induce" a dipole moment into the nonpolar molecule. The magnitude of this effect depends on the strength of the dipole and a fundamental property of the nonpolar molecule called "polarizability." For example, the pi bonding in benzene makes it fairly polarizable. A similar consideration of the spherical averaging described in relation to electrostatic forces results in

$$u_{in} = \frac{-\varepsilon_{in}}{r^6}$$

1.2

Disperse Attraction Forces (Dispersion Forces)

When two nonpolar molecules approach, they may also induce dipoles into each other. Clearly these would be weak, and this is why the forces between the noble gases are so weak. Nevertheless, their dependence on radial distance may be analyzed similarly to the electrostatic forces, and this gives the same form for the attractive forces:

❶ The r^{-6} dependence of attractive forces has a theoretical basis.

$$u_{disp}^{att} = \frac{-\varepsilon_{att}}{r^6}$$

1.3

Repulsive Forces

Another aspect of the collisions between molecules that we have not addressed yet is what happens when the electron clouds begin to interpenetrate significantly. Clearly this would give rise to a strong repulsive force which is generally referred to as a repulsive dispersion force. This force increases rapidly as radial distance decreases, and quickly outweighs the attractive force when the atoms get too close together. Two atoms do not like to be in the same place at the same time. A common and useful way to express this is

$$u_{disp}^{rep} = \frac{\varepsilon_{rep}}{r^{12}}$$

1.4

Specific (Chemical) Forces

What happens when the strength of interaction between two molecules is so strong at certain orientations that it does not make sense to spherically average over it? For instance, it would not make sense to spherically average when two atoms were permanently bound in a specific orientation. But if they were permanently bound together we would call that a chemical reaction and handle it in a different way. An interesting problem arises when the strength of interaction is too strong to be treated entirely by spherically averaging and too weak to be treated as a normal chemical reaction which forms permanent stable chemical species. Clearly, this problem is difficult and it would be tempting to try to ignore it. In fact, most of this course will deal with theories that treat only what can be derived by spherically averaging. But it should be kept in mind that these specific forces are responsible for some rather important properties, especially in the form of hydrogen bonding. Since many systems are aqueous or contain amides or alcohols, ignoring hydrogen bonding entirely would substantially undermine the accuracy of our conceptual foundation. Fortunately, the nature of these forces is such that we can often approximate them as strong association forces to obtain a crude engineering model that is workable.

Examples of Model Potentials

Based on the forms of these electrostatic, induction, and dispersion forces, it should be easy to appreciate the form of the Lennard-Jones potential in Fig. 1.1. Other models of the potential function are possible, such as the square-well potential or the Sutherland potential also shown in Fig. 1.1. These latter potential models represent simplified forms of the Lennard-Jones model that are accurate enough for many applications.

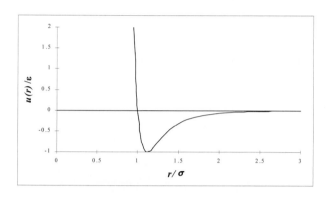

$$u(r) = 4\varepsilon[(\sigma/r)^{12} - (\sigma/r)^6]$$

The Lennard-Jones potential

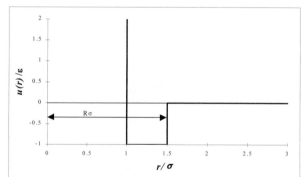

$$u(r) = \begin{cases} \infty & \sigma \geq r \\ -\varepsilon & \sigma < r < R\sigma \\ 0 & r \geq R\sigma \end{cases}$$

The square-well potential for $R = 1.5$.

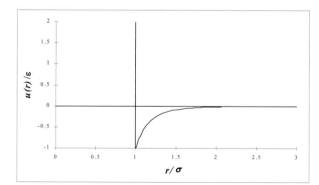

$$u(r) = \begin{cases} \infty & r \leq \sigma \\ -\varepsilon(\sigma/r)^6 & r > \sigma \end{cases}$$

The Sutherland potential.

Figure 1.1 *Schematics of three engineering models for pair potentials on a dimensionless basis.*

The key features of all of these potential models are the representation of the size of the molecule by the parameter σ and the attractive strength (i.e. "stickiness") by the parameter ε. Note that we would need a more complicated potential model to represent the shape of the molecule. Typically, molecules of different shapes are represented by binding together several potentials like those above with each potential site representing one molecular segment. For example, n-butane could be represented by four Lennard-Jones sites that have their relative centers located at distances corresponding to the bond-lengths in n-butane. The potential between two butane molecules would then be the sum of the potentials between each of the individual Lennard-Jones sites on the different molecules. In similar fashion, potential models for very complex molecules can be constructed.

We can gain considerable insight about the thermodynamics of fluids by intuitively reasoning about the relatively simple effects of size and stickiness. For example, a large molecule like buckminsterfullerene would have a larger value for σ than would methane. Water and methane are about the same size, but their difference in boiling temperature indicates a large difference in their stickiness. As you read through this chapter, it should become apparent that water has a higher boiling temperature because it sticks to itself more strongly than does methane. With these simple insights, you should be able to understand the molecular basis for a large number of macroscopic phenomena.

Example 1.1 Intermolecular potentials for mixtures

Our discussion of intermolecular potentials has focused on describing single molecules, but it is actually more interesting to compare and contrast the potential models for different molecules that are mixed together. We can use the square-well potential as the basis for this exercise and focus simply on the size (σ_{ij}) and stickiness (ε_{ij}) of each potential model, where the subscript ij indicates an interaction of molecule i with molecule j. For example, ε_{11} would be the stickiness of molecule 1 to itself, and ε_{12} would be its stickiness to a molecule of type 2. The size parameter for interaction between different molecules is reasonably well represented by $\sigma_{12} = (\sigma_{11} + \sigma_{22})/2$. The estimation of the stickiness parameter for interaction between different molecules requires some intuitive reasoning. For mixtures of hydrocarbons, it is conventional to estimate the stickiness by a geometric mean. To illustrate, sketch on the same pair of axes the potential models for methane and benzene, assuming that the stickiness parameter is given by $\varepsilon_{12} = (\varepsilon_{11} \varepsilon_{22})^{1/2}$.

Solution:

Methane(1) has fewer atoms in it than benzene(2), so we can assume it is smaller. Let's depict this by saying $\sigma_{22} = 2\sigma_{11}$. This means that $\sigma_{12} = 1.5\sigma_{11}$. Similarly, methane's boiling temperature is lower so its stickiness should be smaller in magnitude. Let's depict this by $\varepsilon_{22} = 4\varepsilon_{11}$ and this means $\varepsilon_{12} = 2\varepsilon_{11}$. Thus we obtain Fig. 1.2.

1.2 THE MOLECULAR NATURE OF ENTROPY

To be fair to both of the central concepts of the course, we must mention entropy at this point, in parallel with the mention of energy. Unfortunately, there is no simple analogy that can be drawn like that of the potential energy between the earth and moon. The study of entropy is fairly specific to the study of thermodynamics. The proper development of the subject must await Chapter 3.

What we can say at this point is that entropy has been conceived to account for losses in the prospect of performing useful lost work. The energy can take on many forms and be completely accounted for without contemplating how much energy has been "wasted" by converting work into

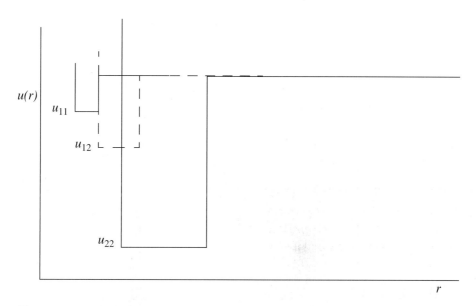

Figure 1.2 *Sketch of intermolecular square-well potential models for a mixture of methane and benzene for R = 1.5 as explained in Example 1.1.*

something like warm water. Entropy accounts for this kind of wastefulness. It turns out that the generation of such wastes can be directly related to the degree of disorder involved in conducting our process. Thus generation of disorder results in lost work. Furthermore, work that is lost by not maintaining order cannot be converted into useful work. To see the difference between energy and entropy, consider the following example. Oxygen and nitrogen are mixed as ideal gases at room temperature and pressure. How much energy is evolved in the mixing process? How much work must be exerted to separate them again? The answer to the first question is that no energy is involved. Ideal gases are point masses with no potential energy to affect the mixing. For the answer to the second question, however, we must acknowledge that a considerable effort would be involved. The minimum amount required is given by the "lost work," which we will discuss in Chapter 3.

1.3   BRIEF SUMMARY OF SEVERAL THERMODYNAMIC QUANTITIES

Internal Energy

It may be somewhat confusing that kinetic and potential energy exist on the macroscopic level and the microscopic level. It is frequently convenient to lump the microscopic energies together and consider them as the ***internal energy*** of the system which is given the symbol U. The remaining potential energy and kinetic energy are the macroscopic properties of the complete system and can be accounted for separately. The internal energy is a function of the temperature and density of the system, and it does not usually change if the entire system is placed on, say, an airplane. This is the convention followed throughout the remainder of Unit I. In Units II and III, we reexamine the molecular potentials as to how they affect the bulk fluid properties. Thus, throughout the remainder of Unit I, when we refer to kinetic and potential energy of a body of fluid as a system, we will be referring to the kinetic energy of the center of mass of the system and the gravitational potential energy of the center of mass of the system.

❶ The sum of microscopic random kinetic energy and intermolecular potential energies are the internal energy.

Work

Work is a familiar term from physics. We know that work is a force acting through a distance. There are several ways forces may interact with the system which all fit under this category. We will discuss the details of how we calculate work and determine its impact on the system in the next chapter.

Density

Density is a measure of the mass per unit volume and may be expressed on a molar basis or a mass basis. In some situations, it is expressed as number of particles per unit volume.

Pressure — An Ultrasimplified Kinetic Theory

Pressure is the force exerted per unit area. We will be concerned primarily with the pressure exerted by the molecules of fluids upon the walls of their containers. For our purposes, the kinetic theory of pressure should provide a sufficient description.

Suppose we have two hard spherical molecules in a container that are bouncing back and forth in the x-direction only and not contacting each other.

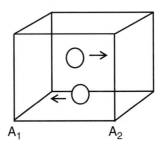

We wish to quantify the forces acting on each wall. Since the particles are only colliding with the walls at A_1 and A_2 in our idealized model, these are the only walls we need consider. Let us assume that particles bounce off the wall with the same speed which they had before striking the wall, but in the opposite direction (a perfectly elastic collision where the wall is perfectly rigid and absorbs no momentum). Thus, the kinetic energy of the particles will be fixed. If u is the initial velocity of the particle (recall that u is a vector quantity and u is a scalar) before it strikes a wall, the change in velocity due to striking the wall is $-2u$. The change in velocity of the particle indicates the presence of interacting forces between the wall and the particle. If we quantify the force necessary to change the velocity of the particle, we will also quantify the forces of the particle on the wall by Newton's third principle. To quantify the force, we may apply Newton's second principle stated in terms of momentum:

The time rate of change of the momentum of a particle is equal to the resultant force acting on the particle and is in the direction of the resultant force.

$$\frac{d\mathbf{p}}{dt} = \mathbf{F} \qquad\qquad 1.5$$

The application of this formula directly is somewhat problematic since the change in direction is instantaneous, and it might seem that the time scale is important. This can be avoided by determining the time-averaged force,[1] $\mathbf{F}_{avg}$ exerted on the wall during time Δt:

---

1. See an introductory physics text for further discussion of time-averaged force.

$$\int_{t^i}^{t^f} \mathbf{F} \, dt = \mathbf{F}_{avg} \, \Delta t = \int_{t^i}^{t^f} \frac{d\mathbf{p}}{dt} \, dt = \Delta \mathbf{p} \qquad 1.6$$

where $\Delta\mathbf{p}$ is the total change in momentum during time Δt. The momentum change for each collision is $-2m\mathbf{u}$ where m is the mass per particle. Each particle will collide with the wall every t_1 seconds, where $t_1 = 2L/u$, where L is the distance between A_1 and A_2. The average force is then

$$\mathbf{F}_{avg} = -2m\mathbf{u} \frac{u}{2L} \qquad 1.7$$

where $\mathbf{u}$ is the velocity before the collision with the wall. Pressure is the force per unit area, and the area of a wall is L^2, thus

$$P = \frac{m}{L^3} (u_1^2 + u_2^2) \qquad 1.8$$

where the subscripts denote the particles.

P is proportional to the number of particles in a volume and to the kinetic energy of the particles.

If the particle motions are generalized to motion in arbitrary directions, collisions with additional walls in the analysis does not complicate the problem dramatically because each component of the velocity may be evaluated independently. To illustrate, consider a particle bouncing around the centers of four walls in a horizontal plane. From the top view, the trajectory would appear as below:

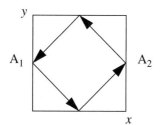

For the same velocity as the first case, the force of each collision would be reduced because the particle strikes merely a glancing blow. The x-component of the force can be related to the magnitude of the velocity by noting that $u_x = u_y$, such that $u = (u_x^2 + u_y^2)^{1/2} = u_x 2^{1/2}$. The time between collisions with wall A_1 would be $4L/(u2^{1/2})$. The formula for the average force in two dimensions then becomes:

$$F_{avg, A_1} = -2mu_x \frac{u\sqrt{2}}{4L} = -2m \frac{u}{\sqrt{2}} \frac{u\sqrt{2}}{4L} = -2mu \frac{u}{2L} \qquad 1.9$$

and the pressure due to two particles that don't collide with each other in two dimensions becomes:

$$P = \frac{m}{2L^3} (u_1^2 + u_2^2) \qquad 1.10$$

The extension to three dimensions is more difficult to visualize, but comparing Eqn. 1.8 to Eqn. 1.10, you should not be surprised to learn that the pressure in three dimensions is:

$$P = \frac{m}{3L^3}(u_1^2 + u_2^2)$$ 1.11

The problem does get more complicated when collisions between particles occur. We can ignore that possibility here because the ideal gases being considered are point masses that do not collide with each other. Computation involving molecular collisions demands a computer and is called molecular dynamics simulation. Despite the need for a computer, the program is fundamentally nothing more than computing the trajectories of two billiard balls colliding, as described briefly in the homework problems.

Temperature

The most reliable definition of temperature is that it is a numerical scale for uniquely ordering the "hotness" of a series of objects.[1] We are guaranteed that a universal scale of temperature can be developed because of the "zeroth law" of thermodynamics: if two objects are in equilibrium with a third, then they are in equilibrium with each other. The zeroth law is a law in the sense that it is a fact of experience that must be regarded as an empirical fact of nature. The significance of the zeroth law is that we can calibrate the temperature of any new object by equilibrating it with objects of known temperature. Temperature is therefore an empirical scale that requires calibration according to specific standards. The Celsius and Fahrenheit scales are in everyday use. The conversions are:

$$(T \text{ in } °C) = \frac{5}{9}((T \text{ in } °F) - 32)$$

When we perform thermodynamic calculations, we must usually use absolute temperature in Kelvin or Rankine. These scales are related by

$$(T \text{ in } K) = (T \text{ in } °C) + 273.15$$

$$(T \text{ in } °R) = (T \text{ in } °F) + 459.67$$

$$(T \text{ in } R) = 1.8 \cdot (T \text{ in } K)$$

❶ Thermodynamic calculations use absolute temperature in °R or K.

This definition is rigorous, but it does not provide the kind of conceptual picture that permits an easy grasp of its significance and how it relates to other important properties. A better conceptual picture is provided by considering that temperature primarily provides a measure of the kinetic energy of a system of molecules. As the molecules move faster, the temperature goes up. The absolute temperature scale has the advantage that the temperature can never be less than absolute zero. This observation is easily understood from the kinetic perspective. The kinetic energy cannot be less than zero; if the molecules are moving, their kinetic energy must be greater than zero. To better understand the connection between temperature and kinetic energy, we can apply the ideal gas law as described on page 15.

❶ Temperature primarily reflects the kinetic energy of the molecules.

1. Denbigh, K., *The Principles of Chemical Equilibrium,* Cambridge University Press, 3rd ed., p. 9, London, 1971.

Heat – Sinks and Reservoirs

Heat is energy in transit between the source from which the energy is coming and a destination toward which the energy is going. When developing thermodynamic concepts, we frequently will assume that our system transfers heat to/from a *sink* or *reservoir*. A heat reservoir is an infinitely large source or destination of heat transfer. The reservoir is assumed to be so large that the heat transfer does not affect the temperature of the reservoir. A sink is a special name sometimes used for a reservoir which can accept heat without a change in temperature. The assumption of constant temperature makes it easier to concentrate on the thermodynamic behavior of the system while making a reasonable assumption about the part of the universe assigned to be the reservoir.

❶ A reservoir is an infinitely large source or destination for heat transfer.

The mechanics of heat transfer are also easy to picture conceptually from the molecular kinetics perspective. In heat conduction, faster moving molecules collide with slower ones, exchanging kinetic energy and equilibrating the temperatures. In this manner, we can imagine heat being transferred from the hot surface to the center of a pizza in an oven until the center of the pizza is cooked. In heat convection, packets of hot mass are circulated and mixed, accelerating the equilibration process. Heat convection is important in getting the heat from the oven flame to the surface of the pizza. Heat radiation, the remaining mode of heat transfer, occurs by an entirely different mechanism having to do with waves of electromagnetic energy emitted from a hot body that are absorbed by a cooler body. Radiative heat transfer is typically discussed in detail during courses devoted to heat transfer.

1.4   BASIC CONCEPTS

The System

A **system** is that portion of the universe which we have chosen to study.

A **closed** system is one in which no mass crosses the system boundaries.

An **open** system is one in which mass crosses the system boundaries. The system may gain or lose mass or simply have some mass pass through it.

System boundaries are established at the beginning of a problem, and simplification of balance equations depends on whether the system is open or closed. *Therefore, the system boundaries should be clearly identified. If the system boundaries are changed, the simplification of the mass and energy balance equations should be performed again, because different balance terms are likely to be necessary.* These guidelines will become more apparent in Chapter 2.

❶ The placement of system boundaries is a key step in problem solving.

The Mass Balance

Presumably, students in this course are familiar with mass balances from an introductory course in material and energy balances. The relevant relation is simply:

$$\begin{bmatrix} \text{rate of mass} \\ \text{accumulation within} \\ \text{system boundaries} \end{bmatrix} = \begin{bmatrix} \text{rate of mass flow} \\ \text{into system} \end{bmatrix} - \begin{bmatrix} \text{rate of mass flow} \\ \text{out of system} \end{bmatrix}$$

1.12

$$\dot{m} = \sum_{inlets} \dot{m}^{in} - \sum_{outlets} \dot{m}^{out}$$

❶ The mass balance.

where $\dot{m} = \dfrac{dm}{dt}$. $\dot{m}^{in}$ and $\dot{m}^{out}$ are the absolute values of mass flow rates entering and leaving, respectively.

We may also write

$$dm = \sum_{inlets} dm^{in} - \sum_{outlets} dm^{out} \qquad 1.13$$

where mass differentials dm^{in} and dm^{out} are *always positive*. When all the flows of mass are analyzed in detail for several subsystems coupled together, this simple equation may not seem to fully portray the complexity of the application. The general study of thermodynamics is similar in that regard. A number of simple relations like this one are coupled together in a way that requires some training to understand. *In the absence of chemical reactions*, we may also write a mole balance by replacing mass with moles in the balance.

Intensive Properties

Intensive properties are those properties which are independent of the size of the system. For example, in a system at equilibrium without internal rigid/insulating walls, the temperature and pressure are uniform throughout the system and are therefore intensive properties. Likewise, mass or mole-specific properties are independent of the size of the system. For example, the molar volume ($[\equiv]$ length^3/mole), mass density ($[\equiv]$ mass/length^3), the specific internal energy ($[\equiv]$ energy/mass) are intensive properties. Notationally in this text, intensive properties are not underlined.

> ❶ The distinction between intensive and extensive properties is key in selecting and using variables for problem solving.

Extensive Properties

Extensive properties depend on the size of the system. For example the volume ($[\equiv]$ length^3) and energy ($[\equiv]$ energy). Extensive properties are underlined, e.g. $\underline{U} = n\, U$, where n is the number of moles and U is molar internal energy.

States and State Properties – The Phase Rule

Two state variables are necessary to specify the state of a *single-phase* pure fluid, i.e., two from the set P, V, T, U. Other state variables to be defined later in the text which also fit in this category are molar enthalpy, molar entropy, molar Helmholtz energy and molar Gibbs energy. *State variables must be intensive properties.* As an example, specifying P and T permits you to find the specific internal energy and specific volume of steam. Note, however, that you need to specify only one variable, the temperature or the pressure, if you want to find the properties of saturated vapor or liquid. This reduction in the needed specifications is referred to as a reduction in the "degrees of freedom." As another example in a ternary, two-phase system, the temperature and the mole fractions of two of the components of the lower phase are state variables (the third component is implicit in summing the mole fractions to unity), but the total number of moles of a certain component is not a state variable because it is extensive. In this example, the pressure and mole fractions of the upper phase may be calculated once the temperature and lower-phase mole fractions have been specified. The number of state variables needed to completely specify the state of a system is given by the Gibbs phase rule:

$$F = C - P + 2 \qquad 1.14$$

where F is the number of state variables needed to specify the state of the system (F is also known as the number of degrees of freedom), C is the number of components, and P is the number of phases. More details on the Gibbs phase rule are given in Chapter 13.

Equilibrium

A system is in equilibrium when there is no driving force for a change of intensive variables within the system.

The Gibbs phase rule provides the number of state variables (intensive properties) to specify the state of the system.

Steady-State Open Systems

The term *steady state* is used to refer to open systems in which the inlet and outlet mass flowrates are invariant with time and there is no mass accumulation. In addition, steady state requires that state variables at all locations are invariant with respect to time. Note that state variables may vary with position. Steady state does not require the system to be at equilibrium. For example, in a heat exchanger operating at steady state with a hot and cold stream, each stream will have a temperature gradient along its length, and there will always be a driving force for heat transfer from the hotter stream to the colder stream.

Steady-state flow is very common in the process industry.

The Ideal Gas Law

Throughout Unit I, we frequently use thermodynamic charts and tables to obtain thermodynamic properties for various compounds. The most prevalent compounds in our study are water and the ideal gas. Properties of water are obtained from the steam tables to be discussed below. Properties of the ideal gas are calculated from the ideal gas model:

$$P\underline{V} = nRT \qquad \text{(ig) 1.15}$$

P-V-T properties of the ideal gas law are shown in Fig. 1.3. The ideal gas model represents many compounds, such as air, nitrogen, oxygen, and even water vapor at temperatures and pressures near ambient conditions. Use of this model simplifies calculations while the concepts of the energy and entropy balances are developed throughout Unit I. This does not imply that the ideal gas model is applicable in all situations. Analysis using more complex fluid models is delayed until Unit II. We rely on thermodynamic charts and tables to obtain properties for fluids which may not be considered ideal gases until Unit II.

The ideal gas law is a model that is not always valid, but gives an initial guess.

One application of the ideal gas law would be timely to promote better understanding of the connection of temperature with kinetic energy. We can apply the ideal gas law in conjunction with Eqn. 1.11 which was derived for a spherical (monatomic) molecule.

$$nRT = P\underline{V} = \frac{m}{3L^3}\left(\sum_{i=1}^{N} u_i^2\right)L^3 \qquad \text{(ig) 1.16}$$

where m is the mass per particle. Rearranging shows how the temperature is related to the average molecular kinetic energy,

$$T = \frac{m}{3R}\frac{\sum_{i=1}^{N} u_i^2}{n} = \frac{2}{3R}\frac{\sum_{i=1}^{N} \frac{m u_i^2}{2}}{N/N_A} = \frac{2N_A}{3R}\left\langle \frac{m u^2}{2} \right\rangle \qquad 1.17$$

where <> brackets denote an average, and N_A denotes Avogadro's number.

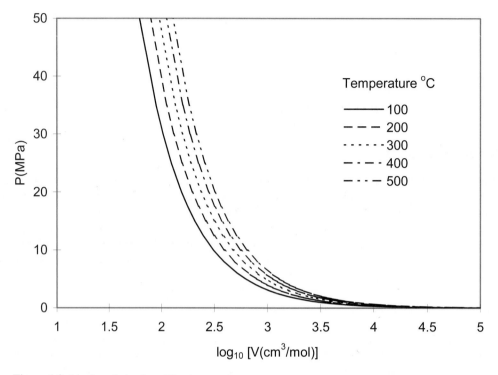

Figure 1.3 *Ideal gas behavior at five temperatures.*

Eqn. 1.17 may seem to be confined to ideal monatomic gases since that was the origin of its derivation, but it is actually applicable to any monatomic classical system, including monatomic liquids and solids. This means that for a pure system of a monatomic ideal gas in thermal equilibrium with a liquid, the average velocities of the molecules are independent of the phase in which they reside. The liquid molecular environment is still different from the gas molecular environment because liquid molecules are confined to move primarily within a much more crowded environment where the potential energies are more significant. Unless a molecule's kinetic energy is sufficient to escape the potential energy, it simply collides with a higher frequency in its local environment. What happens if the temperature is raised such that the liquid molecules can escape the potential energies of its neighbors? We call that "boiling," and the pressure must increase to keep the system in vapor-liquid phase equilibrium. The energy required to promote "boiling" is related to the heat of vaporization. Now you can begin to understand what temperature is and how it relates to other important thermodynamic properties.

For completeness, we may also mention that kinetic energy is the only form of energy for an ideal gas, so the internal energy of a monatomic ideal gas is given by:

$$\underline{U}^{ig} = \frac{Nm\langle u^2 \rangle}{2} = \frac{nN_A m\langle u^2 \rangle}{2} = \frac{3}{2}nRT \qquad \text{(ig) 1.18}$$

The proportionality constant between temperature and internal energy is known as the ideal gas heat capacity at constant volume, denoted C_V. Eqn 1.18 shows that $C_V = 1.5R$ for a monatomic ideal gas. If you refer to the tables of constant pressure heat capacities (C_P) on the end flap of the text and note that $C_P = C_V + R$, you may be surprised by how accurate this ultrasimplified theory actually is for species like helium, neon, and argon at 298 K.

Note that Eqn. 1.18 shows that $U^{ig} = U^{ig}(T)$. In other words, the internal energy depends only on the temperature for an ideal gas. The observation that $U^{ig} = U^{ig}(T)$ is true for any ideal gas, not only for ultrasimplified, monatomic ideal gases. We make the most of this fact in Chapter 5, where we show how to compute changes in energy for any fluid at any temperature and density by systematically correcting the relatively simple ideal gas result.

Real Fluids

The thermodynamic behavior of real fluids differs from the behavior of ideal gases in most cases. Real fluids condense, evaporate, freeze and melt. Characterization of the volume changes and energy changes of these processes is an important skill for the chemical engineer. Many real fluids do behave *as if they are* ideal gases at typical process conditions. Application of the ideal gas law simplifies many process calculations for common gases, *e.g.*, air at room temperature and moderate pressures. However you must always remember that the ideal gas law is an approximation (sometimes an excellent approximation) that must be applied carefully to any fluid. *P-V* behavior of a real fluid (water) and an ideal gas can be compared in Figs. 1.3 and 1.4. The behaviors are presented along *isotherms* (lines of constant temperature) and the deviations from the ideal gas law for water are obvious. Water is one of the most common substances that we work with, and water vapor behaves nearly as an ideal gas at 100°C ($P^{sat} = 0.1014$ MPa), where experimentally the vapor volume is 1.6718 m^3/kg (30,092 cm^3/mol) and by the ideal gas law we may calculate $V = RT/P = 8.314 \cdot 373.15 / 0.1014 = 30,595$ cm^3/mol. However, the state is the normal boiling point, and we are well aware that a liquid phase can co-exist at this state. This is because there is another density of water at these conditions that is also stable.[1]

We will frequently find it convenient to work mathematically in terms of molar density or mass density, which is inversely related to molar volume or mass volume, $\rho = 1/V$. Plotting the isotherms in terms of density yields a *P-ρ* diagram that qualitatively looks like the mirror image of the *P-V* diagram. Density is convenient to use because it always stays finite as $P \rightarrow 0$, whereas *V* diverges. Examples of *P-ρ* diagrams are shown in Fig. 6.1 on page 195.

The conditions where two phases coexist are called *saturation conditions*. The terms *saturation pressure* and *saturation temperature* are used to refer to the state. The volume (or density) is called the saturated volume (or saturated density). Saturation conditions are shown in Fig. 1.4 as the "hump" on the diagram. The hump is called the *phase envelope*. Two phases coexist when the system conditions result in a state *inside* or *on* the envelope. The horizontal lines *inside* the curves are called *tie* lines that show the two volumes (saturated liquid and saturated vapor) that can coexist. The curve labeled "sat'd liquid" is also called the *bubble* line, since it represents conditions where boiling (bubbles) can occur in the liquid. The curve labeled "sat'd vapor" is also called a *dew* line, since it is the condition where droplets (dew) can occur in the vapor. Therefore, *saturation* is a term that can refer to either *bubble* or *dew* conditions. When the total volume of a system results in a system state *on* the saturated vapor line, only an infinitesimal quantity of liquid exists, and the state is indicated by the term *saturated vapor*. Likewise, when a system state is *on* the saturated liquid line, only an infinitesimal quantity of vapor exists, and the state is indicated by term *saturated liquid*. When the total volume of the system results in a system *in between* the saturation vapor and saturation liquid volumes, the system will have vapor and liquid phases coexisting, each phase occupying a finite fraction of the overall system. Note that each isotherm has a unique saturation pressure.

🛑 Real fluids have saturation conditions, bubble points, and dew points.

---

1. This stability is determined by the Gibbs energy and we will defer proof until Chapter 8.

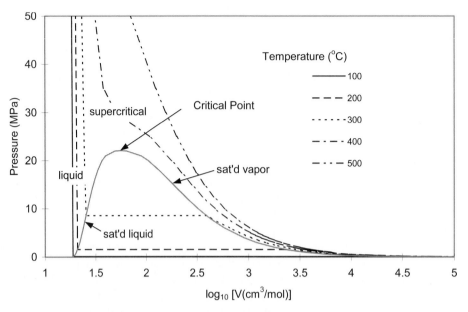

Figure 1.4 *P-V-T behavior of water at the same temperatures used in Fig. 1.3. The plot is prepared from the steam tables in Appendix E.*

This pressure is known as the *saturation* or *vapor pressure*. Although the vapor pressure is often used to characterize a pure liquid's bubble point, recognize that it also represents the dew point for the pure vapor.

Following an isotherm from the right side of the diagram along a path of decreasing volume, the isotherm starts in the vapor region, and the pressure rises as the vapor is isothermally compressed. As the volume reaches the saturation curve at the vapor pressure, a liquid phase begins to form. Notice that further volume decreases do not result in a pressure change until the system reaches the saturated liquid volume, after which further decreases in volume require extremely large pressure changes. Therefore, liquids are often treated as *incompressible* even though the isotherms really do have a finite rather than infinite slope. The accuracy of the incompressible assumption varies with the particular application.

❶ Liquids are quite incompressible.

As we work process problems, we will need to use properties such as the internal energy of a fluid.[1] Properties such as these are available for most common fluids in terms of a table or chart. For steam, both tables and charts are commonly used, and in this section we will introduce the steam tables available in Appendix E.

Steam Tables

When dealing with water, some conventions have developed for referring to the states which can be confusing if the terms are not clearly understood. *Steam* refers to a vapor state, and *saturated steam* is vapor at the dew point. For water, in the two-phase region, the term *wet steam* is used to indicate a vapor + liquid system.

---

1. Calculation of these properties requires mastery of several fundamental concepts as well as application of calculus and will be deferred. We calculate energies for ideal gas in Chapter 2 and for real fluids in Chapter 7.

Steam tables are divided into four tables. The first table presents *saturation* conditions indexed by temperature. This table is most convenient to use when the temperature is known. Each row lists the corresponding saturation values for pressure (vapor pressure), internal energy, volume and two other properties we will use later in the text: enthalpy and entropy. Special columns represent the energy, enthalpy, and entropy of vaporization. These properties are tabulated for convenience although they can be easily calculated by the difference between the saturated vapor value and the saturated liquid value. Notice that the vaporization values decrease as the saturation temperature and pressure increase. The vapor and liquid phases are becoming more similar as the saturation curve is followed to higher temperatures and pressures. At the *critical point*, the phases become identical. Notice in Fig. 1.4 that the two phases become identical at the highest temperature and pressure on the saturation curve, so this is the critical point. For a pure fluid, the critical temperature is the temperature at which vapor and liquid phases are identical on the saturation curve, and is given the notation T_c. The pressure at which this occurs is called the critical pressure, and is given the symbol P_c.

❶ The critical temperature and critical pressure are key characteristic properties of a fluid.

The second steam table organizes saturation properties indexed by pressure, so it is easiest to use when the pressure is known. Like the temperature table, vaporization values are presented. The table duplicates the saturated temperature table, i.e. plotting the saturated volumes from the two tables would result in the same curves. The third steam table is the largest portion of the steam tables, consisting of superheated steam values. *Superheated* steam is vapor above its saturation temperature at the given pressure. The adjective *superheated* specifies that the vapor is above the saturation temperature at the system pressure. The adjective is usually used only where necessary for clarity. The difference between the system temperature and the saturation temperature, $(T - T^{sat})$, is termed the *degrees of superheat*. The superheated steam tables are indexed by pressure and temperature. The saturation temperature is provided at the top of each pressure table so that the superheat may be quickly determined without referring to the saturation tables.

❶ Superheat.

The fourth steam table has liquid-phase property values at temperatures below the critical temperature and above each corresponding vapor pressure. Liquid at these states is sometimes called *subcooled* liquid to indicate that the temperature is below the saturation temperature for the specified pressure. Another common way to describe these states is to identify the system as *compressed* liquid, which indicates that the pressure is above the saturation pressure at the specified temperature. The adjectives *subcooled* and *compressed* are usually only used where necessary for clarity. Notice by scanning the table that pressure has a small effect on the volume and internal energy of liquid water. By looking at the saturation conditions together with the general behavior of Fig. 1.4 in our minds, we can determine the state of aggregation (vapor, liquid or mixture) for a particular state.

❶ Subcooled, compressed.

Example 1.2   Introduction to steam tables

For the following states, specify if water exists as vapor, liquid or a mixture: (a) 110°C and 0.12 MPa; (b) 200°C and 2 MPa; (c) 0.8926 MPa and 175°C.

Solution:

(a) Looking in the saturation temperature table, the saturation pressure at 110°C is 0.143 MPa. Below this pressure, water is vapor (steam).

(b) From the saturation temperature table, the saturation pressure is 1.5549 MPa, therefore water is liquid.

(c) This is a saturation state listed in the saturation temperature table. The water exists as saturated liquid, saturated vapor, or a mixture.

Linear Interpolation

Since the information in the steam tables is tabular, we must interpolate to find values at states that are not listed. To interpolate, we assume the property we desire (e.g., volume, internal energy) varies linearly with the independent variables specified (e.g., pressure, temperature). The assumption of linearity is almost always an approximation, but is a close estimate if the interval of the calculation is small. Suppose we seek the value of volume, V, at pressure, P, and temperature, T, but the steam tables have only values of volume at P_1 and P_2 which straddle the desired pressure value as shown in Fig. 1.5. The two points represent values available in the tables and the solid line represents the true behavior. The dotted line represents a linear fit to the tabulated points.

Linear interpolation is a necessary skill for problem solving using thermodynamic tables.

If we fit a linear segment to the tabulated points, the equation form is $y = mx + b$, where y is the dependent variable (volume in this case), x is the independent variable (pressure in this case), m is the slope $m = \dfrac{\Delta y}{\Delta x} = \dfrac{V_2 - V_1}{P_2 - P_1}$, and b is the intercept. We can interpolate to find V without directly determining the intercept. Since the point we desire to calculate is also on the line with slope m, it also satisfies the equation $m = \dfrac{\Delta y}{\Delta x} = \dfrac{V - V_1}{P - P_1}$. We can equate the two expressions for m to find the interpolated value of V at P.

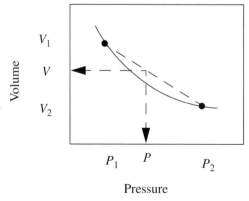

Figure 1.5 *Illustration of linear interpolation.*

There are two quick ways to think about the interpolation. First, since the interpolation is linear, the fractional change in V relative to the volume interval is equal to the fractional change in P relative to the pressure interval. In terms of variables:

$$\frac{V - V_1}{V_2 - V_1} = \frac{P - P_1}{P_2 - P_1}$$

For example, $(V - V_1)$ is 10% of the volume interval $(V_2 - V_1)$, when $(P - P_1)$ is 10% of $(P_2 - P_1)$. We can rearrange this expression to find:

$$V = V_1 + \frac{P - P_1}{P_2 - P_1}(V_2 - V_1) \qquad 1.19$$

If we consider state "1" as the base state, we can think of this expression in words as

$$V = \text{base } V + (\text{fractional change in } P) \cdot (\text{volume interval size})$$

Another way to think of Eqn. 1.19 is by arranging it as:

$$V = V_1 + \frac{V_2 - V_1}{P_2 - P_1}(P - P_1) \qquad 1.20$$

which in words is

$$V = \text{base } V + \text{slope} \cdot (\text{change in } P \text{ from base state})$$

Note that subscripts for 1 and 2 can be interchanged in any of the formulas if desired, provided that *all* subscripts are interchanged.

Example 1.3 Interpolation

Find the volume and internal energy for water at: (a) 5 MPa and 325°C, (b) 5 MPa and 269°C.

Solution:

(a) Looking at the superheated steam table at 5 MPa, we find the saturation temperature in the column heading as 263.9°C; therefore, the state is superheated. Values are available at 300°C and 350°C. Since we are halfway in the temperature interval, by interpolation the desired U and V will also be halfway in their respective intervals (which may be found by the average values):

$$U = (2699.0 + 2809.5)/2 = 2754.3 \text{ kJ/kg}$$

$$V = (0.0453 + 0.0520)/2 = 0.0487 \text{ m}^3\text{/kg}$$

(b) For this state, we are between the saturation temperature (263.9°C) and 300°C, and we apply the interpolation formula

$$U = 2597.0 + \frac{269 - 263.9}{300 - 263.9}(2699.0 - 2597.0) = 2611.4 \text{ kJ/kg}$$

$$V = 0.0394 + \frac{269 - 263.9}{300 - 263.9}(0.0453 - 0.0394) = 0.0402 \text{ m}^3\text{/kg}$$

Double Interpolation

Occasionally, we must perform double or multiple interpolation to find values. The following example illustrates these techniques

Example 1.4 Double interpolation

For water at 160°C and 0.12 MPa, find the internal energy.

Solution: By looking at the saturation tables at 160°C, water is below the saturation pressure, and will exist as superheated vapor, but superheated values at 0.12 MPa are not tabulated in the superheated table. If we tabulate the available values we find

	0.1 MPa	0.12 MPa	0.2 MPa
150°C	2582.9		2577.1
160°C			
200°C	2658.2		2654.6

We may either interpolate the first and third columns to find the values at 160°C, followed by an interpolation in the second row at 160°C, or we may interpolate the first and third rows, followed by the second column. The values found by the two techniques will not be identical because of the non-linearities of the properties we are interpolating. Generally, the more accurate interpolation should be done first, which is over the smaller change in U, which is the pressure interpolation. The pressure increment is 20% of the pressure interval $[(0.12 - 0.1)/(0.2 - 0.1)]$; therefore, interpolating in the first row

$$U = 2582.9 + 0.2 \cdot (2577.1 - 2582.9) = 2581.7 \text{ kJ/kg}$$

and in the third row

$$U = 2658.2 + 0.2 \cdot (2654.6 - 2658.2) = 2657.5 \text{ kJ/kg}$$

and then interpolating between these values, using the value at 150°C as the base value, then

$$U = 2581.7 + \frac{1}{5} \cdot (2657.5 - 2581.7) = 2596.9 \text{ kJ/kg}$$

The final results are tabulated in the shaded cells in the table:

	0.1 MPa	0.12 MPa	0.2 MPa
150°C	2582.9	2581.7	2577.1
160°C		2596.9	
200°C	2658.2	2657.5	2654.5

We also may need to interpolate between values in different tables as shown in the following example.

Example 1.5 Double interpolation using different tables

Find the internal energy for water at 0.12 MPa and 110°C.

Solution: We found in Example 1.2 on page 20 that this is a superheated state. From the superheated table we can interpolate to find the internal energy at 110°C and 0.1 MPa

$$U = 2506.2 + \frac{1}{5}\cdot(2582.9 - 2506.2) = 2521.5 \text{ kJ/kg}$$

At 0.2 MPa, 110°C is not found in the superheated table because the saturation temperature is 120.3°C so the values at this pressure cannot be used. Therefore, we can find the desired internal energy by interpolation using the value above and the saturation value at 110°C and 0.143 MPa from the saturation temperature table.

$$U = 2521.5 + \frac{0.12 - 0.1}{0.143 - 0.1}(2517.7 - 2521.5) = 2519.7 \text{ kJ/kg}$$

Interpolation Program

Please note that an interpolation program is included for HP calculators as described in Appendix P. Occasionally, interpolation must be performed when the T and P are both unknown. Computers or spreadsheets can be helpful as shown in the next example.

<table>
<tr><td colspan="2">

Example 1.6 Double interpolation using Excel

</td></tr>
<tr><td>

</td><td>

Steam undergoes a series of state changes and is at a final state where $U = 2650$ kJ/kg and $V = 0.185$ m³/kg. Find the T and P.

Solution: Scanning the steam tables, the final state is in the range 1.0 MPa $< P <$ 1.2 MPa, 200°C $< T <$ 250°C. The final state requires a double interpolation using U and V. One easy method is to set up the table in Excel. In each of the tables below, the pressure interpolation is performed first in the top and bottom rows, dependent on the pressure variable in the top of the center column, which can be set at any intermediate pressure to start. The temperature interpolation is then entered in the center cell of each table using the temperature variable. The formulas in both tables reference a common temperature variable cell and a common pressure variable cell. Solver is started and T and P are adjusted to make $U = 2650$ kJ/kg subject to the constraint $V = 0.185$ m³/kg. (See Appendix B for Solver instructions.) The converged result is shown at $T = 219.6$°C and $P = 1.17$ MPa.

</td></tr>
</table>

		P^f	
U(kJ/kg) table	P = 1 MPa	1.164752	P = 1.2 MPa
T = 200°C	2622.2	2614.539	2612.9
T^f 219.4486791		2650	
T = 250°C	2710.4	2705.705	2704.7

V(m³/kg) table	P = 1 MPa		P = 1.2 MPa
T = 200°C	0.2060	0.175768	0.1693
		0.185	
T = 250°C	0.2327	0.199502	0.1924

Extrapolation

Occasionally the values we seek are not conveniently between points in the table and we can apply the "interpolation" formulas to extrapolate as shown in Fig. 1.6. In this case, T lies outside the interval. Extrapolation is much less certain than interpolation since we frequently do not know "what curve lies beyond" that we may miss by linear approximation. The formulas used for extrapolation are identical to those used for interpolation. With the steam tables, extrapolation is generally not necessary at normal process conditions and should be avoided if possible.

Phase Equilibrium and Quality

Along the saturation curve in Fig. 1.4 on page 18, there is just one degree of freedom ($F = C - P + 2 = 1 - 2 + 2 = 1$). If we seek saturation, we may choose either a T^{sat} or a P^{sat}, and the other is determined. The vapor pressure increases rapidly with temperature as shown in Fig. 1.7 on page 26. A plot of $\ln P^{sat}$ vs. $\dfrac{1}{T^{sat}}$ is nearly linear and over large intervals, so for accurate interpolations, vapor

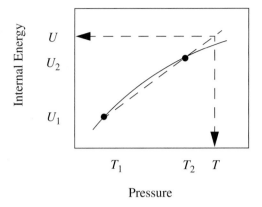

Figure 1.6 *Illustration of linear extrapolation.*

pressure data should be converted to this form before interpolation. However, the steam tables used with this text have small enough intervals that direct interpolation can be applied to P^{sat} and T^{sat} without appreciable error.

The saturation volume values of the steam tables were used to generate the phase diagram of Fig. 1.4 on page 18. Note that as the critical point is approached, the saturation vapor and liquid values approach each other. The same occurs for internal energy and two properties that will be used in upcoming chapters, enthalpy, H, and entropy, S. When a mixture of two phases exists, we must characterize the fraction that is vapor, since the vapor and liquid property values differ significantly.

The mass percentage that is vapor is called the *quality* and given the symbol q. The properties $V, U, H,$ or $S,$ may be represented with a generic variable M. The *overall* value of the state variable M is

> Quality is the vapor mass percentage of a vapor/liquid mixture.

$$M = (1 - q)\, M^L + q M^V \qquad\qquad 1.21$$

which may be rearranged

$$M = M^L + q(M^V - M^L)$$

but $(M^V - M^L)$ is just ΔM^{vap} and for internal energy, enthalpy, and entropy, it is tabulated in columns of the saturation tables. The value of overall M is

$$M = M^L + q\Delta M^{vap} \qquad\qquad 1.22$$

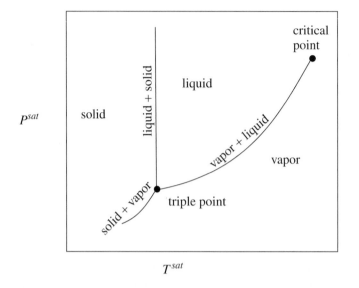

Figure 1.7 *P-T representation of real fluid behavior. Note that only*
vapor and liquid behavior is shown in Fig. 1.4 on page 18.

Example 1.7 Quality calculations

Two kg of water coexists as vapor and liquid at 280°C in a 0.05 m³ rigid container. What is the
pressure, quality, and overall internal energy of the mixture?

Solution:

The overall mass volume is $V = 0.05$ m³/2 kg = 0.025 m³/kg. From the saturation temperature
table, the pressure is 6.417 MPa. Using the saturation volumes at this condition to find q,

$$0.025 = 0.001333 + q\,(0.0302 - 0.0013) \text{ m}^3/\text{kg}$$

which leads to $q = 0.82$. The overall internal energy is

$$U = 1228.33 + 0.82 \cdot 1358.1 = 2342 \text{ kJ/kg}$$

Example 1.8   Constant volume cooling

Steam is initially contained in a rigid cylinder at $P = 30$ MPa and $V = 10^{2.498}$cm³/mole (as shown on Fig. 1.8). The cylinder is allowed to cool to 300°C. What is the pressure, quality, and overall internal energy of the final mixture?

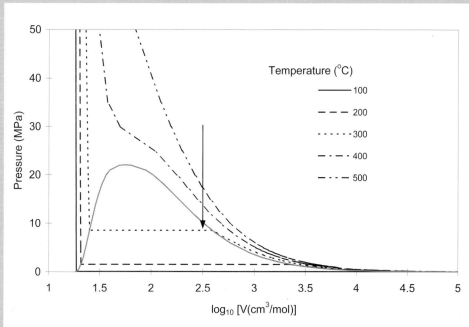

Figure 1.8 *P-V-T behavior of water illustrating a quality calculation.*

Solution:

The overall mass volume is $V = 10^{2.498}$cm³-mole$^{-1}\cdot10^{-6}$(m³/cm³)/(18.02E-3kg/mole) = 0.01747 m³/kg. From the superheated steam table at 30 MPa, the initial temperature is 900°C. When the cylinder is cooled to 300°C, you should notice that there is no pressure in the superheated steam tables that provides a volume of $V = 0.01747$ m³/kg. Look hard, they are all too large. (Imagine yourself looking for this on a test when you are in a hurry.) Now look in the saturated steam tables at 300°C. Notice that the saturated vapor volume is 0.0217 m³/kg. Since that is higher than the desired volume, but it is the lowest vapor volume at this temperature, we must conclude that our condition is somewhere between the saturated liquid and the saturated vapor at a pressure of 8.588 MPa. (When you are in a hurry, it is advisable to check the saturated tables first.)

Using the saturation volumes at 300°C condition to find q,

$$0.01747 = 0.001404 + q\,(0.0217 - 0.001404) \text{ m}^3/\text{kg}$$

which leads to $q = (0.01747 - 0.001404)/(0.0217 - 0.001404) = 0.792$. The overall internal energy is

$$U = 1332.95 + 0.792\cdot1230.67 = 2308 \text{ kJ/kg}$$

1.5   SUMMARY

Years from now you may have some difficulty recalling the details presented in this text. That is nothing to be embarrassed about. On the other hand, the two basic premises outlined in this introductory section are so fundamental to technically educated people that you really should commit them to long-term memory as soon as possible. Formally, we may state our two basic premises as the first and second "laws" of thermodynamics.[1]

First Law: Overall energy is conserved (you can't get something for nothing).

Second Law: Overall entropy changes are greater than or equal to zero (generation of disorder results in lost work).

The first law is further developed in Chapter 2. The concepts of entropy and the second law are developed in Chapters 3 and 4. The exact overall relationship of our two basic premises to these two laws may not become apparent until some time later in this text, but you can refer back to appropriate sections of the text at any time. What you should concentrate on is remembering the basic premises and where to look for the details when the need arises. There are times when the endeavor to apply these simple tasks seems daunting, but the answer appears simple in retrospect, once obtained. By practicing adaptation of the basic principles to many specific problems, you will slowly grasp the appropriate connection between the basic premises and finding the details. Try not to be distracted by the vocabulary or the tedious notation that goes into keeping all the coupled systems classified in textbook fashion. Keep in mind that other students have passed through this and found the detailed analysis to be worth the effort.

> *A theory is the more impressive the greater the simplicity of its premises is, the more different kinds of things it relates, and the more extended is its area of applicability. Therefore the deep impression which classical thermodynamics made upon me.*
>
> *Albert Einstein*

Test Yourself

1. Draw a sketch of the force model implied by the square-well potential, indicating the position(s) where the force between two atoms is zero and the positions where it is nonzero.

2. Explain in words how the pressure of a fluid against the walls of its container is related to the velocity of the molecules.

3. What is it about molecules that requires us to add heat to convert liquids to gases?

4. If the kinetic energy of pure liquid and vapor molecules at phase equilibrium must be the same, and the internal energy of a system is the sum of the kinetic and potential energies, what does this say about the intensive internal energy of a liquid phase compared with the intensive internal energy of the gas phase?

5. Explain the terms "energy," "potential energy," "kinetic energy," "internal energy."

---

1. There is also a "third law" of thermodynamics, as discussed by Denbigh, K., *The Principles of Chemical Equilibrium*, 4th ed., p. 416, Cambridge University Press, London, 1981. We will not be using the third law in this introductory text, however.

6. How is the internal energy of a substance related to the intermolecular pair potentials of the molecules?

7. Are T and P intensive properties? Name two intensive properties and two extensive properties.

8. How many degrees of freedom exist when a pure substance coexists as a liquid and gas?

9. Can an ideal gas condense? Can real fluids that follow the ideal gas law condense?

10. Give examples of bubble, dew, saturation, and superheated conditions. Explain what is meant when wet steam has a quality of 25%.

11. Create and solve a problem that requires double interpolation.

1.6   HOMEWORK PROBLEMS

> *Note: Some of the steam table homework problems involve enthalpy, H, which is defined for convenience using properties discussed in this chapter, $H \equiv U + PV$. The enthalpy calculations can be performed by reading the tabulated enthalpy values from the tables in an analogous manner used for internal energy.*

1.1 In each of the following, sketch your estimates of the intermolecular potentials between the given molecules and their mixture on the same pair of axes.

   (a) Chloroform is about 20% larger than acetone and about 10% stickier, but chloroform and acetone stick to each other much more strongly than they stick to themselves.
   (b) You have probably heard that "oil and water don't mix." What does that mean in molecular terms? Let's assume that oil can be characterized as benzene and that benzene is four times larger than water, but water is 10% stickier than benzene. If the ε_{12} parameter is practically zero, that would represent that the benzene and water stick to themselves more strongly than to each other. Sketch this.

1.2 For each of the states below, calculate the number of moles of ideal gas held in a three liter container.

   (a) $T = 673$ K, $P = 2$ MPa
   (b) $T = 500$ K, $P = 0.7$ MPa
   (c) $T = 450$ K, $P = 1.5$ MPa

1.3 A 5 m^3 tank farm gas storage tank contains methane. The initial temperature and pressure are $P = 1$ bar, $T = 18°C$. Calculate the T and P following each of the successive steps using the ideal gas law.

   (a) 1 m^3 (at standard conditions) is withdrawn isothermally.
   (b) The sun warms the tank to 40°C.
   (c) 1.2 m^3 (at standard conditions) is added to the tank and the final temperature is 35°C.
   (d) The tank cools overnight to 18°C.

1.4 A 5 m^3 outdoor gas storage tank warms from 10°C to 40°C on a sunny day. If the initial pressure was 0.12 MPa at 10°C, what is the pressure at 40°C, and how many moles of gas are in the tank? Use the ideal gas law.

1.5 An automobile tire has a pressure of 255 kPa (gauge) in the summer when the tire temperature after driving is 50°C. What is the wintertime pressure of the same tire at 0°C if the volume of the tire is considered the same and there are no leaks in the tire?

1.6 Calculate the mass density of the following gases at 298 K and 1 bar.

(a) nitrogen
(b) oxygen
(c) air (use average molecular weight)
(d) CO_2
(e) argon

1.7 Calculate the mass of air (in kg) that is contained in a classroom that is 12m x 7m x 3m at 293 K and 0.1 MPa.

1.8 Five grams of the specified pure solvent is placed in a variable volume piston. What is the volume of the pure system when 50% and 75% have been evaporated at: (*i*) 30°C, (*ii*) 50°C? Use the Antoine Equation (Appendix E) to relate the saturation temperature and saturation pressure. Use the ideal gas law to model the vapor phase. Show that the volume of the system occupied by liquid is negligible compared to the volume occupied by vapor.

(a) hexane ($\rho^L = 0.66$ g/cm^3)
(b) benzene ($\rho^L = 0.88$ g/cm^3)
(c) ethanol ($\rho^L = 0.79$ g/cm^3)
(d) water without using the steam tables ($\rho^L = 1$ g/cm^3)
(e) water using the steam tables

1.9 A gasoline spill is approximately 4 liters of liquid. What volume of vapor is created at 1 bar and 293 K when the liquid evaporates? The density of regular gasoline can be estimated by treating it as pure isooctane (2,2,4-trimethylpentane $\rho^L = 0.692$ g/cm^3) at 298 K and 1 bar.

1.10 LPG is a useful fuel in rural locations without natural gas pipelines. A leak during the filling of a tank can be extremely dangerous because the vapor is more dense than air and drifts to low elevations before dispersing, creating an explosion hazard. What volume of vapor is created by a leak of 40L of LPG? Model the liquid before leaking as propane with $\rho^L = 0.24$ g/cm^3. What is the mass density of pure vapor propane after depressurization to 293 K and 1 bar? Compare with the mass density of air at the same conditions.

1.11 The gross lifting force of a balloon is given by $(\rho_{air} - \rho_{gas})V_{ballon}$. What is the gross lifting force (in kg) of a hot air balloon of volume 1.5E6 L, if the balloon contains gas at 100°C and 1 atm? The hot gas is assumed to have an average molecular weight of 32 due to carbon dioxide from combustion. The surrounding air has an average molecular weight of 29 and is at 25°C and 1 atm.

1.12 The gas phase reaction A → 2R is conducted in a 0.1-m^3 spherical tank. The initial temperature and pressure in the tank are 0.05 MPa and 400 K. After species A is 50% reacted, the temperature has fallen to 350 K. What is the pressure in the vessel?

1.13 A gas stream entering an absorber is 20 mol% CO_2 and 80 mol% air. The flowrate is 1 m^3/min at 1 bar and 360 K. When the gas stream exits the absorber, 98% of the incoming CO_2 has been absorbed into a flowing liquid amine stream.

(a) What are the gas stream mass flowrates on the inlet and outlets in g/min?
(b) What is the volumetric flowrate on the gas outlet of the absorber if the stream is at 320 K and 1 bar?

1.14 A permeation membrane separates an inlet air stream, F, (79 mol% N_2, 21 mol% O_2) into a permeate stream, M, and a reject stream, J. The inlet stream conditions are 293 K, 0.5 MPa, and 2 mol/min; the conditions for both outlet streams are 293 K and 0.1 MPa. If the permeate stream is 50 mol% O_2, and the reject stream is 13 mol% O_2, what are the volumetric flowrates (L/min) of the two outlet streams?

1.15 For water at each of the following states, determine the internal energy and enthalpy using the steam tables.

$T(°C)$	$P(MPa)$
(a) 100	0.01
(b) 550	6.25
(c) 475	7.5
(d) 180	0.7

1.16 (a) What size vessel holds 2 kg water at 80°C such that 70% is vapor? What are the pressure and internal energy?
   (b) A 1.6-m^3 vessel holds 2 kg water at 0.2 MPa. What are the quality, temperature, and internal energy?

1.17 Determine the temperature, volume, and quality for one kg of water under the following conditions:

   (a) $U = 3000$ kJ/kg, $P = 0.3$ MPa
   (b) $U = 2900$ kJ/kg, $P = 1.7$ MPa
   (c) $U = 2500$ kJ/kg, $P = 0.3$ MPa
   (d) $U = 350$ kJ/kg, $P = 0.03$ MPa

1.18 Three kg of saturated liquid water are to be evaporated at 60°C.

   (a) At what pressure will this occur at equilibrium?
   (b) What is the initial volume?
   (c) What is the system volume when 2 kg have been evaporated? At this point, what is $\Delta \underline{U}$ relative to the initial state?
   (d) What are $\Delta \underline{H}$ and $\Delta \underline{U}$ relative to the initial state for the process when all three kg have been evaporated?
   (e) Make a qualitative sketch of parts (b) through (d) on a P-V diagram, showing the phase envelope.

1.19 Two kg of water exist initially as a vapor and liquid at 90°C in a rigid container of volume 2.42 m^3.

   (a) At what pressure is the system?
   (b) What is the quality of the system?
   (c) The temperature of the container is raised to 100°C. What is the quality of the system, and what is the pressure? What are $\Delta \underline{H}$ and $\Delta \underline{U}$ at this point relative to the initial state?
   (d) As the temperature is increased, at what temperature and pressure does the container contain only saturated vapor? What is $\Delta \underline{H}$ and $\Delta \underline{U}$ at this point relative to the initial state?
   (e) Make a qualitative sketch of parts (a) through (d) on a P-V diagram, showing the phase envelope.

1.20 A molecular simulation sounds like an exceedingly advanced subject, but it is really quite simple for purely repulsive bodies.[1] To illustrate, write a simple program that computes the time to collision for four purely repulsive disks bouncing in two dimensions around a square box. Let the diameters of your disks, σ, be 0.3 nm, masses be 14 g/mole, and the length of the square box, L, be 0.8 nm. Start the four disks at (0.25L,0.25L) (0.75L,0.25L), (0.25L,0.75L), (0.75L,0.75L) and with initial velocities of $(u,0)$, $(0,u)$, $(u/2^{1/2},-u/2^{1/2})$, $(-u/2^{1/2},-u/2^{1/2})$.

(a) Compute u initially from Eqn. 1.17 assuming a temperature of 298 K.

(b) Write a vector formula for computing the center to center distance between two disks given their velocities, u, and their positions, r_0, at a given time, t_0. Write a similar formula for computing the distance of each disk from each wall.

(c) Noting that energy and momentum must be conserved during a collision, write a vector formula for the changes in velocity of two disks after collision. Write a similar formula for the change in velocity of a disk colliding with a wall.

(d) Each step of your program should compute the time to collision for each disk with each wall, and for disk 1 with disks 2 and 3 and for disk 2 with disk 3. Then find the collision which occurs first and increment the positions of all the disks according to their velocities and the time to collision (Note: $r - r_0 = u(t - t_0)$). Collisions occur when the distance between two disks is equal to one disk diameter.

(e) Compute the directions of the disk(s) after collision then repeat the process for one million collisions.

(f) Compute the "pressure" by accumulating over time the momentum changes per unit length for one of the walls and dividing by the total time. This formula for "pressure" assumes that there is only a single plane where molecules exist, therefore, it is really a tension (τ = force per unit length). What value do you obtain for the quantity $Z = \tau L^2/RT$?

1. Alder, B.J. and Wainwright, T.E., *J. Chem. Phys.*, 31:459 (1959).

CHAPTER

THE ENERGY BALANCE

When you can measure what you are speaking about, and express it in numbers, you know something about it. When you can't measure it, your knowledge is meager and unsatisfactory.

Lord Kelvin

The energy balance is based on the postulate of conservation of energy in the universe, which is known as the "First Law of Thermodynamics." It is a "law" in the same sense as Newton's laws. It is not refuted by experimental observations within a broadly defined range of conditions, but there is no mathematical proof of its validity. Derived from experimental observation, it quantitatively accounts for energy transformations (heat, work, kinetic, potential) and we take the first law as a starting point, a postulate. Facility with computation of energy transformations is a necessary step in developing an understanding of elementary thermodynamics. The first law relates work, heat, and flow to the internal energy, kinetic energy, and potential energy of the system. Therefore, we precede the introduction of the first law with discussion of work and heat.

❶ The energy balance is also known as the *first law of thermodynamics.*

2.1 EXPANSION/CONTRACTION WORK

There is a simple way that a force on a surface may interact with the system to cause expansion/contraction of the system in volume. This is the type of surface interaction which occurs if we release the latch of a piston, and move the piston in/out while holding the cylinder in a fixed location. Note that a moving boundary is not sufficient to distinguish this type of work—there must be movement of the system boundaries *relative to each other*. For expansion/contraction interactions, the size of the system *must change*. This distinction becomes significant when we contrast expansion/contraction work to flow work in Section 2.3.

How can we relate this amount of work to other quantities that are easily measured, like volume and pressure? For a force applied in the *x* direction, the work done on our system is

$$d\underline{W} = F_{applied}\,dx = -F_{system}\,dx$$

where we have used Newton's principle of equal and opposite forces acting on a boundary to relate the applied and system forces. Since it is more convenient to use the system force in calculations, we use the latter form, and drop the subscript with the understanding that we are calculating the work done on the system and basing the calculation on the system force. For a constant force, we may write

$$\underline{W} = -F\Delta x$$

If F is changing as a function of x then we must use an integral of F,

$$\underline{W} = -\int F dx \qquad\qquad 2.1$$

For a fluid acting on a surface of area A, the system force and pressure are related,

$$P = F/A \Rightarrow F = P \cdot A$$

$$\underline{W}_{EC} = -\int PAdx = -\int Pd\underline{V} \qquad\qquad 2.2$$

where the subscript EC refers to expansion/contraction work.

In evaluating this expression, a difficult question of perspective comes up. It would be a trivial question except that it causes major headaches when we later try to keep track of positive and negative signs. The question is essentially this: in the discussion above, is positive work being done on the system, or is negative work being done by the system? When we put heat into the system, we consider it a positive input into the system; therefore, putting work into the system should also be considered as a positive input. On the other hand, the change in volume during compression is negative. The problem is that both perspectives are equally valid—therefore, the choice is arbitrary. Since different people choose differently, there is always confusion about sign conventions. The best we can hope for is to be consistent during our own discussions. We hereby consider work to be positive when performed *on* the system. Thus, energy put into the system is positive. Therefore,

$$\boxed{\underline{W}_{EC} = -\int P \, d\underline{V}} \qquad\qquad 2.3$$

> **⊘** Expansion/Contraction work is associated with a change in *system* size.

where P and $\underline{V}$ are of the *system*. By comparing Eqn. 2.3 with the definitions of work given by Eqns. 2.1 and 2.2, it should be obvious that the $d\underline{V}$ term results from expansion/contraction of the boundary of the system. The P results from the force of the system acting at the boundary. Therefore to use Eqn. 2.3, the pressure in the integral is the pressure *of the system* at the boundary, and the boundary must move. A system which does not have an expanding/contracting boundary does not have expansion/contraction work.[1]

2.2   SHAFT WORK

In a flowing system, we know that a propeller-type device can be used to push a fluid through pipes—this is the basis of a centrifugal pump. Also, a fluid flowing through a pump-like device could cause movement of a shaft—this is the basis for hydroelectric power generation and the water

---

1. Some texts refer to expansion/contraction work as *PV* work. This leads to confusion since we will show in Section 2.3 on page 37 that work associated with flow is *PV*, and the types of work are distinctly different. We have chosen to use the term "expansion/contraction" for work involved in moving boundaries to help avoid this ambiguity.

wheels that powered grain-processing plants in the U.S. in the early 20th century. These are the most commonly encountered forms of shaft work in thermodynamics, but there is another slight variation. Suppose an impeller was inserted into a cylinder containing cookie batter and stirred while holding the piston at a fixed volume. We would be putting work into the cylinder, but the system boundaries would neither expand nor contract. All of these cases exemplify shaft work. The essential feature of shaft work is that work is being added or removed without a change in volume of the system.

Shaft work characterizes the work of turbine or pump.

2.3   WORK ASSOCIATED WITH FLOW

In engineering applications, most problems involve flowing systems. This means that materials typically flow into a piece of equipment then flow out of it, crossing well-defined system boundaries in the process. Thus, we need to introduce an additional characterization of work: the work interaction of the system and surroundings when mass crosses a boundary. For example, when a gas is released out of a tank through a valve, the exiting gas pushes the surrounding fluid, doing work on the surroundings. Likewise, when a tank valve is opened to allow gas from a higher pressure source to flow inward, the surroundings do work on the gas already in the system. We calculate the work in these situations most easily by first calculating the rate at which work is done.

Let us first consider a fluid entering a system as shown in Fig. 2.1. We have $d\underline{W} = Fdx$, and the work interaction of the system is positive since we are pushing fluid into the system. The rate of work is $\underline{\dot{W}} = F\dot{x}$, but $\dot{x}$ is velocity, and $F = P \cdot A$. Further rearranging, recognizing $A\dot{x} = \underline{\dot{V}}$, and that the volumetric flowrate may be related to the mass specific volume and the mass flowrate, $\underline{\dot{V}} = V\dot{m}^{in}$

$$\underline{\dot{W}}_{flow} = PA\dot{x} = P\underline{\dot{V}} = PV\dot{m}^{in} \qquad 2.4$$

Work associated with fluid flowing in/out of boundaries is called *flow* work.

where PV are the properties of the fluid at the point where it crosses the boundary, and $\dot{m}^{in}$ is the absolute value of the mass flowrate across the boundary. When fluid flows out of the system, work is done on the surroundings and the work interaction *of the system* is

$$\underline{\dot{W}}_{flow} = -PA\dot{x} = -P\underline{\dot{V}} = -PV\dot{m}^{out} \qquad 2.5$$

where $\dot{m}^{out}$ is the absolute value of the mass flow across the boundary, and since work is being done on the surroundings, the work interaction of the system is negative. When flow occurs both in and out, the net flow work is the difference:

$$\underline{\dot{W}}_{flow} = (PV)^{in}\dot{m}^{in} - (PV)^{out}\dot{m}^{out} \qquad 2.6$$

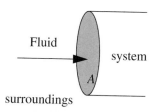

Figure 2.1 *Schematic illustration of flow work.*

where $\dot{m}^{in}$ and $\dot{m}^{out}$ are absolute values of the mass flowrates. For more streams, we simply follow the conventions established, and add inlet streams and subtract outlet streams.

2.4 LOST WORK VS. REVERSIBILITY

In order to properly understand the various characteristic forms that work may assume, we must address an issue which primarily belongs to the upcoming chapter on entropy. The problem is that the generation of disorder reflected by entropy change results in conversion of potentially useful work energy into practically useless thermal energy. If "generation of disorder results in lost work," then operating in a disorderly fashion results in the lost capability to perform useful work, which we abbreviate by the term: "lost work." It turns out that the most orderly manner of operating is a hypothetical process known as a reversible process. Typically, this hypothetical, reversible process is applied as an initial approximation of the real process, and then a correction factor is applied to estimate the results for the actual process. It was not mentioned in the discussion of expansion/contraction work, but we implicitly assumed that the process was performed reversibly, so that all of the work on the system was stored in a potentially useful form. To see that this might not always be the case, and how this observation relates to the term "reversible," consider the problem of stirring cookie batter. Does the cookie batter become unmixed if you stir in the reverse direction? Of course not. The shaft work of stirring has been degraded to effect the randomness of the ingredients. It is impossible to completely recover the work lost in the randomness of this irreversible process. Any real process involves some degree of stirring or mixing, so lost work cannot be eliminated, but we can hope to minimize unnecessary losses if we understand the issue properly.

 Real processes involve "lost work."

Consider a process involving gas enclosed in a piston and cylinder. Let the piston be oriented upward so that an expansion of the gas causes the piston to move upward. Suppose that the pressure in the piston is great enough to cause the piston to move upward when the latch is released. How can the process be carried out so that the expansion process yields the maximum work? First, we know that we must eliminate friction to obtain the maximum movement of the piston.

🖋 Friction results in "lost work."

Friction decreases the work available from a process. Frequently we neglect friction to perform a calculation of maximum work.

If we neglect friction, what will happen when we release the latch? The forces are not balanced. Let us take z as our coordinate in the vertical direction, with increasing values in the upward direction. The forces downward on the piston are the force of atmospheric pressure ($-P_{atm} \cdot A$, where A is the cross-section area of the piston) and the force of gravity ($-m \cdot g$). These forces will be constant throughout movement of the piston. The upward force is the force exerted by the gas ($P \cdot A$). Since the forces are not balanced, the piston will begin to accelerate upward ($F = ma$). It will continue to accelerate until the forces become balanced.[1] However, when the forces are balanced, the piston will have a non-zero velocity. As it continues to move up farther, the pressure inside the piston continues to fall, making the upward force due to the inside pressure smaller than the downward force. This causes the piston to decelerate until it eventually stops. However, when it stops at the top of the travel, it is still not in equilibrium because the forces are again not balanced. It begins to move downward. In fact *in the absence of dissipative mechanisms* we have set up a perpetual motion.[2] The piston would oscillate continuously between the initial state and the state at the top of

---

1. Two other possibilities exist: 1) The piston may hit a stop before it has finished moving upward, a case that will be considered below, or; 2) The piston may fly out of the cylinder if the cylinder is too short, and there is no stop.
2. However, this is not a useful perpetual motion machine because the net effect on the surroundings and the piston is zero at the end of each cycle. If we tried to utilize the motion, we would damp it out.

travel. This would not happen in a real system. One phenomenon which we have failed to consider is viscous dissipation (the effect of viscosity).

Let us consider how velocity gradients dissipate linear motion. Consider two *diatomic* molecules touching each other which both have exactly the same velocity and are traveling in exactly the same direction. Suppose that neither is rotating. They will continue to travel in this direction at the same velocity until they interact with an external body. Now consider the same two molecules in contact, again moving in exactly the same direction, but one moving slightly faster. Now there is a velocity gradient. Since they are touching each other, the fact that one is moving a little faster than the other causes one to begin to rotate clockwise and the other counter-clockwise because of friction as one tries to move faster than the other. Naturally, the kinetic energy of the molecules will stay constant, but the directional velocities are being converted to rotational (directionless) energies. This is an example of viscous dissipation in a shear situation. In the case of the oscillating piston, the viscous dissipation prevents complete transferral of the internal energy of the gas to the piston during expansion, resulting in a stroke that is shorter than a reversible stroke. During compression, viscous dissipation results in a fixed internal energy rise for a shorter stroke than a reversible process. In both expansion and compression, the temperature of the gas at the end of each stroke is higher than it would be for a reversible stroke, and each stroke becomes successively shorter.

> *Velocity gradients* lead to dissipation of directional motion (kinetic energy) into random motion (internal energy) due to the viscosity of a fluid. Frequently, we neglect viscous dissipation to calculate maximum work. A fluid would need to have zero viscosity for this mechanism of dissipation to be non-existent. *Pressure gradients* within a viscous fluid lead to velocity gradients; thus, one type of gradient is associated with the other.

❶ Velocity gradients in viscous fluids lead to lost work.

We can see that friction and viscosity play an important role in the loss of capability to perform useful work in real systems. In our example, these forces will cause the oscillations to decrease somewhat with each cycle until the piston comes to rest. Another possibility of motion which might occur with a piston is interaction with a stop, which limits the travel of the piston. As the piston travels upward, if it hits the stop, it will have kinetic energy which must be absorbed. In a real system, this kinetic energy is converted to internal energy of the piston, cylinder, and gas.

> Kinetic energy is dissipated to internal energy when objects collide inelastically, such as when a moving piston strikes a stop. Frequently we imagine systems where the cylinder and piston can neither absorb nor transmit heat; therefore, the lost kinetic energy is returned to the gas as internal energy.

❶ Inelastic collisions result in lost work.

So far, we have identified three dissipative mechanisms. Additional mechanisms are diffusion along a concentration gradient and heat conduction along a temperature gradient, which will be discussed in Chapter 3. Velocity, temperature, and concentration gradients are always associated with losses of work. If we could eliminate them, we could perform maximum work (but it would require infinite time).

> A process without dissipative losses is called *reversible*. A process is reversible if the system may be returned to a prior state by reversing the motion. We can usually determine that a system is not reversible by recognizing when dissipative mechanisms exist.

❶ A reversible process avoids lost work.

Approaching Reversibility

We can approach reversibility by eliminating gradients within our system. To do this, we can perform motion by differential changes in forces, concentrations, temperatures, etc. Let us consider a piston with a weight on top, at equilibrium. If we slide the weight off to the side, the change in potential energy of the weight is zero, the piston rises so its potential energy increases. If the piston hits a stop, kinetic energy is dissipated. Now let us subdivide the weight into two portions. If we move off one half the weight, the piston strikes the stop with less kinetic energy than before, and in addition, we have now raised half of the weight. If we repeat the process again we could find that we could move increasing amounts of weight by decreasing the weight we initially move off the piston. In the limit, our weight would become like a pile of sand, and we would remove one grain at a time. Since the changes in the system are so small, only infinitesimal gradients would ever develop, and we would approach reversibility. The piston would never develop kinetic energy which would need to be dissipated.

Reversibility by a Series of Equilibrium States

When we move a system differentially, as just discussed, the system is at equilibrium along each step of the process. To test whether the system is at equilibrium at a particular stage, we can imagine freezing the process at that stage and we can ask whether the system would change if we left it at those conditions. If the system would remain static (i.e., not changing) at those conditions, then it must be at equilibrium. Because it is static, we could just as easily go one way as another ⇒ "reversible." Thus, reversible processes are the result of infinitesimal driving forces.

Reversibility by Neglecting Viscosity and Friction

Real processes are not done infinitely slowly. In the previous examples we have used idealized pistons and cylinders for discussion. Real systems can be far from ideal and may have much more complex geometry. For example, projectiles can be fired using gases to drive them, and we need a method to estimate the velocities with which they are projected into free flight. One application of this is the steam catapult used to assist airplanes in becoming airborne from the short flight decks of aircraft carriers. Another application would be determination of the exit velocity of a bullet fired from a gun. These are definitely not equilibrium processes, so how can we begin to calculate the exit velocities? Another case would be the centrifugal pump. The pump works by rapidly rotating a propeller-type device. The pump simply would not work at low speed without velocity gradients! So what do we do in these cases? The answer is that we perform a calculation ignoring viscosity and friction. Then we apply an efficiency factor to calculate the real work done. The efficiency factors are determined empirically from our experience with real systems of a similar nature to the problem at hand. Efficiencies are introduced in Chapter 3. In the remainder of this chapter, we concentrate on the first part of the problem, calculation of reversible work.

❶ Viscosity and friction are frequently ignored for an estimation of optimum work, and an empirical efficiency factor is applied based on experience with similar systems.

Example 2.1 Isothermal compression of an ideal gas

Calculate the work necessary to isothermally perform steady compression of two moles of an ideal gas from 1 to 10 bar and 311 K in a piston. An isothermal process is one at constant temperature. The steady compression of the gas should be performed such that the pressure of the system is always practically equal to the external pressure on the system. In Chapter 3, we will refer to this type of compression as "reversible" compression, and we will make a number of important points in contrasting it to "irreversible" processes.

Solution:

System: closed; Basis: One mole

$$W_{EC} = -\int P dV$$

$$P = \frac{RT}{V} \Rightarrow W_{EC} = -\int_{V_1}^{V_2} \frac{RT}{V} \, dV = -RT \int_{V_1}^{V_2} \frac{dV}{V} = -RT \, \ln(V_2 / V_1) \qquad \text{(ig) 2.7}$$

$$V_2 / V_1 = \frac{RT / P_2}{RT / P_1} = P_1 / P_2 = 1 / 10 \qquad \text{(ig) 2.8}$$

$$W_{EC} = -8.314 \text{ J/gmol-K} \cdot 311 \text{ K } \ln(1/10) = 5954 \text{ J/gmol}$$

$$\underline{W_{EC}} = 2(5954) = 11{,}908 \text{ J}$$

Note: Work is done on the gas since the sign is positive. This is the sign convention set forth in Eqn. 2.3. If the integral for Eqn. 2.3 is always written as shown with the initial state as the lower limit of integration and the P and V properties of the system, the work on the gas will always result with the correct sign.

2.5 PATH PROPERTIES AND STATE PROPERTIES

In the previous example, we have used an *isothermal* path. It is convenient to define other terms which describe pathways concisely. An *isobaric* path is one at constant pressure. An *isochoric* path is one at constant volume. An *adiabatic* path is one without heat transfer.

The heat and work transfer necessary for a change in state are dependent on the pathway taken between the initial and final states. A state property is one that is independent of the pathway taken. For example, when the pressure and temperature of a gas are changed and the gas is returned to its initial state, the net change in temperature, pressure, and internal energy is zero, and these properties are therefore state properties. However, the net work and net heat transfer will not necessarily be zero; their values will depend on the path taken.[1]

> The terms *isothermal, isobaric, isochoric, adiabatic,* describe pathways.

> The work and heat transfer necessary for a change in state are dependent on the pathway taken between the initial and final state.

---

1. Also, it is helpful to recall that heat and work are not properties of the system; therefore, they are not state properties.

Example 2.2   Work as a path function

Consider 1.2 moles of an ideal gas in a piston at 298K and 0.2 MPa and at volume $\underline{V}_1$. The gas is expanded isothermally to twice its original volume, then cooled isobarically to $\underline{V}_1$. It is then heated at constant volume back to T_1. Demonstrate that the net work is non-zero, and that the work depends on the path.

Solution: First sketch the process on a diagram to visualize the process as shown below in Fig. 2.2. Determine the initial volume:

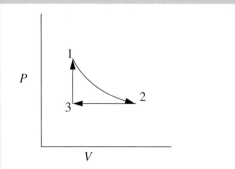

Figure 2.2 *Schematic for Example 2.2.*

$$\underline{V} = nRT/P = \frac{1.2 \text{ moles}}{} \left| \frac{8.314 \text{ cm}^3\text{MPa}}{\text{mole K}} \right| \frac{298 \text{ K}}{0.2 \text{ MPa}} = 14{,}865 \text{ cm}^3 \qquad \text{(ig)}$$

1. isothermally expand that gas

$$\Rightarrow \underline{W}_{EC} = -\int P \, d\underline{V} = -nRT_1 \, \ln(V_2 / V_1) \qquad \text{(ig) 2.9}$$

$$= \frac{1.2 \text{ moles}}{} \left| \frac{8.314 \text{ cm}^3\text{MPa}}{\text{mole K}} \right| 298 \text{ K} \quad \ln(2) = -2060 \text{ J}$$

2. cool isobarically down to V_1

$$\underline{W}_{EC} = -\int_{V_2}^{V_1} P_2 d\underline{V} = -P_2(\underline{V}_1 - \underline{V}_2) = -0.1 \text{ MPa}(-14{,}865 \text{ cm}^3) = 1487 \text{ J} \qquad 2.10$$

3. heat at constant volume back to T_1

$$\Rightarrow \underline{W}_{EC} = 0 \ (because \ d\underline{V} = 0 \ over \ entire \ step)$$

We have returned the system to its original state and all state properties have returned to their initial values. What is the total work done on the system?

$$\underline{W} = \underline{W}_{1\rightarrow 2} + \underline{W}_{2\rightarrow 3} + \underline{W}_{3\rightarrow 1} = -nRT_1 \, \ln(V_2 / V_1) + nP_2(V_2 - V_1) = -573 \text{ J} \qquad \text{(ig) 2.11}$$

Therefore, we conclude that work is a path function, not a state function.

> ### Example 2.2 Work as a path function (Continued)
>
> **Exercise:** If we reverse the path, the work will be different, in fact it will be positive instead of negative (+573.6 J). If we change the path to isobarically expand the gas to double the volume ($\underline{W} = -2973$ J), cool to T_1 at constant volume ($\underline{W} = 0$ J), then isothermally compress to the original volume ($\underline{W} = -2060$ J), the work will be −913 J.
>
> Heat was added and removed during the process which has not been accounted for above. The above process transforms work into heat, and all we have done is computed the amount of work. The important thing to remember is that work is a path function, not a state function.

🛑 Work and heat are path properties.

2.6 HEAT FLOW

A very simple experiment shows us that heat transport is related to energy. If two steel blocks of different temperature are placed in contact with each other, but otherwise insulated from the surroundings, they will come to equilibrium at a common intermediate temperature. The warmer block will be cooled, and the colder block will be warmed.

$$\underline{Q}_{block\,1} = -\underline{Q}_{block\,2}$$

Heat is transferred at a system boundary. Therefore, heat is not a property of the system. It is a form of interaction at the boundary which transfers internal energy. If heat is added to a system for a finite period of time, then the energy of the system increases because the kinetic energy of the molecules is increased. When an object feels hot to our touch, it is because the kinetic energy of molecules is readily transferred to our hand.

Since the rate of heating may vary with time, we must recognize that the total heat flows must be summed (or integrated) over time. In general, we can represent a differential contribution by

$$d\underline{Q} = \dot{\underline{Q}}dt$$

We can also relate the internal energy change and heat transfer for either block in a differential form:

$$d\underline{U} = d\underline{Q} \quad \text{or} \quad \frac{d\underline{U}}{dt} = \dot{\underline{Q}} \qquad \qquad 2.12$$

An idealized system boundary that has no resistance to heat transfer but is impervious to mass is called a *diathermal* wall.

🛑 Diathermal.

2.7 THE CLOSED-SYSTEM ENERGY BALANCE

A closed system is one in which no mass flows in or out of the system, as shown in Fig. 2.3. The introductory sections have discussed heat and work interactions, but we have not yet coupled these to the energy of the system. In the transformations we have discussed, energy can cross a boundary in the form of expansion/contraction work ($-\int PdV$), shaft work (W_S), and heat (Q)[1]. *There are only two ways a closed system can interact with the surroundings*, via heat and work interactions. If we

🛑 A closed system interacts with the surroundings only through heat and work.

---

1. Other possibilities include electric or magnetic fields, or mechanical springs, etc., which we do not address in this text.

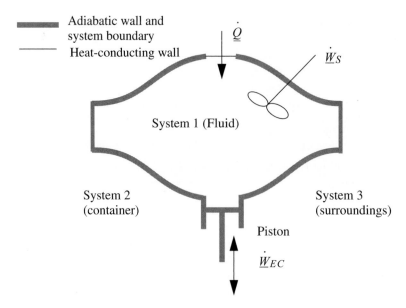

Figure 2.3 *Schematic of a closed system.*

put both these possibilities into one balance equation, then developing the balance for a given application is simply a matter of analyzing a given situation and deleting the balance terms that do not apply. The equation terms can be thought of as a check list.

Experimentally, it is found that if we measure heat and work for a cyclical process which returns to the initial state, we find that the heat and work interactions together always sum to zero. *This is an important result!* This means that, in non-cyclical processes where the sum of heat and work are non-zero, the system has stored or released energy, depending on whether the sum is positive or negative. In fact, by performing enough experiments, we would decide that, in fact, the sum of heat and work interactions in a *closed* system *is* the change in energy of the system! To develop the closed-system energy balance, let us first express the balance in terms of words.

$$
\begin{bmatrix} \text{energy} \\ \text{accumulation within} \\ \text{system boundaries} \end{bmatrix} = \begin{bmatrix} \text{heat flow} \\ \text{into system} \end{bmatrix} + \begin{bmatrix} \text{work done} \\ \text{on system} \end{bmatrix}
\qquad 2.13
$$

Energy within the system is composed of the internal energy (e.g., $\underline{U}$), and the kinetic ($mu^2/2g_c$) and potential energy (mgz/g_c) of the center of mass. For closed systems, the "check list" equation is:

$$
md\left[U + \frac{u^2}{2g_c} + \frac{g}{g_c}z\right] = d\underline{Q} + d\underline{W}_S + d\underline{W}_{EC}
\qquad 2.14
$$

The left-hand side summarizes changes occurring *within* the system boundaries and the right-hand side summarizes changes due to interactions *at* the boundaries. It is a recommended practice to always write the balance in this convention when starting a problem. We will follow this convention throughout our example problems. The kinetic and potential energy of interest in

Eqn. 2.14 is for the center of mass, not the random kinetic and potential energy of molecules about the center of mass. The balance could also be expressed in terms of molar quantities, but if we do so, we need to introduce molecular weight in the potential and kinetic energy terms. Since the mass is constant in a closed system, we may divide the above equation by m,

$$d\left[U + \frac{u^2}{2g_c} + \frac{g}{g_c}z\right] = dQ + dW_S + dW_{EC}$$

2.15

Closed-system balance. The left-hand side summarizes changes *inside* the boundaries, and the right-hand side summarizes interactions *at* the boundaries.

where heat and work interactions are summed for multiple interactions at the boundaries. We can integrate Eqn. 2.15 to obtain

$$\Delta\left(U + \frac{u^2}{2g_c} + \frac{g}{g_c}z\right) = Q + W_S + W_{EC}$$

2.16

We may also express the energy balance in terms of rates of change:

$$\frac{d}{dt}\left[U + \frac{u^2}{2g_c} + \frac{g}{g_c}z\right] = \dot{Q} + \dot{W}_S + \dot{W}_{EC}$$

2.17

where $\dot{Q} = \dfrac{dQ}{dt}$, $\dot{W}_S = \dfrac{dW_S}{dt}$, and $\dot{W}_{EC} = \dfrac{dW_{EC}}{dt}$.

Example 2.3 Internal energy and heat

In Section 2.6 on page 43 we discussed that heat flow is related to the energy of system, and now we have a relation to use to quantify changes in energy. If 2000 J of heat are passed from the hot block to the cold block, how much has the internal energy of each block changed?

Solution: First choose a system boundary. Let us initially place system boundaries around each of the blocks. Let the warm block be *block1* and the cold block be *block2*. Next, eliminate terms which are zero or are not important. The problem statement says nothing about changes in position or velocity of the blocks so these terms can be eliminated from the balance. There is no shaft involved so shaft work can be eliminated. The problem statement doesn't specify the pressure, so it is common to assume that the process is at a constant atmospheric pressure of 0.101 MPa. The cold block does expand slightly when it is warmed, and the warm block will contract; however, since we are dealing with solids, the work interaction is so small that it can be neglected. For example, the blocks together would have to change 10 cm^3 at 0.101 MPa to equal 1 J out of the 2000 J that are transferred.

Example 2.3 Internal energy and heat (Continued)

Therefore, the energy balance for each block becomes:

$$d\left[U + \frac{u^2}{2g_c} + \frac{g}{g_c}z\right] = dQ + dW_S + dW_{EC}$$

We can integrate the energy balance for each block.

$$\Delta \underline{U}_{block1} = \underline{Q}_{block1} \qquad \Delta \underline{U}_{block2} = \underline{Q}_{block2}$$

The magnitude of the heat transfer between the blocks is the same since no heat is transferred to the surroundings, but how about the signs? Let's explore that further. Now, placing the system boundary around both blocks, the energy balance becomes:

$$d\left[U + \frac{u^2}{2g_c} + \frac{g}{g_c}z\right] = dQ + dW_S + dW_{EC}$$

Note that the composite system is considered *isolated* from the surroundings since all heat and work interactions are negligible. Therefore, we have: $\Delta \underline{U} = 0$ or by dividing in subsystems, $\Delta \underline{U}_{block1} + \Delta \underline{U}_{block2} = 0$ which becomes $\Delta \underline{U}_{block1} = -\Delta \underline{U}_{block2}$. Notice that the signs are important in keeping track of which system is giving up heat and which system is gaining heat. In this example, it would be easy to keep track, but other problems will be more complicated, and it is best to develop a good bookkeeping practice of watching the signs. In this example the heat transfer for the initially hot system will be negative, and the heat transfer for the other system will be positive. Therefore, the internal energy changes are $\Delta \underline{U}_{block1} = -2000$ J and $\Delta \underline{U}_{block2} = 2000$ J.

This example has illustrated several important points.

1. It is important to specify exactly where the boundary is for the balance equation that is written.

2. It is possible to subdivide a system into subsystems. The composite system was isolated, but the subsystems were not. Many times problems are more easily solved, or insight is gained by looking at the overall system, even though we can frequently make progress quickly by starting with subsystems. If the subsystem balances look difficult to solve, try an overall balance.

3. It is important to keep track of signs carefully.

4. It is important to recognize when simplifications can be made. Calculation of the work would have certainly been possible, but would not have affected our answer significantly. However, if our two systems would have been gases, then this simplification would have not been reasonable.

Four important points about solving problems.

2.8   THE OPEN-SYSTEM, STEADY-STATE BALANCE

Having established the energy balance for a closed system, and from Section 2.3 the work associated with flowing fluids, let us extend these concepts to develop the energy balance for a steady-state flow system. The term *steady state* means:

1. All state properties throughout the system are invariant with respect to time. The properties may vary with respect to position within the system.

2. The system has constant mass, i.e. the total inlet mass flowrate equals the total outlet mass flowrate, and all flowrates are invariant with respect to time.

3. The center of mass for the system is fixed in space. (This restriction is not strictly required, but will be used throughout this text).

To begin, we write our balance in words, by adding flow to our previous closed-system balance. There are only three ways the surroundings can interact with the system: flow, heat, and work. A schematic of an open steady-state system is shown in Fig. 2.4. In consideration of the types of work encountered in steady-state flow, recognize that expansion/contraction work will usually not be involved so this term is omitted. This is because we typically apply the steady-state balance to systems of rigid mechanical equipment, and there is no change in the size of the system. Therefore, the expansion/contraction work term is set to 0.

> ⓘ Steady-state flow systems are usually fixed size, so $W_{EC} = 0$.

Our balance in words becomes time-dependent since we work with flowrates:

$$\begin{bmatrix} \text{rate of energy} \\ \text{accumulation within} \\ \text{system boundaries} \end{bmatrix} = \begin{bmatrix} \text{energy per unit} \\ \text{mass of fluid at inlet} \end{bmatrix} \begin{bmatrix} \text{mass} \\ \text{flowrate} \\ \text{in} \end{bmatrix} \qquad 2.18$$

$$- \begin{bmatrix} \text{energy per unit} \\ \text{mass of fluid at outlet} \end{bmatrix} \begin{bmatrix} \text{mass} \\ \text{flowrate} \\ \text{out} \end{bmatrix} + \begin{bmatrix} \text{rate of heat flow} \\ \text{into system} \end{bmatrix}$$

$$+ \begin{bmatrix} \text{rate that work} \\ \text{is done on system} \end{bmatrix}$$

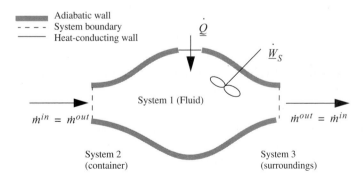

Figure 2.4 *Schematic of a steady-state flow system.*

Again, we follow the convention that the left-hand side quantifies changes *inside* our system. Consider the change of energy inside the system boundary given by the left-hand side of the equation. Due to the restrictions placed on the system by steady-state, there is no accumulation of energy within the system boundaries, so the left-hand side of Eqn. 2.18 becomes 0.

As a result:

$$0 = \sum_{inlets} \left[U + \frac{u^2}{2g_c} + \frac{gz}{g_c} \right]^{in} \dot{m}^{in} - \sum_{outlets} \left[U + \frac{u^2}{2g_c} + \frac{gz}{g_c} \right]^{out} \dot{m}^{out} + \dot{Q} + \dot{W}_S + \dot{W}_{flow} \qquad 2.19$$

where heat and work interactions are summed over all boundaries. The flow work from Eqn. 2.6 may be inserted and summed over all inlets and outlets.

$$0 = \sum_{inlets} \left[U + \frac{u^2}{2g_c} + \frac{gz}{g_c} \right]^{in} \dot{m}^{in} - \sum_{outlets} \left[U + \frac{u^2}{2g_c} + \frac{gz}{g_c} \right]^{out} \dot{m}^{out} + \dot{Q} + \dot{W}_S \qquad 2.20$$

$$+ \sum_{inlets} (PV)^{in} \dot{m}^{in} - \sum_{outlets} (PV)^{out} \dot{m}^{out}$$

and combining flow terms:

$$0 = \sum_{inlets} \left[U + PV + \frac{u^2}{2g_c} + \frac{gz}{g_c} \right]^{in} \dot{m}^{in} - \sum_{outlets} \left[U + PV + \frac{u^2}{2g_c} + \frac{gz}{g_c} \right]^{out} \dot{m}^{out} + \dot{Q} + \dot{W}_S \qquad 2.21$$

Enthalpy

❶ *Enthalpy* is a mathematical property defined for convenience in problem solving.

Note that the quantity $(U + PV)$ arises quite naturally in the analysis of flow systems. Flow systems are very common so it makes sense to define a single symbol which denotes this quantity.

$$H \equiv U + PV$$

Thus, we can tabulate precalculated values of H and save steps in calculations for flow systems. We call H the *enthalpy*.

The open-system, steady-state balance is then,

❶ Open system, steady-state balance.

$$0 = \sum_{inlets} \left[H + \frac{u^2}{2g_c} + \frac{gz}{g_c} \right]^{in} \dot{m}^{in} - \sum_{outlets} \left[H + \frac{u^2}{2g_c} + \frac{gz}{g_c} \right]^{out} \dot{m}^{out} + \dot{Q} + \dot{W}_S \qquad 2.22$$

where the heat and work interactions are summations of the individual heat and work interactions over all boundaries.

Note: Q is positive when the system gains heat energy; W is positive when the system gains work energy; $\dot{m}^{in}$ and $\dot{m}^{out}$ are always positive; and $\dot{m}_{system} = \dot{m}^{in} - \dot{m}^{out}$ is positive when the systems gains mass and zero for steady-state flow. Mass may be replaced with moles in a non-reactive system with appropriate care for unit conversion.

Note that the potential and kinetic energies which are important are for the fluid *entering* and *leaving* the boundaries, *not* for the fluid which is *inside* the system boundaries. When only one inlet and one outlet stream are involved, the steady-state flowrates must be equal, and

$$0 = \left[H + \frac{u^2}{2g_c} + \frac{gz}{g_c} \right]^{in} \dot{m}^{in} - \left[H + \frac{u^2}{2g_c} + \frac{gz}{g_c} \right]^{out} \dot{m}^{out} + \underline{\dot{Q}} + \underline{\dot{W}}_S$$

2.23   ❶ Several common ways the steady-state balance can be written.

When kinetic and potential energy changes are negligible, we may write

$$0 = -\Delta H \dot{m} + \underline{\dot{Q}} + \underline{\dot{W}}_S$$

2.24

where $\Delta H = H^{out} - H^{in}$. We could use molar flowrates for Eqns. 2.22 through 2.24 with the usual care for unit conversions of kinetic and potential energy. For an open steady-state system meeting the restrictions of Eqn. 2.24, we may divide through by the mass flowrate to find

$$0 = -\Delta H + Q + W_S$$

2.25

In common usage, it is traditional to relax the convention of keeping only system properties on the left side of the equation. More simply we often write:

$$\Delta H = Q + W_S$$

2.26

Comment on Notation

In a closed system we use the Δ symbol to denote the change of a property from initial state to final state. In an open, steady-state system, the left-hand side of the energy balance is zero. Therefore, we frequently write Δ as a shorthand notation to combine the first two flow terms on the right-hand side of the balance, with the symbol meaning "outlet relative to inlet" as shown above. You need to learn to recognize which terms of the energy balance are zero or insignificant for a particular problem, whether a solution is for a closed or open system, and whether the Δ symbol denotes "outlet relative to inlet" or "final relative to initial."

❶ Explanation of the use of Δ.

2.9   THE COMPLETE ENERGY BALANCE

An open system which does not meet the requirements of a steady-state system is called an unsteady-state open system as shown in Fig. 2.5. The mass-in may not equal the mass-out, or the state variables (e.g., temperature) may change with time, so the system itself may gain in internal energy, kinetic energy, or potential energy. An example of this is the filling of a tank being heated with a steam jacket. Another example is the inflation of a balloon, where there is mass flow in and the system boundary expands. These considerations lead to a general equation which is applicable to open or closed systems.

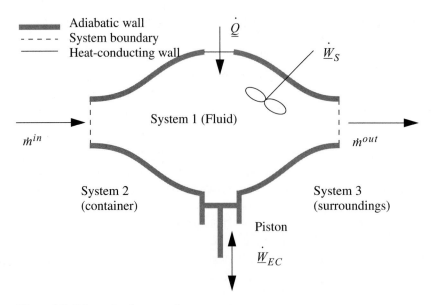

Figure 2.5 *Schematic of a general system.*

❶ Complete Energy Balance.

$$\frac{d}{dt}\left[m\left(U+\frac{u^2}{2g_c}+\frac{gz}{g_c}\right)\right] = \sum_{inlets}\left[H+\frac{u^2}{2g_c}+\frac{gz}{g_c}\right]^{in}\dot{m}^{in} - \sum_{outlets}\left[H+\frac{u^2}{2g_c}+\frac{gz}{g_c}\right]^{out}\dot{m}^{out}$$
$$+\,\dot{\underline{Q}}+\dot{\underline{W}}_{EC}+\dot{\underline{W}}_S \qquad\qquad 2.27$$

where the heat and work interactions are summations of the individual heat and work interactions over all boundaries. We also may write this with the time dependence implied:

$$d\left[m\left(U+\frac{u^2}{2g_c}+\frac{gz}{g_c}\right)\right] = \sum_{inlets}\left[H+\frac{u^2}{2g_c}+\frac{gz}{g_c}\right]^{in}dm^{in} - \sum_{outlets}\left[H+\frac{u^2}{2g_c}+\frac{gz}{g_c}\right]^{out}dm^{out}$$
$$+\,d\underline{Q}+d\underline{W}_{EC}+d\underline{W}_S \qquad\qquad 2.28$$

Note: The signs and conventions are the same as presented following Eqn. 2.22.

Usually, the closed-system or the steady-state equations are sufficient by themselves. But for unsteady-state open systems, the entire equation must be considered. Fortunately, even when the entire energy balance is applied, some of the terms are usually not necessary for a given problem, so fewer terms are usually needed than shown in Eqn. 2.27. With experience, you will recognize which terms apply to a given problem.

2.10   INTERNAL ENERGY, ENTHALPY, AND HEAT CAPACITIES

Before we proceed with more examples, we need to add another thermodynamic tool. Unfortunately, there are no "internal energy" or "enthalpy" meters. In fact, these state properties must be "measured" indirectly by other state properties. The Gibbs phase rule tells us that if two state variables are fixed in a pure single-phase system, then all other state variables will be fixed. Therefore, it makes sense to measure these properties in terms of P, V, and T. In addition, if this relation is developed, it will enable us to find P, V, and/or T changes for a given change in ΔU or ΔH. In Example 2.3, where a warm and cold steel block were contacted, we solved the problem without calculating the change in temperature for each block. However, if we had a relation between U and T, we could have calculated the temperature changes. The relations that we seek are the definitions of the heat capacities.

Constant Volume Heat Capacity

The constant of proportionality between the internal energy change *at constant volume* and the temperature change is known as the constant volume heat capacity. The constant volume heat capacity *is defined by*:

$$ C_V \equiv \left(\frac{\partial U}{\partial T} \right)_V $$

2.29   ❗ Definition of C_V.

Since temperature changes are easily measured, internal energy changes can be calculated once C_V is known. C_V is not commonly tabulated, but, as shown below, it can be easily determined from the constant pressure heat capacity, which is commonly available.

Constant Pressure Heat Capacity

In the last two sections, we have introduced enthalpy, and we can relate the change in enthalpy of a system to temperature in a manner analogous to the method used for internal energy. This relationship will involve a new heat capacity, the *heat capacity at constant pressure*:

$$ C_P \equiv \left(\frac{\partial H}{\partial T} \right)_P $$

2.30   ❗ Definition of C_P

where H is the enthalpy of the system.

The use of two heat capacities, C_V and C_P, forces us to think of constant volume or constant pressure as the important distinction between these two quantities. The important quantities are really internal energy versus enthalpy. You simply must force yourself to remember that C_V refers to changes in U at constant volume, and C_P refers to changes in H at constant pressure.

Relations between Heat Capacities, *U* and *H*

We have said that C_V values are not readily available; therefore, how do we determine internal energy changes? Also, how do we determine enthalpy changes at constant volume or internal energy changes at constant pressure? We will return to the details of these questions in later chapters and handle them rigorously, but the details have been rigorously followed by developers of thermodynamic charts and tables. Therefore, for relating the internal energy or enthalpy to temperature and pressure, a thermodynamic chart or table is preferred. If none is available, or properties are not tabulated in the state of interest, some exact relations and some approximate rules of thumb must be applied.

For an ideal gas

C_P, C_V and relation between ΔU and ΔH for an ideal gas.

$$C_P = C_V + R \quad \Delta H = \Delta U + R\Delta T \quad \text{Exact for ideal gases.} \quad \text{(ig) 2.31}$$

For real gases and for liquids, the relation between C_P and C_V is more complex, and derivatives of P-V-T properties must be used. The background for the relation will be developed in Chapter 5. We will use thermodynamic tables and charts for real gases until these relations are developed. Constant pressure heat capacities for ideal gases are tabulated in Appendix E. Constant volume heat capacities for ideal gases can readily be determined from Eqn. 2.31.

For ideal gases, internal energy and enthalpy are independent of pressure. This relation will be proven in Example 5.8 on page 185. For real gases, the relation is more complex and will be deferred to Unit II. Thermodynamic charts and tables will be used in Unit I. For liquids, it has been experimentally determined that internal energy is only very weakly dependent on pressure below $T_r = 0.75$. In addition, the molar volume is insensitive to pressure below $T_r = 0.75$. We will show in Example 2.4

Pressure dependence of H for condensed phases.

$$\Delta H_T \approx V\Delta P_T \quad \text{Liquids below } T_r = 0.75 \text{ and solids.} \quad 2.32$$

$T_r \equiv \dfrac{T}{T_c}$ is the reduced temperature calculated by dividing the absolute temperature by the critical temperature. The relations for solids and liquids are important because frequently the properties have not been measured, or the measurements available in charts and tables are not available at the pressures of interest. We may then summarize the relations of internal energy and enthalpy with temperature.

Useful formula for relating T, P to U and H in the absence of phase changes.

$$\boxed{\Delta U = \int_{T_1}^{T_2} C_V(T)dT} \quad \text{Ideal gas: exact.} \quad 2.33$$

$$\boxed{\Delta H = \int_{T_1}^{T_2} C_P(T)dT} \quad \begin{array}{l} \text{Ideal gas: exact.} \\ \text{Real gas: valid only if } P = \text{constant.} \end{array} \quad 2.34$$

$$\boxed{\Delta H \approx \int_{T_1}^{T_2} C_P(T)dT + V\Delta P} \quad \begin{array}{l} \text{Liquid below } T_r = 0.75 \text{ or solid: reasonable} \\ \text{approximation.} \end{array} \quad 2.35$$

$$\boxed{\Delta U = \Delta H - \Delta(PV) \approx \Delta H - V\Delta P \approx \int_{T_1}^{T_2} C_P(T)dT} \quad \begin{array}{l} \text{Liquid below } T_r = 0.75 \text{ or solid: reasonable approximation when pressure} \\ \text{change is below several MPa.} \end{array} \quad 2.36$$

Note: These formulas do not account for phase changes which may occur.

Heat capacities are functions of temperature. Recall that the internal energy of a system is simply the potential and random kinetic energy of the molecules, and part of this kinetic energy is the vibrations of the bonds making up the molecules and the random rotations of the molecules. Whenever we assume that the heat capacities are independent of temperature, we are making an approximation because we neglect the effect of temperature on the these properties.[1] For diatomic gases like oxygen, nitrogen, and air, *near room temperature,* the ideal gas heat capacities can be approximated as $C_P = 7/2$ R, and $C_V = 5/2$ R. For a monatomic ideal gas *near room temperature*, the heat capacities may be approximated as $C_P = 5/2$ R and $C_V = 3/2$ R. For polyatomic molecules, the heat capacity generally increases with molecular weight. In this text, ideal gas heat capacity values at 298 K are summarized in the back inside cover of the book, and may be assumed to be independent of temperature over small temperature ranges near room temperature. These approximations fail at high temperatures because increases in the number of vibrational, rotational, and electronic microstates become important. *Whenever we assume heat capacity to be temperature independent in this text, we mark the equation with a (*) symbol near the right margin.* Heat capacities represented as polynomials of temperature are available in Appendix E. Note that an HP program is available for quickly calculating the heat capacity from the polynomial constants. Also, the heat capacity depends on the state of aggregation. For example, water will have different heat capacities as solid (ice), liquid, and vapor (steam).

❶ Whenever we assume heat capacity to be temperature independent in this text, we mark the equation with a (*) symbol near the right margin.

❶ The heat capacity of a substance depends on the state of aggregation.

Example 2.4 Enthalpy of H_2O above its saturation pressure

Determine the enthalpy of H_2O at 20°C and at 50 and 100 MPa.

Solution: The enthalpies can be found using Eqn. 2.32 relative to the saturation state. For gases, we would need to worry about changes in volume with respect to pressure, but liquids are approximately incompressible. Therefore, from the steam tables, for saturated liquid water at 20°C,

$$V^L \approx \text{constant} = 1.002 \text{ cm}^3/\text{g}$$

$$\Rightarrow \Delta H \approx V^L \Delta P = 1.002 \text{ cm}^3/\text{g}(50 \text{ MPa} - 0.00234 \text{ MPa}) = 50.1 \text{ MPa-cm}^3/\text{g for 50 MPa}$$

$$\Rightarrow \Delta H \approx V^L \Delta P = 1.002 \text{ cm}^3/\text{g}(100 \text{ MPa} - 0.00234 \text{ MPa}) = 100.20 \text{ MPa-cm}^3/\text{g for 100 MPa}$$

A convenient way of converting units for these calculations is to multiply and divide by the gas constant, noting its different units. This trick is especially convenient in this case, e.g.:

$$\Delta H = 50.1 \text{ MPa-cm}^3/\text{g} \cdot (8.314 \text{ J/mole-K})/(8.314 \text{ MPa-cm}^3/\text{mole-K}) = 50.1 \text{ kJ/kg}$$

$$\Delta H = 100.20 \text{ MPa-cm}^3/\text{g} \cdot (8.314 \text{ J/mole-K})/(8.314 \text{ MPa-cm}^3/\text{mole-K}) = 100.20 \text{ kJ/kg}$$

Note that the change in enthalpy in kJ/kg is roughly equal to the pressure rise in MPa. That's a convenient rule of thumb. Finally, applying the initial condition of the saturated liquid, we obtain the value of the enthalpy consistent with the steam tables.

at 50 MPa, $\Rightarrow H = 83.95 + 50.1 = 134.05$ kJ/kg

at 100 MPa $\Rightarrow H = 83.95 + 100.20 = 184.15$ kJ/kg.

The value calculated by the NIST steam formulation at 50 MPa is 130 kJ/kg, and at 100 MPa the value is 174 kJ/kg, so the estimation error is about 3% at 50 MPa, and about 6% at 100 MPa.

---

1. In addition, the electronic energies may change due to changes in population of the electronic energy levels.

Relation to Property Tables/Charts

In Section 1.4, we used steam tables to find internal energies of water as liquid or vapor. Tables or charts usually contain enthalpy and internal energy information, which means that these properties can be read from the source for these compounds, eliminating the need to apply Eqns. 2.33–2.36. This is usually more accurate because the pressure dependence of the properties that Eqns. 2.33–2.36 neglect has been included in the table/chart, although the pressure correction method applied in the previous example for liquids is generally accurate enough. Energy and enthalpy changes spanning phase transitions can be determined directly from the tables since energies and enthalpies of phase transitions are implicitly included in tabulated values.

Estimation of Heat Capacities

If heat capacity information cannot be located from appendices in this text or from reference handbooks, it can be estimated by several techniques offered in the *Chemical Engineer's Handbook*[1] and *The Properties of Gases and Liquids*.[2]

Phase Transitions (Liquid-Vapor)

Enthalpies of vaporization are tabulated in Appendix E for many substances at their normal boiling temperatures (their saturation temperatures at 1.01325 bar). In the case of the steam tables, Section 1.4 shows that the energies and enthalpies of vaporization of water are available along the entire saturation curve. Complete property tables for some other compounds are available in the literature; however, most textbooks present charts to conserve space, and we follow that trend. In the cases where tables or charts are available, their use is preferred for phase transitions away from the normal boiling point, although a hypothetical path that passes through the normal boiling point can usually be constructed easily.

The energy of vaporization is more difficult to find than the enthalpy of vaporization. It can be calculated from the enthalpy of vaporization and the P-V-T properties. Since $U = H - PV$,

$$\Delta U^{vap} = \Delta H^{vap} - \Delta(PV)^{vap} = \Delta H^{vap} - (P^{sat}V)^V - (P^{sat}V)^L = \Delta H^{vap} - P^{sat}(V^V - V^L)$$

Far from the critical point, the molar volume of the vapor is much larger than the molar volume of the liquid. Further, at the normal boiling point (the saturation temperature at 1.01325 bar), the ideal gas law is often a good approximation for the vapor volume,

$$\boxed{\Delta U^{vap} \approx \Delta H^{vap} - P^{sat}V^V \approx \Delta H^{vap} - RT^{sat}}$$

(ig) 2.37

❗ Relation between ΔU^{vap} and ΔH^{vap} when the vapor follows the ideal gas law.

Estimation of Enthalpies of Vaporization

If the enthalpy of vaporization cannot be located in the appendices or a standard reference book, it may be estimated by several techniques offered and reviewed in the *Chemical Engineer's Handbook* and *The Properties of Gases and Liquids*. If the vapor pressure is available, the enthalpy of vaporization can be estimated far from the critical point where the ideal gas law is valid for the vapor phase by the Clausius–Clapeyron equation.

1. Perry, R.M., Green, D.W., Maloney, J.O., *Chemical Engineer's Handbook*, 7th ed., McGraw-Hill, New York, (1997).
2. Reid, R.C., Prausnitz, J.M., Poling, B.E., *The Properties of Gases and Liquids*, 4th ed., McGraw-Hill, New York (1987).

$$\Delta H^{vap} \approx -R\frac{d\ln P^{sat}}{d\left(\frac{1}{T}\right)} \tag{ig 2.38}$$

The background for this equation will be developed in Section 8.2. Vapor pressure is often represented by the Antoine equation, $\log P^{sat} = A - B/(T + C)$. If Antoine parameters are available, they may be used to estimate the derivative term of Eqn. 2.38.

$$\frac{d\ln P^{sat}}{d\left(\frac{1}{T}\right)} = \frac{2.3026\,d\log P^{sat}}{d\left(\frac{1}{T}\right)} = \frac{-2.3026B(T+273.15)^2}{(T+C)^2} \tag{ig}$$

where T is in °C, and B and C are Antoine parameters for the common logarithm of pressure. For Antoine parameters intended for other temperature or pressure units, the equation must be carefully converted. The temperature limits for the Antoine parameters must be carefully followed because the Antoine equation does not extrapolate well outside the temperature range where the constants have been fit.

Phase Transitions (Solid-Liquid)

Enthalpies of fusion (melting) are tabulated for many substances at the normal melting temperatures in the appendices as well as handbooks. Internal energies of fusion are not usually available, however the volume change on melting is usually very small, resulting in internal energy changes that are nearly equal to the enthalpy changes

$$\Delta U^{fus} = \Delta H^{fus} - \Delta(PV)^{fus} = \Delta H^{fus} - P(V^L - V^S) \approx \Delta H^{fus} \tag{2.39}$$

ⓘ Relation between ΔU^{fus} and ΔH^{fus}.

Unlike the liquid-vapor transitions, where T^{sat} depends on pressure, the melting (solid-liquid) transition temperature is almost independent of pressure, as illustrated schematically in Fig. 1.7.

Reference States

Notice that our heat capacities do not permit us to calculate absolute values of internal energy or enthalpy; they simply permit us to calculate *changes* in these properties. Therefore, when is internal energy or enthalpy equal to zero—at a temperature of absolute zero? Is absolute zero a reasonable place to assign a reference state from which to calculate internal energies and enthalpies? Actually, we don't usually solve this problem in engineering thermodynamics for the following two reasons:[1] 1) for a gas, there would almost always be at least two phase transitions between room temperature and absolute zero that would require knowledge of energy changes of phase transitions and heat capacities of each phase; and 2) even if phase transitions did not occur, the empirical fit of the heat capacity represented by the constants in the appendices are not valid down to absolute zero! Therefore, for engineering calculations, we arbitrarily set enthalpy *or* internal energy equal to zero at some *convenient* reference state where the heat capacity formula is valid. We calculate changes relative to this state. The *actual* enthalpy or internal energy is certainly not zero, it just makes our reference state location clear. The reference state for water (in the steam tables) is chosen to set enthalpy or

ⓘ Reference states permit the tabulation of *values* for *U, H*.

---

1. In the most detailed calculations, absolute zero *is* used as a reference state to create *some* thermodynamic tables. This is based on a principle known as the Third Law of Thermodynamics, that states that entropy goes to zero for a perfect crystal at absolute temperature. The difficulties in the rigorous calculations are mentioned above, and although the principles are straightforward, the actual calculations are beyond the scope of this book.

internal energy of water equal to zero at the triple point. (Can you verify this in the steam tables? Which property is set to zero, and for which state of aggregation?). If we choose to set the value of enthalpy to zero at the reference state, then $H_R = 0$, and $U_R = H_R - (PV)_R$ where we use subscript R to denote the reference state. Note that U_R and H_R cannot be precisely zero simultaneously at the reference state.

⚠ Reference states are not required for steady-state flow systems with only a few streams, but are recommended when many streams are present.

As you will notice in the following problems, reference states are not necessary when working with a pure fluid in a closed system or in a steady-state flow system. The numerical values of the changes in internal energy or enthalpy will be independent of the reference state.

When multiple components are involved, or many inlet/outlet streams are involved, definition of reference states is *recommended* since flowrates of the inlet and outlet streams will not necessarily match one-to-one. The reference state for each component may be different, so the reference temperature, pressure, and state of aggregation must be clearly designated.

⚠ Reference states are required for unsteady-state changes in U, H when mass accumulation is present.

For unsteady-state open systems that accumulate or lose mass, reference states are *imperative when values of ΔU or ΔH changes of the system or surroundings are calculated* as the numerical values depend on the reference state. It is only when the changes for the system and surroundings are summed together that the reference state drops out.

Ideal Gas Properties

For an ideal gas, we must specify only the reference T and P.[1] An ideal gas cannot exist as a liquid or solid, and this fact completely specifies the state of our system. In addition, we need to set H_R or U_R (but not both!) equal to zero. Since $H = U + PV$, for an ideal gas, $(PV)_R = RT_R$, so $U_R^{ig} = H_R^{ig} - RT_R$.

$$U^{ig} = \int_{T_R}^{T} C_V dT + U_R^{ig} \qquad \text{(ig) 2.40}$$

$$H^{ig} = \int_{T_R}^{T} C_P dT + H_R^{ig} \qquad \text{(ig) 2.41}$$

Example 2.5 Adiabatic compression of an ideal gas in a piston/cylinder

Nitrogen is contained in a cylinder and is compressed adiabatically. The temperature rises from 25°C to 225°C. How much work is performed? Assume that the heat capacity is constant ($C_P/R = 7/2$), and that nitrogen follows the ideal gas law.

Solution: Closed system, system size changes, adiabatic.

$$d\left[U + \frac{u^2}{2g_c} + \frac{g}{g_c}z \right] = d\cancel{Q} + d\cancel{W}_S + dW_{EC}$$

$$\int dU = \int dW_{EC} = W_{EC}$$

$$dU = C_V dT \Rightarrow W_{EC} = \int C_V dT = C_V \Delta T = (C_P - R)\Delta T = 4200 \text{ J/mol} \qquad (*\text{ig})$$

---

1. For calculation of ideal gas U and H, only a reference temperature is required; however, for the entropy introduced in the next chapter, a reference pressure is needed, so we establish the requirement now.

Example 2.6 Transformation of kinetic energy into enthalpy

Water is flowing in a straight horizontal pipe of 2.5 cm ID with a velocity of 6.0 m/s. The water flows into a section where the diameter is suddenly increased. There is no device present for adding or removing energy as work. What is the change in enthalpy of the water if the downstream diameter is 5 cm? If it is 10 cm? What is the maximum enthalpy change for a sudden enlargement in the pipe? How will these changes affect the temperature of the water?

Solution: Designate system boundaries: open system, steady-state:

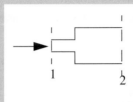

$$0 = \sum_{inlets}\left[H + \frac{u^2}{2g_c} + \frac{gz}{g_c}\right]^{in} \dot{m}^{in} - \sum_{outlets}\left[H + \frac{u^2}{2g_c} + \frac{gz}{g_c}\right]^{out} \dot{m}^{out} + \dot{Q} + \dot{W}_{EC} + \dot{W}_S$$

Simplifying: $\Delta H = \dfrac{-\Delta(u^2)}{2g_c}$

Liquid water is incompressible. Letting A represent cross-sectional area, and letting D represent pipe diameter, $\dot{V} = u_1 A_1 = u_2 A_2 \Rightarrow u_2 = u_1(A_1/A_2)$,

$$\Delta(u^2) = u_1^2\left[\left(\frac{D_1}{D_2}\right)^4 - 1\right]$$

$$\Delta H = \frac{-u_1^2}{2g_c}\left[\left(\frac{D_1}{D_2}\right)^4 - 1\right]$$

$D_2/D_1 = 2 \Rightarrow \Delta H = -6.0^2 \text{ m}^2/\text{s}^2 \ (1\text{J}/1\text{kg-m}^2/\text{s}^2)\ (\frac{1}{2}^4 - 1)/2 = 17 \text{ J/kg}$

$D_2/D_1 = 4 \Rightarrow \Delta H = 18 \text{ J/kg}$

$D_2/D_1 = \infty \Rightarrow \Delta H = 18 \text{ J/kg}$

To calculate the temperature rise, we can choose any method to relate the enthalpy change to temperature since they are both state properties. From the definition of enthalpy, we can calculate the change in temperature at constant pressure, which is numerically the number we need (recognizing that pressure does not affect greatly the enthalpy of condensed phases),

$$\Delta H = C_P \Delta T \tag{*}$$

$$C_P = 4184\frac{\text{J}}{\text{kgK}} \Rightarrow \Delta T = \frac{18.00(\text{J/kg})}{4184(\text{J/(kgK)})} = 0.004\text{K} \tag{*}$$

In this example, the velocity decreases, and enthalpy increases due to greater flow work on the inlet than the outlet. Note that the above result for a liquid does not depend on the pipe containing a sudden enlargement. A gradual taper would give the same result since the results only depend on the initial and final velocities.[1]

2.11   KINETIC AND POTENTIAL ENERGY

Example 2.6 has shown that a large change in kinetic energy resulted in a small change in temperature. Let us explore this relation further by comparing potential, kinetic, and thermal energy changes.

Example 2.7 On the relative magnitude of kinetic, potential, internal energy and enthalpy changes

For a system of 1 kg water, what are the internal energy and enthalpy changes for raising temperature 1°C as a liquid and as a vapor from 24° to 25°C? What are the internal energy enthalpy changes for evaporating from the liquid to the vapor state? How much would the kinetic and potential energy need to change to match the magnitudes of these changes?

Solution: The properties of water and steam can be found from the saturated steam tables, interpolating between 20° and 25°C. For saturated water or steam being heated from 24° to 25°C, and for vaporization at 25°C:

	ΔU(J)	ΔH(J)	ΔU^{vap}(kJ)	ΔH^{vap}(kJ)
water	4184	4184		
steam	1362	1816	2304.3	2441.7

Of these values, the values for ΔU of steam are lowest. How much would kinetic and potential energy of a system have to change to be comparable to 1000J?

Kinetic energy: If $\Delta KE = 1000$ J, and if the kg is initially at rest, then the velocity change must be,

$$\Delta(u^2) = \frac{2(1000\text{J})}{1\text{kg}} \text{ or } \Delta u = 44.7 \text{ m/s}$$

---

1. In a real system, the measured temperature rise will be higher than our calculation presented here. There are irreversibilities introduced by the velocity gradients and swirling in the region of the sudden enlargement that we haven't considered. These losses increase the temperature rise. In fluid mechanics, irreversible losses due to flow are characterized by a quantity known as the *friction factor*. The losses of a valve, fitting, contraction or enlargement can be characterized empirically by the *equivalent length* of straight pipe that would result in the same losses. We will cover these topics further in Section 4.7. From the standpoint of the energy balance, the temperature rise is still small and can be neglected except in the most detailed analysis.

> ### Example 2.7 On the relative magnitude of kinetic, potential, internal energy and enthalpy changes (Continued)
>
> This corresponds to a velocity change of 161 kph (100 mph). A velocity change of this order of magnitude is unlikely in most applications except nozzles (discussed below). Therefore, kinetic energy changes can be neglected in most calculations.
>
> **Potential energy:** If $\Delta PE = 1000$ J, then the height change must be,
>
> $$\Delta z = \frac{1000 \text{ J}}{1\left(9.8066 \frac{\text{N}}{\text{kg}}\right)} = 102 \text{ m}$$
>
> This is equivalent to about one football field in position change. Once again this is very unlikely in most process equipment, so it can usually be ignored relative to heat and work interactions.[1] Further, when a phase change occurs, these changes are even less important relative to heat and work interactions.

Velocity and height changes must be large to be significant in the energy balance when temperature changes also occur.

2.12 ENERGY BALANCES FOR PROCESS EQUIPMENT

Valves and Throttles

A throttling device is used to reduce the pressure of a flowing fluid *without any shaft work or accelerating the fluid significantly.* Throttling is also known as *Joule/Thomson expansion* in honor of the scientists who originally developed its use. An example of a throttle is the kitchen faucet. Industrial valves are modeled as throttles.

$$0 = \left[H + \frac{u^2}{2g_c} + \frac{gz}{g_c}\right]^{in} \dot{m}^{in} - \left[H + \frac{u^2}{2g_c} + \frac{gz}{g_c}\right]^{out} \dot{m}^{out} + \dot{Q} + \dot{W}_{EC} + \dot{W}_S$$

$$\Delta H = 0 \quad \textit{for throttles} \qquad\qquad 2.42$$

Nozzles

Nozzles are specially designed devices utilized for *gas flows* to convert pressure drop into kinetic energy. An example of a nozzle is a booster rocket. Nozzles are also used on the inlets of impulse turbines to convert the enthalpy of the incoming stream to a high velocity before it encounters the turbine blades.[2] Δu is significant for nozzles. A nozzle is a very specially designed device with a specially-tapered neck on the inlet and sometimes the outlet as shown schematically in Fig. 2.6.

---

1. This example has demonstrated that kinetic and potential energy changes of a fluid are usually negligible. Moreover, kinetic and potential energy changes are important in the design of piping networks because the temperature changes *are* negligible. This topic is covered in fluid mechanics courses. Another application where kinetic and potential energy changes are important is in the balance for solids such as projectiles, where again the temperature changes of the solids are negligible and work done on the system is manifested by kinetic and potential energy changes. One example of this application is the steam catapult used to assist planes in take-offs from aircraft carriers. A steam-filled piston + cylinder device is expanded, and the piston drags the plane to a velocity sufficient for the jet engines to lift the plane. While the kinetic and potential energy changes for the *steam* are negligible, the work done by the steam causes important kinetic energy changes in the piston and plane.
2. Turbine design is a specialized topic. Introductions to the actual operation are most readily available in mechanical engineering thermodynamics textbooks, such as *Engineering Thermodynamics*, J.B. Jones, R.E. Dugan, Prentice-Hall, 1996, p. 734–745.

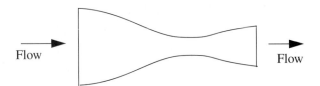

Figure 2.6 *Illustration of a converging-diverging nozzle showing the manner in which inlets and outlets are tapered.*

Nozzles are specially designed for specific applications and specific velocities/pressures of operation. Throttles are much more common in the problems we will address in this text. The meaning of "nozzle" in thermodynamics is much different from the common devices we term "nozzles" in everyday life. Most of the everyday devices we call nozzles are actually throttles. Reducing the energy balance for a nozzle shows:

$$0 = \left[H + \frac{u^2}{2g_c} + \frac{gz}{g_c} \right]^{in} \dot{m}^{in} - \left[H + \frac{u^2}{2g_c} + \frac{gz}{g_c} \right]^{out} \dot{m}^{out} + \cancel{\dot{Q}} + \cancel{\dot{W}_{EC}} + \cancel{\dot{W}_S}$$

$$\Delta H = -\Delta(u^2)/2g_c \quad \text{for nozzles} \qquad 2.43$$

Properly designed nozzles will cause an increase in the velocity of the vapor and a decrease in the enthalpy. A nozzle can be designed to operate nearly reversibly.

Heat Exchangers

Heat exchangers are available in a number of flow configurations. For example, in an industrial heat exchanger, a hot stream flows over pipes that carry a cold stream (or vice versa), and the objective of operation is to cool one of the streams and heat the other. A generic shell-and-tube heat exchanger can be illustrated by a line diagram as shown in Fig. 2.7. Shell-and-tube heat exchangers consist of a shell (or outer sleeve) through which several pipes pass. (Our figure just has one pipe for simplicity.) One of the process streams passes through the shell, and the other passes through the tubes. Stream *A* in our example passes through the shell, and stream *B* passes through the tubes. *The streams are physically separated from each other by the tube walls and do not mix.* Let's suppose that Stream *A* is the hot stream and Stream *B* is the cold stream. In our figure, both streams flow from left to right. This type of flow pattern is called *cocurrent*. The temperatures of the two streams will approach each other as they flow to the right. With this type of flow pattern, we must be careful that the hot stream temperature that we calculate is always higher than the cold stream temperature at every point in the heat exchanger.[1] If we reverse the flow direction of Stream *A,* a *countercurrent* flow pattern results. With a countercurrent flow pattern, the outlet temperature of the cold stream can be higher than the outlet temperature of the hot stream (but still must be lower than the inlet temperature of the hot stream). The hot stream temperature must always be above the cold stream temperature at all points along the tubes in this flow pattern also.

So far, our discussion has assumed that there are no phase transitions occurring in the heat exchanger. If Stream *A* is a hot stream, and Stream *B* is converted from liquid at the inlet to vapor at the outlet, we call the heat exchanger a *boiler* to bring attention to the phase transition occurring

---

1. This may seem like common sense, but sometimes when calculations are performed, it is surprisingly easy to overlook the fact that a valid mathematical result might be physically impossible to obtain.

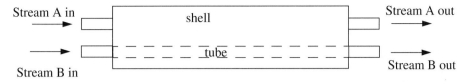

Figure 2.7 *Illustration of a generic heat exchanger with a cocurrent flow pattern. The tube-side usually consists of a set of parallel tubes which are illustrated as a single tube for convenience.*

inside. The primary difference in the operation of a boiler to that of a generic heat exchanger is that the cold stream temperature change might be small or even zero. This is because the phase change will occur isothermally at the saturation temperature of the fluid corresponding to the boiler pressure, absorbing large amounts of heat. In a similar fashion, we could have Stream *A* be cooling water and Stream *B* be an incoming vapor which is condensed. We would call this heat exchanger a *condenser*, to clearly bring attention to the phase change occurring inside. In this case, the temperature change of the hot stream might be small. Another type of heat exchanger that we will use in Chapter 4 is the *superheater*. A superheater takes a vapor that is saturated and superheats it.

There are two more important points to keep in your mind as you perform thermodynamic calculations. For the purposes of this text we will neglect pressure drops in the heat exchangers; the outlet pressure will match the inlet pressure of Stream *A,* and a similar statement applies for Stream *B*. Note that this does not imply that streams *A* and *B* are at the *same* pressure. Also, we neglect heat transfer to or from the surroundings unless specified. Therefore, all heat transfer occurs inside the heat exchanger, not at the boundaries of the heat exchanger and the surroundings.

There are other configurations of heat exchangers such as kettle-type reboilers or plate-and-frame configurations; however, for thermodynamic purposes, only the flow pattern is important, not the details of material construction that lead to the flow pattern. Thus, the shell-and-tube concepts will be adequate for our needs.

The energy balance that we write depends on how we choose our system. Since the streams are physically separated from each other, we may write a balance for each of the streams independently, or we may place the system boundary around the entire heat exchanger and write a balance for both streams. Let us take the system to be Stream *B* and let us suppose that Stream *B* is boiled. In this case, there is just one inlet and outlet. There is no shaft work or expansion/contraction work.[1] If our system is operating at steady-state, the left-hand side of the energy balance is zero,

$$0 = \left[H + \frac{u^2}{2g_c} + \frac{gz}{g_c} \right]^{in} \dot{m}^{in} - \left[H + \frac{u^2}{2g_c} + \frac{gz}{g_c} \right]^{out} \dot{m}^{out} + \underline{\dot{Q}} + \dot{W}_{EC} + \dot{W}_S$$

> ❶ The energy balance for a heat exchanger may be written in several ways.

which simplifies to

$$0 = -\Delta H \dot{m} + \underline{\dot{Q}} \qquad\qquad 2.44$$

---

1. Even though the process fluid is expanding as it evaporates, the system *boundaries* are not expanding. The expansion effects will be automatically included in the energy balance by the enthalpy terms which have the flow work embedded in them.

where $\dot{Q}$ is the rate of heat transfer from the hot stream. On a molar (or mass) basis,

$$0 = -\Delta H + Q \qquad\qquad 2.45$$

If we take the system boundaries to be around the entire heat exchanger, then we have multiple streams, and all heat transfer occurs inside, and we have

$$0 = \sum_{inlets}\left[H + \cancel{\frac{u^2}{2g_c}} + \cancel{\frac{gz}{g_c}}\right]^{in} \dot{m}^{in} - \sum_{outlets}\left[H + \cancel{\frac{u^2}{2g_c}} + \cancel{\frac{gz}{g_c}}\right]^{out} \dot{m}^{out} + \cancel{\dot{Q}} + \cancel{\dot{W}_{EC}} + \cancel{\dot{W}_S}$$

which simplifies to

$$0 = \sum_{inlets} H^{in}\dot{m}^{in} - \sum_{outlets} H^{out}\dot{m}^{out} \qquad\qquad 2.46$$

$$0 = -\Delta H_A \dot{m}_A - \Delta H_B \dot{m}_B \qquad\qquad 2.47$$

Since Eqns. 2.45 and 2.47 look quite different for the same process, it is important that you understand the placement of boundaries and their implications on the balance expression.

Turbine or Expander

A turbine or expander is basically a sophisticated windmill. The term "turbine" implies operation by steam and the term "expander" implies operation by a different process fluid, perhaps a hydrocarbon, although the term "turbine" is used sometimes for both. The objective of operation is to convert the kinetic energy from a gas stream to rotary motion of a shaft to produce work (shaft work). The enthalpy of the high-pressure inlet gas is converted to kinetic energy by special stators (stationary blades) or nozzles *inside the turbine shell*. The high velocity gas drives the rotor. Turbines are designed to be adiabatic, although heat losses can occur. When heat losses are present, they decrease the output that would have otherwise been possible for the turbine. Therefore, when calculations are performed, we assume that turbines or expanders operate adiabatically, unless otherwise noted.

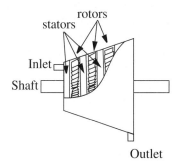

Figure 2.8 *Illustration of a turbine. The rotor (shaft) turns due to the flow of gas. The blades connected to the shell are stationary (stators), and are sometimes curved shapes to perform as nozzles. The stator blades are not shown to make the rotors more clear.*

The energy balance for the turbine only involves the kinetic energy change for the *entering* and *exiting* fluid, not for the changes occurring inside the turbine. Since the nozzles which cause large kinetic energy changes are *inside* the turbine unit, these changes are irrelevant to the balance *around* the unit. Recall from the development of our energy balance that we are only interested in the values of enthalpy, kinetic and potential energy for streams *as they cross the boundaries* of our system. The energy balance for a *steady-state* turbine involving one inlet and one outlet is:

$$0 = \left[H + \cancel{\frac{u^2}{2g_c}} + \cancel{\frac{gz}{g_c}} \right]^{in} \dot{m}^{in} - \left[H + \cancel{\frac{u^2}{2g_c}} + \cancel{\frac{gz}{g_c}} \right]^{out} \dot{m}^{out} + \cancel{\dot{Q}} + \cancel{\dot{W}_{EC}} + \dot{W}_S$$

which becomes

$$0 = (H^{in} - H^{out})\dot{m} + \dot{\underline{W}}_S = -\Delta H \dot{m} + \dot{\underline{W}}_S \qquad 2.48$$

❶ Adiabatic turbine or expander.

and on a mass or molar basis becomes

$$0 = -\Delta H + W_S \qquad 2.49$$

When we calculate values for the ΔH and work, they will be negative values.

Compressors

Compressors can be constructed in a manner qualitatively similar to turbines with stationary vanes (stators). This type of compressor is called an axial compressor. The main differences between turbines and axial compressors are: 1) the details of the construction of the vanes and rotors, which we won't be concerned with; 2) the direction of flow of the fluid; and 3) the fact that we must put work into the compressor rather than obtaining work from it. Thus, the energy balance is the same as the turbine (Eqns. 2.48 and 2.49). When we calculate values for the ΔH and work, they will be positive values, where they were negative values for a turbine. Compressors may also be constructed as reciprocating (piston/cylinder) devices. This modification has no impact on our energy balance, so it remains the same.

Pumps

Pumps are used to move liquids by creating the pressure necessary to overcome the resistance to flow. They are in principle just like compressors, except the liquid will not change density the way a gas does when it is compressed. Again, the energy balance will be the same as a turbine or compressor (Eqns. 2.48 and 2.49). The primary difference we will find in application of the energy balance is that tabulated enthalpies are difficult to find for compressed liquids. Therefore, if we want to calculate the work needed for a pump, we can find it from the energy balance after we have calculated or determined the enthalpy change. Let us find a way to perform this calculation by an example.

❶ Compressors and pumps usually have the same energy balance as turbines.

Example 2.8 The integral representing shaft work

Consider steady state, adiabatic, horizontal operation of a pump, turbine, or compressor. Calculate the integral in terms of P-V-T properties which represents the shaft work.

Solution: As our system, we choose a packet of fluid containing a unit mass as it flows through the equipment. This system is closed. The system, as we have chosen it, does not include a shaft even though it will move past the shaft.[1] Therefore, all work for this closed system is technically expansion/contraction work. We will see that this work from the closed-system perspective is composed of the flow work and shaft work which we have seen from the open-system perspective.

We cannot say exactly what happens to our system at every point, but we can say something about how it starts off and how it finishes up. This observation leads to what is called an integral method of analysis.

System: closed; Basis: packet of mass m. The kinetic energy change will again be small:

Note that this derivation neglects kinetic and potential energy changes.

$$d\left[U + \frac{u^2}{2g_c} + \frac{g}{g_c}z \right] = dQ + dW_S + dW_{EC}$$

Integrating from the inlet (initial) state to the outlet (final) state.

$$U^{out} - U^{in} = W_{EC}$$

We may change the form of the integral representing work via integration by parts:

$$W_{EC} = -\int_{in}^{out} PdV = -[PV]_{in}^{out} + \int_{in}^{out} VdP$$

We recognize the term PV as representing the work done by the flowing fluid entering and leaving the system; it does not contribute to the work output to the turbine. Therefore, the work interaction with the turbine is the remaining integral, $W_S = \int_{in}^{out} VdP$, which can be seen by comparing our result with the open-system energy balance:

$$\Rightarrow [U + PV]^{out} - [U + PV]^{in} = \int_{in}^{out} VdP$$

---

1. If you have trouble seeing this, remember that the system boundaries cannot change throughout a given problem. Since the system boundary doesn't contain the shaft before the packet enters, or after the packet exits, it cannot contain the shaft as it moves through the turbine.

Example 2.8 The integral representing shaft work (Continued)

Recall that $H = U + PV$ and compare with Eqn. 2.49. This means $W_s = \int VdP$ is the work done using the pump, compressor or turbine as the system. When the pressure change is negative, as in proceeding through a turbine or expander, the work done on the system is negative. This is consistent with our sign convention

$$W_S = \int_{in}^{out} V dP \qquad\qquad 2.50$$

Note: The shaft work given by $dW_S = VdP$ is not the same as expansion/contraction work, $dW_{EC} = PdV$. Also, neither is the same as flow work, $d\underline{W}_{flow} = PVdm$.

Now, we usually won't need to evaluate the integral for a compressor or a turbine because we usually can find enthalpy values directly from a table or chart. However, for compressed liquids, these values are harder to find. We may evaluate the integral for a liquid approximately by recognizing that a liquid below $T_r = 0.75$ is incompressible and

$$W_S \approx V^L(P^{out} - P^{in}) = \frac{\Delta P}{\rho}. \qquad\qquad 2.51$$

❶ Shaft work for an adiabatic pump or turbine where kinetic and potential energy changes are small.

❶ Shaft work for an adiabatic *liquid* pump or turbine where kinetic and potential energy changes are small and $T_r < 0.75$ so that the fluid is incompressible.

2.13 STRATEGIES FOR SOLVING PROCESS THERMODYNAMICS PROBLEMS

Before we start several more complicated example problems, it will be helpful to outline the strategies which will be applied. We provide these in a step form to make them easier to use. Many of these steps will seem obvious, but if you run into problems when working a problem, it is usually because one of these steps was omitted or applied inconsistently with system boundaries.

1. Choose system boundaries; decide whether this boundary location will make the system **open or closed.**

2. Identify all given state properties of fluids in system and crossing boundaries. Identify which are invariant with time. **Identify your system as steady or unsteady-state.** (For unsteady-state pumps, turbines or compressors, the accumulation of energy within the device is usually neglected.) For open, steady-state systems, write the mass balance and **solve if possible.**

3. Identify how many state variables are unknown for the system. **Recall that only two state variables are required to specify the state of a pure, single-phase fluid.** The number of unknowns will equal the number of independent equations necessary for a solution. (Remember in a system of known total volume $\underline{V}$, that if n is known, the state variable V is known.)

❶ The phase rule is important in determining the required number of equations.

4. **Write the mass balance and the energy balance. These are the first equations to be used in the solution.** Specify reference states for all fluids if necessary. Simplify energy balance to **eliminate terms which are zero** for the system specified in step 1.[1] Combine the mass balance and the energy balance for open systems.

---

1. Section 2.16 on page 75 "Hints on Application of the Mass and Energy Balances" may be helpful for details regarding how to interpret each term of the balance for new applications.

For unsteady-state problems:

(*a*) Identify whether the individual terms in the energy balance may be integrated directly without combining with other energy balance terms. Often the answer is obtained most easily this way. *This is almost always possible for closed-system problems.*

(*b*) If term-by-term integration of the energy balance is not possible, rearrange the equation to simplify as much as possible before integration.

5. Look for any other information in the problem statement that will provide **additional equations** if unknowns remain. Look for key words such as **adiabatic, isolated, throttling, nozzle, reversible, irreversible.** Using any applicable constraints of throttling devices, nozzles, etc., relate stream properties for various streams to each other and to the system state properties. Constraints on flowrates, heat flow, etc. provide additional equations. With practice, many of these constraints may be recognized immediately before writing the energy balance in steps 3 and 4.

6. Introduce the thermodynamic properties of the fluid (the equation of state). **This provides all equations relating P, V, T, U, H, C_P, C_V.** The information will consist of either 1) the ideal gas approximation; 2) a thermodynamic chart or table; or 3) a volumetric equation of state (which will be introduced in Chapter 6). Using more than one of these sources of information in the same problem may introduce inconsistencies in the properties used in the solution, depending on the accuracy of the methods used.

Combine the thermodynamic information with the energy balance. Work to minimize the number of state variables which remain unknown. Many problems are solved at this point.

7. Do not hesitate to move your system boundary and try again if you are stuck. Do not forget to try an overall balance (frequently, two open systems can be combined to give an overall closed system, and strategy 4*a* can be applied). Make reasonable assumptions.

8. After an answer is obtained, verify assumptions that were made to obtain the solution.

2.14   CLOSED AND STEADY-STATE OPEN SYSTEMS

Example 2.9   Adiabatic, reversible expansion of an ideal gas

Suppose an ideal gas in a piston + cylinder is adiabatically and reversibly expanded to twice its original volume. What will be the final temperature?

Solution: First consider the energy balance. The system will be the gas in the cylinder. The system will be closed. Since a basis is not specified, we can choose 1 mole. Since there is no mass flow, heat transfer or shaft work, the energy balance becomes:

$$d\left[U + \frac{u^2}{2g_c} + \frac{g}{g_c}z\right] = dQ + dW_S + dW_{EC}$$

$$dU = -PdV$$

Example 2.9 Adiabatic, reversible expansion of an ideal gas (Continued)

In this case, as we work down to step 4 in the strategy, we see that we cannot integrate the sides independently since P depends on T. Therefore, we need to combine terms before integrating.

$$C_V dT = -RT\frac{dV}{V} \quad \text{which becomes} \quad \frac{C_V}{T}dT = -\frac{R}{V}dV \qquad \text{(ig) 2.52}$$

The technique that we have performed is called separation of variables. All of the temperature dependence is on the left-hand side of the equation and all of the volume dependence is on the right-hand side. Now, if we assume a constant heat capacity for simplicity, we can see that this integrates to

$$\frac{C_V}{R}\ln\frac{T}{T^i} = \ln\frac{V^i}{V} \qquad (\text{*ig})$$

$$\boxed{\left(\frac{T}{T^i}\right)^{\frac{C_V}{R}} = \frac{V^i}{V}} \qquad (\text{*ig}) \ 2.53$$

> ❶ These boxed equations relate state variables for adiabatic reversible changes of an ideal gas in a closed system.

Although not required, several rearrangements of this equation are useful for other problems. Note that we may insert the ideal gas law to convert to a formula relating T and P. Using $V = RT/P$,

$$\left(\frac{T}{T^i}\right)^{\frac{C_V}{R}} = \frac{T^i P}{P^i T} \qquad (\text{*ig})$$

Rearranging,

$$\left(\frac{T}{T^i}\right)^{\frac{C_V}{R}}\frac{T}{T^i} = \left(\frac{T}{T^i}\right)^{\frac{C_V}{R}+1} = \frac{P}{P^i} \qquad (\text{*ig})$$

which becomes

$$\boxed{\left(\frac{T}{T^i}\right)^{\frac{C_P}{R}} = \frac{P}{P^i}} \qquad (\text{*ig}) \ 2.54$$

Example 2.9 Adiabatic, reversible expansion of an ideal gas (Continued)

We may also insert the ideal gas law into Eqn. 2.53 to convert to a formula relating P and V. Using $T = PV/R$,

$$\left(\frac{PV}{P^i V^i}\right)^{\frac{C_V}{R}} = \frac{V^i}{V} \qquad (\text{*ig})$$

$$\frac{P}{P^i} = \left(\frac{V^i}{V}\right)^{\frac{R}{C_V}} \frac{V^i}{V} = \left(\frac{V^i}{V}\right)^{\frac{R}{C_V}+1} = \left(\frac{V^i}{V}\right)^{\frac{C_P}{C_V}} \qquad (\text{*ig})$$

which may be written

$$\boxed{PV^{\frac{C_P}{C_V}} = \text{const}} \qquad (\text{*ig})\ 2.55$$

Example 2.10 Continuous adiabatic, reversible compression of an ideal gas

Suppose 1 kmole/h of air at 5 bars and 298 K is adiabatically and reversibly compressed in a continuous process to 25 bars. What will be the outlet temperature and power requirement for the compressor in hp?

Solution: Note that air is composed primarily of oxygen and nitrogen and these both satisfy the stipulations for diatomic gases with their reduced temperatures high and their reduced pressures low. In other words, the ideal gas approximation with $C_P/R = 7/2$ is clearly applicable. Next consider the energy balance. The system is the compressor. The system is open. Since it is a steady-state process with no heat transfer, the simplification of the energy balance has been shown in Section 2.8, and the energy balance becomes:

$$\boxed{\Delta H = \cancel{Q} + W_S} \qquad 2.56$$

We can adapt Eqn. 2.50 for an ideal gas as follows:

$$dW_S = dH = VdP$$

In this case, as we work down to step 4 in the strategy, we see that we cannot integrate the sides independently since P depends on T. Therefore, we need to combine terms before integrating.

$$C_P dT = RT\frac{dP}{P} \quad \text{which becomes} \quad \frac{C_P}{T}dT = \frac{R}{P}dP \qquad (\text{ig})\ 2.57$$

Example 2.10   Continuous adiabatic, reversible compression of an ideal gas (Continued)

Once again, we have performed separation of variables. The rest of the derivation is entirely analogous to Example 2.9, and, in fact, the resulting formula is identical.

$$\frac{T_2}{T_1} = \left(\frac{P_2}{P_1}\right)^{\frac{R}{C_P}}$$

(*ig) 2.58

❶ Steady-state adiabatic, reversible processing of an ideal gas results in the same relations as Example 2.9.

Note that this formula comes up quite often as an approximation for both open and closed systems. Making the appropriate substitutions:

$$T_2 = 298\left(\frac{25}{5}\right)^{\frac{2}{7}} = 472 \text{ K}$$

Applying the complete energy balance, $W_S = \Delta H = C_P \Delta T = 3.5 \cdot 8.314 \cdot (472-298) = 5063$ J/mole

At the given flow rate, and reiterating that this problem statement specifies a reversible process:

$W_S^{rev} = 5063$ J/mole·[1000mole/hr]·[1hr/3600sec]·[1hp/(745.7J/s)] = 1.9hp

This kind of calculation is extremely common.

Example 2.11   Continuous, isothermal, reversible compression of an ideal gas

Repeat the compression from the previous example, but consider steady-state isothermal compression. What will be the heat removal rate and power requirement for the compressor in hp?

Solution: First consider the energy balance. The system is the compressor. The system is open. Since it is a steady-state process *with* heat transfer, the energy balance becomes:

$$\Delta H = Q + W_S$$

2.59

Recalling the definition of the heat capacity for an ideal gas,

$$dH = C_P dT = 0$$

Example 2.11 Continuous, isothermal, reversible compression of an ideal gas (Continued)

We can still adapt Eqn. 2.50 for an ideal gas as follows:

$$dW_S = -dQ = VdP$$

In this case, the constant temperature and the ideal gas equation of state provide $V = \dfrac{RT}{P}$ which becomes

$$W_S = -Q = RT\ln\left(\frac{P_2}{P_1}\right) = 8.314(298)\ln 5 = 3987\,\frac{J}{mol} \qquad \text{(ig) 2.60}$$

At the given flow rate, and reiterating that this problem statement specifies a reversible process:

$$W_S^{rev} = 3987 \text{ J/mole}\cdot[1000\text{mole/hr}]\cdot[1\text{hr}/3600\text{sec}]\cdot[1\text{hp}/(745.7\text{J/s})] = 1.5 \text{ hp}$$

Comparing with the previous example, the isothermal compressor requires less work. (Here is a brain teaser. Suppose the process fluid had been steam instead of an ideal gas. How would you have solved the problem then? Note, for steam $V \neq \dfrac{RT}{P}$ and $dH \neq C_P dT$. The answer is given in the next chapter.)

Example 2.12 Heat loss from a turbine

High pressure steam at a rate of 1100 kg/h initially at 3.5 MPa and 350°C is expanded in a turbine to obtain work. Two exit streams leave the turbine. Exiting stream (2) is at 1.5 MPa and 225°C and flows at 110 kg/h. Exiting stream (3) is at 0.79 MPa and is known to be a mixture of saturated vapor and liquid. A fraction of stream (3) is bled through a throttle valve to 0.10 MPa and is found to be 120°C. If the measured output of the turbine is 100 kW, estimate the heat loss of the turbine. Also, determine the quality of the steam in stream (3).

Solution: First draw a schematic. Designate boundaries. Both system A and system B are open steady-state systems.

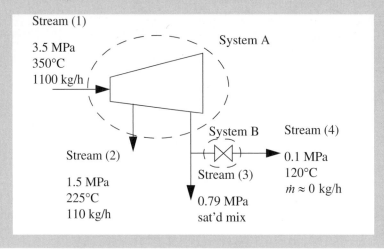

Example 2.12 Heat loss from a turbine (Continued)

The mass balance gives $\dot{m}_3$ = 990 kg/h. Next, determine where the state of the streams are completely specified: streams (1), (2), and (4) are fully specified. Since stream (3) is saturated, the temperature and pressure and specific enthalpies of the saturated vapor and liquid can be found, but the quality needs to be calculated to determine the overall molar enthalpy of the stream. From the steam tables we find H_1 directly. For H_2 we use linear interpolation. The value $H(1.5$ MPa, $225°C)$ is not available directly, so we need to first interpolate at 1.4 MPa between 200°C and 250°C to find $H(1.4$ MPa, $225°C)$ and then interpolate between this value and the value at 1.6 MPa:

$$H(1.4\text{MPa}, 225°C) = \frac{1}{2}(2803.0 + 2927.9) = 2865.5 \text{ kJ/kg}$$

$$H(1.6\text{MPa}, 225°C) = 2792.8 + \frac{23.6}{48.6}(2919.9 - 2792.8) = 2854.5 \text{ kJ/kg}$$

Then to find H_2: $H_2 = 0.5 \cdot (2865.5 + 2854.5) = 2860.0$ kJ/kg. For H_4 we can interpolate in the superheated steam tables:

$$H_4 = 2675.8 + \frac{20}{50}(2776.6 - 2675.8) = 2716.1 \text{ kJ/kg}$$

Recognize system B as a throttle valve; therefore, $H_3 = H_4 = 2716.1$ kJ/kg. We make a table to summarize the results so we can easily find values:

stream	1	2	3, 4
H(kJ/kg)	3104.8	2860.0	2716.1

The energy balance for system A gives, using $\dot{\underline{W}}_S = -100$ kW,

$$0 = H_1\dot{m}_1 - H_2\dot{m}_2 - H_3\dot{m}_3 + \dot{Q} + \dot{\underline{W}}_S = 3104.8(1100) - 2860.0(110) - 2716.1(990) + \dot{Q} + \dot{\underline{W}}_S$$

$$\dot{Q} = \frac{-411{,}741 \text{ kJ}}{h}\bigg|\frac{h}{3600 \text{ s}} + \frac{100 \text{ kJ}}{s} = -14.4 \text{ kJ/s}$$

To find the quality of stream (3), $H_3\dot{m}_3 = H^L\dot{m}^L - H^V\dot{m}^V$,

$$H_3 = H^L\frac{\dot{m}^L}{\dot{m}_3} - H^V\frac{\dot{m}^V}{\dot{m}_3} = H^L + q(\Delta H^{vap})$$

At 0.79 MPa from the sat'd P table, $H^L = 718.5$ kJ/kg and $\Delta H^V = 2049$ kJ/kg.

$$q = \frac{2716.1 - 718.5}{2049} = 0.975$$

2.15   UNSTEADY-STATE OPEN SYSTEMS (OPTIONAL)

Example 2.13   Adiabatic expansion of an ideal gas from a leaky tank

An ideal gas is leaking from an insulated tank. Relate the change in temperature to the change in pressure for gas leaking from a tank. Determine ΔU for the tank.

Solution: Let us choose our system as the gas in the tank at any time. This will be an open unsteady-state system. We have no inlet streams and one outlet stream. The mass balance gives $dn = -dn^{out}$.

We can neglect kinetic and potential energy changes. Although the gas is expanding, the system size remains unchanged, and there is no expansion/contraction work. The energy balance becomes (on a molar basis):

$$d(nU) = H^{in}dn^{in} - H^{out}dn^{out} + dQ + dW_{EC} + dW_S$$

Since the enthalpy of the exit stream matches the enthalpy of the tank, $H^{out} = H$. $d(nU) = -H^{out}dn = Hdn$. Now H depends on temperature, which is changing, so we are not able to apply hint 4a from the problem solving strategy. It will be necessary to combine terms before integrating. By the product rule of differentiation, the left-hand side expands to $d(nU) = ndU + Udn$. Collecting terms in the energy balance,

$$ndU = (H-U)dn$$

Performing some substitutions, the energy balance can be written in terms of T and n,

$$(H-U) = PV = RT; \quad dU = C_V dT; \Rightarrow \frac{C_V}{R}\frac{dT}{T} = \frac{dn}{n} \tag{ig}$$

$$\boxed{\frac{C_V}{R}\ln\frac{T}{T^i} = \ln\frac{n}{n^i}} \tag{*ig}$$

The volume of the tank is constant, ($\underline{V}$ = constant), therefore,

$$\ln\frac{n}{n^i} = \ln\frac{PT^i}{TP^i} = -\ln\frac{T}{T^i} + \ln\frac{P}{P^i} = \frac{C_V}{R}\ln\frac{T}{T^i} \tag{ig}$$

substituting,

$$\left(\frac{C_V}{R} + 1\right)\ln\frac{T}{T^i} = \frac{C_P}{R}\ln\frac{T}{T^i} = \ln\frac{P}{P^i} \tag{*ig}$$

recognizing the relation between C_V and C_P,

$$\boxed{\frac{T}{T^i} = \left(\frac{P}{P^i}\right)^{R/C_P} = \left(\frac{P}{P^i}\right)^{(1-1/\gamma)} \quad \gamma \equiv C_P / C_V (= 1.4 \text{ for ideal diatomic gas})} \tag{*ig} 2.61$$

❗ For fluid *exiting* from an adiabatic tank, the results are the same as a closed system as in Example 2.9.

Example 2.13   Adiabatic expansion of an ideal gas from a leaky tank (Continued)

Through the ideal gas law ($PV = RT$), we can obtain other arrangements of the same formula.

$$V^i / V = \left(P / P^i\right)^{1/\gamma}; P / P^i = \left(V^i / V\right)^{\gamma} = \left(T / T^i\right)^{(1/\gamma)-1}; T / T^i = \left(V^i / V\right)^{\gamma/(\gamma-1)}$$

$$(\text{*ig})\ 2.62$$

The numerical value for the change in internal energy *of the system* depends on the reference state because the reference state temperature will appear in the result:

$$\Delta \underline{U} = n^f(C_V(T^f - T_R) + U_R) - n^i(C_V(T^i - T_R) + U_R) \qquad (\text{*ig})$$

Example 2.14   Adiabatically filling a tank with an ideal gas

Helium at 300 K and 3000 bar is fed into an evacuated cylinder until the pressure in the tank is equal to 3000 bar. Calculate the final temperature of the helium in the cylinder ($C_P/R = 5/2$).

Solution: The system will be the gas inside the tank at any time. The system will be an open, unsteady-state system. The mass balance is $dn = dn^{in}$. The energy balance becomes:

$$d(nU) = H^{in}dn^{in} - H^{out}\cancel{dn^{out}} + \cancel{dQ} + \cancel{dW_{EC}} + \cancel{dW_S}$$

We recognize that H^{in} will be constant throughout the tank filling. Therefore, by hint 4a from the problem solving strategy, we can integrate terms individually. We need to be careful to keep the superscript since the incoming enthalpy is at a different state than the system. The right-hand side of the energy balance can be integrated to give

$$\int_i^f H^{in}dn = H^{in}\int_i^f dn = H^{in}(n^f - n^i) = H^{in}n^f$$

The left-hand side of the energy balance becomes

$$\Delta(Un) = U^f n^f - U^i n^i = U^f n^f$$

Combining the result with the definition of enthalpy:

$$U^f = H^{in} = U^{in} + PV^{in} = U^{in} + RT^{in} \qquad (\text{ig})\ 2.63$$

And with our definition of heat capacity, we can find temperatures:

$$\Delta U = C_V\left(T^f - T^{in}\right) = RT^{in} \Rightarrow T^f = T^{in}\left(R + C_V\right) / C_V = T^{in}C_P / C_V \qquad (\text{*ig})$$

Note that the final temperature is independent of pressure for the case considered here.

Example 2.15 Adiabatic expansion of steam from a leaky tank

An insulated tank contains 500 kg of steam and water at 215°C. Half of the tank volume is occupied by vapor and half by liquid. 25 kg of dry vapor is vented slowly enough that temperature remains uniform throughout the tank. What is the final temperature and pressure?

Solution: There will be some similarities to the solution of Example 2.13 on page 72; however, we can no longer apply the ideal gas law. The energy balance reduces in a similar way, but we note that the exiting stream consists of only vapor; therefore, it is not the overall average enthalpy of the tank:

$$d(mU) = -H^{out}dm^{out} = H^V dm$$

The sides of the equation can be integrated independently if the vapor enthalpy is constant. Looking at the steam table, the enthalpy changes only about 10 kJ/kg out of 2800 kJ/kg (0.3%) along the saturation curve down to 195°C. Let us assume it is constant at 2795 kJ/kg making the integral of the right-hand side simply $H^V \Delta m$. Note that this procedure is equivalent to a numerical integration by trapezoidal rule as given in Appendix B on page 603. Many students forget that analytical solutions are merely desirable, not absolutely necessary. The energy balance then can be integrated using hint 4a on page 66.

$$\Delta \underline{U} = m^f U^f - m^i U^i = 2795(m^f - m^i) = 2795(-25) = -69,875 \text{ kJ}$$

The quantity $m^f = 475$, and $m^i U^i$ will be easy to find, which will permit calculation of U^f. For each m³ of the original saturated mixture at 215°C:

$$\frac{0.5 \text{ m}^3 \text{ vapor}}{} \left| \frac{\text{kg vapor}}{0.0947 \text{m}^3 \text{ vapor}} = 5.28 \text{ kg vapor} \right.$$

$$\frac{0.5 \text{ m}^3 \text{ liquid}}{} \left| \frac{\text{kg liquid}}{0.001181 \text{m}^3 \text{ liquid}} = 423.4 \text{ kg liquid} \right.$$

Therefore:

$$V^i = \frac{\text{m}^3}{423.4 + 5.28 \text{ kg}} = 0.00233 \text{ m}^3/\text{kg}$$

So the tank volume, quality and internal energy are:

$$\underline{V}_T = \frac{500 \text{kg}}{} \left| \frac{\text{m}^3}{428.63} = 1.166 \text{m}^3 \right.$$

$$q^i = 5.28 \text{ kg vapor} / 428.63 \text{ kg} = 0.0123$$

$$U^i = 918.04 + 0.0123(1681.9) = 938.7 \text{ kJ/kg}$$

$$\underline{U}^i = 938.7 \text{ kJ/kg} \cdot 500 \text{ kg} = 469,400 \text{ kJ}$$

Example 2.15 Adiabatic expansion of steam from a leaky tank (Continued)

Then, from the energy balance and mass balance:

$$U^f = (-69{,}875 + 469{,}400)\ \text{kJ} / 475\ \text{kg} = 841.0\ \text{kJ/kg}$$

$$V^f = 1.166\ \text{m}^3 / 475\ \text{kg} = 0.00245\ \text{m}^3/\text{kg}$$

We need to find P^f and T^f which correspond to these state variables. The answer will be along the saturation curve because the overall specific volume is intermediate between saturated vapor and liquid values at lower pressures. We will guess P^f (and the corresponding saturation T^f), find q from V^f, then calculate U^f_{calc} and compare to $U^f = 841.0$ kJ/kg. If U^f_{calc} is too high, the P^f (and T^f) guess will be lowered.

Since $V = V^L + q(V^V - V^L)$,

$$q = \frac{0.00245 - V^L}{V^V - V^L} \quad \text{and from this value of } q,\ U^f_{calc} = U^L + q(U^V - U^L)$$

To guide our first guess, we need $U^L < U^f = 841.0$ kJ/kg. Our first guess is $T^f = 195°C$. Values for the properties from the steam tables are shown in the table below. This initial guess gives $U^f_{calc} = 845$ kJ/kg; no further iteration is necessary. The H^V at this state is 2789, therefore our assumption of $H^{out} \approx$ constant is valid.

state	P(MPa)	T (°C)	V^L	V^V	U^L	ΔU^{vap}	q
initial	2.106	215	0.001181	0.0947	918.04	1681.9	0.0123
guess	1.3988	195	0.001149	0.1409	828.18	1763.6	0.0093

$$P^f = 1.4\ \text{MPa},\ T^f = 195°C,\ \Delta P = 0.7\ \text{MPa},\ \Delta T = 20°C$$

2.16 DETAILS OF TERMS IN THE ENERGY BALANCE (OPTIONAL)

Generally, the strategies discussed in Section 2.13 are sufficient to simplify the energy balance. Occasionally, in applying the energy balance to a new type of system, simplification of the balances may require more detailed analysis of the background leading to the terms and/or details of interactions at boundaries. This section provides an overall summary of the details for the principles covered earlier in this chapter, and it is usually not necessary unless you are having difficulty simplifying the energy balance and need details regarding the meaning of each term.

The universe frequently consists of 3 subsystems, as illustrated in Fig. 2.9. The container (system 2) is frequently combined with system 1 (designated here as system (1 + 2)) or system 3 (designated here as system (2 + 3)). For every balance, all variables are of the *system* for which the balance is written.

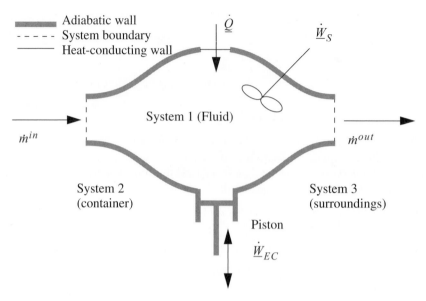

Figure 2.9 *Schematic of a general system.*

non - S.S/ open

$$\frac{d}{dt}\left(\underline{U} + \frac{1}{2}\frac{mu^2}{g_c} + \frac{mgz}{g_c}\right) = \sum_{inlets}\left(H + \frac{1}{2}\frac{u^2}{g_c} + \frac{gz}{g_c}\right)^{in}\dot{m}^{in}$$

$$- \sum_{outlets}\left(H + \frac{1}{2}\frac{u^2}{g_c} + \frac{gz}{g_c}\right)^{out}\dot{m}^{out} + \dot{\underline{Q}} + \dot{\underline{W}}_S + \dot{\underline{W}}_{EC}$$

2.64

Where superscripts "in" and "out" denote properties of the streams which cross the boundaries, which may or may not be equal to properties of the system.

1. Non-zero heat interactions of systems 1 and 3 are not equal unless the heat capacity or mass of system 2 is negligible.

2. H^{out}, H^{in} account for internal energy changes and work done on system *due to flow across boundaries.*

3. $\dot{\underline{W}}_{EC} = \left(-\frac{F_{boundary}}{A_{boundary}}\right)\frac{d\underline{V}_{system}}{dt}$ represents work done on system due to *expansion or con-traction of system size.* $F_{boundary}$ is the absolute value of the *system* force at the boundary.

   (By Newton's third principle, the forces are equal but act in opposite directions at any boundary.) Note that system 1 and system 3 forces are not required to be equal. Unequal forces create movement (acceleration) of any movable barrier (e.g., piston head).

4. $\dot{\underline{W}}_S$ represents the work done *on* the system resulting from *mechanical forces at the surface of the system* except work due to expansion/contraction or mass flow across boundaries. Turbines and compressors are part of system 2; thus are involved with work interactions with the fluid in system 1. Note that piston movement is calculated as $\underline{W}_{EC}$ for systems 1, 3, (1+2), or (2+3), but the movement is calculated as $\underline{W}_S$ for system 2 alone. When a balance for system 2 is considered, the movement of the piston is technically shaft work, even

though no shaft is involved. (The piston is a closed system, and it does not expand or contract when it moves.) As another example of the general definition of shaft work as it relates to forces at the surface of a system, consider the closed system of Fig. 2.3 on page 44 being raised 150 m or accelerated to 75 m/s. There is a work interaction at the surface of the system required for these energy changes even though there is not a "rotating" shaft.

5. Non-zero $\dot{W}_{EC}$ or $\dot{W}_S$ interactions of systems 1 and 3 are not equal unless changes in kinetic and potential energy of the moving portion of system 2 (e.g., piston head for $\underline{W}_{EC}$ or turbine for $\underline{W}_S$) are negligible and the movement is reversible.

6. Frictional forces, if present, *must* be attributed to one of the systems shown above. Irreversibility due to any cause does not require additional energy balance terms because energy is always conserved, even when processes are irreversible.

7. Electrical and magnetic fields have not been included.

8. On the lefthand side of the equation, kinetic and potential energy changes are calculated based on movement of the *center of mass*. In a composite system such as (1+2), they may be calculated for each subsystem and summed.

2.17 SUMMARY

We are trying to be very careful throughout this chapter to anticipate every possibilty that might arise. As a result, the verbiage gets very dense. If we relax the formality, we can summarize most of this chapter casually as follows:

$$\Delta U = Q + W_{EC} \qquad \text{closed systems}$$

$$\Delta H = Q + W_S \qquad \text{open systems}$$

$$d(nU) = H^{in} dn^{in} - H^{out} dn^{out} + d\underline{Q} + d\underline{W}_{EC} + d\underline{W}_S \qquad \text{open, unsteady-state systems}$$

Naturally, it is best to appreciate how these equations result from simplifications. Remember to check the general energy balance for terms that may be significant in exceptional situations.

2.18 PRACTICE PROBLEMS

A. General Reductions of the Energy Balance

The energy balance can be developed for just about any process. Since our goal is to learn how to develop model equations as well as to simply apply them, it is valuable practice to obtain the appropriate energy balance for a broad range of odd applications. If you can deduce these energy balances, you should be well prepared for the more common energy balances encountered in typical chemical engineering processes.

P2.1 A pot of water is boiling in a pressure cooker when suddenly the pressure relief valve becomes stuck, preventing any steam from escaping. System: the pot and its contents after the valve is stuck. (ANS. $d[mU] / dt = \dot{\underline{Q}}$)

P2.2 The same pot of boiling water as above. System: the pot and its contents before the valve is stuck. (ANS. $d[mU] = \dot{m}H^V + \underline{\dot{Q}}$)

P2.3 A gas home furnace has been heating the house steadily for hours. System: the furnace. (ANS. $\Delta\underline{H} = \underline{\dot{Q}}$)

P2.4 A gas home furnace has been heating the house steadily for hours. System: the house and all contents. (ANS. $d[mU]/dt = \underline{\dot{Q}}_{Heat} + \underline{\dot{Q}}_{Loss}$)

P2.5 A child is walking to school when hit by a snowball. He stops in his tracks. System: the child. (ANS. $\Delta[mU + mu^2/2g_c] = m_{snow}[H + u^2/2g_c]_{snow}$)

P2.6 A sealed glass bulb contains a small paddle-wheel. The paddles are painted white on one side and black on the other. When placed in the sun, the paddle wheel begins to turn steadily. System: the bulb and its contents. (ANS. $0 = \Delta\underline{U}$)

P2.7 A sunbather lays on a blanket. At 11:30 A.M., the sunbather begins to sweat. System: the sunbather at 12:00 noon. (ANS. $d[mU]/dt = \dot{m}H^V + \underline{\dot{Q}}$)

P2.8 An inflated balloon slips from your fingers and flies across the room. System: the balloon and its contents. (ANS. $d[mU + mu^2_{balloon}/2g_c]/dt =$

$[H + u^2/2g_c]^{out}\, dm/dt + \underline{\dot{W}}_{EC}$)

B. Numerical problems

P2.9 Consider a block of concrete weighing 1 kg.

(a) How far must it fall to change its potential energy by 1 kJ? (ANS. 100 m)
(b) What would be the value of its velocity at that stage? (ANS. 44.7 m/s)

P2.10 A block of copper weighing 0.2 kg with an initial temperature of 400 K is dropped into 4 kg of water initially at 300 K contained in a perfectly insulated tank. The tank is also made of copper and weighs 0.5 kg. Solve for the change in internal energy of both the water and the block given $C_V = 4.184$ J/g-K for water and 0.380 J/g-K for copper. (ANS. 7480 J, −7570 J)

P2.11 In the preceding problem, suppose that the copper block is dropped into the water from a height of 50 m. Assuming no loss of water from the tank, what is the change in internal energy of the block? (ANS. −7570 J)

P2.12 In the following take $C_V = 5$ and $C_P = 7$ cal/gmol-K for nitrogen gas:

(a) Five moles of nitrogen at 100°C is contained in a rigid vessel. How much heat must be added to the system to raise its temperature to 300°C if the vessel has a negligible heat capacity? (ANS. 5000 cal) If the vessel weighs 80 g and has a heat capacity of 0.125 cal/g-K, how much heat is required? (ANS. 7000 cal)
(b) Five moles of nitrogen at 300°C is contained in a piston/cylinder arrangement. How much heat must be extracted from this system, which is kept at constant pressure, to cool it to 100°C if the heat capacity of the piston and cylinder is neglected? (ANS. 7000 cal)

P2.13 A rigid cylinder of gaseous hydrogen is heated from 300 K and 1 bar to 400 K. How much heat is added to the gas? (ANS. 2080 J/mole)

P2.14 Saturated steam at 660°F is adiabatically throttled through a valve to atmospheric pressure in a steady-state flow process. Estimate the outlet quality of the steam. (ANS. $q = 0.96$)

P2.15 Refer to Example 2.6 about transformation of kinetic energy to enthalpy. Instead of water, suppose N_2 at 1 bar and 298 K was flowing in the pipe. How would that change the answers? In particular, how would the temperature rise change? (ANS. max ~0.001K)

P2.16 Steam at 150 bars and 600°C passes through process equipment and emerges at 100 bars and 700°C. There is no flow of work into or out of the equipment, but heat is transferred.

(a) Using the steam tables, compute the flow of heat into the process equipment per kg of steam. (ANS. 288 kJ/kg)
(b) Compute the value of enthalpy at the inlet conditions, H^{in}, relative to an ideal gas at the same temperature, H^{ig}. Consider steam at 1 bar and 600°C as an ideal gas. Express your answer as $(H^{in} - H^{ig})/RT^{in}$. (ANS. –0.305)

P2.17 A 700 kg piston is initially held in place by a removable latch above a vertical cylinder. The cylinder has an area of 0.1 m²; the volume of the gas within the cylinder initially is 0.1 m³ at a pressure of 10 bar. The working fluid may be assumed to obey the ideal gas equation of state. The cylinder has a total volume of 0.25 m³, and the top end is open to the surrounding atmosphere, at a pressure of 1 bar.

(a) Assume that the piston rises frictionlessly in the cylinder when the latches are removed and the gas within the cylinder is always kept at the same temperature. This may seem like an odd assumption, but it provides an approximate result that is relatively easy to obtain. What will be the velocity of the piston as it leaves the cylinder? (ANS. 13.8 m/s)
(b) What will be the maximum height to which the piston will rise? (ANS. 9.6 m)
(c) What is the pressure behind the piston just before it leaves the cylinder? (ANS. 4 bar)
(d) Now suppose the cylinder was increased in length such that its new total volume is 0.588 m³. What is the new height reached by the piston? (ANS. ~13 m)
(e) What is the maximum height we could make the piston reach by making the cylinder longer? (ANS. ~13 m)

P2.18 A tennis ball machine fires tennis balls at 40 mph. The cylinder of the machine is 1 m long; the installed compressor can reach about 50 psig in a reasonable amount of time. The tennis ball is about 3 inches in diameter and weighs about 0.125 lb$_m$. Estimate the initial volume required in the pressurized firing chamber. [Hint: note the tennis ball machine fires horizontally and the tennis ball can be treated as a frictionless piston. Don't be surprised if iterative solution is necessary and ln $(V_2/V_1) = \ln(1 + \Delta V/V_1)$] (ANS. 390 cm³)

P2.19 A 700 kg piston is initially held in place by a removable latch inside a horizontal cylinder. The totally frictionless cylinder (assume no viscous dissipation from the gas also) has an area of 0.1 m²; the volume of the gas on the left of the piston is initially 0.1 m³ at a pressure of 8 bars. The pressure on the right of the piston is initially 1 bar, and the total volume is 0.25 m³. The working fluid may be assumed to follow the ideal gas equation of state. What would be the highest pressure reached on the right side of the piston and what would be the position of the piston at that pressure? (a) assume isothermal; (b) What is the kinetic energy of the piston when the pressures are equal?[1] (partial ANS. 1.6 bars)

---

1. This problem is reconsidered as an adiabatic process in problem P3.14.

2.19   HOMEWORK PROBLEMS

2.1 Five grams of the specified pure solvent is placed in a variable volume piston. What are the molar enthalpy and total enthalpy of the pure system when 50% and 75% have been evaporated at: (*i*) 30°C, (*ii*) 50°C? Use liquid at 25°C as a reference state.

(a) benzene ($\rho^L = 0.88$ g/cm^3)
(b) ethanol ($\rho^L = 0.79$ g/cm^3)
(c) water without using the steam tables ($\rho^L = 1$ g/cm^3)
(d) water using the steam tables

2.2 Create a table of T, U, H for the specified solvent using a reference state of $H = 0$ for liquid at 25°C and 1 bar. Calculate the table for: (*i*) liquid at 25°C; (*ii*) saturated liquid at 1 bar; saturated vapor at 1 bar; (*iii*) vapor at 110°C and 1 bar. Use the Antoine Equation (Appendix E) to relate the saturation temperature and saturation pressure. Use the ideal gas law to model the vapor phase.

(a) benzene
(b) ethanol
(c) water without using the steam tables
(d) water using the steam tables

2.3 Three moles of an ideal gas (with temperature-independent $C_P = (7/2)R$, $C_V = (5/2)R$) is contained in a horizontal piston/cylinder arrangement. The piston has an area of 0.1 m^2 and mass of 500 g. The initial pressure in the piston is 101 kPa. Determine the heat that must be extracted to cool the gas from 375°C to 275°C at: (a) constant pressure; (b) constant volume.

2.4 One mole of an ideal gas ($C_P = 7R/2$) in a closed piston/cylinder is compressed from $T^i = 100$ K, $P^i = 0.1$ MPa to $P^f = 0.7$ MPa by the following pathways. For each pathway, calculate ΔU, ΔH, Q, and W_{EC}: (a) isothermal; (b) constant volume; (c) adiabatic.

2.5 One mole of an ideal gas ($C_P = 5R/2$) in a closed piston/cylinder is compressed from $T^i = 298$ K, $P^i = 0.1$ MPa to $P^f = 0.25$ MPa by the following pathways. For each pathway, calculate ΔU, ΔH, Q, and W_{EC}: (a) isothermal; (b) constant volume; (c) adiabatic.

2.6 One mole of an ideal gas ($C_P = 7R/2$) in a closed piston/cylinder is expanded from $T^i = 700$ K, $P^i = 0.75$ MPa to $P^f = 0.1$ MPa by the following pathways. For each pathway, calculate ΔU, ΔH, Q, and W_{EC}: (a) isothermal; (b) constant volume; (c) adiabatic.

2.7 One mole of an ideal gas ($C_P = 5R/2$) in a closed piston/cylinder is expanded from $T^i = 500$ K, $P^i = 0.6$ MPa to $P^f = 0.1$ MPa by the following pathways. For each pathway, calculate ΔU, ΔH, Q, and W_{EC}: (a) isothermal; (b) constant volume; (c) adiabatic.

2.8 (a) What is the enthalpy change needed to change 3 kg of liquid water at 0°C to steam at 0.1 MPa and 150°C?
(b) What is the enthalpy change needed to heat 3 kg of water from 0.4 MPa and 0°C to steam at 0.1 MPa and 150°C?
(c) What is the enthalpy change needed to heat 1 kg of water at 0.4 MPa and 4°C to steam at 150°C and 0.4 MPa?
(d) What is the enthalpy change needed to change 1 kg of water of a water-steam mixture of 60% quality to one of 80% quality if the mixture is at 150°C?

(e) Calculate the ΔH value for an isobaric change of steam from 0.8 MPa and 250°C to saturated liquid.

(f) Repeat part (e) for an isothermal change to saturated liquid.

(g) Does a state change from saturated vapor at 230°C to the state 100°C and 0.05 MPa represent an enthalpy increase or decrease? A volume increase or decrease?

(h) In what state is water at 0.2 MPa and 120.21°C? At 0.5 MPa and 151.83°C? At 0.5 MPa and 153°C?

(i) A 0.15 m³ tank containing 1 kg of water at 1 MPa and 179.88°C has how many m³ of liquid water in it? Could it contain 5 kg of water under these conditions?

(j) What is the volume change when 2 kg of H_2O at 6.8 MPa and 93°C expands to 1.6 bar and 250°C?

(k) Ten kg of wet steam at 0.75 MPa has an enthalpy of 22,000 kJ. Find the quality of the wet steam.

2.9 One kg of methane is contained in a piston/cylinder device at 0.8 MPa and 250°C. It undergoes a reversible isothermal expansion to 0.3 MPa. Methane can be considered an ideal gas under these conditions. How much heat is transferred?

2.10 One kilogram of steam in a piston/cylinder device undergoes the following changes of state. Calculate Q and W for each step.

(a) Initially at 350 kPa and 250°C, it is cooled at constant pressure to 150°C.

(b) Initially at 350 kPa and 250°C, it is cooled at constant volume to 150°C.

2.11 In one stroke of a reciprocating compressor, helium is isothermally and reversibly compressed in a piston + cylinder from 298 K and 20 bars to 200 bars. Compute the heat removal and work requirement per mole during each stroke.

2.12 Two moles of nitrogen is initially at 10 bar and 600 K (state 1) in a horizontal piston/cylinder device. It is expanded adiabatically to 1 bar (state 2). It is then heated at constant volume to 600 K (state 3). Finally, it is isothermally returned to state 1. Assume that N_2 is an ideal gas with a constant heat capacity as given in the back endflap of the book. Neglect the heat capacity of the piston/cylinder device. Suppose that heat can be supplied or rejected as illustrated below. Assume each step of the process is reversible.

Hot Reservoir

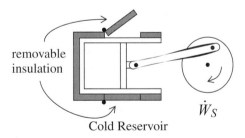

removable insulation

$\dot{W}_S$

Cold Reservoir

(a) Calculate the heat transfer and work done on the gas for each step and overall.

(b) Taking state 1 as the reference state, and setting $U_R^{ig} = 0$, calculate U and H for the nitrogen at each state, and ΔU and ΔH for each step and the overall Q and $\underline{W}_{EC}$.

(c) The atmosphere is at 1 bar and 298 K throughout the process. Calculate the work done on the atmosphere for each step and overall. (Hint: take the atmosphere as the system). How much work is transferred to the shaft in each step and overall?

2.13 One mole of methane gas held within a piston/cylinder, at an initial condition of 600 K and 5 MPa, undergoes the following reversible steps. Use a temperature-independent heat capacity of $C_P = 44$ J/mol-K.

(a) Step 1. The gas is expanded isothermally to 0.2 MPa, absorbing a quantity of heat $\underline{Q}_{H1}$;

Step 2. The gas is cooled at constant volume to 300 K; Step 3. The gas is compressed isothermally to the initial volume; Step 4. The gas is heated to the initial state at constant volume requiring heat transfer $\underline{Q}_{H2}$. Calculate ΔU, Q, W_{EC} for each step and for the cycle. Also calculate the thermal efficiency that is the ratio of total work obtained to heat furnished, $\dfrac{-\underline{W}_{EC, cycle}}{(\underline{Q}_{H1} + \underline{Q}_{H2})}$.

(b) Step1. The gas is expanded to 3.92 MPa isothermally, absorbing a quantity of heat $\underline{Q}_{H1}$;

Step 2. The gas is expanded adiabatically to 0.1 MPa; Step 3. The gas is compressed isothermally to 0.128 MPa; Step 4. The gas is compressed adiabatically to the initial state. Calculate ΔU, Q, W_{EC} for each step and for the cycle. Also calculate the thermal efficiency for the cycle, $\dfrac{-\underline{W}_{EC, cycle}}{\underline{Q}_{H1}}$.

2.14 Air at 30°C and 2MPa flows at steady-state in a horizontal pipeline with a velocity of 25 m/s. It passes through a throttle valve where the pressure is reduced to 0.3 MPa. The pipe is the same diameter upstream and downstream of the valve. What is the outlet temperature and velocity of the gas? Assume air is an ideal gas with a temperature-independent $C_P = 7R/2$, and the average molecular weight of air is 28.8.

2.15 Steam undergoes a state change from 450°C and 3.5 MPa to 150°C and 0.3 MPa. Determine ΔH and ΔU using:

(a) steam table data.
(b) ideal gas assumptions. (Be sure to use the ideal gas heat capacity for water.)

2.16 Argon at 400 K and 50 bar is adiabatically and reversibly expanded to 1 bar through a turbine in a steady process. Compute the outlet temperature and work derived per mole.

2.17 Steam at 500 bar and 500°C undergoes a throttling expansion to 1 bar. What will be the temperature of the steam after the expansion? What would be the downstream temperature if the steam were replaced by an ideal gas, $C_P/R = 7/2$?

2.18 An adiabatic turbine expands steam from 500°C and 3.5 MPa to 200°C and 0.3 MPa. If the turbine generates 750 kW, what is the flow rate of steam through the turbine?

2.19 A steam turbine operates between 500°C and 3.5 MPa to 200°C and 0.3 MPa. If the turbine generates 750 kW and the heat loss is 100 kW, what is the flow rate of steam through the turbine?

2.20 An overall balance around part of a plant involves three inlets and two outlets which only contain water. All streams are flowing at steady-state. The inlets are: 1) liquid at 25°C, $\dot{m} = 54$ kg/min; 2) steam at 1 MPa, 250°C, $\dot{m} = 35$ kg/min; 3) wet steam at 0.15 MPa, 90% quality, $\dot{m} = 30$ kg/min. The outlets are: 1) saturated steam at 0.8 MPa, $\dot{m} = 65$ kg/min; 2) superheated steam at 0.2 MPa and 300°C, $\dot{m} = 54$ kg/min. Two kW of work are being added to the portion of the plant to run miscellaneous pumps and other process equipment, and no work is being obtained. What is the heat interaction for this portion of the plant in kW? Is heat being added or removed?

2.21 (a) A pressure gauge on a high pressure steam line reads 80 bar absolute, but temperature measurement is unavailable inside the pipe. A small quantity of steam is bled out through a valve to atmospheric pressure at 1 bar. A thermocouple placed in the bleed stream reads 400°C. What is the temperature inside the high pressure duct?

(b) Steam traps are common process devices used on the lowest points of steam lines to remove condensate. By using a steam trap, a chemical process can be supplied with so-called *dry* steam, i.e. steam free of condensate. As condensate forms due to heat losses in the supply piping, the liquid runs downward to the trap. As liquid accumulates in the steam trap, it causes a float mechanism to move. The float mechanism is attached to a valve, and when the float reaches a control level, the valve opens to release accumulated liquid, then closes automatically as the float returns to the control level. Most steam traps are constructed in such a way that the inlet of the steam trap valve is always covered with saturated liquid when opened or closed. Consider such a steam trap on a 7 bar (absolute) line that vents to 1 bar (absolute). What is the quality of the stream that exits the steam trap at 1 bar?

2.22 Steam at 550 kPa and 200°C is throttled through a valve at a flowrate of 15 kg/min to a pressure of 200 kPa. What is the temperature of the steam in the outlet state, and what is the change in specific internal energy across the value, $(U^{out} - U^{in})$?

2.23 A common model in distillation column screening is called *constant molar overflow*. In this model, the actual enthalpy of vaporization of a mixture is represented by the average enthalpy of vaporization, which can be assumed to be independent of composition for the purposes of the calculation, $\Delta H^{vap} = (\Delta H_1^{vap} + \Delta H_2^{vap})/2$. A distillation column operates by contacting countercurrent streams of vapor and liquid using internal trays or packing. The actual design of the trays or packing isn't required to perform an energy balance using the constant molar overflow concepts. In the constant molar overflow model for a column with one feed, the column may be represented by 5 sections: a feed section where the feed enters; a *rectifying* section above the feed zone; a condenser above the rectifying section which condenses the vapors and returns a portion of the liquid condensate as *reflux*; a *stripping* section below the feed section; and a *reboiler* that creates that creates vapors from the liquid flowing down the column. A schematic is shown below for the case where the condenser outlet is saturated liquid. According to the assumption of constant molar overflow, the enthalpies of all saturated vapor streams are equal, and the enthalpies of all saturated liquid streams are equal. Streams V_S and V_R are assumed to be saturated vapor unless otherwise noted. Streams B, L_S, L_R and D are assumed to be saturated liquid unless otherwise noted. According to the constant molar overflow model, the vapor and liquid flow rates are constant within the stripping and rectifying sections, and change only at the feed section as determined by mass and energy balances around the feed section. The system to be studied in this problem has an average enthalpy of vaporization of 32 kJ/mol, an average C_P^L of 146 J/mol°C, and an average C_P^V of 93 J/mol°C. Variable names for the various stream flowrates and the heat flowrates are given in the diagram. The feed can be liquid, vapor, or a mixture represented using subscripts to indicate the vapor and liquid flows, $F = F_V + F_L$. The enthalpy flow due to feed can be represented as: for saturated liquid, $F_L H^{satL}$; for saturated vapor, $F_V H^{satV}$; for subcooled liquid, $F_L H^{satL} + F_L C_P^L (T_F - T^{satL})$; superheated vapor, $F_V H^{satV} + F_V C_P^V (T_F - T^{satV})$; mix of vapor and liquid, $F_L H^{satL} + F_V H^{satV}$.

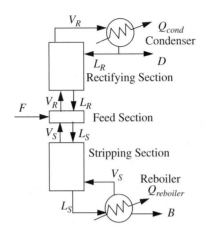

(a) Use a mass balance to show $F_V + V_S - V_R = L_S - L_R - F_L$.

[For parts (b)–(f), use the feed section mass and energy balances to show the desired result.]

(b) For saturated vapor feed, $F_L = 0$. Show $V_R = V_S + F_V$, $L_S = L_R$.

(c) For saturated liquid feed, $F_V = 0$. Show $V_S = V_R$, $L_S = L_R + F_L$.

(d) For subcooled liquid feed, $F_V = 0$. Show $V_R - V_S = F_L C_P (T_F - T^{sat})/\Delta H^{vap}$.

(e) For superheated vapor feed, $F_L = 0$. Show $L_S - L_R = -F_V C_P (T_F - T^{sat})/\Delta H^{vap}$.

(f) For a mixture of saturated liquid and saturated vapor feed, Show $V_R = V_S + F_V$, $L_S = L_R + F_L$.

(g) Use the mass and energy balances around the condenser to relate the condenser duty to the enthalpy of vaporization, for the case of streams L_R and D being saturated liquid.

(h) Use the mass and energy balances around the reboiler to relate the reboiler duty to the enthalpy of vaporization.

(i) In the case of subcooled liquid streams L_R and D, the vapor flow out of the top of the column, more variables are required. V_R' (into the condenser) will be smaller than the rectifying section flowrate V_R. Also the liquid flowrate in the rectifying section, L_R will be larger than the reflux back to the column, L_R'. Using the variables V_R', L_R' to represent the flowrate out of the top of the column and the reflux, respectively, relate V_R to V_R', L_R' and the degree of subcooling $T_{L'} - T^{satL}$.

[For parts (j)–(o), find all other flowrates and heat exchanger duties ($\underline{Q}$ values).]

(j) $F = 100$ mol/hr (saturated vapor), $B = 43$, $L_R/D = 2.23$.

(k) $F = 100$ mol/hr (saturated vapor), $D = 48$, $L_S/V_S = 2.5$.

(l) $F = 100$ mol/hr (saturated liquid), $D = 53$, $L_R/D = 2.5$.

(m) $F = 100$ mol/hr (half vapor, half liquid), $B = 45$, $L_S/V_S = 1.5$.

(n) $F = 100$ mol/hr (60°C subcooled liquid), $D = 53$, $L_R/D = 2.5$.

(o) $F = 100$ mol/hr (60°C superheated vapor), $D = 48$, $L_S/V_S = 1.5$.

2.24 A 0.1 m³ cylinder is initially at a pressure of 10 bar and a temperature of 300 K. The cylinder is emptied by opening a valve and letting pressure drop to 1 bar. What will be the temperature and moles of gas in the cylinder if this is accomplished

(a) Isothermally?

(b) Adiabatically? Neglect heat transfer between the cylinder walls and the gas, and assume an ideal gas with $C_P/R = 7/2$.

2.25 As part of a supercritical extraction of coal, an initially evacuated cylinder is fed with steam from a line available at 20 MPa and 400°C. What is the temperature in the cylinder immediately after filling?

2.26 An adiabatic tank of negligible heat capacity and 1 m³ volume is connected to a pipeline containing steam at 10 bar and 200°C, filled with steam till the pressure equilibrates, and disconnected from the pipeline. How much steam is in the tank at the end of the filling process, and what is its temperature if:

(a) the tank is initially evacuated?
(b) the tank initially contains steam at 1 bar and 150°C?

2.27 A large air supply line at 350 K and 0.5 MPa is connected to the inlet of a well-insulated 0.002 m³ tank. The tank has mass flow controllers on the inlet and outlet. The tank is at 300 K and 0.1 MPa. Both valves are rapidly and simultaneously switched open to a flow of 0.1 mol/min. Model air as an ideal gas with C_P = 29.3 J/mol-K, and calculate the pressure and temperature as a function of time. How long does it take until the tank is within 5 K of the steady-state value?

2.28 Compressed air at room temperature (295 K) is contained in a 20-L tank at 2 bar. The valve is opened and the tank pressure falls slowly and isothermally to 1.5 bar. The frictionless piston-cylinder is isothermal and isobaric (P = 1.5 bar) during movement. The surroundings are at 1 bar. The volumes of the piping and valve are negligible. During the expansion, the piston is pushing on external equipment and doing useful work such that the total resistance to the expansion is equivalent to 1.5 bar. The entire system is then cooled until all of the air is back in the original container. During the retraction of the piston, the piston must pull on the equipment, and the resistance of the external equipment is equivalent to 0.1 bar, so the total force on the piston is less than 1 bar, $1.0 - 0.1 = 0.9$ bar. The valve is closed and the tank is heated back to room temperature. (Air can be considered an ideal gas with a T-independent C_P = 29 J/mol-K.)

Initial and Final State   Intermediate State

(a) The useful work done by the process is the total work done by the piston in the expansion step less than amount of work done on the atmosphere. Calculate the useful work done per expansion stroke in kJ.
(b) Calculate the amount of heat needed during the expansion in kJ. Neglect the heat capacity of the tank and cylinder.
(c) Calculate the amount of cooling needed during the retraction of the cylinder in kJ.

2.29 A well-insulated tank contains 1 mole of air at 2 MPa and 673 K. It is connected via a closed valve to an insulated piston/cylinder device that is initially empty. The piston may be assumed to be frictionless. The volumes of the piping and valve are negligible. The weight of the piston and atmospheric pressure are such that the total downward force can be balanced with gas pressure in the cylinder of 0.7 MPa. The valve between the tank and piston/cylinder is cracked open until the pressure is uniform throughout. The temperature in the tank is found to be 499.6 K. Air can be assumed to be an ideal gas with a temperature-independent heat capacity $C_P = 29.3$ J/molK.

(a) What is the number of moles left in the tank at the end of the process?
(b) Write and simplify the energy balance for the process. Determine the final temperature of the piston/cylinder gas.

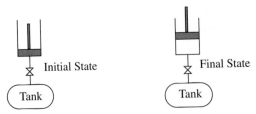

Initial State   Final State

Tank   Tank

2.30 A piston/cylinder has two chambers and includes a spring as illustrated below. The right-hand side contains air at 20°C and 0.2 MPa. The spring exerts a force to the right of $F = 5500x$ N where x is the distance indicated in the diagram, and $x^i = 0.3$ m. The piston has a cross-sectional area of 0.1 m². Assume that the piston/cylinder materials do not conduct heat, and that the piston/cylinder and spring do not change temperature. After oscillations cease, what is the temperature of the air in the right chamber and the final position of the piston, x^f, for the cases listed below? Use a temperature-independent heat capacity for air, $C_P = 7R/2$.

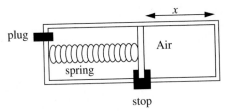

(a) The left chamber is evacuated and the plug remains in place.
(b) The plug is removed so that the left side pressure stays at 0.1MPa throughout the process.

2.31 We wish to determine the final state for the gas in an inflated balloon. Initially, the balloon has a volume of $\underline{V}^i$ at rest. The volume of the balloon is related to the internal pressure by $\underline{V} = a \cdot P + b$, where a and b are constants. The balloon is to be inflated by air from our lungs at P_{lung} and T_{lung} which are known and assumed to remain constant during inflation. Heat transfer through the walls of the balloon can be ignored. The system is defined to be the gas inside the balloon at any time. Starting with the general energy balance, simplify to write the balance in terms of $\{C_P, C_V, T_{lung}, P_{lung}, T^i, T^f, T_R, P_R\}$ and either $\{\underline{V}^i, \underline{V}^f\}$ or $\{P^i, P^f\}$. This will demonstrate that measurement of V^f or P^f is sufficient to determine T^f. Assume air is an ideal gas.

CHAPTER

ENTROPY

$$S = k \, ln \, W$$

L. Boltzmann

3.1 THE CONCEPT OF ENTROPY

We have discussed energy balances and the fact that friction and velocity gradients cause loss of useful work. It would be desirable to determine maximum work output (or minimum work input) for a given process. Our concern for accomplishing useful work inevitably leads to a search for what might cause degradation of our capacity to convert any form of energy into useful work. As an example, isothermally expanding an ideal gas from V^i to $2V^i$ can produce a significant amount of useful work if carried out reversibly, or possibly zero work if carried out irreversibly. If we could understand the difference between these two operations, we would be well on our way to understanding how to minimize wasted energy in many processes. Inefficiencies are addressed by the concept of entropy. Entropy provides a measure of the disorder of a system. As we will see, increased "disorder of the universe" leads to reduced capability for performing useful work.

Entropy may be studied on the microscopic and macroscopic scale. The historical approach to introducing entropy has been to focus on the macroscopic perspective, exploring the implications by repeatedly applying the definition until it becomes routine. Historically, entropy was initially conceived as a means of tracking transformations of heat into work in the context of steam engine design. The term "entropy" was coined by Clausius from the Greek for *transformation*, such that the entropy acts as a measure of the capacity for a body's energy to be transformed into useful work.[1] The macroscopic perspective is especially convenient for solving problems involving heat transfer, as can be easily appreciated by considering its definition,

❶ Entropy is a useful property for determining maximum/minimum work.

Macroscopic definition—Intensive entropy is a state property of the system. For a differential change in state *of a closed simple system* (no internal temperature gradients or composition gradients

---

1. Denbigh, K., *The Principles of Chemical Equilibrium,* 3rd ed. p. 33, Cambridge University Press, NY, NY (1971).

and no internal rigid, adiabatic, or impermeable walls),[1] the differential entropy change of the system is equal to the heat absorbed by the system *along a reversible path* divided by the absolute temperature of the system at the surface where heat is transferred.

$$dS = \frac{dQ_{rev}}{T_{sys}}$$ 3.1

where dS is the entropy change of the system.

While it is possible to rigorously develop the basic theory from purely macroscopic principles,[2] it is difficult to appreciate the manner in which "generation of disorder results in lost work" relates to the purely macroscopic perspective on entropy. For example, how does the macroscopic perspective relate to the entropy generation that occurs when two gaseous species are mixed adiabatically at constant temperature and pressure? The microscopic level is useful for understanding how entropy changes with volume and/or mixing.

Microscopic definition—Entropy is a measure of the molecular disorder of the system. Its value is related to the number of microscopic states available at a particular macroscopic state. Specifically, for a system of fixed energy and number of particles, N,

$$S_i = k\ln(p_i) \text{ or } \Delta S = k\ln\left(\frac{p_2}{p_1}\right)$$ 3.2

where p_i is the number of microstates in the ith macrostate. k is Boltzmann's constant, related to the gas constant, as $Nk = nR$, $k = R/N_A$. We define microstates and macrostates in the next section. To understand how this definition relates to disorder, consider the number of ways that your socks can be arranged in your bedroom. If they are all in one drawer (macrostate 1), there are several possible arrangements that they can assume. But if they can spread anywhere around the room (macrostate 2), there are many more possible arrangements. Therefore, the entropy (and the disorder) of the latter case is greater, and work will be required to return from the latter case to the former.

> *Entropy is a difficult concept to understand, mainly because its influence on physical situations is subtle, forcing us to rely heavily on the mathematical definition. We have ways to try to make some physical connection with entropy, and we will discuss these to give you every opportunity to develop a sense of how entropy changes. Ultimately, you must reassure yourself that entropy is defined mathematically, and like enthalpy, can be used to solve problems even though our physical connection with the property is occasionally less than satisfying.*

❗ The microscopic approach to entropy is discussed first, then the macroscopic approach is discussed.

Since there are two ways to define entropy, each with its own advantages, it would be desirable to understand both, and why they are really equivalent. In Section 3.2, the microscopic definition of entropy is discussed. On the microscopic scale, S is influenced primarily by spatial arrangements (affected by volume), and energetic arrangements (occupation) of energy levels (affected by temperature). First, we clarify the meaning of the microscopic definition by analyzing spatial distributions of molecules. To make the connection between entropy and temperature, we outline how the principles of volumetric distributions extend to energetic distributions.

---

1. A simple system is not acted on also by external force fields or inertial forces.
2. Denbigh, K., *The Principles of Chemical Equilibrium,* 3rd ed. p. 33, Cambridge University Press, NY, NY (1971).

In Section 3.3, we introduce the macroscopic definition of entropy and THE SECOND LAW OF THERMODYNAMICS, and then put the second law in mathematical form with the entropy balance in Section 3.4. Sections 3.5–3.8 are brief sections, but each with key points. In these sections we demonstrate how heat can be converted into work (as in a coal-burning plant). However, the maximum efficiency of the conversion of heat into work is less than 100%, and can be easily calculated using entropy to develop the relationships. We introduce heat engines and heat pumps to demonstrate the concepts of heat-to-work transformations, and present the principles for their application. In Section 3.9 we simplify the entropy balance for common process equipment, and then use the remaining sections to demonstrate applications of the balance. Overall, this chapter provides an understanding of entropy which is essential for Chapter 4 where entropy must be used routinely for process calculations.

3.2   MICROSCOPIC VIEW OF ENTROPY

> *Probability theory is nothing but common sense reduced to calculation.*

> *LaPlace*

First, we need to recognize that there are two ways that disorder of a system can change. First, there is disorder that occurs due to the physical arrangement (distribution) of atoms, and we represent this with the *configurational entropy*.[1] There is also a distribution of kinetic energies of the particles, and we represent this with the *thermal entropy*. For an example of kinetic energy distributions, consider that a system of two particles, one with a kinetic energy of 3 units and the other of 1 unit, is microscopically different from the same system when they both have 2 units of kinetic energy, even when the configurational arrangement of atoms is the same. This second type of entropy is more difficult to implement on the microscopic scale, so we focus on the configurational entropy in this section.[2]

❶ Configurational entropy is associated with spatial distribution. Thermal entropy is associated with kinetic energy distribution.

Entropy and Spatial Distributions: Configurational Entropy

Given *N* molecules and *M* boxes, how can these molecules be distributed among the boxes? Is one distribution more likely than another? Consideration of these issues will clarify what is meant by microstates and macrostates and how entropy is related to disorder. Our consideration will focus on the case of distributing particles between two boxes.

First, let us suppose that we distribute $N = 2$ non-interacting[3] particles in $M = 2$ boxes, and let us suppose that the particles are labeled so that we can identify which particle is in a particular box. There are four ways that we can distribute the labeled particles shown in Fig. 3.1. These arrangements are called *microstates* because the particles are labeled. For two particles and two boxes, there are four possible microstates. However, a macroscopic perspective makes no distinction between which particle is in which box. The only macroscopic characteristic that is important is

❶ Distinguishability of particles is associated with microstates. Indistinguishability is associated with macrostates.

---

1. The term "configurational" is occasionally used in different contexts. We apply the term in the context of K. Denbigh, *The Principles of Chemical Equilibrium*, 4th ed., Cambridge University Press, pp. 54-55. Technically, the configurational entropy includes both the combinatorial contribution discussed here for ideal gases, and the entropy departure function discussed in Unit II. Note that configurational energy is equivalent to the energy departure function of Chapter 7 because the change in energy of spatially rearranging ideal gas particles is zero.
2. The distinctions between these types of entropy are discussed in more detail by K. Denbigh, *The Principles of Chemical Equilibrium*, 4th ed., Cambridge University Press, pg. 353.
3. Non-interacting particles are oblivious to the presence of other particles and the energy is independent of the interparticle separations. In other words, potential energies are ignored.

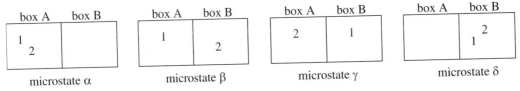

Figure 3.1 *Illustration of configurational arrangements of two particles in two boxes, showing the microstates.*

how many particles are in a box, rather than which particle is in a certain box. For *macrostates*, we just need to keep track of *how many* particles are in a given box, not *which particles* are in a given box. It might help to think about connecting pressure gauges to the boxes. The pressure gauge could distinguish between zero, one, and two particles in a box, but could not distinguish *which* particles are present. Therefore, microstates α and δ are different macrostates because the distribution of particles is different; however, microstates β and γ give the same macrostate. Thus, from our four microstates, we have only three macrostates.

To find out which arrangement of particles is most likely, we apply the "principle of equal *a priori* probabilities." This "principle" states that all microstates of a given energy are equally likely. Since all of the states we are considering for our non-interacting particles are at the same energy, they are all equally likely.[1] From a practical standpoint, we are interested in which macrostate is most likely. The probability of a macrostate is found by dividing the number of microstates in the given macrostate by the total number of microstates in all macrostates as shown in Table 3.1. For our example, the probability of the first macrostate is 1/4 = 0.25. The probability of the evenly distributed state is 2/4 = 0.5. That is, one third of the macrostates possess 50% of the probability. The "most probable distribution" is the evenly distributed case.

What happens when we consider more particles? It turns out that the total number of microstates for N particles in M boxes is M^N so the counting gets tedious. For five particles in two boxes, the calculations are still manageable. There will be two microstates where all the particles

Table 3.1 *Illustration of macrostates for two particles and two boxes.*

Macrostate		# of microstates	Probability of macrostate
# in box A	# in box B		
0	2	1	0.25
1	1	2	0.5
2	0	1	0.25

1. Note that the number of particles and the energy are constant throughout the discussion presented here and the volume is specified at each stage. The constant energy for non-interacting particles means that the temperature will be constant; only the pressure will be reduced at larger volumes because it takes the molecules longer to get around the box and collide with a particular wall. We can think of this as an N, V, U perspective, and we will demonstrate that entropy is maximized at equilibrium within this perspective, but some other quantity might characterize equilibrium if we held other quantities constant.

Table 3.2 *Microstate for the second and third macrostates for five particles distributed in two boxes.*

3.2a One particle in box A		3.2b. Two particles in box A			
box A	box B	box A	box B	box A	box B
1	2,3,4,5	1,2	3,4,5	2,4	1,3,5
2	1,3,4,5	1,3	2,4,5	2,5	1,3,4
3	1,2,4,5	1,4	2,3,5	3,4	1,2,5
4	1,2,3,5	1,5	2,3,4	3,5	1,2,4
5	1,2,3,4	2,3	1,4,5	4,5	1,2,3

are in one box or the other. Let us consider the case of one particle in box A and four particles in box B. Recall that the macrostates are identified by the number of particles in a given box, not by which particles are in which box. Therefore, the five microstates for this macrostate appear as given in Table 3.2a.

The counting of microstates for putting two particles in box A and three in box B is slightly more tedious, and is shown in Table 3.2b. It turns out that there are 10 microstates in this macrostate. The distributions for (three particles in A) + (two in B) and for (four in A) + (one in B) are like the distributions (two in A) + (three in B), and (one in A) + (four in B), respectively. These three cases are sufficient to determine the overall probabilities. There are $M^N = 2^5 = 32$ microstates total summarized in the table below.

Macrostate		# Microstates	Probability of Macrostate
Box A	Box B		
0	5	1	0.0313
1	4	5	0.1563
2	3	10	0.3125
3	2	10	0.3125
4	1	5	0.1563
5	0	1	0.0313

Note now that one-third of the macrostates (two out of six) possess 62.5% of the microstates. Thus, the distribution is now more peaked toward the most evenly distributed states than it was for two particles where one-third of the macrostates possessed 50% of the microstates. This is one of the most important aspects of the microscopic approach. As the number of particles increases, it won't be long before 99% of the microstates are in one-third of the macrostates. The trend will continue,

and increasing the number of particles further will quickly yield 99% of the microstates in that 1/10 of the macrostates. In the limit as N→∞ (the "thermodynamic limit"), virtually all of the microstates are in just a few of the most evenly distributed macrostates, even though the system has a very slight finite possibility that it can be found in a less evenly distributed state. Based on the discussion, and considering the microscopic definition of entropy (Eqn. 3.2), entropy is maximized at equilibrium for a system of fixed energy and total volume.[1]

> ❶ With a large number of particles, the most evenly distributed configurational state is most probable, and the probability of any other state is small.

Generalized Microstate Formulas

To extend the procedure for counting microstates to large values of N ($\sim 10^{23}$), we cannot imagine listing all the possibilities and counting them up. It would require 40 years to simply count to 10^9 if we did nothing but count night and day. We must systematically analyze the probabilities as we consider configurations and develop model equations describing the process.

> ❶ Factorials are a quick tool for counting arrangements.

How do we determine the number of microstates for a given macrostate for large N? For the first step in the process, it is fairly obvious that there are N ways of moving 1 particle to box B, i.e., 1 came first, or 2 came first, . . . , etc., which is what we did to create Table 3.2a. However, counting gets more complicated when we have two particles in a box. Since there are N ways of putting the first particle in the box, and there are $(N-1)$ particles left, then we would be tempted to follow the same logic for the $(N-1)$ remaining particles. For example, with 5 particles, there would then be 5 ways of placing the first particle, then 4 ways of placing the second particle for a total of 20 possible ways of putting 2 particles in box B. One way of writing this would be 5·4, which is equivalent to (5·4·3·2·1)/(3·2·1), which can be generalized to $N!/(N-m)!$, where m is the number of particles we have placed in the first box.[2] ($N!$ is read "N factorial,"[3] and calculated as $N\cdot(N-1)\cdot(N-2)......\cdot 2\cdot 1$). Our formula gives 20 ways, but Table 3.2b shows only 10 ways. What are we missing? Answer: When we count this way, we are implicitly overcounting some microstates. Note in Table 3.2b that although there are two ways that we could put the first particle in box B, the order in which we place them does not matter when we count microstates. Therefore, using $N!/(N-m)!$ implicitly distinguishes between the order in which particles are placed. For counting microstates, the history of how a particular microstate was achieved does not interest us. Therefore, we say there are only 10 *distinguishable* microstates.

It turns out that it is fairly simple to correct for this overcounting. For 2 particles in a box, they could have been placed in the order 1-2, or in the order 2-1, which gives 2 possibilities. For 3 particles, they could have been placed 1-2-3, 1-3-2, 2-1-3, 2-3-1, 3-1-2, 3-2-1, for 6 possibilities. For m particles in a box, without correction of the formula, we will overcount by $m!$. Therefore, we modify the above formula by dividing by $m!$ to correct for overcounting. Finally, the number of microstates for arranging N particles in 2 boxes, with m particles in one of the boxes, is:[4]

$$p_j = \frac{N_j!}{m_j!(N_j - m_j)!} \qquad\qquad 3.3$$

---

1. In an isolated system at constant ($\underline{U}, \underline{V}$), entropy will be generated as equilibrium is approached; $\underline{S}$ will increase and will be maximized at equilibrium. If the system is closed but not isolated, the property which is minimized is determined by the property which is a natural function of the controlled variables: $\underline{H}$ is minimized for constant ($\underline{S},P$); $\underline{A}$ for constant ($T,\underline{V}$); $\underline{G}$ for constant (T,P). A and G will be introduced in future chapters.
2. In statistics, this is called the number of permutations.
3. Factorials are available on the HP48G suggested to accompany this text by the keystrokes MTH, NXT, *softkey* PROB, *softkey* !.
4. The formula for the particles in boxes is an example of a *binomial distribution*, fundamental in the study of probability and statistics. Detailed development of the binomial distribution and the issue of indistinguishability can be found in any textbook or handbook on the subject.

The general formula for M boxes is:[1]

$$p_j = \frac{N_j!}{\displaystyle\prod_{i=1}^{M} m_{ij}!}$$

3.4   ❶ General formula for number of microstates for N particles in M boxes.

m_{ij} is the number of particles in the ith box at the jth macrostate. We will not derive this general formula, but it is a straightforward extension of the formula for two boxes which was derived above. Therefore, with 10 particles, and 3 in the first box, 2 in the second box and 5 in the third box, we have $10!/(3!2!5!) = 3{,}628{,}800/(6{\cdot}2{\cdot}120) = 2520$ microstates for this macrostate.

Example 3.1   Entropy change vs. volume change

Recall the microscopic definition of entropy given by Eqn. 3.2. Let us use it to calculate the entropy change for an ideal gas due to an isothermal change in volume. The statistics we have just derived will apply since an ideal gas consists of non-interacting particles whose energy is independent of nearest neighbors.

Suppose an insulated container, partitioned into two equal volumes, contains N molecules of an ideal gas in one section and no molecules in the other. When the partition is withdrawn, the molecules quickly distribute themselves uniformly throughout the total volume. How is the entropy affected? Let subscript 1 denote the initial state and subscript 2 denote the final state. Here we take for granted that the final state will be evenly distributed.

Solution: Applying Eqn. 3.4, and noting that $0! = 1$,

$$p_1 = \frac{N!}{N!\,0!} = 1; \quad p_2 = \frac{N!}{(N/2)!\,(N/2)!}; \quad \ln(p_2/p_1) = \ln\left(\frac{N!/(N/2)!^2}{1}\right)$$

Substituting into Eqn. 3.2, and recognizing $\ln\left(\frac{N}{2}\right)!^2 = 2\ln\left(\frac{N}{2}\right)!$,

$$\Delta \underline{S} = \underline{S}_2 - \underline{S}_1 = k\,\ln(p_2/p_1) = k\left\{\ln(N!) - 2\ln[(N/2)!]\right\}$$

Stirling's approximation may be used for $\ln(N!)$ when N is large $\Rightarrow \ln(N!) \approx N\ln(N) - N$. The approximation is a mathematical simplification, and not, in itself, related to thermodynamics.

❶ Entropy of a constant temperature system increases when volume increases.

$$\Rightarrow \Delta \underline{S} = k[N\ln(N) - N - 2(N/2)\ln(N/2) + 2(N/2)]$$
$$= k[N\ln(N) - N - N\ln(N) + N\ln(2) + N]$$
$$= kN\ln(2) \Rightarrow \Delta \underline{S} = nR\ln(2)$$

---

1. In statistics this is called the number of combinations.

Example 3.1   Entropy change vs. volume change (Continued)

Therefore, entropy of the system has increased by a factor of $\ln(2)$ when the volume has doubled at constant T.

Suppose the box initially with particles is three times as large as the empty box. In this case the increase in volume will be 33%. Then what is the entropy change? The trick is to imagine four equal size boxes, with three equally filled at the beginning.

$$p_1 = \frac{N!}{\left[\left(\frac{N}{3}\right)!\right]^3 0!} \qquad\qquad p_2 = \frac{N!}{\left[\left(\frac{N}{4}\right)!\right]^4}$$

Using Stirling's approximation

$$4\ln\left[\left(\frac{N}{4}\right)!\right] \approx 4\left[\frac{N}{4}\ln\frac{N}{4} - \frac{N}{4}\right] = N\ln(N/4) - N$$

$$3\ln\left[\left(\frac{N}{3}\right)!\right] \approx 3\left[\frac{N}{3}\ln\frac{N}{3} - \frac{N}{3}\right] = N\ln(N/3) - N$$

Thus, for an isothermal change, (i.e., no redistribution of thermal entropy),

$$\Delta\underline{S} = -k\left\{N\ln\left[\frac{N}{4}\right] - N - N\ln\left[\frac{N}{3}\right] + N\right\} = nR\ln\left[\frac{4}{3}\right]$$

We may generalize the result by noting the pattern with this result and the previous result,

$$\boxed{\Delta S = R\ln\left[\frac{V}{V^i}\right]} \tag{ig 3.5}$$

❶ Formula for *iso-thermal* entropy changes of an ideal gas.

For an isothermal ideal gas, we also may express this in terms of pressure along an isotherm by substituting $V = RT/P$ in Eqn. 3.5

$$\boxed{\Delta S = -R\ln\left[\frac{P}{P^i}\right]} \tag{ig 3.6}$$

Therefore, when the pressure increases isothermally the entropy decreases. Likewise, when the volume decreases isothermally, the entropy decreases.

Example 3.2   Entropy change of mixing ideal gases

Mixing is another important process to which we may apply the statistics that we have developed. We can imagine mixing species as ideal gases then turning on the potential function to compute the changes in properties as the fluid departs from ideal gas behavior. The departure from ideal gas behavior is treated in Unit II, but the mixing process is treated here.

Example 3.2 Entropy change of mixing ideal gases (Continued)

One mole of pure oxygen vapor and three moles of pure nitrogen vapor at the same temperature and pressure are brought into intimate contact and held in this fashion until the nitrogen and oxygen have completely mixed. The resulting vapor is a uniform, random mixture of nitrogen and oxygen molecules. Determine the entropy change associated with this mixing process, assuming ideal-gas behavior.

Solution: Since the T^i and P^i of both ideal gases are the same, then $V^i_{N2} = 3V^i_{O2}$ and $V^i_{tot} = 4V^i_{O2}$. Ideal gas molecules are point masses so the presence of O_2 in the N_2 does not affect anything as long as the pressure is constant. The main effect is that the O_2 now has a larger volume to access and so does N_2.

Recalling the entropy change vs. volume change:

Entropy change for O_2:

$$\Delta \underline{S} = n_{O2}\, R \ln 4 = n_{tot}\, R \left[-x_{O2}\, \ln(0.25) \right] = n_{tot} R \left[-x_{O2}\, \ln(x_{O2}) \right]$$

Entropy change for N_2:

$$\Delta \underline{S} = n_{N2}\, R \ln 4\,/\,3 = n_{tot}\, R \left[-x_{N2}\, \ln(0.75) \right] = n_{tot} R \left[-x_{N2}\, \ln(x_{N2}) \right]$$

Entropy change for total fluid:

$$\Delta \underline{S}_{tot} = -n_{tot} R \left[x_{O2}\, \ln(x_{O2}) + x_{N2}\, \ln(x_{N2}) \right] = -4R(-0.562) = 18.7 \text{ J/K}$$

Note: In general, ideal mixing $\Rightarrow \Delta S^{ig}_{mix}\,/\,R = -\left[\sum x_i\, \ln(x_i) \right]$ 3.7

This is an important result as it gives the entropy change of mixing for non-interacting particles. Therefore, the result applies to ideal gases.[1] **This equation provides the underpinning for much of the discussion of mixtures and phase equilibrium in Unit III.**

In this section we have shown that a system of non-interacting molecules at equilibrium is most likely to be found in the most randomized (distributed) configuration because this is the macrostate with the largest fraction of microstates. In addition, since the entropy is proportional to the logarithm of the number of microstates, the entropy of a state is maximized at equilibrium for a system of fixed energy, $\underline{V}$, and N.

We have also shown that isothermal compression decreases entropy and that isothermal expansion increases entropy. We have shown that mixing increases entropy. These concepts are extremely important in developing an understanding of entropy, but by themselves are not directly helpful in the initial objective of this chapter—that of determining inefficiencies and maximum work. This is because entropy changes *of the system* comprise just one piece of the puzzle we need to solve. We have chosen to develop the microscopic approach first, because it is helpful in understanding how the entropy of a system changes as we extend the concepts to macroscopic calculations.

---

1. Later we will apply this to a broader class of mixtures, ideal solutions, where all interactions and sizes are identical.

Entropy and Temperature Change: Thermal Entropy

The key to understanding thermal entropy is the appreciation that energy is quantized. Thus, there are discrete translational energy levels in which spherical particles may be arranged. These energy levels are analogous to the boxes in the spatial distribution problem. The effect of increasing the temperature is to increase the energy of the molecules and make higher energy levels accessible. To see how this affects the entropy, consider a system of 10 particles. Suppose we are at a low temperature and their total energy is equal to 2. This means one particle could be in energy level 2 and all the others at zero, or two particles could be at energy level one and all the others at zero. These possibilities are very limited. Consider how many more possibilities would be available if we raised the temperature such that the total energy was equal to 25. Clearly, the number of microstates for this macrostate would be larger. Therefore, adding heat increases the temperature which increases the entropy. Unfortunately, this discussion oversimplifies the energy levels for real molecules because the real energy levels are not evenly spaced and counting of actual microstates for a fixed total energy is more tedious. However, the illustration qualitatively illustrates the correct behavior. For this introductory text, it is sufficient to establish the manner in which the microscopic and macroscopic perspectives can be equivalent, and then to calculate this effect macroscopically, as we show in the next section.

As we progress through the thermodynamics of dense fluids, you may notice that practically all of the thermal contributions are relegated to the ideal gas contribution, and then the entire thermal effect appears in the ideal gas heat capacity. If you become curious about the manner in which the heat capacities of polyatomic species differ from those of the spherical particles discussed above, you would be well-advised to take a course dedicated to statistical thermodynamics. The issues of kinetic energy and quantized energy levels are fundamentally the same as described above, but the quantum effects become more significant and complicated because bonded atoms are located very close to each other. The treatment of heat capacity by statistical thermodynamics is particularly interesting because it is a theory[1] which often gives more accurate results than experimental calorimetric measurements. For the purposes of this text, we simply acknowledge that ideal gas heat capacities can be determined and correlated by expressions like polynomials.

In the present day, the subtle relations between entropy and molecular distributions are complex but approachable. Imagine how difficult gaining this understanding must have been for Boltzmann in 1880, before the advent of quantum mechanics. Many people at the time refused even to accept the existence of molecules. Trying to convincingly explain to people the nature and significance of his discoveries must have been extremely frustrating. What we know for sure is that Boltzmann drowned himself in 1903. Try not to take your frustrations with entropy quite so seriously.

3.3   THE MACROSCOPIC DEFINITION OF ENTROPY

In the introduction of the chapter, we alluded to the relation between entropy and maximum process efficiency. We have shown that entropy changes with pressure and temperature. How can we use entropy to help us determine maximum work output or minimum work input? The answer is best summarized by a series of statements.

---

1. This theory also requires experimental spectroscopic measurements, but those are quite different from the calorimetric measurement of enthalpy changes with respect to temperature.

Molar or specific entropy is a *state* property which will assist us in the following ways:

1. Irreversible processes will result in an increase in entropy of the *universe*. (Irreversible processes will result in entropy generation.) Irreversible processes result in loss of capability for performing work.

2. Reversible processes result in no increase in entropy of the *universe*. (Reversible processes result in no entropy generation. This principle will be useful for calculation of maximum work output or minimum work input for a process.)

3. Proposed processes which would result in a decrease of entropy of the universe are impossible. (Impossible processes result in negative entropy generation.)

These three principles are summarized in the **THE SECOND LAW OF THERMODYNAMICS**: reversible processes and/or optimum work interactions occur without entropy generation, and irreversible processes result in entropy generation. The microscopic descriptions in the previous section teach us very effectively about the relation between entropy and disorder, but it is not fair to say that any increase in volume results in a loss of potentially useful work when the entropy of the system increases. (Note that it is the entropy change of the *universe* that determines irreversibility, not entropy change of the *system*.) After all, the only way of obtaining any expansion/contraction work is by a change in volume. To understand the relation between lost work and volume change, we must appreciate the meaning of reversibility, and what types of phenomena are associated with entropy generation. We will explore these concepts in the next sections.

❶ The entropy balance is the SECOND LAW OF THERMO-DYNAMICS.

Entropy Definition (Macroscopic)

Let us define the differential change in entropy of a *closed* simple system by the following equation:

$$dS \quad \frac{dQ_{rev}}{T_{sys}} \qquad\qquad 3.1$$

Recall that we defined enthalpy mathematically also for reasons of convenience. Entropy is much more difficult than enthalpy to conceptualize. In fact, it wasn't obvious until Clausius' famous work with heat engines which we introduce in Section 3.5 on page 110, that he realized that such a state property could be defined. For a change in states, both sides of Eqn. 3.1 may be integrated,

$$\boxed{S = \int_{state\ 1}^{state\ 2} \frac{dQ_{rev}}{T_{sys}}} \qquad\qquad 3.8$$

where:

1. The entropy change on the left-hand side of Eqn. 3.8 is dependent only on states 1 and 2 and not dependent on reversibility. However, to calculate the entropy change via the integral, the *integral* must be evaluated along any *reversible* pathway.

2. T_{sys} is the temperature of the system boundary where heat is transferred. Only if the system boundary temperature is constant along the pathway may this term be taken out of the integral sign.

❶ Entropy is a state property. For a pure single-phase fluid, specific entropy is characterized by two state variables (e.g., T and P).

A change in entropy is completely characterized for a pure single-phase fluid by any other two state variables. It may be surprising that the integral is independent of path since Q is a path-dependent property. The key to understand is that the righthand side integral is independent of path, *as long as the path is reversible*. Thus, a process between two states does not need to be reversible to permit calculation of the entropy change, since we can evaluate it along *any* reversible path of choice. If the *actual* path is reversible, then the actual heat transfer and pathway may be used. If the process is *irreversible*, then any *reversible* path may be constructed for the calculation. This point will be made more clear by later examples.

Note that entropy does not depend directly on the work done on a system. Therefore, it may be used to decouple heat and work in the energy balance for reversible processes. Also for reversible processes, entropy provides a second property that may be used to determine unknowns in a process. Let us investigate some convenient pathways for the evaluation of entropy changes.

❶ Entropy can be used to decouple heat and work.

Calculation of Entropy Changes in Closed Systems

As with enthalpy and internal energy, tables and charts are the preferred sources for entropy information. In the event that thermodynamic charts or tables are unavailable, entropy changes can be easily calculated. Since the integral must be evaluated along a reversible path, let us consider some easy choices of paths. For a closed *reversible* system without shaft work, the energy balance Eqn. 2.15 becomes

$$d\left[U + \frac{u^2}{2g_c} + \frac{g}{g_c}z \right] = dQ_{rev} + dW_S + dW_{EC} \qquad 3.9$$

Inserting Eqn. 2.3

$$[dU + PdV] = dQ_{rev} \qquad 3.10$$

We now consider how this equation may be substituted in the integral of Eqn. 3.8 for calculating entropy changes.

Constant Pressure (Isobaric) Pathway

Many process calculations involve state changes at constant pressure. Recognizing $H = U + PV$, $dH = dU + PdV + VdP$. In the case at hand, dP happens to be zero, therefore Eqn. 3.10 becomes

$$dH = dQ_{rev} \qquad 3.11$$

Since $dH = C_P dT$ at constant pressure, along a *constant-pressure pathway*, substituting for dQ_{rev} in Eqn. 3.8, the entropy change is

$$dS)_P = \frac{C_P}{T} dT \qquad 3.12$$

$$\boxed{\Delta S = \int_{T_1}^{T_2} \frac{C_P}{T} dT} \qquad 3.13$$

❶ Constant pressure.

Constant Volume Pathway

For a constant-volume pathway, Eqn. 3.10 becomes

$$dU = dQ_{rev} \qquad 3.14$$

Since $dU = C_V dT$ along a *constant-volume pathway*, substituting for dQ_{rev} in Eqn. 3.8, the entropy change is

$$\Delta S = \int_{T_1}^{T_2} \frac{C_V}{T} dT$$

3.15   ❶ Constant volume.

Constant Temperature (Isothermal) Pathway

The behavior of entropy at constant temperature is more difficult to generalize in the absence of charts and tables because dQ_{rev} depends on the state of aggregation. For the ideal gas, $dU = 0 = dQ - PdV$, $dQ = RTdV/V$, and plugging into Eqn. 3.8,

$$\Delta S = R \ln \left[\frac{V_2}{V_1} \right] \text{ or } \Delta S = -R \ln \left[\frac{P_2}{P_1} \right]$$

(ig) 3.16   ❶ Isothermal.

For a liquid or solid, the precise relations for detailed calculations will be developed in Chapters 5–7. Looking at the steam tables at constant temperature, entropy is very weakly dependent on pressure for *liquid* water. This result may be generalized to other liquids below $T_r = 0.75$ and also to solids. For *condensed* phases, to a first approximation, entropy can be assumed to be independent of pressure (or volume) at *fixed temperature*.

Adiabatic Pathway

A process that is adiabatic *and reversible* will result in an *isentropic path*. By Eqn. 3.8,

$$\Delta S = \int_{state\ 1}^{state\ 2} \frac{dQ_{rev}}{T_{sys}} = 0 \quad \text{for reversible process only.}$$

3.17   ❶ Adiabatic *and* reversible.

Note that a path that is adiabatic, but not reversible, will not be isentropic. This is because a reversible adiabatic process starting at the same state 1 will not follow the same path, so it will not end at state 2, and reversible heat transfer will be necessary to reach state 2.

Phase Transitions

In the absence of property charts or tables, entropy changes due to phase transitions can be easily calculated. Since equilibrium phase transitions for pure substances occur at constant temperature and pressure, for vaporization

$$\Delta S^{vap} = \int \frac{dQ_{rev}}{T} = \frac{1}{T^{sat}} \int dQ_{rev} = \frac{Q^{vap}}{T^{sat}}$$

where T^{sat} is the equilibrium saturation temperature. Likewise for a solid-liquid transition,

$$\Delta S^{fus} = \frac{Q^{fus}}{T_m}$$

where T_m is the equilibrium melting temperature. Since either transition occurs at constant pressure if along a reversible pathway, we may include Eqn. 3.11, giving

Phase transitions.

$$\Delta S^{vap} = \frac{\Delta H^{vap}}{T^{sat}} \quad \text{and} \quad \Delta S^{fus} = \frac{\Delta H^{fus}}{T_m} \qquad 3.18$$

Now let us examine a process from Chapter 2 that was reversible, and study the entropy change. We will show that the result is the same via two different paths.

Example 3.3 Ideal gas entropy changes in a piston/cylinder

In Example 2.9 on page 66, we derived the temperature change for a closed system adiabatic expansion of an ideal gas. How does the entropy change along this pathway, and what does this example show about changes in entropy with respect to temperature?

Solution: Reexamine the equation $\frac{C_V}{T}dT = -\frac{R}{V}dV$, which may also be written $\frac{C_V}{R}d\ln T = -d\ln V$. We can sketch this path as shown by the diagonal line in Fig. 3.2. Since our path is adiabatic ($dQ = 0$) and reversible, and our definition of entropy is $dS = \frac{dQ_{rev}}{T}$, we expect that this implies that the path is also isentropic (a constant-entropy path). Since entropy is a state property, we can verify this by calculating entropy along the other pathway of the figure consisting of a constant temperature Step A and a constant volume Step B.

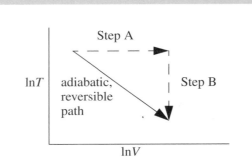

Figure 3.2 *Equivalence of adiabatic path and an alternate path to the same state.*

For the reversible isothermal step we have

$$dU = dQ_{rev} - PdV = 0 \quad \text{or} \quad dQ_{rev} = PdV \qquad \text{(ig)}$$

Thus

$$dS)_T = \frac{dQ_{rev}}{T} = \frac{PdV}{T} \qquad \text{(ig)}$$

Example 3.3 Ideal gas entropy changes in a piston/cylinder (Continued)

Substituting the ideal gas law

$$dS)_T^{ig} = \frac{RTdV}{VT} = R\frac{dV}{V}$$

(ig) 3.19

For the constant volume step, we have

$$dU = dQ_{rev} \quad \text{or} \quad C_V dT = dQ_{rev}$$

Thus

$$dS)_V = \frac{dQ_{rev}}{T} = C_V\frac{dT}{T}$$

3.20

We could replace a differential step along the adiabat with the equivalent differential steps along the alternate pathways; therefore, we can see that the change in entropy is zero.

$$dS_{adiabat}^{ig} = dS)_V + dS)_T^{ig} = \frac{C_V}{T}dT + \frac{R}{V}dV = 0$$

(ig) 3.21

which was shown by the energy balance in Eqn. 2.52, and we verify that the overall expansion is isentropic.

❶ The entropy change along the adiabatic, reversible path is the same as along (step A + step B) illustrating that S is a state property.

The method of subdividing state changes into individual temperature and volume changes can be generalized to any process, not just the adiabatic process of the previous example, giving

$$\Rightarrow dS^{ig} \equiv \frac{C_V}{T}\,dT + \frac{R}{V}\,dV$$

(ig) 3.22

We may integrate Step A and B independently. We also could use temperature and pressure steps to calculate entropy changes resulting in an alternate formula:

$$\Delta S^{ig} = C_V\ln\frac{T}{T^i} + R\ln\frac{V}{V^i} \quad \text{or} \quad \Delta S^{ig} = C_P\ln\frac{T}{T^i} - R\ln\frac{P}{P^i}$$

(*ig) 3.23

❶ Formulas for an ideal gas.

Looking back at Eqn. 3.20, we realize that it does not depend on the ideal gas assumption, and it is a general result,

$$\left(\frac{\partial S}{\partial T}\right)_V = \frac{C_V}{T}$$

3.24

❶ Temperature derivatives of entropy are related to C_P and C_V.

which provides a relationship between C_V and entropy. Looking back at Eqn. 3.12,

$$\left(\frac{\partial S}{\partial T}\right)_P = \frac{C_P}{T}$$

3.25

Example 3.4 Steam entropy changes in a piston/cylinder

Steam at 450°C and 4.5 MPa is held in a piston/cylinder. The cylinder is adiabatically and reversibly expanded to 2.0 MPa. What is the final temperature?

Solution: This is not an ideal gas, but by Eqn. 3.1, the process will be isentropic. From the steam tables, the entropy at the initial state is 6.877 kJ/kgK. At 2 MPa, this entropy will be found between 300°C and 350°C. Interpolating,

$$T = 300 + \frac{6.877 - 6.7684}{6.9583 - 6.7684}(350 - 300) = 329°C$$

Entropy Generation

At the beginning of Section 3.3 on page 96, statement number one declares that irreversible processes generate entropy. Now that some methods for calculating entropy have been presented, this principle can be explored. We discussed in Chapter 2 that friction and velocity gradients result in irreversibilities and thus entropy generation occurs. Entropy generation can also occur during heat transfer, so let us consider that possibility.

Example 3.5 Entropy generation in a temperature gradient

A 500-mL glass of chilled water at 283 K is removed from a refrigerator. It slowly equilibrates to room temperature at 298 K. The process occurs at one bar. Calculate the entropy change of the water ΔS_{water}, the entropy change of the surroundings, ΔS_{surr}, and the entropy change of the universe ΔS_{univ}. Neglect the heat capacity of the container. For liquid water $C_P = 4.184$ J/gK.

Solution:

Water: The system is closed at constant pressure with $T^i = 283$ K and $T^f = 298$ K. We choose any reversible pathway along which to evaluate Eqn. 3.8, a convenient path being constant-pressure heating. Thus, using Eqn. 3.11,

$$d\underline{Q}_{rev} = d\underline{H} = mC_P dT$$

Substituting this into our definition for a change in entropy, and assuming a T-independent C_P,

$$\Delta \underline{S}_{water} = \int_{T^i}^{T^f} \frac{mC_P dT}{T_{sys}} = mC_P \ln \frac{T^f}{T^i} = (500)4.184\ln \frac{298}{283} = 108.0 \, \frac{\text{J}}{\text{K}} \qquad (*)$$

Surroundings: The surroundings also undergo a constant pressure process as a closed system; however, the heat transfer from the glass causes no change in temperature—the surroundings act as a reservoir and the temperature is 298 K throughout the process. The heat transfer of the surroundings is the negative of the heat transfer of the water, so we have

$$\Delta \underline{S}_{surr} = \int \frac{d\underline{Q}_{rev}}{T_{surr}} = -\int \frac{d\underline{Q}_{rev,water}}{T_{surr}} = \frac{-mC_P \Delta T_{water}}{T_{surr}} = \frac{-31380\text{J}}{298\text{K}} = -105.3 \, \frac{\text{J}}{\text{K}} \, (*)$$

Example 3.5 Entropy generation in a temperature gradient (Continued)

Note that the temperature of the surroundings was constant, which simplified the integration.

Universe: For the universe we sum the entropy changes of the two subsystems that we have defined. Summing the entropy change for the water and the surroundings we have

$$\Delta \underline{S}_{univ} = 2.7 \frac{J}{K}$$

Comment. The entropy change for the universe is positive which demonstrates that this is an irreversible process. The irreversibility is caused by the finite temperature driving force. Because entropy is a state property, the integrals that we calculate may be along any reversible pathway, and the time dependence along that pathway is unimportant as we have shown in this example.

> This is an irreversible process because entropy is generated.

Example 3.6 Entropy generation and lost work in a gas expansion

Suppose the expansion of Example 3.3 were carried out adiabatically but irreversibly, such that no work was derived. What would be the final temperature and how much potentially useful work would have been lost?

Solution: Basis: 1 mole

The energy balance gives: $dU = dQ + dW_{EC} = 0 + 0 = C_V dT \quad \Rightarrow \quad dT = 0 \quad \Rightarrow \quad T^f = T^i$

Therefore, the process is isothermal. As for the lost work, a reversible expansion with $T^f = T^i$ would yield

$$W_{rev} = -RT \ln(V/V^i) \tag{ig}$$

When our ability to perform work has been degraded, we do not care whether it was leaving the system or entering the system; therefore, we find that

$$\text{lost work} = |W_{rev}| = |RT \ln (V/V^i)| \tag{ig}$$

Integrating Eqn. 3.19, we find that

$$RT \ln (V/V^i) = T\Delta S \tag{ig}$$

The entropy change would have been zero for an adiabatic, reversible process; thus, any entropy change is due to entropy generation. $\Delta S = S_{gen}$ is the entropy generated *in the system* by conducting the process irreversibly.

Therefore, determination of lost work reduces to finding the difference between the entropy change of the actual process versus the entropy change of the reversible process, and in general, we may write, for a closed system

$$d\underline{S} = \frac{d\underline{Q}}{T} + d\underline{S}_{gen} \tag{3.26}$$

for any process, noting that $d\underline{S}_{gen} \geq 0$.

3.4 THE ENTROPY BALANCE

In Chapter 2 we used the energy balance to track energy changes of the system by the three types of interactions with the surroundings—flow, heat and work. This method was extremely helpful because we could use the balance as a checklist to account for all interactions. Therefore, we present a general entropy balance in the same manner. To solve a process problem we can use an analogous balance approach of starting with an equation including all the possible contributions that might occur and eliminating the balance terms that do not apply for the situation under consideration.

Entropy change within a system boundary will be given by the difference between entropy which is transported in and out, plus entropy changes due to the heat flow across the boundaries, and in addition, since entropy may be generated by an irreversible process, an additional term for entropy generation is added. A general entropy balance is

❶ General entropy balance.

$$\frac{dS}{dt} = \sum_{in} S^{in} \dot{m}^{in} - \sum_{out} S^{out} \dot{m}^{out} + \frac{\dot{Q}}{T_{sys}} + \dot{S}_{gen}$$

3.27

Like the energy balance, the quantity to the left of the equals sign represents the entropy change of the system. The term representing heat transfer should be applied at each location where heat is transferred and the T_{sys} for each term is the system temperature at each boundary where the heat transfer occurs. The heat transfer represented in the general entropy balance is the heat transfer which occurs in the *actual* process. We may simplify the balance for steady-state or closed systems:

❶ Open, steady-state entropy balance.

$$0 = \dot{m}\left(S^{in} - S^{out}\right) + \frac{\dot{Q}}{T} + \underline{\dot{S}}_{gen}$$

3.28

❶ Closed system entropy balance.

$$d\underline{S} = \frac{d\underline{Q}}{T} + d\underline{S}_{gen}$$

3.29

Caution: The entropy balance provides us with an additional equation which may be used in solving thermodynamic problems; however, in irreversible processes, the entropy generation term can usually not be calculated from first principles. Thus, it is an unknown in Eqns. 3.27–3.29. The balance equation is not useful for calculating any other unknowns when $\dot{S}_{gen}$ is unknown. In Example 3.5 the problem would have been difficult if we applied the entropy *balance* to the water or the surroundings independently, because we did not know how to calculate $\dot{S}_{gen}$ for each subsystem. However, we could calculate $\Delta\underline{S}$ for each subsystem along reversible pathways. Summing the entropy changes for the subsystems of the universe, we obtain the entropy change of the universe. Consider the right-hand side of the entropy balance when written for the universe in this example. There is no mass flow in and out of the universe—it all occurs between the subsystems of the universe. In addition, heat flow also occurs between subsystems of the universe, and the first three terms on the right-hand side of the entropy balance are zero. Therefore, the entropy change of the universe is equal to the entropy generated in the universe.

Caution: The criterion for the feasibility of a process is that the entropy generation term must be greater than or equal to zero. The feasibility may not be determined unequivocally by ΔS for the system unless the system is the universe.

As we work examples for irreversible processes, *note that we do not apply the entropy balance to find entropy changes. We always calculate entropy changes by alternative reversible pathways that reach the same states, then we apply the entropy balance to find how much entropy was generated.*

Alternatively, for reversible processes, *we do apply the entropy balance because we set the entropy generation term to zero.*

Let us now apply the entropy balance—first to another heat conduction problem. In Example 3.5 on page 102 we studied an unsteady-state system. Now let us consider steady-state heat transfer to see which entropy balance terms are important in this application. In Example 3.5, we stated that irreversibility was due to a finite-temperature driving force when the glass of water was warmed by the room air, and we indicated that the entropy was generated. In this example, we will show that such heat conduction results in entropy generation because entropy generation occurs where the temperature gradient exists.

Example 3.7 Steady-state entropy generation

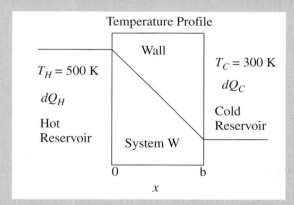

Temperature Profile

Imagine heat transfer occurring between two reservoirs. A steady-state temperature profile for such a system is illustrated above. The entire temperature gradient occurs within the wall. In this ideal case, there is no temperature gradient within either reservoir (therefore, the reservoirs are not a source of entropy generation).

Example 3.7 Steady-state entropy generation (Continued)

Take the universe as three subsystems, hot reservoir, cold reservoir, wall.

> *Note: Keeping track of signs and variables can be confusing when the universe is divided into multiple subsystems. Heat flow on the hot side of the wall will be negative for the hot reservoir, but positive for the wall. As you study this example, keep in mind that we have chosen to use subscript H, C, W to denote properties of the hot reservoir, cold reservoir, and wall, respectively. When we match heat flow at a boundary, we may choose to write it in terms of heat flow for the other subsystem, but with a negative sign.*

For the hot and cold reservoirs, the entropy balance simplifies:

$$\frac{dS}{dt} = \sum_{in} S^{in} \dot{m}^{in} - \sum_{out} S^{out} \dot{m}^{out} + \frac{\dot{Q}}{T_{sys}} + \dot{S}_{gen}$$

The entropy generation term drops out because there is no temperature gradient in the reservoirs. Applying to each reservoir:

$$\frac{dS_H}{dt} = \frac{\dot{Q}_H}{T_H}$$

$$\frac{dS_C}{dt} = \frac{\dot{Q}_C}{T_C} = -\frac{\dot{Q}_H}{T_C}$$

where the heat fluxes are equated in the second equation by the energy balance, which we have not shown.

Now consider the entropy balance for the wall. Entropy is a state property, and since no state properties throughout the wall are changing with time, entropy of the wall is constant, and the left-hand side of our entropy balance is equal to zero. To avoid introducing new variables for the heat flow, we match the heat flow at the boundaries with the heat flow from/to the hot and cold reservoir, and use variables which are already defined (note there are two heat transfer terms because there are two surfaces, and that the entropy generation term is kept because we know there is a temperature gradient):

$$\frac{dS_W}{dt} = 0 = -\frac{\dot{Q}_H}{T_H} + -\frac{\dot{Q}_C}{T_C} + \dot{S}_{gen,W}$$

We may then write for the wall:

$$\dot{S}_{gen,W} = \dot{Q}_H\left(\frac{1}{T_H} - \frac{1}{T_C}\right) \qquad\qquad 3.30$$

and the wall with the temperature gradient is a source of entropy generation.

Example 3.7 Steady-state entropy generation (Continued)

An overall entropy balance gives:

$$\frac{dS_{univ}}{dt} = \frac{dS_H}{dt} + \frac{dS_W}{dt} + \frac{dS_C}{dt} = \dot{S}_{gen} \qquad\qquad 3.31$$

And we see that the wall is the source of entropy generation of the universe, which is positive. Notice that inclusion of the wall is important in accounting for the entropy generation by the entropy balance equations.

⚠ Entropy is generated by a temperature gradient.

We have concluded that heat transfer results in entropy production. How can we transfer heat reversibly? If the size of the gradient is decreased, the right-hand side of Eqn. 3.30 will decrease in magnitude, because for heat conduction through a body:[1]

$$\dot{Q} = Ak\frac{dT}{dx} \quad \text{where } A = \text{cross sectional area}$$

A smaller temperature gradient will decrease the rate of production of entropy, but from a practical standpoint, will require a longer time to transfer a fixed amount of heat. If we wish to transfer heat reversibly from two reservoirs at finitely different temperatures, we can use a heat engine, which will be introduced below. In addition to transferring the heat reversibly, use of a heat engine will permit generation of work.

> **Summary:** *This example has shown that boundaries (walls) between systems can generate entropy. In this example, entropy was not generated in either reservoir because no temperature profile existed. The entropy generation occurred within the wall.*

⚠ Entropy may be generated at system boundaries.

Example 3.8 Reversible work between heat reservoirs, lost work

Reconsider Example 3.7, but let us introduce a reversible heat engine to replace the wall as the mode for transfer of the heat. At this point, do not be concerned about the actual configuration of the engine. Accept that such a device can be imagined and operated in such a manner that it does not generate entropy. We will show how this can be done in Section 3.5 on page 110. Let us also consider a batch process rather than steady-state heat flow. The analysis is very similar, but now we deal with quantities of change rather than rates of change.

A reversible heat engine absorbs 1000 J at 500°C produces work and discards heat at 100°C. The hot and cold reservoirs are also at 500°C and 100°C, respectively. What is the change in entropy of the heat source and the heat sink, and what is the total entropy change resulting from the process? How much work is produced?

---

1. This relation is known as Fourier's Law and is studied in heat-transfer courses.

Example 3.8 Reversible work between heat reservoirs, lost work (Continued)

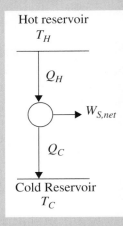

Hot reservoir
T_H

Q_H

$W_{S,net}$

Q_C

Cold Reservoir
T_C

To perform the heat transfer, an engine will absorb heat from the hot reservoir, process it, reject some heat to the cold reservoir, then return to the initial state ready to accept another batch of heat. Thus, a heat engine is left in the same *state* at the end of cycle as it was at the beginning.

Solution: Entropy Balance: $\Delta \underline{S}_{univ} = 0$ (reversible)

First for the engine, since it is left in the same *state* where it starts, $\Delta S_{engine} = 0$. Recall that entropy is a *state* property, and, therefore, at the end of an engine cycle, the entropy has returned to the initial value.

$$\Delta \underline{S}_{univ} = \Delta \underline{S}_H + \Delta \underline{S}_C + \Delta \underline{S}_{engine} = \Delta \underline{S}_H + \Delta \underline{S}_C + 0 = 0$$

> This example applies the entropy balance to the reservoirs and heat engine individually.

Looking next at the reservoirs:

$$\Delta \underline{S}_H = \underline{Q}_H / T_H = -1000 \text{ J/ } 773 \text{ K} = -1.29 \text{ J/K}$$

Since the process is reversible, $\Delta \underline{S}_{univ} = \Delta \underline{S}_H + \Delta \underline{S}_C = \underline{Q}_H/T_H + \underline{Q}_C/T_C = 0$

$$\Rightarrow \Delta \underline{S}_C = 1.29 \text{ J/K} = \underline{Q}_C/T_C \Rightarrow \underline{Q}_C = 481 \text{ J}$$

(This process is different from Example 3.7 because $Q_H \neq -Q_C$)

Energy Balance (on engine): $\underline{W}_{S,net} = -\underline{Q}_{net} = -(\underline{Q}^{in} + \underline{Q}^{out}) = -(-\underline{Q}_H - \underline{Q}_C)$
$\Rightarrow \underline{W}_{S,net} = -(1000 - 481) = -519 \text{ J}$

Note in Example 3.8 that we could have written an entropy balance on the engine. Since the engine is a closed system which does not generate entropy, and it is left in the same state as it started, the entropy balance for the heat engine is:

> Entropy balance applied to a heat engine.

$$\Delta \underline{S} = 0 = \frac{\underline{Q}^{in}}{T_H} + \frac{\underline{Q}^{out}}{T_C} = -\frac{\underline{Q}_H}{T_H} - \frac{\underline{Q}_C}{T_C} \Rightarrow \underline{Q}_C = -\underline{Q}_H\left(\frac{T_C}{T_H}\right) \qquad 3.32$$

which matches what we have written above, but is a different way of looking at the same process. We may write the work in terms of the heat flow and temperatures by an energy balance around the engine:

$$-\underline{W}_{S,net} = \underline{Q}_H + \underline{Q}_C = \underline{Q}_H\left(1-\frac{T_C}{T_H}\right) = T_C \cdot \underline{Q}_H\left(\frac{1}{T_C} - \frac{1}{T_H}\right) \qquad 3.33$$

where $\underline{Q}_H$ and $\underline{Q}_C$ are for the engine (the negative of the values for the reservoirs). Compare this to Eqns. 3.30 and 3.31, the results of the steady-state entropy generation example. If we had run the heat transfer process irreversibly in Example 3.8, then the universe would have lost a quantity of work equal to $-T_C\underline{S}_{gen}$. Note that T_C, the colder temperature of our engine, is important in relating the entropy generation to the lost work. Sometimes this is referred to as the temperature at which the work was lost. Also note that we have chosen to operate the heat engine at temperatures which match the reservoir temperatures. This is arbitrary, but is required to obtain the maximum amount of work. The heat engine may be reversible without this constraint, but the entire process will not be reversible. These details will be clarified later in Section 3.7 on page 113.

Example 3.9 Entropy change of quenching

A carbon-steel engine casting [C_P = 0.5 kJ/kg°C] weighing 100 kg and having a temperature of 700 K is heat-treated to control hardness by quenching in 300 kg of oil [C_P = 2.5 kJ/kg°C] initially at 298 K.

If there are no heat losses from the system, what is the change in entropy of: (a) the casting; (b) the oil; (c) both considered together; (d) Is this process reversible?

Solution: Unlike the previous examples, there are no reservoirs, and the casting and oil will both change temperature. The final temperature of the oil and the steel casting is found by an energy balance. Let T^f be the final temperature in K.

Energy Balance: The total change in energy of the oil and steel is zero.

Heat lost by casting:

$$\underline{Q} = 100\,(0.5)\,(700 - T^f)$$

Heat gained by oil:

$$\underline{Q} = 300\,(2.5)\,(T^f - 298) \implies T^f = 323.1 \text{ K}$$

Entropy Balance: The entropy change of the universe will be the sum of the entropy changes of the oil and casting. We will not use the entropy balance directly except to note that $\Delta\underline{S}_{univ} = \underline{S}_{gen}$.

We can calculate the change of entropy of the casting and oil by any reversible pathway which begins and ends at the same states. Consider an isobaric path:

$$\text{Using the macroscopic definition} \implies \Delta\underline{S} = \int\frac{dQ}{T} = m\int\frac{C_P}{T}\,dT = mC_P\,\ln\left(\frac{T_2}{T_1}\right) \qquad (*)$$

(a) Change in entropy of the casting:

$$\Delta\underline{S} = 100\,(0.5)\,\ln[323.1/700] = -38.7 \text{ kJ/K} \qquad (*)$$

> ## Example 3.9 Entropy change of quenching (Continued)
>
> (b) Change in entropy of the oil (the oil bath is of finite size and will change temperature as heat is transferred to it):
>
> $$\Delta \underline{S} = 300\,(2.5)\,\ln[323.1/298] = 60.65 \text{ kJ/K} \qquad\qquad (*)$$
>
> (c) Total entropy change: $\underline{S}_{gen} = \Delta\,\underline{S}_{univ} = 60.65 - 38.7 = 21.9 \text{ kJ/K}$
>
> (d) $\underline{S}_{gen} > 0$; therefore irreversible; compare the principles with Example 3.5 on page 102 to note the similarities. The difference is that both subsystems changed temperature.

❶ Compare with Example 3.5 on page 102.

3.5   THE CARNOT ENGINE

❶ The Carnot cycle is *one* method of constructing a heat engine.

In Example 3.8 on page 107, we have proposed an engine which is able to perform heat transfer between two systems with finite temperature differences in an internally reversible manner, and simultaneously obtain work. To perform the thermodynamic calculations for the process, we did not need to know *how* the engine was constructed. You may be wondering how such a mechanical device could be constructed. In this section, we develop the Carnot cycle to demonstrate the principles which could be used to construct such a device. The Carnot engine has been shown to provide the highest *thermal efficiency* of any feasible engine. The thermal efficiency is the ratio of the work obtained to the heat supplied. The definition of thermal efficiency is

❶ Thermal efficiency.

$$\eta \equiv \frac{work\ output}{heat\ input} = -\frac{\dot{W}_{S,net}}{\dot{\underline{Q}}_H} \qquad\qquad 3.34$$

Note: In some of the previous examples and discussions, we have used subscripts H and C to refer to heat flows of the hot and cold reservoirs and $Q_H < 0$, $Q_C > 0$, where the reservoirs are our systems. However, now we consider only the heat engine in our analysis. Q_H and Q_C will now denote heat flows for the engine, and $Q_H > 0$, $Q_C < 0$.

A schematic of a Carnot heat engine is shown in Fig. 3.3. Heat is transferred from the hot reservoir to the heat engine. The heat engine extracts some work from this energy and rejects the remaining energy to the cold reservoir.

We may write an energy balance around the heat engine itself, as in the discussion following Example 3.8, giving the equation

$$-\dot{\underline{W}}_{S,net} = \dot{\underline{Q}}_H + \dot{\underline{Q}}_C \quad\Rightarrow\quad \eta = -\frac{\dot{\underline{W}}_{S,net}}{\dot{\underline{Q}}_H} = 1 + \frac{\dot{\underline{Q}}_C}{\dot{\underline{Q}}_H} \qquad\qquad 3.35$$

Note the change in sign relative to Eqn. 3.33 because the heat terms belong to the engine. Let us consider the steps which comprise the Carnot engine cycle.

Caution: Many of the following derivations apply only to heat reservoirs that do not change temperature when heat is added or removed. Analysis of reversible heat transfer in systems of finite size is also possible, but temperature will change as heat is transferred in a finite system.

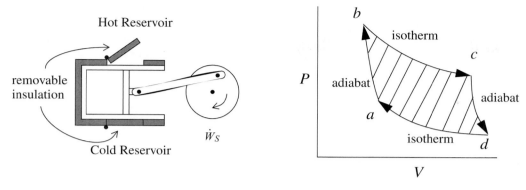

Figure 3.3 *Schematic of the Carnot Engine, and the Carnot P-V cycle when using a gas as the process fluid.*

The Carnot engine consists of the steps illustrated in Fig. 3.3. Consider a piston in the vicinity of both a hot reservoir and a cold reservoir. The insulation on the piston may be removed to transfer heat from the hot reservoir during one step of the process, and also removed from the cold side to transfer heat to the cold reservoir during another step of the process. Carnot conceived of the cycle consisting of the steps shown schematically on the *P-V* diagram beginning from point *a*. Between points *a* and *b*, the gas undergoes an adiabatic compression. From point *b* to *c*, the gas undergoes an isothermal expansion, absorbing heat from the hot reservoir. From point *c* to *d*, the gas undergoes an adiabatic expansion. From point *d* to *a*, the gas undergoes an isothermal compression, rejecting heat to the cold reservoir. The work obtainable by this process is work done by the gas which is given by the $\int P dV$ over the whole cycle. This work is represented by the area within the sketched cycle on the *P-V* diagram.

The heat transferred and work performed in the various steps of the process are summarized in the table below. For this calculation we assume the gas within the piston follows the ideal gas law with temperature-independent heat capacities. We calculate reversible changes in the system; thus, we neglect temperature and velocity gradients within the gas (or we perform the process very slowly so that these gradients do not develop).

> *Note: The temperatures T_H and T_C here refer to the hot and cold temperatures of the gas, which are not required to be equal to the temperatures of the reservoirs for the Carnot engine to be reversible. Later we will show that if these temperatures do equal the reservoir temperatures, the work is maximized.*

Table 3.3 *Illustration of Carnot cycle calculations for an ideal gas.[a]*

Step	Type	$\underline{Q}$		$\underline{W}$		$\underline{\Delta S}$
a → b	adiabat	0		$nC_V(T_H - T_C)$	(*ig)	0
b → c	isotherm	$\underline{Q}_H = nRT_H \ln \dfrac{V_c}{V_b}$	(ig)	$-nRT_H \ln \dfrac{V_c}{V_b}$	(ig)	$\dfrac{Q_H}{T_H}$
c → d	adiabat	0		$nC_V(T_C - T_H)$	(*ig)	0
d → a	isotherm	$\underline{Q}_C = nRT_C \ln \dfrac{V_a}{V_d}$	(ig)	$-nRT_C \ln \dfrac{V_a}{V_d}$	(ig)	$\dfrac{Q_C}{T_C}$

a. *The Carnot cycle calculations are shown here for an ideal gas. There are no requirements that the working fluid is an ideal gas.*

The Carnot engine operates on a cycle that we have arbitrarily chosen to begin in state a and return to state a. Since the state of the fluid is returned at the end of the cycle, all of our defined thermodynamic *state* quantities are the same at the end of the cycle as at the beginning. A net change in internal energy, enthalpy, and entropy for the cycle will be zero for the fluid. Therefore, we may write

$$\Delta \underline{S} = 0 = \frac{Q_H}{T_H} + \frac{Q_C}{T_C}$$

The same equation can be derived by starting with the entropy balance for a Carnot engine operating at steady state. This results in

$$\underline{\dot{S}} = 0 = \frac{\dot{Q}_H}{T_H} + \frac{\dot{Q}_C}{T_C} \qquad 3.36$$

These last two equations are simply direct applications of the entropy balance to the reversible heat engine. We may rearrange either equation to

$$\frac{Q_C}{Q_H} = -\frac{T_C}{T_H} \qquad 3.37$$

The thermal efficiency of the Carnot cycle is given by inserting the ratio of temperatures into Eqn. 3.35

$$\boxed{\eta = -\frac{W_{S,net}}{\underline{Q}_H} = 1 + \frac{Q_C}{Q_H} = 1 - \frac{T_C}{T_H}} \qquad 3.38$$

Note from Eqn. 3.38 that we cannot achieve $\eta = 1$ unless the temperature of the hot reservoir becomes infinite or the temperature of the cold reservoir approaches 0 K. Such reservoir temperatures are not practical for real applications. For real processes, we typically operate between the temperature of a furnace and the temperature of cooling water. For a normal power-plant cycle, these temperatures might be 600 K for the hot reservoir and 300 K for the cold reservoir, so the maximum thermal efficiency for the process is near 50%. Most real power plants operate with thermal efficiencies closer to 30–35% due to inherent inefficiencies in real processes.

The work summarized in Table 3.3 represents work done by the gas in each step of the cycle. This work is equal to the work done on the shaft plus the expansion/contraction work done on the atmosphere for each step. By summing the work terms for the entire cycle, the net work done on the atmosphere in a complete cycle is zero since the net atmosphere volume change is zero. Therefore, the work represented by the shaded portion of the *P-V* diagram is the useful work transferred to the shaft.

3.6　CARNOT HEAT PUMP

In the above section, we have demonstrated a device capable of reversibly transferring heat and simultaneously obtaining work. Since work is obtained, we call the device a Carnot engine. Heat flow from the colder reservoir to the warmer reservoir will not occur without adding work, but the Carnot cycle can be reversed with work added, and results in a heat pump. In essence, the device is performing the same role as a refrigerator or freezer, where heat is removed from the cold chamber and transferred to the warmer surroundings. However, the objective for the use of a heat pump may actually be heating; a heat pump may be used for home heating in the winter where the outdoor surroundings serve as the cold reservoir, and the heat is pumped to the warmer house. For a given transfer of heat from the cold reservoir, a Carnot heat pump requires the minimum amount of work for any conceivable process. We therefore use the Carnot heat pump as a benchmark for comparing real processes, in the same way that thermal efficiency serves for engines. The *coefficient of performance, COP,* is the ratio of the work required to the heat transferred from the cold reservoir. (You may wish to think of it as the ratio of cooling obtained to net work).

$$COP \equiv \frac{\dot{\underline{Q}}_C}{\underline{\dot{W}}_{S,net}} \qquad\qquad 3.39$$

Writing an entropy balance around the Carnot heat pump,

$$\dot{\underline{S}} = 0 = \frac{\dot{\underline{Q}}_H}{T_H} + \frac{\dot{\underline{Q}}_C}{T_C} \qquad \Rightarrow \frac{\dot{\underline{Q}}_H}{\dot{\underline{Q}}_C} = -\frac{T_H}{T_C} \qquad\qquad 3.40$$

Introducing the energy balance, $\underline{\dot{W}}_{S,net} = -(\dot{\underline{Q}}_C + \dot{\underline{Q}}_H)$, which can be rearranged,

$$\frac{\underline{\dot{W}}_{S,net}}{\dot{\underline{Q}}_C} = -\left(1 + \frac{\dot{\underline{Q}}_H}{\dot{\underline{Q}}_C}\right) = \left(\frac{T_H}{T_C} - 1\right) ,$$

$$\boxed{COP \equiv \frac{\dot{\underline{Q}}_C}{\underline{\dot{W}}_{S,net}} = \left(\frac{T_H}{T_C} - 1\right)^{-1}} \qquad\qquad 3.41 \quad \textbf{❶} \; \text{Coefficient of performance.}$$

3.7   INTERNAL REVERSIBILITY

A *process* may be irreversible due to interactions at the boundaries (such as discussed in Example 3.7 on page 105) even when the system is reversible. Such a process is called *internally reversible*. Such a *system* has no entropy generation *inside* the system boundaries. We have derived equations for Carnot engines and heat pumps, assuming that the devices will operate between temperatures that match the reservoir temperatures. While such restrictions are not necessary for internal reversibility, we show here that the work is maximized in a Carnot engine at these conditions and minimized in a Carnot heat pump. Note that in development of the Carnot devices, the only temperatures of concern are the operating temperatures at the hot and cold portions of the cycle. In the following illustrations, the internally reversible engine or pump operates between T_H, T_C, and the reservoir temperatures are T_2 and T_1.

Heat Engine

A schematic for a Carnot engine is shown below. Heat is being transferred from the reservoir at T_2

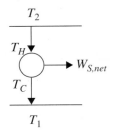

to the reservoir at T_1, and work is being obtained as a result. In order for heat transfer to occur between the reservoirs and the heat engine in the desired direction, we must satisfy $T_2 \geq T_H > T_C \geq T_1$, and since the efficiency is given by Eqn. 3.38, for maximum efficiency (maximum work), T_C should be as low as possible and T_H as high as possible, i.e., set $T_H = T_2$, $T_C = T_1$.

❶ The operating temperatures of a reversible heat engine or heat pump are not necessarily equal to the surrounding's temperatures; however, the optimum work interactions occur if they match the surrounding's temperature because matching the temperatures eliminates the finite temperature driving force that generates entropy.

Heat Pump

A schematic for a Carnot heat pump is shown below. Heat is being transferred from a reservoir at T_1

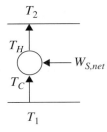

to the reservoir at T_2, and work is being supplied to achieve the transfer. In order for heat transfer to occur between the reservoirs and the heat engine in the desired direction, we must satisfy $T_2 \leq T_H > T_C \leq T_1$. Since the COP is given by Eqn. 3.41, for maximum COP (minimum work), T_C should be as high as possible and T_H as low as possible, i.e., set $T_C = T_1$, $T_H = T_2$. *Therefore, optimum work interactions occur when the Carnot device operating temperatures match the surrounding temperatures.* We will use this feature in future calculations without special notice.

3.8   MAXIMUM/MINIMUM WORK IN REAL PROCESS EQUIPMENT

Our analysis of the Carnot devices supports statement number two at the beginning of Section 3.3. We have seen that work is maximized/minimized when the entropy generation is zero. Analysis of other processes would verify this useful conclusion. Work is *lost* by processes which generate entropy. If a device is not *internally reversible,* work will be lost within the device. Also, even if the device is internally reversible, work may be lost by irreversible interactions with the surroundings. Therefore, in setting up and solving problems to find maximum/minimum work, the objectives must be clear as to whether the system is internally reversible or whether the entire process is reversible. When we apply the entropy balance to a reversible process, the term representing entropy generation is zero.

In Chapter 2, both velocity gradients and friction were discussed as phenomena that lead to irreversibilities. Indeed, entropy is generated by both these phenomena as well as by heat conduction along temperature gradients discussed in this chapter. Considering factors which affect reversibility (generation of entropy), you may have challenged yourself to consider a practical way to transfer heat with only infinitesimal temperature differences, or move fluid with only infinitesimal velocity gradients. You are probably convinced that such a process would not be practical. Indeed, the rate at which heat is transferred increases as the temperature driving forces increase, and we need finite temperature differences to transfer heat practically. Likewise, a pump will have large velocity gradients. Although we can measure changes in other properties by which we can calculate entropy changes arising from irreversibilities, an *a priori* prediction of lost work (lw) or entropy generated ($\underline{S}_{gen}$) is extremely difficult and generally impractical. Direct evaluation of lost work in process equipment, such as turbines and compressors, is far beyond routine calculation and determined by empirical experience. It may seem that all of the effort to characterize reversible processes will be difficult to relate to real processes.

However, we can use practical experience to relate real processes to the idealized reversible processes. Therefore, in analyzing or designing processes involving operations of this nature, it is often necessary to approximate the real situation with a reversible one in which (lw) = 0 (no entropy generation). Past experience with many devices, such as compressors and turbines, often permits the engineer to relate performance under hypothetical reversible conditions to actual operation under real conditions. The relation is usually expressed by means of an efficiency factor. For devices such as pumps and compressors which utilize work from the surroundings, efficiency is defined as

$$\text{pump or compressor efficiency} = \left| \frac{W'}{W} \times 100\% \right|$$

3.42   ❶ Primes are used to denote reversible processes.

where the ' denotes the reversible work. This notation will be used throughout the text when irreversible and reversible calculations are performed in the same problem. For turbines and other expansion devices that supply work to the surroundings, the definition is inverted to give

$$\text{turbine or expander efficiency} = \left| \frac{W}{W'} \times 100\% \right|$$

3.43

For *adiabatic* pumps, compressors, turbines or expanders, the work terms may be calculated from the reversible and irreversible enthalpy changes by application of the energy balance.

The strategy is to first calculate the work involved in a reversible process, then apply an efficiency which is empirically derived from previous experience with similar equipment. The outlet **pressure** of an irreversible adiabatic turbine or pump is always at the same pressure as a reversible device, but the enthalpy is always higher for the same inlet state. *This means that if the outlet of the reversible adiabatic device is a single phase, the outlet of the irreversible adiabatic device will be at a higher temperature. If the outlet of the reversible adiabatic device is a two-phase mixture, the quality for the irreversible adiabatic device will be higher or the outlet could potentially be a single phase.*

3.9 ENTROPY BALANCE FOR PROCESS EQUIPMENT

Before moving on to more examples, it will be helpful to consider the entropy balance for common steady-state process equipment, because understanding these balances is a key step for the calculation of reversible heat and work interactions.

Adiabatic Turbine, Compressor, Pump

The entropy balance for a steady-state *adiabatic* device is:

$$\frac{dS}{dt} = \sum_{in} S^{in} \dot{m}^{in} - \sum_{out} S^{out} \dot{m}^{out} + \frac{\dot{Q}}{T_{sys}} + \dot{S}_{gen} \qquad 3.44$$

The lefthand side drops out because the system is at steady-state. If the device is reversible, $\dot{S}_{gen}$ is zero. Further, these devices typically have a single inlet or outlet,[1] and $\dot{m}^{in} = \dot{m}^{out}$, thus,

🛑 Adiabatic reversible turbine, compressor, pump.

$$S^{out} = S^{in} \qquad 3.45$$

Therefore, if we know the inlet state, we can find S^{in}. The outlet pressure is generally given, so for a pure fluid, the outlet state is completely specified by the two state variables S^{out} and P^{out}. We then use thermodynamic relations to find the other thermodynamic variables at this state, and use the energy balance at this state to find W_{rev}.

Heat Exchanger

The entropy balance for a standard two-stream heat exchanger is also given by Eqn. 3.44. Since the unit is at steady-state, the lefthand side is zero. Since the entropy balance is written around the *entire* heat exchanger, there is no heat transfer *across* the system boundaries (in the absence of heat loss), so the heat-transfer term is eliminated. Since heat exchangers operate by conducting heat across tubing walls with finite temperature driving forces, we would expect the devices to be irreversible. Indeed, if the inlet and outlet states are known, the flow terms may be evaluated, thus permitting calculation of entropy generation.

We also may perform "paper" design of ideal heat exchangers that operate reversibly. If we set the entropy generation term equal to zero, we find that the inlet and outlet states are constrained. Since there are multiple streams, the temperature changes of the streams are coupled to satisfy the entropy balance. In order to construct such a reversible heat exchanger, the device would need to be impracticably large to only have small temperature gradients.

Throttle Valves and Nozzles

Steady-state throttle valves are usually assumed to be adiabatic, so they follow Eqn. 3.44. Since there is almost always a pressure drop through the valve, the entropy on the outlet will be larger than the entropy on the inlet, indicating that entropy is generated. The entropy increase for a pressure drop across a valve will be large for a gas, and it is generally small, but non-zero, for liquids. It is important to recall also that liquid streams near saturation may flash as they pass through throttle valves, which also results in large entropy changes.

Steady-state nozzles can be designed to operate nearly reversibly; therefore, we may assume $\dot{S}_{gen} = 0$, and Eqn. 3.45 applies.

---

1. In multistage units, the stages may be considered individually.

3.10   CHARTS INCLUDING ENTROPY

Turbines, compressors and pumps occur so frequently that we need convenient tools to aid in process calculations. Visualization of the state change is possible by plotting entropy on charts. This technique also permits the charts to be used directly in the process calculations. One common representation is the *T-S* chart shown in Fig 3.4. The phase envelope appears as a fairly symmetrical hump. A reversible turbine, compressor or pump creates state changes along a vertical line on these coordinates. Lines of constant enthalpy and pressure are also shown on these diagrams, as sketched in the figure. Volumes are also usually plotted, but lie so close to the pressure lines that they are not illustrated in the figure here to assure clarity.

Visualizing state changes on charts will be helpful when using tables or computers for physical properties.

   P-H diagrams are also useful; they are used frequently for refrigeration processes. The phase envelope tends to lean to the right because the enthalpies of vapor and liquid are both increasing along the saturation curve until the critical point is approached, where the vapor-phase enthalpy decreases due to significant non-idealities. Lines of constant entropy on these plots are slightly diagonal with a positive slope as shown in Fig. 3.5*a*. For some hydrocarbons and halogenated compounds, the phase envelope can lean more sharply than the isentropic lines as shown in Fig. 3.5*b*. A reversible compressor will operate along a line of constant entropy.

   Another convenient representation of entropy is the *H-S* diagram (Mollier diagram). In this diagram, lines of constant pressure are diagonal, and isotherms have a downward curvature as in Fig 3.6. The saturation curve is quite skewed.

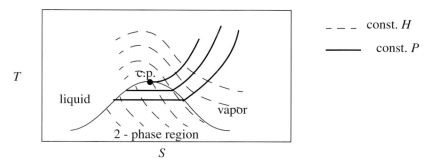

Figure 3.4 *Illustration of a T-S diagram showing lines of constant pressure and enthalpy.*

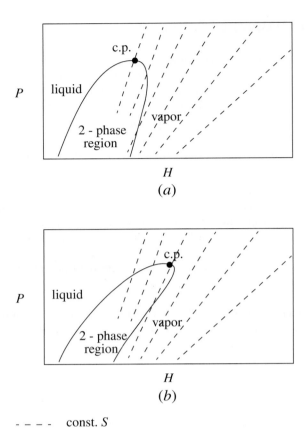

(a)

(b)

- - - -   const. S

Figure 3.5 *Illustration of a P-H diagram showing (a) lines of constant entropy for a species where the saturation curve leans less than isentropes (e.g., water) and (b) Illustration of a P-H diagram showing lines of constant entropy for a species where the saturation curve leans more than isentropes (e.g., hexane).*

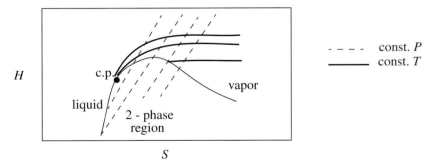

- - - -   const. P
————   const. T

Figure 3.6 *Illustration of an H-S diagram showing lines of constant entropy.*

3.11   TURBINE CALCULATIONS

For a reversible adiabatic turbine, the entropy balance in Section 3.9 shows that the outlet entropy will be equal to the inlet entropy. For an irreversible turbine, the outlet entropy will be greater than the inlet entropy. We may now visualize the state change on diagrams sketched in Section 3.10. For example, on a *T-S* diagram, the performance of a turbine can be visualized as shown in Fig. 3.7 below. Note that the isobars are important in sketching the behavior because the *outlet pressure will be the same* for the reversible and irreversible turbines, but the outlet enthalpies (not shown) and entropies will be different.

> ❗ The outlet entropy of an *irreversible* adiabatic turbine will be greater than the outlet entropy of a *reversible* adiabatic turbine with the same outlet pressure.

Steam Quality Calculations

A common problem encountered when adiabatically reducing the pressure of real fluids like steam, methane, or Freon is the formation of a vapor-liquid mixture. Since the thermodynamic properties change dramatically depending on the mass fraction that is vapor (the quality), it is important to know how to calculate that fraction. The calculation procedure will differ from the case shown in Fig. 3.7 where the outlets for the reversible and irreversible cases are both one phase. Since the reversible adiabatic turbine is isentropic, the line representing the reversible process will be vertical. As shown in Fig. 3.8, if the upstream entropy is less than the saturated vapor entropy at the outlet pressure, the *reversible* process ends up *inside the liquid-vapor region,* to the left of the saturated vapor curve. In this case, we must perform a quality calculation to determine the fraction vapor. Since the real turbine will have an outlet state of higher entropy, due to entropy generation, the

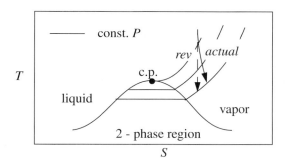

Figure 3.7 *Illustration of a reversible and actual (irreversible) turbine on a T-S diagram.*

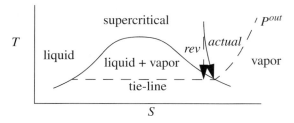

Figure 3.8 *Illustration of need for quality calculation on turbine outlet where the actual outlet is saturated steam.*

The actual outlet state might be in the one-phase region when the reversible outlet state is in the two-phase region.

outlet state can lie inside the phase envelope, on the saturation curve, or outside the phase envelope, depending on the proximity of the reversible outlet state to the saturation curve and also depending on the turbine or expander efficiency. When students first confront this problem they often ask, "How do I know when I need to do a quality calculation?" The calculation will be required if the *inlet* entropy is less than the saturation entropy at the *outlet* pressure as illustrated in the figure. A quality calculation may also be required for the actual state, if the actual enthalpy turns out to be less than the saturation enthalpy at the outlet pressure.

Example 3.10 Turbine efficiency

A test made on a stand-by turbine power unit produced the following results. With steam supplied to the turbine at 1.35 MPa and 375°C, the discharge from the turbine at 0.01 MPa was saturated vapor only. Determine the efficiency, the lost work, and the effective temperature at which the lost work was lost.

Solution: Assuming the turbine to be properly designed, we can neglect heat transfer and kinetic and potential energy terms.

Basis: The flowrate is not specified; therefore, we solve on the basis of 1 kg. A sketch of the reversible and actual pathway is shown in Fig. 3.8. The reversible pathway will fall inside the phase envelope since the actual process outlet is saturated steam. Since the reversible adiabatic turbine is isentropic, the line representing the performance will be vertical.

For an adiabatic turbine, the energy balance gives $\Delta H = W_s$. Under these circumstances,

$$\eta = \frac{\Delta H}{\Delta H'} = \frac{W_S}{W_S'}$$

Performing a double interpolation at the inlet:

T(°C)	H (kJ/kg)			S (kJ/kg-K)		
	1.2 MPa	1.35MPa	1.4MPa	1.2 MPa	1.35MPa	1.4MPa
350	3154.2	3151.1	3150.1	7.2139	7.1569	7.1379
375		3205			7.2401	
400	3261.3	3258.9	3258.1	7.3793	7.3233	7.3046

Initially, $H_{1.35}^{375C} = 3205$ kJ/kg; $S_{1.35}^{375C} = 7.2401$ kJ/kg-K

Using primes to denote the reversible process, the energy balance gives:

$$W_{act} = \Delta H = H_{0.01MPa}^{sat} - 3205 = 2584.3 - 3205 = -620.7 \text{ kJ/kg}$$

Example 3.10 Turbine efficiency (Continued)

Entropy balance: $\Delta S' = 0$ (reversible) and $P^f = 0.01\text{MPa}$

$$7.2401 = q'(8.1488) + (1 - q')(0.6492)$$

$$\Rightarrow q' = \frac{(7.2401 - 0.6492)}{(8.1488 - 0.6492)} = 0.8788$$

$$H^f = 0.8788(2583.9) + (1 - 0.8788)(191.8) = 2294.0 \text{ kJ/kg}$$

$$W = 2294.0 - 3205 = -911 \text{ kJ/kg}$$

$$\eta = \frac{-620.7}{-911} = 0.68$$

Lost work $= -911 - (-620.7) = -290.3$ kJ/kg. For saturated steam at the outlet, $S^f = 8.1488$ kJ/kg-K

$$S_{gen} = (S^f - 7.2401) = 8.1488 - 7.2401 = 0.91 \text{ kJ/kg-K}$$

$$T_0 = \frac{|\text{lost work}|}{S_{gen}} = \frac{290}{0.91} = 318.7 \approx (273.15 + 45.8) = T^{sat} \text{ at exit condition}$$

That is, work was lost at the temperature at which it was rejected downstream.

3.12 MULTISTAGE TURBINES

Commonly, turbines are staged for several reasons that we will explore in Chapter 4. Generally, some steam is drawn off at intermediate pressures for other uses. The important point that needs to be stressed now is that the convention used for characterizing efficiency is important. Consider the three-stage turbine shown in Fig. 3.9 and the schematic that represents the overall reversible path and the actual path. The isobars on the *H-S* diagram for water curve slightly upward, and are spaced

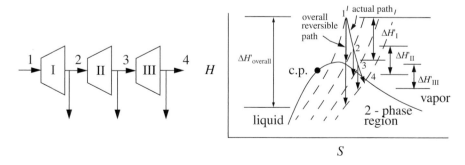

Figure 3.9 *Illustration that overall efficiency of an adiabatic turbine will be higher than the efficiency of the individual stages.*

slightly closer together at the bottom of the diagram than at the top as shown in Figure 3.9. The overall efficiency will be given by $\eta_{overall} = \dfrac{H_4 - H_1}{\Delta H'_{overall}}$, and the efficiency of an individual stage

will be given by $\eta_i = \dfrac{H_{i+1} - H_i}{\Delta H'_i}$. If we consider the reversible work as $\Delta H'_{overall}$, that quantity

will be smaller than $\displaystyle\sum_{stages} \Delta H'_i$. In fact, because the isobar spacing is increasing to the right of the diagram, the vertical drop between any isobars on the line marked as the overall reversible path will be smaller than the vertical drop between the same two isobars starting along the actual path (except for the very first turbine). Therefore, the efficiency calculated for the overall system will be higher than the efficiency for the individual stages. This comparison does not imply that staging turbines improves their performance. The difference in efficiencies is due to differences in what is considered to be the basis for the reversible calculation. The cautionary note to retain from this discussion is that the distinction between overall or individual efficiencies is important when communicating the performance of a staged turbine system.

> ❗ Overall turbine efficiency will be greater than stage efficiencies for the same total work output.

3.13 PUMPS AND COMPRESSORS

An irreversible adiabatic pump or compressor will generate entropy. If these devices are reversible, they will be isentropic. Examples of both are shown in Fig. 3.10. The calculation procedures are generally straightforward. Consider the case where the inlet state and the outlet pressure is known. First, the reversible outlet state is determined based on the isentropic condition, and the enthalpy at the reversible state is known. The efficiency can then be used to determine the actual outlet enthalpy, using Eqn. 3.44. The typical energy balance for a pump compressor is $\Delta H = Q + W_S$.

Multistage Compression

During adiabatic compression of vapors, the temperature rises. This can cause equipment problems if the temperature rise or compression ratio (P^{out}/P^{in}) is too large. To address this problem, interstage cooling is used to lower the gas temperature between compression stages. Such operations are common when high pressures need to be reached. A schematic of a compressor with interstage cooling is shown in Fig. 3.11.

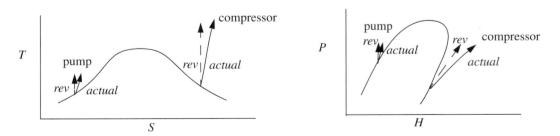

Figure 3.10 *Illustrations of pathways for reversible and irreversible pumps and compressors. The P-H diagram is for a system like Fig. 3.5a.*

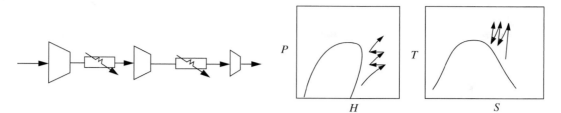

Figure 3.11 *Illustration of a multistage compression and the corresponding P-H diagram. On the P-H diagram, the compressors appear as the curves of increasing pressure and the heat exchangers are the horizontal lines at constant pressure.*

3.14   STRATEGIES FOR APPLYING THE ENTROPY BALANCE

When solving thermodynamic problems, usually the best approach is to begin with the mass and energy balances. In this chapter, we have introduced some new terms which can specify additional constraints when used in the problem statement, e.g., "isentropic," "reversible," "internally reversible," "irreversible," "thermal efficiency," "mechanical efficiency."

❶ New key words have been defined that specify constraints.

The entropy balance is useful to calculate maximum work available from a process or to evaluate reversibility. The entropy balance should be introduced with care because it is often redundant with the energy balance when simplified with information from step 5 from the strategies of Section 2.13. (For example, the entropy balance applied to Example 2.9 on page 66 results in the same simplified equation as the energy balance). *In general, if the pressures and temperatures of the process are already known, the entropies at each point, and the entropy changes, can be determined without direct use of the entropy balance. However, if either the pressure or temperature is unknown for a process, the entropy balance may be the key to the solution.*

❶ In some cases, *T* and *P* are known, so *S* can be determined without the entropy balance in a pure system.

Before beginning more examples, it is also helpful to keep in mind those processes which generate entropy. This is important because, in the event that such processes arise, the entropy-generation term cannot be set to zero unless we modify the process to eliminate the source of the generation. Entropy is generated by the following processes:

1. Heat conduction along a temperature gradient.
2. Diffusion along a concentration gradient.
3. Mixing of substances of different composition.
4. Adiabatic mixing at constant system volume of identical substances initially at different molar entropies due to (T, P) differences.
5. Velocity gradients within equipment. This is accounted for in pipe flow by the friction factor developed in textbooks on fluid flow.
6. Friction.
7. Electrical resistance.

In an open system, irreversibilities are always introduced when streams of different temperatures are mixed at constant pressure (item 4 above) because we could have obtained work by operating a

● Problem statements will rarely explicitly point out entropy generation, so you will need to look for causes.

heat engine between the two streams to make them isothermal before mixing. If the streams are isothermal, but of different composition, mixing will still generate entropy (e.g., see Eqn. 3.7 on page 95), and we have not yet devised a general method to obtain work from motion on this molecular scale.

As we work examples, recall the comments from Section 3.4 which we repeat here:

> *For* irreversible processes, *note that we* do not *apply the entropy balance to find entropy changes. We always calculate the entropy change by an alternate reversible pathway that reaches the final state, then we apply the entropy balance to find how much entropy was generated, or we find the reversible work, apply an efficiency factor and identify the final state via the energy balance.*
>
> *Alternately, for* reversible processes, *we* do *apply the entropy balance because we set the entropy generation term to zero.*

3.15 ADDITIONAL STEADY-STATE EXAMPLES

We have already considered many important examples. Probably the most important applications for process engineering calculations are the compressor, pump, and turbine calculations discussed above. However, before we leave the topic of entropy, we offer some more examples of important applications of the entropy balance.

Example 3.11 Heat pump analysis

Suppose your family is considering replacement of their furnace with a heat pump. Work is necessary in order to "pump" the heat from the low outside temperature up to the inside temperature. The best you could hope for is if the heat pump acts as a reversible heat pump between a heat source (outdoors in this case) and the heat sink (indoors). The average winter temperature is 4°C, and the building is to be maintained at 21°C. The coils outside and inside for transferring heat are of such size that the temperature difference between the fluid inside the coils and the air is 5°C. We generally refer to this as the "approach temperature." What would be the maximum cost of electricity in ($/kW-h) for which the heat pump would be competitive with conventional heating where a fuel is directly burned for heat. Consider the cost of fuel as $7.00 per 10^9J, and electricity as $0.10 per kW-h. Consider only energy costs.

Solution: The entropy balance gives $\dot{\underline{S}}_{pump} = 0$, using Eqn. 3.40 with the Carnot heat pump COP, Eqn. 3.41

$$\dot{\underline{W}}_{S,net} = \dot{\underline{Q}}_C\left(\frac{T_H}{T_C} - 1\right) = \dot{\underline{Q}}_H\left(\frac{-T_C}{T_H}\right)\left(\frac{T_H}{T_C} - 1\right) = \dot{\underline{Q}}_H\left(\frac{T_C - T_H}{T_H}\right)$$

$$\dot{\underline{W}}_{S,net} = \dot{\underline{Q}}_H\frac{(-1-26)}{(26+273.15)} = -\frac{27}{299.15}\dot{\underline{Q}}_H$$

Example 3.11 Heat pump analysis (Continued)

Letting $\dot{Q}_H$ be the heating requirement in kW,

Heat pump operating cost = $(0.09) \cdot \dot{Q}_H \cdot (\theta \text{ h}) \cdot [x \text{ \$/kW-h}]$; where θ is an arbitrary time and x is the cost.

Direct heating operating cost = $\dot{Q}_H \cdot (3600 \text{ s/h}) \cdot (\theta \text{ h}) \cdot (\$7/10^6 \text{ kJ})$

For the maximum cost of electricity for competitive heat pump operation, let heat pump cost = direct heating cost at the break even point.

$$\Rightarrow (0.09) \cdot \dot{Q}_H \cdot (\theta \text{ h}) \cdot [x \text{ \$/kW-h}] = \dot{Q}_H \cdot (3600 \text{ s/h}) \cdot (\theta \text{ h}) \cdot (\$7/10^6 \text{ kJ}) \qquad x = \$0.28/\text{kW-h}$$

Since the actual cost of electricity is given as \$0.10/kW-h it might be worthwhile if the heat pump is reversible and does not break down. (Purchase, installation and maintenance costs have been assumed equal in this analysis although the heat pump is more complex with more moving parts.)

Example 3.12 Entropy in a heat exchanger

A heat exchanger for cooling a hot hydrocarbon liquid uses 10 kg/min of cooling H_2O which enters the exchanger at 25°C. Five kg/min of hot oil enters at 300°C and leaves at 150°C and has an average specific heat of 2.51 kJ/kg-K.

(a) Demonstrate that the process is irreversible as it operates now;

(b) Assuming no heat losses from the exchanger, calculate the maximum work which could be obtained if we replaced the heat exchanger with a Carnot device which replaced the water stream and transferred heat to the surroundings at 25°C

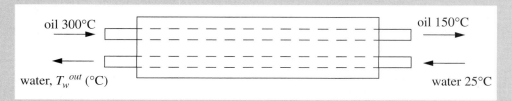

Solution: (a) System is heat exchanger (open system in steady-state flow)

Energy balance:

$$\Delta H_{oil}\dot{m}_{oil} + \Delta H_{water}\dot{m}_{water} = 0$$

$$10 \cdot 4.184 \, (T_{water}^{out} - 25) + 5(2.51)(150 - 300) = 0 \qquad T_{water}^{out} \Rightarrow 70°C$$

Example 3.12 Entropy in a heat exchanger (Continued)

Entropy balance

$$\Delta S_{oil}\dot{m}_{oil} + \Delta S_{water}\dot{m}_{water} = \dot{S}_{gen}$$

$$dS_i = dQ/T = C_P\,dT/T \Rightarrow \Delta S_i = C_{Pi}\ln(T_i^{\,out}/T_i^{\,in}) \tag{*}$$

$$\Delta S_{oil} = C_P\ln(T^{out}/T^{in}) = (2.51)\ln(423.15/573.15) = -0.7616 \text{ kJ/kgK} \tag{*}$$

$$\Delta S_{water} = C_P\ln(T^{out}/T^{in}) = (4.184)\ln(343.15/298.15) = 0.5881 \text{ kJ/kgK} \tag{*}$$

$$\Delta S_{oil}\dot{m}_{oil} + \Delta S_{water}\dot{m}_{water} = \underline{\dot{S}}_{gen} = 5\cdot(-0.7616) + 10\cdot0.5881 = 2.073 \text{ kJ/K-min}$$

(not reversible)

(b) The modified process is represented by the schematic shown below. To simplify analysis, the overall system boundary will be used.

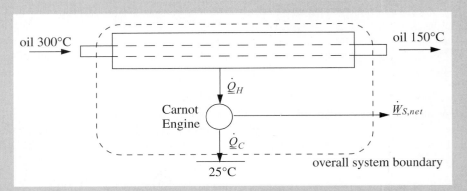

By an energy balance around the overall system,

$$0 = \dot{n}(H^{in} - H^{out}) + \underline{\dot{Q}}_C + \underline{\dot{W}}_S$$

We can only solve for the enthalpy term,

$$\dot{n}(H^{in} - H^{out}) = \dot{n}C_P(T^{in} - T^{out}) = 5(2.51)(300 - 150) = 1882.5 \text{ kJ/min}$$

Since heat and work are both unknown, we need another equation. Consider the entropy balance, which since it is a reversible process, $\underline{\dot{S}}_{gen} = 0$ gives

$$0 = (S^{in} - S^{out})\dot{n} + \frac{\underline{\dot{Q}}_C}{T}, \qquad \underline{\dot{Q}}_C = 298.15\cdot(-0.7616)\cdot5 = -1135 \text{ kJ/min}$$

Now inserting these results into the overall energy balance gives the work,

$$\underline{\dot{W}}_{S,net} = -1883. + 1135. = -748 \text{ kJ/min}$$

3.16   UNSTEADY-STATE OPEN SYSTEMS (OPTIONAL)

Example 3.13 Entropy change in a leaky tank

Consider air (an ideal gas) leaking from a tank. How does the entropy of the gas in the tank change? Use this perspective to develop a relation between T^f and P^f and compare it to the expression we obtained previously by the energy balance.

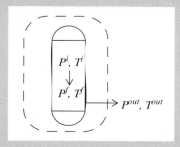

Solution:

m-Balance: $dn = -\, dn^{out}$

S-Balance:

$$\frac{d(nS)}{dt} = -S^{out}\frac{dn^{out}}{dt}$$

$$\Rightarrow ndS + Sdn = -S^{out}dn^{out}$$

But physically, we know that the leaking fluid is at the same state as the fluid in the tank; therefore, the S-balance becomes $ndS + S\!dn = -S^{out}dn^{out}$, or $\Delta S = 0$.

For an ideal gas with a constant heat capacity:

$$\Delta S = C_V \ln(T_2 / T_1) + R \ln(V_2 / V_1) = 0 \qquad (\text{*ig})$$

$$= C_V \ln(T_2 / T_1) + R \ln\!\left(\frac{T_2 P_1}{T_1 P_2}\right) = (C_V + R)\ln(T_2 / T_1) - R\ln(P_2 / P_1) \qquad (\text{*ig})$$

$$\Rightarrow \qquad T_2 / T_1 = (P_2 / P_1)^{R/Cp} \qquad (\text{*ig})$$

Compare with Example 2.13 on page 72. The entropy balance and energy balance in this case are not independent. Either can be used to derive the same result. This also shows that our analysis in Example 2.13 was assumed to be reversible.

Illustration that the energy and entropy balances may not be independent.

Example 3.14 An ideal gas leaking through a turbine
(unsteady-state)

A portable power supply consists of a 28-liter bottle of compressed helium, charged to 13.8 MPa at 300 K, connected to a small turbine. During operation, the He drives the turbine continuously until the pressure in the bottle drops to 0.69 MPa. The turbine exhausts at 0.1 MPa. Neglecting heat transfer, calculate the maximum possible work from the turbine. Assume helium to be an ideal gas with $C_P = 20.9$ J/mol-K.

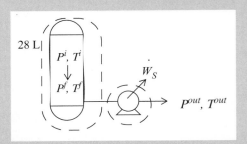

Consider a balance on the *tank only*. The result of the balance will match the result of Example 3.13.

Writing an entropy balance for a reversible adiabatic *turbine only*,

$$(S^{out} - S^{in})\dot{n} = 0 \;\;\Rightarrow\; \Delta S = 0$$

which shows that the turbine also does not change the molar entropy. Thus, the molar entropy of the exiting fluid is the same as the entropy in the tank, which is identical to the molar entropy at the start of the process. Therefore, the molar entropy and the pressure of the exiting gas are fixed. Since only two intensive properties fix all other intensive properties for a pure fluid, the exiting temperature is also fixed. The relation for an ideal gas along a reversible adiabat gives:

$$T^{out} = T^i\,(P^{out}/P^i)^{R/C_P} = 42.3 \text{ K} \tag{*ig}$$

$$\text{Likewise: } T^f = T^i\,(P^f/P^i)^{R/C_P} = 91.1 \text{ K} \tag{*ig}$$

Solution by overall energy balance:

$d(nU) = H^{out}\,dn + d\underline{W}_S$ and H^{out} is fixed since T^{out}, P^{out} are fixed; therefore, we may apply hint 4a from Section 2.13.

Integrating this expression

$$n^f U^f - n^i U^i = H^{out}(n^f - n^i) + \underline{W}_S$$

Rearranging:

$$\underline{W}_S = n^f(U^f - H^{out}) - n^i(U^i - H^{out}) \tag{3.46}$$

Determining variables in the equation,

$$n^f = P^f\,\underline{V}/RT^f;\; n^f = 25.5\text{gmol};\; n^i = 154.9\text{gmol} \tag{ig}$$

> ### Example 3.14 An ideal gas leaking through a turbine (unsteady-state) (Continued)
>
> Choose reference temperature, $T_R \equiv 300$ K, $\Rightarrow$ setting $U_R = 0$, then since $H_R = U_R + (PV)_R$, and since our fluid is an ideal gas, $C_V = C_P - R = 20.9 - 8.314 = 12.586$ J/molK
>
> $$H_R = (PV)_R = RT_R = R(300) \qquad \text{(ig) 3.47}$$
>
> Note: $\Rightarrow H(T) = C_P(T - T_R) + H_R = C_P(T - 300) + R \cdot (300) \qquad (*\text{ig})$
>
> $$H^{out} = -2892 \text{ J/mol} \qquad (*\text{ig})$$
>
> $$U^f = C_V(T - T_R) + U_R = -2629 \text{ J/mol}; \ U^i = 0 \qquad (*\text{ig})$$
>
> Now, plugging into Eqn. 3.46
>
> $$\underline{W}_S = 25.5(-2629 + 2892) - 154.9(0 + 2892)$$
>
> $$\Rightarrow \underline{W}_S = -441,200 \text{ J}$$

🛈 Illustration using a reference state.

3.17   THE ENTROPY BALANCE IN BRIEF

In this section, we refer to a division of the universe into the same three subsystems described in the Section 2.13 on page 65 "Strategies for Solving Process Thermodynamics Problems."

$$\frac{d\underline{S}}{dt} = \sum_{inlets} S^{in} \dot{n}^{in} - \sum_{outlets} S^{out} \dot{n}^{out} + \sum_{surfaces} \frac{\dot{Q}}{T} + \underline{\dot{S}}_{gen}$$

a. T is the *system* temperature at the location where Q is transferred.

b. S^{in}, S^{out} are state variables, and *any* pathway may be used to calculate the change from inlet to outlet. The pathway for calculation does not need to be the pathway for the actual process.

c. $\underline{\dot{S}}_{gen}$ represents entropy generation due to irreversibilities *within* the system, e.g., internal heat transfer or conduction, viscous dissipation or mixing of streams of differing composition, or degradation of mechanical energy to thermal energy. Entropy generation at system boundaries is not included in the balance.

d. Entropy generation may occur at container walls. The entropy generation of the universe must be calculated by summing $\underline{\dot{S}}_{gen}$ for all three subsystems, not just system 1 and system 3.

3.18   SUMMARY

The concept of entropy is important. It enables an analysis of reversible processes. The primary impact for pure-fluid applications is that compressors and turbines can be analyzed using empirical efficiencies. On a larger scale, however, you should appreciate the limitations of the conversion of heat into work. Also, the entropy of mixing will be of major importance in Unit III in the discussion of mixtures.

3.19 PRACTICE PROBLEMS

P3.1 Call placement of a particle in box A, "heads" and placement in box B, "tails." Given one particle, there are two ways of arranging it, H or T. For two particles, there four ways of arranging them, {HH,HT,TH,TT}.We can treat the microstates by considering each particle in order. For example, {H T H H} means the first particle is in box A, the second in box B, the third in box A, and the fourth in box A.

(a) List and count the ways of arranging three particles. Now consider four particles. What is the general formula for the number of arrangements versus the number of particles? (ANS. 2^N)

(b) How many arrangements correspond to having two particles in box A and one in Box B? What is the probability of {2H,1T}? (ANS. 3/8)

(c) How many arrangements correspond to {2H,2T}. {3H,2T}. {4H,2T}. {3H,3T}? (ANS. $N!/[(N-m)!m!])$

(d) List the macrostates and corresponding number of microstates for an eight-particle, two box system. What portion of all microstates are parts of either 5:3, 4:4, or 3:5 macrostates?

(e) What is the change of entropy in going from a 5:3 macrostate to a 4:4 macrostate? Use Stirling's approximation to estimate the change of entropy in going from a distribution of 50.1% of 6.022E23 in box A to a distribution of 50.001%? From 50.001% to 50.000%? (ANS. 3.08E–24J/K)

P3.2 20 molecules are contained in a piston + cylinder at low pressure. The piston moves such that the volume is expanded by a factor of 4 with no work produced of any kind. Compute $\Delta S/k$. (ANS. 23.19)

P3.3 15 molecules are distributed as 9:4:2 between equal-sized boxes A:B:C, respectively. The partitions between the boxes are removed, and the molecules distribute themselves evenly between the boxes. Compute $\Delta S/k$. (ANS. 11.23)

P3.4 Rolling two dice (six sided cubes with numbers between 1 and 6 on each side) is like putting two particles in six boxes. Compute $\Delta S/k$ for going from double sixes to a four and three. (ANS. 0.693)

P3.5 Estimate the change in entropy when one mole of nitrogen is compressed by a piston in a cylinder from 300 K and 23 liters/gmole to 400 K and 460 liters/gmol. ($C_P = 7/2R$) (ANS. 1.07 kJ/kgK)

P3.6 Steam at 400°C and 10 bars is left in an insulated 10 m³ cylinder. The cylinder has a small leak, however. Compute the conditions of the steam after the pressure has dropped to 1 bar. What is the change in the specific entropy of the steam in the cylinder? Is this a reversible process? The mass of the cylinder is 600 kg, and its heat capacity is 0.1 cal/g-K. Solve the problem with and without considering the heat capacity of the cylinder. (ANS. (a)~120°C; (b) 350°C)

P3.7 A mixture of 1CO:2H$_2$ is adiabatically and continuously compressed from 5 atm and 100°F to 100 atm and 1100°F. Hint: for this mixture, $C_P = x_1 C_{P1} + x_2 C_{P2}$.

(a) Estimate the work of compressing 1 ton/h of the gas. ($C_P = 7/2R$) (ANS. 1.3E6 Btu/h)

(b) Determine the efficiency of the compressor. (ANS. 76%)

P3.8 An adiabatic compressor is used to continuously compress nitrogen ($C_P/R = 7/2$) from 2 bar and 300 K to 15 bar. The compressed nitrogen is found to have an outlet temperature of 625 K. How much work is required (kJ/kg)? What is the efficiency of the compressor? (ANS. 9.46kJ/mol, 72%)

P3.9 An adiabatic compressor is used to continuously compress low pressure steam from 0.8 MPa and 200°C to 4.0 MPa and 500°C in a steady-state process. What is the work required per kg of steam through this compressor? Compute the efficiency of the compressor. (ANS. 606 J/g, 67%)

P3.10 An adiabatic turbine is supplied with steam at 2.0 MPa and 600°C and the steam exhausts at 98% quality and 24°C. Compute the work output per kg of steam. Compute the efficiency of the turbine. (ANS. 1.2E3 kJ, 85%)

P3.11 An adiabatic compressor has been designed to continuously compress 1 kg/s of saturated vapor steam from 1 bar to 100 bar and 1100°C. Estimate the power requirement of this compressor in horsepower. Determine the efficiency of the compressor. (ANS. 3000hp, 60%)

P3.12 Ethylene gas is to be continuously compressed from an initial state of 1 bar and 20°C to a final pressure of 18 bar in an adiabatic compressor. If compression is 70% efficient compared with an isentropic process, what will be the work requirement and what will be the final temperature of the ethylene? Assume the ethylene behaves as an ideal gas with $C_P = $ 44 J/mol-K (ANS. 13.4 kJ/mol, 596K)

P3.13 Steam is produced at 30 bar and some unknown temperature. A small amount of steam is bled off and goes through an adiabatic throttling valve to 1 bar. The temperature of the steam exiting the throttling valve is 110°C. What is the value of the specific entropy of the steam before entering the throttle? (ANS. 5.9736 J/g-K)

P3.14 Suppose the expansion in problem P2.19 was completely adiabatic instead of isothermal and $C_P = 7$ cal/(gmol-K). How would the height of the piston be affected? Must we generate heat or consume heat to maintain isothermal operation? (ANS. decrease, generate)

3.20   HOMEWORK PROBLEMS

3.1 An ideal gas, with temperature independent $C_P = (7/2)R$, at 15°C and having an initial volume of 60 m³, is heated at constant pressure ($P = 0.1013$ MPa) to 30°C by transfer of heat from a reservoir at 50°C. Calculate ΔS_{gas}, $\Delta S_{heat\ reservoir}$, $\Delta S_{universe}$. What is the irreversible feature of this process?

3.2 Steam undergoes a state change from 450°C and 3.5 MPa to 150°C and 0.3 MPa. Determine ΔH and ΔS using:

(a) steam table data
(b) ideal gas assumptions (Be sure to use the ideal gas heat capacity for water).

3.3 The following problems involve one mole of an ideal monatomic gas, $C_P = 5R/2$, in a variable volume piston/cylinder with a stirring paddle, an electric heater and a cooling coil through which refrigerant can flow (see figure). The piston is perfectly insulated. The piston contains 1 gmole of gas. Unless specified, the initial conditions are: $T^i = 25°C$, $P^i = 5$ bar.

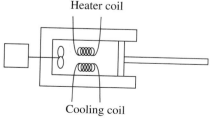

Heater coil

Cooling coil

(a) Status: Heater on; cooler off; paddle off; piston fixed. Five kJ are added by the heater. Find ΔU, ΔS, ΔP and ΔT.

(b) Status: Heater off; cooler off; paddle off; piston moveable. What reversible volume change will give the same temperature rise as in part (a)? Also find ΔU, ΔS and ΔP.

(c) Status: Heater off; cooler off; paddle on; piston fixed. What shaft work will give the same ΔU, ΔS as part (a)?

(d) Status: Heater off; cooler off; paddle on; piston fixed. The stirring motor is consuming 55 watts and is 70% efficient. What rate is the temperature changing? At what initial rates are U and S changing?

(e) Status: Heater unknown; cooler unknown; paddle off; piston free. We wish to perform a reversible isothermal compression until the volume is half of the initial volume. If the volume is decreasing at 2.0 cm^3/s, at what rate should we heat or cool? Express your answer in terms of the instantaneous volume. What is the total heat transfer necessary?

3.4 When a compressed gas storage tank fails, the resulting explosion occurs so rapidly that the gas cloud can be considered adiabatic and assumed to not mix appreciably with the surrounding atmosphere. Consider the failure of a 2.5-m^3 air storage tank initially at 15 bar. Atmospheric pressure is 1 bar, $C_P = 7R/2$.

(a) Calculate the work done on the atmosphere.

(b) A detonation of one kg of TNT releases about 4.5 MJ of work. Calculate the equivalent mass of TNT that performs the same work.

3.5 Work problem 3.4 but consider a steam boiler that fails. The boiler is 250 L in size, operating at 4 MPa, and half full of liquid.

3.6 An isolated chamber with rigid walls is divided into two equal compartments, one containing gas at 600 K and 1 MPa and the other evacuated. The partition between the two compartments ruptures. Compute the final T, P, and ΔS for:

(a) an ideal gas with $C_P/R = 7/2$

(b) steam.

3.7 An isolated chamber is divided into two equal compartments, one containing gas and the other evacuated. The partition between the two compartments ruptures. At the end of the process, the temperature and pressure are uniform throughout the chamber.

(a) If the filled compartment initially contains an ideal gas at 25 MPa and 650 K, what is the final temperature and pressure in the chamber? What is ΔS for the process? Assume a constant heat capacity of $C_P/R = 4.041$.

(b) If the filled chamber initially contains steam at 25 MPa and 650 K, what is the final temperature and pressure in the chamber? What is ΔS for the process? (Use the steam tables).

3.8 Consider the wintertime heating of a house with a furnace compared to addition of Carnot heat engines/pumps. To compensate for heat losses to the surroundings, the house is maintained at a constant temperature T_{house} by a constant rate of heat transfer, $\dot{Q}_{house}$. The furnace operates at a constant temperature T_F, and with direct heat transfer, the heat required from the furnace, $\dot{Q}_F$ is equal to $\dot{Q}_{house}$.

(a) Instead of direct heat transfer, if we utilize the surroundings, at T_S, as an additional heat source and include heat pump technology, $\dot{Q}_F$ may be reduced by generating work from a heat engine operating between T_F and T_S, then applying that work energy to a heat pump operating between T_S and T_{house}. Given that $T_F = 800$ K, $T = 293$ K, $T_S = 265$ K, and $\dot{Q}_{house} = 40$ kJ/h, determine $\dot{Q}_F$ utilizing heat pump technology. No other sources of energy may be used.

(b) Another option is to run a heat engine between T_F and T_{house} and the heat pump between T_S and T_{house}. Compare this method with part (a).

3.9 An ideal gas enters a valve at 500 K and 3 MPa at a steady-state rate of 3 mol/min. It is throttled to 0.5 MPa. What is the rate of entropy generation? Is the process irreversible?

3.10 SO_2 vapor enters a heat exchanger at 100°C and at a flowrate of 45 mole/h. If heat is transferred to the SO_2 at a rate of 1,300 kJ/h, what is the rate of entropy transport in the gas at the outlet relative to the inlet in kJ/K/h given by $(S^{out} - S^{in})\dot{n}$?

3.11 An ideal gas stream (Stream A), $C_P = 5R/2$, 50 mole/h, is heated by a steady-state heat exchanger from 20°C to 100°C by another stream (Stream B) of another ideal gas, $C_P = 7R/2$, 45 mole/h, which enters at 180°C. Heat losses from the exchanger are negligible.

(a) For concurrent flow in the heat exchanger, calculate the molar entropy changes $(S^{out} - S^{in})$ for each stream, and $\dot{S}_{gen}$ for the heat exchanger.
(b) For countercurrent flow in the heat exchanger, calculate the molar entropy changes $(S^{out} - S^{in})$ for each stream, and $\dot{S}_{gen}$ for the heat exchanger. Comment on the comparison of results from parts (a) and (b).

3.12 An inventor has applied for a patent on a device that is claimed to utilize 1 mole/min of air (assumed to be an ideal gas) with temperature independent $C_P = (7/2)R$ which enters at 500 K and 2 bar, and leaves at 350 K and 1 bar. The process is claimed to produce 2000 J/min of work and to require an undisclosed amount of heat transfer with a heat reservoir at 300 K. Should the inventor be issued a patent on this device?

3.13 Two streams of air are mixed in a steady-state process shown below. Assume air is an ideal gas with a constant heat capacity $C_P = 7R/2$.

(a) What is the temperature of the stream leaving the tank?
(b) What is the rate of entropy generation within the tank?
(c) If we duplicated the stream conditions (temperatures, pressures, and flowrates) with an internally reversible process, what is the maximum rate at which work could be obtained? If desirable, you are permitted to transfer heat to the surroundings at the surroundings' temperature of 295 K.

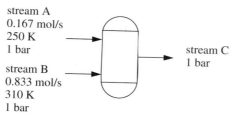

stream A
0.167 mol/s
250 K
1 bar

stream B
0.833 mol/s
310 K
1 bar

stream C
1 bar

3.14 Air is flowing at steady-state through a 5-cm diameter pipe at a flowrate of 0.35 mole/min at $P = 5$ bar and T = 500 K. It flows through a throttle valve and exits at 1 bar. Assume air is an ideal gas with $C_P = 29.1$ J/moleK. If the throttle valve was replaced by a reversible steady-state flow device to permit exactly the same state change for the air in this steady-state process, at what rate could work could be obtained? Heat transfer, if desired, can occur with the surroundings at 298 K, which may be considered a reservoir.

3.15 A common problem in the design of chemical processes is the steady-state compression of gases from a low pressure P_1 to a much higher pressure P_2. We can gain some insight about optimal design of this process by considering adiabatic reversible compression of ideal gases with stagewise intercooling. If the compression is to be done in two stages, first compressing the gas from P_1 to P^*, then cooling the gas at constant pressure down to the compressor inlet temperature T_1, and then compressing the gas to P_2, what should the value of the intermediate pressure be to accomplish the compression with minimum work?

3.16 Steam flowing at steady-state enters a turbine at 400°C and 7 MPa. The exit is at 0.275 MPa. The turbine is 85% efficient. What is the quality of the exiting stream? How much work is generated per kg of steam?

3.17 A steam turbine inlet is to be 4 MPa. The outlet of the adiabatic turbine is to operate at 0.01 MPa, and provide saturated steam. The turbine has an efficiency of 85%. Determine the superheat which is required on the turbine inlet, and the work produced by the turbine.

3.18 Steam is fed to an adiabatic turbine at 4 MPa and 500°C. It exits at 0.1 MPa.

   (a) If the turbine is reversible, how much work is produced per kg of steam?
   (b) If the turbine is 80% efficient, how much work is produced per kg of steam?

3.19 Methane is compressed in a steady-state adiabatic compressor (87% efficient) to 0.4 MPa. What is the required work per mole of methane in kJ? If the flow is to be 17.5 kmol/h, how much work must be furnished by the compressor (in kW)? What is the rate of entropy generation (in kJ/K/h)? (a) the inlet is at 0.1013 MPa and −240°F; (b) the inlet is 0.1013 MPa and 200 K.

3.20 Methane is to be compressed from 0.05 MPa and −120°F to 5 MPa in a two-stage compressor. In between adiabatic, reversible stages, a heat exchanger returns the temperature to −120°F. The intermediate pressure is 1.5 MPa.

   (a) What is the work required (kJ/kg) in the first compressor?
   (b) What is the temperature at the exit of the first compressor (°C)?
   (c) What is the cooling requirement (kJ/kg) in the interstage cooler?

3.21 (a) A steam turbine in a small electric power plant is designed to accept 5000 kg/h of steam at 60 bar and 500°C and exhaust the steam at 1 bar. Assuming that the turbine is adiabatic and reversible, compute the exit temperature of the steam and the power generated by the turbine.
   (b) If the turbine in part (a) is adiabatic but only 80% efficient, what would be the exit temperature of the steam? At what rate would entropy be generated within the turbine?
   (c) One simple way to reduce the power output of the turbine in part (a) (100% efficient) is by adjusting a throttling valve that reduces the turbine inlet steam pressure to 30 bar. Compute the steam temperature to the turbine, the rate of entropy generation, and the power output of the turbine for this case. Is this a thermodynamically efficient way of reducing the power output? Can you think of a better way?

3.22 A steady stream (1000kg/hr) of air flows through a compressor, entering at (300 K, 0.1 MPa) and leaving at (425 K, 1 MPa). The compressor has a cooling jacket where water flows at 1500 kg/hr and undergoes a 20 K temperature rise. Assuming air is an ideal gas, calculate the work furnished by the compressor, and also determine the minimum work required for the same state change of air.

3.23 Steam is used in the following adiabatic turbine system to generate electricity.

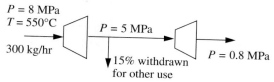

$P = 8$ MPa
$T = 550°C$

$P = 5$ MPa

300 kg/hr

$P = 0.8$ MPa

15% withdrawn
for other use

(a) How much work (in kJ/h) is generated by the first turbine which is 80% efficient?
(b) How much work (in kJ/h) is generated by the second turbine which is 80% efficient?
(c) Steam for the turbines is generated by a boiler. How much heat must be supplied to the boiler (not shown) which has 300 kg/h of flow? The stream entering the boiler is $T = 170°C$, $P = 8$ MPa. The stream exiting the boiler matches the inlet to the first turbine.

3.24 Propane is to be compressed from 0.4 MPa and 360 K to 4 MPa using a two-stage compressor. An interstage cooler returns the temperature of the propane to 360 K before it enters the second compressor. The intermediate pressure is 1.2 MPa. Both adiabatic compressors have a compressor efficiency of 80%.

(a) What is the work required in the first compressor per kg of propane?
(b) What is the temperature at the exit of the first compressor?
(c) What is the cooling requirement in the interstage cooler per kg of propane?

3.25 Liquid nitrogen is useful for medical purposes and for research laboratories. Determine the minimum shaft work needed to liquefy nitrogen initially at 298 K and 0.1013 MPa and ending with saturated liquid at the normal boiling point, 77.4 K and 0.1013 MPa. The heat of vaporization at the normal boiling point is 5.577 kJ/mol, and the surroundings are at 298 K. The constant pressure heat capacity of gaseous nitrogen can be assumed to be independent of temperature at $7/2R$ for the purpose of this calculation.

(a) Consider nitrogen entering a flow device at 1 mol/min. Give shaft work in kW.
(b) Consider nitrogen in a piston/cylinder device. Give the work in kJ per mole liquefied.
(c) Compare the minimum shaft work for the two processes. Is one of the processes more advantageous than the other on a molar basis?

3.26 Propane flows into a steady-state process at 0.2 MPa and 280 K. The final product is to be saturated liquid propane at 300 K. Liquid propane is to be produced at 1000 kg/h. The surroundings are at 295 K. Using a propane property chart, determine the rate of heat transfer and minimum work requirement if the process is to operate reversibly?

3.27 Propane (1000 kg/hr) is to be liquefied following a two stage compression. The inlet gas is to be at 300 K and 0.1 MPa. The outlet of the adiabatic compressor I is 0.65 MPa, and the propane enters the interstage cooler where it exits at 320 K, then adiabatic compressor II raises the propane pressure to 4.5 MPa. The final cooler lowers the temperature to 320 K before it is throttled adiabatically to 0.1 MPa. The adiabatic compressors have an efficiency of 80%.

(a) Determine the work required by each compressor.

(b) If the drive motors and drive linkages are together 80% efficient, what size motors are required?

(c) What cooling is required in the interstage cooler and the final cooler?

(d) What percentage of propane is liquefied, and what is the final temperature of the propane liquid?

3.28 A heat exchanger operates with the following streams: Water in at 20°C, 30 kg/hr; water out at 70°C; Organic in at 100°C, 41.8 kg/hr; organic out at 40°C.

(a) What is the maximum work that could be obtained if the flowrates and temperatures of the streams remain the same, but heat transfer is permitted with the surroundings at 298 K. ($C_{P\,water}$ = 4.184 kJ/(kgK), $C_{P\,organic}$ = 2.5 kJ/(kgK)).

(b) What is the maximum work that could be obtained by replacing the heat exchanger with a reversible heat transfer device, where the inlet flowrates and temperatures are to remain the same, the organic outlet temperature remains the same, and no heat transfer with the surroundings occurs?

3.29 Presently, benzene vapors are condensed in a heat exchanger using cooling water. The benzene (100 kmol/h) enters 0.1013 MPa and 120°C, and exits at 0.1013 MPa and 50°C. Cooling water enters at 10°C and exits at 40°C.

(a) What is the current demand for water (kg/h)?

(b) To what flowrate could the water demand be lowered by introducing a reversible heat transfer device? The temperature rise of water is to remain the same. What work could be obtained from the new heat transfer device?

3.30 A Hilsch vortex tube is an unusual device that takes an inlet gas stream and produces a hot stream and a cold stream without moving parts. A high pressure inlet stream (A) enters towards one end of the tube. The cold gas exits at outlet B on the end of the tube near the inlet where the port is centered in the end cap. The hot stream exits at outlet C on the other end of the tube where the exit is a series of holes or openings around the outer edge of the end cap.

The tube works in the following way. The inlet stream A enters tangent to the edge of the tube, and swirls as it cools by expansion. Some of the cool fluid exits at port B. The remainder of the fluid has high kinetic energy produced by the volume change during expansion, and the swirling motion dissipates the kinetic energy back into internal energy so the temperature rises before the gas exits at port C.

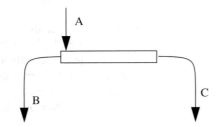

Inlet A is at 5 bar and 310 K and 3.2 mol/min. Outlet B is at 1 bar and 260 K. Outlet C is at 1 bar and 315 K. The tube is insulated and the fluid is air with $C_P = 7R/2$.

(a) Determine the flowrates of streams B and C.

(b) Determine $\underline{S}_{gen}$ for the Hilsch tube.

(c) Suppose a reversible heat engine is connected between the outlet streams B and C which is run to produce the maximum work possible. The proposed heat engine may only exchange heat between the streams and not with the surroundings as shown. The final temperature of streams B and C will be $T_{B'}$ as they exit the apparatus. What is $T_{B'}$?

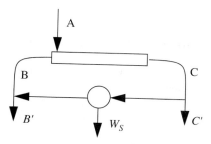

(d) What work output is possible in W? What is $\dot{S}_{gen}$ for the entire system including the tube plus the heat engine?

(e) Suppose that instead of using the heat engine, streams B and C were mixed directly with each other to form a single outlet stream. What would this temperature be, and how does it compare with $T_{B'}$ from part (c)? Calculate $\dot{S}_{gen}$ and compare it with $\dot{S}_{gen}$ from part (c). What do you conclude from the comparison?

3.31 Airplanes are launched from aircraft carriers by means of a steam catapult. The catapult is a well-insulated cylinder that contains steam, and is fitted with a frictionless piston. The piston is connected to the airplane by a cable. As the steam expands, the movement of the piston causes movement of the plane. A catapult design calls for 270 kg of steam at 15 MPa and 450°C to be expanded to 0.4 MPa. How much work can this catapult generate during a single stroke? Compare this to the energy required to accelerate a 30,000-kg aircraft from rest to 350 km per hour.

3.32 Methane gas is contained in a 0.65-m³ gas cylinder at 6.9 MPa and 300 K. The cylinder is vented rapidly until the pressure falls to 0.5 MPa. The venting occurs rapidly enough that heat transfer between the cylinder walls and the gas can be neglected, as well as between the cylinder and the surroundings. What is the final temperature and the final number of moles of gas in the tank immediately after depressurization, assuming the expansion within the tank is reversible and:

(a) methane is considered to be an ideal gas with $C_P/R = 4.298$
(b) methane is considered to be a real gas with properties given by a property chart.

3.33 A thermodynamically interesting problem is to analyze the fundamentals behind the product called "fix-a-flat." In reality, this product is a 500-ml can that contains a volatile compound under pressure, such that most of it is liquid. Nevertheless, we can make an initial approximation of this process by treating the contents of the can as an ideal gas. If the initial temperature of both the compressed air and the air in the tire is 300 K, estimate the initial pressure in the compressed air can necessary to reinflate one tire from 1 bar to 3 bar. Also, estimate the final air temperature in the tire and in the can. For the purposes of this calculation you may assume: air is an ideal gas with $C_P/R = 7/2$, the tire does not change its size or shape during the inflation process, the inner tube of the tire has a volume of 40,000 cm³. We will reconsider this problem with liquid contents, after discussing phase equilibrium in a pure fluid.

3.34 Wouldn't it be great if a turbine could be put in place of the throttle in problem 3.33? Then you could light a small bulb during the inflation to see what you were doing at night. How much energy (J) could possibly be generated by such a turbine if the other conditions were the same as in problem 3.33?

3.35 A 1-m^3 tank is to be filled using N$_2$ at 300 K and 20 MPa. Instead of throttling the N$_2$ into the tank, a reversible turbine is put in line to get some work out of the pressure drop. If the pressure in the tank is initially zero, and the final pressure is 20 MPa, what will be the final temperature in the tank? How much work will be accomplished over the course of the entire process? (Hint: consider the entropy balance carefully.)

3.36 An insulated cylinder is fitted with a freely floating piston, and contains 1 lb$_m$ of steam at 120 psia and 90% quality. The space above the piston, initially 1 ft^3, contains air at 300 K to maintain the pressure on the steam. Additional air is forced into the upper chamber, forcing the piston down and increasing the steam pressure until the steam has 100% quality. The final steam pressure is 428 psia, and the work done on the steam is 91 Btu, but the air above the steam has not had time to exchange heat with the piston, cylinder, or surroundings. The air supply line is at 700 psia and 300 K. What is the final temperature of the air in the upper chamber?

3.37 Two well-insulated tanks are attached as shown in the figure below. The tank volumes are given in the figure. There is a mass-flow controller between the two tanks. Initially, the flow controller is closed. At $t = 0$, the mass flow controller is opened to a flow of 0.1 mol/s. After a time of 500 seconds, what are the temperatures of the two tanks? Neglect the heat capacity of the tanks and piping. No heat transfer occurs between the two tanks. (After 500 seconds, the pressure in the left tank is still higher than the pressure in the right tank). The working fluid is nitrogen and the ideal gas law may be assumed. The ideal gas heat capacity $C_P = 7/2 \cdot R$ may be assumed to be independent of T.

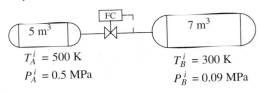

$$T_A^i = 500 \text{ K} \qquad\qquad T_B^i = 300 \text{ K}$$
$$P_A^i = 0.5 \text{ MPa} \qquad\qquad P_B^i = 0.09 \text{ MPa}$$

3.38 Two storage tanks (0.1 m^3 each) contain air at 2 bar. They are connected across a small reversible compressor. The tanks, connecting lines, and compressor are immersed in a constant temperature bath at 280 K. The compressor will take suction from one tank, compress the gas, and discharge it to the other tank. The gas is at 280 K at all times. Assume that air is an ideal gas with $C_P = 29.3$ J/molK.

(a) What is the minimum work interaction required to compress the gas in one tank to 3 bar?

(b) What is the heat interaction with the constant temperature bath?

3.39 A constant pressure air supply is connected to a small tank (A) as shown in the figure below. With valves B and C, the tank can be pressurized or depressurized. The initial conditions are $T = 300$ K, $P = 1.013$ bar, $C_P = 29.3$ J/(mol·K). Consider the system adiabatic.

(a) The tank is pressurized with valve B open and valve C closed. What is the final temperature of the tank? Neglect the heat capacity of the tank and valves.

(b) Taking the system as the tank plus the valves, what is the entropy change of the system due to pressurization? What is the entropy change of the air supply reservoir? What is the entropy change of the universe? Use a reference state of 300 K and 1.013 bar.

(c) During depressurization with valve B closed and valve C open, how does the molar entropy entering valve C compare with the molar entropy leaving? What is the temperature of the tank following depressurization?

Air supply   B   A   C   Atmosphere
9 bar $\underline{V} = 1m^3$ 1.013 bar
292 K   295 K

3.40 The pressurization of problem 3.39 is performed by replacing the inlet valve with a reversible device that permits pressurization that is internally reversible. The system is to remain adiabatic with respect to heat transfer of the surroundings.

(a) What is the final temperature of the tank?
(b) How much work could be obtained?

3.41 A 2m³ tank is at 292 K and 0.1 MPa and it is desired to pressurize the tank to 3 MPa. The gas is available from an infinite supply at 350 K and 5 MPa connected to the tank via a throttle valve. Assume that the gas follows the ideal gas law with a constant heat capacity of $C_P = 29$ J/(molK).

(a) Modeling the pressurization as adiabatic, what is the final temperature in the tank and the final number of moles when the pressure equals 3 MPa?
(b) Identify factors included in the idealized calculation of part (a) that contribute to irreversibilities.
(c) Identify factors neglected in the analysis of part (a) that would contribute to irreversibilities in a real process.
(d) If the pressurization could be performed reversibly, the final temperature might be different from that found in part (a). Clearly outline a procedure to calculate the temperature indicating that enough equations are provided for all unknowns. Also clearly state how you would use the equations. Additional equipment is permissible provided that the process remains adiabatic with regard to heat transfer to the surroundings.
(e) In part (d), would work be added, removed, or not involved in making the process reversible? Provide equations to calculate the work interaction.

3.42 Two gas storage tanks are interconnected through a isothermal expander. Tank 1 ($\underline{V} = 1$ m³) is initially at 298 K and 30 bar. Tank 2 ($\underline{V} = 1$ m³) is initially at 298 K and 1 bar. Reversible heat transfer is provided between the tanks, the expander and the surroundings at 298 K. What is the maximum work that can be obtained from the expander when isothermal flow occurs from Tank 1 to Tank 2?

CHAPTER 4

THERMODYNAMICS OF PROCESSES

There cannot be a greater mistake than that of looking superciliously upon practical applications of science. The life and soul of science is its practical application.

Lord Kelvin (William Thompson)

In the first three chapters, we have concentrated on application of the first and second law to simple systems (e.g. turbine, throttle) and the constraints imposed by the second law should be clear. In this section, we show how the analyses we have developed for one or two operations at a time can be put together in complex processes. In this way, we provide several specific examples of ways that operations can be connected to create power cycles, refrigeration cycles, and liquefaction cycles. We can consider these processes as paradigms for some general observations about energy and entropy constraints.

4.1 THE CARNOT CYCLE

Consider a steam turbine as discussed in relation to the energy and entropy balances. A turbine requires high-pressure steam at the inlet and rejects lower pressure steam or wet steam. Suppose we would like to recycle the turbine outlet stream to a boiler and regenerate it. Then we must somehow raise it to a pressure high enough to get it back into our high pressure boiler. If we try to just compress the low pressure steam from the turbine outlet, we will require the same work that we obtained so that is not a good choice. Suppose instead that we condense most of it to water and raise it to pressure by compressing the liquid (which requires little work). How much work could we possibly get out of this process per unit of heat input? The quantity that we use to characterize the work out vs. the heat input is known as the thermal efficiency (η) defined by Eqn. 3.34,

$$\eta \equiv -W_{S,net}/Q_H$$

4.1 ❶ General formula for thermal efficiency.

> ***Note:*** *the thermal efficiency is not the same as the turbine or compressor efficiency. Thermal efficiency refers to the over-all efficiency in use of heat to generate work for an over-all process which may include turbines or compressors with their own efficiencies.*

We saw in the last chapter how a Carnot cycle could be set up using an ideal gas as a working fluid. We could plot this cycle in *P-V* or *T-S* coordinates.

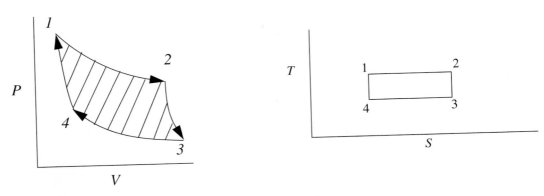

Figure 4.1 *Illustration of a Carnot cycle on P-V and T-S coordinates.*

In this case, the area inside the *P-V* cycle represents the work done by the gas in one cycle, and the area enclosed by the *T-S* path is equal to the net intake of energy as heat by the gas in one cycle.

The Carnot cycle has two advantages over any other cycle. First, the cycle operates at the highest temperature available for as long as possible. In contrast, other cycles may only reach the highest temperature for a short time. So the Carnot cycle is making maximal use of the high temperature reservoir, and a similar argument holds regarding the low temperature reservoir. The second advantage of the Carnot cycle is that it assumes that all the processes are reversible. As a result of these two advantages, it is impossible to construct an engine that operates in a cycle that is more efficient than a reversible Carnot engine. It turns out that it is impossible to make full use of the advantages of the Carnot cycle in practical applications, as discussed below, but the Carnot cycle is so simple that it provides a useful estimate for checking results from calculations regarding other cycles.

The Carnot cycle given below shows one way of carrying out reversible steps cyclically. Note that the turbine. compressor and all heat transfer must be reversible. Since this entire process is reversible: $\Delta \underline{\dot{S}}_{gen} = 0$. But we have shown in Section 3.5,

$$-\underline{\dot{W}}_{S,net} = \underline{\dot{Q}}_H + \underline{\dot{Q}}_C = \underline{\dot{Q}}_H \left(1 - \frac{T_C}{T_H}\right) \qquad 4.2$$

Therefore, $\eta = -W_{S,net}/Q_H = (T_H - T_C)/T_H$. We can control thermal efficiency by controlling the temperature difference. In order to perform the isothermal expansion and compression, we could utilize phase transitions, where the volume changes dramatically at constant temperature. Such a cycle is represented below.

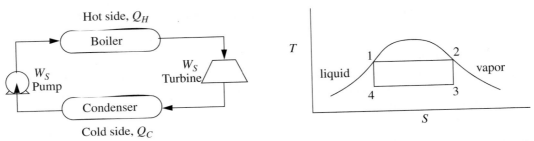

Figure 4.2 *Illustration of a possible flowsheet for a Carnot cycle compared to the T-S diagram.*

4.2   THE RANKINE CYCLE

When water is used as the working fluid, the Carnot cycle is impractical for two reasons: 1) compressing a partially condensed fluid is much more complex than compressing an entirely condensed liquid; 2) low-quality (wet) steam damages turbine blades by rapid erosion due to water droplets. Therefore, most power plants are based on modifications of the Rankine cycle depicted below. The vapor is superheated before entering the turbine. Note in Fig. 4.3 that the superheater between the boiler and the turbine is not drawn, and only a boiler is shown. In actual power plants, separate boilers and superheaters are usually used; however, for the sake of simplicity in our discussions the boiler/superheater steam generator combination will be represented by a single boiler in the schematic.

> ❶ Most plants will have separate boilers and superheaters. We show just a boiler for simplicity.

The cycle in Fig. 4.3 is idealized from a real process because the inlet to the pump is considered saturated. In a real process, it will be subcooled to avoid difficulties (cavitation[1]) in pumping. In fact, real processes will have temperature and pressure changes along the piping between individual components in the schematic, but these changes will be considered negligible in the Rankine cycle and all other processes discussed in the chapter, unless otherwise stated. These simplifications allow focus on the most important concepts, but the simplifications must be reconsidered in a detailed process design.

In Fig. 4.3, state 4' is the outlet state for a reversible adiabatic turbine. *We will use the prime to denote a reversible outlet state throughout this chapter.* State 4 is the actual outlet state which is calculated by applying the efficiency to the enthalpy change.

> ❶ The prime denotes a reversible outlet state.

$$\Delta H_{3 \to 4} = \eta \Delta H_{3 \to 4'} \qquad\qquad 4.3$$

Because a real turbine always generates entropy, state 4 will always be to the right of 4' on a *T-S* diagram. State 4 and 4' can be inside or outside the phase envelope. Efficiencies are greater if state 4 is slightly inside the phase envelope because the enthalpy change will be larger for the same pressure drop due to the large enthalpy of vaporization; however, to avoid turbine blade damage, quality is kept above 90% in most cases.

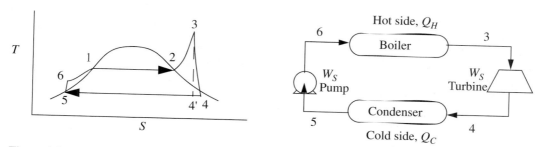

Figure 4.3 *Rankine cycle.*

---

1. Cavitation occurs when vapor bubbles form in the inlet line of a liquid pump, and the bubbles prevent the pump from drawing the liquid into the pump cavity.

Example 4.1 Rankine cycle

A power plant uses the Rankine cycle. The maximum desired temperature in the boiler is 500°C. If the turbine is reversible, and the outlet of the turbine is saturated vapor at $P = 0.025$ MPa, determine:

(a) The operating pressure of the boiler;

(b) The thermal efficiency;

(c) The circulation rate to provide 1 MW net power output.

Solution: We will refer to Figure 4.3 for stream numbers. The recommended method for solving process problems is to establish a table to record values as they are determined. *In this text we will show values in the tables in shaded cells if they have been determined by balance calculations.* The turbine outlet can be read from the temperature table without interpolation. Unshaded cells refer to properties determined directly from the problem statement.

state	T (°C)	P (MPa)	H (kJ/kg)	S (kJ/kgK)	V (m³/kg)
4', sat V	65	0.025	2617.5	7.8296	
5, sat L	65	0.025	272.12	0.8937	0.00102
3	500	0.8	3481.3	~7.83	
6			273		
2, sat V	170.4	0.8	2768.3	6.6616	

We begin by determining the process locations where two state variables are known, and working forwards or backwards from those locations. In this case, state 4' is known. (Since the turbine is reversible, state 4 = state 4'). Let us work backwards around the Rankine cycle.

(a) State 3 is determined most quickly from the T-S diagram by following the line of constant entropy up to 500°C. It may also be found by scanning the steam tables to find this entropy value at 500°C. Since the boiler operates at 500°C, the pressure is 0.8 MPa. The enthalpy at this state will be needed to calculate turbine output later, so it is recorded, $H_3 = 3481.3$ kJ/kg.

(b) We must calculate $\dot{Q}_{boiler}$ and $\dot{W}_{net}$. The energy balance gives $\dot{Q}_{boiler} = (H_3 - H_6)$. The enthalpy at state 6 is determined by calculating the adiabatic work input by the pump to increase the pressure from state 5. For the pump, we have

$$\Delta H_{pump} = W_{S,pump} = \int V dP \approx V \Delta P.$$

$$\Delta H_{pump} = \frac{1020 \text{ cm}^3 (0.8 - 0.025)\text{MPa}}{\text{kg}} \left| \frac{1\text{J}}{\text{cm}^3\text{MPa}} \right| \frac{\text{kJ}}{10^3\text{J}} = 0.79 \text{ kJ/kg}$$

Example 4.1   Rankine cycle (Continued)

Thus, the work of the pump is extremely small, $H_6 = 272.12 + 0.79 = 273$ kJ/kg. To find $W_{S,turbine}$, $W'_{S,turbine} = (H_4' - H_3) = (2617.5 - 3481.3) = -863.8$ kJ/kg   and $W'_{S,net} = -863.8 + 0.79 = -863.0$ kJ/kg.

$$\eta = \frac{-W'_{S,net}}{Q_{boiler}} = \left(\frac{863}{3481.3 - 273}\right) = 0.27$$

(c) for 1MW capacity, $\dot{W}'_S = \dot{m}W'_S$, the circulation rate is

$$\dot{m} = \frac{-1000 \text{ kJ}}{\text{s}} \left| \frac{\text{kg}}{-863 \text{ kJ}} \right| \frac{3600 \text{ s}}{\text{h}} = 4171 \text{ kg/h}$$

Example 4.2   Two-phase turbine output

Suppose we modify the above process to use the outlet from a boiler at 500°C (the same temperature), but at 1.6 MPa. If the output of the turbine remains at 0.025 MPa and 65°C and the turbine is 80% efficient, what is the turbine outlet enthalpy?

Solution: We may enter the known values in a table. Again referring to Fig. 4.3 for stream numbers, we may enter the values in the unshaded cells.

state	T (°C)	P (MPa)	H (kJ/kg)	S (kJ/kgK)
3	500	1.6	3472.6	7.5409
4'	65	0.025	2519	7.5409
4	65	0.025	2710	

The reversible outlet state is in the two-phase region because the entropy for saturated vapor at 0.025 MPa is $S^{satV} = 7.8296$ kJ/kgK, and the turbine outlet entropy is less. To find the quality for state 4', based on the outlet entropy, interpolating between the saturated liquid and saturated vapor values, $S = S^L + q\Delta S^{vap}$,

$$S_3 = S_4' = 7.5409 = 0.8937 + q'\cdot(6.9359) \text{ kJ/kgK}$$

which gives $q' = 0.958$. The enthalpy in this state is $H_4' = H^L + q'\Delta H^{vap}$

$$H_4' = 272.12 + 0.958\cdot2345.4 = 2519 \text{ kJ/kg}$$

which gives $W_S' = H_4' - H_3 = -953.6$ kJ/kg. Since the turbine is 80% efficient, $W_{S,turbine} = 0.8W_{S,turbine}' = H_4 - H_3$, which gives $H_4 = 3472.6 - 0.8(953.6) = 2710$ kJ/kg. The actual quality of turbine outlet is higher than q'; in fact, we find that H^V at 0.025 MPa is 2617.5 kJ/kg. Therefore, the actual outlet state is superheated, even though the reversible outlet state is two-phase.

⊗ The irreversible turbine outlet can be one-phase when the reversible outlet is two-phase.

4.3   RANKINE MODIFICATIONS

Two modifications of the Rankine cycle are in common use to improve the efficiency. A Rankine cycle with reheat increases the boiler pressure but keeps the maximum temperature approximately the same. The maximum temperatures of the boilers are limited by corrosion concerns. This modification uses a two-stage turbine with reheat in-between. An illustration of the modified cycle is shown in Fig. 4.4.

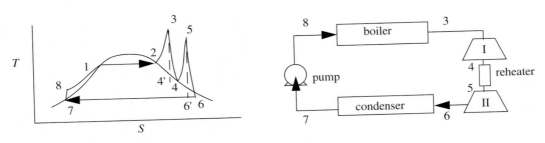

Figure 4.4 *Rankine cycle with reheat.*

Example 4.3 Rankine with reheat

Consider a modification of Example 4.1. If we limit ourselves to a 500°C boiler and reheat, we can develop a new cycle to investigate any improvement in efficiency and circulation rate. Let us operate a cycle utilizing two reversible turbines with saturated vapor exit streams. Let the feed to the first turbine be steam at 433°C and 6 MPa. Let the feed to the second stage be 0.8 MPa and 500°C. Determine the improvement in efficiency and circulation rate relative to Example 4.1.

Solution: Refer to Fig. 4.4 for stream numbers. First, let us find state 3. Interpolating at 6MPa between 400 and 450°C,

$$H_3 = 3178.2 + 0.66(3302.9 - 3178.2) = 3261 \text{ kJ/kg}$$

$$S_3 = 6.5432 + 0.66(6.7219 - 6.5432) = 6.6611 \text{ kJ/kgK}$$

Upon expansion through the first reversible turbine, we find state 4' is the same as state 2 in the previous example. This means there is a correspondence of states 4' through 7 with states 2 through 5 of Example 4.1. We can therefore fill out almost the entire table directly.

state	T (°C)	P (MPa)	H (kJ/kg)	S (kJ/kgK)	V (m³/kg)
6', sat V	65	0.025	2617.5	~7.87	
7, sat L	65	0.025	272.12	0.8937	0.00102
5	500	0.8	3481.3	7.8692	
8			278.2		
4', sat V	170.4	0.8	2768.3	6.6616	
3	433	6	3261	6.6611	

Example 4.3   Rankine with reheat (Continued)

$$\Delta H_{pump} = W_{S,pump} = 1020\frac{(6 - 0.025)}{1000} = 6.09 \text{ kJ/kg}$$

$H_8 = 272.12 + 6.09 = 278.2$ kJ/kg. Thus, the pump work is still small, but larger than before. Heat is now required in both the boiler and reheater, and work comes from both turbines.

$$Q_H = (H_3 - H_8) + (H_5 - H_4') = (3261 - 278.2) + (3481.3 - 2768.5) = 3695.6 \text{ kJ/kg}$$

$$W_{S,turbine} = (H_4' - H_3) + (H_6' - H_5) = (2768.3 - 3261) + (2617.5 - 3481.3) = -1356.5 \text{ kJ/kg}$$

$$\eta = -W_{S,net}/Q_H = \frac{1356.5 - 6.1}{3695.6} = 0.365, \quad \dot{m} = \frac{(-1000)(3600)}{(1349)} = 2669 \text{ kg/h}$$

The efficiency has improved by $\frac{(0.365 - 0.27)}{0.27} \times 100\% = 35\%$, and the circulation rate has been decreased by 36%.

> ❶ Reheat improves thermal efficiency.

Another common modification of the Rankine cycle is a regenerative cycle using feedwater preheaters. A portion of high-pressure steam is used to preheat the water as it passes from the pump back to the boiler. A schematic of such a process is shown in Fig. 4.5 using closed feedwater preheaters. The economic favorability increases until about 5 preheaters are used, then the improvements are not worth the extra cost. Three preheaters are more common. As the condensate from each preheater enters the next preheater, it throttles through a valve to the next lower pressure and partially or totally vaporizes as it throttles. It is also common to withdraw some steam from turbine outlets for process use and heating. One type of implementation is illustrated in Fig. 4.6. Often in actual processes, open feedwater preheaters are used. In an open feedwater preheater, all of the incoming streams mix. The advantage of this preheater is that dissolved oxygen in the returning condensate can be removed by heating, and if provision is made to vent the non-condensables from the open feedwater preheater, it may serve as a *deaerator*. A system with an open feedwater preheater is shown in Fig 4.7.

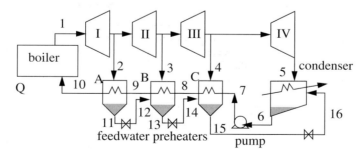

Figure 4.5 *Regenerative Rankine cycle using closed feedwater preheaters.*

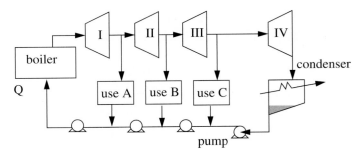

Figure 4.6 *Rankine cycle with side draws for process steam. Pumps and/or throttles may be used in returning process steam to the boiler.*

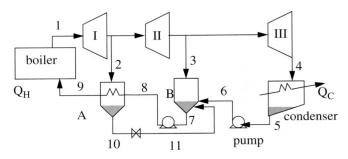

Figure 4.7 *Schematic for a system with a closed feedwater preheater, A, and an open feedwater preheater, B.*

Example 4.4 Regenerative Rankine cycle

Steam exits a boiler/superheater at 500°C and 5 MPa. A process schematic is shown in Fig. 4.8. Known process conditions are given in the table. Find the net power output per kg of flow in stream 1.

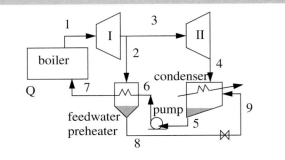

Figure 4.8 *Regenerative Rankine cycle for Example 4.4.*

Example 4.4   Regenerative Rankine cycle (Continued)

Stream	T(°C)	P(MPa)	H(kJ/kg)	S(kJ/kg-K)
1	500	5	3434.7	6.9781
2	300	1	3051.6	7.1246
4		0.1	2615.7	7.2000
5			417.5	
6		5	422.7	
7		5	745.0	
8			762.5	

Solution: Often, one of the key steps in working a problem involving a regenerative cycle is to solve for the fraction of each flow diverted rather than solving for the individual flowrates. The flowrate of streams 7, 6, and 5 are equal to the flowrate of stream 1, and we may write the energy balance around the feedwater preheater using the mass flowrate of stream 1 together with the mass flowrate of stream 2, $\dot{m}_1(H_6 - H_7) + \dot{m}_2(H_2 - H_8) = 0$. Dividing by $\dot{m}_1$ and plugging in enthalpy values $(423 - 745) + \dfrac{\dot{m}_2}{\dot{m}_1}(3052 - 763) = 0$, which gives $\dfrac{\dot{m}_2}{\dot{m}_1} = 0.14$.

The net work is given by

$$\dot{W}_{S,net} = \dot{m}_1(W_{S,I}) + \dot{m}_3(W_{S,II}) + \dot{m}_1(W_{S,pump})\text{, and on the basis of one kg from the boiler/}$$

superheater, $\dfrac{\dot{W}_{S,net}}{\dot{m}_1} = (W_{S,I}) + \left(1 - \dfrac{\dot{m}_2}{\dot{m}_1}\right)(W_{S,II}) + (W_{S,pump})$, and using enthalpies to calculate the work of each turbine,

$$\frac{\dot{W}_{S,net}}{\dot{m}_1} = (3052 - 3435) + (1 - 0.14)(2616 - 3052) + (423 - 418) = -753\frac{kJ}{kg\ stream\ 1}$$

Solving for flowrate ratios in regenerative cycles can be helpful when the total flowrate is unknown.

4.4   REFRIGERATION

Suppose we were to operate the Carnot cycle in reverse. Instead of taking work out, put work in. Then we raise temperature of our working fluid through compression, we cool it to liquefy at this high pressure, we drop the pressure isentropically to obtain a cooler two-phase mixture, and boil off the rest of the liquid at the lower temperature. For this process, we would be concerned with the amount of heat removed from the low temperature source for a given amount of work; this ratio is called the *coefficient of performance, COP*. For a Carnot cycle, we already have the necessary formula:

❶ COP for Carnot cycle.

$$COP = \frac{\dot{Q}_C}{\dot{W}_{S,net}} \Rightarrow$$ 4.4

$$For\ Carnot:\ COP = \left(\frac{T_H}{T_C} - 1\right)^{-1}$$ 4.5

Ordinary Vapor Compression Cycle

❶ The ordinary vapor compression cycle is the most common refrigeration cycle.

The Carnot cycle is not practical for refrigeration for much the same reasons as discussed for power production. Therefore, most refrigerators operate on the ordinary vapor-compression cycle which is shown in Fig. 4.9.

As we did with the Rankine cycle, we will make some simplifications that would have to be reevaluated in a detailed calculation. Again, we neglect pressure losses in piping. We assume that the vapor is saturated at the inlet to the compressor, and that the outlet of the condenser is saturated liquid. Thus, saturated vapor enters the compressor and exits heated above the condenser temperature, then cools in the condenser until it condenses to a saturated liquid. In the cyclic process, the saturated liquid is passed through a throttle valve at constant enthalpy and exits as a two-phase mixture. The evaporator is assumed to be isothermal, and accepts heat at the colder temperature to complete the vaporization. The OVC cycle is frequently evaluated using a *P-H* diagram as shown in Fig. 4.10.

COP for ordinary vapor compression cycle

$$COP = Q_C/W_{S,net};\ Q_C = (H_2 - H_1)$$

$$Energy\ balance:\ W_{S,net} = \Delta H_{2 \to 3} = (H_3 - H_2)$$

❶ COP for ordinary vapor compression cycle.

$$\Rightarrow COP = (H_2 - H_1)/(H_3 - H_2) = (H_2 - H_4)/(H_3 - H_2)$$ 4.6

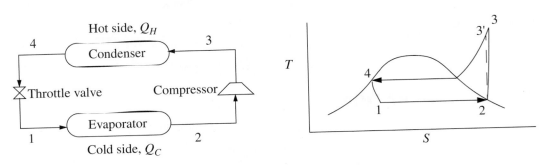

Figure 4.9 *OVC refrigeration cycle process schematic and T-S diagram.*

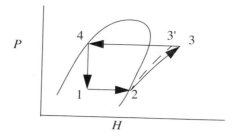

Figure 4.10 *OVC refrigeration cycle plotted on the more commonly used P-H diagram. State numbers correspond to Fig. 4.9.*

Example 4.5 Refrigeration by vapor compression cycle

An industrial freezer room is to be maintained at −15°C by a cooling unit utilizing refrigerant HFC-134*a* as the working fluid. The evaporator coils should be maintained at −20°C to assure efficient heat transfer. Cooling water is available at 10°C from an on-site water well and can be warmed to 25°C in the condenser. The refrigerant temperature exiting the condenser is to be 30°C. The cooling capacity of the freezer unit is to be 120,000 BTU/h (126,500 kJ/h). 12,650 kJ/h is known as one *ton* of refrigeration because it is approximately the cooling rate required to freeze one ton of water at 0°C to one ton of ice at 0°C in 24 h. So this refrigerator represents a 10 ton refrigerator. As a common frame of reference, typical home air conditioners are about 2-3 tons, but they typically weigh less than 100 kg. Calculate the COP and recirculation rate (except part (a)) for the industrial freezer in the following cases:

> ❗ 12,650 kJ/h is known as a *ton* of refrigeration capacity.

(a) Carnot cycle.

(b) Ordinary vapor compression cycle with a reversible compressor.

(c) Vapor compression cycle with the throttle valve replaced with an expander.

(d) Ordinary vapor compression cycle for which compressor is 80% efficient.

Solution: We will refer to Fig 4.9 for identifying state by number. The operating temperatures of the refrigeration unit will be

$$T_H = T_4 = 303 \text{ K} \qquad T_C = T_2 = 253 \text{ K}$$

(a) Carnot Cycle

$$COP = \frac{T_C}{T_H - T_C} = \frac{253}{303 - 253} = 5.06$$

(b) Ordinary VC cycle with reversible compressor

We will create a table to summarize results. Values determined from balances are shown in shaded cells. Other valves are from the HFC-134a chart in Appendix E. State 2 is a convenient place to start since it is a saturated vapor and the temperature is known. $T_2 = -20°C$, from the chart, $H_2^{satV} = 386.5$ kJ/kg, $S_2^{satV} = 1.7414$ kJ/kg-K. The condenser outlet (state 4) is taken as saturated liquid at 30°C, so the pressure of the condenser will be $P_4^{sat}(30°C) = 0.77$ MPa, and $H_4 = 241.5$ kJ/kg, $S_4 = 1.1428$ kJ/kg-K.

Example 4.5 Refrigeration by vapor compression cycle (Continued)

State	T (K)	P (MPa)	H (kJ/kg)	S (kJ/kg-K)
1	253	0.13	241.5	
1' (for part (c))	253	0.13	235.0	1.1428
2, satV	253	0.13	386.5	1.7414
3'	316	0.77	422	1.7414
4	303	0.77	241.5	1.1428

If the process is reversible, the entropy at state 3' will be the same as S_2. Finding H_3' from $S_3' = 1.7414$ kJ/kg-K and $P_3 = 0.77$ MPa, using the chart, $H_3' = 422$ kJ/kg. Note that the pressure in the condenser, not the condenser temperature, fixes the endpoint on the isentropic line from the saturated vapor.

$$COP = \frac{Q_C}{W_{S,net}} = \frac{(H_2 - H_1)}{(H_3' - H_2)} = \frac{(H_2 - H_4)}{(H_3' - H_2)} = \frac{386.5 - 241.5}{422 - 386.5} = 4.08$$

The required circulation rate is

$$\dot{Q} = 120,000 = \dot{m}Q_C = \dot{m}(H_2 - H_1) = \dot{m}(386.5 - 241.5), \text{ which gives } \dot{m} = 828 \text{ kg/h.}$$

(c) VC cycle with turbine expansion

The throttle valve will be replaced by a reversible expander. Therefore, $S_1' = S_4 = 1.1428$ kJ/kg-K. The saturation values at 253 K are $S^{satL} = 0.8994$ kJ/kg-K and $S^{satV} = 1.7414$ kJ/kg-K; therefore, $S_1' = 1.1428 = q' \cdot 1.7414 + (1 - q')0.8994$, which gives $q' = 0.289$.

Then using the saturated enthalpy values and the quality: $H_1' = 235.0$ kJ/kg. In order to calculate the COP, we must recognize that we are able to recover some work from the expander, given by $H_1' - H_4$

$$COP = \frac{Q_C}{W_{S,net}} = \frac{(H_2 - H_1')}{(H_3' - H_2) + (H_1' - H_4)} = \frac{386.5 - 235.0}{(422 - 386.5) + (235.0 - 241.5)} = 5.22$$

> ## Example 4.5 Refrigeration by vapor compression cycle (Continued)
>
> The increase in COP requires a significant increase in equipment complexity and cost, since a two-phase expander would probably have a short life due to erosion of turbine blades by droplets.
>
> ### (d) like (b) but with irreversible compressor
>
> States 1, 2, and 4 are the same as in *b*. The irreversibility simply changes state 3.
>
> $$COP = \frac{Q_C}{W_{S,net}} = \frac{(H_2 - H_1)}{(H_3 - H_2)} = 0.8\frac{(H_2 - H_4)}{(H_3' - H_2)} = 0.8(4.08) = 3.26$$

Note: Choice of refrigerant dictated by

1. Toxicity (Freon R-12 depletes ozone and has been phased out; Freon R-22 is being phased out).

2. Vapor pressure ~ atmospheric at T_{evap}. Consequently, the driving force for leakage will be small.

3. Vapor pressure not too high at T_H so that the operating pressure is not too high; high pressure increases compressor and equipment costs.

4. High heat of vaporization per unit mass.

5. Small C_P/C_V of vapor to minimize temperature rise and work of compressor.

6. High heat transfer coefficient in vapor and liquid.

Flash Chamber Intercooling

When the temperature difference between the condenser and evaporator is increased, the compressor must span larger pressure ranges. If the compression ratio becomes too large, interstage cooling can be used to increase efficiency, as shown in Fig. 4.11. Because the process temperatures are usually below cooling water temperatures, a portion of the condensed refrigerant stream can be flashed to provide the interstage cooling. The interstage cooler is sometimes called an economizer.

> Flash chamber intercooling is a common method of increasing COP.

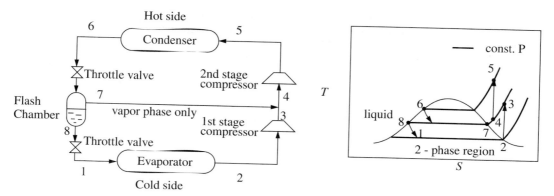

Figure 4.11 *Flash chamber intercooling.*

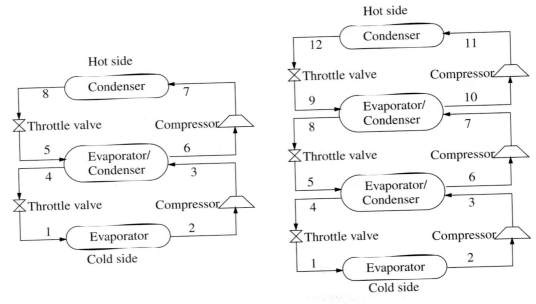

Figure 4.12 *Binary cycle (left) and three-cycle cascade (right) refrigeration cycles. The refrigerants do not mix in the evaporator/condensers.*

Cascade Refrigeration

In order to span extremely large temperature ranges, a single refrigerant cannot be used, because the condenser must always be below the critical temperature, and the evaporator must always be above the freezing temperature. Therefore, to span extremely large ranges, *binary* vapor cycles or *cascade* vapor cycles are used. In a binary cycle, a refrigerant with a normal boiling point below the coldest temperature is used on the cold cycle, and a refrigerant that condenses at a moderate pressure is used on the hot cycle. The two cycles are coupled at the condenser of the cold cycle and the evaporator of the hot cycle as shown in Fig 4.12. Because the heat of vaporization is coupled to the saturation temperature for any refrigerant, usually the operating temperatures are selected, and the circulation rates are determined for each cycle. Certainly, there are many variables to optimize in a process design of this type. For extremely large ranges, such as for cryogenic processing of liquefied gases, cascade refrigeration can be used with multiple cycles. For example, for the liquefaction of natural gas, the three cycles might be ammonia, ethylene, and methane.

Cascade refrigeration is used to reach cryogenic temperatures.

4.5   LIQUEFACTION

We have encountered liquefaction since our first quality calculation in dealing with turbines. In refrigeration, throttling or isentropic expansion results in a partially liquid stream. The point of a liquefaction process is simply to recover the liquid part as the primary product.

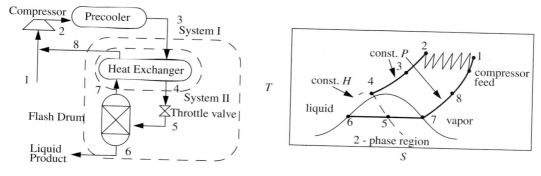

Figure 4.13 *Linde liquefaction process schematic. The system boundaries shown on the left are used in Example 4.6.*

Linde Liquefaction

The Linde process works by throttling high-pressure vapor. The Joule-Thomson coefficient, $\left(\dfrac{\partial T}{\partial P}\right)_H$, must be such that the gas cools on expansion,[1] and the temperature must be low enough and the pressure high enough to assure that the expansion will end in the two-phase region. Since less than 100% is liquefied, the vapor phase is returned to the compressor, and the liquid phase is withdrawn. Multistage compression is usually used in the Linde liquefaction process to achieve the required high pressures. An example of the process pathways on a *T-S* diagram is shown in Fig. 4.13. The actual state of the gas entering the multistage compressor depends on the state of the feed.

Example 4.6   Liquefaction of methane by the Linde process

Methane is to be liquefied in a simple Linde process. The feed and recycle are mixed, compressed to 60 bar, and precooled to 300 K. The vapor then passes through a heat exchanger for additional cooling before being throttled to 1 bar. The unliquefied fraction leaves the separator at the saturation temperature, and passes through the heat exchanger, then exits at 295 K. (a) What fraction of the gas is liquefied in the process; and (b) what is the temperature of the high-pressure gas entering the throttle valve?

Solution: The schematic is shown in Fig. 4.13. To solve this problem, first recognize that states 3, 6, 7 and 8 are known. State 3 is at 300K and 60 bar; state 6 is saturated liquid at 1 bar; state 7 is saturated vapor at 1 bar; and state 8 is at 295 K and 1 bar. Use the furnished methane chart from Appendix E.

(a) The System I energy balance is: $H_3 - [qH_8 + (1-q)H_6] = 0$

$$\Rightarrow q = \frac{H_3 - H_6}{H_8 - H_6} = \frac{H(60, 300) - H(1, satL)}{H(1, 295) - H(1, satL)} = \frac{1130 - 284}{1195 - 284} = 0.9286 \Rightarrow 7.14\% \text{ liquefied}$$

(b) The energy balance for System II is: $H_4 - H_3 = -q(H_8 - H_7) = -0.9286(1195 - 796.1)$
$$= -370.5 \Rightarrow H_4 = 780$$

$$\Rightarrow H_4 = 780 \text{ @ } 60 \text{ bar} \quad \Rightarrow \text{chart gives} -95°F = 203 \text{ K}$$

---

1. This criteria can be evaluated by looking at the dependence of temperature on pressure at constant enthalpy on a thermodynamic chart or table. In Chapter 5, we introduce principles for calculating this derivative from *P, V, T* properties.

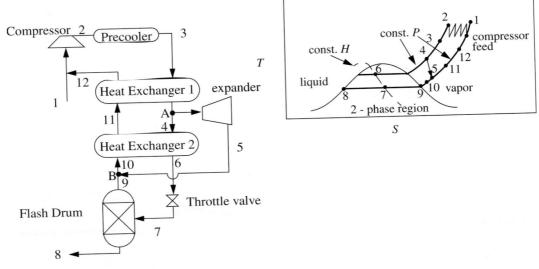

Figure 4.14 *The Claude liquefaction process.*

Claude Liquefaction

The throttling process between states 4 and 5 in the Linde process is irreversible. To improve this, a reversible expansion is desirable; however, since the objective is to liquefy large fractions of the inlet stream, turbines are not practical because they cannot handle low-quality mixtures. One compromise, the Claude liquefaction, is to expand a portion of the high-pressure fluid in an expander under conditions that avoid the two-phase region, as shown in Fig. 4.14. Only a smaller fraction of the compressed gas enters the irreversible throttle valve, so the over-all efficiency can be higher but more sophisticated equipment is required.

4.6   INTERNAL COMBUSTION ENGINES

Steam is not the only working fluid that can be used in a power producing cycle. A common alternative is to use air, mixed with a small amount of fuel that is burned. The heat of combustion provides energy to heat the gas mixture before it does work in an expansion step. A major benefit of using air is that a physical loop is not necessary; we can imagine the atmosphere as the recycle loop. This approach forms the basis for internal combustion engines like lawn mowers, jet engines, diesels, and autos.

Combustion Gas Turbine (Brayton Cycle)

The typical approach for analysis of air standard cycles is illustrated by the Brayton Cycle in Fig. 4.15. To understand the cycle, the basic idea is to write the balances for each step individually, then sum them up.

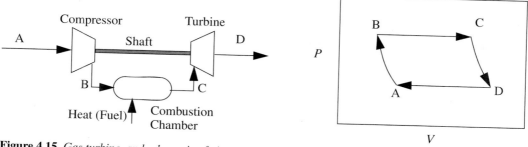

Figure 4.15 *Gas turbine, and schematic of air-standard Brayton cycle.*

Example 4.7   Air-standard Brayton cycle thermal efficiency

Analyze the air-standard Brayton cycle to find the thermal efficiency and the dependence of temperature changes on pressure changes across the compressor and turbine.

Solution:

Basis: Assume the working fluid is an ideal gas, and that the moles of gas produced by combustion are small relative to the total moles of gas throughput. Refer to Fig. 4.15 for stream labels.

System: closed packet of 1 mole of fluid passing through system.

$$A{\rightarrow}B \text{ compression:} \quad W_{S,AB} = \Delta H = C_P(T_B - T_A) \tag{*ig}$$

$$C{\rightarrow}D \text{ expansion:} \quad W_{S,CD} = \Delta H = C_P(T_D - T_C) \tag{*ig}$$

$$B{\rightarrow}C \text{ heat:} \quad Q = \Delta H = C_P(T_C - T_B) \tag{*ig}$$

$$\eta = \frac{W_{S,AB} + W_{S,CD}}{-Q_{BC}} = \frac{-C_P\left[(T_B - T_A) + (T_D - T_C)\right]}{C_P(T_C - T_B)} + 1 - \frac{(T_C - T_B)}{(T_C - T_B)} = 1 - \left[\frac{T_D - T_A}{T_C - T_B}\right] \tag{*ig}$$

Substituting relations for adiabatic reversible ideal gases and defining the specific heat ratio $\gamma\ (\equiv C_P/C_V)$:

$$(T_D - T_A) = \left[\frac{P_A}{P_B}\right]^{\frac{(\gamma-1)}{\gamma}} (T_C - T_B) \Rightarrow \eta = 1 - \left[\frac{P_A}{P_B}\right]^{\frac{(\gamma-1)}{\gamma}} \tag{*ig) 4.7}$$

Therefore, efficiency is controlled by compression ratio.

Turbofan Jet Engines

Many of the principles of the gas turbine are also present in turbofan jet aircraft engines. The engine also introduces fuel in a combustor. Like the gas turbine, the engine has a compressor system run by using some of the energy from the hot gases. The major difference is that, since the objective is to produce thrust, only a couple of turbine stages are used to drive the air compression system as shown in Fig. 4.16. The gases exiting the turbine are at relatively high temperature and pressure, can be further heated in an afterburner, and create thrust by exiting the engine at high velocity through an exhaust nozzle. Afterburners consume fuel rapidly but are necessary for the highest thrust, and are turned on/off as needed. Afterburners are used almost exclusively on military aircraft and the Concorde.

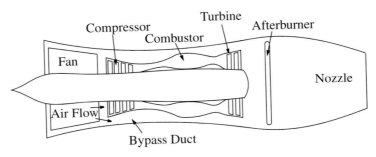

Figure 4.16 *Schematic of a typical turbofan jet engine.*

There are two other features of a turbofan jet engine: 1) a large fan precedes the compressor on the same rotating shaft; 2) 40 to 97% of the low-pressure gas from the fan bypasses the compressor. Therefore, the overall zones in a turbofan jet engine are, from front to rear: the fan; the compressor with a surrounding bypass duct; the combustor that uses the high temperature/high pressure air from the compressor; a turbine, driven by the combustion gases to power the compressor; an afterburner to further heat the bypass gas and turbine outlet gases; an exhaust nozzle.

The Otto Cycle

Other kinds of air-based engines can be easily imagined. The Otto cycle describes the modern automobile engine. The actual *P-V* relations in an Otto engine follow a complex path of intake, compression, ignition/combustion (so rapid that $\Delta V = 0$), expansion/work, and exhaust. Since the processes that occur in the Otto engine are extremely complex, a semiquantitative model has been developed. It is referred to as the air-standard Otto cycle.

Example 4.8 Thermal efficiency of the Otto engine

Determine the thermal efficiency of the air-standard Otto cycle as a function of the specific heat ratio $\gamma\ (= C_P/C_V)$ and the compression ratio $r \equiv V_1/V_2$.

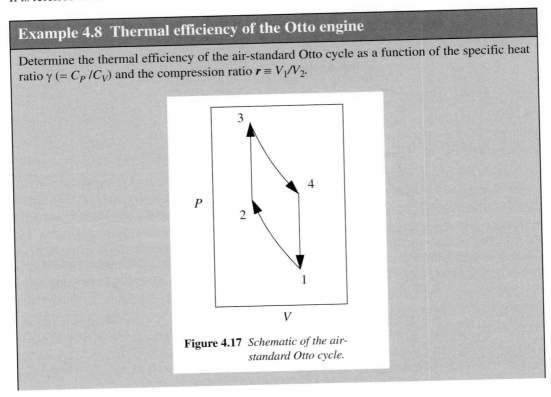

Figure 4.17 *Schematic of the air-standard Otto cycle.*

> ### Example 4.8 Thermal efficiency of the Otto engine (Continued)
>
> **Solution:**
>
> $$\text{Basis: model as ideal gas, } Q_H = C_V(T_3 - T_2) \qquad (\text{*ig})$$
>
> $$Q_C = C_V(T_1 - T_4) \qquad (\text{*ig})$$
>
> $$-W_{S,net} = Q_H + Q_C = C_V(T_3 - T_2 + T_1 - T_4) \qquad (\text{*ig})$$
>
> $$\eta = C_V(T_3 - T_2 + T_1 - T_4)/[C_V(T_3 - T_2)] + 1 - (T_3 - T_2)/(T_3 - T_2) = 1 + (T_1 - T_4)/(T_3 - T_2)(\text{*ig})$$
>
> $$\frac{T_2}{T_1} = \left(\frac{V_1}{V_2}\right)^{R/C_V}; \frac{T_4}{T_3} = \left(\frac{V_3}{V_4}\right)^{R/C_V} = \left(\frac{V_2}{V_1}\right)^{R/C_V} \qquad (\text{*ig})$$
>
> $$\Rightarrow T_4 = T_3\, r^{-R/C_V}; T_1 = T_2\, r^{-R/C_V} \qquad (\text{*ig})$$
>
> $$\eta = 1 - r^{-R/C_V} = 1 - r^{(1-\gamma)} \qquad (\text{*ig})\ 4.8$$

The Diesel Engine

The Diesel Engine is similar to the Otto engine except that the fuel is injected after the compression and the combustion occurs relatively slowly. This necessitates "fuel-injectors," but has the advantage that higher compression ratios can be obtained without concern of pre-ignition (ignition at the wrong time in the cycle). Pre-ignition occurs when the temperature rise due to compression goes past the spontaneous ignition temperature of the fuel, creating an annoying pinging and knocking sound, and reducing the efficiency of the cycle. Pre-ignition does not occur in a diesel engine because there is no fuel during compression. Some spontaneous ignition temperatures are given below.

Spontaneous ignition temperatures (°C) of sample hydrocarbons

Isooctane	447
Benzene	592
Toluene	568
n-Octane	240
n-Decane	232
n-Hexadecane	230
Methanol	470
Ethanol	392

Diesel fuel is much like decane and hexadecane and burns without a spark. Auto fuel is much like isooctane and benzene and, ideally, will not ignite until the spark goes off. But all fuels are mixtures, and if the gasoline contains enough n-octane instead of isooctane, then undesirable pre-ignition will occur.

The thermodynamics of the air-standard Diesel engine can be analyzed like the Otto Cycle. But, instead of rapid combustion at constant volume, the Diesel engine has relatively slow combustion at constant pressure.

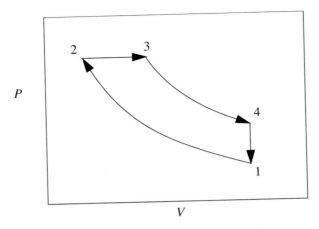

Figure 4.18 *Schematic of air-standard Diesel cycle.*

Example 4.9 Thermal efficiency of a Diesel engine

Develop an expression for the thermal efficiency of the air-standard diesel cycle as a function of the compression ratio $rc = V_1/V_2$ and the expansion ratio $re = V_4/V_3$. Assume the working fluid is an ideal gas, and the volume effect of moles of gas generated is small relative to the effect of heating from combustion.

Solution: This process is a little more complicated than the Otto cycle because of the heat addition at constant pressure. Since the volume is changing, the energy balance is:

$$Q_H + W_{S,combustion} = dU \Rightarrow \Delta U - W_{S,combustion} = \Delta U + P_H(V_3 - V_2) = \Delta(U + PV) = \Delta H$$

$$\Rightarrow Q_H = \Delta H = C_P(T_3 - T_2) \tag{*ig}$$

$$Q_C = C_V(T_1 - T_4) \tag{*ig}$$

$$-W_{S,net}/Q_H = (Q_H + Q_C)/Q_H = 1 + \frac{C_V(T_1 - T_4)}{C_P(T_3 - T_2)} = 1 + \frac{1}{\gamma}\left[\frac{T_1}{T_3 - T_2} - \frac{T_4}{T_3 - T_2}\right] \tag{*ig}$$

$$\frac{T_3 - T_2}{T_1} = \frac{T_3}{T_1} - \frac{T_2}{T_1} = \frac{P_3 V_3}{P_1 V_1} - rc^{\gamma-1} = rc^{\gamma}/re - rc^{\gamma-1} \tag{*ig}$$

$$\frac{T_3 - T_2}{T_4} = \frac{T_3}{T_4} - \frac{T_2}{T_4} = re^{\gamma-1} - \frac{P_2 V_2}{P_4 V_4} = re^{\gamma-1} - \frac{P_3 V_2}{P_4 V_4} = re^{\gamma-1} - re^{\gamma}/rc \tag{*ig}$$

$$\eta = 1 + \frac{1}{\gamma}\left\{\frac{1}{rc^{\gamma}/re - rc^{\gamma-1}} - \frac{1}{re^{\gamma-1} - re^{\gamma}/rc}\right\}$$

$$= 1 + \frac{1}{\gamma}\left\{\frac{re}{rc^{\gamma} - re\cdot rc^{\gamma-1}} - \frac{rc}{rc\cdot re^{\gamma-1} - re^{\gamma}}\right\} \tag{*ig 4.9}$$

4.7 FLUID FLOW

Consider the general flow system of Fig. 4.19a in which work and heat are transferred and the fluid undergoes changes in kinetic and potential energy. Recognize that the compressor or pump in the schematic could be replaced with an expander or turbine. Rather than deriving an integral equation between points 1 and 4 in the schematic, let us consider a balance over a differential element at steady-state as shown in Fig. 4.19b where the possibility of heat and work transfer are permitted. The steady-state balance for a single stream becomes:[1]

$$0 = \lim_{dL \to 0} \left\{ \left[H + \frac{u^2}{2g_c} + \frac{gz}{g_c} \right]^{in} - \left[H + \frac{u^2}{2g_c} + \frac{gz}{g_c} \right]^{out} \right\} + dQ + dW_S$$

For a differential element, $H^{out} = H^{in} + dH$, and

$$\left(\frac{u^2}{2g_c} \right)^{out} = \left(\frac{u^2}{2g_c} \right)^{in} + d\left(\frac{u^2}{2g_c} \right), \quad \left(\frac{gz}{g_c} \right)^{out} = \left(\frac{gz}{g_c} \right)^{in} + d\left(\frac{gz}{g_c} \right)$$

The differential balance becomes

$$0 = -dH - d\left(\frac{u^2}{2g_c} \right) - d\left(\frac{gz}{g_c} \right) + dQ + dW_S$$

Recognizing $d\left(\dfrac{u^2}{2g_c} \right) = \dfrac{udu}{g_c}$, and over practical distances g is constant, resulting in $d\left(\dfrac{gz}{g_c} \right) = \dfrac{g}{g_c} dz$.

$$(dH - dQ) + \frac{udu}{g_c} + \frac{g}{g_c} dz = dW_S$$

The entropy balance for the differential system of Fig. 4.19b,

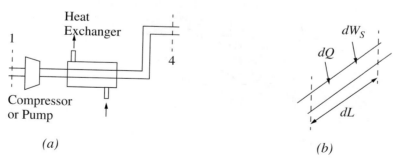

(a)   (b)

Figure 4.19 (a) Schematic of a general overall system, and (b) differential balance for an arbitrary piece of overall system.

---

1. Note that the velocity in this equation is a mean velocity.

$$0 = \lim_{dL \to 0} \{S^{in} - S^{out}\} + \frac{dQ}{T} + dS_{gen}$$

and $S^{out} = S^{in} + dS$,

$$TdS - TdS_{gen} = dQ$$

Combining with the energy balance,

$$(dH - TdS) + TdS_{gen} + \frac{udu}{g_c} + \frac{g}{g_c}dz = dW_S$$

Noting the fundamental relation for enthalpy, $dH = TdS + VdP$,

$$VdP + TdS_{gen} + \frac{udu}{g_c} + \frac{g}{g_c}dz = dW_S$$

Although H and S are state properties, TdS is path-dependent, therefore VdP is path-dependent and must be evaluated along the actual path. The term TdS_{gen} represents the losses due to viscosity, and is often called the lost work, lw.

⚠ Lost work is due to entropy generation.

$$\boxed{TdS_{gen} \equiv d(lw)}$$

4.10

The energy balance becomes

⚠ The differential energy balance.

$$\boxed{VdP + d(lw) + \frac{udu}{g_c} + \frac{g}{g_c}dz = dW_S}$$

4.11

The treatment of $d(lw)$ depends on the device. Typically, the differential balance is integrated over the individual pieces of equipment in Fig. 4.19a in the three categories: 1) pipes or fittings, $(lw)_f$; 2) pumps or compressors, $(lw)_p$; 3) expanders or turbines, $(lw)_t$.

Pipes

The frictional losses in pipes can be predicted by an empirical variable f known as the Fanning friction factor, and the losses due to fittings or sudden cross-section changes can be handled separately

$$d(lw)_f = \frac{2fu^2}{g_c D}dL + d(lw)_{fittings}$$

4.12

where D is the diameter of the pipe and $d(lw)_{fittings}$ is the lost work due to fittings and sudden cross-section changes. The friction factor is relatively insensitive to considerable temperature changes. Work producing/generating devices are not present in flow through pipes, $dW_S = 0$,

⚠ General differential flow in a pipe.

$$VdP + \frac{2fu^2}{g_c D}dL + d(lw)_{fittings} + \frac{udu}{g_c} + \frac{g}{g_c}dz = 0$$

4.13

For an incompressible fluid in a pipe, the density changes will be negligible. Integrating term by term, where ρ is the mass density,

$$\boxed{\frac{\Delta P}{\rho} + \frac{2fu^2 L}{g_c D} + (lw)_{fittings} + \frac{\Delta u^2}{2g_c} + \frac{g}{g_c}\Delta z = 0}$$

4.14   ❗ Flow of an in-compressible fluid.

For a compressible fluid such as a gas, velocity and density can change appreciably with pressure and temperature. The mass flowrate is constant at steady state,

$$\dot{m} = \frac{Au}{V} = AG$$

4.15

where A is the cross-sectional area of the pipe, $G \equiv u/V = u\rho$, and G is called the mass velocity and is constant when A is constant. Using $u = G/\rho$, $du = (-G/\rho^2)d\rho$. Substituting into Eqn. 4.13 for a horizontal pipe without fittings or sudden cross-section changes,

$$\frac{dP}{\rho} + \frac{2fG^2}{g_c\rho^2 D}dL - \frac{G^2}{g_c\rho^3}d\rho = 0, \quad \rho dP + \frac{2fG^2}{g_c D}dL - \frac{G^2}{g_c}\frac{d\rho}{\rho} = 0$$

For an ideal gas,

$$(MW)\int\frac{P}{RT}dP + \frac{2fG^2 L}{g_c D} - \frac{G^2}{g_c}\ln\frac{\rho_2}{\rho_1} = 0$$

(ig)

where (MW) is the molecular weight. If the flow is isothermal,

$$\boxed{\frac{(MW)(P_2^2 - P_1^2)}{2RT} + \frac{2fG^2 L}{g_c D} - \frac{G^2}{g_c}\ln\frac{P_2}{P_1} = 0}$$

(ig) 4.16   ❗ Isothermal flow of an ideal gas in a horizontal pipe of constant size and no fittings.

The Fanning friction factor, f, depends on the properties of the fluid and the size and roughness of the pipe. The friction factor is most easily characterized in terms of the dimensionless Reynolds number, $Re = Du\rho/\mu$, where D is the diameter of the pipe, ρ is the fluid mass density, and μ is the fluid viscosity. The relationships for smooth pipes are:

$$f = \frac{16}{Re} \text{ for } Re < 2100$$

4.17

$$f = 0.0014 + \frac{0.125}{Re^{0.32}} \text{ for } 3000 < Re < 3.\text{E }6$$

4.18

Below $Re = 2100$ the flow is laminar; above 4000 the flow is turbulent. The range of Reynolds number between 2100 and 4000 is known as the transition region where the friction factor correlations are somewhat uncertain because the flow is in between the laminar and turbulent flow regimes. Fluid mechanics textbooks can be consulted for calculating flows in rough pipes, fittings, sudden contractions or expansions, non-circular pipes, or non-isothermal conditions.

Pumps or Compressors

The lost work due to pump/compressor irreversibility, $(lw)_p$, is incorporated into the energy balance by the pump/compressor efficiency

$$\eta = \frac{W_S - (lw)_p}{W_S} \quad \text{where } W_S > 0, (lw)_p > 0 \qquad 4.19$$

W_S is the actual work transferred by the pump/compressor. Therefore, the terms $W_S - (lw)_p$ in the energy balance can be replaced by ηW_S. For a flow system including piping for an *incompressible* fluid, Eqn. 4.14 becomes

❗ Pumping of an incompressible fluid.

$$\boxed{\frac{\Delta P}{\rho} + \frac{2fu^2 L}{g_c D} + (lw)_{fittings} + \frac{\Delta u^2}{2g_c} + \frac{g}{g_c}\Delta z = \eta W_S} \qquad 4.20$$

By analogy, an *isothermal* pump or compressor for a *compressible* fluid will result in ηW_S on the right-hand side of Eqn. 4.16.

Turbines or Expanders

The lost work due to turbine/expander irreversibilities $(lw)_t$, is incorporated by the turbine/expander efficiency

$$\eta = \frac{W_S}{W_S - (lw)_t} \quad \text{where } W_S < 0, (lw)_t > 0. \qquad 4.21$$

W_S is the actual work interaction of the fluid with the turbine/expander. Therefore the terms $W_S - (lw)_t$ in the energy balance may be replaced with W_S/η. By analogy with Eqn. 4.20 for an *incompressible* fluid

❗ Turbine for an incompressible fluid.

$$\boxed{\frac{\Delta P}{\rho} + \frac{2fu^2 L}{g_c D} + (lw)_{fittings} + \frac{\Delta u^2}{2g_c} + \frac{g}{g_c}\Delta z = \frac{W_S}{\eta}} \qquad 4.22$$

A similar modification may be made to Eqn. 4.16 for an *isothermal* turbine/expander for a compressible fluid.

4.8 PROBLEM-SOLVING STRATEGIES

❗ A review of common assumptions and hints.

As you set up more complex problems, use the strategies in Section 2.13 on page 65, and incorporate the energy balances developed in Section 2.12 on page 59 for valves, nozzles, heat exchangers, turbines, and pumps and entropy balances developed in Section 3.9 on page 116 for turbines, compressors, and heat pumps/engines as you work through step 5 of the strategies. A stream that exits a condenser is assumed to exit as saturated liquid unless otherwise specified. Likewise, a stream that is vaporized in a boiler is assumed to exit saturated unless otherwise specified; however, recall that in a Rankine cycle the steam is always superheated, and we omit the superheater in schematics for simplicity. Read problem statements carefully to identify the outlet states of turbines. Outlets of turbines *are not required to be saturated or in the two-phase region,* although operation in this manner is common. In a multistage turbine, only the last stages will be near saturation unless reheat is used.

Unless specified, pressure drops are considered negligible in piping and heat exchangers as a first approximation. Throttle valves are assumed to be adiabatic unless otherwise stated, and they are always irreversible and *do* have an important pressure drop unless otherwise stated. Recognize that the entropy balances for throttle valves or heat exchangers are usually not helpful since, in practical applications, these devices are inherently irreversible and generate entropy.

Recognize that the energy balance must be used often to find mass flow rates. In order to do this for open steady-state flow systems, the enthalpies for all streams must be known, in addition to the mass flow rates for all but one stream. You can try moving the system boundary as suggested in step 7 of the strategy in Section 2.13 on page 65 to search for balances that satisfy these conditions. Mass flowrates can be found using the entropy balance also, but this is not done very often, since the entropy balance is useful only if the process is internally reversible or if the rate of entropy generation is known (i.e., no irreversible heat exchangers or throttle valves or irreversible turbines/compressors inside the system boundary).

Basically, the energy and entropy balance and the *P-V-T* relation are the only equations that always apply. While we have shown common simplifications, there are always new applications that can arise, and it is wise to learn the principles involved in simplifying the balance to a given situation.

4.9   PRACTICE PROBLEMS

P4.1 An ordinary vapor compression cycle is to operate a refrigerator on HFC-134a between −40°C and 40°C (condenser temperatures). Compute the coefficient of performance and the heat removed from the refrigerator per day if the power used by the refrigerator is 9000 J per day. (ANS. 1.76)

P4.2 An ordinary vapor compression cycle is to be operated on methane to cool a chamber to −260°F. Heat will be rejected to liquid ethylene at −165°F. The temperatures in the condenser and evaporator are −160°F and −280°F. Compute the coefficient of performance. (ANS. 0.86)

P4.3 A simple Rankine cycle is to operate on steam between 200°C and 99.6°C, with saturated steam exhausting from the turbine. What is the maximum possible value for its thermodynamic efficiency? (ANS. 7.8%)

P4.4 An adiabatic turbine is supplied with steam at 300 psia and 550°F that exhausts at atmospheric pressure. The quality of the exhaust steam is 95%.

(a) What is the efficiency of the turbine? (ANS. 76%)
(b) What would be the thermodynamic efficiency of a Rankine cycle operated using this turbine at these conditions? (ANS. 17%)

4.10   HOMEWORK PROBLEMS

4.1 A steam power plant operates on the Rankine cycle according to the specified conditions below. Using stream numbering from Fig 4.3 on page 143, for each of the options below, determine:

(a) The work output of the turbine per kg of steam;
(b) The work input of the feedwater pump per kg of circulated water;
(c) The flowrate of steam required;
(d) The heat input required in the boiler;
(e) the thermal efficiency

Option	$T_3(°C)$	P_3(MPa)	P_4(MPa)	$\eta_{turbine}$	η_{pump}	plant capacity
i	600	10	0.01	0.8	0.75	80MW
ii	600	8	0.01	0.8	0.75	80MW
iii	400	4	0.01	0.83	0.95	75MW
iv	500	4	0.02	0.85	0.8	80MW
v	600	4	0.02	0.85	0.8	80MW

4.2 A steam power plant operates on the Rankine cycle with reheat, using the specified conditions below. Using stream numbering from Fig. 4.4 on page 146, for each of the options below, determine:

(a) The work output of each turbine per kg of steam;
(b) The work input of the feedwater pump per kg of circulated water;
(c) The flowrate of steam required;
(d) The heat input required in the boiler and reheater;
(e) the thermal efficiency.

Option	$T_3(°C)$	P_3(MPa)	P_4(MPa)	$T_5(°C)$	P_6(MPa)	$\eta_{turbines}$	η_{pump}	plant capacity
i	450	5	0.4	400	0.01	0.8	0.75	80MW
ii	400	5	0.6	450	0.01	0.8	0.75	80MW
iii	400	4	0.5	500	0.01	0.9	0.95	75MW
iv	500	4	0.8	500	0.01	0.85	0.8	80MW
v	600	4	1.2	600	0.01	0.85	0.8	80MW

4.3 A modified Rankine cycle using a single feedwater preheater as shown in Fig. 4.8 on page 148 has the following characteristics:

(a) the inlet to the first turbine is at 500°C and 0.8 MPa;
(b) the feedwater preheater reheats the recirculated water so that stream 7 is 140°C, and steam at 0.4 MPa is withdrawn from the outlet of the first turbine to perform the heating;
(c) the efficiency of each turbine and pump is 79%;
(d) the output of the plant is to be 1 MW;
(e) the output of the second turbine is to be 0.025 MPa.

Determine the flowrates of streams 1 and 8 and the quality of stream 9 *entering* the condenser (after the throttle valve). Use the stream numbers from Fig. 4.8 to label streams in your solution.

4.4 A modified Rankine cycle uses reheat and one closed feedwater preheater. The schematic is a modification of Fig. 4.8 on page 148 obtained by adding a reheater between the T-joint and turbine II. Letting stream 3 denote the inlet to the reheater, and stream 3a denote the inlet to the turbine, the conditions are given below. The plant capacity is to be 80 MW. Other constraints are: the efficiency of each turbine stage is 85%; the pump efficiency is

80%; the feedwater leaving the closed preheater is 5°C below the temperature of the condensate draining from the bottom of the closed preheater. For the options below, calculate:

(a) the flowrate of stream 1;
(b) the thermal efficiency of the plant;
(c) the size of the feedwater pump (kW).

Options:

(i) $T_1 = 500°C$, $P_1 = 4$ MPa, $P_2 = 0.8$ MPa, $T_{3a} = 500°C$, $P_4 = 0.01$ MPa.
(ii) $T_1 = 600°C$, $P_1 = 4$ MPa, $P_2 = 1.2$ MPa, $T_{3a} = 600°C$, $P_4 = 0.01$ MPa.

4.5 A regenerative Rankine cycle utilized the schematic of Fig. 4.7 on page 148. Conditions are: stream 1, 450°C, 3 MPa; stream 2, 250°C, 0.4 MPa; stream 3, 150°C, 0.1 MPa; stream 4, 0.01 MPa; stream 9, 140°C, $H = 592$ kJ/kg.

(a) Determine the pressures for streams 5,6,8,9,10.
(b) Determine $\dot{m}_2 / \dot{m}_1$.
(c) Determine the enthalpies of streams 5 and 6 if the pump is 80% efficient.
(d) Determine the efficiency of turbine stage I.
(e) Determine the output of turbine stage III per kg of stream 4 if the turbine is 80% efficient.
(f) Determine $\dot{m}_3 / \dot{m}_1$
(g) Determine the work output of the system per kg of stream 1 circulated.

4.6 A regenerative Rankine cycle uses one open feedwater preheater and one closed feedwater preheater. Using the stream numbering from Fig. 4.7 on page 148, and the specified conditions below, the plant capacity is to be 75 MW. Other constraints are: the efficiency of each turbine stage is 85%; the pump efficiencies are 80%; the feedwater leaving the closed preheater is 5°C below the temperature of the condensate draining from the bottom of the closed preheater. For the options below, calculate:

(a) the flowrate of stream 1;
(b) the thermal efficiency of the plant;
(c) the size of the feedwater pumps (kW);
(d) and the overall efficiency of the multistage turbine.

Options:

(i) the conditions are: $T_1 = 500°C$, $P_1 = 4$ MPa, $P_2 = 0.7$ MPa, $P_3 = 0.12$ MPa, $P_4 = 0.02$ MPa.
(ii) the conditions are: $T_1 = 600°C$, $P_1 = 4$ MPa, $P_2 = 1.6$ MPa, $P_3 = 0.8$ MPa, $P_4 = 0.01$ MPa.

4.7 A regenerative Rankine cycle uses three closed feedwater preheaters. Using the stream numbering from Fig. 4.5 on page 147, and the specified conditions below, the plant capacity is to be 80 MW. Other constraints are: the efficiency of each turbine stage is 88%; the pump efficiency is 80%; the feedwater leaving each preheater is 5°C below the temperature of the condensate draining from the bottom of each preheater. For the options below, calculate:

(a) the flowrate of stream 1;
(b) the thermal efficiency of the plant;
(c) the size of the feedwater pump (kW);
(d) and the overall efficiency of the multistage turbine.

Options:

(*i*) the conditions are: $T_1 = 700°C$, $P_1 = 4$ MPa, $P_2 = 1$ MPa, $P_3 = 0.3$ MPa, $P_4 = 0.075$ MPa, $P_5 = 0.01$ MPa.

(*ii*) the conditions are: $T_1 = 750°C$, $P_1 = 4.5$ MPa, $P_2 = 1.2$ MPa, $P_3 = 0.4$ MPa, $P_4 = 0.05$ MPa, $P_5 = 0.01$ MPa.

4.8 An ordinary vapor compression refrigerator is to operate with refrigerant HFC-134a with evaporator and condenser temperatures at −20°C and 35°C. Assume the compressor is reversible.

(a) Make a table summarizing the nature (e.g., saturated, superheated, temperature, pressure, and *H)* of each point in the process.
(b) Compute the coefficient of performance for this cycle and compare it to the Carnot cycle value.
(c) If the compressor in the cycle were driven by a 1 hp motor, what would be the tonnage rating of the refrigerator? Neglect losses in the motor.

4.9 An ordinary vapor compression refrigeration cycle using HFC-134a is to operate with a condenser at 45°C and an evaporator at −10°C. The compressor is 80% efficient.

(a) Determine the amount of cooling per kg of HFC-134a circulated.
(b) Determine the amount of heat rejected per kg of HFC-134a circulated.
(c) Determine the work required per kg of HFC-134a circulated, and the COP.

4.10 An ordinary vapor compression cycle using propane operates at temperatures of 240 K in the cold heat exchanger, and 280 K in the hot heat exchanger. How much work is required per kg of propane circulated if the compressor is 80% efficient? What cooling capacity is provided per kg of propane circulated? How is the cooling capacity per kg of propane affected by lowering the pressure of the hot heat exchanger, while keeping the cold heat exchanger pressure the same?

4.11 The low-temperature condenser of a distillation column is to be operated using a propane refrigeration unit. The evaporator is to operate at −20°C. The cooling duty is to be 10,000,000 kJ/hr. The compressor is to be a 2-stage compressor with an adiabatic efficiency of 80% (each stage). The compression ratio for each stage is to be the same. The condenser outlet is to be at 50°C. Refer to Fig. 4.9 on page 150 for stream numbers.

(a) Find the condenser, evaporator, and compressor interstage pressures.
(b) Find the refrigerant flowrate through each compressor.
(c) Find the work input required for each compressor.
(d) Find the cooling rate needed in the condenser.

4.12 Solve problem 4.11 using an economizer at the intermediate pressure and referring to Fig. 4.11 on page 153 for stream numbers.

4.13 A refrigeration process with interstage cooling uses refrigerant HFC-134a, and the outlet of the condenser is to be saturated liquid at 40°C. Refer to Fig 4.11 on page 153 for stream numbers in your solution. The pressure of the flash chamber and the intermediate pressure between compressors is to be 290 kPa. The evaporator is to operate at −20°C and the outlet is to be saturated vapor. The flowrate of stream 1 is 23 kg/h. The flash chamber may be considered adiabatic. The compressors may be considered to be 80% efficient. Attach the *P-H* chart with your solution.

(a) What is the work input required to the first compressor in kJ/h?
(b) What are the flowrates of streams 7 and 6?
(c) What is the enthalpy of stream 4?

4.14 A refrigeration process with interstage cooling uses refrigerant HFC-134a. The outlet of the condenser is to be saturated liquid at 40°C. The evaporator is to operate at −20°C, and the outlet is saturated vapor. The economizer is to operate at 10°C. Refer to Fig. 4.11 on page 153 for stream numbers in your solution.

(a) Determine the required flowrate of stream 1 if the cooling capacity of the unit is to be 8250 kJ/h.
(b) Determine the pressure of stream 3, and the work required by the first compressor if it has an efficiency of 85%.
(c) What are the flowrates of streams 7 and 6?
(d) What is the enthalpy of stream 4?
(e) Determine the work required by the second compressor (85% efficient) and the COP.

4.15 The Claude liquefaction process is to be applied to methane. Using the schematic of Fig. 4.14 on page 156 for stream numbering, the key variables depend on the fraction of stream 3 that is liquefied, $\dot{m}_8/\dot{m}_3$, and the fraction of stream 3 that is fed through the expander, $\dot{m}_5/\dot{m}_3$. Create a table listing all streams from low to high stream numbers. Fill in the table as you complete the problem sections. Attach a *P-H* diagram with your solution.

(a) Write a mass balance for the system boundary encompassing all equipment except the compressor and precooler.
(b) Write an energy balance for the same boundary described in part (a), and show

$$\frac{\dot{m}_8}{\dot{m}_3} = \frac{(H_3 - H_{12}) + (\dot{m}_5/\dot{m}_3)W_{S\,\text{expander}}}{(H_8 - H_{12})}$$

(c) Stream 3 is to be 300 K and 3 MPa, stream 4 is to be 280 K and 3 MPa, stream 12 is to be 290 K, 0.1 MPa, and the flash drum is to operate at 0.1 MPa. The expander has an efficiency of 91%. The fraction liquefied is to be $\dot{m}_8/\dot{m}_3 = 0.15$. Determine how much flow to direct through the expander, $\dot{m}_5/\dot{m}_3$. What affect would increasing the flow through the expander have on the fraction liquefied if all state variables remained fixed?
(d) Find the enthalpies of streams 3-12, and the temperatures and pressures.

4.16 A Brayton gas turbine typically operates with only a small amount of fuel added so that the inlet temperatures of the turbine are kept relatively low because of material degradation at higher temperatures, thus the flowing streams can be modeled as only air. Refer to Fig. 4.15 on page 157 for stream labels. Consider a Brayton cycle modeled with air under the following conditions: $T_A = 298$ K, $P_A = P_D = 0.1$ MPa, $P_B = 0.6$ MPa, $T_C = 973$ K. The efficiency of the turbine and compressor are to be 85%. Consider air as an ideal gas stream with $C_P = 0.79 \cdot C_{P,N2} + 0.21 \cdot C_{P,O2}$. Determine the thermal efficiency, heat required, and the net work output per mole of air assuming:

(a) the heat capacities are temperature-independent at the values at 298 K.
(b) the heat capacities are given by the polynomials in Appendix E.

4.17 The thermal efficiency of a Brayton cycle can be increased by adding a regenerator as shown in the schematic below. Consider a Brayton cycle using air under the following conditions: $T_A = 298$ K, $P_A = P_E = P_F = 0.1$ MPa, $P_B = 0.6$ MPa, $T_D = 973$ K, $T_F = 563$ K. The efficiency of the turbine and compressor are to be 85%. Consider air as an ideal gas stream with $C_P = 0.79 \cdot C_{P,N2} + 0.21 \cdot C_{P,O2}$, and assume the molar flows of B and E are equal. Determine the thermal efficiency, heat required, and the net work output per mole of air assuming:

(a) the heat capacities are temperature-independent at the values at 298 K.
(b) the heat capacities are given by the polynomials in Appendix E.

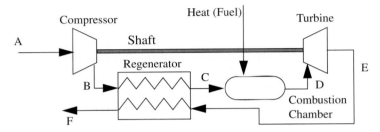

4.18 Consider the air-standard Otto cycle of Example 4.8 on page 158. At the beginning of the compression stroke, $P_1 = 95$ kPa, $T_1 = 298$ K. Consider air as an ideal gas stream with $C_P = 0.79 \cdot C_{P,N2} + 0.21 \cdot C_{P,O2}$. If the compression ratio is 6, determine T_2, T_4, and the thermal efficiency, if $T_3 = 1200$ K and:

(a) the heat capacities are temperature-independent at the values at 298 K.
(b) the heat capacities are given by the polynomials in Appendix E.

4.19 A hexane ($\rho \approx 0.66$ kg/L, $\mu = 3.2$ E-3 g/(cm·s)) storage tank in the chemical plant tank farm is 250 m from the 200 L solvent tank that is to be filled in 3 min. A pump is located at the base of the storage tank at ground level. The storage tank is large enough so that the liquid height doesn't change significantly when 200 L are removed. The bends and fittings in the pipe contribute lost work equivalent to 15 m of additional pipe. The pump and motor are to be sized based on a storage tank liquid level of 0.3 m above ground level to assure adequate flowrate when the storage tank is nearly empty. Find the required power input to the pump and motor.

(a) The pipe is to be 2.5 cm in diameter and the outlet is to be 10 m above ground level. The pump efficiency is 85%, the motor efficiency is 90%.
(b) The pipe is to be 3.0 cm in diameter and the outlet is to be 8.5 m above ground level. The pump efficiency is 87%, the motor efficiency is 92%.
(c) Determine the time required to fill the solvent tank using the pump and motor sized in part (a) if the storage tank liquid level is 6.5 m above ground.
(d) Answer part (c) except determine the filling time for part (b).

GENERALIZED ANALYSIS OF FLUID PROPERTIES

Forming an intermediate state between liquids, in which we assume no external pressure, and gases, in which we omit molecular forces, we have the state in which both terms occur. As a matter of fact, we shall see further on, that this is the only state which occurs in nature.

van der Waals (1873, ch2)

In Unit I we focused predominately on a relatively small number of pure fluids. But the number of chemical compounds encountered when considering all possible applications is vast, and new compounds are being invented and applied every day. Imagine how many charts and tables would be necessary to have properties available for all those compounds. Then imagine how many charts would be necessary to represent the properties of all the conceivable mixtures of those compounds. Clearly, we cannot address all problems by exactly the same techniques as applied in Unit I. We must still use the energy and entropy balance, but we need to be able to represent the physical properties of pure compounds and mixtures in some condensed form, and we desire to predict physical properties based on very limited data.

As one might expect, an excellent shorthand is offered by the language of mathematics. When we sought values in the steam tables, we noticed that specification of any two variables was sufficient to determine the variable of interest (e.g, S or H). This leads to an excellent application of the calculus of two variables. Changes in each value of interest may be expressed in terms of changes in whatever other two variables are most convenient. It turns out that the most convenient variables mathematically are temperature and density, and that the most convenient variables experimentally are temperature and pressure.

There is a limit to how condensed our mathematical analysis can be. That limit is dictated by how much physical insight is required to represent the properties of interest to the desired accuracy. With no physical insight, we can simply measure the desired values, but that is impractical. With maximum physical insight, we can represent all the properties purely in terms of their fundamental electronic structure as given by the periodic table and their known molecular structure. The current state-of-the-art lies between these limits, but somewhat closer to the fundamental side. By developing a sophisticated analysis of the interactions on the molecular scale, we can show that three carefully selected parameters characterizing physical properties are generally sufficient to characterize properties to the accuracy necessary in most engineering applications. This analysis leads to the equation of state, which is then combined with the necessary mathematics to provide methods for computing and predicting physical properties of interest. The development of van der Waals' equation of state provides an excellent case study in the development of engineering models based on insightful physics and moderately clever extrapolation. Note that before van der Waals the standard conception was that the vapor phase was represented by what we now refer to as an ideal gas, and the liquid was considered to be an entirely different species. Van der Waals' analysis led to a unification of these two conceptions of fluids that also set the stage for the analysis of interfaces and other inhomogeneous fluids. Van der Waals' approach lives on in modern research on inhomogeneous fluids.

CLASSICAL THERMODYNAMICS— GENERALIZATION TO ANY FLUID

5

When I first encountered the works of J.W. Gibbs, I felt like a small boy who had found a book describing the facts of life.

T. von Karmann

When someone refers to "classical thermodynamics" with no context or qualifiers, they are generally referring to a subtopic of physical chemistry which deals with the mathematical transformations of energy and entropy in fluids. These transformations are subject to several constraints owing to the nature of state functions. This field was developed largely through the efforts of J.W. Gibbs during the late 1800's. Our study focuses on three aspects of the field:

1. the fundamental property relation: $dU(S,V) = TdS - PdV$;

2. development of general formulas for property dependence on measurable variables, e.g., temperature and pressure dependence of U, H;

3. phase equilibrium: e.g., quality and composition calculations.

The fundamental property relation provides a very general connection between the energy balance and the entropy balance. It relates state functions to each other mathematically such that no specific physical situation is necessary when considering how the energy and entropy change. That is, it tells how one variable changes with respect to some other variables that we may know something about. We implicitly applied this approach for solving several problems involving steam, determining the properties upstream of a throttle from the pressure and enthalpy, for instance. In this chapter we focus intensely on understanding how to transform from one set of variables to another as preparation for developing general formulas for property dependence on measured variables.

Through "classical thermodynamics," we can generalize our insights about steam to any fluid at any conditions. All engineering processes simply involve transitions from one set of conditions to another. To get from one state to another, we must learn to develop our own paths. It does not matter what path we take, only that we can compute the changes for each step and add them up. In Chapter 6, we present the insights that led van der Waals to formulate his equation of state, enabling the estimation of any fluid's pressure given the density and temperature. In Chapter 7, we show the paths that are convenient for applying equations of state to estimate the thermodynamic properties for any fluid at any state based on a minimal number of experimental measurements.

Finally, a part of property estimation involves calculating changes of thermodynamic properties upon phase transitions so that they may be used in process calculations (e.g., formation of condensate during expansion through a turbine and characterization of the quality). The generalized analysis of phase changes requires the concept of phase equilibrium and an understanding of how the equilibrium is affected by changes in temperature and pressure. The skills developed in Chapter 5 will be required in a slightly different form to analyze the thermodynamics of phase transitions in Chapter 8.

5.1 THE FUNDAMENTAL PROPERTY RELATION

There is one equation that can be said to underlie all the other equations that we will be discussing in this chapter. It is essentially the energy balance. The only special feature that we add in this section is that we eliminate any references to specific physical situations. Transforming to a purely mathematical realm, we are free to apply multivariable calculus at will, rederiving every problem in whatever terms that seem most convenient at the time. Some of these relatively convenient forms appear frequently throughout the text so we present them here as clear implications of the fundamental property relation changes in U.

The Fundamental Property Relation for *dU* in Simple Systems

We restrict our treatment here to systems with no internal temperature gradients, rigid impermeable walls, and no external fields. These restrictions comprise what we refer to as *simple* systems. This is not a strong restriction, however. Any system can be treated as a sum of simple systems. Our goal is to transform the energy balance from extensive properties like heat and work to intensive (state) properties like density, temperature, and specific entropy. For this purpose, we may imagine any convenient physical path, recognizing that the final result will be independent of path as long as it simply relates state properties.

The energy balance for a closed simple system is

$$d(U + E_K + E_P) = dQ + dW_S + dW_{EC}$$

5.1

where E_K and E_P are the intensive kinetic and potential energies of the center of mass of the system. Eliminating all surface forces except those that cause expansion or contraction, because a simple system has no gradients or shaft work and neglecting E_K and E_P changes by taking the system's center of mass as the frame of reference

$$dU = dQ - PdV$$

5.2

Emphasizing the neglect of gradients, the reversible differential change between states is

$$dU_{rev} = dQ_{rev} - (PdV)_{rev}$$

5.3

but, by definition,

$$d\underline{S} = \frac{d\underline{Q}_{rev}}{T_{sys}} \qquad \Rightarrow T_{sys}\, d\underline{S} = d\underline{Q}_{rev} \qquad (T_{sys}\text{-system temperature where } Q \text{ transferred})$$

Since the system is simple, for the process to be internally reversible, the temperature must be uniform throughout the system (no gradients). So the system temperature has a single value throughout. On a molar basis, the fundamental property relation for dU is $dU_{rev} = T\,dS_{rev} - (P\,dV)_{rev}$. Noting that U, S, and V are all state functions, not path functions, we recognize that this relation applies whether the path is reversible or not. Consequently, the subscripts have been dropped.

$$dU = TdS - PdV$$

5.4   ❗ dU for a closed simple system.

The significance of this relation is that changes in one state variable, dU, have been related to changes in two other state variables, dS and dV. Therefore, the physical problem of relating heat flow and volume changes to energy changes has been transformed into a purely mathematical problem of the calculus of two variables. This transformation liberates us from having to think of a physical means of attaining some conversion of energy—instead if we know changes in S and V, we can apply some relatively simple rules of calculus.

Auxiliary Relations for Convenience Properties

Because dU is most simply written as a function of S and V it is termed a *natural function* of S and V. We can express changes of internal energy in terms of other state properties (such as $\{P, T\}$ or $\{T, V\}$), but when we do so, the expression always involves additional derivatives. We will show this in more detail in Example 5.9 on page 185. We also should explore the natural variables for the convenience properties.

We have defined enthalpy, $H \equiv U + PV$. Therefore, $dH = dU + PdV + VdP = TdS - PdV + PdV + VdP$,

$$dH = TdS + VdP$$ which shows H is a natural function of S and P[1]

5.5   ❗ dH for a reversible, closed simple system. Enthalpy is convenient when heat and pressure are manipulated.

Enthalpy is termed a *convenience* property because we have specifically *defined* it to be useful in problems where reversible heat flow and pressure are manipulated. By now you have become so used to using it that you may not stop to think about what the enthalpy really is. If you look back to our introduction of enthalpy, you will see that we defined it in an arbitrary way when we needed a new tool. The fact that it relates to the heat transfer in a constant-pressure closed system, and relates to the heat transfer/shaft work in steady-state flow systems, is a result of our careful choice of its definition.

We may want to control T, V for some problems, particularly in statistical mechanics, where we create a system of particles and want to change the volume (intermolecular separation) at fixed temperature. Situations like this also arise quite often in our studies of pistons and cylinders. Since U is not a natural function of T, V, such a state property is convenient. Therefore, we define *Helmholtz energy* $A \equiv U - TS$. Therefore, $dA = dU - TdS - SdT = TdS - PdV - TdS - SdT$,

$$dA = -SdT - PdV$$ which shows A is a natural function of T and V

5.6   ❗ Helmholtz energy is convenient when T, V are manipulated.

---

1. The mathematical manipulation used here is called a Legendre transformation. It involves addition of a conjugate pair like PV or TS, then differentiation and simplification. Details are given by J.W. Tester and M. Modell, *Thermodynamics and Its Applications,* 3rd Ed, Prentice Hall, Upper Saddle River, N.J., 1996.

For systems constrained by constant T and V, the equilibrium occurs when the derivative of the Helmholtz energy is zero. The other frequently used convenience property is *Gibbs energy* $G \equiv U - TS + PV = A + PV = H - TS$. Therefore, $dG = dH - TdS - SdT = TdS + VdP - TdS - SdT$.

⬤ Gibbs energy is convenient when *T*, *P* are manipulated.

$$\boxed{dG = -SdT + VdP}$$ which shows G is a natural function of T and P^1 5.7

The Gibbs energy is used specifically in phase equilibria problems where temperature and pressure are controlled. We find that for systems constrained by constant T and P, the equilibrium occurs when the derivative of the Gibbs energy is zero ($\Rightarrow$ driving forces sum to zero and Gibbs energy is minimized). These Helmholtz and Gibbs energies are such that an increase in entropy detracts from our other energies, $A = U - TS$, $G = U - TS + PV$. In other words, increases in entropy detract from increases in energy. These are sometimes called the *free energies*.

Often, students' first intuition is to expect that internal energy is minimized at equilibrium. But some deeper thought shows that equilibrium based purely on internal energy would eventually reach a state where all atoms are at the minimum of their potential wells with respect to each other. All the world would be a solid block. More interesting phenomena are only possible over a narrow range of conditions where the spreading generated by entropic driving forces balances the compaction generated by energetic driving forces. A greater appreciation for how this balance occurs should be developed over the next several chapters.

5.2 DERIVATIVE RELATIONS

⬤ Generalized expression of *U* and *H* as functions of variables like *T* and *P* are desired. Further, the relations should use *P-V-T* properties and heat capacities.

In the first four chapters, we analyzed processes using either the ideal gas law to describe the fluid or a thermodynamic chart or table. We have not yet addressed what to do in the event that a thermodynamic chart/table is not available for a compound of interest and the ideal gas law is not valid for our fluid. To meet this need, it would be ideal if we could express U or H in terms of other state variables such as P,T. In fact, we did this for the ideal gas in Eqns. 2.29 through 2.31. Unfortunately, such an expression is more difficult to derive for a real fluid. The required manipulations have been done for us when we look at a thermodynamic chart or table. These charts and tables are created by utilizing the P-V-T properties of the fluid, together with their derivatives to calculate the values for H, U, S which you see tabulated in the charts and tables. We explore the details of how this is done in Chapter 7 after discussing the equations of state used to represent the P-V-T properties of fluids in Chapter 6. The remainder of this chapter exploits primarily mathematical tools necessary for the manipulations of derivatives to express them in terms of measurable properties. By the term *measurable properties*, we mean:

1. P-V-T and partial derivatives involving only P-V-T;

2. C_P and C_V which are known functions of temperature at low pressure (in fact, C_P and C_V are special names for derivatives of entropy).

3. S is acceptable if it is not a derivative constraint or within a derivative term. S can be calculated once the state is specified.

Recall that the Gibbs' phase rule specifies for a pure single-phase fluid that any state variable is a function of any two other state variables. For convenience, we could write internal energy in terms

---

1. Note that the restrictions in derivation of fundamental relation for dU also apply for Eqns. 5.5–5.7.

of $\{P,T\}$, $\{V,T\}$ or any other combination. In fact, we have already seen that the internal energy is a natural function of $\{S,V\}$

$$dU = T\,dS - P\,dV$$

In real processes, this form is not the easiest to apply since $\{V,T\}$ and $\{P,T\}$ are more often manipulated than $\{S,V\}$. Therefore, what we seek is something of the form:

$$dU = f(P,\ V,\ T,\ C_P,\ C_V)\,dV + g(P,\ V,\ T,\ C_P,C_V)\,dT \qquad 5.8$$

The problem we face now is determining the functions $f(P,\ V,\ T,\ C_P,\ C_V)$ and $g(P,\ V,\ T,\ C_P,C_V)$. The only way to understand how to find the functions is to review multivariable calculus, then apply the results to the problem at hand. If you find that you need additional background to understand the steps applied here, try to understand whether you seek greater understanding of the mathematics or the thermodynamics. The mathematics generally involve variations of the chain rule. The thermodynamics pertain more to choices of preferred variables into which the final results should be transformed. Keep in mind that the development here is very mathematical, but the ultimate goal is to express U, H, A, G in terms of measurable properties.

First, let us recognize that we have a set of state variables $\{T, S, P, V, U, H, A, G\}$ that we desire to interrelate. Further, we know from the phase rule that specification of any two of these variables will completely specify all others for a pure, simple system (i.e., we have two degrees of freedom). *The relations developed in this section are applicable to pure simple systems, the relations are entirely mathematical, and proofs do not lie strictly within the confines of "thermodynamics."* The first four of the state properties in our set $\{T, S, P, V\}$ are the most useful subset experimentally, so this is the subset we frequently choose to use as the controlled variables. Therefore, if we know the changes of any two of these variables, we will be able to determine changes in any of the others, including U, H, A, G. Let's say we want to know how U changes with any two properties which we will denote symbolically as x and y. The way we express this mathematically is:

❗ The principles that we apply use multivariable calculus.

$$dU = (\partial U/\partial x)_y\,dx + (\partial U/\partial y)_x\,dy \qquad 5.9$$

where x and y are any two other variables from our set of properties. We also could write

$$dT = (\partial T/\partial x)_y\,dx + (\partial T/\partial y)_x\,dy \qquad 5.10$$

where x and y are any properties except T. The structure of the mathematics provides a method to determine how all of these properties are coupled. We could extend the analysis to all combinations of variables in our original set. As we will see in the remainder of the chapter, there are some combinations which are more useful than others. In the upcoming chapter on equations of state, some very specific combinations will be required.

Basic Identities

Frequently as we manipulate derivatives we obtain derivatives of the following forms which should be recognized.

$$\left(\frac{\partial x}{\partial y}\right)_z = \frac{1}{\left(\dfrac{\partial y}{\partial x}\right)_z} \qquad 5.11$$

!Basic Identities.

$$\left(\frac{\partial x}{\partial y}\right)_x = 0 \quad \text{and} \quad \left(\frac{\partial x}{\partial y}\right)_y = \infty \qquad\qquad 5.12$$

$$\left(\frac{\partial x}{\partial x}\right)_y = 1 \qquad\qquad 5.13$$

Triple Product Rule

Suppose $F = F(x,y)$, then

$$dF = (\partial F/\partial x)_y\, dx + (\partial F/\partial y)_x\, dy \qquad\qquad 5.14$$

Consider what happens when $dF = 0$ (i.e., at constant F). Then,

$$0 = \left(\frac{\partial F}{\partial x}\right)_y\left(\frac{\partial x}{\partial y}\right)_F + \left(\frac{\partial F}{\partial y}\right)_x \Rightarrow \left(\frac{\partial x}{\partial y}\right)_F = \frac{-\left(\frac{\partial F}{\partial y}\right)_x}{\left(\frac{\partial F}{\partial x}\right)_y} = \frac{-\left(\frac{\partial x}{\partial F}\right)_y}{\left(\frac{\partial y}{\partial F}\right)_x} \text{ or}$$

!Triple product rule.

$$\left(\frac{\partial x}{\partial y}\right)_F\left(\frac{\partial y}{\partial F}\right)_x\left(\frac{\partial F}{\partial x}\right)_y = -1 \qquad\qquad 5.15$$

Two Other Useful Relations

First, for any partial derivative involving three variables, say x, y, and F, we can interpose a fourth variable z using the *chain rule*:

!Chain rule interposing a variable.

$$\left(\frac{\partial x}{\partial y}\right)_F = \left(\frac{\partial x}{\partial z}\right)_F\left(\frac{\partial z}{\partial y}\right)_F \qquad\qquad 5.16$$

Another useful relation is found by a procedure known as the *expansion rule*. The details of this expansion are usually not covered in introductory calculus texts:

!The expansion rule.

$$\left(\frac{\partial F}{\partial w}\right)_z = \left(\frac{\partial F}{\partial x}\right)_y\left(\frac{\partial x}{\partial w}\right)_z + \left(\frac{\partial F}{\partial y}\right)_x\left(\frac{\partial y}{\partial w}\right)_z \qquad\qquad 5.17$$

Recall that we started with a function $F = F(x,y)$. If you look closely at the expansion rule, it provides a method to evaluate a partial derivative $\left(\frac{\partial F}{\partial w}\right)_z$ in terms of $\left(\frac{\partial x}{\partial w}\right)_z$ and $\left(\frac{\partial y}{\partial w}\right)_z$. Thus, we have transformed the calculation of a partial derivative of F to partial derivatives of x and y. This relation is particularly useful in manipulation of the fundamental relations S, U, H, A, G when one of these properties is substituted for F, and the natural variables are substituted for x and y. We will demonstrate this in Examples 5.1 and 5.2. Look again at Eqn. 5.17. It *looks* like we have taken the differential expression of Eqn. 5.14 and divided through the differential terms by dw and constrained to constant z, but this procedure violates the rules of differential operators. *What we have actually done is not nearly this simple.*[1] However, looking at the equation this way provides a fast way to remember a complicated-looking expression.

Exact Differentials

In this section, we apply calculus to the fundamental properties. Our objective is to derive relations known as Maxwell's relations. We begin by reminding you that we can express any state property in terms of any other two state properties. For a function which is only dependent on two variables, we can obtain the following differential relation, called in mathematics an *exact differential.*

$$U = U(S,V) \Rightarrow dU = (\partial U/\partial S)_V \, dS + (\partial U/\partial V)_S \, dV \qquad 5.18$$

Developing the ability to express any state variable in terms of any other two variables from the set $\{P, T, V, S\}$ as we have just done is very important. But the equation looks a little formidable. However, the fundamental property relationship says:

$$dU = TdS - PdV \qquad 5.19$$

Comparison of the above equations shows that:

$$T = (\partial U/\partial S)_V \quad \text{and} \quad -P = (\partial U/\partial V)_S \qquad 5.20$$

This means that these unusual derivatives are really properties that are familiar to us. Likewise, we can learn something about formidable-looking derivatives from enthalpy:

$$H = H(S,P) \Rightarrow dH = (\partial H/\partial S)_P dS + (\partial H/\partial P)_S \, dP \qquad 5.21$$

But the result of the fundamental property relationship is:

$$dH = TdS + VdP$$

Comparison shows that:

$$T = (\partial H/\partial S)_P \quad \text{and} \quad V = (\partial H/\partial P)_S$$

Now, we see that a definite pattern is emerging, and we could extend the analysis to Helmholtz and Gibbs energy. We can, in fact, derive relations between certain second derivatives of these relations. Since the properties U, H, A, G are state properties of only two other state variables, the differentials we have given in terms of two other state variables are known mathematically as exact differentials; we may apply properties of exact differentials to these properties. We show the features here; for details consult an introductory calculus textbook. Consider a general function of two variables: $F = F(x,y)$, and

$$dF = \left(\frac{\partial F}{\partial x}\right)_y dx + \left(\frac{\partial F}{\partial y}\right)_x dy$$

For an exact differential, differentiating with respect to x we can define some function M:

$$M \equiv (\partial F/\partial x)_y = M(x,y) \qquad 5.22$$

Similarly differentiating with respect to y:

$$N \equiv (\partial F/\partial y)_x = N(x,y) \qquad 5.23$$

Exact differentials.

1. Leithold, L., *The Calculus with Analytical Geometry*, 3rd ed., p. 929, Harper & Rowe, 1976.

Taking the second derivative and recalling from multivariable calculus that the order of differentiation should not matter:

$$\frac{\partial^2 F(x, y)}{\partial x \partial y} = \frac{\partial}{\partial x}\left(\left(\frac{\partial F(x, y)}{\partial y}\right)_x\right)_y = \frac{\partial}{\partial y}\left(\left(\frac{\partial F(x, y)}{\partial x}\right)_y\right)_x = \frac{\partial^2 F(x, y)}{\partial y \partial x} \qquad 5.24$$

$$\left(\frac{\partial N}{\partial x}\right)_y = \left(\frac{\partial M}{\partial y}\right)_x \qquad 5.25$$

● Euler's Law.

This simple observation is known as Euler's Law. To apply Euler's Law, recall the total differential of enthalpy considering $H = H(S,P)$:

$$dH = (\partial H/\partial S)_P\, dS + (\partial H/\partial P)_S\, dP = TdS + VdP \qquad 5.26$$

Considering second derivatives:

$$\frac{\partial^2 H}{\partial S \partial P} = \frac{\partial^2 H}{\partial P \partial S} = \left[\frac{\partial}{\partial S}\left(\frac{\partial H}{\partial P}\right)_S\right]_P = \left[\frac{\partial}{\partial P}\left(\frac{\partial H}{\partial S}\right)_P\right]_S \qquad 5.27$$

$$\Rightarrow \left(\frac{\partial V}{\partial S}\right)_P = \left(\frac{\partial T}{\partial P}\right)_S \qquad 5.28$$

A similar derivation applied to each of the other thermodynamic functions yields the equations known as Maxwell's Relations.

Maxwell's Relations

$$dU = TdS - PdV \Rightarrow -(\partial P/\partial S)_V = (\partial T/\partial V)_S \qquad 5.29$$

● Maxwell's
Relations.

$$dH = TdS + VdP \Rightarrow (\partial V/\partial S)_P = (\partial T/\partial P)_S \qquad 5.30$$

$$dA = -SdT - PdV \Rightarrow (\partial P/\partial T)_V = (\partial S/\partial V)_T \qquad 5.31$$

$$dG = -SdT + VdP \Rightarrow -(\partial V/\partial T)_P = (\partial S/\partial P)_T \qquad 5.32$$

Example 5.1 Pressure dependence of H

Derive the relation for $\left(\dfrac{\partial H}{\partial P}\right)_T$ and evaluate the derivative for water at 20°C where $\left(\dfrac{\partial V}{\partial T}\right)_P = 2.07\times10^{-4}$ cm³/g-K and $\left(\dfrac{\partial V}{\partial P}\right)_T = -4.9\times10^{-5}$ cm³/g-bar, $\rho = 0.998$ g/cm³.

Solution: $dH = TdS + VdP$. Applying the expansion rule,

$$\left(\frac{\partial H}{\partial P}\right)_T = T\left(\frac{\partial S}{\partial P}\right)_T + V\left(\frac{\partial P}{\partial P}\right)_T^{\ 1} = T\left(\frac{\partial S}{\partial P}\right)_T + V$$

Example 5.1 Pressure dependence of H (Continued)

by a Maxwell relation, the entropy derivative may be replaced

$$\left(\frac{\partial H}{\partial P}\right)_T = -T\left(\frac{\partial V}{\partial T}\right)_P + V$$

Plugging in values,

$$\left(\frac{\partial H}{\partial P}\right)_T = -293.15(2.07 \times 10^{-4}\,\text{cm}^3/\text{g-K}) + 1.002$$

$$= -0.061 + 1.002$$

Therefore, within 6% at room temperature, $\left(\frac{\partial H}{\partial P}\right)_T \approx V$ as used in Chapter 2.

Example 5.2 Entropy change with respect to T at constant P

Evaluate $(\partial S/\partial T)_P$ in terms of C_P, C_V, T, P, V, and their derivatives.

Solution: Recall, $dH = TdS + VdP$

Applying the expansion rule, Eqn. 5.17, we find,

$$\left(\frac{\partial H}{\partial T}\right)_P = T\left(\frac{\partial S}{\partial T}\right)_P + V\left(\frac{\partial P}{\partial T}\right)_P \qquad\qquad 5.33$$

Applying the basic identity of Eqn. 5.12 to the second term on the right-hand side,

$$\left(\frac{\partial H}{\partial T}\right)_P = T\left(\frac{\partial S}{\partial T}\right)_P$$

But the definition of the left-hand side is given by Eqn. 2.30: $C_P \equiv \left(\frac{\partial H}{\partial T}\right)_P$

Therefore, $(\partial S/\partial T)_P = C_P/T$, which we have seen before as Eqn. 3.25, and we have found that the constant-pressure heat capacity is related to the constant-pressure derivative of entropy with respect to temperature. An analogous analysis of U at constant V will result in a relation between the constant-volume heat capacity and the derivative of entropy with respect to temperature at constant V. That is, Eqn. 3.24

$$(\partial S/\partial T)_V = C_V/T$$

Example 5.3 Entropy as a function of T and P

Derive a general relation for entropy changes of any fluid with respect to temperature and pressure in terms of C_P, C_V, P, V, T and their derivatives.

Solution: First, since we choose T, P to be the controlled variables, applying Eqn. 5.14

$$dS = (\partial S/\partial T)_P \, dT + (\partial S/\partial P)_T \, dP \qquad 5.34$$

but $(\partial S/\partial T)_P = C_P/T$ as derived above, and Maxwell's Relations show that

$$(\partial S/\partial P)_T = -(\partial V/\partial T)_P.$$

$$\Rightarrow dS(T,P) = C_P \, dT/T - (\partial V/\partial T)_P \, dP \qquad 5.35$$

This is a useful expression. It is ready for application, given an equation of state which describes $V(T,P)$.

Note that expressions similar to Eqn. 5.35 can be derived for other thermodynamic variables in an analogous fashion. These represent powerful short-hand relations that can be used to solve many different process-related problems. In addition to Eqn. 5.35, some other useful expressions are:

$$dS(T,V) = C_V \, dT/T + (\partial P/\partial T)_V \, dV \qquad 5.36$$

$$dS(P,V) = C_V \, (\partial T/\partial P)_V \, dP/T + C_P \, (\partial T/\partial V)_P \, dV/T \qquad 5.37$$

$$dH(T,P) = C_P \, dT + [V - T(\partial V/\partial T)_P]dP \qquad 5.38$$

$$dU(T,V) = C_V \, dT + [T(\partial P/\partial T)_V - P]dV \qquad 5.39$$

> ❶ A summary of some useful relations.

Handwritten margin notes:
$A = U - TS$
$G = H - TS$
$H = U + PV$
$dU = TdS - PdV$
$= Q + W_{ec}$

The Importance of Derivative Manipulations

One may wonder, "What is so important about the variables C_P, C_V, P, V, T, and their derivatives?" The answer is that the equation of state is written in terms of these fundamental properties. In fact, it provides the link between P, V, and T. So knowing an equation for $P = P(V,T)$, we can solve for all the derivatives and add up all the changes. As for C_P and C_V, we should actually be very careful about specifying when we are referring to C_P of a real fluid or the C_P^{ig} of an ideal gas, but the distinction is frequently not made clear in literature. Occasionally, students forget that there is a difference, but clearly the heat capacity of liquid water is different from that of the ideal gas. Since this lack of indistinction is fairly universal, the best approach is simply to warn you about it. You need to recognize from the context of the source whether C_P is for the ideal gas or something else.

Two measurable derivatives are commonly used to discuss fluids properties, the isothermal compressibility and the isobaric coefficient of thermal expansion. The isothermal compressibility is

> ❶ Isothermal compressibility.

Handwritten margin note: $C_P = \left(\dfrac{\partial H}{\partial T}\right)_P$

$$\kappa_T = \frac{-1}{V}\left(\frac{\partial V}{\partial P}\right)_T = \frac{1}{\rho}\left(\frac{\partial \rho}{\partial P}\right)_T \qquad 5.40$$

The isobaric coefficient of thermal expansion is

> ❶ Isobaric coefficient of thermal expansion.

Handwritten margin note: $C_V = \left(\dfrac{\partial U}{\partial T}\right)_V$

$$\alpha_P = \frac{1}{V}\left(\frac{\partial V}{\partial T}\right)_P = \frac{-1}{\rho}\left(\frac{\partial \rho}{\partial T}\right)_P \qquad 5.41$$

Example 5.4 Entropy change for an ideal gas

As a sample application, revisit the dependence of entropy on T and P, further develop the entropy formula of Eqn. 5.35, and evaluate it for an ideal gas.

Solution: The derivative $(\partial V/\partial T)_P$ is required.

$$V = RT/P \;\Rightarrow\; (\partial V/\partial T)_P = R/P \tag{ig}$$

$$dS = \frac{C_P}{T}dT - \frac{R}{P}dP = C_P d\ln T - R d\ln P \tag{ig}$$

Assuming C_P is independent of T and integrating,

$$\Delta S = C_P \ln(T_2/T_1) - R\ln(P_2/P_1) \tag{*ig}$$

Once again, we arrive at Eqn. 3.23, but this time, it is easy to recognize the necessary changes for applications to real gases. That is, we must simply replace the P-V-T relation by a more realistic equation of state.

Example 5.5 Entropy change for a simple non-ideal gas

Derive the pressure and temperature dependence of entropy for a fluid that follows the simple equation of state $V = RT/P + (a + bT)$, where a and b are constants.

Solution:

$$(\partial V/\partial T)_P = R/P + b$$

Inserting into Eqn. 5.35,

$$dS = \frac{C_P}{T}dT - \frac{R}{P}dP - b\,dP = C_P d\ln T - R d\ln P - b\,dP$$

Assuming C_P is independent of T and integrating,

$$\Delta S = C_P \ln(T_2/T_1) - R\ln(P_2/P_1) - b\Delta P \tag{*}$$

Hints on Manipulating Partial Derivatives

1. Learn to recognize $\left(\dfrac{\partial S}{\partial T}\right)_V$ and $\left(\dfrac{\partial S}{\partial T}\right)_P$ as being related to C_P and C_V.

   Useful hints on manipulating derivatives.

2. If a derivative involves entropy, enthalpy, or Helmholtz or Gibbs energy being held constant, e.g., $\left(\dfrac{\partial T}{\partial P}\right)_H$, bring it inside the parenthesis using the triple product relation (Eqn. 5.15). Then apply the expansion rule (Eqn. 5.17) to eliminate immeasurable quantities. The expansion rule is very useful when F is U, H, A or G.

3. When a derivative involves $\{T, S, P, V\}$ only, look to apply a Maxwell relation.

4. When nothing else seems to work, apply the Jacobian method.[1] The Jacobian method will always result in derivatives with the desired independent variables.

---

1. This method is covered in optional Section 5.3.

Example 5.6 Application of the triple product relation

Evaluate $(\partial S/\partial V)_A$ in terms of C_P, C_V, T, P, and V. Your answer may include absolute values of S if it is not a derivative constraint or within a derivative term.

Solution: This problem illustrates a typical situation where the triple product rule is helpful because the Helmholtz energy is held constant (hint #2). It is easiest to express changes in the Helmholtz energies as changes in other variables.

Applying the triple product rule:

$$(\partial S/\partial V)_A = -(\partial A/\partial V)_S / (\partial A/\partial S)_V$$

Applying the expansion rule twice, $dA = -PdV - SdT \Rightarrow (\partial A/\partial V)_S = -P - S(\partial T/\partial V)_S$ and $(\partial A/\partial S)_V = 0 - S(\partial T/\partial S)_V$. Recalling Eqn. 3.24 and converting to measurable derivatives

$$\Rightarrow \left(\frac{\partial T}{\partial S}\right)_V = \frac{T}{C_V} \text{ and } (\partial T/\partial V)_S = -\left(\frac{\partial T}{\partial S}\right)_V\left(\frac{\partial S}{\partial V}\right)_T = -\frac{T}{C_V}\left(\frac{\partial P}{\partial T}\right)_V$$

Substituting

$$\Rightarrow \left(\frac{\partial S}{\partial V}\right)_A = \frac{-PC_V}{ST} + \left(\frac{\partial P}{\partial T}\right)_V$$

Example 5.7 $\left(\frac{\partial U}{\partial V}\right)_T$ for an ideal gas

Derive an expression for $\left(\frac{\partial U}{\partial V}\right)_T$ in terms of measurable properties. Evaluate for the ideal gas.

Solution: Beginning with the fundamental relation for dU,

$$dU = TdS - PdV$$

Applying the expansion rule

$$\left(\frac{\partial U}{\partial V}\right)_T = T\left(\frac{\partial S}{\partial V}\right)_T - P\left(\frac{\partial V}{\partial V}\right)_T \qquad 5.42$$

Using a Maxwell relation and a basic identity

$$\left(\frac{\partial U}{\partial V}\right)_T = T\left(\frac{\partial P}{\partial T}\right)_V - P \qquad 5.43$$

For an ideal gas, $P = RT/V$

$$\left(\frac{\partial P}{\partial T}\right)_V = \frac{R}{V}; \quad \left(\frac{\partial U}{\partial V}\right)_T^{ig} = \frac{RT}{V} - P = 0 \qquad \text{(ig)}$$

Thus, internal energy of an ideal gas does not depend on volume (or pressure) at a given T. (We will reevaluate this derivative in Chapter 6 for a real fluid.)

Example 5.8 Volumetric dependence of C_V for ideal gas

Determine how C_V depends on volume (or pressure) by deriving an expression for $\left(\dfrac{\partial C_V}{\partial V}\right)_T$.
Evaluate the expression for an ideal gas.

Solution: From Eqn. 3.24

$$C_V = T\left(\frac{\partial S}{\partial T}\right)_V$$

By the chain rule:

$$\left(\frac{\partial C_V}{\partial V}\right)_T = \left(\frac{\partial S}{\partial T}\right)_V \left(\frac{\partial T}{\partial V}\right)_T + T\frac{\partial}{\partial V}\left[\left(\frac{\partial S}{\partial T}\right)_V\right]_T$$

Changing the order of differentiation:

$$\left(\frac{\partial C_V}{\partial V}\right)_T = T\frac{\partial}{\partial T}\left[\left(\frac{\partial S}{\partial V}\right)_T\right]_V = T\left(\frac{\partial^2 P}{\partial T^2}\right)_V$$

For an ideal gas, $P = RT/V$, we have $\left(\dfrac{\partial P}{\partial T}\right)_V$ in Example 5.7,

$$\frac{\partial}{\partial T}\left[\left(\frac{\partial P}{\partial T}\right)_V\right]_V = \frac{\partial}{\partial T}\left(\frac{R}{V}\right)_V = 0 \qquad \text{(ig) 5.44}$$

Thus, heat capacity of an *ideal* gas does not depend on volume (or pressure) at a fixed temperature. (We will reevaluate this derivative in Chapter 6 for a real fluid.)

Example 5.9 Master equation for an ideal gas

Derive a master equation for calculating changes in U for an ideal gas in terms of $\{V, T\}$.
Solution:

$$dU = \left(\frac{\partial U}{\partial V}\right)_T dV + \left(\frac{\partial U}{\partial T}\right)_V dT$$

Applying results of the previous examples

$$dU = C_V dT + \left[T\left(\frac{\partial P}{\partial T}\right)_V - P\right]dV \qquad 5.45$$

Notice that this expression is more complicated than the fundamental property relation in terms of $\{S, V\}$. As we noted earlier, this is why $\{S, V\}$ are the natural variables for dU, rather than $\{T,V\}$ or any other combination. For an ideal gas, we can use the results of Example 5.7 to find:

$$dU^{ig} = C_V dT \qquad \text{(ig) 5.46}$$

Example 5.10 Relating C_P to C_V

Derive a general formula to relate C_P and C_V.

Solution: Start with an expression that already contains one of the desired derivatives (e.g., C_V) and introduce the variables necessary to create the second derivative (e.g., C_P). Beginning with Eqn. 5.36,

$$dS = \frac{C_V}{T}dT + \left(\frac{\partial P}{\partial T}\right)_V dV$$

and using the expansion rule with T at constant P,

$$\left(\frac{\partial S}{\partial T}\right)_P = \frac{C_V}{T}\left(\frac{\partial T}{\partial T}\right)_P^{1} + \left(\frac{\partial P}{\partial T}\right)_V\left(\frac{\partial V}{\partial T}\right)_P, \text{ where the left-hand side is } \frac{C_P}{T}.$$

$$C_P = C_V + T\left(\frac{\partial P}{\partial T}\right)_V\left(\frac{\partial V}{\partial T}\right)_P$$

Exercise: Verify that the last term simplifies to R for an ideal gas.

5.3 ADVANCED TOPICS (OPTIONAL)

Hints for Remembering the Auxiliary Relations

Auxiliary relations can be easily written by memorizing the fundamental relation for dU and the natural variables for the other properties. Note that $\{T,S\}$ and $\{P,V\}$ always appear in pairs, and each pair is a set of conjugate variables. A Legendre transformation performed on internal energy among conjugate variables changes the dependent variable and the sign of the term involving the conjugate variables. For example, to transform P and V, the product PV is added to U, resulting in Eqn. 5.5. To transform T and S, the product TS is subtracted. $A = U - TS$, $dA = dU - TdS - SdT = -SdT - PdV$. The pattern can be easily seen in the following table. Note that $\{T,S\}$ always appear together, and $\{P,V\}$ always appear together, and the sign changes upon transformation.

Table 5.1 *Fundamental and Auxiliary Property Relations*

	Natural Variables	**Legendre Transformation**	**Transformed variable sets**
$dU = T\,dS - P\,dV$	$U(S,V)$		---------
$dH = T\,dS + V\,dP$	$H(S,P)$	$H = U + PV$	$\{V,P\}$
$dA = -S\,dT - P\,dV$	$A(T,V)$	$A = U - TS$	$\{S,T\}$
$dG = -S\,dT + V\,dP$	$G(T,P)$	$G = U - TS + PV$	$\{S,T\}, \{V,P\}$

Jacobian Method of Derivative Manipulation

A partial derivative may be converted to derivatives of measurable properties with any two desired independent variables from the set $\{P, V, T\}$. Jacobian notation can be used to manipulate partial derivatives, and there are several useful rules for manipulating derivatives with the notation. The Joule-Thomson coefficient is a derivative that indicates how temperature changes upon pressure change at fixed enthalpy, $\left(\dfrac{\partial T}{\partial P}\right)_H$, which is written in Jacobian notation as $\left(\dfrac{\partial T}{\partial P}\right)_H = \dfrac{\partial(T, H)}{\partial(P, H)}$. Note how the constraint of constant enthalpy is incorporated into the notation. The rules for manipulation of the Jacobian notation are:

1. Jacobian notation represents a determinant of partial derivatives,

$$\frac{\partial(K, L)}{\partial(X, Y)} = \left(\frac{\partial K}{\partial X}\right)_Y \left(\frac{\partial L}{\partial Y}\right)_X - \left(\frac{\partial K}{\partial Y}\right)_X \left(\frac{\partial L}{\partial X}\right)_Y = \begin{vmatrix} \left(\dfrac{\partial K}{\partial X}\right)_Y & \left(\dfrac{\partial K}{\partial Y}\right)_X \\[2ex] \left(\dfrac{\partial L}{\partial X}\right)_Y & \left(\dfrac{\partial L}{\partial Y}\right)_X \end{vmatrix} \qquad 5.47$$

The Jacobian is particularly simple when the numerator and denominator have a common variable,

$$\frac{\partial(K, L)}{\partial(X, L)} = \left(\frac{\partial K}{\partial X}\right)_L \qquad 5.48$$

which is a special case of Eqn. 5.47.

2. When the order of variables in the numerator or denominator are switched, the sign of the Jacobian changes. Switching the order of variables in both the numerator and denominator results in no sign change due to cancellation. Consider switching the order of variables in the numerator,

$$\frac{\partial(K, L)}{\partial(X, Y)} = -\frac{\partial(L, K)}{\partial(X, Y)} \qquad 5.49$$

3. The Jacobian may be inverted.

$$\frac{\partial(K, L)}{\partial(X, Y)} = \left[\frac{\partial(X, Y)}{\partial(K, L)}\right]^{-1} = \frac{1}{\dfrac{\partial(X, Y)}{\partial(K, L)}} \qquad 5.50$$

4. Additional variables may be interposed. When additional variables are interposed, it is usually convenient to invert one of the Jacobians.

$$\frac{\partial(K, L)}{\partial(X, Y)} = \frac{\partial(K, L)}{\partial(B, C)}\frac{\partial(B, C)}{\partial(X, Y)} = \frac{\dfrac{\partial(K, L)}{\partial(B, C)}}{\dfrac{\partial(X, Y)}{\partial(B, C)}} \qquad 5.51$$

Manipulation of Derivatives

Before manipulating derivatives, the desired independent variables are selected. The selected independent variables will be held constant outside the derivatives in the final formula. The general procedure is to interpose the desired independent variables, rearrange as much as possible to obtain Jacobians with common variables in the numerator and denominator, write the determinant for any Jacobians without common variables, then use Maxwell relations, the expansion rule, etc., to simplify the answer.

1. If the starting derivative already contains both the desired independent variables, the result of Jacobian manipulation is redundant with the triple product rule. The steps are: 1) write the Jacobian; 2) interpose the independent variables; 3) rearrange to convert to partial derivatives.

Example: Convert $\left(\dfrac{\partial T}{\partial P}\right)_H$ to derivatives that use T and P as independent variables.

$$\left(\frac{\partial T}{\partial P}\right)_H = \frac{\partial(T, H)}{\partial(P, H)} = \frac{\partial(T, H)}{\partial(T, P)}\frac{\partial(T, P)}{\partial(P, H)} = \frac{\dfrac{\partial(H, T)}{\partial(P, T)}}{-\dfrac{\partial(H, P)}{\partial(T, P)}} = \frac{-\left(\dfrac{\partial H}{\partial P}\right)_T}{\left(\dfrac{\partial H}{\partial T}\right)_P} = \frac{-\left(\dfrac{\partial H}{\partial P}\right)_T}{C_P}$$

and the numerator can be simplified using the expansion rule as presented in Example 5.1.

2. If the starting derivative has just one of the desired independent variables, the steps are: 1) write the Jacobian; 2) interpose the desired variables; 3) write the determinant for the Jacobian without a common variable; 4) rearrange to convert to partial derivatives.

Example: Find a relation for the adiabatic compressibility, $\kappa_S = -\dfrac{1}{V}\left(\dfrac{\partial V}{\partial P}\right)_S$ in terms of derivatives using T, P as independent variables.

$$\left(\frac{\partial V}{\partial P}\right)_S = \frac{\partial(V, S)}{\partial(P, S)} = \frac{\partial(V, S)}{\partial(P, T)}\frac{\partial(P, T)}{\partial(P, S)} = \left[\left(\frac{\partial V}{\partial P}\right)_T\left(\frac{\partial S}{\partial T}\right)_P - \left(\frac{\partial V}{\partial T}\right)_P\left(\frac{\partial S}{\partial P}\right)_T\right]\left(\frac{\partial T}{\partial S}\right)_P$$

Now, including a Maxwell relation as we simplify the second term in square brackets, and then combining terms:

$$\left(\frac{\partial V}{\partial P}\right)_S = \left(\frac{\partial V}{\partial P}\right)_T + \frac{T}{C_P}\left(\frac{\partial V}{\partial T}\right)_P^2$$

$$\kappa_S = -\frac{1}{V}\left(\left(\frac{\partial V}{\partial P}\right)_T + \frac{T}{C_P}\left(\frac{\partial V}{\partial T}\right)_P^2\right) = \kappa_T - \frac{T}{VC_P}\left(\frac{\partial V}{\partial T}\right)_P^2$$

3. If the starting derivative has neither of the desired independent variables, the steps are: 1) write the Jacobian; 2) interpose the desired variables; 3) write the Jacobians as a quotient and write the determinants for both Jacobians; 4) rearrange to convert to partial derivatives.

Example: Find $\left(\dfrac{\partial S}{\partial V}\right)_U$ in measurable properties using P and T as independent variables.

$$\frac{\partial(S, U)}{\partial(V, U)} = \frac{\partial(S, U)}{\partial(P, T)}\frac{\partial(P, T)}{\partial(V, U)} = \frac{\dfrac{\partial(S, U)}{\partial(P, T)}}{\dfrac{\partial(V, U)}{\partial(P, T)}}$$

Writing the determinants for both Jacobians:

$$\frac{\left(\frac{\partial S}{\partial P}\right)_T\left(\frac{\partial U}{\partial T}\right)_P - \left(\frac{\partial U}{\partial P}\right)_T\left(\frac{\partial S}{\partial T}\right)_P}{\left(\frac{\partial V}{\partial P}\right)_T\left(\frac{\partial U}{\partial T}\right)_P - \left(\frac{\partial U}{\partial P}\right)_T\left(\frac{\partial V}{\partial T}\right)_P}$$

Now, using the expansion rule for the derivatives of U, and also introducing Maxwell relations,

$$\frac{-\left(\frac{\partial V}{\partial T}\right)_P\left[C_P - P\left(\frac{\partial V}{\partial T}\right)_P\right] + \frac{C_P}{T}\left[T\left(\frac{\partial V}{\partial T}\right)_P + P\left(\frac{\partial V}{\partial P}\right)_T\right]}{\left(\frac{\partial V}{\partial P}\right)_T\left[C_P - P\left(\frac{\partial V}{\partial T}\right)_P\right] + T\left(\frac{\partial V}{\partial T}\right)_P^2 + P\left(\frac{\partial V}{\partial P}\right)_T\left(\frac{\partial V}{\partial T}\right)_P} =$$

$$\frac{P\left(\frac{\partial V}{\partial T}\right)_P^2 + C_P\frac{P}{T}\left(\frac{\partial V}{\partial P}\right)_T}{T\left(\frac{\partial V}{\partial T}\right)_P^2 + C_P\left(\frac{\partial V}{\partial P}\right)_T} = \frac{P}{T}$$

The result is particularly simple. We could have derived this directly if we had recognized that S and V are the natural variables for U. Therefore $dU = TdS - P\,dV = 0$, $TdS\big|_U = -PdV\big|_U$, $T\left(\frac{\partial S}{\partial V}\right)_U = P$, $\left(\frac{\partial S}{\partial V}\right)_U = \frac{P}{T}$. However, the exercise demonstrates the procedure and power of the Jacobian technique even though the result will usually not simplify to the extent of this example.

5.4 SUMMARY

We have seen in this chapter that calculus provides powerful tools to permit us to calculate changes in immeasurable properties in terms of other measurable properties. The ability to perform these manipulations easily is one sign that you have a mastery of the calculus necessary for understanding how we exploit these tools in the development of general methods for calculating thermodynamic properties from P-V-T data.

Due to all the interrelations between all the derivatives, there is usually more than one way to derive a useful result. This can be frustrating to the novice; however, patience in attacking the problems, and attacking a problem from different angles help you visualize the structure of the calculus in your mind. Patience in developing these tools is rewarded with a mastery of the relations that permit quick insight into the easiest way to solve problems.

Test Yourself

1. What are the restrictions necessary to calculate one state property in terms of only two other state variables?

2. When integrating Eqn. 5.46, under what circumstances may C_V be taken out of the integral?

3. May Eqn. 5.46 be applied to a condensed phase?

4. Is the heat capacity different for liquid acetone than for acetone vapor?

5. Can the tabulated heat capacities be used in Eqn. 5.46 for gases at high pressure?

5.5   HOMEWORK PROBLEMS

5.1 Express in terms of P, V, T, C_P, C_V, and their derivatives. Your answer may include absolute values of S if it is not a derivative constraint or within a derivative.

(a) $(\partial G/\partial P)_T$
(b) $(\partial P/\partial A)_V$
(c) $(\partial T/\partial P)_S$
(d) $(\partial H/\partial T)_U$
(e) $(\partial T/\partial H)_S$
(f) $(\partial A/\partial V)_P$
(g) $(\partial T/\partial P)_H$
(h) $(\partial A/\partial S)_P$
(i) $(\partial S/\partial P)_G$

5.2 (a) Derive $\left(\dfrac{\partial H}{\partial P}\right)_T$ and $\left(\dfrac{\partial U}{\partial P}\right)_T$ in terms of measurable properties.

(b) $dH = dU + d(PV)$ from the definition of H. Apply the expansion rule to show the difference between $\left(\dfrac{\partial H}{\partial P}\right)_T$ and $\left(\dfrac{\partial U}{\partial P}\right)_T$ is the same as the result from part (a).

5.3 Express $\left(\dfrac{\partial H}{\partial V}\right)_T$ in terms of α_P and/or κ_T.

5.4 Express the adiabatic compressibility, $\kappa_S = -\dfrac{1}{V}\left(\dfrac{\partial V}{\partial P}\right)_S$, in terms of measurable properties.

5.5 (a) Prove $\left(\dfrac{\partial P}{\partial T}\right)_S = \dfrac{C_P}{TV\alpha_P}$.

(b) For an ideal gas along an adiabat, $(P/P^i) = (T/T^i)^{C_P/R}$. Demonstrate that this equation is consistent with the expression from part (a).

5.6 Determine the difference $C_P - C_V$ for the following liquids using the data provided near 20°C.

Liquid	MW	ρ (g/cm³)	$10^3 \alpha_P$ (K⁻¹)	$10^6 \kappa_T$ (bar⁻¹)
(a) Acetone	58.08	0.7899	1.487	111
(b) Ethanol	46.07	0.7893	1.12	100
(c) Benzene	78.12	0.87865	1.237	89
(d) Carbon disulfide	76.14	1.258	1.218	86
(e) Chloroform	119.38	1.4832	1.273	83
(f) Ethyl ether	74.12	0.7138	1.656	188
(g) Mercury	200.6	13.5939	0.18186	3.95
(h) Water	18.02	0.998	0.207	49

5.7 A rigid container is filled with liquid acetone at 20°C and 1 bar. Through heat transfer at constant volume, a pressure of 100 bar is generated. $C_P = 125$ J/mol·K. (Other properties of acetone are given in problem 5.6.) Provide your best estimate of:

(a) the temperature rise;

(b) ΔS, ΔU, ΔH;

(c) the heat transferred per mole.

5.8 In Chapter 2, internal energy of condensed phases was stated to be more weakly dependent on pressure than enthalpy. This problem evaluates that statement.

(a) Derive $\left(\frac{\partial H}{\partial P}\right)_T$ and $\left(\frac{\partial U}{\partial P}\right)_T$ in terms of measurable properties.

(b) Evaluate $\left(\frac{\partial H}{\partial P}\right)_T$ and compare the magnitude of the terms contributing to $\left(\frac{\partial H}{\partial P}\right)_T$ for the fluids listed in problem 5.6.

(c) Evaluate $\left(\frac{\partial U}{\partial P}\right)_T$ for the fluids listed in problem 5.6 and compare with the values of $\left(\frac{\partial H}{\partial P}\right)_T$.

5.9 The fundamental internal energy relation for a rubber band is $dU = TdS - FdL$ where F is the system force, which is negative when the rubber band is in tension. The applied force is given by $F_{applied} = k(T)(L - L_o)$ where $k(T)$ is positive and increases with increasing temperature. The heat capacity at constant length is given by $C_L = \alpha(L) + \beta(L) \cdot T$. Stability arguments show that $\alpha(L)$ and $\beta(L)$ must provide for $C_L \geq 0$.

(a) Show that temperature should increase when the rubber band is stretched adiabatically and reversibly.

(b) Prof. Lira in his quest for scientific facts hung a weight on a rubber band and measured the length in the laboratory at room temperature. When he hung the rubber band with the same weight in the refrigerator, he noticed that the length of the rubber band had changed. Did the length increase or decrease?

(c) The heat capacity at constant force is given by

$$C_F = T\left(\frac{\partial S}{\partial T}\right)_F$$

Derive a relation for $C_F - C_L$ and show whether this difference is positive, negative, or zero.

(d) The same amount of heat flows into two rubber bands, but one is held at constant tension and the other at constant length. Which has the largest increase in temperature?

(e) Show that the dependence of $k(T)$ on temperature at constant length is related to the dependence of entropy on length at constant temperature. Offer a physical description for the signs of the derivatives.

5.10 Express the follwing it terms of $U, H, S, G,$ and their derivatives.

$$\left(\frac{\partial(A/RT)}{\partial T}\right)_V$$

CHAPTER

6

ENGINEERING EQUATIONS OF STATE FOR *PVT* PROPERTIES

I am more than ever an admirer of van der Waals.

Lord Rayleigh (1891)

From Chapter 5, it is obvious that we can calculate changes in U, S, H, A, and G by knowing changes in any two variables from the set $\{P\text{-}V\text{-}T\}$ plus C_P or C_V. This chapter introduces the various ways available for quantitative prediction of the $P\text{-}V\text{-}T$ properties we desire in a general case. The method of calculation of thermodynamic properties like U, H, etc. requires the use of *departure functions* which will be the topic of the next chapter. The development of the departure functions is a relatively straightforward application of Maxwell's relations. What is less straightforward is the logical development of a connection between P, V, and T. We introduced the concept in Chapter 1 that the pressure, temperature, and density (i.e., V^{-1}) are connected through intermolecular interactions. We must now apply that concept to derive quantitative relationships that are applicable to any fluid at any conditions, not simply to ideal gases. You will see that making the connection between P, V, and T hinges on the transition from the molecular-scale pressure and energy to the macroscopic pressure and energy. This transition is most obvious for the energy equation, but the pressure equation is clearly analogous.

Making the transition from the molecular scale to the macroscopic represents the final piece in our conceptual puzzle. The overall puzzle has to do with energy, entropy, and equilibrium, as we have reiterated several times. We drew the connection between the microscopic and macroscopic scales for entropy during our introduction to entropy. For energy, however, we have left a gap that you may not have noticed. We discussed the molecular energy in Chapter 1, but we did not discuss the macroscopic implications. We discussed the macroscopic implications of energy in Chapter 2, but we did not discuss the molecular basis. It is time to fill that gap, and in doing so, to finalize the conceptual framework of the entire course.

From one perspective, the purpose of the material preceding this chapter was to explain the need for making the transition from the molecular scale to the macroscopic scale. The purpose of the material following this chapter is to demonstrate the reduction to practice of this conceptual framework in several different contexts. So in many ways, this chapter represents the conceptual kernel for all thermodynamic modeling.

6.1   EXPERIMENTAL MEASUREMENTS

The preferred method of obtaining *P-V-T* properties is from experimental measurements of the desired fluid or fluid mixture. We spend most of the text discussing theories, but you should never forget the precious value of experimental data. Experimental measurements beat theories every time. The problem with experimental measurements is that they are expensive, relative to pushing a few buttons on a computer.

> ❶ The basic procedure for calculating properties involves using derivatives of *P,V,T* data.

To illustrate the difficulty of measuring all properties experimentally, consider the following case. One method to determine the *P-V-T* properties is to control the temperature of a container of fluid, change the volume of the container in carefully controlled increments, and carefully measure the pressure. The required derivatives are then calculated by numerical differentiation of the data obtained in this manner. It is also possible to make separate measurements of the enthalpy by carefully adding measured quantities of heat and determining changes in *P*, *V*, and *T*. These measurements can be cross-referenced for consistency with the estimated changes as determined by applying Maxwell's relations to the *P-V-T* measurements. Imagine what a daunting task this approach would present when considering all fluids and mixtures of interest. It should be understandable that detailed measurements of this type have been made for relatively few compounds. Water is the most completely studied fluid, and the steam tables are a result of this study. Ammonia, carbon dioxide, refrigerants, and light hydrocarbons have also been quite thoroughly studied. The charts which have been used in earlier chapters are results of these careful measurements. Equations of state permit correlation and extrapolation of experimental data that can be much more convenient and more broadly applicable than the available charts.

An experimental approach is naturally impractical for all substances due to the large number of fluids needing to be characterized. The development of equations of state is the engineering approach to describing fluid behavior for prediction, interpolation, and extrapolation of data using the fewest number of adjustable parameters possible for the desired accuracy. Typically, when data are analyzed today, they are fitted with elaborate equations (typically complicated embellishments of the equations of state discussed in this chapter) before determination of interpolated values or derivatives. The charts are then generated from the fitted results of the equation of state.

As a summary of the experimental approach to equations of state, some review of the historical development of *P-V-T* measurements may be beneficial. First, it should be recalled that early measurements of *P-V-T* relations laid the foundation for modern physical chemistry. Knowing the densities of gases in bell jars led to the early characterizations of molecular weights, molecular formulas, and even the primary evidence for the existence of molecules themselves. At first, it seemed that gases like nitrogen, hydrogen, and oxygen were non-condensable and something quite different from liquids like water or wood alcohol (methanol). As technology advanced, however, experiments were performed at higher temperatures and pressures. Carbon dioxide was a very common compound in the early days (known as "carbonic acid" to van der Waals), and it soon became apparent that it showed a high degree of compressibility. Experimental data were carefully measured in 1871 for carbon dioxide ranging to 110 bars, and these data were referenced extensively by van der Waals. Carbon dioxide is especially interesting because it has some very "peculiar" properties that are exhibited near room temperature and at high pressure. At 31°C and about 70 bars, a very small change in pressure can convert the fluid from a gas-like density to a liquid density. Van der Waals showed that the cause of this behavior is the balance between the attractive forces from the intermolecular potential being accentuated at this density range and the repulsive forces being accentuated by the high velocity collisions at this temperature. This "peculiar" range of conditions is known as the critical region. The precise temperature, pressure, and density where the vapor and the liquid become indistinguishable is called the critical point. Above the critical point, there is no longer an abrupt change in the density with respect to pressure while holding temperature constant. Instead, the balance between forces leads to a single-phase region spanning vapor-like densities and liquid-like densities. With the work of van der Waals, researchers began to recognize that the behavior was not "peculiar," and that all substances have critical points.[1]

> ❗ Fortunately, *P,V,T* behavior of fluids follows the same trends for all fluids. All fluids have a critical point.

6.2 THREE-PARAMETER CORRESPONDING STATES

If we plot the P versus ρ behavior of several different fluids, we find some remarkably similar trends. As shown in Fig. 6.1 below, both methane and pentane show the saturated vapor density approaching the saturated liquid density as the temperature increases. Compare these figures to Fig. 1.4 on page 20, and note that the P versus ρ figure is qualitatively a mirror image of the P versus V figure. The isotherms are shown in terms of the *reduced temperature*, $T_r \equiv T/T_c$. Saturation densities are the values obtained by intersection of the phase envelope with horizontal lines drawn at the saturation pressures. The isothermal compressibility $\equiv -\dfrac{1}{V}\left(\dfrac{\partial V}{\partial P}\right)_T = \dfrac{1}{\rho}\left(\dfrac{\partial \rho}{\partial P}\right)_T$ is infinite, and its reciprocal is zero, at the critical point (e.g., 191 K and 4.6 MPa for methane). It is also worth noting that the critical temperature isotherm exhibits an inflection point at the critical point. This means that $(\partial^2 P/\partial \rho^2)_T = 0$

> ❗ The isothermal compressibility is infinite at the critical point.

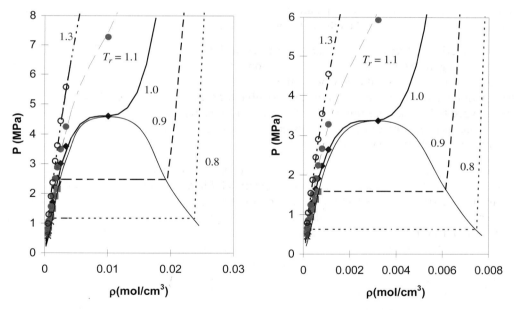

Figure 6.1 *Comparison of the PρT behavior of methane (left) and pentane (right) demonstrating the qualitative similarity which led to corresponding states treatment of fluids. The lines are calculated with the Peng-Robinson equation to be discussed later. The phase envelope is an approximation sketched through the points available in the plots. The smoothed experimental data are from Brown, G.G., Sounders Jr., M., Smith, R.L., Ind. Eng. Chem., 24, 513 (1932). Although not shown, the Peng-Robinson equation is not particularly accurate for modeling liquid densities.*

---

1. Naturally, some compounds decompose before their critical point is reached, or like carbon or tungsten they have such a high melting temperature that such a measurement is impossible even at the present time.

at the critical point as well as $(\partial P/\partial\rho)_T = 0$. The principle of corresponding states asserts that all fluid properties are similar if expressed properly in reduced variables.

Although the behaviors in Fig. 6.1 are globally similar, when researchers superposed the P-V-T behaviors based on only T_c and P_c, they found the superposition was not sufficiently accurate. For example, one way of comparing the behavior of fluids is to plot the *compressibility factor Z*. The compressibility factor is defined as

! The compress-
ibility factor.

$$Z \equiv \frac{PV}{RT}$$

6.1

> *Note: The compressibility factor **is not the same as** the isothermal compressibility. The similarity in names can frequently result in confusion as you first learn the concepts.*

The compressibility factor has a value of one when a fluid behaves as an ideal gas, but will be non-unity when the pressure increases. By plotting the data and calculations from Fig 6.1 as a function of *reduced temperature* $T_r = \dfrac{T}{T_c}$, and *reduced pressure*, $P_r = \dfrac{P}{P_c}$, the plot of Fig 6.2 results.

Clearly, another parameter is needed to accurately correlate the data. Note that the vapor pressure for

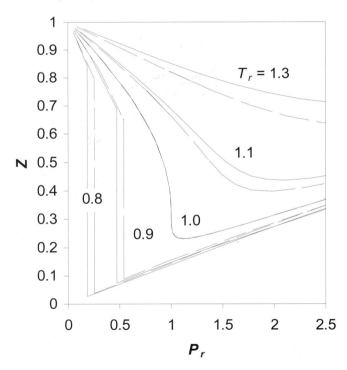

Figure 6.2 *The Peng-Robinson lines from Fig. 6.1 plotted in terms of the reduced pressure at $T_r = 0.8$, 0.9, 1.0, 1.1 and 1.3, demonstrating that critical temperature and pressure alone are insufficient to accurately represent the P-V-T behavior. Dashed lines are for methane, solid lines for pentane. The figure is intended to make an illustrative point. Accurate calculations should use the compressibility factor charts developed in the next section.*

methane and pentane differ on the compressibility factor chart as indicated by the vertical lines on the subcritical isotherms. The same behavior is followed by other fluids. For example, the vapor pressures for six compounds are shown in Fig 6.3, and although they are all nearly linear, the slopes are different. In fact, we may characterize this slope with a third parameter, known as the acentric factor, ω. The *acentric factor* is a parameter which helps to specify the vapor pressure curve which, in turn, correlates the rest of the thermodynamic variables.[1]

$$\omega \equiv -1 - \log_{10}\left(\frac{P^{sat}}{P_c}\right)\Bigg|_{T_r = 0.7} \equiv \text{acentric factor} \qquad 6.2$$

> Critical temperature and pressure are insufficient characteristic parameters by themselves. The acentric factor serves as a third important parameter.

Note: The specification of T_c, P_c, and ω provides two points on the vapor pressure curve. T_c and P_c specify the terminal point of the vapor pressure curve. ω specifies a vapor pressure at a reduced temperature of 0.7. The acentric factor was first introduced by Pitzer et al.[2] Its definition is arbitrary in that, for example, another reduced temperature could have been chosen for the definition. The definition above gives values of $\omega \sim 0$ for spherical molecules like argon, xenon, neon, krypton, and methane. Deviations from zero *usually* derive from deviations in spherical symmetry. Nonspherical molecules are "not centrally symmetric," so they are "acentric." In general, there is no

> The acentric factor is a measure of the slope of the vapor pressure curve plotted as ln P^{sat} vs. 1/T.

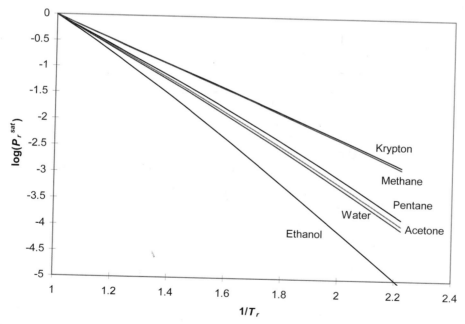

Figure 6.3 *Reduced vapor pressures plotted as a function of reduced temperature for six fluids demonstrating that the shape of the curve is not highly dependent on structure, but that the primary difference is the slope as given by the acentric factor.*

---

1. The significance of the vapor pressure curve in determining the thermodynamic properties can be readily appreciated if you consider the difference between a vapor enthalpy and a liquid enthalpy. The detailed consideration of vapor pressure behavior is treated in Chapter 8.
2. K.S. Pitzer, D.Z. Lippmann, R.F. Curl, Jr., C.M. Huggins, and D.E. Petersen, *J. Am. Chem. Soc.*, 77:3427,3433 (1955).

direct theoretical connection between the acentric factor and the shape of the intermolecular potential. Rather, the acentric factor provides a convenient experimental vapor pressure which can be correlated with the shape of the intermolecular potential in an *ad hoc* manner. It is convenient in the sense that its value has been experimentally determined for a large number of compounds and that knowing its value permits a significant improvement in the accuracy of our engineering equations of state.

6.3   GENERALIZED COMPRESSIBILITY FACTOR CHARTS

P-V-T behavior can be generalized in terms of T_c, P_c, and ω. The original correlation was presented by Pitzer, and is given in the form

Pitzer
Correlation.

$$Z = Z^0 + \omega Z^1$$ 6.3

where tables or charts summarized the values of Z^0 and Z^1 at reduced temperature and pressure. The broad availability of computers and programmable calculators is making this approach somewhat obsolete, but it is worthwhile to illustrate this approach, if only for the benefit of visualizing the trends. Fig. 6.4[1] may be applied for most hydrocarbons. The plot of Z^0 represents the behavior of a fluid that would have an acentric factor of 0, and the plot of Z^1 represents the quantity $Z|_{\omega = 1} - Z|_{\omega = 0}$, which is the correction factor for a hypothetical fluid with an acentric factor of 1. By perusing the table on the endsheet of this book, you will notice that most fluids fall between these ranges so that the charts may be used for interpolation.

Eqn. 6.3 can be applied to any fluid once T_r, P_r, and ω are known. It should be noted, however, that this graphical approach is rarely used in current practice since computer programs are more conveniently written in terms of the equations of state as demonstrated in Section 6.5 and the homework.

Example 6.1   Application of the generalized charts

Estimate the specific volume in cm^3/g for carbon dioxide at 310 K and (a) 8 bars (b) 75 bars by the compressibility factor charts and compare to the experimental values[2] of 70.58 and 3.90 respectively.

Solution: $\omega = 0.228$ and $T_r = 310/304.2 = 1.02$ for both cases (a) and (b), so,
   (a) $P_r = 8/73.82 = 0.108$; from the charts, $Z^0 = 0.96$ and $Z^1 = 0$, so $Z = 0.96$

$V = ZRT/(P \cdot MW) = (0.96 \cdot 83.14 \cdot 310)/(8 \cdot 44) = 70.29$, within 0.4% of the experimental value.

   (b) $P_r = 75/73.82 = 1.016 \approx 1.02$; Note that the compressibility factor is extremely sensitive to temperature in the critical region. To obtain a reasonable degree of accuracy in reading the charts, we must interpolate between the reduced temperatures of 1.0 and 1.05 which we can read with more confidence.

At $T_r = 1.0$, $Z^0 = 0.22$ and $Z^1 = -0.08$ so $Z = 0.22 + 0.228 \cdot (-0.08) = 0.202$

At $T_r = 1.05$, $Z^0 = 0.58$ and $Z^1 = 0.03$ so $Z = 0.58 + 0.228 \cdot (0.03) = 0.587$

1. This is the Lee-Kesler equation, Lee, B. I., M.G. Kesler *AIChE J.*, **21**:510 (1975).
2. Vargaftik, N.B., *Handbook of Physical Properties of Liquids and Gases,* 2nd ed., Hemisphere, New York (1975).

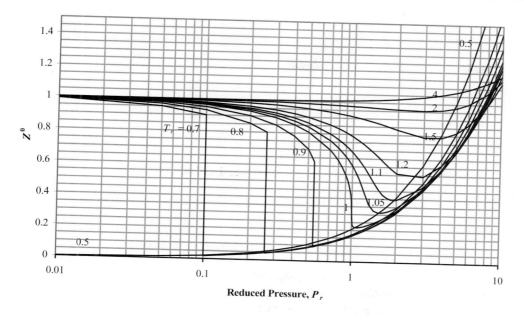

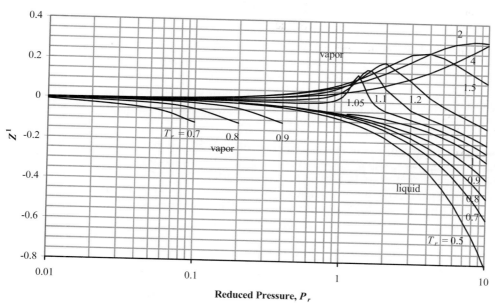

Figure 6.4 *Generalized charts for estimating the compressibility factor.* (Z^0) *applies the Lee-Kesler equation using* $\omega = 0.0$, *and* (Z^1) *is the correction factor for a hypothetical compound with* $\omega = 1.0$. *Note the semilog scale.*

Example 6.1 Application of the generalized charts (Continued)

Interpolating, $Z = 0.202 + (0.587 - 0.202) \cdot 2/5 = 0.356$

$V = ZRT/(P \cdot MW) = (0.356 \cdot 8.314 \cdot 310)/(7.5 \cdot 44) = 2.78$, giving 29% error relative to the experimental value.

It should be noted that the relative error encountered in this example is somewhat exaggerated relative to most conditions because the Z-charts are highly non-linear in the critical region used in this problem. Since the compressibility factor charts essentially provide a "linear interpolation" between Z values for $\omega = 0$ and $\omega = 1$, the error is large in the critical region. If the reduced temperature had been slightly higher, (e.g., $T_r = 1.1$), then the relative error would have been roughly 1%, as demonstrated in the homework problems. It would be a simple matter to specify conditions that would make the chart look much more reliable, but then students might tend to err liberally rather than conservatively. For better reliability, computer methods provide proper alternatives, and these are easily applied on any modern engineering calculator. Example 6.4 on page 208 will demonstrate in detail the validity of this perspective.

6.4   THE VIRIAL EQUATION OF STATE

At low reduced pressure, deviations from ideal gas behavior are sufficiently small that we can write our equation of state as explicit in a power series with respect to density. That is,

$$Z = 1 + B\rho + C\rho^2 + D\rho^3 + \ \dots \qquad\qquad 6.4$$

where B, C, and D are the second, third, and fourth virial coefficients. This can be considered an expansion in powers of ρ. Coefficients C and D are rarely applied because this power series is not very accurate over a broad range of conditions. The most common application of the virial equation of state is to truncate it after the second virial coefficient and to limit the range of application appropriately. It provides a simple equation which still has a reasonable number of viable applications. It has become common usage to refer to the equation truncated after the second virial coefficient as *the* virial equation, even though we know that it is really a specialized form. We, too, will follow this common usage. Furthermore, the truncated form may alternatively be expressed as $Z = Z(P,T)$. Hence, we often refer to *the* virial equation as:

$$Z = 1 + B(P/RT) \qquad\qquad 6.5$$

> ❶ Virial Equation. *B* is known as the second virial coefficient, and it is a measure of the slope of the *Z* chart isotherms in the linear region.

where B is a function of T. Note that Eqn. 6.5 indicates that Z varies linearly with pressure along an isotherm. Look back at Fig. 6.4 and notice that the region in which linear behavior occurs is limited, but in general, the approximation can be used at higher reduced pressures when the reduced temperature is higher. The virial equation can be generalized in reduced coordinates as given by Eqns. 6.6–6.9.[1] Eqn. 6.10 checks for restriction of the calculation to the linear Z region.

---

1. Smith, J.M., Van Ness, H.C., *Introduction to Chemical Engineering Thermodynamics,* 3rd ed., p. 87, McGraw-Hill, New York, 1975.

$$Z = 1 + (B^0 + \omega B^1)P_r/T_r \quad \text{or} \quad Z = 1 + BP/RT \tag{6.6}$$

$$\text{where } B(T) = (B^0 + \omega B^1)RT_c/P_c \tag{6.7}$$

$$B^0 = 0.083 - 0.422/T_r^{1.6} \tag{6.8}$$

$$B^1 = 0.139 - 0.172/T_r^{4.2} \tag{6.9}$$

$$\text{Subject to } T_r > 0.686 + 0.439P_r \text{ or } V_r > 2.0 \tag{6.10}$$

The temperature dependence of the slope of the Z lines is not sufficiently represented by 1/T, so the temperature dependence of B in Eqns. 6.8 and 6.9 is required. The virial equation is limited in its range of applicability, but it has the advantage of simplicity. Its simplicity is especially advantageous when illustrating derivations of real-fluid behavior for the first time and extending thermodynamic relations to vapor mixtures. Unfortunately, it does *not* apply to liquids, and many interesting results in thermodynamics appear in the study of liquids. To develop a global perspective applicable to gases and liquids, we must consider the physics of fluids in a more sophisticated manner. The simplest form which still permits this level of sophistication is the cubic equation, discussed in the following section.

Example 6.2 Application of the virial equation

Estimate the specific volume in cm^3/g for carbon dioxide at 310 K and (a) 8 bars (b) 75 bars by the virial equation and compare to the experimental values of 70.58 and 3.90 respectively.

Solution: $\omega = 0.228$ and $T_r = 310/304.2 = 1.02$ for both cases (a) and (b), so,

$B^0 = 0.083 - 0.422/1.02^{1.6} = -0.326$

$B^1 = 0.139 - 0.172/1.02^{4.2} = -0.0193$

$B(T)P_c/RT_c = (B^0 + \omega B^1) = (-0.326 + 0.228 \cdot (-0.0193)) = -0.3304$

(a) $P_r = 8/73.82 = 0.108$; so $Z = 1 + (B^0 + \omega B^1)P_r/T_r = 1 - 0.3304 \cdot 0.108/1.02 = 0.965$

$V = ZRT/(P \cdot MW) = (0.965 \cdot 83.14 \cdot 310)/(8 \cdot 44) = 70.66$, within 0.1% of the experimental value.

(b) $P_r = 75/73.82 = 1.016$; applying Eqn. 6.10, $0.686 + 0.439 \cdot 1.016 = 1.13 > T_r = 1.02$. Therefore, the virial equation should not be applied using only the second virial coefficient.

There is an adaptation of the form of the virial series which should be mentioned before concluding this discussion. It should not seem surprising that the inclusion of extra adjustable parameters in the form of the virial series is an extremely straightforward task—just add higher order terms to the series. In many cases, exponential terms are also included as in Eqn. 6.11 on page 202. In this way, it is possible to fit the P-V-T behavior of the liquid as well as the vapor to a reasonable degree of accuracy. It turns out that the mathematical foundation for the series expansion in this way is tenuous, however. Reading "the fine print" in discussions of series expansions like Taylor series shows that such an approach is only applicable to "analytic" functions. At present, there is a general acceptance that the behavior of real fluids is "non-analytic" at the critical point. This means that application of such a series expansion above the critical density and below the critical temperature is without a rigorous mathematical basis. Nevertheless, engineers occasionally invoke the motto that "we can fit the shape of an elephant with enough adjustable parameters." It is in this spirit that empirical equations like the Benedict-Webb-Rubin equation are best appreciated. One particular modification of the Benedict-Webb-Rubin form is given below. It is the form that Lee and Kesler[1] developed to render the Pitzer correlation in terms of computer-friendly equations. The Lee-Kesler equation was used to generate Fig. 6.4.

$$Z = 1 + \frac{B}{V_r} + \frac{C}{V_r^2} + \frac{D}{V_r^5} + \frac{E_0}{T_r^3 V_r^2}\left(E_1 + \frac{E_2}{V_r^2}\right)\exp\left(-\frac{E_2}{V_r^2}\right) \qquad 6.11$$

Twelve parameters are used to specify the temperature dependence of B, C, D, E_0, E_1, and E_2 for each compound. Readers are directed to the original article for the exact values of the parameters as part of the homework.

6.5   CUBIC EQUATIONS OF STATE

The acronym EOS will be used to mean equation of state.

To apply the relationships that we can develop for relating changes in properties to C_P, C_V, P, T, V, and their derivatives, we need really general relationships between P, V, and T. These relationships are dictated by the equation of state (EOS). Constructing an equation of state with a firm physical and mathematical foundation requires considering how the intermolecular forces are affecting the energy and pressure in a fluid. As the fluid becomes dense, we know that the molecules will be closer together on the average, and such closeness will give rise to a potential energy contribution and to an increase in the contribution of attractive forces. A common practical implication of this attractive energy is the heat of vaporization of a boiling liquid. But how can we make a quantitative connection between molecular forces and macroscopic properties? A firm understanding of this physical and mathematical foundation is necessary to understand the extensions to multicomponent mixtures and multiphase equilibria.

A proper derivation would provide a mathematical connection between the microscopic potential and the macroscopic properties. We will lay the groundwork for such a rigorous derivation later in the chapter. For introductory purposes, however, we would like to see how some typical equations look and how to use them in conjunction with the theorem of corresponding states.

The van der Waals Equation of State

One of the most influential equations of state has been the van der Waals (1873) equation. Even the most successful engineering equations currently used are only minor variations on the theme originated by van der Waals. The beauty of his argument is that detailed knowledge of the molecular interactions is not necessary. Simply by noting that there are two characteristic molecular quantities (ε and σ) and two characteristic macroscopic quantities (T_c and P_c), he was able to infer a simple equation that captured the key features of each fluid through the principle of corresponding states. His final equation expressed the attractive energy in terms of a parameter which he referred to as a, and the size parameter b, but the choice of symbols was arbitrary. The key feature to recognize at this stage is that there are at least two parameters in all the equations and that these can be determined by matching experimental data.

---

1. Lee, B.I., Kesler, M.G., *AIChE J.* 21:510 (1975).

The resulting equation of state is:

$$Z = 1 + \frac{b\rho}{(1-b\rho)} - \frac{a\rho}{RT} = \frac{1}{(1-b\rho)} - \frac{a\rho}{RT}$$

6.12 ❶ van der Waals EOS.

where ρ = molar density = $n/\underline{V}$.

> *Note: Common engineering practice is to use ρ to denote intensive density. We follow that convention here, using ρ as molar density. Advanced chemistry and physics books and research publications frequently use ρ as number density $N/\underline{V} = nN_A/\underline{V}$.*

❶ ρ will be used to denote molar density.

The exact manner of determining the values for the parameters a and b is discussed in a later section. (See "Matching the Critical Point" on page 220.)

$$a \equiv \frac{27}{64}\frac{R^2 T_c^2}{P_c} \qquad ; \qquad b \equiv 0.125 R\frac{T_c}{P_c}$$

6.13

We may write the equation of state as $Z = 1 + Z^{rep} + Z^{attr}$, where Z^{rep} represents the deviations from the ideal gas law due to repulsive interactions; Z^{attr} represents the deviations due to attractive interactions. For the van der Waals equation,

$$Z = 1 + \frac{b\rho}{1-b\rho} - \frac{a\rho}{RT}$$

6.14 ❶ The van der Waals equation written in the form $Z = 1 + Z^{rep} + Z^{attr}$.

where the second and third terms on the right-hand side are Z^{rep} and Z^{attr}, respectively. There are two key features of the van der Waals equation. First, the repulsive term accounts for the asymptotic divergence of the compressibility factor at the close-packed density. The divergence occurs because increasing the pressure cannot increase the density as close packing is approached. Second, the roles of attractive forces increase as the temperature decreases. The impacts of attractive forces increase at low temperature because the kinetic energy can no longer overwhelm the potential attractions at low temperature. As we have discussed, this eventually leads to condensation. The discussion in Section 6.8 provides a better understanding of the theoretical basis of the van der Waals equation.

The Peng-Robinson Equation of State

Since the time of van der Waals (1873), many approximate equations of state have been proposed. For the most part, these have been semi-empirical corrections to "a = constant," and most have taken the form $a = a(T)$. One of the most successful examples of this approach is that of Peng and Robinson (1976). We refer to this equation many times throughout this text and use it to demonstrate many central themes in thermodynamic theory as well as useful applications.

The Peng-Robinson equation of state (EOS) is given by:

❶ The Peng-Robinson EOS. Note that a is a temperature-dependent parameter, not a constant. Note the dependence on the acentric factor.

$$P = \frac{RT\rho}{(1-b\rho)} - \frac{a\rho^2}{1+2b\rho-b^2\rho^2} \qquad \text{or} \qquad Z = \frac{1}{(1-b\rho)} - \frac{a}{bRT}\cdot\frac{b\rho}{1+2b\rho-b^2\rho^2}$$

6.15

where ρ = molar density = $n/\underline{V}$

$$a \equiv a_c \alpha; \quad a_c \equiv 0.45723553 \frac{R^2 T_c^2}{P_c} \qquad\qquad b \equiv 0.07779607 R \frac{T_c}{P_c} \qquad 6.16$$

$$\alpha \equiv \left[1 + \kappa\left(1 - \sqrt{T_r}\right)\right]^2 \quad \kappa \equiv 0.37464 + 1.54226\omega - 0.26993\omega^2 \qquad 6.17$$

and a relation that will be useful later:

$$\frac{da}{dT} = \frac{-a_c \kappa \sqrt{\alpha T_r}}{T} \qquad 6.18$$

T_c, P_c, and ω are reducing constants according to the principle of corresponding states.[1] Writing the equation of state as $Z = 1 + Z^{rep} + Z^{attr}$,

$$Z = 1 + \frac{b\rho}{1 - b\rho} - \frac{a}{bRT} \cdot \frac{b\rho}{1 + 2b\rho - b^2\rho^2} \qquad 6.19$$

Comparison of the Peng-Robinson equation to the van der Waals equation shows one obvious similarity; the repulsive term is the same. There are some differences: temperature dependence of the attractive parameter a is incorporated in the Peng-Robinson equation; dependence of a on the acentric factor is introduced; and the density dependence of Z^{attr} is altered. The manner in which these extra details were added was almost entirely empirical; different equations were tried until one was found which seemed to fit the data most accurately while retaining cubic behavior. (Many equations can be tried in 103 years.) There is not much to say about this empirical approach beyond the importance of including the acentric factor. The main reason for the success of the Peng-Robinson equation is that it is primarily applied to vapor-liquid equilibria and that the representation of vapor-liquid equilibria is strongly influenced by the more accurate representation of vapor pressure implicit in the inclusion of the acentric factor. Since the critical point and the acentric factor characterize the vapor pressure fairly accurately, it should not be surprising that the Peng-Robinson equation accurately represents vapor pressure.

One caution is given. The differences in accuracy between various equations of state are subtle enough that one equation may be most accurate for one narrow range of applications, while another equation of state is most accurate over another range. This may require a practicing engineer to adopt an equation of state other than the Peng-Robinson equation for his specific application. Nevertheless, the treatment of the Peng-Robinson equation presented here is entirely analogous to the treatment required for any equation of state. If you understand this treatment, you should have no problem adapting. A brief review of several thermodynamic models commonly encountered in chemical process simulations is given in Appendix D.

> *Note: The variables a and b are used throughout equation of state literature (and this text) to denote equation of state parameters. The formulas or values of these parameters for a given equation of state cannot be used directly with any other equation of state.*

🚫 There are a lot of EOSs. We focus on the Peng-Robinson to illustrate the concepts.

🚫 The variables a and b are commonly used in EOSs. Do not interchange the formulas.

---

1. The number of significant figures presented in Eqn. 6.16 is important in reproducing the universal value of $Z_c = 0.307$ predicted by the Peng-Robinson equation of state.

6.6   SOLVING THE EQUATION OF STATE FOR Z

In most applications we are given a pressure and temperature and asked to determine the density and other properties of the fluid. This becomes slightly difficult because the equation involves terms of density that are to the third power ("cubic") even when simplified as much as possible. Standard methods for solutions to cubic equations can be applied. The equation can be made dimensionless prior to application of the solution method. By noting that

$$b\rho \equiv B/Z \qquad\qquad 6.20$$

$$B \equiv bP/RT \qquad\qquad 6.21$$

$$Z \equiv P/\rho RT \qquad\qquad 6.22$$

$$A \equiv aP/R^2T^2 \qquad\qquad 6.23$$

Dimensional analysis is an important engineering tool. Here we make the EOS dimensionless so that it can be solved in a generalized way.

the Peng-Robinson equation of state becomes

$$\boxed{Z = \frac{1}{(1 - B/Z)} - \frac{A}{B} \cdot \frac{B/Z}{1 + 2B/Z - (B/Z)^2}} \qquad\qquad 6.24$$

Do not confuse the EOS parameters A and B with other uses of the variables. The intended use is almost always clear.

> ***Note:*** *The variable A is also used elsewhere in the text to denote Helmholtz energy. The variable B is used elsewhere in the text to represent the second virial coefficient. The context of the variable usage should make the meaning of the variable clear. We choose to use A and B as reduced equation of state parameters for consistency with equation of state discussions in the literature.*

Rearranging the dimensionless Peng-Robinson equation yields a cubic function in Z that must be solved for vapor, liquid, or fluid roots

$$Z^3 - (1 - B)Z^2 + (A - 3B^2 - 2B)Z - (AB - B^2 - B^3) = 0 \qquad\qquad 6.25$$

Naming this function $F(Z)$, we can plot $F(Z)$ versus Z to gain some understanding about its roots as shown in Fig. 6.5. Considering the case when $P = P^{sat}$, we see that the larger root of $F(Z)$ will be the vapor root and will be the value of Z for saturated vapor. The smallest root will be the liquid root and will be the value of Z for saturated liquid. The middle root corresponds to a condition that violates thermodynamic stability, and cannot be found experimentally; the derivative of volume with respect to pressure must always be negative in a real system, and this root violates that condition. Below the critical temperature, when $P > P^{sat}$, the fluid will be a compressed liquid, and the vapor root is less stable. Below the critical temperature, when $P < P^{sat}$, the fluid will be a superheated vapor and the liquid root is less stable. When $T > T_c$, we have a supercritical fluid which can only have a single root but it may vary continuously between a "vapor-like" or "liquid-like" densities and compressibility factors.

Methods of Solving the Cubic Equation

Solution of the equation of state in terms of Z is greatly preferred over solution for V. The value of Z often falls between 0 and 1. (See Fig. 6.4 on page 199). V often varies from 50-100 cm^3/mole for liquids to near infinity for gases as P approaches zero. It is much easier to solve for roots over the smaller variable range using the compressibility factor Z. There are two basic approaches to solving

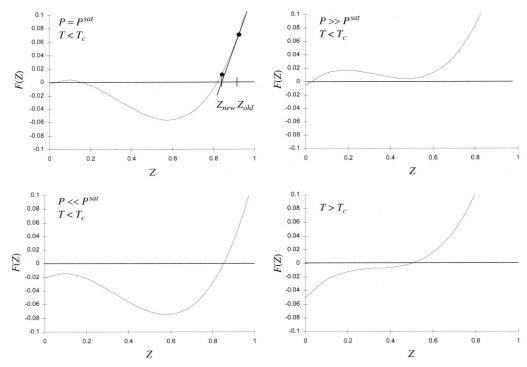

Figure 6.5 *Comparison of behavior of cubic in Z for the Peng-Robinson equation of state at several conditions.*

cubic equations of state. First, we may use an iterative method. One such method is the iterative Newton-Raphson method. Another method is to solve analytically. A computer or calculator program is helpful in either solution method.

Iterative Method

Suppose we have made an initial guess Z_{old} which gives a value F_{old}, as shown in the upper left graph on Fig. 6.5. We are seeking a value of Z that results in $F = 0$. If F_{old} is the current value, and if we use the derivative of F as a linear approximation of the function behavior, then $0 = m \cdot Z_{new} + b$, (where the slope m can be calculated analytically from Eqn. 6.25 as $dF/dZ = (3Z^2 - 2(1 - B)Z + (A - 3B^2 - 2B)$. Since the current point is on the same line, we may also write $F_{old} = m \cdot Z_{old} + b$. Taking the difference we get $0 - F_{old} = m \cdot (Z_{new}-Z_{old}) + (b-b)$ or rearranging, $-F_{old}/m + Z_{old} = Z_{new}$. Since $m = (dF/dZ)$, we have $Z_{new} = Z_{old} - F/(dF/dZ)$. The procedure can be repeated until the answer is obtained. A summary of steps is:

1. Guess $Z_{old} = 1$ or $Z_{old} = 0$ and compute $F_{old}(Z_{old})$.

2. Compute dF/dZ

3. $Z_{new} = Z_{old} - F/(dF/dZ)$

4. If $|\Delta Z/Z_{new}| < 1.E - 5$, print the value of Z_{new} and stop.

5. Compute $F_{new}(Z_{new})$ and use this as F_{old}. Return to step 3 until step 4 terminates.

Note that an initial guess of $Z = 0$ converges on the smallest real root. An initial guess of $Z = 1$ almost always converges on the largest real root. At very high reduced pressures, an initial guess greater than one is sometimes required since the compressibility factor can exceed one (see Fig. 6.4 on page 199).

Analytical Solution

The other choice we have for solution of the cubic is to analytically obtain the roots. This method is detailed in Appendix B. The method varies depending on whether one or three roots exist at the pressure of interest. A spreadsheet is provided with the text to obtain the solution (PREOS.xls).

HP PrI uses a built-in analytical procedure.

PRPURE and PRMIX include cubic subroutines.

PREOS.xls uses the procedures from Appendix B and shows the intermediate calculations.

Example 6.3 Solution of the Peng-Robinson equation for molar volume

Find the molar volume predicted by the Peng-Robinson equation of state for argon at 105.6 K and 4.98 bar.

Solution: Use PREOS.xls. The critical data are entered from the table on the endflap of the text. The spreadsheet is shown in Fig. 6.6. The answers are given for the three-root region, whereas the cells for the one-root region are labeled #NUM! by EXCEL. This means that we are in the three-root region at these conditions of temperature and pressure. Many of the intermediate calculations are also illustrated in case you want to write your own program some day. The answers are 27.8, 134, and 1581 cm^3/mole. The lower value corresponds to the liquid volume and the upper value corresponds to the vapor.

Peng-Robinson Equation of State (Pure Fluid)   Spreadsheet protected, but no password used.

Properties

Gas	T_c (K)	P_c (MPa)	ω	R(cm^3MPa/molK)
Argon	150.86	4.898	-0.004	8.314

Current State		Roots		
T (K)	105.6	Z	V	fugacity
P (MPa)	0.498		cm^3/gmol	MPa
answers for three		0.896744	1580.931	0.451039
root region		0.076213	134.3613	
		0.015743	27.75473	0.450754
& for 1 root region		#NUM!	#NUM!	#NUM!

Stable Root has a lower fugacity

Intermediate Calculations		
T_r	0.699987	a (MPa cm^6/gmol^2)
P_r	0.101674	165065.2
κ	0.368467	b (cm^3/gmol)
α	1.123999	19.92155
fugacity ratio		A 0.106644
	1.000633	B 0.0113

To find vapor pressure, or saturation temperature, see cell A28 for instructions

Solution to Cubic $Z^3 + a_2Z^2 + a_1Z + a_0 = 0$

a_2	a_1	a_0	p	q
-0.9887	0.083661	-0.00108	-0.24218	-0.0451

R = q^2/4 + p^3/27 =   -1.8E-05
If Negative, three unequal real roots,
If Positive, one real root

Method 1 - For region with one real root

P	Q	Root to equation in x
#NUM!	#NUM!	#NUM!

Method 2 - For region with three real roots

m	3q/pm	3*θ_1	θ_1	Roots to equation in x		
0.568251	0.983041	0.184431	0.061477	0.567177	-0.25335	-0.31382

Figure 6.6 *Sample output from PREOS.xls as discussed in Example 6.3*

Prl

PREOS.xls

PENG

Example 6.4 Application of the Peng-Robinson equation

Estimate the specific volume in cm^3/g for carbon dioxide at 310 K and (*a*) 8 bars (*b*) 75 bars by the Peng-Robinson equation and compare to the experimental values of 70.58 and 3.90 respectively.[1]

Solution: $\omega = 0.228$ and $T_r = 310/304.2 = 1.02$ for both cases (a) and (b), so,

(a) $P_r = 8/73.82 = 0.108$; so $Z = 0.961$

$V = ZRT/(P \cdot MW) = (0.961 \cdot 83.14 \cdot 310)/(8 \cdot 44) = 70.37$, within 0.3% of the experimental value.

(b) $P_r = 75/73.82 = 1.016$; so $Z = 0.496$

$V = ZRT/(P \cdot MW) = (0.496 \cdot 83.14 \cdot 310)/(73.82 \cdot 44) = 3.93$, giving 0.8% error relative to the experimental value.

Note the equation of state is much more reliable than reading the compressibility factor charts in Example 6.1 on page 198.

Determining Stable Roots

When a cubic equation provides one real root, we characterize the solution as being in the *one-root region*. Cubic equations of state will provide three real roots over a range of pressures along any isotherm below the critical temperature in the *three-root region* as shown in Fig. 6.7. Note that above the critical temperature the isotherms do not have "humps" and only one root is predicted. Below the critical temperature, one root or three roots can exist depending on the pressure. This section discusses the circumstances when three real roots are found. As mentioned in Example 6.3, the smallest root *usually* corresponds to a liquid state, and the largest root usually corresponds to the vapor state.[2] However, when three real roots are found over a range of pressures for a given temperature, these roots *do not* indicate vapor + liquid coexistence at all these conditions. Vapor and liquid phases *coexist* only at the vapor pressure—above the vapor pressure, the liquid root is most stable—below the vapor pressure, the vapor root is most stable. The most *stable* root represents the phase that will exist *at equilibrium*. When three roots exist, the most stable root has the lower Gibbs energy or fugacity. At phase equilibrium, the Gibbs energy and fugacity of the roots will be equal. (Fugacity will be described in Chapter 8, but we can begin to use the calculated values before we discuss the calculation procedures completely.) When three roots exist, the center root is thermodynamically *unstable* because the derivative of pressure with respect to volume is positive, which violates our common sense, and is shown to be thermodynamically unstable in advanced thermodynamics texts. Physically, we can understand the meaning of the unstable root as follows. Imagine placing Lennard-Jones spheres all in a row such that the force pulling on each molecule from one direction is exactly balanced by the force in the other direction. Next, imagine placing similar rows perpendicular to that one until you obtain the desired density at the desired temperature. Physically and mathematically, this configuration is conceivable, but what will happen when one of these atoms moves? The perfect balance will be destroyed, and a large number of atoms will cluster together to form a liquid. The atoms that do not form a liquid will remain in the form of a low density

The *stable* root represents the phase that will exist *at equilibrium*. The stable root has the lower Gibbs energy or fugacity.

1. Vargaftik, N.B., *Handbook of Physical Properties of Liquids and Gases,* Hemisphere Publishers, New York, 1975.
2. However, the Peng-Robinson equation does give small real roots at high pressures, and the smallest real root is not always the liquid root. See problem 6.12.

vapor. Although this discussion has discussed only the energy of interactions, the entropy is also important. It is actually a balance of enthalpy and entropy that results in phase equilibrium, as we will discuss in Chapter 8.

A simple way to visualize the conditions that lead to vapor-liquid equilibrium along an isotherm is to consider the *P-V* diagram illustrated in Fig. 6.7. The "humps" are characteristic of the cubic equation isotherm shape below the critical temperature. We will show in Chapter 8 that the condition for equilibrium between vapor and liquid roots occurs when the horizontal line on the *P-V* diagram is positioned such that the area enclosed above the line is exactly equal to the area enclosed below the line. Even though the enclosed areas have different shapes, you can imagine moving this line up and down until it looks like the areas are equal. *The dots in the figure are the predictions of the saturated liquid and vapor volumes, and form the phase envelope.* The parts of the isotherms that are between the saturated vapor and saturated liquid roots are either *metastable* or unstable. To determine whether a given point is metastable or unstable, look for the point where the isotherm reaches a maximum or a minimum. Points between the liquid root and the minimum are considered metastable liquids; points between the maximum and the vapor root are considered metastable vapors. The metastable state can be experimentally obtained in careful experiments. Under clean conditions, it is possible to experimentally heat a liquid above its boiling point to obtain superheated liquid. Likewise, under clean conditions, it is possible to experimentally obtain subcooled vapor. However, a metastable state is easily disrupted by vibrations or nucleation sites, e.g., provided by a boiling chip or dust, and once disrupted, the state will decay rapidly to the equilibrium state. We will discuss more details about characterizing proper fluid roots when we treat phase equilibrium in a pure fluid.

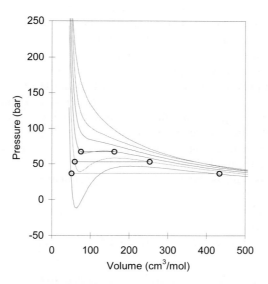

Figure 6.7 *Illustration of the prediction of isotherms by the Peng-Robinson equation of state for CO_2 ($T_c = 304.2$ K) at 275 K, 290 K, 300 K, 310 K, 320 K and 350 K. Higher temperatures result in a high pressure for a given volume. The "humps" are explained in the text. The calculated vapor pressures are 36.42 bar at 275 K, 53.2 bar at 290 K, and 67.21 bar at 300 K.*

6.7 IMPLICATIONS OF REAL FLUID BEHAVIOR

There is one implication of non-ideal fluid behavior that should be clear from the equations presented above: real fluids behave differently from ideal gases. How differently? An example provides the most straightforward answer to that question.

Example 6.5 Derivatives of the Peng-Robinson equation

Determine $\left(\dfrac{\partial P}{\partial T}\right)_V$, $\left(\dfrac{\partial C_V}{\partial V}\right)_T$, and $\left(\dfrac{\partial U}{\partial V}\right)_T$ for the Peng-Robinson equation.

Solution: Differentiating the Peng-Robinson equation,

$$\left(\frac{\partial P}{\partial T}\right)_V = \frac{R\rho}{1 - b\rho} - \frac{\rho^2}{1 + 2b\rho - b^2\rho^2}\frac{da}{dT}$$

which approaches the ideal gas limit: $\displaystyle\lim_{\rho \to 0}\left(\frac{\partial P}{\partial T}\right)_V = R\rho = \frac{R}{V}$

$$\left(\frac{\partial C_V}{\partial V}\right)_T = T\left(\frac{\partial^2 P}{\partial T^2}\right)_V = \frac{-T\rho^2}{1 + 2b\rho - b^2\rho^2}\frac{d^2 a}{dT^2} = \frac{-\rho^2}{1 + 2b\rho - b^2\rho^2}\frac{a_C\kappa}{2}\left(\frac{\kappa}{T_C} + \frac{\sqrt{\alpha T_r}}{T}\right)$$

which approaches the ideal gas limit at low density.

$$\left(\frac{\partial U}{\partial V}\right)_T = T\left(\frac{\partial P}{\partial T}\right)_V - P = \frac{\rho^2}{1 + 2b\rho - b^2\rho^2}\left[a - \frac{da}{dT}\right] = \frac{\rho^2 a_C}{1 + 2b\rho - b^2\rho^2}\left[\alpha + \kappa\sqrt{\alpha T_r}\right]$$

which also approaches the ideal gas limit at low density.

We have shown that C_V depends on volume. To calculate a value of C_V, first we determine C_V^{ig} $= C_P^{ig} - R$, where C_P^{ig} is the heat capacity tabulated in Appendix E. Then, at a given $\{P,T\}$, the equation of state is solved for ρ. The resulting density is used as the limit in the following integrals, noting as $V \to \infty$, $\rho \to 0$, and $dV = -d\rho/\rho^2$:

$$C_V - C_V^{ig} = \int_{\infty}^{V}\left(\frac{\partial C_V}{\partial V}\right)_T dV = \left(\frac{d^2 a}{dT^2}\right)\int_0^{\rho}\frac{T\rho^2}{1 + 2b\rho - b^2\rho^2}\frac{d\rho}{\rho^2} = \frac{T}{2\sqrt{2}b}\left(\frac{d^2 a}{dT^2}\right)\ln\left[\frac{1 + \left(1 + \sqrt{2}\right)b\rho}{1 + \left(1 - \sqrt{2}\right)b\rho}\right]$$

where $\left(\dfrac{d^2 a}{dT^2}\right) = \dfrac{a_c\kappa}{2T_c^2 T_r}\left[\kappa + \sqrt{\dfrac{\alpha}{T_r}}\right]$

6.8 THE MOLECULAR THEORY BEHIND EQUATIONS OF STATE

In the previous sections we alluded to the equation of state as an empirical equation that may have appeared by magic. In this section, we attempt to de-mystify the origins behind equations of state by systematically recounting the current outlook on equation of state development. At this point in the text, it may seem like overkill to develop so much theory to justify such simple equations. As empirical equations go, equations of state are not much more difficult to accept than, say, Newton's laws of motion.

On the other hand, we know from our introductory discussion that the macroscopic properties must be derived from the molecular interactions through Newton's laws. So Newton's laws are different from a macroscopic equation of state; they are more fundamental. Furthermore, our goal is to understand molecular behavior in a way that will enable us to make quick judgments about which kinds of interactions are favorable and which are likely to lead to complications. We cannot really accomplish that without building a bridge from the molecular scale to the macroscopic scale. It might be possible to postpone the molecular development until we begin treating mixtures, where the interactions are so complicated that some systematic development is really demanded. On the other hand, we can introduce the basic concepts now in ways that are relevant but not overwhelmingly complicated. Then you will have some time to assimilate them before we deal with mixtures.

The key to bridging the gap from the molecular scale to the macroscopic is to consider the average number of molecules at each distance from the center of an average molecule. To get the configurational internal energy,[1] multiply this average number of molecules by the amount of potential energy at that distance and integrate over all distances. To get the pressure, multiply this average number of molecules by the amount of force per unit area at that distance and integrate over all distances. These considerations give rise to the energy equation and the pressure equation and, most importantly, to the definition of this "average number of molecules." The average number of molecules at a particular distance from an average molecule is characterized by the "radial distribution function" which is discussed in detail below.

The Energy Equation

The ideal gas continues to be an important concept, because it is a convenient reference fluid. If we calculate the internal energy of a real gas, it is easiest to calculate the internal energy of the real fluid relative to an ideal gas at the same temperature and pressure. This way, the kinetic energy of the gas is included in the ideal gas internal energy, and we calculate the contribution to internal energy due to the intermolecular potentials of the real gas.

$$U - U^{ig} = \frac{N_A \rho}{2} \int_0^\infty N_A u \; g(r) \; 4\pi r^2 \; dr \qquad\qquad 6.26$$

where u is the pair potential and $g(r) \equiv$ the radial distribution function defined by Eqn. 6.32. This equation can be written in dimensionless form as:

$$\frac{U - U^{ig}}{RT} = \frac{N_A \rho}{2} \int_0^\infty \frac{N_A u}{RT} g(r) 4\pi r^2 dr \qquad\qquad 6.27$$

The Pressure Equation

We also may choose to solve for the pressure of our real fluid. Once again it is convenient to use the ideal gas as our reference fluid and calculate the pressure of the real fluid relative to the ideal gas

---

1. The configurational energy is that energy due solely to the intermolecular separations, hence the adjective *configurational*.

law. Since intermolecular force is the derivative of the intermolecular potential, we note the derivative of the intermolecular potential in the following equation.

$$P = \rho RT - \frac{\rho^2 N_A^2}{6} \int_0^\infty r\left(\frac{du}{dr}\right) g(r) \; 4\pi r^2 \; dr \qquad\qquad 6.28$$

This equation can also be written in dimensionless form, recalling the definition of the compressibility factor:

$$\frac{P}{\rho RT} = 1 - \frac{\rho N_A}{6} \int_0^\infty \frac{N_A r}{RT}\left(\frac{du}{dr}\right) g(r) \; 4\pi r^2 \; dr \qquad\qquad 6.29$$

Note in both the energy equation and the pressure equation, that our integral extends from 0 to infinity. Naturally, we never have a container of infinite size. How can we represent a real fluid this way? Look again at the form of the intermolecular potentials in Chapter 1. At long molecular distances, the pair potential and the derivative of the pair potential both go to zero. Long distances on the molecular scale are 4 to 5 molecular diameters (on the order of nanometers), and the integrand is practically zero outside this distance. Therefore, we may replace the dimensions of our container with infinity, and obtain the same numerical result. This substitution makes a single equation valid for all containers of any size greater than a few molecular diameters.[1]

An Introduction to the Radial Distribution Function

As a prelude to a general description of atomic distributions, it may be helpful to review the structure of crystal lattices like those in body-centered cubic (bcc) metals, as shown in Fig. 6.8. Such a lattice possesses long-range order due to repetitive arrangements of the unit cell in three dimensions. This close-packed arrangement of atoms gives a single value for the density, and the density correlates with many of the macroscopic properties of the material (e.g., strength, ductility). These are some of the key considerations fundamental to materials science, and more details are given in common texts on the subject. One goal of introducing the radial distribution function is to generalize the concept of atomic arrangements so that non-lattice fluids can be included.

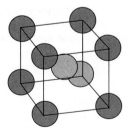

Figure 6.8 *The body centered cubic unit cell.*

---

1. The exception to this discussion occurs very near the critical point, but addressing this problem is beyond the scope of this text.

The distribution of atoms in a bcc crystal is fairly easy to understand, but how can we address the distribution of atoms in a fluid? For a fluid, the positions of the atoms around a central atom are less well-defined than in a crystal. To get started, think about the simplest fluid, an ideal gas.

The Fluid Structure of an Ideal Gas

Consider a fluid of point particles surrounding a central particle. What is the number of particles in a given volume element surrounding the central particle? Since they are point particles, they do not influence one another. This means that the number of particles is simply related to the density.

$$dN_V = N \frac{d\underline{V}}{\underline{V}} \qquad \text{(ig) 6.30}$$

where dN_V is the number of particles in the volume element

N is the total number of particles in the total volume

$\underline{V}$ is the total volume

$d\underline{V}$ is the size of the volume element

$dN_V = N_A \rho \, d\underline{V}$

If we would like to know the number of particles within some spherical neighborhood of our central particle then,

$$d\underline{V} = 4\pi \, r^2 \, dr$$

where r is the radial distance from our central particle

$$N_c = \int_0^{N_c} dN_V = N_A \int_0^{R_o} \rho \, 4\pi \, r^2 dr \qquad \text{(ig) 6.31}$$

where R_o defines the range of our spherical neighborhood.

N_c is the number of particles in the neighborhood (coordination number)

The Fluid Structure of a Low Density Hard-Sphere Fluid

Now consider the case of atoms which have a finite size. In this case, the number of particles within a given neighborhood is strongly influenced by the range of the neighborhood. If the range of the neighborhood is less than two atomic radii, or one atomic diameter, then the number of particles in the neighborhood is zero (not counting the central particle). Outside the range of one atomic diameter, the exact variation in the number of particles is difficult to anticipate *a priori*. You can anticipate it, however, if you think about the way cars pack themselves into a parking lot. We can express these insights mathematically by defining a "weighting factor" which is a function of the radial distance. The weighting factor takes on a value of zero for ranges less than two atomic radii, and for larger ranges, we can consider its behavior undetermined as yet.

❶ The hard-sphere fluid has been studied extensively to represent spherical fluids.

Then we may write

$$N_c = N_A \rho \int_0^{R_o} g(r)\, 4\pi r^2 \, dr \qquad\qquad 6.32$$

where $g(r)$ is our average "weighting function," called the radial distribution function. The radial distribution function is the number of atomic centers located in a spherical shell from r to $r + dr$ from each other, divided by the volume of the shell and the bulk number density.

This is a lot like algebra. It helps us to organize what we do know and what we do not know. The next task is to develop some insights about the behavior of this weighting factor so that we can make some engineering approximations.

As a first approximation, we might assume that atoms outside the range of two atomic radii do not influence each other. Then the number of particles in a given volume element goes back to being proportional to the size of the volume element, and the radial distribution function has a value of one for all r greater than one diameter. The approximation that atoms outside the atomic diameter do not influence each other is reasonable at low density. An analogy can be drawn between the problem of molecular distributions and the problem of parking cars. When the parking lot is empty, cars can be parked randomly at any position, as long as they are not parked on top of each other. Recalling the relation between a random distribution and a flat radial distribution function, Fig. 6.9 should seem fairly obvious at this point.

The Structure of a BCC Lattice

Far from the low-density limit, we approach close-packing. The ultimate in close-packing is a crystal lattice. Let's clarify what is meant by the radial distribution function of a lattice. The radial distribution function of a bcc lattice can be deduced from knowledge of N_c and the defining relation for $g(r)$.

$$N_c = \rho N_A \int_0^{R_O} g(r)\, 4\pi r^2 \, dr \qquad\qquad 6.33$$

If we assume that the atoms in a crystal are located in specific sites, and no atoms are out of their sites, then $g(r)$ must be zero everywhere except at a site. For a body-centered cubic crystal these sites are at $r = \{\sigma, 1.15\sigma, 1.6\sigma,...\}$ so that $g(r)$ looks like a series of spikes. In the parking lot analogy, the best way of parking the most cars is to assign specific regular spaces with regular space between, as shown in Fig. 6.10.

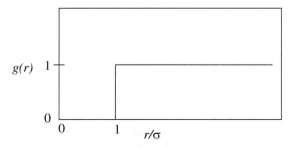

Figure 6.9 *The radial distribution function for the low density hard-sphere fluid.*

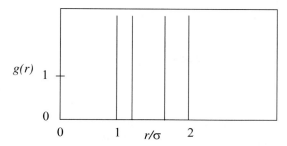

Figure 6.10 *The radial distribution function for the bcc hard-sphere fluid.*

The Fluid Structure of High Density Hard-Sphere Fluids

The distributions of atoms in a fluid are most conveniently referred to as the fluid's *structure*. The structures of these simple cases clarify what is meant by structure in the context that we will be using, but the behavior of a dense liquid illustrates why this concept of structure is necessary. Dense-liquid behavior is something of a hybrid between the low-density fluid and the solid lattice. At large distances, atoms are too far away to influence each other and the radial distribution function approaches unity because the increase in neighbors becomes proportional to the size of the neighborhood. Near the atomic diameter, however, the central atom influences its neighbors to surround it in "layers" in an effort to approach the close packing of a lattice. Thus, the value of the radial distribution function is large very close to one atomic diameter. Because liquids lack the long-range order of crystals, the influence of the central atom on its neighbors is not as well defined as in a crystal, and we get smeared peaks and valleys instead of spikes. Returning to the parking lot analogy once again, the picture of liquid structure is considerably more realistic than the assumption of a regular lattice structure. There are no "lines" marking the proper "parking spaces" in a real fluid. If a few individuals park out of line, the regularity of the lattice structure is disrupted, and it becomes impossible to say what the precise structure is at 10 or 20 molecular diameters. It is true, however, that the average parking around any particular object will be fairly regular for a somewhat shorter range, and the fluid structure in Fig. 6.11 reflects this by showing sharp peaks and valleys at short range and an approach to a random distribution at long range.

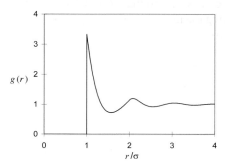

Figure 6.11 *The radial distribution function for the hard-sphere fluid at a packing fraction of $b\rho = 0.4$.*

The Structure of Fluids in the Presence of Attractions and Repulsions

As a final case, consider the influence of a square-well potential (presented in Section 1.1) on its neighbors. The range $r < \sigma$ is off-limits, and the value of the radial distribution function there is still zero. But what about the radial distribution function at low density for the range where the attractive potential is influential? We would expect some favoritism for atoms inside the attractive range, $\sigma < r < R\sigma$, since that would release energy. How much favoritism? The rigorous derivation is beyond our current scope, but it turns out to be simply related to the energy inherent in the potential function.

$$\lim_{\rho \to 0} g(r) = \exp\left[-\frac{u(r)}{kT}\right] \qquad\qquad 6.34$$

❗ The low-density limit of the radial distribution function is related to the pair potential.

This exponential function, known as a Boltzmann distribution, accounts for the off-limits range and the attractive range as well as the no-influence ($r > R\sigma$) range. Referring to the parking lot analogy again, imagine the distribution around a coffee and doughnut vending truck early in the morning when the parking is nearly empty. Many drivers would be attracted by such a prospect and would naturally park nearby, if the density was low enough to permit it.

As for the radial distribution function at high density, we expect packing effects to dominate and attractive effects to subordinate because attaining a high density is primarily affected by efficient packing. At intermediate densities, the radial distribution function will be some hybrid of the high and low density limits, as shown in Fig. 6.12.

A mathematical formalization of these intuitive concepts is presented in several texts, but the difficulty of such a rigorous treatment is beyond the scope of our introductory presentation. For our purposes, we would simply like to understand that the number of particles around a central particle has some character to it that depends on the temperature and density, and that representing this temperature and density dependence in some way will be necessary in analyzing the energetics of how molecules interact. In other words, we would like to appreciate that something called "fluid structure" exists, and that it is described in detail by the "radial distribution function." This appreciation will be of use again when we extend these considerations to the energetics of mixing. Then we will

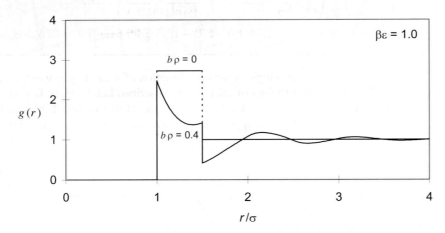

Figure 6.12 *The square-well fluid ($R = 1.5$) at zero density and at a packing fraction of $b\rho = 0.4$. The variable $\beta \equiv \frac{1}{kT}$.*

develop expressions that can be used to predict partitioning of components between various phases (e.g., vapor-liquid equilibria).

Example 6.6 Deriving your own equation of state

Appendix B shows how the following equation can be derived to relate the macroscopic equation of state to the microscopic properties characterized in terms of the square-well potential for $R = 1.5$.

$$Z = 1 + \frac{4\pi N_A \rho \sigma^3}{6} \{g(\sigma^+) - 1.5^3[1 - \exp(-\varepsilon/kT)]g(1.5\sigma^-)\} \qquad 6.35$$

Apply this result to develop your own equation of state with a radial distribution function of the form:

$$g(x) = \frac{\exp(-u/kT)}{(1 - 2b\rho/x^6)\{1 + 2Sb\rho/x^6\}} \qquad 6.36$$

where $x = r/\sigma$, $b = \pi N_A \sigma^3/6$, and S is the "Student" parameter. You pick a number for S, and this will be your equation of state. Evaluate your equation of state at $\varepsilon/kT = 0.5$ and $b\rho = 0.4$.

Solution:

At first glance, this problem may look outrageously complicated, but it is actually quite simple. We only need to evaluate the radial distribution function at $x = 1$ and $x = 1.5$ and insert these two results into Eqn. 6.35.

$$g(\sigma^+) = \frac{\exp(\varepsilon/kT)}{(1 - 2b\rho)\{1 + 2Sb\rho\}} \qquad 6.37$$

$$g(1.5\sigma^-) = \frac{\exp(\varepsilon/kT)}{(1 - 0.18b\rho)\{1 + 0.18Sb\rho\}} \qquad 6.38$$

$$Z = 1 + \frac{4b\rho}{1 - 2b\rho}\frac{\exp(\varepsilon/kT)}{1 + Sb\rho} - \frac{13.5b\rho[\exp(\varepsilon/kT) - 1]}{(1 - 0.18b\rho)(1 + 0.18Sb\rho)} \qquad 6.39$$

Supposing $S = 3$, $Z(0.5,0.4) = 1 + 4\cdot0.4\cdot1.648/(0.2\cdot1.2) - 13.5\cdot0.4\cdot0.648/(0.92\cdot1.216) = 8.86$.

Congratulations! You have just developed your own equation of state. If you want to apply it, you may want to match its properties to the critical point as described below. Have fun with it and feel free to experiment with different approximations for the radial distribution function. Hansen and McDonald[1] describe several systematic approaches to developing such approximations if you would like to know more.

The Virial Equation

The second virial coefficient can be easily derived using the concepts presented in this section, together with a little more mathematics. Advanced chemistry and physics texts customarily derive the virial equation as an expansion in density,

---

1. Hansen, J.P., McDonald, I.R., *Theory of Simple Liquids*, Academic Press, 1986.

$$Z = 1 + B\rho + C\rho^2 + D\rho^3 + \ldots \tag{6.40}$$

The result of the advanced derivation is that each virial coefficient can be expressed exactly as an integral over the intermolecular interactions characterized by the potential function. Even at the introductory level we can illustrate this approach for the second virial coefficient. Comparing the virial equation at low-density, $Z = 1 + B\rho$, with Eqn. 6.29, we can see that the second virial coefficient is related to the radial distribution function at low density. Inserting the low-density form of the radial distribution function as given by Eqn. 6.34, and subsequently integrating by parts (the topic of homework problem 6.23), we find

$$B = 2\pi N_A \int_0^\infty \left(1 - \exp\left(-\frac{u}{kT}\right)\right) r^2 dr \tag{6.41}$$

This relationship is particularly valuable, because experimental virial coefficient data may be used to obtain parameter values for pair potentials.

The van der Waals Equation of State

As we discussed in previous sections, the predecessor of modern cubic equations of state is the equation of van der Waals. The beauty of his argument is that detailed knowledge of the radial distribution function is not necessary, only the kind of general knowledge of its existence as described in the preceding section. Van der Waals started with the pressure equation and reasoned that the integral could be broken into two parts: a repulsive part and an attractive part.

$$\frac{P}{\rho RT} = 1 - \frac{\rho N_A^2}{6RT} \int_0^\sigma r\left(\frac{du}{dr}\right) g(r) \ 4\pi r^2 dr \ - \ \frac{\rho N_A^2}{6RT} \int_\sigma^\infty r\left(\frac{du}{dr}\right) g(r) \ 4\pi r^2 dr \tag{6.42}$$

Each integral can now be analyzed separately. Mathematically, analysis of the first integral is difficult because the number of atoms in the range 0 to σ is zero except at $r = \sigma$, and the derivative of the potential is zero except at $r = \sigma$, where it is basically equal to infinity. The proper mathematical analysis is presented in Appendix B. The result of that analysis is that the only quantity of interest in the repulsive range is the value of the radial distribution function at $r = \sigma$, i.e., the "contact value" of the radial distribution function with the conclusion,

$$\frac{P^{HS}}{\rho RT} = 1 + 4b\rho g^{HS}(\sigma) \tag{6.43}$$

Technically, the contact value of the radial distribution function depends on the strength of the attractive forces, as well as the repulsive forces. But we can identify the repulsive contribution to the contact value as the part that remains when the temperature approaches infinity. The attractive forces have no impact on the fluid structure in that limit. Then, we can treat the first part of the integral just as if it was simply for the hard-sphere potential. Considering the low-density limit as outlined above, the value of the radial distribution function at contact must approach unity at low densities. The high-density limiting value for the radial distribution function at contact must diverge at the close-packing limit, for the same reasons that the pressure diverges. The only way to characterize this divergence is to perform computer simulations based on the hard-sphere potential. This has been done[1] and the result is:

$$g(\sigma) \sim 1/(1 - 1.5b\rho) \text{ (high density limit for hard spheres)} \qquad 6.44$$

Unfortunately, this result was about 100 years too late for van der Waals' benefit. Even so, the approximate reasoning applied by van der Waals has turned out to be remarkably accurate. He missed the precise values of the close-packed density and the second virial coefficient for hard-sphere fluids, but the qualitative trends were correct. The fact that the van der Waals equation still provides the basis of modern engineering equations of state is a testimonial to the utility of simple physical reasoning in engineering analysis. With a little effort, you can learn from van der Waals' example. Making the substitution into Eqn. 6.43, we have:

$$\frac{-\rho N_A^2}{6RT} \int_0^\sigma r\left(\frac{du}{dr}\right) g(r)\, 4\pi\, r^2\, dr \;\approx\; 4b\rho g(\sigma) = \frac{4b\rho}{(1 - 1.5b\rho)} \sim \frac{b\rho}{(1 - b\rho)} \qquad 6.45$$

As for the second integral of Eqn. 6.42, this basically represents the attractive contribution to the pressure. Consider, for example, what happens to this integral when u is given by the Sutherland potential. Then,

$$\frac{\rho N_A^2}{6RT} \int_\sigma^\infty r\left(\frac{du}{dr}\right) g(r)\, 4\pi r^2 = \frac{\rho N_A^2 \sigma^3 \varepsilon}{6RT} \int_1^\infty x\left(\frac{d(u/\varepsilon)}{dx}\right) g(x)\, 4\pi\, x^2\, dx \text{ where } x \equiv r/\sigma \qquad 6.46$$

The integral on the right hand side is independent of the particular substance of interest because the only way of distinguishing different substances in the Sutherland potential is by different values of σ and ε. By factoring the σ^3 and ε out of the integral, we obtain an integral which can be applied universally to any substance. Van der Waals did not have information available about $g(r)$; therefore, he made the approximation that the value of this integral was some universal constant independent of T and ρ. This may seem somewhat crude since we know that $g(r)$ changes significantly with respect to density, but the way that $g(r)$ oscillates about unity leads to a weak density dependence for the integral. When this universal constant is factored in with σ^3 and ε, a single, substance-dependent constant is obtained,

$$a \equiv \frac{4\pi\, N_A \sigma^3 N_A \varepsilon}{6} \left\{ \frac{(g(\sigma) - g^{HS}(\sigma))}{N_A \varepsilon} + \int_{1^+}^\infty x\left(\frac{du/\varepsilon}{dx}\right) g(x)\, x^2\, dx \right\} \qquad 6.47$$

The resulting equation of state is:

$$\boxed{Z = 1 + \frac{b\rho}{(1 - b\rho)} - \frac{a\rho}{RT} = \frac{1}{(1 - b\rho)} - \frac{a\rho}{RT}} \qquad 6.48$$

This is the equation presented in Section 6.5. At the present time, we can use molecular simulations to investigate the inaccuracies of the attractive term as well as the repulsive term. These simulations show that the temperature and density dependence of the attractive part of the van der Waals equation of state are actually quite accurate. The primary sources of inaccuracy are in the repulsive term as discussed above.

---

1. Tobochnik, J., Chapin, P.M., J. Chem. Phys., 88:5824 (1988).

We conclude these theoretical developments with the comment that this same analysis will be repeated in the treatment of mixtures. At that time, it should become apparent that the extension to mixtures is primarily one of accounting; the conceptual framework is identical. The sooner you master the concepts of separate contributions from repulsive forces and attractive forces, the sooner you will master your understanding of fluid behavior from the molecular scale to the macro scale.

6.9 MATCHING THE CRITICAL POINT

The capability of a relatively simple equation to represent the complex physical phenomena illustrated in Figs. 6.7–6.12 is a tribute to the genius of van der Waals. His method for characterizing the difference between subcritical and supercritical fluids was equally clever. He recognized that, at the critical point,

$$\left(\frac{\partial P}{\partial \rho}\right)_T = 0 \quad \text{and} \quad \left(\frac{\partial^2 P}{\partial \rho^2}\right)_T = 0 \quad \text{at} \quad T_c, P_c$$

You can convince yourself that this is true by looking at the P versus ρ plots of Fig. 6.1 on page 195. From this observation, we obtain two equations that characterize the equation of state parameters a and b in terms of the critical constants T_c and P_c. In principle, this is all we need to say about this problem. In practice, however, it is much simpler to obtain results by recognizing another key feature of the critical point: the vapor and liquid roots are exactly equal at the critical point (and the spurious middle root is also equal). We can apply this latter insight by specifying that $(Z - Z_c)^3 = 0 = Z^3 - 3Z_cZ^2 + 3Z_c^2 Z - Z_c^3 = Z^3 - a_2Z^2 + a_1Z - a_0$. (See Appendix B). Equating the coefficients of these polynomials gives three equation in three unknowns: Z_c, A_c, B_c.

Example 6.7 Critical parameters for the van der Waals equation

Apply the above method to determine the values of Z_c, A_c, and B_c for the van der Waals equation.

Solution: Rearranging the equation in terms of A_c and B_c we have:

$$0 = Z^3 - (1 + B_c)Z^2 + A_cZ - A_cB_c = 0 = Z^3 - 3Z_cZ^2 + 3Z_c^2 Z - Z_c^3$$

By comparison, $Z_c = (1 + B_c)/3$; $A_c = 3Z_c^2$; $A_cB_c = Z_c^3$;

Substituting in favor of B_c, we have: $3Z_c^2 B_c = Z_c^2 (1 + B_c)/3$

Cancelling the Z_c^2 and solving we have $B_c = 1/8 = 0.125$. The other equations then give $Z_c = 0.375$ and $A_c = 27/64$.

6.10 SUMMARY AND CONCLUDING REMARKS

The simple physical observations and succinct mathematical models set forth in this chapter provide powerful tools for current chemical applications and provide an excellent example of model development that we would all do well to emulate. This chapter has illustrated applications of physical reasoning, dimensional analysis, asymptotic analysis, and parameter estimation that have set the standard for many modern engineering developments.

Furthermore, the final connection has been drawn between the molecular level and the macroscopic scale. In retrospect, the microscopic definition of entropy is relatively simple. It follows naturally from the elementary statistics of the binomial distribution. The qualitative description of molecular interaction energy is also simple; it was discussed in the introductory chapter. Last, but not least, the macroscopic description of energy is easy to understand; it gives the macroscopic energy balance. What is not so simple is the connection of the qualitative description of molecular energies with the macroscopic energy balance. This is the significant development of this chapter. Having complete descriptions of the molecular and macroscopic energy and entropy, all the "pieces to the puzzle" are now in our hands. What remains is to put the pieces together. This final step requires a fair amount of mathematics, but it is largely a straightforward application of tools that are readily available from elementary courses in calculus.

6.11   PRACTICE PROBLEMS

P6.1   For $T_r < 1$ and $P_r \approx P_r^{sat}$, the Peng-Robinson equation of state will have three roots corresponding to compressibility factors between zero and ten. The smallest root will be the compressibility factor of the liquid. The largest root will be the compressibility factor of the vapor and the middle root will not have physical significance. This gives us a general method for finding the compressibility factor of any fluid obeying the Peng-Robinson equation. For the iterative method, use an initial guess of $Z = 0$ to find the liquid roots and $Z = 1$ to find the vapor roots of methane at the following conditions:

T_r	P_r
0.9	0.55
0.8	0.26
0.7	0.10
0.6	0.03

Compare to experimental data from N.B Vargaftik. *Handbook of Physical Properties of Liquids and Gases*, 2nd ed., New York, NY: Hemisphere, 1975.

T_r	0.9	0.8	0.7	0.6
Z^L	0.0908	0.0413	0.0166	0.0050
Z^V	0.6746	0.8124	0.9029	0.9608

P6.2 (a) Estimate the value of the compressibility factor, Z, for neon at $P_r = 30$ and $T_r = 15$.
   (b) Estimate the density of neon at $P_r = 30$ and $T_r = 15$. (ANS. 1.14, 0.25 g/cc)

P6.3 Above the critical point or far from the saturation curve,[1] only one real root to the cubic equation exists. If we are using Newton's method, we can check how many phases exist by trying the two different initial guesses and seeing if they both converge to the same root. If they do, then we can assume that only one real root exists. Find the compressibility factors for methane at the following conditions, and identify whether they are vapor, liquid, or supercritical fluid roots. Compare your results to Z charts.

T_r	P_r	Z	Phase
0.9	0.77		
0.9	0.05		
1.1	2.00		

---

1. Fig. 6.7 on page 209 shows that three roots will exist at all pressures below P^{sat} when the reduced temp is low, but over a limited range near $T_r = 1$.

When Newton's method is applied with an initial guess of zero, erratic results are obtained at these conditions. Explain what is happening, and why, by plotting $F(Z)$ versus Z for each iteration.

P6.4 A rigid vessel is filled to one half its volume with liquid methane at its normal boiling point (111 K). The vessel is then closed and allowed to warm to 77°F. Calculate the final pressure using the Peng-Robinson equation. (ANS. 33.8 MPa)

P6.5 $4 \, m^3$ of methane at 20°C and 1 bar is roughly equivalent to 1 gal of gasoline in an automotive engine of ordinary design. If methane were compressed to 200 bar and 20°C, what would be the required volume of a vessel to hold the equivalent of 10 gal of gasoline? (ANS. 16 L)

P6.6 A carbon dioxide cylinder has a volume of $0.15 \, m^3$ and is filled to 100 bar at 38°C. The cylinder cools to 0°C. What is the final pressure in the cylinder and how much more CO_2 can be added before the pressure exceeds 100 bar? If you add that much CO_2 to the cylinder at 0°C, what will the pressure be in the cylinder on a hot, 38°C day? What will happen if the cylinder can stand only 200 bar? [Hint: $\log (P_r^{sat}) \approx (7(1 + \omega)/3) (1 - 1/T_r)$] (ANS. 3.5 MPa, 38 MPa, boom!)

6.12   HOMEWORK PROBLEMS

6.1 The compressibility factor chart provides a quick way to assess when the ideal gas law is valid. For the following fluids, what is the minimum temperature in K where the fluid has a gas phase compressibility factor greater than 0.95 at 30 bar?

(a) nitrogen
(b) carbon dioxide
(c) ethanol

6.2 A container having a volume of 40 L contains one of the following fluids at the given initial conditions. After a leak, the temperature and pressure are remeasured. For each option, determine the kilograms of fluid lost due to the leak, using:

(a) compressibility factor charts
(b) the Peng-Robinson equation

Options:

(i)   methane $T^i = 300$ K, $P^i = 100$ bar, $T^f = 300$ K, $P^f = 50$ bar
(ii) propane $T^i = 300$ K, $P^i = 50$ bar, $T^f = 300$ K, $P^f = 0.9$ bar
(iii) n-butane $T^i = 300$ K, $P^i = 50$ bar, $T^f = 300$ K, $P^f = 10$ bar

6.3 From experimental data it is known that at moderate pressures the volumetric equation of state may be written as $PV = RT + B \cdot P$, where the second virial coefficient B is a function of temperature only. Data for methane are given by Dymond and Smith (1969) as:[1]

T(K)	120	140	160	180	200	250	300	350	400	500	600
B(cm^3/mole)	−284	−217	−169	−133	−107	−67	−42	−27.0	−15.5	−0.5	8.5

1. Dymond, J.H., Smith, E.B., *The Virial Coefficients of Pure Gases and Mixtures,* Oxford University Press, New York (1969).

(a) Identify the Boyle temperature (the temperature at which $B = 0$) and the inversion temperature (the temperature at which $(\partial T/\partial P)_H = 0$) for gaseous methane. [Hint: Plot B versus T^{-1} and regress a trendline, then differentiate analytically.]

(b) Plot these data versus T^{-1} and compare to the curve generated from Eqn. 6.7. Use points without lines for the experimental data and lines without points for the theoretical curve.

6.4 Data for hydrogen are given by Dymond and Smith (1969) as:

T(K)	19	25	30	40	50	75	100	150	200	300	400
B(cm^3/mole)	−164	−111	−85	−54	−35	−12	−1.9	7.1	11.3	14.8	15.2

(a) Plot these data versus T^{-1} and compare to the results from the generalized virial equation (Eqn. 6.7). Suggest a reason that this specific compound does not fit the generalized equation very accurately. Use points without lines for the experimental data and lines without points for the theoretical curve.

(b) Use the generalized virial equation to speculate whether a small leak in an H_2 line at 300 bar and 298 K might raise the temperature of H_2 high enough to cause it to spontaneously ignite.

6.5 Estimate the liquid density (g/cm^3) of propane at 298 K and 10 bar. Compare the price per kilogram of propane to the price per kilogram of regular gasoline assuming the cost of 5 gal of propane for typical gas grills is roughly \$20. The density of regular gasoline can be estimated by treating it as pure isooctane (2,2,4-trimethylpentane $\rho = 0.692$ g/cm^3) at 298 K and 1 bar.

6.6 N.B. Vargaftik (1975)[1] lists the following experimental values for the specific volume of isobutane at 175°C. Compute theoretical values and their percent deviations from experiment by

(a) the generalized charts
(b) the Peng-Robinson equation

P (atm)	10	20	35	70
V (cm^3/g)	60.7	27.79	13.36	3.818

6.7 Evaluate $(\partial P/\partial V)_T$ for the equation of state:

$$P = RT/(V - b)$$

6.8 Evaluate $(\partial P/\partial T)_V$ for the equation of state:

$$P = RT/(V - b) + a/T^{3/2}$$

1. Vargaftik, N.B., *Handbook of Physical Properties of Liquids and Gases,* 2nd ed., Hemisphere, New York (1975).

6.9 Evaluate $\left(\dfrac{\partial P}{\partial T}\right)_V$ for the Redlich-Kwong equation of state

$$P = \frac{RT}{V-b} - \frac{a}{T^{1/2}V(V+b)}\,,\text{ where } a \text{ and } b \text{ are temperature-independent parameters.}$$

6.10 (a) The derivative $(\partial V/\partial T)_P$ is tedious to calculate by implicit differentiation of an equation of state such as the Peng-Robinson equation. Show that calculus permits us to find the derivative in terms of derivatives of pressure, which are easy to find, and provide the formula for this equation of state.

  (b) Using the Peng-Robinson equation, calculate the isothermal compressibility of ethylene for saturated vapor and liquid at the following conditions: $\{T_r = 0.7, P = 0.414\text{ MPa}\}$; $\{T_r = 0.8, P = 1.16\text{ MPa}\}$; $\{T_r = 0.9, P = 2.60\text{ MPa}\}$.

6.11 Plot P_r versus ρ_r for the Peng-Robinson equation with $T_r = [0.7, 0.9, 1.0]$, showing both vapor and liquid roots in the two-phase region. Assume $\omega = 0.040$ as for N_2. Include the entire curve for each isotherm, as illustrated in Fig. 6.1 on page 195. Also show the horizontal line that connects the vapor and liquid densities at the saturation pressure. Use lines without points for the theoretical curves. Estimate T_r^{sat} by $\log(P_r^{sat}) = 2.333(1 + \omega)$ $(1 - 1/T_r^{sat})$.

6.12 When cubic equations of state give three real roots for Z, usually the smallest root is the liquid root and the largest is the vapor root. However, the Peng-Robinson equation can give real roots at high pressure that differ from this pattern. To study this behavior, tabulate all the roots found for the specified gas and pressures. As the highest pressures are approached at this temperature, is the fluid a liquid or gas? Which real root (smallest, middle, or largest) represents this phase at the highest pressure, and what are the Z values at the specified pressures?

  (a) ethylene at 250 K and 1, 3, 10, 100, 150, 170, 175, and 200 MPa
  (b) hexane at 400 K and 0.2, 0.5, 1, 10, 100, 130 and 150 MPa

6.13 Within the two-phase envelope, one can draw another envelope representing the limits of supercooling of the vapor and superheating of liquid that can be observed in the laboratory; along each isotherm these are the points for which $(\partial P/\partial \rho)_T = 0$. Obtain this envelope for the Peng-Robinson equation, and plot it on the same figure as generated in problem 6.11. This is the spinodal curve. The region between the saturation curve and the curve just obtained is called the metastable region of the fluid. Inside the spinodal curve, the fluid is unconditionally unstable. The saturation curve is called the binodal curve. Outside, the fluid is entirely stable. It is possible to enter the metastable region with hot water by heating at atmospheric pressure in a very clean flask. Sooner or later, the superheated liquid becomes unstable, however. Describe what would happen to your flask of hot water under these conditions and a simple precaution that you might take to avoid these consequences.

6.14 Develop a spreadsheet that computes the values of the compressibility factor as a function of reduced pressure for several isotherms of reduced temperature using the Lee-Kesler (1975) equation of state (*AIChE J.*, 21:510). A tedious but straightforward way to do this is to tabulate reduced densities from 0.01 to 10 in the top row and reduced temperatures in the first column. Then, enter the Lee-Kesler equation for the compressibility factor of the simple fluid in one of the central cells and copy the contents of that cell to all other cells in the table. Next, copy that entire table to a location several rows lower. Replace the contents of the new cells by the relation $P_r = Z \cdot \rho_r \cdot T_r$. You now have a set of reduced pressures corresponding to a set of compressibility factors for each isotherm, and these can be plotted to reproduce the chart in the chapter, if you like. Copy this spreadsheet to a new one, and change the values of the B, C, D, and E parameters to correspond to the reference fluid. Finally, copy the simple fluid worksheet to a new worksheet, and replace the contents of the compressibility factor cells by the formula: $Z = Z_0 + \omega(Z_{ref} - Z_0)/\omega_{ref}$, where the Z_{ref} and Z_0 refer to numbers in the cells of the other worksheets.

6.15 The Soave-Redlich-Kwong equation[1] is given by:

$$P = \frac{RT\rho}{(1 - b\rho)} - \frac{a\rho^2}{1 + b\rho} \qquad \text{or} \qquad Z = \frac{1}{(1 - b\rho)} - \frac{a}{bRT} \cdot \frac{b\rho}{1 + b\rho} \qquad 6.49$$

where ρ = molar density = $n/\underline{V}$

$$a \equiv a_c\alpha; \quad a_c \equiv 0.42748\frac{R^2 T_c^2}{P_c} \qquad\qquad b \equiv 0.08664\frac{RT_c}{P_c} \qquad 6.50$$

$$\alpha \equiv \left[1 + \kappa\left(1 - \sqrt{T_r}\right)\right]^2 \qquad \kappa \equiv 0.480 + 1.574\omega - 0.176\omega^2 \qquad 6.51$$

T_c, P_c, and ω are reducing constants according to the principle of corresponding states. Solve for the parameters at the critical point for this equation of state (a_c, b_c, and Z_c) and list the next five significant figures in the sequence 0.08664.......

6.16 Show that $B_c = bP_c/RT_c = 0.07780$ for the Peng-Robinson equation by setting up the cubic equation for B_c analogous to the van der Waals equation and solving analytically as described in Appendix B.

6.17 Determine the values of ε/kT_c, Z_c and b_c in terms of T_c and P_c for the equation of state given by

$$Z = \frac{1 + 2b\rho}{1 - 2b\rho} - Fb\rho$$

where $F = \exp(\varepsilon/kT) - 1$.

6.18 Consider the equation of state

$$Z = 1 + \frac{4c\eta}{1 - 1.9\eta} - \frac{a\eta}{bRT}$$

where $\eta = b/V$.

(a) Determine the relationships between a, b, c and T_c, P_c, Z_c.
(b) What practical restrictions are there on the values of Z_c that can be modeled with this equation?

---

1. Soave, G., *Chem. Eng. Sci.*, 27:1197 (1972).

6.19 The ESD equation of state[1] is given by

$$\frac{PV}{RT} = 1 + \frac{4\langle c\eta \rangle}{1 - 1.9\eta} - \frac{9.5\langle qY\eta \rangle}{1 + 1.7745\langle Y\eta \rangle}$$

$\eta = b\rho$, c is a "shape parameter" which represents the effect of non-sphericity on the repulsive term, $q = 1 + 1.90476(c - 1)$, and Y is a temperature-dependent function whose role is similar to the temperature dependence of the a parameter in the Peng-Robinson equation.

(a) Use the methods of Appendix B to fit b and Y to the critical point for ethylene using $c = 1.3$.

(b) Use the methods of Appendix B to fit c, b, and Y at the critical point for ethylene using the experimental value of the critical compressibility factor.

6.20 The discussion in the chapter focuses on the square-well fluid, but the same reasoning is equally applicable for any model potential function. Illustrate your grasp of this reasoning with some sketches analogous to those in the chapter.

(a) Sketch the radial distribution function vs. radial distance for a low density Lennard-Jones (LJ) fluid. Describe in words why it looks like that.
(b) Repeat the exercise for the high density LJ fluid. Also sketch on the same plot the radial distribution function of a hard-sphere fluid at the same density. Compare and contrast the hard-sphere fluid to the LJ fluid at high density.

6.21 Suppose that a reasonable approximation to the radial distribution function is

$$g(x) = \begin{cases} 0 & r < \sigma \\ \dfrac{1}{1 - 2b\rho/x^6} + \dfrac{2.5F}{(1 + x^6 Fb\rho)} & r \geq \sigma \end{cases}$$

where $x = r/\sigma$, $F = \exp(\varepsilon/kT) - 1$, $b = \pi N_A \sigma^3/6$. Derive an expression for the equation of state of the square-well fluid based on this approximation. Evaluate the equation of state at $\rho N_A \sigma^3 = 0.6$ and $\varepsilon/kT = 1$.

6.22 Suppose that a reasonable approximation for the radial distribution function is $g(r) = 0$ for $r < \sigma$, and

$$g(r) = \frac{\exp(-u(r)/kT}{1 - b\rho}$$

for $r \geq \sigma$, where u is the square-well potential and $b = \pi N_A \sigma^3/6$. Derive an equation of state for the square-well fluid based on this approximation.

1. Elliott, J.R., Suresh, S.J., Donohue, M.D., *Ind. Eng. Chem. Res.* 29:1476 (1990).

6.23 The truncated virial equation (density form) is $Z = 1 + B\rho$. According to Eqn. 6.29, the virial coefficient is given by

$$B = -\frac{2}{3}\frac{\pi N_A}{kT}\int_0^\infty \left(\frac{du}{dr}\right)g(r)r^3 dr$$

where the low pressure limit of $g(r)$ given by Eqn. 6.34 is to be used. Another commonly cited equation for the virial coefficient is Eqn. 6.41. Show that the two equations are equivalent by the following steps:

(a) Beginning with $B = -\frac{2}{3}\frac{\pi N_A}{kT}\int_0^\infty \left(\frac{du}{dr}\right)g(r)r^3 dr$, insert the low pressure limit for $g(r)$, and simplify as much as possible.

(b) Integrate by parts to obtain

$$\frac{1}{3}\int_0^\infty d(r^3 \exp(-u/kT)) = \int_0^\infty \exp(-u/kT)r^2 dr - \frac{1}{3kT}\int_0^\infty \left(\frac{du}{dr}\right)g(r)r^3 dr$$

(c) Show that the left hand side of the answer to part b may be written as $\int_0^\infty r^2 dr$ for a physically realistic pair potential. Then combine integrals to complete the derivation of Eqn. 6.41.

6.24 The virial coefficient can be related to the pair potential by Eqn. 6.41.

(a) Derive the integrated expression for the second virial coefficient in terms of the square-well potential parameters ε/k, σ, R.
(b) Fit the parameters to the experimental data for argon.[1]

T(K)	B (cm^3/mole)	T(K)	B (cm^3/mole)
85	−251	400	−1
100	−183.5	500	7
150	−86.2	600	12
200	−47.4	700	15
250	−27.9	800	17.7
300	−15.5	900	20

(c) Fit the parameters to the experimental data for propane.[1]

T(K)	B (cm^3/mole)	T(K)	B (cm^3/mole)
250	−584	350	−276
260	−526	375	−238
270	−478	400	−208
285	−424	430	−177
300	−382	470	−143
315	−344	500	−124
330	−313	550	−97

1. Dymond, J.H., Smith, E. B. *The Virial Coefficients of Pure Gases and Mixtures.* New York, NY: Oxford University Press, 1980.

6.25 One suggestion for a simple pair potential is the triangular potential

$$u(r) \; = \; \infty \;\; \text{for } r < \sigma \text{ and } u(r) = 0 \text{ for } r > R\sigma$$

$$u(r) \; = \; -\varepsilon\left[\frac{(r/\sigma) - R}{1 - R}\right] \text{ for } \sigma < r < R\sigma$$

Derive the second virial coefficient and fit the parameters σ, ε, and R to the virial coefficient data for argon tabulated in problem 6.24.

CHAPTER

DEPARTURE FUNCTIONS

All the effects of nature are only the mathematical consequences of a small number of immutable laws.

LaPlace

Maxwell's relations make it clear that changes in any one variable can be represented as changes in some other pair of variables. In chemical processes, we are often concerned with the changes of enthalpy and entropy as functions of temperature and pressure. As an example, recall the operation of a reversible turbine between some specified inlet conditions of T and P and some specified outlet pressure. Using the techniques of Unit I, we typically determine the outlet T and q which match the upstream entropy then solve for the change in enthalpy. Applying this approach to steam should seem quite straightforward at this stage. But what if our process fluid is a new refrigerant or a multicomponent natural gas, for which no thermodynamic charts or tables exist? How would we analyze this process? In such cases, we need to have a general approach that is applicable to any fluid. A central component of developing this approach is the ability to express changes in variables of interest in terms of variables which are convenient (via Maxwell's relations). The other important consideration is the choice of "convenient" variables. Experimentally, P and T are preferred; however, V and T are easier to use with cubic equations of state.

The equation of state describes the effects of pressure on our system properties, including the low pressure limit of the ideal gas law. However, integration of properties over pressure ranges is relatively complicated because general equations of state express changes in thermodynamic variables as functions of density, not pressure. Recall that our engineering equations of state are typically of the pressure-explicit form:

$$P = P(V,T) \qquad\qquad 7.1$$

and the cubic equations of state typically cannot be rearranged to the volume explicit form:

$$V = V(P,T) \qquad\qquad 7.2$$

Therefore, development of thermodynamic properties based on $\{V, T\}$ will be consistent with the most widely used forms of equations of state, and deviations from ideal gas behavior will be

> ❗ Experimentally, P and T are usually specified. However, equations of state are typically density (volume) dependent.

expressed with the density-dependent formulas for departure functions in Sections 7.1-7.5. In Section 7.6, we present the pressure-dependent form useful for the virial equation. In Section 7.7, we show how reference states are used in tabulating thermodynamic properties.

7.1 THE DEPARTURE FUNCTION PATHWAY

Suppose we desire to calculate the change in U in a process which changes state from (V_L, T_L) to (V_H, T_H). Now, it may seem unusual to pose the problem in terms of T and V, since we stated above that our objective was to use T and P. The choice of T and V as variables is because we must work with equations of state that are functions of volume. The volume corresponding to any pressure is rapidly found by the methods of Chapter 6. We have two obvious pathways for calculating a change in U using $\{V, T\}$ as state variables as shown in Figure 7.1. Pathway A consists of an isochoric step followed by an isothermal step. Pathway B consists of an isothermal step followed by an isochoric step. Naturally, since U is a state function, ΔU for the process is the same by either pathway. Recalling the relation for $dU(T,V)$, ΔU may be calculated by either

Path *A:*

$$\Delta U = \int C_V\big|_{V_L} dT + \int \left[T\left(\frac{\partial P}{\partial T}\right)_V - P \right]\bigg|_{T_H} dV \qquad 7.3$$

or Path *B:*

$$\Delta U = \int C_V\big|_{V_H} dT + \int \left[T\left(\frac{\partial P}{\partial T}\right)_V - P \right]\bigg|_{T_L} dV \qquad 7.4$$

We have previously shown, in Example 6.5 on page 210, that C_V depends on volume for a real fluid. Therefore, even though we could insert the equation of state for the integrand of the second integral, we must also estimate C_V by the equation of state for at least one of the volumes, using the results of Example 6.5. Not only is this tedious, but estimates of C_V by equations of state tend to be less reliable than estimates of other properties.

To avoid this calculation, we devise an equivalent pathway of three steps. First, we calculate the difference in internal energy between the real fluid and an ideal gas at the same $\{P, T\}$ known as

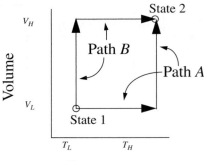

Figure 7.1 *Comparison of two alternate pathways for calculation of a change of state.*

the *departure function*. Then, we calculate the changes in state for an ideal gas. Then, we add the difference in the final state between the real fluid and an ideal gas. Being careful with signs of the terms, we may combine the calculations for the desired result:

$$\Delta U = U_2 - U_1 = (U_2 - U_2^{ig}) + (U_2^{ig} - U_1^{ig}) - (U_1 - U_1^{ig}) \qquad 7.5$$

The calculation can be generalized to any fundamental property from the set $\{U, H, A, G, S\}$, using the variable M to denote the property

$$\Delta M = M_2 - M_1 = (M_2 - M_2^{ig}) + (M_2^{ig} - M_1^{ig}) - (M_1 - M_1^{ig}) \qquad 7.6$$

The steps can be seen graphically in Fig. 7.2.

> ❶ Departure functions permit us to use the ideal gas calculations that are easy, and incorporate a departure property value for the initial and final state.

Note that all the ideal gas terms in Eqns. 7.5 and 7.6 cancel to yield the desired property difference. A common mistake is to get the sign wrong on one of the terms in these equations. Make sure that you have the terms in the right order by checking for cancelation of the ideal gas terms. The advantage of this pathway is that all temperature calculations are done in the ideal gas state where:

$$C_P^{ig} = C_V^{ig} + R$$
$$dU^{ig} = C_V^{ig} dT \qquad 7.7$$

and the ideal gas heat capacities are pressure- (and volume-) independent (see Example 5.8 on page 185).

To derive the formulas to be used in calculating the values of enthalpy, internal energy, and entropy for real fluids, we must apply our fundamental property relations once and our Maxwell's relations once.

7.2   INTERNAL ENERGY DEPARTURE FUNCTION

Fig. 7.3 schematically compares a real gas isotherm and an ideal gas isotherm at identical temperatures. Note that at a given $\{T, P\}$ the volume of the real fluid is V, and the ideal gas volume is $V^{ig} = RT/P$. Similarly, the ideal gas pressure is not equal to the true pressure when we specify $\{T, V\}$. Note that we

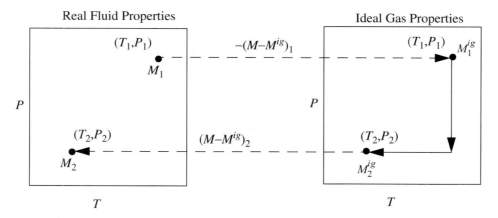

Figure 7.2 *Illustration of calculation of state changes for a generic property M using departure functions where M is U, H, S, G, or A.*

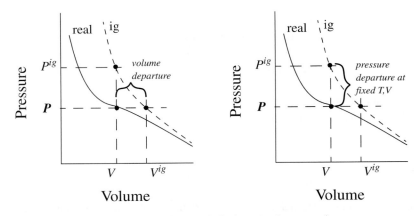

Figure 7.3 *Comparison of real fluid and ideal gas isotherms at the same temperature, demonstrating the departure function, and the departure function at fixed T,V.*

may characterize the departure from ideal gas behavior in two ways: 1) at the same $\{T, V\}$; or 2) at the same $\{T, P\}$. We will find it convenient to use both concepts, but we need nomenclature to distinguish between the two departure characterizations. When we refer to the departure of the real fluid property and the same ideal gas property at the same $\{T, P\}$, we will call it the *departure function, and use the notation* $U - U^{ig}$. When we compare the departure at the same $\{T,V\}$ we will call it the *departure function at fixed T,V, and designate it as* $(U - U^{ig})_{TV}$.[1]

The departure for property M is at fixed T and P, and is given by $(M-M^{ig})$. The departure at fixed T,V is also useful (particularly in Chapter 10) and is denoted by $(M-M^{ig})_{TV}$.

To calculate the change in internal energy along an isotherm for the real fluid, we write:

$$U(T, V) - U(T, \infty) = \int_{\infty}^{V} dU = \int_{\infty}^{V} \left(\frac{\partial U}{\partial V}\right)_T dV \qquad 7.8$$

For an ideal gas:

$$U^{ig}(T, V) - U^{ig}(T, \infty) = \int_{\infty}^{V} dU = \int_{\infty}^{V} \left(\frac{\partial U}{\partial V}\right)_T^{ig} dV \qquad 7.9$$

Since the real fluid approaches the ideal gas at infinite volume, we may take the difference in these two equations to find the departure function at fixed T,V:

$$(U - U^{ig})_{TV} = \int_{\infty}^{V} \left[\left(\frac{\partial U}{\partial V}\right)_T - \left(\frac{\partial U}{\partial V}\right)_T^{ig} \right] dV \qquad 7.10$$

We have obtained a calculation with the real fluid in our desired state (T, P, V); however, we are referencing an ideal gas at the same volume rather than the same pressure. To see the difference, consider methane at 250 K, 10 MPa, and 139 cm^3/mole. The volume of the ideal gas should be

---

1. Generalization of the ideal gas state is also possible beyond the two choices discussed here. For a discussion, see *The Properties of Gases and Liquids*, 4th ed. R.C. Reid, J.M. Prausnitz, B.E. Poling, McGraw-Hill, 1987.

$V^{ig} = 8.314 \cdot 250/10 = 208$ cm^3/mole. To obtain the departure function denoted by $(U - U^{ig})$ (which is referenced to an ideal gas at the same pressure), we must add a correction to change the ideal gas state to match the pressure rather than the volume. Note in Fig. 7.3 that the real state is the same for both departure functions—the difference between the two departure functions has to do with the volume used for the ideal gas part of the calculation. The result is

$$U - U^{ig} = (U - U^{ig})_{TP} = (U - U^{ig})_{TV} - (U^{ig}_{TP} - U^{ig}_{TV}) = (U - U^{ig})_{TV} - \int_V^{V^{ig}} \left(\frac{\partial U}{\partial V}\right)_T^{ig} dV \quad 7.11$$

We have already solved for $(\partial U^{ig}/\partial V)_T$ (see Example 5.7 on page 184), and found that it is equal to zero. We are fortunate in this case because the internal energy of an ideal gas does not depend on the volume. When it comes to properties involving entropy, however, the dependency on volume will require careful analysis. Then the systematic treatment developed above will be quite valuable.

$$\left(\frac{\partial U}{\partial V}\right)_T = T\left(\frac{\partial P}{\partial T}\right)_V - P \text{ and } \left(\frac{\partial U}{\partial V}\right)_T^{ig} = 0 \qquad 7.12$$

Making these substitutions, we have

$$\boxed{U - U^{ig} = \int_\infty^V \left[T\left(\frac{\partial P}{\partial T}\right)_V - P\right] dV} \qquad 7.13$$

If we transform the integral to density, the resulting expression is easier to integrate for a cubic equation of state. Recognizing $dV = -d\rho/\rho^2$, and as $V \to \infty$, $\rho \to 0$, thus,

$$\boxed{\frac{U - U^{ig}}{RT} = \int_0^\rho \left[\frac{P}{\rho RT} - \frac{1}{\rho R}\left(\frac{\partial P}{\partial T}\right)_\rho\right] \frac{d\rho}{\rho} = -\int_0^\rho T\left(\frac{\partial Z}{\partial T}\right)_\rho \frac{d\rho}{\rho}} \qquad 7.14$$

The above equation applies the product rule in a way that may not be obvious at first:

$$T\left(\frac{\partial Z}{\partial T}\right)_V = \frac{T}{\rho RT}\left(\frac{\partial P}{\partial T}\right)_V - \frac{PT}{R\rho T^2}\left(\frac{\partial T}{\partial T}\right)_V = \frac{1}{\rho R}\left(\frac{\partial P}{\partial T}\right)_V - Z \qquad 7.15$$

We now have a compact equation to apply to any equation of state. Knowing $Z = Z(T,\rho)$, we simply differentiate once, cancel some terms, and integrate. This a perfect sample application of the multivariable calculus which should be familiar at this stage in the curriculum. More important, we have developed a systematic approach to solving for any departure function. The steps are:

1. Take the derivative with respect to volume.
2. Take the difference between the real fluid and the ideal gas.
3. Integrate the difference from infinite volume (where the real fluid and the ideal gas are the same).
4. Correct the ideal gas reference state.
5. Substitute the expressions from classical thermodynamics.
6. Rearrange in terms of density and compressibility factor to make it more compact.

Some of these steps could have been omitted for the internal energy, because $(\partial U^{ig}/\partial V)_T = 0$. To see the importance of all the steps, consider the entropy departure function.

7.3   ENTROPY DEPARTURE FUNCTION

To calculate the entropy departure, adapt Eqn. 7.11,

$$S - S^{ig} = (S - S^{ig})_{TV} - \int_V^{V^{ig}} \left(\frac{\partial S}{\partial V}\right)_T^{ig} dV \qquad 7.16$$

Inserting the integral for the departure at fixed $\{T, V\}$, we have (using a Maxwell relation),

$$S - S^{ig} = \int_\infty^V \left[\left(\frac{\partial S}{\partial V}\right)_T - \left(\frac{\partial S}{\partial V}\right)_T^{ig}\right] dV - \int_V^{V^{ig}} \left(\frac{\partial S}{\partial V}\right)_T^{ig} dV = \int_\infty^V \left[\left(\frac{\partial P}{\partial T}\right)_V - \left(\frac{\partial P}{\partial T}\right)_V^{ig}\right] dV - \int_V^{V^{ig}} \left(\frac{\partial P}{\partial T}\right)_V^{ig} dV$$

$$7.17$$

Since $\left(\frac{\partial P}{\partial T}\right)_V^{ig} = \frac{R}{V}$ (note that this is not zero whereas the analogous equation for energy was zero),

$$S - S^{ig} = \int_\infty^V \left[\left(\frac{\partial P}{\partial T}\right)_V - \frac{R}{V}\right] dV + R \ln \frac{V}{V^{ig}} \qquad 7.18$$

Recognizing $V^{ig} = RT/P$, $V/V^{ig} = PV/RT = Z$,

$$\boxed{\frac{S - S^{ig}}{R} = \int_\infty^V \left[\frac{1}{R}\left(\frac{\partial P}{\partial T}\right)_V - \frac{1}{V}\right] dV + \ln Z = \int_0^\rho \left[-T\left(\frac{\partial Z}{\partial T}\right)_\rho - (Z - 1)\right] \frac{d\rho}{\rho} + \ln Z} \qquad 7.19$$

Note the $\ln(Z)$ term on the end of this equation. It arises from the change in ideal gas $S_{TP}^{ig} - S_{TV}^{ig}$. Changes in states like this may seem pedantic and arcane, but they turn out to be subtle details that often make a big difference. In Example 6.3 on page 207, we determined vapor and liquid roots for Z. The vapor root was close to unity so $\ln(Z)$ would make little difference in that case. For the liquid root, however, $Z = 0.016$, and $\ln(Z)$ makes a substantial difference (the equivalent of a nearly 100-fold change in pressure). These arcane details surrounding the subject of state specification are the thermodynamicist's curse. We will see that the importance of state specification reappears in the discussion of mixtures and in the discussion of reacting systems.

7.4   OTHER DEPARTURE FUNCTIONS

❶ The departures for U and S are the building blocks from which the other departures can be written by combining the relations derived in the previous sections.

The remainder of the departure functions may be derived from the first two and the definitions:

$$H = U + PV \Rightarrow \frac{H - H^{ig}}{RT} = \frac{U - U^{ig}}{RT} + \frac{PV - RT}{RT} = \frac{U - U^{ig}}{RT} + Z - 1$$

$$A = U - TS \Rightarrow \frac{A - A^{ig}}{RT} = \frac{U - U^{ig}}{RT} - \frac{S - S^{ig}}{R}$$

$$7.20$$

where we have used $PV^{ig} = RT$ for the ideal gas in the enthalpy departure. Using $H - H^{ig}$ just derived,

$$\frac{G - G^{ig}}{RT} = \frac{H - H^{ig}}{RT} - \frac{S - S^{ig}}{R} \qquad 7.21$$

7.5   SUMMARY OF DENSITY-DEPENDENT FORMULAS

Formulas for departures at fixed T,P are listed below. These formulas are useful for an equation of state written most simply as $Z = f(T, \rho)$. For treating cases where an equation of state is written most simply as $Z = f(T,P)$, see Section 7.6.

$$\frac{\left(U - U^{ig}\right)}{RT} = \int_0^\rho -T\left[\frac{\partial Z}{\partial T}\right]_\rho \frac{d\rho}{\rho} \qquad 7.22$$

$$\frac{\left(S - S^{ig}\right)}{R} = \int_0^\rho \left[-T\left[\frac{\partial Z}{\partial T}\right]_\rho - (Z - 1)\right]\frac{d\rho}{\rho} + \ln Z \qquad 7.23$$

$$\frac{\left(H - H^{ig}\right)}{RT} = \int_0^\rho -T\left[\frac{\partial Z}{\partial T}\right]_\rho \frac{d\rho}{\rho} + Z - 1 \qquad 7.24$$

$$\frac{\left(A - A^{ig}\right)}{RT} = \int_0^\rho \frac{(Z - 1)}{\rho} d\rho - \ln Z \qquad 7.25$$

$$\frac{\left(G - G^{ig}\right)}{RT} = \int_0^\rho \frac{(Z - 1)}{\rho} d\rho + (Z - 1) - \ln Z \qquad 7.26$$

Useful formulas at fixed T,V include:

$$\frac{\left(A - A^{ig}\right)_{TV}}{RT} = \int_0^\rho \frac{(Z - 1)}{\rho} d\rho \qquad 7.27$$

$$\frac{\left(S - S^{ig}\right)_{TV}}{R} = \int_0^\rho \left[-T\left[\frac{\partial Z}{\partial T}\right]_\rho - (Z - 1)\right]\frac{d\rho}{\rho} \qquad 7.28$$

The tasks that remain are to select a particular equation of state, take the appropriate derivatives, make the substitutions, develop compact expressions, and add up the change in properties. The good news is that many years of engineering research have yielded several preferred equations of state (see Appendix D) which can be applied generally to any application with a reasonable degree of accuracy. For the purposes of the text, we have chosen the Peng-Robinson equation to illustrate the principles of calculating properties. This means that our applications in this course will almost always apply the Peng-Robinson equation of state over and over. However, many applications require higher accuracy; new equations of state are being developed all the time. This means that it is necessary for each student to know where the final equations come from and how to adapt them to new situations as they come along. The derivations are somewhat tedious, however, and it may be helpful to see how simple their eventual application is before getting overwhelmed with the mathematics. Hence, the first example below simply shows how to obtain results based on the computer programs furnished with the text. The second example explores the subtleties of reference states through a computational exercise. The subsequent examples finally confront the derivation of the departure function formulas that appear in the computer programs for the Peng-Robinson equation.

HP DEPFUN (see Prl)

TI PENG

PREOS.xls, or PRPURE

Example 7.1 Enthalpy and entropy departures from the Peng-Robinson equation

Propane gas undergoes a change of state from an initial condition of 5 bar and 105°C to 25 bar and 190°C. Compute the change in enthalpy and entropy.

Solution: For propane, $T_c = 369.8$ K; $P_c = 4.249$ MPa; $\omega = 0.152$. The heat capacity coefficients are given by $A = -4.224$, $B = 0.3063$, $C = -1.586E-4$, $D = 3.215E-8$. We may use the HP48 program PRI, the TI program PENG, or the spreadsheet PREOS.xls. If we select the spreadsheet, we can use the PROPS page to calculate thermodynamic properties. (On the calculator we enter the DEPFUN menu key.) We extract the following results:

For State 2:

T (K)	463.15	Z	V	$H-H^{ig}$	$U-U^{ig}$	$S-S^{ig}$
P (MPa)	2.5		cm^3/gmol	J/mol	J/mol	J/molK
& for 1 root region		0.889085	1369.414	-1489.42	-1062.33	-2.29176

For State 1:

T (K)	378.15	Z	V	$H-H^{ig}$	$U-U^{ig}$	$S-S^{ig}$
P (MPa)	0.5		cm^3/gmol	J/mol	J/mol	J/molK
& for 1 root region		0.957398	6020.002	-400.399	-266.461	-0.70805

Ignoring the specification of the reference state for now (refer to Example 7.6 on page 245 to see how to apply the reference state approach), divide the solution into the three steps described in Section 7.1: I. Departure Function; II. Ideal gas; III. Departure function.

The overall solution path for $H_2 - H_1$ is

$$\Delta H = H_2 - H_1 = (H_2 - H_2^{ig}) + (H_2^{ig} - H_1^{ig}) - (H_1 - H_1^{ig})$$

Similarly, for $S_2 - S_1 =$

$$\Delta S = S_2 - S_1 = (S_2 - S_2^{ig}) + (S_2^{ig} - S_1^{ig}) - (S_1 - S_1^{ig})$$

Example 7.1 Enthalpy and entropy departures from the Peng-Robinson equation (Continued)

The three steps that make up the overall solution are covered individually.

Step I. Departures at state 2 from the spreadsheet:

$$(H_2 - H_2^{ig}) = -1490 \text{ J/mol}$$

$$(S_2 - S_2^{ig}) = -2.292 \text{ J/mol-K}$$

Step II. State change for ideal gas:

$$H_2^{ig} - H_1^{ig} = \int_{T_1}^{T_2} C_P dT = \int_{T_1}^{T_2} (A + BT + CT^2 + DT^3) dT =$$

$$= A(T_2 - T_1) + \frac{B}{2}(T_2^2 - T_1^2) + \frac{C}{3}(T_2^3 - T_1^3) + \frac{D}{4}(T_2^4 - T_1^4)$$

$$= -4.224(463.15 - 378.15) + \frac{0.3063}{2}(463.15^2 - 378.15^2) +$$

$$\frac{-1.586 \times 10^{-4}}{3}(463.15^3 - 378.15^3) + \frac{3.215 \times 10^{-8}}{4}(463.15^4 - 378.15^4) = 8405 \text{ J/mol}$$

$$S_2^{ig} - S_1^{ig} = \int_{T_1}^{T_2} \frac{C_P}{T} dT - R\ln\frac{P_2}{P_1}; \quad \int_{T_1}^{T_2} \frac{C_P}{T} dT = \int_{T_1}^{T_2} \frac{(A + BT + CT^2 + DT^3)}{T} dT$$

$$S_2^{ig} - S_1^{ig} = A\ln\left(\frac{T_2}{T_1}\right) + B(T_2 - T_1) + \frac{C}{2}(T_2^2 - T_1^2) + \frac{D}{3}(T_2^3 - T_1^3) - R\ln\frac{P_2}{P_1}$$

$$= -4.224\ln\frac{463.15}{378.15} + 0.3063(463.15 - 378.15) + \frac{-1.586 \times 10^{-4}}{2}(463.15^2$$

$$-378.15^2) + \frac{3.215 \times 10^{-8}}{3}(463.15^3 - 378.15^3) - 8.314\ln\frac{25}{5} = 6.613 \text{ J/mol-K}$$

Step III. Departures at state 1 from the spreadsheet:

$$(H_1 - H_1^{ig}) = -400 \text{ J/mole}$$

$$(S_1 - S_1^{ig}) = -0.708 \text{ J/mole-K}$$

The total changes may be obtained by summing the steps of the calculation.

$$\Delta H = -1490 + 8405 + 400 = 7315 \text{ J/mole}$$

$$\Delta S = -2.292 + 6.613 + 0.708 = 5.029 \text{ J/mole-K}$$

Example 7.2   Real entropy in an engine

A properly-operating internal combustion engine requires a spark plug when it is operating. The cycle involves adiabatically compressing the fuel-air mixture and then introducing the spark. Assume that the fuel-air mixture in an engine enters the cylinder at 0.08 MPa and 20°C and is adiabatically and reversibly compressed in the closed cylinder until its volume is 1/7 the initial volume. Assuming that no ignition has occurred at this point, determine the final T and P, as well as the work needed to compress each mole of air-fuel mixture. You may assume that C_V^{ig} for the mixture is 32 J/mole-K (independent of T), and that the gas obeys the equation of state

$$PV = RT + aP$$

where a is a constant with value $a = 187$ cm^3/mole. Do not assume that C_V is independent of ρ.

Solution: This example helps to understand the difference between departure functions at fixed T and V and departure functions at fixed T and P. The equation of state in this case is simple enough that it can be applied either way. It is valuable to note how the $\ln(Z)$ term works out. Fixed T and V is convenient since the volume change is specified in this example, and we cover this as Method I, and then use fixed T and P as Method II.

First, we need to rearrange our equation of state in terms of $Z = f(T, \rho)$. This rearrangement may not be immediately obvious.[1] Note that dividing all terms by RT gives $PV/RT = 1 + aP/RT$. Note that $V\rho = 1$. Multiplying the last term by $V\rho$, $Z = 1 + aZ\rho$ which rearranges to

$$Z = \frac{1}{1 - a\rho}$$

Also, we find the density at the two states using the equation of state,

$$\rho = \frac{P}{RT + aP} \Rightarrow \rho_1 = 3.257\text{E-}5 \text{ gmole/cm}^3 \Rightarrow \rho_2 = 2.280\text{E-}4 \text{ gmole/cm}^3$$

Method I. In terms of fixed T and V.

$$\left(\frac{\partial Z}{\partial T}\right)_\rho = 0; \quad Z - 1 = \frac{1}{1 - a\rho} - \frac{1 - a\rho}{1 - a\rho} = \frac{a\rho}{1 - a\rho}$$

$$\frac{\left(S - S^{ig}\right)_{TV}}{R} = \int_0^\rho \left[-T\left[\frac{\partial Z}{\partial T}\right]_\rho - (Z - 1)\right]\frac{d\rho}{\rho} = -a\int_0^\rho \frac{d\rho}{1 - a\rho} = \ln(1 - a\rho)$$

$$S_2 - S_1 = (S - S^{ig})_{TV,2} + (S_2^{ig} - S_1^{ig}) - (S - S^{ig})_{TV,1}$$

$$= R \cdot [\ln(1 - 187 \cdot 2.28\text{E-}4) + \{C_V/R\ln(T_2/T_1) + \ln(V_2/V_1)\} - \ln(1 - 187 \cdot 3.257\text{E-}5)]$$

$$\Delta S/R = 0 = -0.04357 + 32/8.314 \cdot \ln(T_2/293.15) - \ln(7) + 0.00611 = 0 \Rightarrow T_2 = 490.8 \text{ K}$$

---

1. We provide a homework problem later with this same problem statement, but use the form $Z = f(T, P)$, which requires less rearrangement.

Example 7.2   Real entropy in an engine (Continued)

Method II. In terms of T and P

$$\frac{(S - S^{ig})}{R} = \int_0^\rho \left[-T\left[\frac{\partial Z}{\partial T}\right]_\rho - (Z - 1) \right] \frac{d\rho}{\rho} + \ln Z$$

$$= -a \int_0^\rho \frac{d\rho}{1 - a\rho} + \ln(Z) = \ln(1 - a\rho) + \ln\left[\frac{1}{1 - a\rho}\right] = 0$$

Since the departure is zero, it drops out of the calculations.

$$S_2 - S_1 = S_2^{ig} - S_1^{ig} = C_P \ln(T_2 / T_1) - R \ln (P_2 / P_1)$$

However, since we are given the final volume, we need to calculate the final pressure. Note that we cannot insert the ideal gas law into the second term even though we are performing an ideal gas calculation. We must use the pressure ratio for our real gas.

$$\Delta S = C_P \ln(T_2 / T_1) - R \ln\left[\frac{RT_2}{V_2 - a} \Big/ \frac{RT_1}{V_1 - a}\right] = (C_P - R) \ln(T_2 / T_1) - R \ln\left(\frac{V_1 - a}{V_2 - a}\right)$$

Now, if we rearrange, we can show that the result is the same as Method I:

$$\Delta S = C_V \ln(T_2 / T_1) + R \ln(V_2 / V_1) + R \ln\left(\frac{1 - a\rho_2}{1 - a\rho_1}\right)$$

$$= R \ln(1 - a\rho_2) + C_V \ln(T_2 / T_1) + R \ln(V_2 / V_1) - R \ln(1 - a\rho_1)$$

This is equivalent to the equation obtained by Method I and $T_2 = 490.8$ K

Finally, $P_2 = \dfrac{RT_2}{V_2 - a} = \dfrac{8.314(490.8)}{1 / 2.28 \times 10^{-4} - 187} = 0.972$ MPa   and

$$W = \Delta U = \left(U - U^{ig}\right)_2 + C_V \Delta T - \left(U - U^{ig}\right)_1 = 0 + C_V \Delta T - 0 = 6325 \text{ J/mole}$$

❶ The final result of this example is incorporated into the programs Prl, PENG, PRPURE, PREOS.xls.

Example 7.3 Enthalpy departure for the Peng-Robinson equation

Obtain a general expression for the enthalpy departure function of the Peng-Robinson equation.

Solution: In the previous example we were able to obtain both pressure-explicit and density-explicit equations. Therefore, we could solve the problem two different ways. For the Peng-Robinson equation, we can only solve one way.

$$Z = \frac{1}{(1 - b\rho)} - \frac{a\rho/RT}{(1 + 2b\rho - b^2\rho^2)}$$

$$-T\left(\frac{\partial Z}{\partial T}\right)_\rho = \frac{\rho T/R}{(1 + 2b\rho - b^2\rho^2)}\left[\frac{-a}{T^2} + \frac{1}{T}\left[\frac{da}{dT}\right]\right]$$

where

$$\frac{da}{dT} = -\frac{a_C\kappa\sqrt{\alpha T_r}}{T}$$

$$-T\left(\frac{\partial Z}{\partial T}\right)_\rho = \frac{b\rho}{1 + 2b\rho - b^2\rho^2}\left[\frac{-a}{bRT} - \frac{a_c\kappa\sqrt{\alpha T_r}}{bRT}\right]$$

We introduce $F(T_r)$ as a shorthand.

$$-T\left(\frac{\partial Z}{\partial T}\right)_\rho \equiv \frac{b\rho}{(1 + 2b\rho - b^2\rho^2)}F(T_r)$$

Also $B \equiv bP/RT \Rightarrow b\rho = B/Z$ and $A \equiv aP/R^2T^2 \Rightarrow a/bRT = A/B$. Note that the integration is simplified by integration over $b\rho$. (see Eqn. B.25)

$$\int_0^{b\rho} -T\left(\frac{\partial Z}{\partial T}\right)_\rho \frac{d(b\rho)}{b\rho} = \int_0^{b\rho} \frac{b\rho}{(1 + 2b\rho - b^2\rho^2)}F(T_r)\frac{d(b\rho)}{b\rho} =$$

$$\frac{F(T_r)}{\sqrt{8}}\left[\ln\left(\frac{1 - \sqrt{2}}{1 + \sqrt{2}}\right)\left(\frac{b\rho(1 + \sqrt{2}) + 1}{b\rho(1 - \sqrt{2}) + 1}\right)\right]_0^{b\rho} = \frac{F(T_r)}{\sqrt{8}}\ln\left[\frac{1 + (1 + \sqrt{2})b\rho}{1 + (1 - \sqrt{2})b\rho}\right]$$

$$\frac{B}{Z} = \frac{\frac{bP}{RT}}{\frac{P}{\rho RT}} = b\rho \Rightarrow \int_0^{b\rho} -T\left(\frac{\partial Z}{\partial T}\right)_\rho \frac{d(b\rho)}{b\rho} = \frac{F(T_r)}{\sqrt{8}}\ln\left[\frac{Z + (1 + \sqrt{2})B}{Z + (1 - \sqrt{2})B}\right]$$

$$\boxed{\frac{(H - H^{ig})}{RT} = Z - 1 + \frac{1}{\sqrt{8}}\ln\left[\frac{Z + (1 + \sqrt{2})B}{Z + (1 - \sqrt{2})B}\right]\left[\frac{-a}{bRT} - \frac{a_c\kappa\sqrt{\alpha T_r}}{bRT}\right]}$$

$$\boxed{= Z - 1 - \ln\left[\frac{Z + (1 + \sqrt{2})B}{Z + (1 - \sqrt{2})B}\right]\frac{A}{B\sqrt{8}}\left[1 + \frac{\kappa\sqrt{T_r}}{\sqrt{\alpha}}\right]} \qquad 7.29$$

Example 7.4   Gibbs departure for the Peng-Robinson equation

Obtain a general expression for the Gibbs energy departure function of the Peng-Robinson equation.

$$Z = \frac{1}{(1 - b\rho)} - \frac{a\rho/RT}{(1 + 2b\rho - b^2\rho^2)}$$

Solution:

The answer is obtained by evaluating Eqn. 7.26. The argument for the integrand is

$$Z - 1 = \frac{1}{1 - b\rho} - \frac{1 - b\rho}{1 - b\rho} - \frac{a\rho/RT}{(1 + 2b\rho - b^2\rho^2)} = \frac{b\rho}{(1 - b\rho)} - \frac{a\rho/RT}{(1 + 2b\rho - b^2\rho^2)}$$

Evaluating the integral, noting again the change in integration variables,

$$\int_0^{b\rho} (Z - 1) \frac{d(b\rho)}{b\rho} = \int_0^{b\rho} \frac{d(b\rho)}{(1 - b\rho)} + \frac{a}{bRT} \int_0^{b\rho} \frac{d(b\rho)}{(1 + 2b\rho - b^2\rho^2)}$$

$$= -\ln(1 - b\rho) - \frac{a}{bRT\sqrt{8}} \ln\left[\frac{1 + (1 + \sqrt{2})b\rho}{1 + (1 - \sqrt{2})b\rho}\right]$$

Making the result dimensionless,

$$\boxed{\frac{(G - G^{ig})}{RT} = Z - 1 - \ln(Z - B) - \frac{A}{B\sqrt{8}} \ln\left[\frac{Z + (1 + \sqrt{2})B}{Z + (1 - \sqrt{2})B}\right]} \qquad 7.30$$

7.6   PRESSURE-DEPENDENT FORMULAS

Occasionally, our equation of state is difficult to integrate to obtain departure functions using the formulas from Section 7.5. This is because the equation of state is more easily arranged and integrated in the form $Z = f(T,P)$.

These formulas are useful for an equation of state written most simply as $Z = f(T,P)$. For treating cases where an equation of state is written most simply as $Z = f(T, \rho)$, see Section 7.5. We omit derivations and leave them as a homework problem. The two most important departure functions at fixed T,P are:

$$\left(\frac{H - H^{ig}}{RT}\right) = -\int_0^P T\left(\frac{\partial Z}{\partial T}\right)_P \frac{dP}{P} \qquad 7.31$$

$$\left(\frac{S - S^{ig}}{R}\right) = -\int_0^P \left[(Z - 1) + T\left(\frac{\partial Z}{\partial T}\right)_P\right] \frac{dP}{P} \qquad 7.32$$

The other departure functions can be derived from these. Note the mathematical similarity between P in the pressure dependent formulas and ρ in the density-dependent formulas.

Example 7.5 Application of pressure-dependent formulas in compression of methane

Methane gas undergoes a continuous throttling process from upstream conditions of 40°C and 20 bars to a downstream pressure of 1 bar. What is the gas temperature on the downstream side of the throttling device? An expression for the molar heat capacity of methane in the ideal gas state is

$$C_P = 19.25 + 0.0523\,T + 1.197\text{E-}5\,T^2 - 1.132\text{E-}8\,T^3;\ T\,[\equiv]\,\text{K};\ C_P\,[\equiv]\,\text{J/mol–K}$$

The virial equation of state may be used for methane at these conditions:

$$Z = 1 + BP/RT = 1 + (B^0 + \omega B^1)P_r/T_r$$

where $B^0 = 0.083 - 0.422/T_r^{1.6}$ and $B^1 = 0.139 - 0.172/T_r^{4.2}$

Solution: Recognizing B is a function of temperature, performing the integrations,

$$\left(\frac{H - H^{ig}}{RT}\right) = \frac{P}{R}\left(\frac{B}{T} - \frac{dB}{dT}\right)$$

$$\left(\frac{S - S^{ig}}{R}\right) = -\frac{P}{R}\frac{dB}{dT}$$

We can easily show

$$\frac{dB^0}{dT_r} = \frac{0.6752}{T_r^{2.6}} \qquad \frac{dB^1}{dT_r} = \frac{0.7224}{T_r^{5.2}}$$

Substituting the relations for B^0 and B^1 into the formulas for the departure functions for a pure fluid, we get

$$\left(\frac{H - H^{ig}}{RT}\right) = -P_r\left[\frac{1.0972}{T_r^{2.6}} - \frac{0.083}{T_r} + \omega\left(\frac{0.8944}{T_r^{5.2}} - \frac{0.139}{T_r}\right)\right] \qquad 7.33$$

$$\left(\frac{S - S^{ig}}{R}\right) = -P_r\left[\frac{0.675}{T_r^{2.6}} + \omega\frac{0.722}{T_r^{5.2}}\right] \qquad 7.34$$

For the initial state, 1

$$\left(\frac{H - H^{ig}}{RT}\right) = -0.110$$

$$(H - H^{ig})_1 = -287\ \text{J/mole}$$

Assuming a small temperature drop, $C_P \approx 36$ J/mole-K

> **Example 7.5   Application of pressure-dependent formulas in compression of methane (Continued)**
>
> For a throttle, $\Delta H = 0 \Rightarrow (H - H^{ig})_2 + 36(T_2 - 40) + 287 = 0$
>
> Trial and error at state 2 where $P = 1$ bar, $T_2 = 35°C \Rightarrow -13 + 36(35 - 40) + 287 = 94$
>
> $$T_2 = 30°C \Rightarrow -13 + 36(30 - 40) + 287 = -87$$
>
> Interpolating, $T_2 = 35 + (35 - 30)/(94 + 87)(-94) = 32.4°C$

7.7   REFERENCE STATES

If we wish to calculate state changes in a property, then the reference state is not important, and all reference state information drops out of the calculation. However, if we wish to generate a chart or table of thermodynamic properties, or compare our calculations to a thermodynamic table/chart, then designation of a reference state becomes essential. Also, if we need to solve unsteady-state problems, the reference state is important because the answer may depend on the reference state as shown in Example 2.13 on page 72. The quantity $H_R - U_R = (PV)_R$ is non-zero, and although we may substitute $(PV)_R = RT_R$ for an ideal gas, for a real fluid we must use $(PV)_R = Z_R RT_R$, where Z_R has been determined at the reference state. We also may use a real fluid reference state or an ideal gas reference state. Whenever we compare our calculations with a thermodynamic chart/table, we must take into consideration any differences between our reference state and that of the chart/table. Therefore, to specify a reference state for a real fluid, we need to specify:

   Pressure

   Temperature

In addition we must specify the state of aggregation at the reference state from one of the following:

   1. ideal gas

   2. real gas

   3. liquid

   4. solid

Further, we set $S_R = 0$, and either (*but not both*) of U_R and H_R to zero. The principle of using a reference state is shown in Fig. 7.4 and is similar to the calculation outlined in Fig. 7.2 on page 231.

Ideal Gas Reference States

For an ideal gas reference state, to calculate a value for enthalpy, we have

$$H = (H - H^{ig})_{T,P} + \int_{T_R}^{T} C_P \, dT + H_R^{ig} \qquad 7.35$$

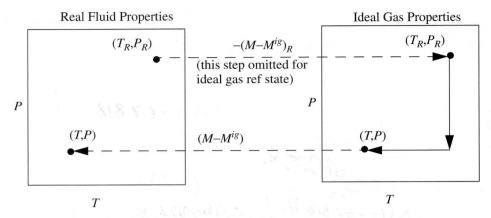

Figure 7.4 *Illustration of calculation of state changes for a generic property M using departure functions where M is U, H, S, G, or A. The calculations are an extension of the principles used in Fig. 7.2 where the initial state is designated as the reference state.*

where the quantity in parenthesis is the departure function from Section 7.5 or 7.6 and H_R^{ig} may be set to zero. An analogous equation may be written for the internal energy. Because entropy of the ideal gas depends on pressure, we must include a pressure integral for the ideal gas:

$$S = (S - S^{ig})_{T,P} + \int_{T_R}^{T} \frac{C_P}{T} \, dT - R \ln \frac{P}{P_R} + S_R^{ig} \qquad 7.36$$

where the reference state value, S_R^{ig}, may be set to zero. From these results we may calculate other properties using relations from Section 5.1, $G = H - TS$, $A = U - TS$, $U = H - PV$.

Real Fluid Reference State

For a real fluid reference state, to calculate a value for enthalpy, we adapt the procedure of Eqn. 7.5.

$$H = (H - H^{ig})_{T,P} + \int_{T_R}^{T} C_P \, dT - (H - H^{ig})_R + H_R \qquad 7.37$$

For entropy:

$$S = (S - S^{ig})_{T,P} + \int_{T_R}^{T} \frac{C_P}{T} \, dT - R \ln \frac{P}{P_R} - (S - S^{ig})_R + S_R \qquad 7.38$$

Changes in State Properties

Changes in state properties are independent of the reference state, or reference state method. To calculate changes in enthalpy, we have the analogy of Eqn. 7.5

$$\Delta H = (H - H^{ig})_{T_2, P_2} + \int_{T_1}^{T_2} C_P dT - (H - H^{ig})_{T_1, P_1} \qquad 7.39$$

To calculate entropy changes:

$$\Delta S = (S - S^{ig})_{T_2,P_2} + \int_{T_1}^{T_2} \frac{C_P}{T} \, dT - R \ln \frac{P_2}{P_1} - (S - S^{ig})_{T_1,P_1} \qquad 7.40$$

Example 7.6 Enthalpy and entropy from the Peng-Robinson equation

UHSG (See Prl)

PREOS.xls,
PRPURE.

Propane gas undergoes a change of state from an initial condition of 5 bar and 105°C to 25 bar and 190°C. Compute the change in enthalpy and entropy. What fraction of the total change is due to the departure functions at 190°C?

Solution: For propane, $T_c = 369.8$ K; $P_c = 4.249$ MPa; $\omega = 0.152$. The heat capacity coefficients are given by $A = -4.224$, $B = 0.3063$, $C = -1.586$E-4, $D = 3.215$E-8. We may use the HP48 program PRI, the TI program PENG, or the spreadsheet PREOS.xls. If we select the spreadsheet, we can use the "Props" page to specify the critical constants and heat capacity constants. The reference state is specified on the companion spreadsheet "Ref State." An arbitrary choice for the reference state is the liquid at 230K and 0.1MPa. Returning to the PROPS worksheet and specifying the desired temperature and pressure gives the thermodynamic properties for V,U,H, and S. (On the HP calculator, press REF to enter the critical constants and set the reference state, PVTF to calculate the compressibility factor, and UHSG to calculate the property values at the P and T used for the PVTF program.)

State	T(K)	P(MPa)	V(cm³/mole)	U(J/mole)	H(J/mole)	S(J/mole-K)
2	463.15	2.5	1369	33478	36901	109.15
1	378.15	0.5	6020	26576	29586	104.13

The changes in the thermodynamic properties are $\Delta H = 7315$J/mole and $\Delta S = 5.024$J/mole-K, identical to the more tediously determined values of Example 7.1 on page 236. The purpose of computing the fractional change due to departure functions is to show that we understand the roles of the departure functions and how they fit into the overall calculation. For the enthalpy, the appropriate fraction of the total change is 20%, for the entropy, 46%.

Example 7.7 Liquefaction revisited

UHSG (See Prl)

PREOS.xls,
PRPURE.

Reevaluate the liquefaction of methane considered in Example 4.6 on page 155 utilizing the Peng-Robinson equation. Previously the methane chart was used. Natural gas, assumed here to be pure methane, is liquefied in a simple Linde process. Compression is to 60 bar, and precooling is to 300 K. The separator is maintained at a pressure of 1.013 bar and unliquefied gas at this pressure leaves the heat exchanger at 295 K. What fraction of the methane entering the heat exchanger is liquefied in the process?

Example 7.7   Liquefaction revisited (Continued)

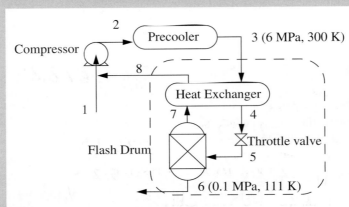

Figure 7.5 *Linde liquification schematic.*

Solution: The solution is easily obtained by using PREOS.xls. When running PREOS, we must specify the temperature of the flash drum which is operating at the saturation temperature at 1.013 bar. This is specified as the boiling temperature for now (111 K).[1]

Before we calculate the enthalpies of the streams, a reference state must be chosen. A convenient choice is the enthalpy of the inlet stream (Stream 3, 6 MPa and 300 K). The results of the calculations from PREOS are summarized in Fig. 7.6.

State 8

Current State		Roots		Stable Root has a lower fugacity			
T (K)	295	Z	V	fugacity	H	U	S
P (MPa)	0.1013		cm³/gmol	MPa	J/mol	J/mol	J/molK
& for 1 root region		0.9976741	24156.108	0.101064	883.5669	-1563.45	35.86805

State 6

Current State		Roots		Stable Root has a lower fugacity			
T (K)	111	Z	V	fugacity	H	U	S
P (MPa)	0.1013		cm³/gmol	MPa	J/mol	J/mol	J/molK
answers for three		0.9666276	8806.4005	0.09802	-4736.62	-5628.7	6.758321
root region		0.0267407	243.61908		-6972.95	-6997.63	-26.6614
		0.0036925	33.640222	0.093712	-12954.3	-12957.7	-66.9014

Figure 7.6 *Summary of enthalpy calculations for methane as taken from the file PREOS.xls.*

The fraction liquefied is calculated by the energy balance:

$m_3 H_3 = m_8 H_8 + m_6 H_6$; then incorporating the mass balance: $H_3 = (1 - m_6/m_3)H_8 + (m_6/m_3)H_6$

Fraction liquefied $= m_6/m_3 = (H_3 - H_8)/(H_6 - H_8) = (0 - 883)/(-12,954 - 883) = 0.064$, or 6.4% liquefied. This is in good agreement with the value obtained in Example 4.6 on page 155.

---

1. It is actually fairly easy to compute the boiling temperature from the equation of state in a self-consistent fashion. The computation requires a consideration of phase equilibrium in a pure fluid, however. This topic is taken up in the following chapter. Practically speaking, there is no significant difference between the experimental saturation temperature and the one determined from the Peng-Robinson equation for this fluid.

Example 7.8 Adiabatically filling a tank with propane (optional)

Propane is available from a reservoir at 350 K and 1 MPa. An evacuated cylinder is attached to the reservoir manifold, and the cylinder is filled adiabatically until the pressure is 1 MPa. What is the final temperature in the cylinder?

Solution: The critical properties, acentric factor and heat capacity constants, are entered on the "Props" page of PREOS.xls. On the "Ref State" page, the reference state is arbitrarily selected as the real vapor at 298 K and 0.1 MPa, and $H_R = 0$. At the reservoir condition, propane is in the one-root region with $Z = 0.888$, $H = 3290$ J/mol, $U = 705$ J/mol, and $S = -7.9766$ J/mol-K. The same type of problem has been solved for an ideal gas in Example 2.14 on page 73; however, in this example the ideal gas law cannot be used. The energy balance reduces to $U^f = H^{in}$, where $H^{in} = 3290$ J/mol. The answer is easily found by using Solver to adjust the temperature on the "Props" page until $U = 3290$ J/mol. The converged answer is 381 K.

Current State		Roots		Stable Root has a lower fugacity			
T (K)	381.364292	Z	V	fugacity	H	U	S
P (MPa)	1		cm³/gmol	MPa	J/mol	J/mol	J/molK
& for 1 root region		0.9153071	2902.13	0.920297	6192.235	3290	-0.03906

7.8 GENERALIZED CHARTS FOR THE ENTHALPY DEPARTURE

As in the case of the compressibility factor, it is often useful to have a visual idea of how generalized properties behave. Fig. 7.7 on page 248 is analogous to the compressibility factor charts from the previous chapter except that the formula for enthalpy is $(H - H^{ig}) = (H - H^{ig})^0 + \omega(H - H^{ig})^1$. Note that one primary influence in determining the liquid enthalpy departure is the heat of vaporization. Also, the subcritical isotherms shift to liquid behavior at lower pressures when the saturation pressures are lower. The enthalpy departure function is somewhat simpler than the compressibility factor in that the isotherms do not cross each other. Note that the temperature used to make the departure dimensionless is T_c. A sample calculation for propane at 463.15 K and 2.5 MPa gives $H^{ig} - H = [0.45 + 0.152(0.2)] (8.314) 369.8 = 1480$ J/mole compared to 1489.2 from the Peng-Robinson equation.

7.9 SUMMARY

The study of departure functions often gives students the most difficulty in this course. That's understandable since it involves simultaneous application of physics and the multivariable calculus. This may be the first instance in which students have applied these subjects in combination to such an extent. On the other hand, this application shows that you studied those subjects for a reason.

When you get beyond the technical details, however, it seems obvious that there must be some difference between an ideal gas and a real fluid. As the accountants for energy movements, we need to be aware of such contributions. Our method is to first add up all the contributions as if everything behaved like an ideal gas, then to compute and add up all the departures from ideal gas behavior. We apply this over and over again. The calculations are greatly facilitated by computers such that

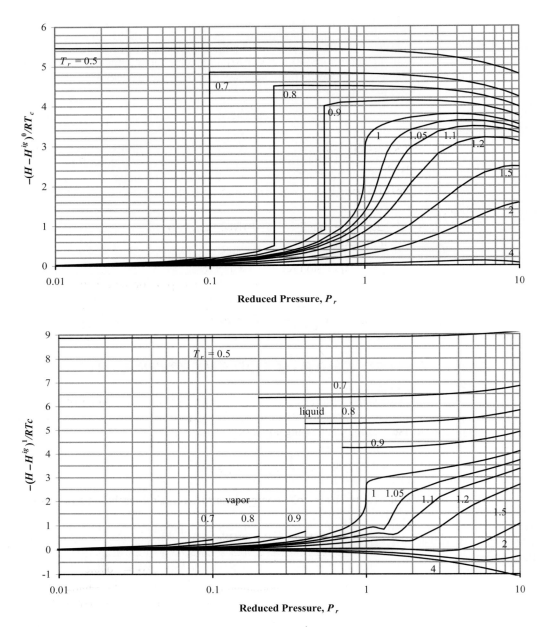

Figure 7.7 *Generalized charts for estimating the $(H - H^{ig})/RT_c$ using the Lee-Kesler equation of state. $(H - H^{ig})^0/RT_c$ uses $\omega = 0.0$, and $(H - H^{ig})^1/RT_c$ is the correction factor for a hypothetical compound with $\omega = 1.0$. Divide by reduced temperature to obtain the enthalpy departure function.*

the minimum requirement is the knowledge of what calculation is required and which buttons to push. If you practice, the button-pushing should quickly become second nature. Then you can turn your attention to developing a better understanding of the subtleties underlying the equations inside the computer programs.

7.10   PRACTICE PROBLEMS

P7.1 Develop an expression for the Gibbs energy departure function based on the Redlich-Kwong (1958) equation of state

$$Z = 1 + \frac{b\rho}{1 - b\rho} - \frac{a\rho}{RT^{3/2}(1 + b\rho)}$$

(ANS. $(G - G^{ig})/RT = -\ln(1 - b\rho) - a\ln(1 + b\rho)/(bRT^{3/2}) + Z - 1 - \ln Z$)

P7.2 For certain fluids, the equation of state is given by: $Z = 1 - b\rho/T_r$

Develop an expression for the enthalpy departure function for fluids of this type. (ANS. $-2b\rho/T_r$)

P7.3 In our discussion of departure functions we derived Eqn. 7.14 for the internal energy departure for any equation of state.

(a) Derive the analogous expression for $(C_V - C_V^{ig})/R$.
(b) Derive an expression for $(C_V - C_V^{ig})/R$ in terms of a, b, ρ, T for the equation of state

$$Z = 1 + \frac{b\rho}{1 + b\rho} - \rho[\exp(a/T) - 1]$$

(ANS. (a) $\int_0^\rho (-2T\left(\frac{\partial Z}{\partial T}\right)_\rho - T^2\left(\frac{\partial^2 Z}{\partial T^2}\right)_\rho)\frac{d\rho}{\rho}$; (b) $a^2\rho T^{-2}\exp(a/T)$)

P7.4 Even in the days of van der Waals, the second virial coefficient for square-well fluids ($R = 1.5$) was known to be: $B/b = 4 + 9.5\ [\exp(N_A\varepsilon/RT) - 1]$. Noting that $e^x \sim 1 + x + x^2/2 + \ldots$, this observation suggests the following equation of state:

$$Z = 1 + \frac{4b\rho}{1 - b\rho} - \frac{9.5N_A\varepsilon}{RT}b\rho$$

Derive an expression for the Helmholtz energy departure function for this equation of state. (ANS. $-4\ln(1 - b\rho) - 9.5N_A\varepsilon b\rho/kT$)

P7.5 Making use of the Peng-Robinson equation, calculate ΔH, ΔS, ΔU, and ΔV for 1-gmole of 1,3-butadiene when it is compressed from 25 bar and 400 K to 125 bar and 550 K. (ANS. $\Delta H = 12{,}570$ J/mol; $\Delta S = 17.998$ J/mol-K; $\Delta U = 11{,}690$ J/mol; $\Delta V = -640.8$ cm^3/mol)

P7.6 Ethane at 425 K and 100 bar initially is contained in a 1-m^3 cylinder. An adiabatic, reversible turbine is connected to the outlet of the tank and exhausted to atmosphere at 1 bar absolute.

(a) Estimate the temperature of the first gas to flow out of the turbine. (ANS. 185 K)
(b) Estimate the rate of work per mole at the beginning of this operation. (ANS. 8880 J/mol)

P7.7 Ethylene at 350°C and 50 bar is passed through an adiabatic expander to obtain work and exits at 2 bar. If the expander has an efficiency of 80%, how much work is obtained per mole of ethylene, and what is the final temperature of the ethylene? How does the final temperature compare with what would be expected from a reversible expander? (ANS. 11 kJ/mole, 450 K versus 404 K)

P7.8 A Rankine cycle is to operate on methanol. The boiler operates at 200°C (P^{sat} = 4.087 MPa), and a superheater further heats the vapor. The turbine outlet is saturated vapor at 0.1027 MPa, and the condenser outlet is saturated liquid at 65°C (P^{sat} = 0.1027 MPa). What is the maximum possible value for the cycle thermal efficiency ($\eta = -W/Q_H$)? (ANS. 26%)

P7.9 An alternative to the pressure equation route from the molecular scale to the macroscopic scale is through the energy equation (Eqn. 6.28). The treatment is similar to the analysis for the pressure equation, but the expression for the radial distribution function must now be integrated over the range of the potential function.

(a) Suppose that $u(r)$ is given by the square-well potential ($R = 1.5$) and $g(r) = 10 - 5(r/\sigma)$ for $r > \sigma$. Evaluate the internal energy departure function where $\rho\sigma^3 = 1$ and $\varepsilon/kT = 1$. (ANS. -5.7π)

(b) Suppose that the radial distribution function at intermediate densities can be reasonably represented by: $g \sim (1 + 2(\sigma/r)^2)$ at all temperatures. Derive an expression for the attractive contribution to the compressibility factor for fluids that can be accurately represented by the Sutherland potential. (ANS. $3\pi\rho N_A\sigma^3 N_A\varepsilon/RT$)

7.11 HOMEWORK PROBLEMS

7.1 What forms does the derivative $(\partial C_V/\partial V)_T$ have for a van der Waals' gas and Redlich-Kwong gas? (The Redlich-Kwong equation is given in Problem P7.1) Comment on the results.

7.2 (a) Derive the enthalpy and entropy departure functions for a van der Waals' fluid.
(b) Derive the formula for the Gibbs energy departure.

7.3 In Example 7.2 we wrote the equation of state in terms of $Z = f(T, \rho)$. The equation of state is also easy to rearrange in the form $Z = f(T, P)$. Rearrange the equation in this form, and apply the formulas from Section 7.6 to resolve the problem using departures at fixed T and P.

7.4 Derive the integrals necessary for departure functions for U, G, and A for an equation of state written in terms of $Z = f(T, P)$ using the integrals provided for H and S in Section 7.6.

7.5 The Soave-Redlich-Kwong equation is presented in problem 6.15. Derive expressions for the enthalpy and entropy departure functions in terms of this equation of state.

7.6 A gas has a constant-pressure ideal-gas heat capacity of 15R. The gas follows the equation of state:

$$Z = 1 + \frac{aP}{RT}$$

over the range of interest, where $a = -1000$ cm^3/mole.

(a) Show that the enthalpy departure is of the form

$$\frac{H - H^{ig}}{RT} = \frac{aP}{RT}$$

(b) Evaluate the enthalpy change for the gas as it undergoes the state change:

$$T_1 = 300 \text{ K}, P_1 = 0.1 \text{ MPa}, T_2 = 400 \text{ K}, P_2 = 10 \text{ MPa}$$

7.7 Derive the integrated formula for the Helmholtz energy departure for the virial equation (Eqn. 6.7), where B is dependent on temperature only. Express your answer in terms of B and its temperature derivative.

7.8 A gas is to be compressed in a steady-state flow reversible isothermal compressor. The inlet is to be 300 K and 1 MPa and the gas is compressed to 20 MPa. Assume that the gas can be modeled with equation of state:

$$PV = RT - \frac{a}{T}P + bP$$

where $a = 385.2$ cm^3-K/mol, $b = 15.23$ cm^3/mol. Calculate the required work per mole of gas.

7.9 Using the Peng-Robinson equation, estimate the change in entropy (J/mole-K) for raising n-butane from a saturated liquid at 271 K and 1 bar to a vapor at 353 K and 10 bar. What fraction of this total change is given by the departure function at 271 K? What fraction of this change is given by the departure function at 353 K?

7.10 Suppose we would like to establish limits for the rule $T_2 = T_1(P_2/P_1)^{R/C_P}$ by asserting that the estimated T_2 should be within 5% of the one calculated using the departure functions. For $\omega = 0$ and $T_r = [1,10]$ at state 1, determine the values of P_r where this assertion holds valid by using the Peng-Robinson equation as the benchmark.

7.11 A piston contains 2 moles of propane vapor at 425 K and 8.5 MPa. The piston is taken through several state changes along a pathway where the total work done by the gas is 2 kJ. The final state of the gas is 444 K and 3.4 MPa. What is the change, ΔH, for the gas predicted by the Peng-Robinson equation and how much heat is transferred? Note: A reference state is optional; if one is desired, use vapor at 400 K and 0.1 MPa.

7.12 An 1-m^3 isolated chamber with rigid walls is divided into two compartments of equal volume. The partition permits transfer of heat. One side contains a nonideal gas at 5 MPa and 300 K and the other side contains a perfect vacuum. The partition is ruptured, and after sufficient time for the system to reach equilibrium, the temperature and pressure are uniform throughout the system. The objective of the problem statements below is to find the final T and P.

The gas follows the equation of state:

$$\frac{PV}{RT} = 1 + \left(b - \frac{a}{T}\right)\frac{P}{RT}$$

where $b = 20$ cm^3/ mole; $a = 40{,}000$ cm^{3}K/mole; and $C_p = 41.84 + 0.084T$(K) J/molK

(a) Set up and simplify the energy balance and entropy balance for this problem.
(b) Derive formulas for the departure functions required to solve the problem.
(c) Determine the final P and T.

7.13 P-V-T behavior of a simple fluid is found to obey the following equation of state given in problem 7.12.

(a) Derive a formula for the enthalpy departure for the fluid.
(b) Determine the enthalpy departure at 20 bar and 300 K.
(c) What value does the entropy departure have at 20 bar and 300 K?

7.14 N.B. Vargaftik[1] (1975) lists the following experimental values for the enthalpy departure of isobutane at 175°C. Compute theoretical values and their percent deviations from experiment by

(a) the generalized charts
(b) the Peng-Robinson equation

Table 7.1 *Enthalpy departure of isobutane at 175°C*

P (atm)	10	20	35	70
$H-H^{ig}$ (J/g)	−15.4	−32.8	−64.72	−177.5

7.15 *n*-pentane is to be heated from liquid at 298 K and 0.01013 MPa to vapor at 360 K and 0.3 MPa. Compute the change in enthalpy using the Peng-Robinson equation of state. If a reference state is desired, use vapor at 310 K, 0.103 MPa, and provide the enthalpy departure at the reference state.

7.16 For each of the fluid state changes below, perform the following calculations using the Peng-Robinson equation: (a) prepare a table and summarize the molar volume, enthalpy, entropy for the initial and final states; (b) calculate ΔH and ΔS for the process; (c) compare with ΔH and ΔS for the fluid modeled as an ideal gas. Specify your reference states.

(i) Propane vapor at 1 bar and 60°C is compressed to a state of 125 bar and 250°C.
(ii) Methane vapor at −40°C and 0.1013 MPa is compressed to a state of 10°C and 7 MPa.

7.17 1 m^3 of CO_2 initially at 150°C and 50 bar is to be isothermally compressed in a frictionless piston/cylinder device to a final pressure of 300 bar. Calculate the volume of the compressed gas, ΔU, the work done to compress the gas, and the heat flow on compression assuming:

(a) CO_2 is an ideal gas
(b) CO_2 obeys the Peng-Robinson equation of state

7.18 Solve problem 7.17 for an adiabatic compression.

7.19 Consider problem statement 2.29 using benzene as the fluid rather than air and eliminating the ideal gas assumption. Use the Peng-Robinson equation. For the same initial state,

(a) The final tank temperature will not be 499.6 K. What will the temperature be?
(b) What is the number of moles left in the tank at the end of the process?
(c) Write and simplify the energy balance for the process. Determine the final temperature of the piston/cylinder gas.

7.20 Solve problem 7.19 using *n*-pentane.

---

1. See Vargaftik reference homework problem 6.6.

7.21 A tank is divided into two equal halves by an internal diaphragm. One half contains argon at a pressure of 700 bar and a temperature of 298 K, and the other chamber is completely evacuated. Suddenly, the diaphragm bursts. Compute the final temperature and pressure of the gas in the tank after sufficient time has passed for equilibrium to be attained. Assume that there is no heat transfer between the tank and the gas, and that argon:

(a) is an ideal gas
(b) obeys the Peng-Robinson equation

7.22 The diaphragm of the preceding problem develops a small leak instead of bursting. If there is no heat transfer between the gas and tank, what is the temperature and pressure of the gas in each tank after the flow stops? Assume that argon obeys the Peng-Robinson equation.

7.23 A practical application closely related to the above problem is the use of a compressed fluid in a small can to reinflate a flat tire. Let's refer to this product as "Fix-a-flat." Suppose we wanted to design a fix-a-flat system based on propane. Let the can be 500 cm^3 and the tire be 40,000 cm^3. Assume the tire remains isothermal and at low enough pressure for the ideal gas approximation to be applicable. The can contains 250 g of saturated liquid propane at 298 K and 10 bar. If the pressure in the can drops to 0.85 MPa, what is the pressure in the tire and the amount of propane remaining in the can? Assuming that 20 psig is enough to drive the car for a while, is the pressure in the tire sufficient? Could another tire be filled with the same can?

7.24 Ethylene at 30 bar and 100°C passes through a throttling valve and heat exchanger and emerges at 20 bar and 150°C. Assuming that ethylene obeys the Peng-Robinson equation, compute the flow of heat into the heat exchanger per mole of ethylene.

7.25 In the final stage of a multistage, adiabatic compression, methane is to be compressed from −75°C and 2 MPa to 6 MPa. If the compressor is 76% efficient, how much work must be furnished per mole of methane, and what is the exit temperature? How does the exit temperature compare with that which would result from a reversible compressor? Use the Peng-Robinson equation.

7.26 (a) Ethane at 280 K and 1 bar is continuously compressed and cooled to 310 K and 75 bar. Compute the change in enthalpy per mole of ethane using the Peng-Robinson equation.
(b) Ethane is expanded through an adiabatic, reversible expander from 75 bar and 310 K to 1 bar. Estimate the temperature of the stream exiting the expander and the work per mole of ethane using the Peng-Robinson equation. (Hint: Is the exiting ethane vapor, liquid, or a little of each? The saturation temperature for ethane at 1 bar is 184.3 K)

7.27 Our space program requires a portable engine to generate electricity for a space station. It is proposed to use sodium ($T_c = 2300$ K; $P_c = 195$ bar; $\omega = 0$; $C_P/R = 2.5$) as the working fluid in a customized form of a "Rankine" cycle. The high-temperature stream is not super-heated before running through the turbine. Instead, the saturated vapor ($T = 1444$ K, $P^{sat} = 0.828$ MPa) is run directly through the (100% efficient, adiabatic) turbine. The rest of the Rankine cycle is the usual. That is, the outlet stream from the turbine passes through a condenser where it is cooled to saturated liquid at 1155 K (this is the normal boiling temperature of sodium), which is pumped (neglect the pump work) back into the boiler.

(a) Estimate the quality coming out of the turbine.
(b) Compute the work output per unit of heat input to the cycle, and compare it to the value for a Carnot cycle operating between the same T_H and T_C.

7.28 Find the minimum shaft work (in kW) necessary to liquefy n-butane in a steady-state flow process at 0.1 MPa pressure. The saturation temperature at 0.1 MPa is 271.7 K. Butane is to enter at 12 mol/min and 0.1 MPa and 290 K and to leave at 0.1 MPa and 265 K. The surroundings are at 298 K and 0.1 MPa.

7.29 The enthalpy of normal liquids changes nearly linearly with temperature. Therefore, in a single pass countercurrent heat exchanger for two normal liquids, the temperature profiles of both fluids are nearly linear. However, the enthalpy of a high pressure gas can be nonlinearly related to temperature because the constant pressure heat capacity becomes very large in the vicinity of the critical point. For example, consider a countercurrent heat exchanger to cool a CO_2 stream entering at 8.6 MPa and 115°C. The outlet is to be 8.6 MPa and 22°C. The cooling is to be performed using a countercurrent stream of water that enters at 10°C. Use a basis of 1 mol/min of CO_2.

(a) Plot the CO_2 temperature (°C) on the ordinate versus $\underline{H}$ on the abscissa, using $H = 0$ for the outlet state as the reference state.

(b) Since $d\underline{H}_{water}/dx = d\underline{H}_{CO2}/dx$ along a differential length, dx, of countercurrent of heat exchanger, the corresponding plot of T versus $\underline{H}$ for water (using the inlet state as the reference state) will show the water temperature profile for the stream that contacts the CO_2. The water profile must remain below the CO_2 profile for the water stream to be cooler than the CO_2. If the water profile touches the CO_2 profile, the location is known as a pinch point and the heat exchanger would need to be infinitely big. What is the maximum water outlet temperature that can be feasibly obtained for an infinitely sized heat exchanger?

(c) Approximately what water outlet temperature should be used to assure a minimum approach temperature for the two streams of approximately 10°C?

7.30 Estimate C_P, C_V and the difference $C_P - C_V$ in (J/mol-K) for liquid n-butane from the following data from Starling, K.E.[1]

T(°F)	P(psia)	V(ft^3/lb)	H (BTU/lb)	U(BTU/lb)
20	14.7	0.02661	−780.22	−780.2924302
40	1400	0.02662	−765.05	−771.9507097
0	14.7	0.02618	−791.24	−791.3112598

7.31 Estimate C_P, C_V, and the difference $C_P - C_V$ in (J/mol-K) for saturated n-butane liquid at 298 K n-butane as predicted by the Peng-Robinson equation of state. Repeat for saturated vapor.

7.32 An alternative to the pressure equation route from the molecular scale to the macroscopic scale is through the energy equation (Eqn. 6.26). The treatment is similar to the analysis for the pressure equation, but the expression for the radial distribution function must now be integrated over the range of the potential function. Suppose that the radial distribution function can be reasonably represented by:

$$g = 0 \text{ for } r < \sigma$$

$$g \sim 1 + \rho N_A \sigma^6 \varepsilon/(r^3 kT) \text{ for } r > \sigma$$

---

1. *Fluid Thermodynamic Properties for Light Petroleum Substances*, Houston, TX: Gulf Publishing 1973.

at all temperatures and densities. Derive an expression for the internal energy departure function for fluids that can be accurately represented by:

(a) the square-well potential with $R = 1.5$
(b) the Sutherland potential

Evaluate each of the above expressions at $\rho N_A \sigma^3 = 0.6$ and $\varepsilon/kT = 1$.

7.33 Starting with the pressure equation as shown in Chapter 6, evaluate the internal energy departure function at $\rho N_A \sigma^3 = 0.6$ and $\varepsilon/kT = 1$ by performing the appropriate derivatives and integrations of the equation of state obtained by applying

$$g = 0 \text{ for } r < \sigma$$

$$g \sim 1 + \rho N_A \sigma^6 \varepsilon/(r^3 kT) \text{ for } r > \sigma$$

at all temperatures and densities,

(a) the square-well potential with $R = 1.5$
(b) the Sutherland potential
(c) Compare these results to those obtained in problem 7.32 and explain why the numbers are not identical.

PHASE EQUILIBRIUM IN A PURE FLUID

One of the principal objects of theoretical research is to find the point of view from which it can be expressed with greatest simplicity.

J.W. Gibbs (1881)

The problem of phase equilibrium is distinctly different from "(In–Out) = Accumulation." The fundamental balances were useful in describing many common operations like throttling, pumping, compressing, etc. and, fundamentally, they provide the basis for understanding all processes. But the balances make a relatively simple contribution in solving problems of phase equilibrium—so much so that they are largely ignored when simple questions like "How many phases are present?" take primary significance.

The general problem of phase equilibrium has a broad significance which begins to distinguish chemical thermodynamics from more generic thermodynamics. If we only care about steam, then it makes sense to concentrate on the various things we can do with steam and to use the steam tables for any properties we need. But, if our interest is in a virtually infinite number of chemicals and mixtures, then we need something more generally applicable than a steam table. Since our interest is chemical thermodynamics, we must deal extensively with property estimations. The determination of phase equilibrium is one of the most important and difficult estimations to make. Ability to understand, model, and predict phase equilibria is necessary for designing industrial separation processes. Typically, these operations comprise one of the most significant components of the capital cost of plant facilities, and require knowledgeable engineers to design, maintain, and troubleshoot them.

In such separation processes, the variables which are controlled in most separation processes are the temperature and pressure. Thus, when we approach the modeling of the phase behavior, we should seek thermodynamic properties which are natural functions of these two properties. In our earlier discussions of convenience properties, the Gibbs energy was shown to be such a function:

$$G \equiv U + PV - TS \qquad dG = -S\,dT + V\,dP \qquad 8.1$$

As a defined mathematical property, the Gibbs energy will remain abstract, in the same way that enthalpy and entropy are difficult to conceptualize. However, our need for a natural function of P and T requires the use of this property.

❶ Phase equilibrium at fixed T and P is most easily understood using G, which is a natural function of P, T.

257

8.1   CRITERIA FOR EQUILIBRIUM

As an introduction to the constraint of phase equilibrium, let us consider an example. A piston-cylinder contains both propane liquid and vapor at −12°C. The piston is forced down a specified distance. Heat transfer is provided to maintain isothermal conditions. What is the final pressure in this vessel?

This is a trick question. As long as two phases are present for a single component and the temperature remains constant, then the system pressure remains fixed at the vapor pressure. The molar volume of vapor phase and liquid phase also stay constant since they are state properties. However, as the total system volume is changed, the quantity of liquid increases, and the quantity of vapor decreases. First of all, we are working with a closed system where $n = n^L + n^V$. For the whole system: $\underline{V} = n^L V^{satL} + n^V V^{satV} = n \cdot V^{satL} + q \cdot n \cdot (V^{satV} - V^{satL})$ and since V^{satL} and V^{satV} are fixed and $V^{satL} < V^{satV}$, a decrease in $\underline{V}$ causes a decrease in q.[1]

Since the temperature and pressure from beginning to end are constant as long as two phases exist, applying Eqn. 8.1 shows that the change in Gibbs energy of each phase of the system from beginning to end must be zero, $dG^L = dG^V = 0$.

For the whole system:

$$\underline{G} = n^L G^L + n^V G^V, \text{ by the product rule} \Rightarrow d\underline{G} = n^L dG^L + n^V dG^V + G^L dn^L + G^V dn^V \qquad 8.2$$

But by the mass balance, $dn^L = -dn^V$ which reduces Eqn. 8.2 to $0 = G^L - G^V$ or

$$G^L = G^V \qquad\qquad 8.3$$

❶ Gibbs energy is the key property for characterizing phase equilibria.

This is a very significant result. In other words, $G^L = G^V$ is a constraint for phase equilibrium. None of our other thermodynamic properties, *U, H, S, A*, will be equivalent in both phases. If we specify phase equilibrium must exist and one additional constraint (e.g. *T*), then all of our other state properties of each phase are fixed and can be determined by the equation of state and heat capacities.

Only needing to specify one variable at saturation to compute all state properties should not come as a surprise, based on our experience with the steam tables. The constraint of $G^L = G^V$ is simply a mathematical way of saying "saturated." The advantage of the mathematical expression is that it yields a specific equality applicable to many chemicals. Essentially, for a pure component, the Gibbs energy in one phase must equal the Gibbs energy in another phase at equilibrium. This powerful insight leads us to the answers of many more difficult and significant questions concerning phase equilibrium.

8.2   THE CLAUSIUS-CLAPEYRON EQUATION

We can apply these concepts of equilibrium to obtain a remarkably simple equation for the vapor-pressure dependence on temperature at low pressures. As a "point of view of greatest simplicity," the Clausius-Clapeyron equation is an extremely important example. Suppose we would like to find

---

1. Once the system volume is decreased below a volume where $\underline{V} < nV^{satL}$, we are compressing a liquid, and the pressure could become quite high. We would need to compute how high using an equation of state. An analogous discussion could be developed for expansion of system volume showing that only vapor will exist for $\underline{V} > nV^{satV}$. The key to notice is that values of q are only physically meaningful in the range $0 < q < 1$.

the slope of the vapor pressure curve, dP^{sat}/dT. Since we are talking about vapor pressure, we are constrained by the requirement that the Gibbs energies of the two phases remain equal as the temperature is changed. Thus, if the Gibbs energy in the vapor phase changes, the Gibbs energy in the liquid phase must change by the same amount. Thus,

$$dG^L = dG^V$$

Rewriting the fundamental property relation $\Rightarrow V^V\,dP^{sat} - S^V\,dT = V^L\,dP^{sat} - S^L\,dT$ and rearranging,

$$\Rightarrow (V^V - V^L)dP^{sat} = (S^V - S^L)dT \qquad\qquad 8.4$$

Entropy is a difficult property to measure. Let us use a fundamental property to substitute for entropy. By definition of G: $G^V = H^V - TS^V = H^L - TS^L = G^L$

$$\boxed{S^V - S^L = \Delta S^{vap} = \frac{(H^V - H^L)}{T} = \frac{\Delta H^{vap}}{T}} \qquad\qquad 8.5$$

Substituting Eqn. 8.5 in for $S^V - S^L$ in Eqn. 8.4, we have the *Clapeyron equation* which is valid for pure fluids along the saturation line:

$$\boxed{\frac{dP^{sat}}{dT} = \frac{\Delta H^{vap}}{T(V^V - V^L)}} \qquad\qquad 8.6 \quad \text{❶ Clapeyron equation.}$$

> *Note: This general form of Clapeyron equation can be applied to any kind of phase equilibrium including solid-vapor and solid-liquid equilibria by substituting the alternative sublimation or fusion properties into Eqn. 8.6; we derived the current equation based on vapor-liquid equilibria.*

Several simplifications can be made in the application of the equation to vapor pressure. To write the equation in terms of Z^V and Z^L, we multiply both sides by T^2 and divide both sides by P^{sat}

$$\frac{T^2}{P^{sat}}\frac{dP^{sat}}{dT} = \frac{\Delta H^{vap}}{R(Z^V - Z^L)}$$

We then use calculus to change the way we write the Clapeyron equation:

$$\frac{dP^{sat}}{P^{sat}} = d\ln P^{sat} \quad \text{and} \quad d\!\left(\frac{1}{T}\right) = -\frac{dT}{T^2}$$

Combining the results, we have an alternative form of the Clapeyron equation

$$\boxed{d\ln P^{sat} = \frac{-\Delta H^{vap}}{R(Z^V - Z^L)}d\!\left(\frac{1}{T}\right)} \qquad\qquad 8.7 \quad \text{❶ Clapeyron equation.}$$

For a gas far from the critical point at "low" reduced temperatures, $Z^V - Z^L \approx Z^V$. In addition, for vapor pressures near 1 bar, where ideal gas behavior is approximated, $Z^V \approx 1$, resulting in the *Clausius-Clapeyron equation*,

❗ Clausius-Clapeyron equation.

$$d\ln P^{sat} = \frac{-\Delta H^{vap}}{R}d\left(\frac{1}{T}\right)$$

(ig) 8.8

Example 8.1 Clausius-Clapeyron equation near or below the boiling point

Derive an expression based on the Clausius-Clapeyron equation to predict vapor-pressure dependence on temperature.

Solution: If we assume that ΔH^{vap} is fairly constant in some range near the boiling point, integration of each side of the Clausius-Clapeyron can be performed from the boiling point to another state on the saturation curve, which yields

$$\ln\left[\frac{P^{sat}}{P_R^{sat}}\right] = \frac{-\Delta H^{vap}}{R}\left[\frac{1}{T} - \frac{1}{T_R}\right]$$

8.9

where P_R^{sat} is 0.1013 MPa and T_R is the normal boiling temperature. This result may be used in a couple of different ways: (1) we may look up ΔH^{vap} so we can calculate P^{sat} at a new temperature T; (2) we may use two vapor pressure points to calculate ΔH^{vap} and subsequently apply method (1) to determine other P^{sat} values.

One vapor pressure point is commonly available through the acentric factor, which is the reduced vapor pressure at a reduced temperature of 0.7.[1] That means, we can apply the definition of the acentric factor to obtain a value of the vapor pressure relative to the critical point.

8.3 SHORTCUT ESTIMATION OF SATURATION PROPERTIES

We found that the Clausius-Clapeyron equation leads to a simple, two-constant equation for the vapor pressure at low temperatures. What about higher temperatures? Certainly, the assumption of ideal gases used to derive the Clausius-Clapeyron is not valid as the vapor pressure becomes large at high temperature; therefore, we need to return to the Clapeyron equation. If $\Delta H^{vap}/\Delta Z^{vap}$ was constant over a wide range of temperature, then we could recover this simple form. Obviously, ΔZ^{vap} is not constant; as we approach the critical point, the vapor and liquid volumes get closer together until they eventually become equal and $\Delta Z^{vap} \to 0$. However, the enthalpies of the vapor and liquid approach each other at the critical point, so it is possible that $\Delta H^{vap}/\Delta Z^{vap}$ may be approximately constant. To analyze this hypothesis, let us plot the experimental data in the form of Eqn. 8.7 *assuming* that $\Delta H^{vap}/\Delta Z^{vap}$ is constant. A constant slope would confirm a constant value of $\Delta H^{vap}/\Delta Z^{vap}$. A plot is shown for two fluids in Fig. 8.1.

The conclusion is that setting $\Delta H/\Delta Z$ equal to a constant is a reasonable approximation, especially over the range of $0.5 < T_r < 1.0$. The plot for ethane shows another nearly linear region

1. Frequently, this temperature is near the normal boiling temperature.

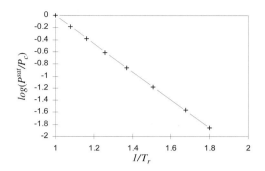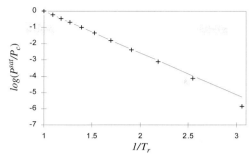

Figure 8.1 *Plot to evaluate Clausius-Clapeyron for calculation of vapor pressures*
at high pressures, argon (left) and ethane (right).

for $1/T_r > 2$ (temperatures below the normal boiling temperature), with a different slope and intercept. Integrating the Clapeyron expression for vapor pressure, we obtain,

$$\ln\left(\frac{P^{sat}}{P_R}\right) = \frac{-\Delta H^{vap}}{R\Delta Z^{vap}}\left(\frac{1}{T} - \frac{1}{T_R}\right)$$

8.10

❗ The plot of $\ln P^{sat}$ vs. $1/T$ is nearly linear.

Example 8.2 Vapor pressure interpolation

What is the value of the pressure in a piston/cylinder at −12°C (261.2 K) with vapor and liquid propane present? Use only the boiling temperature (available from a handbook), critical properties, and acentric factor to determine the answer.

Solution: We will use the boiling point and the vapor pressure given by the acentric factor to determine $\frac{-\Delta H^{vap}}{R\Delta Z^{vap}}$ for Eqn. 8.10, and then use the boiling temperature with $\frac{-\Delta H^{vap}}{R\Delta Z^{vap}}$ to determine the desired vapor pressure. First, let us use the acentric factor to determine the vapor pressure value at $T_r = 0.7$. For propane, $T_c = 369.8$ K, $P_c = 4.249$ MPa, $\omega = 0.152$. Solving for the vapor pressure in terms of MPa by rearranging the definition of the acentric factor, $\left(P^{sat}\big|_{Tr=0.7}\right)$ $= P_c \cdot 10^{(-(1 + 0.152))} = 0.2994$ MPa.[1] The temperature corresponding to this pressure is $T = T_r \cdot T_c = 0.7 \cdot 369.8 = 258.9$ K. The CRC handbook lists the normal boiling temperature of propane as −42°C = 231.2 K. Using these two vapor pressures in Eqn. 8.10:

$$\ln(0.2994/0.1013) = -\Delta H^{vap}/(R\Delta Z^{vap})(1/258.9 - 1/231.2) \Rightarrow -\Delta H^{vap}/(R\Delta Z) = -2342 \text{ K}$$

Therefore, using the boiling point and the value of $-\Delta H^{vap}/(R\Delta Z^{vap})$,

$$P^{sat}(261.2 \text{ K}) = 0.1013 \text{ MPa} \cdot \exp[-2342(1/261.2 - 1/231.2)] = 0.324 \text{ MPa}$$

The calculation is in good agreement with the experimental value.

---

1. Could we use the Clausius-Clapeyron equation at this condition? Since the Clausius-Clapeyron equation requires the ideal gas law, the P^{sat} value must be low enough for the ideal gas law to be followed. The deviations at this state can be quickly checked with the virial equation, $P_r = 0.07$, $T_r = 0.7$, $B^0 = -0.664$, $B^1 = -0.630$, $Z = 0.924$; therefore, the Clausius-Clapeyron equation should probably not be used.

Since the linear relationship of Eqn. 8.10 applies over a broad range of temperatures, we can derive an approximate general estimate of the saturation pressure based on the critical point as the reference and acentric factor as a second point on the vapor pressure curve.

Setting $P_R = P_c$ and $T_R = T_c$,

$$\ln(P_r^{sat}) \approx \frac{-\Delta H^{vap}}{R \Delta Z^{vap} T_c} \left(\frac{T_c}{T} - \frac{T_c}{T_c} \right) = \frac{\Delta H^{vap}}{R \Delta Z^{vap} T_c} \left(1 - \frac{1}{T_r} \right)$$

We convert to common logarithms because they are easier to use to visualize orders of magnitude for estimating quickly.

$$\log\left(P_r^{sat}\right) = \frac{1}{2.303} \frac{\Delta H}{R \Delta Z T_c} \left(1 - \frac{1}{T_r} \right) \equiv A\left(1 - \frac{1}{T_r} \right)$$

Relating this equation to the the acentric factor defined by Eqn. 6.2,

$$\log_{10} P_r^{sat} \Big|_{T_r = 0.7} \equiv -(\omega + 1) = A\left(1 - \frac{1}{0.7} \right) = -\frac{3}{7} A$$

$$A = \frac{7}{3}(1 + \omega)$$

which results in a shortcut vapor pressure equation,

$$\log_{10} P_r^{sat} \approx \frac{7}{3}(\omega + 1)\left(1 - \frac{1}{T_r} \right) \qquad\qquad 8.11$$

> ❗ Shortcut vapor pressure equation. Use care with the shortcut equation below $T_r = 0.5$.

> *The shortcut vapor pressure equation must be regarded as an approximation for rapid estimates. The approximation is generally good above $T_r = 0.5$; however, when the vapor pressure is below approximately 2 bar, the percent error can become large. Always keep in mind that it is estimating based on the critical pressure which is generally 40–50 bar.*

Example 8.3 Application of the shortcut vapor pressure equation

Use the shortcut vapor pressure equation to calculate the vapor pressure of propane at –12°C, and compare the calculation with the results from Example 8.2.

Solution: For propane at –12°C, $T_r = 261.2/369.8 = 0.7063$, $P^{sat}(-12°C) \approx$

$$P_c \cdot 10^{\frac{7(1 + \omega)}{3}\left(1 - \frac{1}{0.7063} \right)} = 0.324 \text{ MPa}$$

This is in excellent agreement with the result of Example 8.2, with considerably less effort.

Example 8.4 General application of the Clapeyron equation

Liquid butane is pumped to a vaporizer as a saturated liquid under a pressure of 1.88 MPa. The butane leaves the exchanger as a wet vapor of 90 percent quality and at essentially the same pressure as it entered. From the following information, estimate the heat load on the vaporizer per gram of butane entering.

For butane, $T_c = 425.2$ K; $P_c = 3.797$ MPa; $\omega = 0.193$; Use the shortcut method to estimate the temperature of the vaporizer, and the Peng-Robinson equation to determine the enthalpy of vaporization.

Solution:

To find the T at which the process occurs:[1]

$$\log_{10}(P_r^{sat}) \approx \frac{7}{3}(1 + \omega)\left(1 - \frac{1}{T_r}\right) \Rightarrow T_r^{sat} = 0.90117 , T = 383.2 \text{ K}$$

First, we use the Peng-Robinson equation to find departure functions for each phase, and subsequently determine the heat of vaporization at 383.2 K and 1.88 MPa,

$$\frac{H^V - H^{ig}}{RT} = -0.9949; \frac{H^L - H^{ig}}{RT} = -5.256;$$

Therefore, $\Delta H^{vap} = (-0.9949 + 5.256)8.314\cdot0.90117\cdot425.2 = 13,575$ J/mol

Since the butane enters as saturated liquid and exits at 90% quality, an energy balance gives

$Q = 0.9\cdot13,575 = 12,217$ J/mol $\cdot$1mol/58g $= 210.6$ J/g

Alternatively, we could have used the shortcut equation another way by comparing the Clapeyron and shortcut equations:

Clapeyron: $\ln(P^{sat}) = -\Delta H^{vap}/RT(Z^V - Z^L) + \Delta H^{vap}/RT_c(Z^V - Z^L) + \ln P_c$

Shortcut: $\ln(P^{sat}) = 2.3025\frac{7}{3}(1 + \omega)\left(1 - \frac{1}{T_r}\right) + \ln P_c$

Comparing, we find: $\dfrac{\Delta H^{vap}}{R\Delta Z^{vap}} = 2.3025\frac{7}{3}(1 + \omega)T_c = 2725$ K

Therefore, using the Peng-Robinson at 383.3 K and 1.88 MPa to determine compressibility factor values:

$$\Delta H^{vap} = 2725R(Z^V - Z^L) = 2725(8.314)(0.6744 - 0.07854) = 13,500 \text{ J/mol}$$

which would give a result in good agreement with the first approach.

---

1. In principle, since we are asked to use the Peng-Robinson equation for the rest of the problem, we could have used it to determine the saturation temperature also, but we were asked to use the shortcut method. The use of equations of state to calculate vapor pressure is discussed in Section 8.10.

The Antoine Equation

The simple form of the shortcut vapor-pressure equation is extremely appealing, but there are times when we desire greater accuracy than such a simple equation can provide. One obvious alternative would be to use the same form over a shorter range of temperatures. With a different slope and intercept, an excellent fit could be obtained. To extend the range of applicability slightly, one modification is to introduce an additional adjustable parameter in the denominator of the equation. The resulting equation is referred to as the Antoine equation:

$$\log_{10} P^{sat} = A - B/(T + C) \qquad 8.12$$

❶ Antoine equation. Use with care outside the stated parameter temperature limits, and watch use of log, ln, and units carefully.

where T is conventionally in Celsius. Values of coefficients for the Antoine equation are widely available, notably in the compilations of vapor-liquid equilibrium data by Gmehling and coworkers.[1] The Antoine equation provides accurate correlation of vapor pressures over a narrow range of temperatures, *but a strong caution must be issued about applying the Antoine equation outside the stated temperature limits; it does not extrapolate well.* If you use the Antoine equation, you should be sure to report the temperature limits as well as the values of coefficients with every application. Antoine coefficients for some compounds are summarized in Appendix E and within the Microsoft® Excel workbook Actcoeff.xls.

8.4 CHANGES IN GIBBS ENERGY WITH PRESSURE

We have seen that the Gibbs energy is the key property which must be used to characterize phase equilibria. In the previous section, we have used Gibbs energy in the derivation of useful relations for vapor pressure. For our discussions here, we have been able to relate the two phases of a pure fluid to each other, and the actual calculation of *values* of the Gibbs energy were not needed. However, extension to general phase equilibria in the next chapters will require a capability to calculate departures of Gibbs energies of individual phases, sometimes using different techniques of calculation for each phase.

❶ *Values* for Gibbs energy departures are needed for further generalization of phase equilibria.

By observing the mathematical behavior of Gibbs energy for fluids derived from the above equations, some sense may be developed for how pressure affects Gibbs energy, and the property becomes somewhat more tangible. Beginning from our fundamental relation, $dG = -SdT + VdP$, the effect of pressure is most easily seen at constant temperature.

❶ Starting point for many derivations.

$$\boxed{dG = VdP \text{ (const. } T)} \qquad 8.13$$

Eqn. 8.13 is the basic equation used as a starting point for derivations used in phase equilibrium. In actual applications the appearance of the equation may differ, but it is useful to recall that most derivations originate with the variation of Eqn. 8.13. To evaluate the change in Gibbs energy, we simply need the *P-V-T* properties of the fluid. These *P-V-T* properties may be in the form of tabulated data from measurements, or predictions from a generalized correlation or an equation of state. For a change in pressure, Eqn. 8.13 may be integrated

$$G_2 - G_1 = \int_{P_1}^{P_2} VdP \text{ (const. } T) \qquad 8.14$$

---

1. Gmehling, J., *Vapor-liquid Equilibrium Data Collection*, Frankfort, Germany, DECHEMA, 1977-.

Methods for calculating Gibbs energies and related properties differ for gases, liquids, and solids. Each type of phase will be covered in a separate section to make the distinctions of the calculation methods more clear. Before proceeding to those analyses, however, we consider a problem which arises in the treatment of Gibbs energy at low pressure. This problem motivates the introduction of the term "fugacity" which takes the place of the Gibbs energy in the presentation in the following sections.

Gibbs Energy in the Low Pressure Limit

The calculation of ΔG is illustrated in Fig. 8.2, where the shaded area represents the integral. The slope of a G versus P plot at constant temperature is equal to the molar volume.

For a real fluid, the ideal-gas approximation is valid only at low pressures. The volume is given by $V = ZRT/P$; thus

$$dG = RT Z \frac{dP}{P} \qquad \qquad 8.15$$

which permits use of generalized correlations or volume-explicit equations of state to represent Z at any T and P.[1] Of course, we may also use Eqn. 8.13 directly, using an equation of state to calculate V. Both techniques will be shown later, but first the qualitative aspects of the calculations will be illustrated.

For an ideal gas, we may substitute $Z = 1$ into Eqn. 8.15 to obtain

$$dG^{ig} = RT \frac{dP}{P} = RT \, d\ln P \qquad \qquad \text{(ig) 8.16}$$

$$\Delta G^{ig} = \int_{P_1}^{P_2} \frac{RT}{P} \, dP = RT \, \ln \frac{P_2}{P_1} \qquad \qquad \text{(ig) 8.17}$$

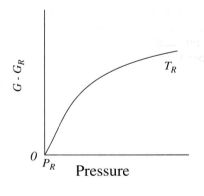 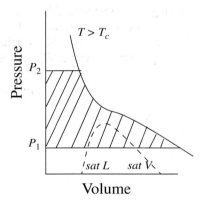

Figure 8.2 *Schematic of dependence of G on pressure for a real fluid at T_R, and an isothermal change on a P-V diagram for a change from P_1 to P_2.*

---

1. Naturally, the accuracy of our calculation is dependent on the accuracy of predicting Z, so we must use an accurate equation of state or correlation.

Both dG and dG^{ig} become infinite as pressure approaches zero. This means that both Eqns. 8.15 and 8.16 are difficult to use directly at low pressure. However, as a real fluid state approaches zero pressure, Z will approach the ideal gas limit and dG approaches dG^{ig}. Thus the difference $dG - dG^{ig}$ will remain finite, and goes to zero as P goes to zero.

However,

$$dG - dG^{ig} = d(G - G^{ig})$$

which is simply the change in departure function. Therefore, we combine Eqns. 8.15 and 8.16 and write:

Differential form of the Gibbs departure.

$$d(G - G^{ig})/RT = (Z - 1)/P \, dP \qquad 8.18$$

This relates the departure function to the P-V-T properties in a way that we have seen before. If you look back to Eqn. 7.26, that equation looks different because we are integrating over volume rather than pressure, but they are really related. We use this departure to define a new state property, *fugacity*, to describe phase behavior. We reserve further discussion of pressure effects in gases for the following sections, where fugacity and Gibbs energy can be considered simultaneously. The generalized treatment by departure functions is also discussed there.

8.5 FUGACITY AND FUGACITY COEFFICIENT

In principle, all pure-component, phase-equilibrium problems could be solved using Gibbs energy. However, historically, an alternative property has been applied in chemical engineering calculations, the *fugacity*. The fugacity has one advantage over the Gibbs energy in that its application to mixtures is a straightforward extension of its application to pure fluids. It also has some empirical appeal because the fugacity of an ideal gas equals the pressure and the fugacity of a liquid equals the vapor pressure under common conditions, as we will show in Section 8.8. The vapor pressure was the original property used for characterization of phase equilibrium by experimentalists.

Fugacity can be directly related to measurable properties under the correct conditions.

The forms of Eqns. 8.16 and 8.15 are similar, and the simplicity of Eqn. 8.16 is appealing. G.N. Lewis *defined* fugacity by

$$dG \;=\; VdP \;\equiv\; RT \, d\ln f \qquad 8.19$$

and comparing to Eqn. 8.16, we see that

Fugacity and fugacity coefficient are convenient ways to quantify the Gibbs departure.

$$d(G - G^{ig})/RT = d \ln (f/P) \qquad 8.20$$

Integrating from low pressure, at constant temperature, we have for the left-hand side:

$$\frac{1}{RT}\int_0^P d(G - G^{ig}) \;=\; \frac{1}{RT}\left[(G - G^{ig})\Big|_P - (G - G^{ig})\Big|_{P=0} \right] \;=\; \frac{(G - G^{ig})}{RT}$$

because $(G - G^{ig})$ approaches zero at low pressure. Integrating the right-hand side of Eqn. 8.20, we have

$$\ln\left(\frac{f}{P}\right)\Bigg|_P - \ln\left(\frac{f}{P}\right)\Bigg|_{P=0}$$

To complete the definition of fugacity, we define the low pressure limit,

$$\lim_{P \to 0}\left(\frac{f}{P}\right) = 1 \qquad \qquad 8.21$$

and we define the ratio f/P to be the *fugacity coefficient*, φ.

$$\frac{(G - G^{ig})}{RT} = \ln\left(\frac{f}{P}\right) = \ln\varphi \qquad \qquad 8.22$$

> ❶ Fugacity has units of pressure, and the fugacity coefficient is dimensionless.

The fugacity coefficient is simply another way of characterizing the Gibbs departure function at a fixed T, P. For an ideal gas, the fugacity will equal the pressure, and the fugacity coefficient will be unity. For representations of the P-V-T data in the form $Z = f(T,P)$ (like the virial equation of state), the fugacity coefficient is evaluated from an equation of the form:

$$\boxed{\frac{(G - G^{ig})}{RT} = \ln\left(\frac{f}{P}\right) = \ln\varphi = \frac{1}{RT}\int_0^P (V - V^{ig})dP = \int_0^P \frac{(Z - 1)}{P}dP} \qquad 8.23$$

or the equivalent form for P-V-T data in the form $Z = f(T,V)$, which is essentially Eqn. 7.26:

$$\boxed{\frac{(G - G^{ig})}{RT} = \ln\left(\frac{f}{P}\right) = \int_0^\rho \frac{(Z - 1)}{\rho}d\rho + (Z - 1) - \ln Z} \qquad 8.24$$

which is the form used for cubic equations of state.

A graphical interpretation of the fugacity coefficient can be seen in Fig. 8.3. The integral of Eqn. 8.23 is represented by the negative value of the shaded region between the real gas isotherm and the ideal gas isotherm. The fugacity coefficient is a measure of non-ideality. *Under most common conditions, the fugacity coefficient is less than one.* At very high pressures, the fugacity coefficient can become greater than one.

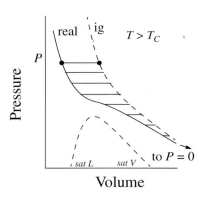

Figure 8.3 *Illustration of RT* ln φ *as a departure function.*

> *In practice, we do not evaluate the fugacity of a substance directly. Instead, we evaluate the fugacity coefficient, and then calculate the fugacity by*
>
> $$f = \varphi P \qquad\qquad 8.25$$

8.6 FUGACITY CRITERIA FOR PHASE EQUILIBRIA

We began the chapter by showing that Gibbs energy was equivalent in phases at equilibrium. Here we show that equilibrium may also be described by equivalence of fugacities. Since

$$G^L = G^V \qquad\qquad 8.3$$

We may subtract G^{ig} from both sides and divide by RT giving

$$\frac{(G^L - G^{ig})}{RT} = \frac{(G^V - G^{ig})}{RT} \qquad\qquad 8.26$$

Substituting Eqn. 8.22

$$\ln\!\left(\frac{f^L}{P}\right) = \ln\!\left(\frac{f^V}{P}\right)$$

which becomes

$$\boxed{f^L = f^V} \qquad\qquad 8.27$$

Therefore, calculation of fugacity and equating in each phase becomes the preferred method of calculating phase equilibria. In the next few sections, we discuss the methods for calculation of fugacity of gases, liquids and solids.

8.7 CALCULATION OF FUGACITY (GASES)

The principle of calculation of the fugacity coefficient is the same by all methods—Eqn. 8.23 or 8.24 is evaluated. The methods *look* considerably different, usually because the *P-V-T* properties are summarized differently.

Equations of State

Equations of state are the dominant method used in process simulators because the EOS can be solved rapidly by computer. We consider two equations of state, the virial equation and the Peng-Robinson equation. We also consider the generalized compressibility factor charts as calculated with the Lee-Kesler equation.

1. The Virial Equation

The virial equation may be used to represent the compressibility factor in the *low-to-moderate pressure region* where Z is linear with pressure at constant temperature. Eqn. 6.10 should be used to evaluate the appropriateness of the virial coefficient method. Substituting $Z = 1 + BP/RT$, or $Z - 1 = BP/RT$ into Eqn. 8.23,

$$\ln \varphi = \int_0^P \frac{B}{RT} dP = \frac{BP}{RT}$$ 8.28

Thus,

$$\ln \varphi = \frac{BP}{RT}$$ 8.29   ❶ The virial equation for gases.

Writing the virial coefficient in reduced temperature and pressure

$$\ln \varphi = \frac{P_r}{T_r}(B^0 + \omega B^1)$$ 8.30

where B^0 and B^1 are given in Eqns. 6.8 and 6.9 on page 201.

2. The Peng-Robinson Equation

Cubic equations of state are particularly useful in the petroleum and hydrocarbon-processing industries because they may be used to represent both vapor and liquid phases. Chapter 6 discussed how equations of state may be used to represent the volumetric properties of gases. The integral of Eqn. 8.23 is difficult to use for pressure-explicit equations of state; therefore, it is solved in the form of Eqn. 8.24. The integral is evaluated analytically by methods of Chapter 7. In fact, the result of Example 7.4 on page 241 is $\ln \varphi$ according to the Peng-Robinson equation.

$$\ln \varphi = -\ln(Z - B) - \frac{A}{B\sqrt{8}} \ln\left[\frac{Z + (1 + \sqrt{2})B}{Z + (1 - \sqrt{2})B}\right] + Z - 1$$ 8.31   ❶ Peng-Robinson equation.

To apply, the technique is analogous to the calculation of departure functions. At a given P, T, the cubic equation is solved for Z, and the result is used to calculate φ and then fugacity is calculated, $f = \varphi P$. This method has been programmed into PREOS.xls for the computer, and PrI for the HP calculator.

Below the critical temperature, equations of state may also be used to predict vapor pressure, saturated vapor volume, and saturated liquid volume, as well as liquid volumetric properties. While Eqn. 8.31 can be used to calculate fugacity coefficients for liquids, the details of the calculation will be discussed in the next section. Note again that Eqn. 8.24 is closely related to Eqn. 7.26 as used in Example 7.4 on page 241.

3. Generalized Charts

Properties represented by generalized charts may help to visualize the magnitudes of the fugacity coefficient in various regions of temperature and pressure. To use the generalized chart, we write

$$\ln \varphi = \int_0^P (Z - 1)\frac{dP}{P} = \frac{G - G^{ig}}{RT} = \int_0^\rho (Z - 1)\frac{d\rho}{\rho} + Z - 1 - \ln Z$$ 8.32   ❶ Generalized charts.

The Gibbs energy departure chart can be generated from the Lee-Kesler equation by specifying a particular value for the acentric factor. The charts are for the correlation $\ln \varphi = \ln \varphi^0 + \omega \ln \varphi^1$. The entropy departure can also be estimated by combining Fig. 8.4 with Fig. 7.7,

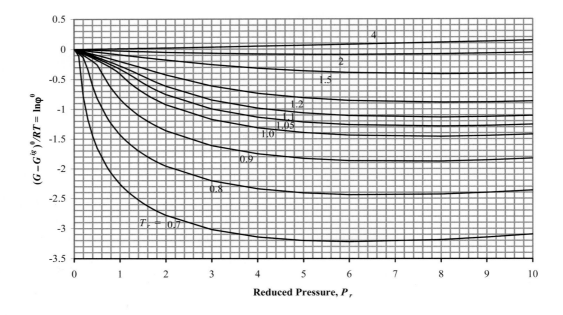

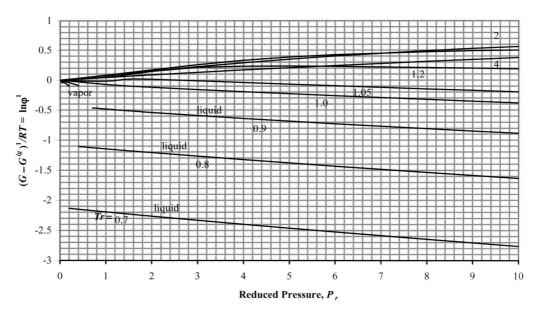

Figure 8.4 *Generalized charts for estimating the Gibbs departure function using the Lee-Kesler equation of state.* $(G - G^{ig})^0/RT$ *uses* $\omega = 0.0$, *and* $(G - G^{ig})^1/RT$ *is the correction factor for a hypothetical compound with* $\omega = 1.0$.

$(S - S^{ig})/R = [(H - H^{ig})/RT_c]/T_r - (G - G^{ig})/RT$. Fig. 8.4 can be useful for hand calculation, if you do not have a calculator. However, the accuracy and convenience of using a calculator should not be discounted. A sample calculation for propane at 463.15 and 2.5 MPa gives

$$\ln(f/P) = \frac{G - G^{ig}}{RT} = -0.1 + 0.152\,(0.05) = -0.09,$$ compared to the value of -0.112 from the Peng-Robinson equation.

8.8   CALCULATION OF FUGACITY (LIQUIDS)

To introduce the calculation of fugacity for liquids, consider Fig. 8.5. The shape of an isotherm below the critical temperature differs significantly from an ideal-gas isotherm. Such an isotherm is illustrated which begins in the vapor region at low pressure, intersects the phase boundary where vapor and liquid coexist and then extends to higher pressure in the liquid region. Point *A* represents a vapor state, point *B* represents saturated vapor, point *C* represents saturated liquid, and point *D* represents a liquid.

We showed in Section 8.6 on page 268 that

$$f_C = f_B = f^{sat} \qquad\qquad 8.33$$

Note that we have designated the fugacity at points *C* and *B* equal to f^{sat}. This notation signifies a saturation condition, and as such, does not require a distinction between liquid or vapor. Therefore, we may refer to points *B* or *C* as saturated vapor or liquid interchangeably when we discuss fugacity. The calculation of the fugacity at point *B* (saturated vapor) is also adequate for calculation of the fugacity at point *C*, the fugacity of saturated liquid. Calculation of the saturation fugacity may be carried out by any of the methods for calculation of vapor fugacities from the above section. Methods differ slightly on how the fugacity is calculated between points *C* and *D*. There are two primary methods for calculating this fugacity change. They are the Poynting method and the equation of state method.

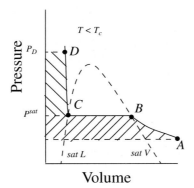

Figure 8.5 *Schematic for calculation of Gibbs energy and fugacity changes at constant temperature for a pure liquid.*

Poynting Method

The Poynting method applies Eqn. 8.19 between saturation (points *B*, *C*) and point *D*. The integral is

$$RT \ln \frac{f_D}{f^{sat}} = \int_{P^{sat}}^{P_D} V dP \qquad 8.34$$

Since liquids are fairly incompressible for $T_r < 0.9$, the volume is approximately constant over the interval of integration, and may be removed from the integral, with the resulting Poynting correction becoming

❶ Poynting correction.

$$\frac{f}{f^{sat}} = \exp\left(\frac{V^L(P - P^{sat})}{RT}\right) \qquad 8.35$$

The fugacity is then calculated by

❶ Poynting method for liquids.

$$\boxed{f = \varphi^{sat} P^{sat} \exp\left(\frac{V^L(P - P^{sat})}{RT}\right)} \qquad 8.36$$

Saturated liquid volume can be estimated within a few percent error using the Rackett equation

$$V^{satL} = V_c Z_c^{(1 - T_r)^{0.2857}} \qquad 8.37$$

The Poynting correction, Eqn. 8.35, is essentially unity for many compounds near room *T* and *P*; thus, it is frequently ignored.

❶ Frequent approximation.

$$\boxed{f \approx \varphi^{sat} P^{sat}} \qquad 8.38$$

Equation of State Method

Calculation of liquid fugacity by the equation of state method uses Eqn. 8.24 just as for vapor. To apply the Peng-Robinson equation of state, we can use Eqn. 8.31. *The only significant consideration is that the liquid compressibility factor must be used.* To understand the mathematics of the calculation, consider the isotherm shown in Fig. 8.6. When $T_r < 1$, the equation of state predicts an isotherm with "humps" in the vapor/liquid region. Surprisingly, these swings can encompass a range of negative values of the pressure near *C'* (although not shown in our example). The exact values of these negative pressures are not generally taken too seriously, however, because they occur in a region of the *P-V* diagram that is unimportant for routine calculations. Since the Gibbs energy from an equation of state is given by an integral of the volume with respect to pressure, the quantity of interest is represented by an integral of the humps. The downward and upward humps tend to cancel each other in generating that integral. This observation gives rise to the equal area rule for computing saturation conditions to be discussed in Section 8.10 where we will show that the shaded area above line $\overline{BC}$ is equal to the shaded area below, and that the pressure where the line is located represents the saturation condition (vapor pressure). With regard to fugacity calculations, it is sufficient simply to note that these humps are in fact integrable, and easily computed by the same formula derived for the vapor fugacity by an equation of state.

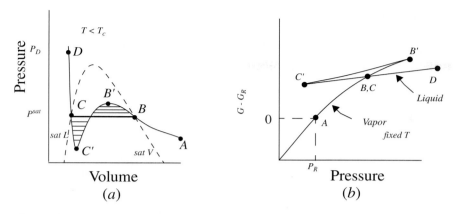

Figure 8.6 *Schematic illustration of the prediction of an isotherm by a cubic equation of state. Compare with Fig. Fig. 8.5 on page 271. The figure on the right shows the calculation of Gibbs energy relative to a reference state. The fugacity will have the same qualitative shape.*

Figure 8.6*b* shows that the molar Gibbs energy is indeed continuous as the fluid transforms from the vapor to the liquid. The Gibbs energy first increases according to Eqn. 8.14 based on the vapor volume. Note that the volume and pressure changes are both positive, so the Gibbs energy relative to the reference value is monotonically increasing. During the transition from vapor to liquid, the "humps" lead to the triangular region associated with the name of "van der Waals loop." Then the liquid behavior takes over and Eqn. 8.14 comes back into play, this time using the liquid volume. Note that the isothermal pressure derivative of the Gibbs energy is not continuous. Can you develop a simple expression for this derivative in terms of P, V, T, C_P, C_V and their derivatives? Based on your answer to the preceding question, would you expect the change in the derivative to be a big change or a small change?

8.9   CALCULATION OF FUGACITY (SOLIDS)

Fugacities of solids are calculated using the Poynting method, with the exception that the volume in the Poynting correction is the volume of the solid phase.

$$f^S = \varphi^{sat} P^{sat} \exp\left(\frac{V^S(P - P^{sat})}{RT}\right)$$

8.39   ❶ Poynting method for solids.

Any of the methods for vapors may be used for calculation of φ^{sat}. P^{sat} is obtained from thermodynamic tables. Equations of state are generally not applicable for calculation of solid phases because they are used only to represent liquid and vapor phases. However, they may be used to calculate the fugacity of a vapor phase in equilibrium with a solid, given by $\varphi^{sat}P^{sat}$. As for liquids, the Poynting correction may be frequently set to unity with negligible error.

$$f \approx \varphi^{sat} P^{sat}$$

8.40   ❶ Frequent approximation.

8.10   SATURATION CONDITIONS FROM AN EQUATION OF STATE

The only thermodynamic specification that is required for determining the saturation temperature or pressure is that the Gibbs energies (or fugacities) of the vapor and liquid be equal. *This involves finding the pressure or temperature where the vapor and liquid fugacities are equal.* The interesting part of the problem comes in computing the saturation condition by iterating on the temperature or pressure.

PREOS.xls.

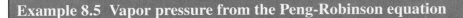
Example 8.5   Vapor pressure from the Peng-Robinson equation

Use the Peng-Robinson equation to calculate the normal boiling point of methane.

Solution: Vapor pressure calculations are available with either PRPURE.EXE or PREOS.xls. The spreadsheet is more illustrative in showing the steps to the calculation. Computing the saturation temperature or pressure in EXCEL is rapid using the Solver tool in Excel.

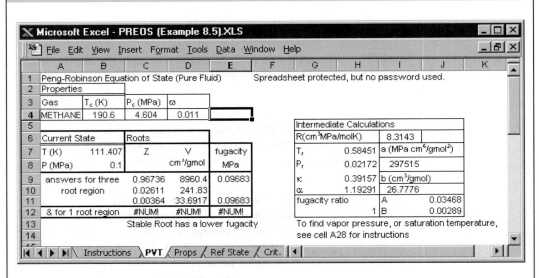

Figure 8.7 *Example of PREOS.XLS used to calculate vapor pressure.*

On the spreadsheet shown in Fig. 8.7, cell H12 is included with the fugacity ratio of the two phases; the cell can be used to locate a saturation condition. Initialize Excel by entering the desired P in cell B8, in this case 0.1 MPa. Then, adjust the temperature to provide a guess in the two-phase (three-root) region. Then, instruct Solver to set the cell for the fugacity ratio (H12) to a value of one by adjusting temperature (B7), subject to the constraint that the temperature (B7) is less than the critical temperature. For methane the solution is found to be 111.4 K which is very close to the experimental value used in Example 7.7 on page 245. Saturation pressures can also be found by adjusting pressure at fixed temperature.

Phase equilibria involves finding the state where $f^L = f^V$.

Fugacity and P-V isotherms for CO_2 as calculated by the Peng-Robinson equation are shown in Fig. 8.8 and Fig. 6.7 on page 209. Fig. 8.8 shows more clearly how the shape of the isotherm is

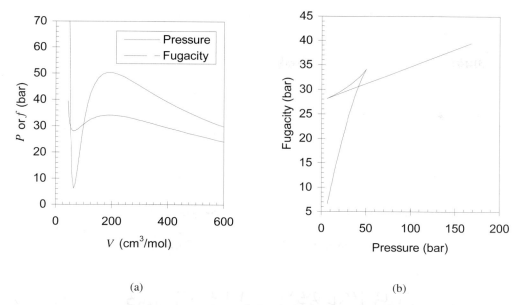

(a)   (b)

Figure 8.8 *Predictions of the Peng-Robinson equation of state for CO_2: (a) Prediction of the P-V isotherm and fugacity at 280 K; (b) Plot of data from Fig (a) as fugacity versus pressure, showing the crossover of fugacity at the vapor pressure. Several isotherms for CO_2 are shown in Fig. 6.7 on page 209.*

related to the fugacity calculation. Note that the fugacity of the liquid root at pressures between B and B' of Fig. 8.6 is lower than the fugacity of the vapor root in the same range, and thus is more stable because the Gibbs energy is lower. Analagous comparisons of vapor and liquid roots at pressures between C and C' show that vapor is more stable.

Just as the vapor pressure estimated by the shortcut vapor pressure equation is less than 100% accurate, the vapor pressure estimated by an equation of state is less than 100% accurate. For example, the Peng-Robinson equation tends to yield about 5% average error over the range $0.4 < T_r < 1.0$. This represents a significant improvement over the shortcut equation. The van der Waals equation, on the other hand, yields much larger errors in vapor pressure. One problem is that the van der Waals equation offers no means of incorporating the acentric factor to fine-tune the characterization of vapor pressure. But the problems with the van der Waals equation go much deeper, as illustrated in the example below.

Example 8.6 Acentric factor for the van der Waals equation

To clarify the problem with the van der Waals equation with respect to phase-equilibrium calculations, it is enlightening to compute the reduced vapor pressure at a reduced temperature of 0.7. Then we can apply the definition of the acentric factor to characterize the vapor pressure behavior of the van der Waals equation. If the acentric factor computed by the van der Waals equation deviates significantly from the acentric factor of typical fluids of interest, then we can quickly assess the magnitude of the error by applying the shortcut vapor-pressure equation. Perform this calculation and compare the resulting acentric factor to those on the inside covers of the book.

Adapting PREOS.xls to a different equation of state.

Example 8.6 Acentric factor for the van der Waals equation (Continued)

Solution: The computations for the van der Waals equation are very similar to those for the Peng-Robinson equation. We simply need to derive the appropriate expressions for a_0, a_1, and a_2, that go into the analytical solution of the cubic equation: $Z^3 + a_2 Z^2 + a_1 Z + a_0 = 0$.

Adapting the procedure for the Peng-Robinson equation given in Section 6.6 on page 205, we can make Eqn. 6.13 dimensionless:

$$Z = \frac{1}{1 - b\rho} - \frac{a\rho}{RT} = \frac{1}{1 - B/Z} - \frac{A}{Z} \qquad 8.41$$

where the dimensionless parameters are given by Eqns. 6.21-6.24; $A = (27/64)\, P_r/T_r^2$; $B = 0.125\, P_r/T_r$. After writing the cubic in Z, the coefficients can be identified $a_0 = -AB$; $a_1 = A$; $a_2 = -(1 + B)$. For the calculation of vapor pressure, the fugacity coefficient for the van der Waals' equation is quickly derived as:

$$\ln\left(\frac{f}{P}\right) = \int_0^\rho \frac{Z - 1}{\rho}\, d\rho + Z - 1 - \ln Z = -\ln(Z - B) - A/Z + Z - 1 \qquad 8.42$$

Substituting these relations in place of their equivalents in PREOS.xls, the problem is ready to be solved. Since we are not interested in any specific compound, we can set $T_c = 1$ and $P_c = 1$, $T_r = 0.7$. Setting an initial guess of $P_r = 0.1$, Solver gives the result that $P_r = 0.20046$. The definition of the acentric factor gives

$$\omega = -\log(P_r) - 1 = -\log(0.20046) - 1 = -0.302$$

Comparing this value to the acentric factors listed in the table on the endflap, the only compound that even comes close is hydrogen, for which we rarely calculate fugacities at $T_r < 1$. This is the most significant shortcoming of the van der Waals equation. This shortcoming becomes most apparent when attempting to correlate phase-equilibria data for mixtures. Then it becomes very clear that accurate correlation of the mixture phase equilibria is impossible without accurate characterization of the pure component phase equilibria, and thus the van der Waals equation will not be useful for *quantitative* calculations.

The Equal Area Rule

As noted above, the swings in the P-V curve give rise to a cancellation in the area under the curve that becomes the free energy/fugacity. A brief discussion is helpful to develop an understanding of how the saturation pressure and liquid and vapor volumes are determined from such an isotherm.

To make this analysis quantitative, it is helpful to recall the formulas for the Gibbs departure functions, noting that the Gibbs departure for the vapor and liquid phases are equal (Eqn. 8.26).

$$\frac{G^L - G^V}{RT} = \frac{G^L - G^{ig}}{RT} - \frac{G^V - G^{ig}}{RT} = \int_{\rho^V}^{\rho^L} \frac{Z - 1}{\rho} d\rho + Z^L - Z^V - \ln(Z^L / Z^V)$$

$$\frac{G^L - G^V}{RT} = -\int_{V^V}^{V^L} \left(\frac{P}{RT} - \frac{1}{V}\right) dV + \frac{1}{RT}\left(PV^L - PV^V\right) - \ln\left(\frac{PV^L / RT}{PV^V / RT}\right)$$

$$\frac{G^L - G^V}{RT} = \frac{1}{RT}\left(PV^L - PV^V\right) - \int_{V^V}^{V^L} \left(\frac{P}{RT}\right) dV + \int_{V^V}^{V^L} \left(\frac{1}{V}\right) dV - \ln\left(\frac{V^L}{V^V}\right)$$ 8.43

$$\frac{G^L - G^V}{RT} = \frac{-1}{RT}\left\{-[PV]_{V^V}^{V^L} + \int_{V^V}^{V^L} PdV\right\}$$

In the final equation, the second term on the right-hand side braces represents the area under the isotherm, and the first term on the right-hand side represents the rectangular area described by drawing a horizontal line at the saturation pressure from the liquid volume to the vapor volume in Fig. 8.6a. Since this area is subtracted from the total inside the braces, the shaded area above a vapor pressure is equal to the shaded area below the vapor pressure for each isotherm. This method of computing the saturation condition is very sensitive to the shape of the P-V curve in the vicinity of the critical point and can be quite useful in estimating saturation properties at near-critical conditions.

8.11   SUMMARY

We began this chapter by introducing the need for Gibbs energy for calculating phase equilibria in pure fluids because it is a natural function of temperature and pressure. We also introduced fugacity, which is a convenient property to use instead of Gibbs energy because it resembles the vapor pressure more closely. We also showed that the fugacity coefficient is directly related to the deviation of a fluid from ideal gas behavior, much like a departure function (see Eqns. 8.23, 8.24). This principle of characterization of non-ideality extends into the next chapter where we consider non-idealities of mixtures. In fact, much of the pedagogy presented in this chapter finds its significance in the following chapters, where the phase equilibria of mixtures become much more complex.

Methods for calculating fugacities were introduced using charts and equations of state. (In the homework problems, we offer illustration of how tables may also be used.) Liquids and solids were considered in addition to gases, and the Poynting correction was introduced for calculating the effect of pressure on condensed phases.

Table 8.1 *Pure component fugacities can be calculated by the techniques summarized below:*

Gases	Liquids	Solids

1. Ideal Gas Law
2. Equation of State
   a. Virial Equation ($V_r \geq 2$)
   b. Cubic Equation
	1. Poynting Method[a]
	2. Equation of State
		1. Poynting Method[a]

a. *The saturation fugacity may be determined by any of the methods for gases, and the Poynting correction is omitted near the vapor pressure.*

8.12   TEMPERATURE EFFECTS ON *G* AND *f* (OPTIONAL)

The effect of temperature at fixed pressure is

$$\left(\frac{\partial G}{\partial T}\right)_P = -S \qquad\qquad 8.44$$

The Gibbs energy change with temperature is then dependent on entropy. Gibbs energy will decrease with increasing temperature. Since the entropy of a vapor is higher than the entropy of a liquid, the Gibbs energy will change more rapidly with temperature for vapor. Similar statements are valid comparing liquids and solids.[1]

8.13   PRACTICE PROBLEMS

P8.1 Carbon dioxide ($C_P = 38$ J/mol-K) at 1.5 MPa and 25°C is expanded to 0.1 MPa through a throttle valve. Determine the temperature of the expanded gas. Work the problem as follows:

   (a) assuming the ideal gas law (ANS. 298 K)
   (b) using the Peng-Robinson equation (ANS. 278 K, sat L + V)
   (c) using a CO_2 chart, noting that the triple point of CO_2 is at −56.6C and 5.2 atm, and has a heat of fusion, ΔH^{fus} of 43.2 cal/g. (ANS. 194 K, sat S + V)

P8.2 Consider a stream of pure carbon monoxide at 300 bar and 150 K. We would like to liquefy as great a fraction as possible at 1 bar. One suggestion has been to expand this high-pressure fluid across a Joule-Thompson valve and take what liquid is formed. What would be the fraction liquefied for this method of operation? What entropy is generated per mole processed? Use the Peng-Robinson equation. Provide numerical answers. Be sure to specify your reference state. (Assume $C_P = 29$ J/mol-K for a quick calculation.) (ANS. 32% liquefied)

P8.3 An alternative suggestion for the liquefaction of CO discussed above is to use a 90% efficient adiabatic turbine in place of the Joule-Thomson valve. What would be the fraction liquefied in that case? (ANS. 60%)

P8.4 At the head of a methane gas well in western Pennsylvania, the pressure is 250 bar, and the temperature is roughly 300 K. This gas stream is similar to the high-pressure stream exiting the precooler in the Linde process. A perfect heat exchanger (approach temperature of zero) is available for contacting the returning low-pressure vapor stream with the incoming high-pressure stream (similar to streams 3-8 of Example 7.7 on page 245). Compute the fraction liquefied using a throttle if the returning low pressure vapor stream is 30 bar. (ANS. 30%)

---

1. Frequently, we arbitrarily set $S_R = 0$ and either H_R or $U_R = 0$ at our reference states. For consistency in our calculations, $G_R = H_R - T_R S_R$. As a result, the calculated value of S at a given state depends on our current state relative to the reference sate. Calculated S values may be positive or negative due to our choice of setting $S_R = 0$, and Gibbs energy thus calculated may increase or decrease with temperature. Entropy does not actually go to zero except for a perfect crystal at absolute zero, and entropy of all substances at practical conditions is positive. The fact that our calculations result in negative numbers for S is purely a result of our choice of setting $S_R = 0$ at our reference state (to avoid more difficult calculation of the actual value relative to a perfect crystal at absolute zero). See third law of thermodynamics in Subject Index.

8.14   HOMEWORK PROBLEMS

8.1 The heat of fusion for the ice-water phase transition is 335 kJ/kg at 0°C and 1 bar. The density of water is $1g/cm^3$ at these conditions and that of ice is 0.915 g/cm^3. Develop an expression for the change of the melting temperature of ice as a function of pressure. Quantitatively explain why ice skates slide along the surface of ice for a 100 kg hockey player wearing 10 cm x 01 cm blades. Can it get too cold to ice skate? Would it be possible to ice skate on other materials such as solid CO_2?

8.2 Thermodynamics tables and charts may be used when both H and S are tabulated. Since $G = H - TS$, at constant temperature, $\Delta G = RT \ln(f_2/f_1) = \Delta H - T\Delta S$. If state 1 is at low pressure where the gas is ideal, then $f_1 = P_1$, $RT \ln(f_2/P_1) = \Delta H - T\Delta S$, where the subscripts indicate states. Use this method to determine the fugacity of steam at 400 °C and 15 MPa. What value does the fugacity coefficient have at this pressure?

8.3 This problem reinforces the concepts of phase equilibria for pure substances.

   (a) Use steam table data to calculate the Gibbs energy of 1 kg saturated steam at 150°C, relative to steam at 150°C and 50 kPa (the reference state). Perform the calculation by plotting the volume data and graphically integrating. Express your answer in kJ. (Note: Each square on your graph paper will represent [pressure·volume] corresponding to the area, and can be converted to energy units.)
   (b) Repeat the calculations using the tabulated enthalpies and entropies. Compare your answer to part (a).
   (c) The saturated vapor from part (a) is compressed at constant T and 1/2 kg condenses. What is the total Gibbs energy of the vapor liquid mixture relative to the reference state of part (a)? What is the total Gibbs energy relative to the same reference state when the mixture is completely condensed to form saturated liquid?
   (d) What is the Gibbs energy of liquid water at 600 kPa and 150°C relative to the reference state from part (a)? You may assume that the liquid is incompressible.
   (e) Calculate the fugacities of water at the states given in parts (a) and (d). You may assume that $f = P$ at 50 kPa.

8.4 Derive the formula for fugacity according to the van der Waals equation.

8.5 Use the result of problem 8.4 to calculate the fugacity of ethane at 320 K and at a molar volume of 150 cm^3/mole. Also calculate the pressure in bar.

8.6 Calculate the fugacity of ethane at 320 K and 70 bar using:

   (a) generalized charts
   (b) the Peng-Robinson equation

8.7 CO_2 is compressed at 35°C to a molar volume of 200 cm^3/gmole. Use the Peng-Robinson equation to obtain the fugacity in MPa.

8.8 Use the generalized charts to obtain the fugacity of CO_2 at 125 °C and 220 bar.

8.9 Estimate the fugacity of pure *n*-pentane (C_5H_{12}) at 97°C and 7 bar by utilizing the virial equation.

8.10 Calculate the fugacity of pure *n*-octane vapor as given by the virial equation at 580 K and 0.8 MPa.

8.11 Develop tables for H, S, and Z for N_2 over the range $P_r = [0.5, 1.5]$ and $T_r = [T_r^{sat}, 300 \text{ K}]$ according to the Peng-Robinson equation. Use the saturated liquid at 1 bar as your reference condition for $H = 0$ and $S = 0$.

8.12 Develop a P-H chart for saturated liquid and vapor n-butane in the range $T = [260, 340]$ using the Peng-Robinson equation. Show constant S lines emanating from saturated vapor at 260 K, 300 K, and 340 K. For an ordinary vapor compression cycle, what would be the temperature and state leaving an adiabatic, reversible compressor if the inlet was saturated vapor at 260 K? (Hint: This is a tricky question.)

8.13 Compare the Antoine and shortcut vapor-pressure equations for temperatures from 298 K to 500 K. (Note in your solution where the equations are extrapolated.) For the comparison, use a plot of $\log_{10} P^{sat}$ versus $1/T$ except provide a separate plot of P^{sat} versus T for vapor pressures less than 0.1 MPa.

(a) n-hexane
(b) acetone
(c) methanol
(d) 2-propanol
(e) water

8.14 Compare the Peng-Robinson vapor pressures to the experimental vapor pressures (represented by the Antoine constants) for the species listed in problem 8.13.

8.15 Carbon dioxide can be separated from other gases by preferential absorption into an amine solution. The carbon dioxide can be recovered by heating at atmospheric pressure. Suppose pure CO_2 vapor is available from such a process at 80°C and 1 bar. Suppose the CO_2 is liquefied and marketed in 43-L laboratory gas cylinders that are filled with 90% (by mass) liquid at 295 K. Explore the options for liquefaction, storage, and marketing via the following questions. Use the Peng-Robinson for calculating fluid properties. Submit a copy of the H-U-S table for each state used in the solution.

(a) Select and document the reference state used throughout your solution.
(b) What is the pressure and quantity (kg) of CO_2 in each cylinder?
(c) A cylinder marketed as specified needs to withstand warm temperatures in storage/ transport conditions. What is the minimum pressure that a full gas cylinder must withstand if it reaches 373 K?
(d) Consider the liquefaction process via: 1) compression of the CO_2 vapor from 80 °C, 1 bar to 6.5 MPa in a single adiabatic compressor ($\eta = 0.8$); 2) the compressor is followed by cooling in a heat exchanger to 295 K and 6.5 MPa. Determine the process temperatures and pressures and the total work and heat transfer requirement for each step.
(e) Consider the liquefaction via: 1) compression of the CO_2 vapor from 80°C, 1 bar to 6.5 MPa in a two-stage compressor with interstage cooling. Each stage ($\eta = 0.8$) operates adiabatically. The interstage pressure is 2.5 MPa, and the interstage cooler returns the CO_2 temperature to 295 K; 2) the two-stage compressor is followed by cooling in a heat exchanger to 295 K and 6.5 MPa. Determine all process temperatures and pressures and the total work and heat transfer requirement for each step.
(f) Calculate the minimum work required for the state change from 80°C, 1 bar to 295 K, 6.5 MPa with heat transfer to the surroundings at 295 K. What is the heat transfer required with the surroundings?

8.16 A three-cycle cascade refrigeration unit is to use methane (cycle 1), ethylene (cycle 2), and ammonia (cycle 3). The evaporators are to operate at: cycle 1, 115.6 K; cycle 2, 180 K; cycle 3, 250 K. The outlet of the compressors are to be: cycle 1, 4 MPa; cycle 2, 2.6 MPa; cycle 3, 1.4 MPa. Use the Peng-Robinson equation to estimate fluid properties. Use stream numbers from Fig. 4.12 on page 154. The compressors have efficiencies of 80%.

   (a) Determine the circulation rate for cycle 2 and cycle 3 relative to the circulation rate in cycle 1.
   (b) Determine the work required in each compressor per kg of fluid in the cycle.
   (c) Determine the condenser duty in cycle 3 per kg of flow in cycle 1.
   (d) Suggest two ways that the cycle could be improved.

8.17 Work problem 6.18, then obtain an expression for the fugacity. Then modify the PREOS.xls spreadsheet for this equation of state. Determine the value of c (+/− 0.5) that best represents the vapor pressure of the specified compound below. Use the shortcut vapor pressure equation to estimate the experimental vapor pressure for the purposes of this problem.

   (a) CO_2
   (b) ethane
   (c) ethylene
   (d) propane
   (e) hexane

FLUID PHASE EQUILIBRIA IN MIXTURES

We have already encountered the phase equilibrium problem in our discussions of pure fluids. In Unit I, we were concerned with the quality of the steam. In Unit II, we developed generalized relations for the vapor pressure. These analyses enable us to estimate both the conditions when a liquid/vapor phase transition occurs and the ratio of vapor to liquid. In Unit III, we are not only concerned with the ratio of vapor to liquid, but also with the ratio of each component in the liquid to that in the vapor. These ratios may not be the same because all components are not equally soluble in all phases. These issues arise in a number of applications (e.g., distillation or extraction) that are extremely common. Unfortunately, prediction of the desired properties to the required accuracy is challenging. In fact, *no currently available method is entirely satisfactory*, even for the limited types of phase equilibria commonly encountered in the chemical processing industries. Nevertheless, the available methods do provide an adequate basis for correlating the available data and for making modest extrapolations, and the methods can be successfully applied to process design. Understanding of the difference between modest extrapolations and radical predictions is facilitated by a careful appreciation of the underlying theory as developed from the molecular level. Developing this understanding is strongly encouraged as a means of avoiding extrapolations that are unreasonable. This unit relies on straightforward extensions of the concepts of energy, entropy, and equilibrium to provide a solid background in the molecular thermodynamics of non-reactive mixtures. The final unit of the text, Unit IV, will treat reactive systems.

CHAPTER

9

INTRODUCTION TO MULTICOMPONENT SYSTEMS

What we obtain too cheap we esteem too lightly.

Thomas Paine

Superficially, the extension of pure component concepts to mixtures may seem simple. In fact, this is a significant problem of modern science which impacts phase transitions in semi- and superconductors, polymer solutions and blends, alloy materials, composites, and biochemistry. Specialists in each of these areas devote considerable effort to the basic problem of group interactions between molecules. From a thermodynamic perspective, these different research efforts are very similar. The specific types of molecules differ, and the pair potential models may be a little different, but the radial distribution function and the energy equation still provide the connection from the molecular scale to the macro scale.

Our coverage of multicomponent systems consists of: (1) a very brief extension to mixtures of the mathematical and physical principles; and (2) an introductory description of a few common methods for reducing these principles to practice. This description is merely introductory because learning the specific methods is basically what distinguishes the polymer scientist from the ceramicist. Such specialized study will be greatly facilitated by having an appreciation of the type of molecular interactions that are most influential in each situation. Of similar importance is the ability to analyze thermodynamic data such that insights about key aspects of processes are easily ascertained.

9.1 PHASE DIAGRAMS

Before we delve into the details of calculating phase equilibria, let us introduce elementary concepts of common vapor-liquid phase diagrams. For a pure fluid, vapor-liquid equilibrium occurs with only one degree of freedom, $F = C - P + 2 = 1 - 2 + 2 = 1$. At one atmosphere pressure, vapor-liquid equilibria will occur at only one temperature—the normal boiling point temperature. However, with a binary mixture, we have two degrees of freedom, $F = 2 - 2 + 2 = 2$. For a system with fixed pressure, phase compositions and temperature can both be varied over a finite range when two phases coexist. Experimental data for experiments of this type are usually presented as a function of

T and composition on a plot known as a *T-x-y* diagram, such as that shown qualitatively in Fig. 9.1. At fixed temperature, we may vary pressure and composition in a binary mixture and obtain data to create a *P-x-y* diagram as shown also in Fig. 9.1. The region where two phases coexist is shown by the area enclosed by the curved lines on either plot and is known as the *phase envelope*.[1] On the *T-x-y* diagram, the vapor region is at the top (raising *T* at fixed *P* causes vaporization of liquid). On the *P-x-y* diagram, the vapor region is at the bottom (lowering *P* at fixed *T* causes vaporization of liquid). Note that the intersections of the phase envelope with the ordinate scales at the pure component compositions give the pure component saturation temperatures on the *T-x-y* diagram, and the pure component vapor (saturation) pressures on the *P-x-y* diagram. Therefore, significant information about the shape of the diagram can often be deduced with a single mixture data point when combined with the pure component end points. Qualitatively, the shape of the *P-x-y* diagram can be found by inverting the *T-x-y*, and vice versa.[2] Customarily, for binary systems in the separations literature, the more volatile component composition is plotted along the abscissa in mole fraction or percent.

The lower curve on the *T-x-y* diagram is next to the liquid region, and it is known as the *bubble line*. The upper curve is next to the vapor region, and is known as the *dew line*. The two lines meet at the axes if the conditions are below the critical pressure of both components. At a given composition, the temperature along the bubble line is the temperature where an infinitesimal bubble of vapor coexists with liquid. Thus, at an over-all composition of 50 mole% *A*, the system of Fig. 9.1 at fixed pressure is 100% liquid below 300 K at the pressure of the diagram. As the temperature is raised, the overall composition is constrained to follow the vertical dashed line constructed on the diagram, and the first vapor bubble forms at the intersection of the bubble line at 300 K at point *a,* which is known as the *bubble-point temperature* for a 50 mole% mixture at the system pressure.

> ❶ The ability to quickly read phase diagrams is an essential skill.

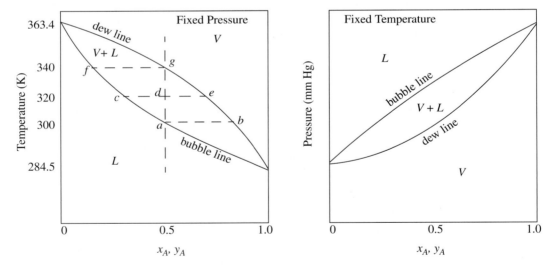

Figure 9.1 *Illustration of T-x-y (left) and P-x-y (right) diagrams.*

---

1. There are many variations of the diagrams, and this discussion is meant to introduce only the most commonly encountered types of diagrams. More complex diagrams are introduced gradually, and discussed in depth in Chapter 13. These diagrams are actually cross-sections of three dimensional diagrams which are discussed in Chapter 13.
2. This is a convenient manipulation to visualize both diagrams if only one diagram is available.

Phase compositions at a given P and T may be found by reading the compositions from intersections of the bubble and dew lines with horizontal lines constructed on the diagram, such as the dashed line at 300 K. For our example at the bubble temperature, the liquid phase will be 50 mole% A because the first bubble of vapor has not yet caused a measurable change in the liquid composition. The vapor phase composition coexisting at the bubble point temperature will be 80 mole% A (point b). As the temperature is increased to 320 K, the overall mixture is at point d, the liquid phase will be 30 mole% A (point c), and the vapor phase will be 70 mole% A (point e). Suppose we start an experiment with a 50 mole% mixture at 350 K, where the mixture is 100% vapor. As the temperature is lowered, the *dew-point temperature* is encountered at 340 K for the 50 mole% mixture at system pressure (point g), and the first drop of liquid is formed which is about 20 mole percent A (point f). Note that the bubble and dew temperatures are composition-dependent. For example, the bubble temperature of a 30 mole% mixture is 320 K (point c), and the bubble temperature of a 20 mole% mixture is 340 K (point f). Similar discussion could be presented for the dew temperatures. The bubble and dew-point discussions could also be presented on the pressure diagram, but in this case we would refer to the bubble and dew *pressures*. The horizontal dotted lines connecting coexisting compositions are *tie lines*.

When we speak of composition in a two-phase mixture, we must be clear about which phase we are discussing. We use x to denote a liquid phase mole fraction, y to denote a vapor phase mole fraction, and z to denote an overall mole fraction.[1] For the example, we have been discussing using a 50 mole% mixture: at the bubble point of 300 K we have $z_A = 0.5$, $x_A = 0.5$, $y_A = 0.8$; at 320 K we have $z_A = 0.5$, $x_A = 0.3$, $y_A = 0.7$; at 340 K we have $z_A = 0.5$, $x_A = 0.20$, $y_A = 0.5$. At 320 K, the system is in the two-phase region, and we may use the compositions of the vapor and liquid phases, together with an overall mass balance, to calculate the fraction of the overall mixture that is vapor or liquid. If the initial number of moles is denoted by F, and it separates into L moles of liquid and V moles of vapor, the over-all mole balance is $F = L + V$, which can be written $1 = L/F + V/F$. The A component balance is $z_A F = y_A V + x_A L$, which can be written $z_A = y_A \cdot V/F + x_A \cdot L/F$. Combining the two balances to eliminate V/F, the percentage that is liquid will be

$$\frac{L}{F} = \frac{z_A - y_A}{x_A - y_A} \qquad\qquad 9.1$$

which is simply given by line segment lengths, $\dfrac{\overline{de}}{\overline{ce}}$. Likewise the fraction that is vapor may be calculated

$$\frac{V}{F} = \frac{x_A - z_A}{x_A - y_A} \qquad\qquad 9.2$$

which is given by line segment lengths, $\dfrac{\overline{cd}}{\overline{ce}}$. These balance equations are frequently called the *lever* ❶ The lever rule.

rule. Note that the two fractions sum to one, $\dfrac{L}{F} + \dfrac{V}{F} = 1$.

---

1. Note that when at 100% liquid, $x = z$ and at 100% vapor, $y = z$. In some cases, such as formulas for an equation of state, we discuss a generic phase which may be liquid *or* vapor, and thus use either x or y.

9.2 CONCEPTS

Generalization of pure-component principles to multicomponent systems is fairly straightforward. The only additional complexity is that we must consider how the properties change with respect to small changes in the amounts of individual components. For a pure fluid, the natural properties were simply a function of two state variables. In multicomponent mixtures, these energies and the entropy also depend on composition.

 These equations extend the use of calculus from Chapter 5 to composition variables.

$$e.g.,\ d\underline{U}(T, P, n_1, n_2, \dots n_i) = \left(\frac{\partial \underline{U}}{\partial P}\right)_{T,n} dP + \left(\frac{\partial \underline{U}}{\partial T}\right)_{P,n} dT + \sum_i \left(\frac{\partial \underline{U}}{\partial n_i}\right)_{P,T,n_{j \neq i}} dn_i \qquad 9.3$$

$$e.g.,\ d\underline{G}(T, P, n_1, n_2, \dots n_i) = \left(\frac{\partial \underline{G}}{\partial P}\right)_{T,n} dP + \left(\frac{\partial \underline{G}}{\partial T}\right)_{P,n} dT + \sum_i \left(\frac{\partial \underline{G}}{\partial n_i}\right)_{P,T,n_{j \neq i}} dn_i \qquad 9.4$$

Note that these equations follow the rules developed in Chapter 5. Each term on the right-hand side consists of a partial derivative with respect to one variable, with all other variables held constant. The summation is simply a shorthand method to avoid writing a term for each component. The subscript $n_{j \neq i}$ means that the moles of all components except i are held constant. In other words, for a ternary system, $\left(\frac{\partial \underline{U}}{\partial n_1}\right)_{P,T,n_{j \neq i}}$ means the partial derivative of $\underline{U}$ with respect to n_1 while holding P, T, n_2 and n_3 constant. Eqn. 9.4 is more useful than Eqn. 9.3 for phase equilibria where P and T are manipulated because the Gibbs energy is a natural function of P and T. At constant moles and composition of material, the mixture must follow the same constraints as a pure fluid. That is, the state is dependent on only two state variables *if we keep the composition constant*.

$$\Rightarrow (\partial \underline{G}/\partial P)_{T,n} = \underline{V} \quad \text{and} \quad (\partial \underline{G}/\partial T)_{P,n} = -\underline{S};$$

These complicated-looking derivatives are really fundamental properties, therefore we can rewrite Eqn. 9.4 as

$$d\underline{G} = \underline{V}dP - \underline{S}dT + \sum_i (\partial \underline{G}/\partial n_i)_{T,P,n_{j \neq i}} dn_i \qquad 9.5$$

The quantity $(\partial \underline{G}/\partial n_i)_{T,P,n_{j \neq i}}$ tells us how the total Gibbs energy of the mixture changes with an infinitesimal change in the number of moles of species i, when the number of moles of all other species fixed, and at constant P and T. The quantity $(\partial \underline{G}/\partial n_i)_{T,P,n_{j \neq i}}$ will become very important in our later discussion so we call it the *chemical potential*, and give it a symbol.

Chemical potential.

$$\mu_i \equiv (\partial \underline{G}/\partial n_i)_{T,P,n_{j \neq i}} \qquad 9.6$$

Partial Molar Properties

Another name for the special derivative of Eqn. 9.6 is the partial molar Gibbs energy. We may generalize the form of the derivative and apply it to other properties. For any *extensive* thermodynamic property $\underline{M}$, we may write $\left(\frac{\partial \underline{M}}{\partial n_i}\right)_{T,P,n_{j \neq i}} \equiv \overline{M}_i$. Note that T, P, and $n_{j \neq i}$ are always held constant. This derivative is called the partial molar property, where the overbar indicates a partial molar quantity, i.e., for $\underline{V}$, $(\partial \underline{V}/\partial n_i)_{T,P,n_{j \neq i}}$ is called the partial molar volume and given the symbol $\overline{V}_i$. A special mathematical result of the differentiation is that we may write at constant temperature and pressure:

$$\underline{M} = \sum_i n_i \overline{M}_i \quad \text{or} \quad M = \sum_i x_i \overline{M}_i$$

9.7 ❶ Partial molar quantities provide a mathematical way to assign the overall mixture property according to composition expressed in moles or mole fractions.

As a result, we may write

$$G = \sum_i x_i \overline{G}_i = \sum_i x_i \mu_i$$

9.8

We will return to these basic equations as we develop more relations for mixtures.

Equilibrium Criteria

For equilibrium at constant T and P, the Gibbs energy is minimized and mathematically the minimum means $d\underline{G} = 0$. Therefore, Eqn. 9.5 is equal to 0 at a minimum, since dT and dP are zero, and for a closed system all dn_i are zero. Thus,

$$d\underline{G} = 0 \qquad \text{at equilibrium, for constant } T \text{ and } P$$

9.9

This equation applies to whatever system we define. Suppose we define our system to consist of two components (e.g. EtOH + H_2O) distributed between two phases (e.g. vapor and liquid), $d\underline{G} = d\underline{G}^L + d\underline{G}^V = 0$, and at constant T and P, the moles may redistribute between the two phases,

$$\mu_1^L dn_1^L + \mu_2^L dn_2^L + \mu_1^V dn_1^V + \mu_2^V dn_2^V = 0$$

but if component 1 leaves the liquid phase then it must enter the vapor phase (and similarly for component 2) because the overall system is closed.

$$\Rightarrow dn_1^L = -dn_1^V \quad \text{and} \quad dn_2^L = -dn_2^V$$

$$\Rightarrow \left(\mu_1^V - \mu_1^L\right) dn_1^V + \left(\mu_2^V - \mu_2^L\right) dn_2^V = 0$$

The only way to make this equal to zero in general is:

$$\mu_1^V = \mu_1^L \quad \text{and} \quad \mu_2^V = \mu_2^L$$

9.10 ❶ The chemical potential of each component must be the same in each phase at equilibrium.

Setting the chemical potentials and T and P in each of the phases equal to each other provides a set of constraints (simultaneous equations) which may be solved for phase compositions provided we know the dependency of the chemical potentials on the phase compositions. Suppose the functions

$\underline{G}^L (x,T,P)$ and $\underline{G}^V (y,T,P)$ are available for a binary system. Then $\left(\dfrac{\partial \underline{G}^L}{\partial n_1}\right)_{T,P,n_2} = \left(\dfrac{\partial \underline{G}^V}{\partial n_1}\right)_{T,P,n_2}$;

$$\left(\frac{\partial \underline{G}^L}{\partial n_2}\right)_{T,P,n_1} = \left(\frac{\partial \underline{G}^V}{\partial n_2}\right)_{T,P,n_1} ; x_2 = 1 - x_1; y_2 = 1 - y_1;$$ which gives 4 equations with 4 unknowns

(x_1, x_2, y_1, y_2) that we can solve, in principle.[1] The first two equations are simply the equivalency of chemical potentials in the two phases.

Chemical Potential of a Pure Fluid

In Chapter 8, we showed the equilibrium constraint for a pure fluid is equality of the specific Gibbs energy in each the phases (*c.f.* Eqn. 8.3). How does this relate to Eqn. 9.10?

$$\mu_i \equiv (\partial(nG)/\partial n_i)_{T,P} = G(\partial n/\partial n_i)_{T,P} + n(\partial G/\partial n_i)_{T,P}$$

For a pure fluid, there is only one component, so $dn_i = dn$, and since $G(T,P)$ is intensive, then $n(\partial G/\partial n)_{T,P} = 0$. Also $(\partial n/\partial n_i)_{T,P} = 1$ by Eqn. 5.13. Therefore,

$$\mu_{i,pure} = G \qquad\qquad 9.11$$

❗ The chemical potential of a pure fluid is simply the molar Gibbs energy.

That is, the chemical potential of a pure fluid is simply the molar Gibbs energy. Pure components can be considered as a special case of the same general statement of the equilibrium constraint.

Fugacity

We introduced fugacity in the last chapter. The chemical potential constraint is sufficient for solving any phase equilibrium problem, but the most popular approach for actual computations makes use of the concept of fugacity. Fugacity is basically a quantity that has come to be recognized as more "user-friendly" than "chemical potential" or "partial molar Gibbs energy." It may not sound user-friendly, but it makes sense if one studies the roots of the term. Fuga- comes from a root referring to flight. The suffix -ity comes from a root meaning "character." Thus, fugacity was invented to mean flight-character, commonly called "escaping tendency" and is best thought of relative to liquid solutions. In an equimolar liquid solution, the component with the higher fugacity ("escaping tendency") will be more prevalent in the vapor phase.[2]

We note that the chemical potential is applicable to pure fluids as well as to components in mixtures. Therefore, let us generalize our pure component relations to become relations for components in mixtures: at constant T, we defined $RTd\ln f \equiv dG$ (Eqn. 8.19) which can be generalized to mixtures as

$$RT\, d \ln \hat{f}_i \equiv d\mu_i \qquad\qquad 9.12$$

❗ Fugacity is another way to express the chemical potential that is used more widely in engineering than chemical potential.

where $\hat{f}_i$ is the fugacity of component i in a mixture and μ_i is the chemical potential of the component. In the limit as the composition approaches a pure composition, these properties become

---

1. Even though $\mu^L = \mu^V$ at equilibrium, the dependency of μ^V on composition will be quite different from the dependency of μ^L on composition because the molecules are arranged very differently.

2. The concept of fugacity becomes useful when we begin to discuss phase equilibrium in mixtures. In that case, it is conceivable that we could have some supercritical component dissolved in the liquid phase despite its high escaping tendency, (e.g. CO_2 in soda pop at 100°F). The possibility of a component that cannot be a liquid still dissolving in a liquid requires a very general concept of escaping tendency because the pure-component vapor pressure does not exist at those conditions. The definition of fugacity provides us with that general concept.

equal to the pure component values. In a manner analogous to the chemical potential, the fugacity of a component becomes the pure component value as the composition approaches purity of that component.

Equality of Fugacities as Equilibrium Criteria

The equality of chemical potentials at equilibrium can easily be reinterpreted in terms of fugacity in a manner analogous to our methods for pure components from Eqn. 9.10

$$\mu_i^V(T, P) = \mu_i^L(T, P)$$

9.13

By integrating Eqn. 9.12 as a function of composition at fixed T from a state of pure i to a mixed state, we find

$$\mu_i^V - \mu_{i, pure} = RT \ln \frac{\hat{f}_i^V}{f_i}$$

9.14

where $\mu_{i, pure}$ and f_i are for the pure fluid at the system temperature. Writing an analogous expression for the liquid phase, and equating the chemical potentials using Eqn. 9.13, we find

$$\mu_i^V - \mu_i^L = RT \ln\left[\hat{f}_i^V / \hat{f}_i^L\right] = 0$$

9.15

$$\hat{f}_i^V = \hat{f}_i^L \quad \text{at equilibrium.}$$

9.16

Eqn. 9.13 or 9.16 becomes the starting point for all phase-equilibrium calculations. Therefore, we need to develop the capability to calculate chemical potentials or fugacities of components in vapor, liquid, and solid mixtures. Here we briefly introduce the framework for calculating the fugacities of components before we begin the direct calculations.

Ideal Gases

Properties of ideal gases are relatively easy to calculate. The ideal gas is also a convenient starting point to introduce the calculation of mixture properties and component fugacities. Since ideal gas molecules do not have intermolecular potentials, the internal energy consists of entirely kinetic energy. When components are mixed at constant temperature and pressure, the internal energy is simply the sum of the component internal energies, which can be written

$$U^{ig} = \sum_i y_i U_i^{ig} \quad \text{or} \quad \underline{U}^{ig} = \sum_i n_i U_i^{ig}$$

(ig) 9.17

The total volume of a mixture is related to the number of moles

$$\underline{V}^{ig} = \left(RT \sum_i n_i\right) / P$$

(ig) 9.18

Combing U and V to obtain the definition of H, $H = U + PV$ and using Eqns. 9.17 and 9.18

$$\underline{H}^{ig} = \sum_i n_i U_i^{ig} + RT \sum_i n_i = \sum_i n_i (U_i^{ig} + RT) = \sum_i n_i H_i \qquad \text{(ig) 9.19}$$

Therefore the enthalpy of a mixture is given by the sum of the enthalpies of the components at the same temperature and pressure. On a molar basis,

$$H^{ig} = \sum_i y_i H_i^{ig} \qquad \text{(ig) 9.20}$$

Entropy for an ideal gas mixture is more complicated because, as shown in Chapter 3, even systems of fixed total energy have an entropy change associated with mixing due to the distinguishability of the components. The entropy of an ideal gas is calculated by the sum of the entropies of the components plus the entropy change of mixing as given in Chapter 3,

$$\underline{S}^{ig} = \sum_i n_i S_i^{ig} + \Delta \underline{S}_{mix} = \sum_i n_i S_i^{ig} - R \sum_i n_i \ln y_i \quad \text{or} \quad S^{ig} = \sum_i y_i S_i^{ig} - R \sum_i y_i \ln y_i \qquad \text{(ig) 9.21}$$

The Gibbs energy and the fugacity will be at the core of phase equilibria calculations. The Gibbs energy of an ideal gas is obtained from the definition, $G \equiv H - TS$. Using H^{ig} an S^{ig} from above,

$$\underline{G}^{ig} \equiv \underline{H}^{ig} - T \underline{S}^{ig} = \sum_i n_i H_i^{ig} - T\left(\sum_i n_i S_i^{ig} - R \sum_i n_i \ln y_i\right) = \sum_i n_i G_i^{ig} + RT \sum_i n_i \ln y_i \qquad \text{(ig) 9.22}$$

The chemical potential of a component is given by Eqn. 9.6 and taking the derivative of Eqn. 9.22,

$$\mu_i^{ig} = (\partial \underline{G}^{ig} / \partial n_i)_{T, P, n_{j \neq i}} = G_i^{ig} + RT\left(\partial \left(\sum_i n_i \ln y_i\right) / \partial n_i\right)_{T, P, n_{j \neq i}} \qquad \text{(ig) 9.23}$$

The derivative is most easily seen by expanding the logorithm before differentiation, $\ln y_i = \ln n_i - \ln n$. Then,

$$\left(\partial \left(\sum_i n_i \ln n_i - \ln n \sum_i n_i\right) / \partial n_i\right)_{T, P, n_{j \neq i}} = \ln n_i + 1 - 1 - \ln n = \ln y_i \qquad 9.24$$

therefore

$$\mu_i^{ig} = G_i^{ig} + RT \ln y_i \qquad \text{(ig) 9.25}$$

By Eqn. 9.14, using Eqn. 9.25 and Eqn. 9.11,

$$\mu_i^{ig} - \mu_{i, pure}^{ig} = RT \ln \frac{\hat{f}_i^{ig}}{f_{i, pure}^{ig}} = RT \ln y_i \quad \text{or} \quad \hat{f}_i^{ig} = y_i f_{i, pure}^{ig} \qquad \text{(ig) 9.26}$$

Since by Eqn. 8.22, $f_{pure}^{ig} = P$,

$$\boxed{\hat{f}_i^{ig} = y_i P}$$

(ig) 9.27   ❶ The fugacity of a component in an ideal-gas mixture is particularly simple, it is equal to the $y_i P$, the partial pressure!

Therefore the fugacity of an ideal gas component is simply its partial pressure. This makes the ideal gas fugacity easy to quantify rapidly. One of the goals of the calculations that will be pursued in Chapter 10 is the quantification of the deviations of the fugacity from ideal gas values quantified by the component fugacity coefficient.

Activity, Activity Coefficient, and Fugacity Coefficient

The fugacity may be calculated via one of three characterizations, the activity, activity coefficient, or fugacity coefficient. To understand these relations, it is first helpful to visualize the framework in which they are developed. The framework is shown in terms of a Venn diagram in Fig. 9.2. You will develop skills in using these relations as you work forward through the material. It may be helpful to refer back to this section periodically to keep some perspective on the relations between the various approaches for modeling phase equilibria. We will follow this philosophy in future chapters as we develop the appropriate relations by presenting modifications of Fig. 9.2 as we introduce the relations that we use.

Ideal gas behavior is the simplest type of mixture behavior because the particles are non-interacting and is shown in the center of Fig. 9.2. Clearly, ideal gas behavior is not followed by all mixtures, therefore ideal gases are a subset of real mixtures, and the relations can only be applied under

❶ Carets (^) are used to denote component properties in mixtures for f and φ. When working with μ, the context of the calculation must be carefully followed.

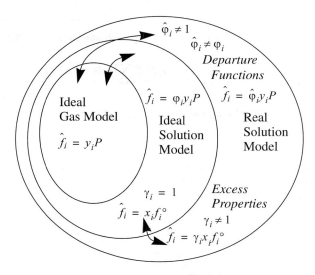

Figure 9.2 *Schematic of the relations between different fluid models. Ideal gases are a subset of ideal mixtures, which in turn are a subset of real mixtures. Departure functions (fugacity coefficients) characterize deviations from ideal-gas behavior, and excess properties (activity coefficients) characterize deviations from ideal-solution behavior.*

stringent conditions. For example, ideal gases cannot condense because they have no attractive forces. Therefore, if all fluids were ideal gases, there would be no liquids, and VLE would not occur.[1] The fugacity of a component in an ideal-gas mixture is particularly simple; it is equal to y_iP, the partial pressure. In the treatment of pure fluids, we defined the fugacity coefficient to quantify the deviation from ideal-gas behavior for a pure fluid. Likewise, we define a component *fugacity coefficient* to quantify the deviations from component behavior in ideal-gas mixtures. The fugacity coefficient is

> The fugacity coefficient characterizes deviations from the ideal gas component fugacity.

$$\hat{\varphi}_i \equiv \frac{\hat{f}_i}{y_iP}$$

9.28

We may calculate, in principle, the fugacity coefficient of a gas or liquid by developing appropriate solution models of the Gibbs energy. If you look closely at Eqn. 9.28, you notice that the component fugacity coefficient is just a measure of the deviation of the component fugacity from the partial pressure. The fugacity of a vapor-phase component will be calculated by determining the value of the component fugacity coefficient, and then applying:

> Equation of state methods calculate the fugacity coefficient, then multiply by the partial pressure.

$$\boxed{\hat{f}_i = y_i\hat{\varphi}_iP}$$

9.29

When $\hat{\varphi}_i$ is one, the component behaves as if it is in an ideal-gas mixture.

There is a model of solutions, *ideal solutions*, intermediate between ideal gases and real mixtures. Ideal solutions are composed of very similar molecules which can be modeled rather easily. For example, *liquid n-propane* and *liquid n-butane* are certainly not ideal gases. However their molecular structures are very similar,[2] and when they are mixed, a propane-butane interaction will be almost like a propane-propane or butane-butane interaction; thus, there is negligible energy change for the mixture relative to the pure components. Since they are very similar in size, the entropic effect of mixing is analogous to that of mixing distinguishable marbles of the same size, and the entropy of mixing is the same as it would be for ideal gases. As shown in Fig. 9.2, the component fugacity is proportional to f_i^o or f_i. Deviations from ideal solution behavior are quantified by the activity and activity coefficient given by

> Activity is commonly used with electrolyte solutions and reacting systems.

$$\text{"Activity"} \quad a_i = \frac{\hat{f}_i}{f_i^o};$$

9.30

$$\text{"Activity coefficient"} \quad \gamma_i = \frac{\hat{f}_i}{x_if_i^o}$$

9.31

> Activity coefficients are commonly used for highly non-ideal solutions.

where f_i^o is the value of the fugacity at *standard state*.[3] The most common standard state is the pure component at the same temperature and pressure as the system of interest. In that case our standard state fugacity is simply the pure component fugacity which we introduced in Chapter 8, $f_i^o = f_i$, where f_i is at the same T as $\hat{f}_i$. Note that the defining equation for activity coefficient can be rearranged to become an equation for calculating the fugacity of a component in a mixture. If the fugacity of interest was for an ideal gas, then γ_i would be equal to unity (from Fig. 9.2 we see that ideal

---

1. However, it is fortunate that we can frequently model the vapor phase *as if it is an ideal gas.*
2. They are members of a homologous series; their structures differ by only one repeat unit.
3. See *standard state* and *reference state* in the glossary for comparing the similarities and differences between these states.

gases are a subset of ideal solutions), and $f_i^{\,\circ}$ would be P, and we would have the formula for the partial pressure. The non-ideality of gases may be characterized by activity and activity coefficients in this manner; however, the procedure is used most frequently for liquids. The "gamma approach" is an extension of the principles used for pure components discussed in the last chapter, Eqn. 8.36.

$$\hat{f}_i^L \;=\; \gamma_i x_i f_i^{\,\circ} = \gamma_i x_i \varphi_i^{sat} P_i^{sat} \exp\!\left(\frac{V_i^L(P_i - P_i^{sat})}{RT}\right)$$

9.32 ❶ The full form of the gamma approach for calculating fugacity of a liquid component.

If the vapor pressure is moderate, we can make a first approximation by including P_i^{sat} in place of $f_i^{\,\circ}$. Then we obtain

$$\hat{f}_i^L \;=\; \gamma_i x_i P_i^{sat}$$

9.33 ❶ The form of the fugacity formula after common simplifications.

Recall that the starting point for calculating VLE will be Eqn. 9.16. Any assumed model for the vapor phase from Fig. 9.2 may be matched with any assumed model for the liquid phase; however, the vapor phases are more likely to approach ideal approximations than the liquid phases for a given mixture. The ideal-solution approximation is sometimes called the *Lewis/Randall rule*, or the *Lewis fugacity rule*. The ideal-solution approximation can also be used *with a hypothetical pure fugacity and it is then known as Henry's Law.* Henry's Law is often used for gases above their critical point, where vapor pressure is undefined. The formulas for the various approximations are shown in Table 9.1.

Within the remainder of this chapter and the next chapters, methods will be developed to determine $\hat{f}_i$ from experimental data, and methods will be developed to predict and/or correlate the component fugacity values. Suppose that experiments have been conducted in a binary system that have led to the values of $\hat{f}_1$ in Fig. 9.3 plotted as points and connected with the smooth curve. The fugacity of component 1 follows Henry's law at low concentrations, and follows the Lewis/Randall rule at high concentrations. The behavior of the component 2 fugacity is not plotted, but it will be a *qualitative* mirror image of the behavior of the plotted component 1 fugacity; the component 2 fugacity will approach zero as x_2 approaches 0, and the values of the ideal solution lines at $x_2 = 1$ will be h_2 and f_2. The fugacities of both components in this example will lie above their respective Lewis/Randall ideal solution lines, and the mixture is described as having *positive deviations* from the ideal solution. If the fugacity curves of the components lie below the Lewis/Randall lines, the mixture is described as having *negative deviations* from the ideal solution. The convention for characterizing the deviations as positive or negative generally refers to the deviations from the Lewis/Randall rule rather than Henry's law.

Table 9.1 *Summary of approximations that may be made for calculating component fugacities.*

	Vapor	**Liquid**
Ideal Gas	$y_i P$	- - -
Ideal Solution	$y_i \varphi_i^V P$	$x_i f_i^{\,\circ}$ where usually $f_i^{\,\circ} = f_i = \varphi_i^{sat} P_i^{sat} \exp[V_i^L(P - P_i^{sat})/RT]$ or $\varphi_i^L P$ or $f_i^{\,\circ} = h_i$
Real Solution	$y_i \;_i P$	$x_i \hat{\varphi}_i^L P$ *or* $x_i \gamma_i f_i$ where $f_i = \varphi_i^{sat} P_i^{sat} \exp[V_i^L(P - P_i^{sat})/RT]$

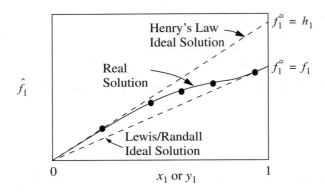

Figure 9.3 *Schematic representation of the fugacity of component 1 in a binary mixture.*

If the plotted real solution fugacity represents a vapor phase, it is characterized using Eqn. 9.18 and approaches the value $f_1 = \varphi_1 P$ as y_1 approaches 1. If the plotted real solution fugacity represents a liquid phase, it can be characterized by Eqns. 9.32 or 9.33. For either vapors or liquids, the deviations from the ideal solution behaviors depend on composition, so $\hat{\varphi}_i$ and/or γ_i values depend on composition. Before we handle the general case of treating solutions that deviate from ideal solutions, we will introduce calculations for systems that follow ideal solution behavior.

9.3   IDEAL SOLUTIONS

Sometimes the simplest analysis deserves more consideration than it receives. Ideal solutions can be that way. For an ideal solution, there are no synergistic effects of the components being mixed together; each component operates independently. Thus, there is no energy change from mixing and no volume change. Examples of ideal solutions are ideal gases and family members like benzene + toluene, n-butanol + n-pentanol, n-pentane + n-hexane. Since $H \equiv U + PV$, the enthalpy of the mixture will simply be the sum of the pure component enthalpies times the number of moles of that component.

$$H^{is} = \sum_i x_i H_i$$

9.34

Comparing Eqn. 9.34 with Eqn. 9.7, we see that

$$\overline{H}_i^{is} = H_i$$

9.35

Let us define the enthalpy of mixing as the enthalpy of the mixture relative to the enthalpies of the pure components before mixing.

$$\Delta H_{mix} \equiv H - \sum_i x_i H_i$$

9.36

Therefore, for an ideal solution,

$$\boxed{\Delta H^{is}_{mix} = 0}$$

9.37   ❗ The enthalpy of mixing is zero for an ideal solution.

The volume of mixing is the volume of the mixture relative to the volumes of the components before mixing.

$$\Delta V_{mix} \equiv V - \sum_i x_i V_i$$

9.38

An ideal solution will have a zero volume of mixing, so in a manner analogous to the component partial molar enthalpies,

❗ The volume of mixing is zero for an ideal solution.

$$\overline{V}^{is}_i = V_i$$

9.39

As for the entropy change of mixing, the loss of order due to mixing is unavoidable, even for ideal solutions. During our consideration of the microscopic definition of entropy, we derived a general expression for the ideal entropy change of mixing, Eqn. 3.7.

$$\boxed{\Delta S^{is}_{mix} / R = -\sum_i x_i \ln(x_i)}$$

9.40   ❗ The entropy of mixing is *non-zero* for an ideal solution.

Although we derived Eqn. 3.7 for mixing ideal gases, it also provides a reasonable approximation for mixing liquids of equal-sized molecules. The reason is that changes in entropy are related to the change in accessible volume. Even though a significant volume is occupied by the molecules themselves in a dense liquid, the void space in one liquid will be very similar to the void space in another liquid if the molecules of each liquid are similar in size. That means that the accessible volume for each component doubles when we mix equal parts of two equal-sized components. That is essentially the same situation that we had when mixing ideal gases.

Given the effect of mixing on these two properties, we can derive the effect on other thermodynamic properties.

$$\boxed{\frac{\Delta G^{is}_{mix}}{RT} = \frac{\Delta H^{is}_{mix}}{RT} - \frac{\Delta S^{is}_{mix}}{R} = \sum_i x_i \ln(x_i)}$$

9.41

The general relationship for ΔG_{mix} gives a relationship for fugacity. We can extend our definition of the enthalpy of mixing to the Gibbs energy of mixing,

$$\boxed{\Delta G_{mix} \equiv G - \sum_i x_i G_i}$$

9.42

But by using Eqns. 9.8 and 9.14,

$$\frac{\Delta G_{mix}}{RT} = \sum_i x_i \frac{(\mu_i - G_i)}{RT} = \sum_i x_i \ln\left[\frac{\hat{f}_i}{f_i}\right]$$

9.43

Thus, comparing Eqns. 9.43 and 9.41, for an ideal solution, $\dfrac{\Delta G_{mix}^{is}}{RT} = \sum_i x_i \ln(x_i) = \sum_i x_i \ln\left[\dfrac{\hat{f}_i}{f_i}\right]$.

By comparing the relations in the logarithms, we obtain the *Lewis/Randall* rule:

❶ Lewis/Randall rule for component fugacity in an ideal solution.

$$\Rightarrow \hat{f}_i^{is} / f_i = x_i \quad \Rightarrow \hat{f}_i^{is} = x_i f_i \qquad\qquad 9.44$$

Application of the Ideal-Solution Approximation to Vapor-Liquid Equilibrium (VLE)

We would like to generate a method for calculating $K_i \equiv y_i / x_i$ for each component. This "equilibrium K-ratio" will prove to be very useful in solving our system of nonlinear simultaneous equations. By our equilibrium constraint,

$$\hat{f}_i^V = \hat{f}_i^L \qquad\qquad 9.45$$

By our ideal solution approximation in both phases, the equilibrium criteria becomes

$$y_i f_i^V = x_i f_i^L \qquad\qquad 9.46$$

Now we need to substitute the expressions for f_i^V and f_i^L that we developed in Chapter 8. The fugacity of the pure vapor comes from Eqn. 8.25

$$f_i^V = \varphi_i^V P \qquad\qquad 9.47$$

The fugacity of the liquid comes from Eqn. 8.36

$$f_i^L = \varphi_i^{sat} P_i^{sat} \exp\left(\frac{V_i^L (P - P_i^{sat})}{RT}\right) \qquad\qquad 9.48$$

Combining Eqns. 9.46–9.48

$$y_i \varphi_i^V P = x_i \varphi_i^{sat} P_i^{sat} \exp\left(\frac{V_i^L (P - P_i^{sat})}{RT}\right)$$

Writing in terms of the K_i ratio,

$$K_i = \frac{y_i}{x_i} = \frac{P_i^{sat}}{P} \left[\frac{\varphi_i^{sat} \exp[V_i^L (P - P_i^{sat})/RT]}{\varphi_i^V}\right]$$

Note: *at reasonably low pressures,*

$$\frac{\varphi_i^{sat}}{\varphi_i} \approx 1, \text{ and } \exp[V_i^L(P - P_i^{sat})/RT] \approx 1$$

$$\boxed{K_i = \frac{P_i^{sat}}{P}} \quad \text{or} \quad \boxed{y_iP = x_iP_i^{sat}}$$

9.49   ❶ Raoult's law.

Some systems which follow Raoult's law are shown in Fig. 9.4 and Fig. 9.5. We reiterate that Raoult's law is only valid for molecularly similar components; ideal behavior is demonstrated in the figures because of careful selection of molecularly similar binary pairs. If we mixed methanol from Fig. 9.4*a* and benzene from Fig. 9.5*c*, the resulting system would be very non-ideal. We will revisit this example before we leave this chapter. Note that the bubble line on the *P-x-y* diagrams is nearly linear for all the systems. Also note that the *T-x-y* and *P-x-y* diagram shapes are related qualitatively by inverting one of the diagrams.

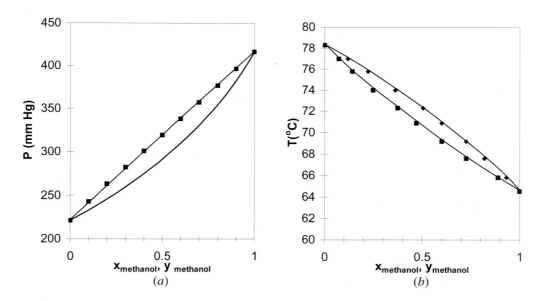

(a)   (b)

Figure 9.4 *a, b. Phase Behavior of the methanol-ethanol system. Left figure at 50°C. Right figure at 760 mm Hg. (P-x-y from Schmidt, G.C., Z. Phys.Chem. **121**, 221(1926), T-x-y from Wilson, A., Simons, E.L., Ind. Eng. Chem., **44**, 2214(1952)).*

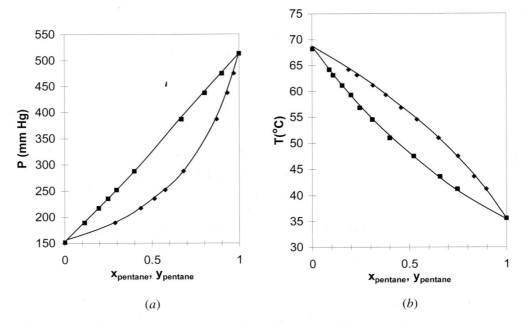

(a)   *(b)*

Figure 9.5 *(a); (b) Phase Behavior of the pentane-hexane system. Left figure at 25°C. Right figure at 750 mm Hg.(P-x-y from Chen, S.S., Zwolinski, B.J., J.Chem. Soc. Faraday Trans., 70, 1133(1974), T-x-y from Tenn, F.G, Missen, R.W., Can. J. Chem. Eng., 41, 12(1963)).*

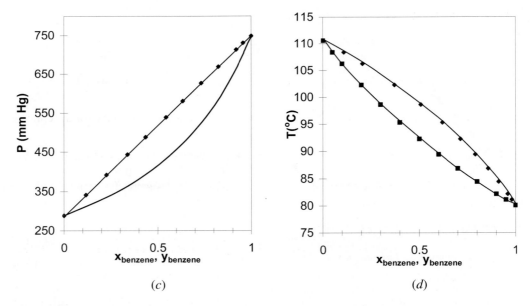

(c)   *(d)*

Figure 9.5 *(c); (d) Phase Behavior of the benzene-toluene system. Left figure at 79.7°C. Right figure at 760 mm Hg.(P-x-y from Rosanoff, M.A., et al., J. Am.Chem.Soc. 36, 1993(1914), T-x-y from Delzenne A.O., Ind. Eng. Chem., Chem. Eng. Data Series. 3, 224(1958)).*

Shortcut Estimation of VLE *K*-ratios

Given Raoult's law and recalling our shortcut procedure for estimating vapor pressures, it is very useful to consider combining these to give a simple and quick method for estimating ideal solution *K*-ratios. Substituting, we obtain the shortcut *K*-ratio,

$$K_i = \frac{P_i^{sat}}{P} \approx \frac{P_{c,i} \, 10^{\frac{7}{3}(1+\omega)\left(1 - \frac{1}{T_{r,i}}\right)}}{P}$$

9.50

> **Shortcut K-ratio** for Raoult's law. Not only is this expression restricted to Raoult's law, but it is also subject to the restrictions of Eqn. 8.11.

Since the ideal solution model is somewhat crude anyway, it is not unreasonable to apply the above equation as a first approximation for any ideal solution when rapid approximations are needed.[1]

9.4   VAPOR-LIQUID EQUILIBRIUM (VLE) CALCULATIONS

Classes of VLE Calculations

Depending on the information provided, one may perform one of several types of calculations to assess the vapor-liquid partitioning. These are: bubble-point pressure (BP), dew-point pressure (DP), bubble-point temperature (BT), dew-point temperature (DT), and isothermal flash (FL). The specifications of the information required and the information to be computed are tabulated below. Also shown in the table are indications of the relative difficulty of each calculation.[2] The best approach to understanding the calculations is to gain experience by plotting phase envelopes in various situations.

> The classes of VLE calculations presented here will be used through the remainder of the text, so the concepts are extremely important.

Principles of Calculations

The principles of approaching one of the calculations shown in Table 9.2 are based on the information available in the second column of the table. For a bubble calculation, all the x_i are known, and we find the y_i by solving for the condition where $\sum_i y_i = 1$. For dew calculations, all y_i are known, and we solve for the condition where $\sum_i x_i = 1$. For a flash calculation, we solve for the condition where $\sum_i x_i - \sum_i y_i = 0$. The information in Table 9.2 is rigorous. The method used to calculate K_i is model-dependent. For Raoult's law, we rearrange our objective function in terms of variables which are known using Eqn. 9.49.

---

1. An interesting problem arises when we must calculate the VLE *K*-ratio for a component in the liquid phase but above its critical temperature. Carbon dioxide ($T_c = 31°C$) in soda pop on a 32°C day would be a common example. Since the saturation pressure of CO_2 does not exist above the critical temperature and pure CO_2 cannot condense, we might consider that CO_2 wouldn't exist in the liquid phase and that Raoult's law might indicate an infinite value for the *K*-ratio. Experience tells us that this component will exist in the liquid phase over a portion of the composition range, but not near pure CO_2 liquid-phase compositions. What is interesting is that extrapolated vapor pressures in the above formula give reasonably accurate results at small liquid phase concentrations of non-condensable components. Of course, it is more accurate for components that are only slightly above their critical temperature, because then the extrapolation is slight. But it is surprisingly accurate at more stringent conditions.

2. If a supercritical component is present in significant quantity, the user must beware, because the shortcut *K*-ratio may falsely predict a liquid phase due to extrapolation of the vapor pressure. Calculations with supercritical components are best done with an equation of state as shown in Example 10.9 on page 340.

Table 9.2 *Summary of the types of phase equilibria calculations. This table is independent of the VLE model.*

Type	Information known	Information computed	Criteria	Convergence
BP	$T, x_i = z_i$	P, y_i	$\sum_i y_i = \sum_i K_i x_i = 1$	Easiest
DP	$T, y_i = z_i$	P, x_i	$\sum_i x_i = \sum_i y_i / K_i = 1$	Not bad
BT	$P, x_i = z_i$	T, y_i	$\sum_i y_i = \sum_i K_i x_i = 1$	Difficult
DT	$P, y_i = z_i$	T, x_i	$\sum_i x_i = \sum_i y_i / K_i = 1$	Difficult
FL	P, T, z_i	$x_i, y_i, L/F$	$\sum_i z_i (1 - K_i)/(K_i + (L/F)(1 - K_i)) = 0$	Most difficult

Methods for Binary Raoult's Law

Note that as we begin applications, some of the equations are model-dependent.

For a *bubble-pressure* calculation, writing $\sum_i y_i = 1$, or $\sum_i K_i x_i = 1$, $\sum_i \dfrac{P_i^{sat}}{P} x_i = 1$ and rearranging:

$$1 = \frac{P_1^{sat}}{P} x_1 + \frac{P_2^{sat}}{P} x_2 \qquad\qquad 9.51$$

$$P = x_1 P_1^{sat} + x_2 P_2^{sat} \qquad\qquad 9.52$$

where no iterations are required because temperature, and therefore vapor pressures, are known. Note from this equation that it is now obvious that Raoult's law will predict a linear bubble pressure line on a P-x-y diagram, because $x_2 = 1 - x_1$, substituting into Eqn. 9.52, $P = x_1 P_1^{sat} + (1 - x_1)P_2^{sat} = x_1(P_1^{sat} - P_2^{sat}) + P_2^{sat}$.

For a *dew-pressure* calculation, writing $\sum_i x_i = 1$, or $\sum_i \dfrac{y_i}{K_i} = 1$ and rearranging:

$$\frac{y_1 P}{P_1^{sat}} + \frac{y_2 P}{P_2^{sat}} = 1 \qquad\qquad 9.53$$

which may be rearranged and solved without iteration, because the vapor pressures are fixed at the specified temperature:

$$P = \frac{1}{\dfrac{y_1}{P_1^{sat}} + \dfrac{y_2}{P_2^{sat}}} \qquad\qquad 9.54$$

For a *bubble-temperature* calculation, writing $\sum_i y_i = 1$, or $\sum_i K_i x_i = 1$, and rearranging:

$$P = x_1 P_1^{sat} + x_2 P_2^{sat}$$

9.55

To solve this equation, it is necessary to iterate on temperature (which changes P_i^{sat}) until P equals the specified pressure.

For a *dew-temperature* calculation, writing $\sum_i x_i = 1$, or $\sum_i \frac{y_i}{K_i} = 1$, and rearranging:

$$P = \frac{1}{\dfrac{y_1}{P_1^{sat}} + \dfrac{y_2}{P_2^{sat}}}$$

9.56

To solve this equation, it is necessary to iterate on temperature (which changes P_i^{sat}) until P equals the specified pressure.

Flash drums[1] are frequently used in chemical processes. For an isothermal drum, the temperature and pressure of the drum are fixed. Consider that a feed stream is liquid which becomes partially vaporized after entering the drum. For a binary system, a flash calculation may be avoided by plotting the *P-x-y* diagram between the dew and bubble pressures and reading the vapor and liquid compositions from the graph. Application of the lever rule permits calculation of the total fraction that is vapor; however, this method requires a lot of calculations to generate the diagram if it is not already available. Also, plotting the curves is slower than a direct calculation.

> *Bubble and dew calculations at the overall composition are recommended first to assure the flash drum is between the bubble and dew pressures at the given temperature. (These calculations are easier than the flash calculation and may save you from doing it if you are outside the phase envelope.)*

The flash equations are easily derived by modification of the overall and component balances used in the development of the lever rule in Section 9.1. If z is the feed composition and L/F is the liquid-to-feed ratio, then $V/F = 1 - L/F$, and the component balance is $z_i = x_i \frac{L}{F} + y_i \frac{V}{F}$. Substituting for V/F from the overall balance, and using $y_i = x_i K_i$, therefore $z_i = x_i \left[\frac{L}{F} + K_i \left(1 - \frac{L}{F} \right) \right]$, which becomes

$$x_i = \frac{z_i}{K_i + \dfrac{L}{F}(1 - K_i)}$$

9.57

---

1. The procedures developed here are for isothermal flash calculations (heat transfer occurs between the surroundings and the flash drum). For adiabatic calculations, the energy balance must also be used to determine that fraction vaporized and the temperature of the flash drum. Adiabatic calculations are much more tedious.

using $y_i = x_i K_i$, we may just multiply Eqn. 9.57 by K_i to obtain y_i.

$$y_i = \frac{z_i K_i}{K_i + \frac{L}{F}(1 - K_i)} \qquad\qquad 9.58$$

The obvious thing to do at this point is to iterate to find the L/F ratio which satisfies $\sum_i x_i = 1$. But the flash problem is different from the dew and bubble point problems because we must solve $\sum_i y_i = 1$ also. Fortunately, a reliable method has been developed to solve this problem. The reliable way to satisfy both these constraints simultaneously is to solve $\left(\sum_i x_i - \sum_i y_i\right) = 0$. We define this difference to be the objective function, $\sum_i D_i$, and iterate until the sum approaches zero. For a binary system

$$\sum_i x_i - \sum_i y_i = (x_1 - y_1) + (x_2 - y_2) = \frac{z_1(1 - K_1)}{K_1 + (L/F)(1 - K_1)} + \frac{z_2(1 - K_2)}{K_2 + (L/F)(1 - K_2)} \qquad 9.59$$

$$\equiv D_1 + D_2 = 0$$

For Raoult's law, K_i is only a function of temperature and pressure, both of which are fixed for an isothermal flash calculation. Therefore, in Eqn. 9.59, the only unknown is L/F. The value of Eqn. 9.59 always decreases as L/F increases; therefore, we search for the value of L/F which satisfies the equation. Then, $V/F = 1 - L/F$ from our basis. Note that $0 < L/F < 1$ for a physically realistic answer. Outside this range, we are either above the bubble pressure or below the dew pressure. One other point should be made before leaving the flash calculation. The name of the flash procedure implies that it is applicable only for flashing liquids, but in fact, the procedure is also valid for partial condensation of a vapor or for any number of incoming vapor and/or liquid streams. To apply the procedure, the overall composition of the components, z_i, and total feed flowrate, F, just need to be known before the procedure is started.

Multicomponent VLE Calculations

Extending our equations to multicomponent systems is straightforward. For a bubble calculation we have:

$$1 = \frac{\sum_i x_i P_i^{sat}}{P} = \sum_i x_i K_i \qquad\qquad 9.60$$

For dew calculation we have:

$$1 = P\sum_i \frac{y_i}{P_i^{sat}} = \sum_i \frac{y_i}{K_i} \qquad\qquad 9.61$$

These equations may be used for bubble or dew-pressure calculations without iterations. For bubble or dew-point temperatures, iteration is required. A first guess may be obtained from one of the following formulas:

$$T = \sum_i x_i T_i^{sat}$$

$$T = \frac{\sum_i y_i T_{r,i} T_i^{sat}}{\sum_i y_i T_{c,i}} \qquad 9.62$$

But these are somewhat inaccurate guesses which require subsequent iteration.[1]

For isothermal flash calculations, the general formula is:

$$\sum_i x_i - \sum_i y_i = \sum_i \frac{z_i(1 - K_i)}{K_i + (L \, / \, F)(1 - K_i)} \equiv \sum_i D_i = 0 \qquad 9.63$$

Iterative Calculations by Excel

The Solver tool in Excel is useful for performing iterative calculations. Many of the following examples summarize detailed calculations to illustrate fully the iterative procedure. In practice, the detailed calculations can be performed rapidly on a spreadsheet. Appendix P summarizes the use of Solver and the methods for successive substitution.

Example 9.1 Bubble and dew temperatures and isothermal flash of ideal solutions

The overhead from a distillation column is to have the following composition:

	z(Overhead)
Propane	0.23
Isobutane	0.67
n-Butane	0.10
Total	1.00

---

1. Note that Eqns. 9.60–9.62 can be used to estimate the bubble and dew points regardless of whether the components are supercritical or whether vapor and liquid phases are indeed possible. We will see in the discussion of equations of state that mixtures can have critical points, too, and this leads to a number of subtle complexities.

Example 9.1 Bubble and dew temperatures and isothermal flash of ideal solutions (Continued)

A schematic of the top of a distillation column is shown below. The overhead stream in relation to the column and condenser is shown where V represents vapor flow and L represents liquid flow. In an ideal column, the vapor leaving each tray is in phase equilibrium with the liquid leaving the same tray. If the cooling water to the condenser is turned off, only vapor product will be obtained. If cooling water is provided, and the vapor product stream is turned off, the condenser will be known as a total condenser, and only liquid product will be collected. If both vapor and liquid product are collected, the condenser will provide some additional separation, and operates as a partial condenser. In an ideal partial condenser, the vapor and liquid leave in phase equilibrium with each other.

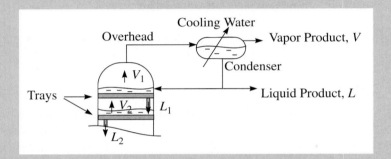

(a) Calculate the temperature at which the condenser must operate in order to condense the overhead product completely at 8 bars.

(b) Assuming the overhead product vapors are in equilibrium with the liquid on the top plate of the column, calculate the temperature of the overhead vapors and the composition of the liquid on the top plate when operating at the pressure of part *a*.

(c) The vapors are condensed by a partial condenser operating at 8 bar and 320 K. What fraction of liquid is condensed?

Solution: Use the shortcut estimates of the K-ratios

It is easy to create an Excel spreadsheet to do these calculations quickly, (*e.g.,* FLSHR.xls can be modified).

HP BPIS (see Prl) does Raoult's law calculations using the shortcut vapor pressure equation.

(a) The maximum temperature is the bubble point temperature.

	Guess $T = 310$ K		Guess $T = 320$ K	
	K_i	y_i	K_i	y_i
C_3	1.61	0.370	2.03	0.466
iC_4	0.616	0.413	0.80	0.536
nC_4	0.438	0.044	0.58	0.058
		0.827		1.061

$$\Rightarrow T = 310 + \left(\frac{1.000 - 0.827}{1.061 - 0.827}\right)(320 - 310) = 317 \text{ K}$$

Example 9.1 Bubble and dew temperatures and isothermal flash of ideal solutions (Continued)

(b) The saturated vapor is at its dew point temperature.

	Guess $T = 325$ K		Guess $T = 320$ K	
	K_i	x_i	K_i	x_i
C_3	2.26	0.102	2.03	0.113
iC_4	0.905	0.740	0.80	0.838
nC_4	0.658	0.152	0.58	0.172
		0.994		1.123

$$T = 325 + \left(\frac{1.00 - 0.994}{1.123 - 0.994}\right)(320 - 325) = 324.8 \text{ K}$$

(c) Recognize that the solution will involve an isothermal flash calculation.

We begin by noting that the specified temperature is between 317 and 324 K, so vapor-liquid equilibrium is indeed possible. We seek a solution for Eqn 9.63 at 320 K:

	z_i	K_i	Guess $L/F = 0.5$ D_i	Guess $L/F = 0.6$ D_i	Guess $L/F = 0.77$ D_i
C_3	0.23	2.03	−0.1564	−0.1678	−0.1915
iC_4	0.67	0.80	0.1489	0.1457	0.1405
nC_4	0.10	0.58	0.0532	0.0505	0.0465
			0.0457	0.0284	−0.0045

$$\frac{L}{F} = 0.77 + \left(\frac{0. + 0.0045}{0.0284 + 0.0045}\right)(0.6 - 0.77) = 0.7467$$

$L/F = 0.75$, applying Eqns. 9.57 and 9.58,

$$\Rightarrow \{x_i\} = \{0.1829, 0.7053, 0.1117\} \text{ and } \{y_i\} = \{0.3713, 0.5642, 0.0648\}.$$

Note: *The flash problem converges much more slowly than the bubble and dew point calculations, so the third iteration is necessary.*

9.5 EMISSION MODELING

Hydrocarbon emission monitoring is an important aspect of environmentally conscious chemical manufacturing and processing. The U.S. Environmental Protection Agency (EPA) has published guidelines[1,2,3] on the calculations of emissions of volatile organic compounds (VOCs), and VOC emissions are monitored in the U.S. Most of the emission models apply the ideal gas law and Raoult's law and thus the calculation methods are easily applied. While many of the mixtures represented with these techniques are not accurately modeled for phase equilibria by Raoult's law, the method is suitable as a first approximation for emission calculations. This section explores emission calculations for batch processes. Batch processes are common in specialty chemical manufacture. In most cases, air or an inert gas such as nitrogen is present in the vapor space (also known as the *head space*). In some cases, the inert head space gas flows through the vessel, and is called a *purge* or *sweep* gas. These gases typically have negligible solubilities in the liquid phase and are thus considered *noncondensable*. There are several common types of unit operations encountered with VOC emissions, which will be covered individually.

Filling or Charging

During filling of a tank with a volatile component, gas is displaced from the head space. The displaced gas is assumed to be saturated with the volatile components as predicted with Raoult's law and the ideal gas law. Initially in the head space, $n_{head}^i = (P\underline{V}_{head}^i)/(RT)$ and after filling, $n_{head}^f = (P\underline{V}_{head}^f)/(RT)$, where the subscript *head* indicates the head space. The volume of liquid charged is equal to the volume change of the head space. The mole fractions of the VOC components are determined by Raoult's law, and the noncondensable gas makes up the balance of the head space. The moles of VOC emission from the tank is estimated by $y_i(n_{head}^i - n_{head}^f)$ for each VOC.

Purge Gas (Liquid VOC present)

When a purge (sweep) gas flows through a vessel containing a liquid VOC, the effluent will contain VOC emissions. At the upper limit, the vessel effluent is assumed to be saturated with VOC as predicted by Raoult's law. For VOC component i,

$$\Delta n_i = \Delta n_{sweep} \cdot k_m(y_i/y_{nc}) \qquad 9.64$$

where $y_{nc} = 1 - \sum_i y_i$, where the sum is over VOCs only. The variable k_m is the saturation level, and is set to one for the assumption of saturation and adjusted lower if justified when the purge gas is known to be unsaturated. The flow of noncondensables Δn_{sweep} can be related to a volumetric flow of purge gas using the ideal gas law,

$$\Delta n_{sweep} = \dot{n}t = \frac{P\dot{\underline{V}}_{sweep}}{RT}t \qquad 9.65$$

Purge Gas (No Liquid VOC present)

Vessels need to be purged for changeover of reactants or before performing maintenance. After draining all liquid, VOC vapors remain in the vessel at the saturation level present before draining. The typical assumption upon purging is that the vessel is well-mixed. A mole balance on the VOC

1. OAQPS, Control of Volatile Organic Compound Emissions from Batch Processes–Alternate Control Techniques Information Document, EPA-450/R-94-020, Research Triangle Park, NC 27711, February 1994.

2. OAQPS, Control of Volatile Organic Emissions from Manufacturing of Synthesized Pharmaceutical Products, EPA-450/2-78-029, December 1978.

3. U.S. E.P.A., Compilation of Air Pollution Emission Factors-Volume 1, (1993) EPA Publication AP-42.

gives $dn_i/dt = -y_i \dot{n}_{sweep}$, dividing by y_i the equation becomes $dn_i/y_i dt = -\dot{n}_{sweep}$. The left-hand side can subsequently be written $n_{tank} dy_i/y_i dt = (P\underline{V}_{tank})/(RT_{tank}) \cdot (dy_i)/(y_i dt)$, and the right-hand side can be written $(P\dot{\underline{V}}_{sweep})/(RT_{sweep})$. When the sweep gas and tank are at the same temperature, which is usually a valid case, the equation rearranges to $(dy_i)/y_i = (-\dot{\underline{V}}_{sweep}/\underline{V}_{tank})dt$, which integrates to

$$y_i^f = y_i^i \exp\left(\frac{-\dot{\underline{V}}_{sweep}}{\underline{V}_{tank}} t\right) \tag{9.66}$$

The emissions are calculated by

$$\Delta n_i = n_{tank}(y_i^i - y_i^f) \tag{9.67}$$

Heating

During a heating process, emissions arise because the vapors in the head space must expand as the temperature rises. Since vapor pressure increases rapidly with increasing temperature, VOC concentrations in the vapor phase increase also. Detailed calculations of emissions during heating are somewhat tedious, so an approximation is made; the emission of each VOC is based on the arithmetic average of the molar ratio of VOC to noncondensable gas at the beginning and the end of the heating multiplied by the total moles of noncondensable gas leaving the vessel. At the beginning of the heating, representing the VOC with subscript i and the noncondensables with subscript nc, the ratio of interest is $(n_i/n_{nc})^i = (y_i/y_{nc})^i = (y_i P/y_{nc} P)^i$, and at the end $(n_i/n_{nc})^f = (y_i/y_{nc})^f = (y_i P/y_{nc} P)^f$. The emission of VOC component i is calculated as

$$\Delta n_i = \frac{\Delta n_{nc}}{2}\left[\left(\frac{y_i}{y_{nc}}\right)^i + \left(\frac{y_i}{y_{nc}}\right)^f\right] \tag{9.68}$$

where $y_{nc} = 1 - \sum_i y_i$, and the sum is over VOCs only. The value of Δn_{nc} is given by

$$\Delta n_{nc} = \frac{V_{head}}{R}\left[\left(\frac{\left(1 - \sum_i y_i\right)P}{T}\right)^i - \left(\frac{\left(1 - \sum_i y_i\right)P}{T}\right)^f\right] + \Delta n_{sweep} \tag{9.69}$$

where the summations are over VOCs only and Δn_{sweep} is the total moles of noncondensable that are swept (purged) through the vessel during heating and is set to zero when purging is not used. Eqns. 9.68 and 9.69 can overestimate the emissions substantially if the tank approaches the bubble point of the liquid because y_{nc} approaches zero, and then calculations are more accurately handled by a more tedious integration. The integration can be approximated by using the method presented here over small temperature steps and summing the results.

Depressurization

Three assumptions are made to model depressurizations: the pressure is decreased linearly over time; air leakage into the vessel is negligible; the process is isothermal. The relationship is then the same as Eqn. 9.68, where Δn_{nc} is calculated by Eqn. 9.69 using $\Delta n_{sweep} = 0$.

Other Operations

Other operations involve condensers, reactors, vacuum vessels, solids drying, and tank farms. Condensers are commonly used for VOC recovery; however, the VOCs have a finite vapor pressure even at condenser temperatures and the emissions can be calculated by using Eqns. 9.64 and 9.65 at the condenser temperature. Reactors may convert or produce VOCs and, in a vented reactor, the emissions can be calculated by adapting one of the above techniques, keeping aware that generation of gas causes additional vapor displacement and possible temperature rise due to reaction. Vacuum units and solids drying operations are also direct adaptations of the methods above. Tank farm calculations are more detailed and empirical. Tank emission calculations depend on factors such as the climate and the paint color of the tank. Fixed roof tanks must breathe as they warm during the day due to sunlight, and then cool during the night hours. There are also working losses due to routine filling of the tanks as covered above. Although heating and cooling in a tank with a static level can be treated by the methods presented above, when the levels are also changing due to usage, EPA publications are recommended for these more tedious calculation procedures.

9.6 NON-IDEAL SYSTEMS

We have seen how easy the ideal-solution calculations can be. Unfortunately, Raoult's law is accurate for only a few of the systems you will actually encounter in practice. Examples of the phase diagrams which deviate from Raoult's law are shown in Fig. 9.6 and Fig. 9.7. There are several features of these diagrams that need to be discussed before proceeding with the next chapters. First of all, the Raoult's law bubble lines are shown as dotted lines in the P-x-y diagrams, so that deviations are obvious. Note again that the phase diagrams of each pair can be qualitatively related by inverting one diagram of the pair.

❶ Positive and negative deviations from Raoult's law.

In Fig. 9.6 the bubble line lies above the Raoult's law line, and these systems are said to have positive deviations from Raoult's law. Positive deviations occur when the components in the mixture would prefer to be near molecules of their own type rather than near molecules of the other component. The 2-propanol + water system has vapor pressures that are close to each other. As a result, the positive deviations are large enough to cause the phase envelope to close at a composition known as the azeotropic composition. Recalling that the dew and bubble lines represent coexisting compositions at equilibrium, this means that at the azeotropic point $x_i = y_i$. When an azeotrope forms in a system with positive deviations, the azeotrope is a maximum on the P-x-y diagram and a minimum on the T-x-y diagram. Since the vapor and liquid phases are equivalent at the azeotrope, a flash drum or distillation column will not be able to separate a mixture of the azeotropic composition. The methanol + 3-pentanone system has significantly different vapor pressures for the components, and the deviations from ideality are not large enough to create azeotrope formation.

In Fig. 9.7 the systems have negative deviations from Raoult's law because the bubble line lies below the Raoult's line. Similar azeotrope behavior is found in these systems if the vapor pressures are close to each other or the deviations are large. When an azeotrope forms in a system with negative deviations, the azeotrope is a minimum on the P-x-y diagram and a maximum on the T-x-y diagram. Azeotropic compositions for both positive and negative deviating systems will depend on temperature and pressure, however the dependencies are usually weak unless large pressure or temperature changes are made.

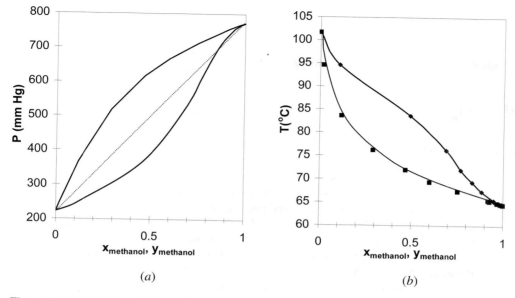

(a)   (b)

Figure 9.6 *(a); (b) Phase Behavior of the methanol + 3-pentanone system. Left figure at 65°C. Right figure at 760 mm Hg. (T-x-y from Glukhareva, M.I., et al. Zh. Prikl. Khim. (Leningrad) **49**, 660(1976), P-x-y calculated from fit of T-x-y.)*

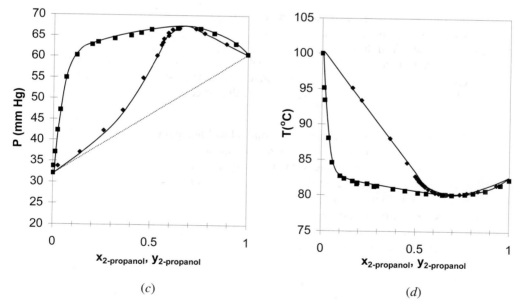

(c)   (d)

Figure 9.6 *(c); (d) Phase Behavior of the 2-propanol + water system. Left figure at 30°C. Right figure at 760 mm Hg. (T-x-y from Wilson, A., Simons, E.L., Ind. Eng. Chem., **44**, 2214(1952), P-x-y from Udovendo, V.V., and Mazanko, T.F., Zh. Fiz. Khim., **41**, 1615(1967).)*

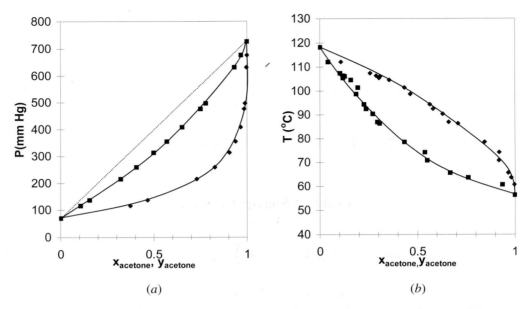

Figure 9.7 *(a); (b) Phase Behavior of the acetone + acetic acid system. Left figure at 55°C. Right figure at 760 mm Hg. (T-x-y from York, R., Holmes, R.C., Ind. Eng. Chem., **34**, 345(1942), P-x-y from Waradzin, W., Surovy, J., Chem. Zvesti **29**, 783(1975).)*

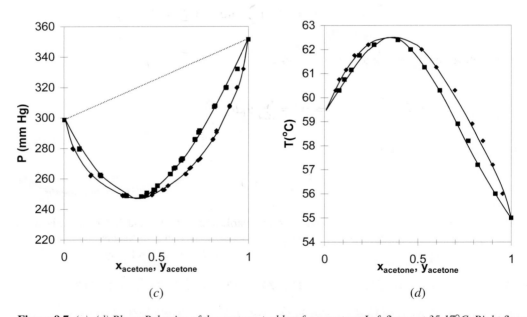

Figure 9.7 *(c); (d) Phase Behavior of the acetone + chloroform system. Left figure at 35.17°C. Right figure at 732 mm Hg. (T-x-y from Soday, F., Bennett, G.W., J.Chem. Educ., **7**, 1336(1930), P-x-y from Zawidzki, V.J., Z. Phys. Chem., **35**, 129(1900).)*

Two Modeling Approaches

There are two main approaches to modeling non-ideal fluids. They differ in the way they treat the liquid phase. The first approach is to use an equation of state. These calculations are complicated enough to require a computer. In this approach, we will use the equation of state to calculate liquid fugacities. This approach is discussed in Chapter 10. The other approach is to use deviations from ideal solution behavior, using activity coefficients, which we will cover in Chapter 11.

9.7   ADVANCED TOPICS (OPTIONAL)

Gibbs-Duhem Equation

A useful expression known as the Gibbs-Duhem equation results when we carefully examine the implications of Eqn. 9.6 at constant T and P. $\Rightarrow$

$$dG = \underline{V}dP - \underline{S}dT + \sum_i \mu_i dn_i \qquad 9.70$$

Suppose we keep the composition the same but add more material.

$$\Rightarrow \Delta\underline{G} = \underline{G}^i + G\,\Delta n \ or \ d\underline{G} = G\,dn$$

The amount of each component added is $dn_i = n_i /n\, dn$. Substituting,

$$G\,dn = \sum_i (\mu_i n_i)\,dn/n = (dn/n)\sum_i \mu_i n_i \Rightarrow \underline{G} = \sum_i \mu_i n_i \qquad 9.71$$

Differentiate according to product rule $\Rightarrow dG = \sum_i \mu_i dn_i + \sum_i n_i d\mu_i$

But, at constant T and P, $\qquad \Rightarrow dG = \sum_i \mu_i dn_i$

Therefore, we must conclude:

$$\sum_i n_i d\mu_i = 0 \qquad 9.72$$

This is called the *Gibbs-Duhem equation* and is applicable at constant T and P. We find this relation useful in testing data for experimental errors (inconsistent data), and also for development of theories for the Gibbs energy of a mixture, since our model must follow this relation.

Extension of Fundamental Relations to Open Systems

This section shows that the Eqn. 9.3 can be derived starting from the flow equations from earlier chapters. For an open *simple* system:

$$d\underline{U} = \sum_{inlets} H^{in} dn^{in} - \sum_{outlets} H^{out} dn^{out} + d\underline{Q} - Pd\underline{V} \qquad 9.73$$

$$dS = \sum_{inlets} S^{in} dn^{in} - \sum_{out} S^{out} dn^{out} + \frac{dQ}{T_{sys}} + d S_{gen} \qquad 9.74$$

where $dS_{gen} = 0$ for an internally reversible process. Elimination of dQ in first balance provides

$$dU = \sum_{inlets} (H^{in} - T_{sys} S^{in}) dn^{in} - \sum_{outlets} (H^{out} - T_{sys} S^{out}) dn^{out} + T_{sys} dS - PdV \qquad 9.75$$

Note that this relation cannot be applied to a simple system if an entering stream temperature is different from the system temperature because of the introduction of irreversibilities. Therefore, $T_{sys} = T$ for a simple system since T is uniform, and $H^{out} = H$, $S^{out} = S$

$$dU = TdS - PdV + \sum_{inlets} G^{in} dn^{in} - \sum_{outlets} G^{out} dn^{out} \qquad 9.76$$

Open, reversible simple system, no shaft work, $T^{in} = T^{out} = T_{system}$

The flow terms can be subdivided into the molar component flows if we introduce the partial molar quantities. The resulting expressions for open systems are:

Closed	Open	Natural Variables
$dU = TdS - PdV$	$dU = TdS - PdV + \sum_i \mu_i dn_i$	S, V
$dH = TdS + VdP$	$dH = TdS + VdP + \sum_i \mu_i dn_i$	S, P
$dA = SdT - PdV$	$dA = -SdT - PdV + \sum_i \mu_i dn_i$	T, V
$dG = SdT + VdP$	$dG = -SdT + VdP + \sum_i \mu_i dn_i$	T, P

9.8   SUMMARY AND CONCLUDING REMARKS

The concepts in this chapter are relatively simple but far-reaching. A simple extension of the chain rule to multicomponent systems led to the equilibrium constraint for multicomponent multiphase equilibria. A simple application of the entropy of mixing derived in Chapter 3 led to the ideal solution model. This simple solution model enabled us to demonstrate the computational procedures for bubble points, dew points, and flashes.

In the remaining chapters of Unit III we proceed in a manner that is extremely similar. We propose a solution model and apply the equilibrium constraint to derive an expression for the K-ratios. Then we follow exactly the same computational procedures as developed here. The primary difference is the increasing level of sophistication incorporated into each solution model. Thus, the chapters ahead focus increasingly on the detailed description of the molecular interactions and the impacts of assumptions on the accuracy of the resulting solution models.

9.9   PRACTICE PROBLEMS

P9.1 The stream from a gas well consists of 90 mol% methane, 5 mol% ethane, 3 mol% propane, and 2 mol% n-butane. This stream is flashed isothermally at 233 K and 70 bar. Use the shortcut K-ratio method to estimate the L/F fraction and liquid and vapor compositions. (ANS. $L/F = 0.181$)

P9.2 An equimolar mixture of n-butane and n-hexane at pressure P is isothermally flashed at 373 K. The liquid-to-feed ratio is 0.35. Use the shortcut K-ratio method to estimate the pressure and liquid and vapor compositions. (ANS. $P = 0.533$ MPa, $x_{C6} = 0.78$)

P9.3 A mixture of 25 mol% n-pentane, 45 mol% n-hexane, and 30 mol% n-heptane is flashed isothermally at 353 K and 2 bar. Use the shortcut K-ratio method to estimate the L/F fraction and liquid and vapor compositions. (ANS. $L/F = 0.56$)

P9.4 A mixture containing 15 mol% ethane, 35 mol% propane, and 50 mole% n-butane is isothermally flashed at 9 bar and temperature T. The liquid-to-feed ratio is 0.35. Use the shortcut K-ratio method to estimate the pressure and liquid and vapor compositions. (ANS. 319.4 K, $x_{C4} = 0.74$)

9.10   HOMEWORK PROBLEMS

9.1 For a separations process it is necessary to determine the VLE compositions of a mixture of ethyl bromide and n-heptane at 30°C. At this temperature the vapor pressure of pure ethyl bromide is 0.7569 bar, and the vapor pressure of pure n-heptane is 0.0773 bar. Calculate the bubble pressure and the composition of the vapor in equilibrium with a liquid containing 47.23 mol% ethyl bromide assuming ideal solution behavior. Compare the calculated pressure to the experimental value of 0.4537 bar.

9.2 Benzene and ethanol form azeotropic mixtures. Prepare a y-x and a P-x-y diagram for the benzene-ethanol system at 45°C assuming the mixture is ideal. Compare the results with the experimental data of Brown and Smith, *Austral. J. Chem.* 264 (1954).(P in bars)

x_e	0	0.0374	0.0972	0.2183	0.3141	0.4150	0.5199	0.5284	0.6155	0.7087	0.9591	1.000
y_e	0	0.1965	0.2895	0.3370	0.3625	0.3842	0.4065	0.4101	0.4343	0.4751	0.8201	1.000
P	0.2939	0.3613	0.3953	0.4088	0.4124	0.4128	0.41	0.4093	0.4028	0.3891	0.2711	0.2321

9.3 The following mixture of hydrocarbons is obtained as one stream in a petroleum refinery on a mole basis: 5% ethane, 10% propane, 40% n-butane, 45% isobutane. Assuming the shortcut K-ratio model: (a) compute the bubble point of the mixture at 5 bar; (b) compute the dew point of the mixture at 5 bar; (c) find the amounts and compositions of the vapor and liquid phases that would result if this mixture were to be isothermally flash vaporized at 30°C from a high pressure to 5 bar.

9.4 Consider a mixture of 50 mol% n-pentane and 50 mol% n-butane at 14 bar.

(a) What is the dew point temperature? What is the composition of the first drop of liquid?
(b) At what temperature is the vapor completely condensed if the pressure is maintained at 14 bar? What is the composition of the last drop of vapor?

9.5 A 50 mol% mixture of propane(1) and *n*-butane(2) enters an isothermal flash drum at 37°C. If the flash drum is maintained at 0.6 MPa, what fraction of the feed exits as liquid? What are the compositions of the phases exiting the flash drum? Work the problem in the following two ways:

(a) Use Raoult's Law (use the Peng-Robinson equation to calculate pure component vapor pressures).

(b) Assume ideal mixtures of vapor and liquid. (Use the Peng-Robinson equation to obtain f^{sat} for each component.)

9.6 Above a solvent's flash point temperature, the vapor concentration in the headspace is sufficient that a spark will initiate combustion; therefore, extreme care must be exercised to avoid ignition sources. Calculate the vapor phase mole fraction for the following liquid solvents using flash points listed, which were obtained from the manufacturer's material safety data sheets (MSDS). The calculated vapor concentration is an estimate of the lower flammability limit (LFL). Assume that the headspace is an equilibrium mixture of air and solvent at 760 mmHg. The mole fraction of air dissolved in the liquid solvent is negligible for this calculation.

(a) methane, −187.8°C   (d) hexane, −21.7°C   (g) toluene, 4.4°C
(b) propane, −104.5°C   (e) ethanol, 12.7°C   (h) *m*-xylene, 28.8°C
(c) pentane, −48.9°C   (f) 2-butanone, −5.6°C   (i) ethyl acetate, −4.5°C

9.7 The flash point of a liquid or liquid mixture is the temperature above which the vapor mole fraction in surrounding air exceeds the lower flammability limit (LFL), and a flame will propagate across the surface of the liquid. The flashpoint of an aqueous binary mixture can be estimated using Raoult's law (providing Raoult's law is a valid approximation for the mixture). The flashpoint is determined by the vapor phase concentration as discussed in problem 9.6. The LFL for ethanol is 4.3%, and for 2-propanol, 2.0%. (Note: These problems are revisited again, modeling the solutions as non-ideal solutions in Chapter 11).

(a) Use Raoult's law to estimate the flashpoint temperature at 760 mmHg for 70 mol% and 90 mol% of the following solvents in liquid water: (*i*) ethanol; (*ii*) 2-propanol.

(b) Use Raoult's law to estimate the liquid phase mol% concentration of the solvent that would give a flashpoint of 22°C at 760 mmHg for: (*i*) ethanol; (*ii*) 2-propanol.

9.8 Solvent vessels must be purged before maintenance personnel enter in order to assure that: (1) sufficient oxygen is available for breathing; (2) vapor concentrations are below the flashpoint; (3) vapor concentrations are below the Occupational Safety and Health Association (OSHA) limits if breathing apparatus is not to be used. Assuming that a 8-m^3 fixed-roof solvent tank has just been drained of all liquid and that the vapor phase is initially saturated at 22°C, estimate the length of purge necessary with 2 m^3/min of gas at 0.1 MPa and 22°C to reach the OSHA 8-hr exposure limit.[1]

(a) hexane 500 ppm
(b) 1-butanol 100 ppm
(c) chloroform 50 ppm
(d) ethanol 1000 ppm
(e) toluene 200 ppm

---

1. OSHA may change these limits at any time.

9.9 Benzyl chloride is manufactured by the thermal or photochemical chlorination of toluene. The chlorination is usually carried out to no more than 50% toluene conversion to minimize the benzal chloride formed. Suppose reactor effluent emissions can be modeled ignoring by-products, and the effluent is 50 mol% toluene and 50 mol% benzyl chloride. Estimate the emission of toluene and benzyl chloride (moles of each) when an initially empty 4-m³ holding tank is filled with the reactor effluent at 30°C and 0.1 MPa.

9.10 A pharmaceutical product is crystallized and washed with absolute ethanol. A 100-kg batch of product containing 10% ethanol by weight is to be dried to 0.1% ethanol by weight by passing 0.2 m³/min of 50°C nitrogen through the dryer. Estimate the rate (mol/min) that ethanol is removed from the crystals, assuming that ethanol exerts the same vapor pressure as if it were pure liquid. Based on this assumption, estimate the residence time for the crystals in the dryer. The dryer operates at 0.1 MPa and the vapor pressure of the pharmaceutical is negligible.

9.11 This problem explores emissions during heating of hexane(1) and toluene(2) in a tank with a fixed roof that is vented to the atmosphere through an open pipe in the roof. Atmospheric pressure is 760 mmHg. The tank volume is 2000 L, but the maximum operating liquid level is 1800 L. Determine the emissions of each VOC (in g) when the tank is heated.

(a) The liquid volume is 1800 L, $x_1 = 0.5$, $T^i = 10°C$, $\Delta T = 15°C$.
(b) The liquid volume is 1800 L, $x_1 = 0.5$, $T^i = 25°C$, $\Delta T = 15°C$.
(c) The liquid volume is 1500 L, $x_1 = 0.5$, $T^i = 25°C$, $\Delta T = 15°C$.
(d) The liquid volume is 1800 L, $x_1 = 1.0$, $T^i = 25°C$, $\Delta T = 15°C$.
(e) Explain why the ratio [(emission of toluene in part (b))/(emission of toluene in part (a))] is different from the corresponding ratio of hexane emissions.

9.12 The flash point is described in problems 9.6 and 9.7. For a mixture, the flashpoint can be estimated by the temperature where $\Sigma(y_i/LFL_i) = 1$. Use Raoult's law to estimate the flash point temperature for the following equimolar liquid mixtures in an air atmosphere at 750 mmHg total pressure.

(a) pentane (LFL = 1.5%) and hexane (LFL = 1.2%)
(b) methanol (LFL = 7.3%) and ethanol (LFL = 4.3%)
(c) benzene (LFL = 1.3%) and toluene (LFL = 1.27%)

CHAPTER
10

PHASE EQUILIBRIA IN MIXTURES BY AN EQUATION OF STATE

The whole is simpler than the sum of its parts.

J.W. Gibbs

Suppose it was required to estimate the vapor-liquid *K*-ratio of carbon monoxide in a mixture at room temperature. For an initial guess, we might assume it follows ideal-solution behavior. It is a relatively simple molecule and much like nitrogen (e.g. weak polar moments, no hydrogen donors). But we cannot use Raoult's law because the required temperature is well-above the critical temperature. If we wanted to use an activity coefficient approach applied to the liquid phase, we would need to define a standard state of pure CO at the conditions of interest, but pure CO does not exist as a liquid at the conditions of interest. Clearly, the obvious ways of considering such a solution are not so straightforward to apply. The most straightforward approach is by consideration of the equation of state.

When we extend equations of state to mixtures, the basic form of the equations do not change. The fluid properties of the mixture are written in terms of the same equation of state parameters as for the pure fluids; however, the equation of state parameters are functions of composition. The equations we use to incorporate compositional dependence into the mixture constants are termed the *mixing rules*.

❶ The composition dependence is introduced into an equation of state by *mixing rules* for the parameters. The basic equation form does not change.

As a preliminary introduction, we will accept these mixing rules based on little justification. That way we can quickly sample the broad range of applications that are made possible through the specification of the mixing rules. For a full development of the origin of the mixing rules, it is necessary to return to the molecular level and reconstruct the mixtures with reference to their intermolecular potentials. Such an analysis also lays the groundwork for development of the activity coefficient models, because they can be derived from slight variations on the equation of state derivations. The introduction to the basic foundation for the treatment of mixtures is presented in Section 11.9 after the importance of the detailed treatment of composition effects has been established through performing a large number of examples.

Sample Mixing Rules for the Virial Equation of State

The virial equation mixing rule is derived theoretically using statistical mechanics.[1] Since the second virial coefficient represents two-body interactions, the second virial coefficient of a mixture represents the summation of these interactions over the pairwise interactions in the mixture. In a binary mixture for example, there are three pairwise interactions in the mixture: 1-1, 2-2, and 1-2 interactions.

$$B = \sum_i \sum_j y_i y_j B_{ij}$$

10.1

which for a binary mixture becomes

$$B = y_1^2 B_{11} + 2y_1 y_2 B_{12} + y_2^2 B_{22}$$

10.2

where the pure component virial coefficients are indicated by identical subscripts on the virial coefficients. In other words, B_{11} represents the contribution of 1-1 interactions, B_{22} represents 2-2 interactions, B_{12} represents the contribution of 1-2 interactions, and the mixing rule provides the mathematical method to sum up the contributions of the interactions. B_{12} is called the *cross coefficient*, indicating that it represents two-body interactions of unlike molecules. In the above sum, it is understood that B_{12} is equivalent to B_{21}. *The cross coefficient B_{12} is not the virial coefficient for the mixture.*

> **Combining rules** are used to quantify the parameters that represent unlike molecule interactions.

To obtain the cross coefficient, B_{12}, we must create a *combining rule* to propose how the cross coefficient depends on the properties of the pure components 1 and 2. For the virial coefficient, the relationship between the pair potential and the virial coefficient was given in Chapter 8. However, this level of rigor is rarely used in engineering applications. Rather, *combining rules* are created to use the corresponding state correlations developed for pure components in terms of T_{c12} and P_{c12}. The combining rules used to determine the values of the cross coefficient critical properties are:

$$T_{c12} = (T_{c1} T_{c2})^{1/2} (1 - k'_{12})$$

10.3

> **Binary interaction parameters** are used to adjust the combining rule to better fit experimental data, if available.

The parameter k'_{12} is an adjustable parameter (called the binary interaction parameter) to force the combining rules to more accurately represent the cross coefficients found by experiment.[2] However, in the absence of experimental data, it is customary to set $k'_{12} = 0$.

$$V_{c12} = \left(\frac{V_{c1}^{1/3} + V_{c2}^{1/3}}{2} \right)^3$$

10.4

$$Z_{c12} = \tfrac{1}{2}(Z_{c1} + Z_{c2})$$

10.5

and

$$\omega_{12} = \tfrac{1}{2}(\omega_1 + \omega_2)$$

10.6

---

1. The derivation is too detailed to present here; however in Section 10.1, we offer a rationale for the form of Eqn. 10.1
2. Reid, R, Prausnitz, J.M., Poling, B. *The Properties of Gases and Liquids,* page 133, McGraw-Hill (1987).

The first three of these combining rules lead to:

$$P_{c12} = Z_{c12}RT_{c12}/V_{c12}$$

10.7

Then, T_{c12}, P_{c12}, and ω_{12} are used in the virial coefficient correlation presented in Chapter 6 to obtain B_{12} (Eqns. 6.6–6.10) which is subsequently incorporated into the equation for the mixture. If Z_c (or V_c) is not available, it may be estimated using $Z_c = 0.291 - 0.08\,\omega$. The virial equation for a binary mixture is implemented on the spreadsheet VIRIALMX.xls furnished with the text.

Example 10.1 The virial equation for vapor mixtures

Calculate the molar volume for a 60 mole% mixture of neopentane(1) in CO_2(2) at 310 K and 0.2 MPa.

VIRIALMX.xls

Solution: The conditions are entered in the spreadsheet VIRIALMX.xls, with the following results:

X Microsoft Excel - Virialmx.xls

File Edit View Insert Format Tools Data Window Help

	A	B	C	D	E	F	G	H	I	J
2		Virial Equation for a Mixture								
3										
4			T (K)		310					
5			P (MPa)		0.2					
6			kij		0					
7										
8		Compound	T_C (K)	P_C (MPa)	ω	V_C (cm³/mol)	Z_C	Tr	Pr	criteria
9		Neopentane	433.8	3.199	0.196	303.28	0.269	0.7146	0.0625	0.00117
10		CO2(2)	304.2	7.382	0.228	93.87	0.274	1.0191	0.0271	0.32117
11		(1)-(2)	363.27	4.59	0.212	178.62	0.2715	0.8534	0.0436	
12										
13										
14		Compound	B^0	B^1	BP_C/RT_C	B_{ij}(cm³/mol)		B_{ij} matrix		
15		Neopentane	-0.6394321	-0.5663837	-0.7504433	-846.06		-846.06	-330.50	
16		CO2(2)	-0.3264383	-0.0198832	-0.3309717	-113.39		-330.50	-113.39	
17		(1)-(2)	-0.460867	-0.195774	-0.5023711	-330.50				

Introduction   **Virial Mix**

The original spreadsheet is modified slightly for this solution. Cells J9 and J10 are programmed with a rearranged form of Eqn. 6.10, $T_r - 0.686 - 0.439P_r$, and if these cells are positive, then the virial equation is suitable. The critical volume is calculated from T_c, P_c, and Z_c. Cells F15-F17 list the virial coefficients for neopentane, CO_2, and the cross coefficient, respectively.

The virial coefficient for the mixture is given by Eqn. 10.1,

$$B = 0.6^2 \cdot (-846.06) + 2(0.6)(0.4)(-330.5) + 0.4^2 \cdot (-113.39) = -481.36 \text{ cm}^3/\text{mol}$$

$$V = RT/P + B = 8.314 \cdot 310/0.2 - 481.36 = 12,405 \text{ cm}^3/\text{mol}$$

The volumetric behavior of the mixture depends on composition. The mixture volume differs from an ideal solution, $V^{is} = \sum_i y_i V_i$. The difference $V - V^{is}$ is called the excess volume, V^E.

The molar volume of pure neopentane is

$$V = RT/P + B = 8.314 \cdot 310/0.2 - 846.1 = 12,041 \text{ cm}^3/\text{mol}$$

Example 10.1 The virial equation for vapor mixtures (Continued)

The molar volume of pure CO_2 is

$$V = RT/P + B = 8.314 \cdot 310/0.2 - 113.4 = 12{,}773 \text{ cm}^3/\text{mol}$$

The molar volume of an ideal solution of a 60 mole% neopentane mixture is

$$V^{is} = 0.6(12{,}041) + 0.4(12{,}773) = 12{,}334 \text{ cm}^3/\text{mol}$$

and the excess volume is

$$V^E = 12{,}405 - 12{,}334 = 71.2 \text{ cm}^3/\text{mol}.$$

The molar volume and excess volume can be determined across the composition range by changing y's in the formulas.

10.1   A SIMPLE MODEL FOR MIXING RULES

To understand the terms in equations of state for mixtures, we must reconsider the molecular perspective which provides the basis of the equation of state and extend that perspective to mixtures. This extension is most easily approached by reconsidering van der Waals' equation.

$$Z = \frac{1}{1 - b\rho} - \frac{a\rho}{RT}$$

Cubic equations of state use two parameters and require a mixing rule for each parameter. We must simply define b and a for the mixture. The parameter b represents the close-packed volume. For most mixtures of roughly equal-sized molecules, the volumes mix ideally at high density. Therefore, $b = \sum_i x_i b_i$ is reasonable[1] . As for a, we must carefully consider how this term relates to the internal energy of mixing because the Gibbs energy of mixing will be closely related. The departure function indicates that:

$$\frac{U - U^{ig}}{RT} = -\frac{a\rho}{RT}$$

The parameter a should represent the interactions due to each type of molecular interaction in the fluid mixture.

In a binary mixture there are three types of interactions for molecules (1) and (2). First, a molecule can interact with itself (1-1 or 2-2 interactions), or it can interact with a molecule of the other type (a 1-2 interaction). In a random fluid, the probability of finding a (1) molecule is the fraction of (1) atoms, x_1. The probability of a 1-1 interaction is a *conditional probability*. A conditional probability is the probability of finding a second interacting molecule of a certain type given the first is a certain type. For independent events, a conditional probability is calculated by the product of the individual probabilities. Therefore the probability of a 1-1 interaction is x_1^2. By similar arguments, the probability of a 2-2 interaction is x_2^2 . The probability of a 1-2 interaction is $x_1 x_2$ and the

---

1. The variable x is customarily used as a generic composition variable for the mixing rule applied to vapor or liquid roots.

probability of a 2-1 interaction is also x_1x_2.[1] If the attractive interactions are characterized by a_{11}, a_{22}, and a_{12}, the mixing rule for a is given by:

$$U - U^{ig} = -a\rho = -\rho(x_1^2 a_{11} + 2x_1 x_2 a_{12} + x_2^2 a_{22})$$ 10.8

This kind of mixing rule is called a quadratic mixing rule, because all cross-products of the compositions are included. It represents a fairly obvious initial approximation to the way mixing should be represented. We will see that most theories before 1964 relied on this basic perspective, including regular-solution theory and Flory-Huggins theory as well as equations of state. After 1964, researchers began to realize that the distributions of molecules around each other might not be entirely random, as implied by the simple form of the conditional probabilities assumed above. Instead, non-random distributions might give rise to local compositions which are different from the bulk compositions. One way of expressing this possibility is to use Eqn. 10.8 but to admit that the a_{ij} may be dependent on composition. To see how this composition dependence arises naturally from the analysis of intermolecular energies, it is necessary to formally extend the energy equation to mixtures. The extension is straightforward, but we reserve it for the chapter on non-ideal solution models, following the introduction to local-composition models. We can accomplish quite a lot with a simple "random mixing" approximation. It is best to explore the limits of this simple approximation before introducing any more complexity than necessary.

Mixing Rules for Cubic Equations of State

From the above discussion the customary mixing rules for cubic equations of state are:

$$a = \sum_i \sum_j x_i x_j a_{ij} \quad \text{and} \quad b = \sum_i x_i b_i$$ 10.9

Note the mathematical similarity of the mixing rule for a with the mixing rule used for the virial coefficient. All of the compositional dependence of the equation of state is incorporated into the two relations. A combining rule is not necessary for the b term, however the a term does require a combining rule. The customary combining rule is

$$a_{ij} = (a_i a_j)^{1/2}(1 - k_{ij})$$ 10.10

where k_{ij} is referred to as a *binary interaction parameter*. This is similar to the form of the geometric mean rule for critical temperatures used for virial coefficients. The adjustable parameter k_{ij} is used to adjust the combining rule to fit experimental data more closely. Technically, this just transfers our ignorance into the adjustable parameter k_{ij}. Values for k_{ij} for various binary combinations are tabulated in the literature.[2]

In the absence of experimental data or literature values for k_{ij}, we may make a first-order approximation by letting $k_{ij} = 0$. This approximation serves our purpose nicely, because the equation of state approach then requires no more information than the ideal solution approach (T_c, P_c, ω, T, P, x, y), but it offers the possibility of more realistic representation of the phase diagram

---

1. The actual probability for a 1-1 interaction is $\dfrac{N_1(N_1-1)}{N\,(N-1)}$, but when N is large it is equal to x_1^2. Likewise, for a

2-2 the probability is $\dfrac{N_2(N_2-1)}{N\,(N-1)}$. For a 1-2 it is $\dfrac{N_1}{N}\dfrac{N_2}{(N-1)}$, which is effectively x_1x_2.

2. Reid, R., Prausnitz, J.M., Poling, B. *The Properties of Gases and Liquids,* page 83, McGraw-Hill (1987).

because of the more fundamental molecular basis. We can demonstrate this improved accuracy by considering some examples.

10.2 FUGACITY AND CHEMICAL POTENTIAL FROM AN EOS

In an equation of state approach, the equation of state is applied, in principle, to all fluid phases at all densities and temperatures. The effect of changing the composition is given by the *mixing rules* on the equation of state parameters (e.g., a and b in the Peng-Robinson equation). In terms of the Venn diagram presented in the preceding chapter, we have the schematic shown in Fig. 10.1.

Once the expression for the Gibbs energy departure is obtained for the mixture, that expression is differentiated to yield the chemical potential and then the fugacity of a given component.

$$\frac{\mu_i(T, P) - \mu_i^{ig}(T, P)}{RT} = \left(\frac{\partial G/RT}{\partial n_i}\right)_{T,P,n_{j\neq i}} - \left(\frac{\partial \underline{G}^{ig}/RT}{\partial n_i}\right)_{T,P,n_{j\neq i}} = \left(\frac{\partial(\underline{G} - \underline{G}^{ig})/RT}{\partial n_i}\right)_{T,P,n_{j\neq i}}$$

10.11

The fugacity coefficient is calculated by

$$\boxed{\ln\left(\frac{\hat{f}_i}{y_i P}\right) = \frac{(\mu_i - \mu_i^{ig})}{RT} = \left(\frac{\partial(\underline{G} - \underline{G}^{ig})/(RT)}{\partial n_i}\right)_{T,P,n_{j\neq i}}}$$

10.12

Review of the concepts from Section 9.2 may help put the approaches in context. Keep in mind that the objective is still to perform bubble, dew and flash calculation, but after relaxing the ideal solution assumption.

General form of fugacity coefficient in a mixture useful for EOSs of the form $Z(T,P)$.

Another name for the ideal solution model is the Lewis fugacity rule.

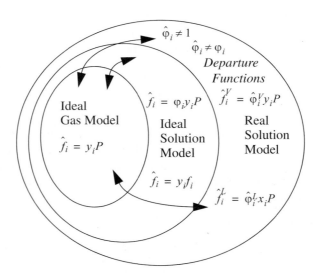

Figure 10.1 *Schematic showing the equation of state approach to modeling fugacities of components. Departure function (fugacity coefficient) methods are used for both the vapor and liquid phase. Superscripts are used to distinguish the fugacity coefficients of each phase. Liquid-phase compositions are conventionally denoted by x_i and vapor-phase compositions by y_i.*

For the virial equation, we have

$$\frac{G - G^{ig}}{RT} = \frac{BP}{RT}$$

Differentiation leads to the form: for a binary,

$$\boxed{\ln\hat{\varphi}_1 = [2y_1 B_{11} + 2y_2 B_{12} - B]P/RT} \quad \boxed{\ln\hat{\varphi}_2 = [2y_1 B_{12} + 2y_2 B_{22} - B]P/RT}$$

10.13　❶ Fugacity coefficient for virial equation of state.

which will be shown in more detail later. The fugacity coefficient of a component in a mixture may be directly determined at a given T and P by evaluating the virial coefficients at the temperature, then using this equation to calculate the fugacity coefficient.

Differentiation of the Gibbs departure function is difficult for a pressure-explicit equation of state like the Peng-Robinson equation of state. The difficulty arises because the Gibbs departure function derived in Chapter 8 is written in terms of volume and temperature rather than pressure, and differentiation at constant pressure as required by Eqn. 10.12 is difficult. As in the case of pure fluids, classical thermodynamics provides the means to solve this problem. Instead of differentiating the Gibbs departure function, we differentiate the Helmholtz departure function. Recalling,

$$d\underline{G} = \underline{V}dP - \underline{S}dT + \sum_i (\partial \underline{G}/\partial n_i)_{T,P,n_{j\neq i}} \, dn_i$$

and noting,

$$d\underline{A} = -Pd\underline{V} - \underline{S}dT + \sum_i \left(\frac{\partial \underline{A}}{\partial n_i}\right)_{T,\underline{V},n_{j\neq i}} dn_i$$

we also use $A = G - PV$, or $d\underline{A} = d\underline{G} - d(PV)$

$$d\underline{A} = d\underline{G} - Pd\underline{V} - \underline{V}dP = \underline{V}dP - \underline{S}dT + \sum_i \left(\frac{\partial \underline{G}}{\partial n_i}\right)_{T,P,n_{j\neq i}} dn_i - Pd\underline{V} - \underline{V}dP$$

$$\Rightarrow -Pd\underline{V} - \underline{S}dT + \sum_i \left(\frac{\partial \underline{A}}{\partial n_i}\right)_{T,\underline{V},n_{j\neq i}} dn_i = -Pd\underline{V} - \underline{S}dT + \sum_i \left(\frac{\partial \underline{G}}{\partial n_i}\right)_{T,P,n_{j\neq i}} dn_i$$

Equating coefficients of dn_i

$$\left(\frac{\partial \underline{A}}{\partial n_i}\right)_{T,\underline{V},n_{j\neq i}} = \left(\frac{\partial \underline{G}}{\partial n_i}\right)_{T,P,n_{j\neq i}} = \mu_i(T,P)$$

10.14

Note that T, P, and V identify the same conditions for the real fluid. Therefore, when we evaluate the departure, the ideal gas state must be corrected from V^{ig} to V,

$$\frac{\mu_i(T,P) - \mu_i^{ig}(T,P)}{RT} = \left(\frac{\partial \underline{A}/RT}{\partial n_i}\right)_{T,\underline{V},n_{j\neq i}} - \left(\frac{\partial \underline{A}^{ig}/RT}{\partial n_i}\right)_{T,\underline{V}^{ig},n_{j\neq i}}$$

$$= \left(\frac{\partial(\underline{A}-\underline{A}^{ig})_{TV}/RT}{\partial n_i}\right)_{T,\underline{V},n_{j\neq i}} - \ln Z \qquad 10.15$$

where the notation $(A - A^{ig})_{TV}$ denotes a departure function at the same T, V, which is the integral of Eqn. 7.27. The last term, $\ln Z$, represents the correction of the ideal gas Helmholtz energy from V to V^{ig}. Careful inspection of the true form on the integral leading to $\ln Z$ will convince you that differentiation does not change this term, and only the integral must be differentiated.

Therefore, the fugacity coefficient is calculated using

$$\ln\left(\frac{\hat{f}_i}{y_i P}\right) = \frac{(\mu_i - \mu_i^{ig})}{RT} = \left(\frac{\partial(\underline{A}-\underline{A}^{ig})_{TV}/(RT)}{\partial n_i}\right)_{T,\underline{V},n_{j\neq i}} - \ln Z \qquad 10.16$$

To apply this, consider the Peng-Robinson equation as an example.

$$\frac{\left(A - A^{ig}\right)_{TV}}{RT} = -\ln(1 - B/Z) - \frac{A}{B\sqrt{8}}\ln\left(\frac{Z + (1+\sqrt{2})B}{Z + (1-\sqrt{2})B}\right)$$

$$= -\ln(1 - b\rho) - \frac{a}{bRT\sqrt{8}}\ln\left(\frac{1 + (1+\sqrt{2})b\rho}{1 + (1-\sqrt{2})b\rho}\right)$$

By extending the method of reducing the equation of state parameters developed in Eqns. 6.21 and 6.23, $A^V = \sum_i \sum_j y_i y_j A_{ij}$ and $B^V = \sum_i y_i B_i$, where $A_{ij} = \sqrt{A_{ii}A_{jj}}\,(1 - k_{ij})$. Then, differentiation as we will show in Example 10.5 on page 334, yields for a binary system

$$\ln\left(\frac{\hat{f}_i^V}{y_i P}\right) = \frac{B_i}{B^V}\left(Z^V - 1\right) - \ln\left(Z^V - B^V\right) - $$

$$\frac{A^V}{B^V\sqrt{8}}\ln\left(\frac{Z^V + (1+\sqrt{2})B^V}{Z^V + (1-\sqrt{2})B^V}\right)\left(\frac{2(y_1 A_{i1} + y_2 A_{i2})}{A^V} - \frac{B_i}{B^V}\right) \qquad 10.17$$

Note: $A^L = \sum_i \sum_j x_i x_j A_{ij}$ and $B^L = \sum_i x_i B_i$ and the derivation of the liquid fugacity coefficient would be analagous:

$$\ln\left(\frac{\hat{f}_i^L}{x_i P}\right) = \frac{B_i}{B^L}\left(Z^L - 1\right) - \ln\left(Z^L - B^L\right) -$$

$$\frac{A^L}{B^L\sqrt{8}} \ln\left(\frac{Z^L + (1+\sqrt{2})B^L}{Z^L + (1-\sqrt{2})B^L}\right)\left(\frac{2(x_1 A_{i1} + x_2 A_{i2})}{A^L} - \frac{B_i}{B^L}\right)$$

10.18

As we saw in the case of equations of state for pure fluids, there is no fundamental reason to distinguish between the vapor and liquid phases except by the size of Z. The equation of state approach encompasses both liquids and vapors very simply.

To obtain an expression for K_i, it is convenient to use the fugacity coefficients of a mixture as defined by Eqn. 9.28

$$\left(\frac{\hat{f}_i^V}{y_i P}\right) \equiv \hat{\varphi}_i^V \quad \text{and} \quad \left(\frac{\hat{f}_i^L}{x_i P}\right) \equiv \hat{\varphi}_i^L$$

Recalling that $\hat{f}_i^V = \hat{f}_i^L$ at equilibrium, we find that

$$K_i = \frac{\hat{\varphi}_i^L}{\hat{\varphi}_i^V} \quad \text{or} \quad y_i \hat{\varphi}_i^V P = x_i \hat{\varphi}_i^L P$$

10.19

> Eqns. 10.19 provide the primary equations for VLE via equations of state.

Given K_i for all i, it is straightforward to solve VLE problems using the same procedures as for ideal solutions.

> *Eqns. 10.19 provides the primary equations for VLE via equations of state. These equations are implemented by iteration procedures summarized in Appendix C. Only the bubble method will be presented in the chapter in detail. Although cubic equations can represent both vapor and liquid phases, note that the virial equation cannot be used for liquid phases.*

Bubble Pressure Method

For a bubble pressure calculation, the T and all x_i are known as shown in Table 9.2 on page 302. Like the simple calculation performed in the last chapter, the criterion for convergence is $\sum_i y_i = 1$ which needs to be expressed in terms of variables for the current method. Rearranging

10.19, this sum becomes $\sum_i \frac{x_i \hat{\varphi}_i^L P}{\hat{\varphi}_i^V P} = \sum_i x_i K_i = 1$. Unlike the calculation in the last chapter, we

cannot explicitly solve for pressure because all $\hat{\varphi}_i^L$ and $\hat{\varphi}_i^V$ depend on pressure. Additionally, all

$\hat{\varphi}_i^V$ depend on composition of the vapor phase, which is not exactly known until the problem is solved. Typically, we use Raoult's law with the shortcut vapor pressure equation for the first guesses of y_i and P. From these values, we determine all K_i and check the sum of y values. If the sum is greater than one, the pressure guess is increased, if less than one, the pressure guess is decreased. A complete flowchart and example will be discussed in Section 10.4, but for now, let us explore the methods for calculating the fugacity coefficients.

Example 10.2 K-values from the Peng-Robinson equation

KVAL may be helpful in following this example.

PRFUG.xls may be helpful in following this example.

The bubble point pressure of an equimolar nitrogen (1) + methane (2) system is to be calculated by the Peng-Robinson equation and compared to the shortcut K-ratio estimate at 100 K. The shortcut K-ratio estimate will be used as an initial guess: $P = 0.4119$ MPa, $y_{N_2} = 0.958$. Apply the formulas for the fugacity coefficients to obtain an estimate of the K-values for nitrogen and methane and evaluate the sum of the vapor mole fractions based on this initial guess.

Solution:

	$T_c(K)$	$P_c(MPa)$	ω
N_2	126.1	3.394	0.040
CH_4	190.6	4.604	0.011

The spreadsheet PRFUG.xls may be used for the calculations. From the shortcut calculation, $P = 0.4119$ MPa at 100 K. Applying Eqn. 6.21 and 6.23 for the pure component parameters we have:

For N_2: $A_{11} = 0.09686$; $B_1 = 0.011906$

For CH_4: $A_{22} = 0.18242$; $B_2 = 0.013266$

By the square-root combining rule Eqn. 10.10: $A_{12} = 0.13293$

Based on the vapor composition of the shortcut estimate at $y_1 = 0.958$, the mixing rule gives $A^V = 0.099913$; $B^V = 0.01196$; Solving the cubic for the vapor root gives $Z^V = 0.9059$.

$$\ln\left(\frac{\hat{f}_i^V}{y_i P}\right) = \frac{B_i}{B^V}\left(Z^V - 1\right) - \ln\left(Z^V - B^V\right)$$

$$- \frac{A^V}{B^V \sqrt{8}} \ln\left(\frac{Z^V + (1 + \sqrt{2})B^V}{Z^V + (1 - \sqrt{2})B^V}\right)\left(\frac{2(y_1 A_{i1} + y_2 A_{i2})}{A^V} - \frac{B_i}{B^V}\right)$$

Then

$$\ln\hat{\varphi}_1^V = \frac{0.011906}{0.01196}(0.9059 - 1) - \ln(0.9059 - 0.01196) -$$

$$\frac{0.099913}{0.01196 \cdot 2.8284} \ln\left(\left(\frac{0.9059 + 2.414 \cdot 0.01196}{0.9059 - 0.4142 \cdot 0.01196}\right) \cdot\right.$$

$$\left.\left(\frac{2(0.958 \cdot 0.9686 + 0.042 \cdot 0.13293)}{0.099913} - \frac{0.011906}{0.01196}\right)\right)$$

$$= -0.08756, \quad \hat{\varphi}_1^V = 0.9162$$

Example 10.2　*K*-values from the Peng-Robinson equation　(Continued)

Many of the terms are the same for the methane in the mixture:

$$\ln \hat{\varphi}_2^V = \frac{0.013266}{0.01196}(0.9059 - 1) - \ln(0.9059 - 0.01196) -$$

$$\frac{0.099913}{0.01196 \cdot 2.8284} \ln\left(\left(\frac{0.9059 + 2.414 \cdot 0.01196}{0.9059 - 0.4142 \cdot 0.01196}\right) \cdot \right.$$

$$\left.\left(\frac{2(0.958 \cdot 0.13293 + 0.042 \cdot 0.18242)}{0.099913} - \frac{0.013266}{0.01196}\right)\right)$$

$$= -0.16571, \quad \hat{\varphi}_2^V = 0.8473$$

To save some tedious calculations, the liquid formulas have already been applied to obtain:

$\hat{\varphi}_1^L = 1.791; \quad \hat{\varphi}_2^L = 0.0937.$　Determining　the　*K*　values,　$K_1 = \dfrac{\hat{\varphi}_1^L}{\hat{\varphi}_1^V} = \dfrac{1.791}{0.9162} = 1.955,$

$$K_2 = \frac{\hat{\varphi}_2^L}{\hat{\varphi}_2^V} = \frac{0.0937}{0.8473} = 0.1106$$

$y_1 = 0.5 \cdot 1.955 = 0.978; \; y_2 = 0.5 \cdot 0.1106 = 0.055; \; \displaystyle\sum_i y_i = 1.033.$

A higher guess for *P* would be appropriate for the next iteration in order to make the *K*-values smaller. $\hat{\varphi}_1^L$ and $\hat{\varphi}_2^L$ would need to be evaluated at the new pressure. The calculations are obviously tedious. K_i calculations are available via the KVAL HP48G program (without seeing intermediate calculations), or they are possible in Excel by first copying the *Fugacities* sheet on PRFUG.xls, using one sheet for liquid and the other for vapor, and then referencing cells on one of the sheets to calculate the K_i. To carry out the subsequent iterations to convergence, bubble calculations are available in the PC program PRMIX.EXE and the HP program PRMIX. More details on the entire procedure will follow in Section 10.4.

10.3　DIFFERENTIATION OF MIXING RULES

Since a compositional derivative is necessary to obtain the partial molar quantities, and the compositions are present in summation terms, we must understand the procedures for differentiation of the sums. Since *all of the compositional dependence* is embedded in these terms, if we understand how these terms are handled, we can then apply the results to *any* equation of state. Only three types of sums appear in most forms of equations of state, which have been introduced above. The first type of derivative we will encounter is of the form:

$$\left(\frac{\partial nb}{\partial n_k}\right)_{T,\underline{V},n_{j\neq k}}$$

10.20

❶ Since the compositional dependence is within the mixing rule, if we understand how to differentiate the general mixing rules, then we can easily apply them to the models that use them.

where $b = \displaystyle\sum_i y_i b_i$. For a binary $nb = n_1 b_1 + n_2 b_2$, and *k* will be encountered once in the sum, whether $k = 1$ or $k = 2$, thus:

$$\left(\frac{\partial nb}{\partial n_1}\right)_{T,\underline{V},n_2} = b_1 \quad \text{and} \quad \left(\frac{\partial nb}{\partial n_2}\right)_{T,\underline{V},n_1} = b_2 \qquad \text{10.21}$$

and the general result is

$$\left(\frac{\partial nb}{\partial n_k}\right)_{T,\,\underline{V},\,n_{j\neq k}} = b_k \qquad \text{10.22}$$

The second type of derivative which we will encounter is of the form

$$\left(\frac{\partial n^2 a}{\partial n_k}\right)_{T,\underline{V},n_{j\neq k}} \qquad \text{10.23}$$

$n^2 a$ may be written as $n^2 \displaystyle\sum_{i=1}^{n}\sum_{j=1}^{n} x_i x_j a_{ij}$. For a binary mixture, $n_1^2 a_{11} + 2n_1 n_2 a_{12} + n_2^2 a_{22}$. Taking the appropriate derivative,

$$\left(\frac{\partial n^2 a}{\partial n_1}\right)_{T,\underline{V},n_2} = 2n_1 a_{11} + 2n_2 a_{12} \quad \text{and} \quad \left(\frac{\partial n^2 a}{\partial n_2}\right)_{T,\underline{V},n_1} = 2n_1 a_{12} + 2n_2 a_{22} \qquad \text{10.24}$$

The general result is

$$\left(\frac{\partial n^2 a}{\partial n_k}\right)_{T,\,\underline{V},\,n_{j\neq k}} = 2\sum_{j} n_j a_{jk} \qquad \text{10.25}$$

For the virial equation, we need to differentiate a function that will look like:

$$\left(\frac{\partial nB}{\partial n_k}\right)_{T,P,n_{j\neq k}} = \left(\frac{\partial\left(\frac{1}{n}\right)\left(\sum_i \sum_j n_i n_j B_{ij}\right)}{\partial n_k}\right)_{T,P,n_{j\neq k}} \qquad \text{10.26}$$

Differentiation by the product rule gives

$$\frac{1}{n}\left(\frac{\partial\left(\sum_i \sum_j n_i n_j B_{ij}\right)}{\partial n_k}\right)_{T,P,n_{j\neq k}} - \frac{\sum_i \sum_j n_i n_j B_{ij}}{n^2} \qquad \text{10.27}$$

The double sum in the derivative is n^2B which we have evaluated in equivalent form in Eqn 10.24. The second term is just B given by Eqn. 10.1. Therefore we have for a binary system

$$\left(\frac{\partial nB}{\partial n_1}\right)_{T,P,n_2} = \left(\frac{\partial\left(\frac{1}{n}\right)\left(\sum_i\sum_j n_i n_j B_{ij}\right)}{\partial n_1}\right)_{T,P,n_2} = 2y_1 B_{11} + 2y_2 B_{12} - B \quad \text{and}$$

$$\left(\frac{\partial nB}{\partial n_2}\right)_{T,P,n_1} = \left(\frac{\partial\left(\frac{1}{n}\right)\left(\sum_i\sum_j n_i n_j B_{ij}\right)}{\partial n_2}\right)_{T,P,n_1} = 2y_1 B_{12} + 2y_2 B_{22} - B \qquad 10.28$$

The general result is,

$$\left(\frac{\partial\left(\frac{1}{n}\right)\left(\sum_i\sum_j n_i n_j B_{ij}\right)}{\partial n_k}\right)_{T,P,n_{j\neq k}} = 2\sum_j y_j B_{jk} - B \qquad 10.29$$

Example 10.3 Fugacity coefficient from the virial equation

For moderate deviations from the ideal-gas law, a common method is to use the virial equation given by:

$$Z = 1 + BP/RT$$

where $B = \sum_i\sum_j y_i y_i B_{ij}$

Develop an expression for the fugacity coefficient.

Solution: For the virial equation, we have the result of Eqn. 8.28

$$\frac{G - G^{ig}}{RT} = \ln\varphi = \frac{BP}{RT}$$

Example 10.3 Fugacity coefficient from the virial equation (Continued)

Applying Eqn. 10.12

$$\left(\frac{\partial (G - G^{ig})/RT}{\partial n_k}\right)_{T,P,n_{j\neq k}} = \frac{P}{RT}\left(\frac{\partial \left[n\sum_i \sum_j y_i y_j B_{ij}\right]}{\partial n_k}\right)_{T,P,n_{j\neq k}}$$

the argument we need to differentiate looks like $n\sum_i \sum_j y_i y_j B_{ij} = \frac{1}{n}\sum_i \sum_j n_i n_j B_{ij}$.

Differentiation has been performed in Eqn. 10.28, which we can generalize:

$$\ln \frac{\hat{f}_k}{y_k P} = \left(2\left(\sum_j y_j B_{jk}\right) - B\right)\frac{P}{RT} \qquad\qquad 10.30$$

which has been shown earlier for a binary in Eqn. 10.13.

Example 10.4 Fugacity coefficient for van der Waals equation

Van der Waals' equation of state provides a simple but fairly accurate representation of key equation of state concepts for mixtures. The main manipulations developed for this equation are the same for other equations of state but the algebra is a little simpler. Recalling van der Waals' equation from Chapter 6,

$$Z = \frac{1}{1 - b\rho} - \frac{a\rho}{RT} = 1 + \frac{b\rho}{1 - b\rho} - \frac{a\rho}{RT}$$

where $a = \sum_i \sum_j y_i y_i a_{ij}$

$b = \sum_i y_i b_i$

Develop an expression for the fugacity coefficient.

Solution:

$$\ln(\hat{\varphi}_k) = \left(\frac{\partial (\underline{A} - \underline{A}^{ig})_{TV}/RT}{\partial n_k}\right)_{T,\underline{V},n_{k\neq i}} - \ln Z$$

Example 10.4 Fugacity coefficient for van der Waals equation (Continued)

Note that we must apply Eqn. 7.27 because the differentiation indicated above is performed at constant volume, not constant pressure,

$$\frac{(A - A^{ig})_{TV}}{RT} = \int_0^{b\rho} (Z - 1) \frac{d(b\rho)}{b\rho} = \int_0^{b\rho} \left(\frac{b\rho}{1 - b\rho} - \frac{a}{bRT} b\rho \right) \frac{d(b\rho)}{b\rho} = -\ln(1 - b\rho) - \frac{a}{bRT} b\rho$$

$$\frac{(\underline{A} - \underline{A}^{ig})_{TV}}{RT} = -n \ln(1 - b\rho) - \frac{an^2}{\underline{V}RT}$$

Apply Eqn. 10.16, but instead of differentiating directly, use Eqn. 5.16,

$$\left(\frac{\partial(term)}{\partial n_k} \right)_{T, \underline{V}, n_{k \neq i}} = \left(\frac{\partial(term)}{\partial(b\rho)} \right)_{T, \underline{V}, n_{k \neq i}} \left(\frac{\partial(b\rho)}{\partial n_k} \right)_{T, \underline{V}, n_{k \neq i}} \quad \text{or}$$

$$\left(\frac{\partial(term)}{\partial n_k} \right)_{T, \underline{V}, n_{k \neq i}} = \left(\frac{\partial(term)}{\partial(n^2 a)} \right)_{T, \underline{V}, n_{k \neq i}} \left(\frac{\partial(n^2 a)}{\partial n_k} \right)_{T, \underline{V}, n_{k \neq i}}$$

$$\left(\frac{\partial(\underline{A} - \underline{A}^{ig})_{TV}/RT}{\partial n_k} \right)_{T, \underline{V}, n_{k \neq i}} = -\ln(1 - b\rho) + \frac{n}{1 - b\rho} \left(\frac{\partial b\rho}{\partial n_k} \right)_{T, \underline{V}, n_{k \neq i}} - \frac{1}{\underline{V}RT} \left(\frac{\partial n^2 a}{\partial n_k} \right)_{T, \underline{V}, n_{k \neq i}}$$

$$b\rho = \frac{nb}{\underline{V}} \Rightarrow \left(\frac{\partial nb/\underline{V}}{\partial n_k} \right)_{T, \underline{V}, n_{i \neq k}} = \frac{b_k}{\underline{V}}$$

$$\ln(\hat{\varphi}_k) = -\ln(1 - b\rho) + \frac{b_k \rho}{1 - b\rho} - \frac{2 \sum_j n_j a_{kj}}{\underline{V}RT} - \ln Z$$

$$= -\ln(1 - b\rho) + \frac{b_k \rho}{1 - b\rho} - \frac{2\rho \sum_j x_j a_{kj}}{RT} - \ln Z$$

10.31

$$b\rho \equiv \frac{B}{Z}; \quad \frac{a}{bRT} \equiv \frac{A}{B}; \quad \frac{a_{jk}}{a} \equiv \frac{A_{jk}}{A}; \quad \frac{b_k}{b} \equiv \frac{B_k}{B}$$

$$\boxed{\ln(\hat{\varphi}_k) = -\ln(Z - B) + \frac{B_k}{Z - B} - \frac{2 \sum_j x_j A_{kj}}{Z}}$$

10.32

Example 10.5 Fugacity coefficient from the Peng-Robinson equation

The Peng-Robinson equation is given by

$$Z = \frac{1}{1 - b\rho} - \frac{a\rho}{RT} \frac{1}{\left(1 + 2b\rho - b^2\rho^2\right)}$$

where $a = \displaystyle\sum_i \sum_j y_i y_j a_{ij}$;

$b = \displaystyle\sum_i y_i b_i$

Develop an expression for the fugacity coefficient.

Solution:

$$\ln(\hat{\varphi}_k) = \left(\frac{\partial (\underline{A} - \underline{A}^{ig})_{TV} / RT}{\partial n_k}\right)_{T, \underline{V}, n_{k \neq i}} - \ln Z$$

From our integration for the pure fluid,

$$\frac{(A - A^{ig})_{TV}}{RT} = -\ln(1 - b\rho) - \frac{a}{bRT\sqrt{8}} \ln\left(\frac{1 + (1 + \sqrt{2})b\rho}{1 + (1 - \sqrt{2})b\rho}\right)$$

$$\frac{(\underline{A} - \underline{A}^{ig})_{TV}}{RT} = -n \ln(1 - b\rho) - \frac{an^2}{nbRT\sqrt{8}} \left\{\ln\left[1 + (1 + \sqrt{2})b\rho\right] - \ln\left[1 + (1 - \sqrt{2})b\rho\right]\right\}$$

The next steps look intimidating. Basically, they apply the same procedure for differentiation as the last example,

$$\left(\frac{\partial (\underline{A} - \underline{A}^{ig})_{TV} / RT}{\partial n_k}\right)_{T, V, n_{k \neq i}} = -\ln(1 - b\rho) + \frac{n}{1 - b\rho}\left(\frac{\partial b\rho}{\partial n_k}\right)$$

$$- \frac{an^2}{nbRT\sqrt{8}} \left\{\frac{(1 + \sqrt{2})\left(\frac{\partial b\rho}{\partial n_k}\right)}{1 + (1 + \sqrt{2})b\rho} - \frac{(1 - \sqrt{2})\left(\frac{\partial b\rho}{\partial n_k}\right)}{1 + (1 - \sqrt{2})b\rho}\right\}$$

$$- \ln\left[\frac{1 + (1 + \sqrt{2})b\rho}{1 + (1 - \sqrt{2})b\rho}\right]\left[\frac{\left(\frac{\partial an^2}{\partial n_k}\right)}{nbRT\sqrt{8}} - \frac{an^2}{RT\sqrt{8}}\frac{\left(\frac{\partial nb}{\partial n_k}\right)}{(nb)^2}\right]$$

$$\ln(\hat{\varphi}_k) = -\ln(1 - b\rho) - \ln Z + \frac{b_k\rho}{1 - b\rho} - \frac{ab_k\rho}{bRT\sqrt{8}}\left\{\frac{(1 + \sqrt{2})}{1 + (1 + \sqrt{2})b\rho} - \frac{(1 - \sqrt{2})}{1 + (1 - \sqrt{2})b\rho}\right\}$$

$$- \frac{a}{bRT\sqrt{8}} \ln\left[\frac{1 + (1 + \sqrt{2})b\rho}{1 + (1 - \sqrt{2})b\rho}\right]\left[\frac{2\displaystyle\sum_j x_j a_{jk}}{a} - \frac{b_k}{b}\right]$$

Example 10.5 Fugacity coefficient from the Peng-Robinson equation

Note a simplification that is not obvious:

$$\frac{b_k\rho}{1-b\rho} - \frac{ab_k\rho}{bRT\sqrt{8}}\left[\frac{(1+\sqrt{2})}{1+(1+\sqrt{2})b\rho} - \frac{(1-\sqrt{2})}{1+(1-\sqrt{2})b\rho}\right] =$$

$$\frac{b_k}{b}\left\{\frac{b\rho}{1-b\rho} - \frac{ab\rho}{bRT\sqrt{8}}\left[\frac{1}{1+2b\rho - b^2\rho^2}\right]\right\} = \frac{b_k}{b}\{Z-1\}$$

$$\ln(\hat{\varphi}_k) = -\ln(1-b\rho) - \ln Z + \frac{b_k}{b}\{Z-1\} - \frac{a}{bRT\sqrt{8}}\ln\left[\frac{1+(1+\sqrt{2})b\rho}{1+(1-\sqrt{2})b\rho}\right]\left[\frac{2\sum_j x_j a_{jk}}{a} - \frac{b_k}{b}\right]$$

$$b\rho \equiv \frac{B}{Z}; \quad \frac{a}{bRT} \equiv \frac{A}{B}; \quad \frac{a_{jk}}{a} \equiv \frac{A_{jk}}{A}; \quad \frac{b_k}{b} \equiv \frac{B_k}{B}$$

$$\ln(\hat{\varphi}_k) = -\ln(Z-B) + \frac{B_k}{B}\{Z-1\} - \frac{A}{B\sqrt{8}}\ln\left[\frac{Z+(1+\sqrt{2})B}{Z+(1-\sqrt{2})B}\right]\left[\frac{2\sum_j x_j A_{jk}}{A} - \frac{B_k}{B}\right] \quad 10.33$$

which has been shown in Eqns. 10.17-10.18 for a binary.

10.4 VLE CALCULATIONS BY AN EQUATION OF STATE

At the end of Section 10.2, the bubble pressure method was briefly introduced to show how the fugacity coefficients are incorporated into a VLE calculation, without concentrating on the details of the iterations. Section 10.3 offered derivations of formulas for the fugacity coefficients that were presented without proof at the beginning of the chapter. Now, it is time to turn to the applied engineering objective: calculation of phase equilibria. Refer again to Table 9.2 on page 302, that lists the types of routines that are needed and the convergence criteria. Note that Table 9.2 is independent of the model used for calculating VLE. As an example of the iteration procedure for cubic equations of state, the bubble pressure flowsheet is presented in Fig. 10.2. The flowsheet puts detail to the procedure discussed superficially in Example 10.2 and immediately preceding the example. Flowsheets for bubble temperature, dew, and flash are available in Appendix C. As with ideal solutions, the bubble pressure routine is the easiest to apply, so we cover it in detail in the following examples.

❗ The engineering objective is to use equations of state for bubble, dew, and flash calculations.

❗ Flowsheets for bubble temperature, dew, and flash routines are in Appendix C.

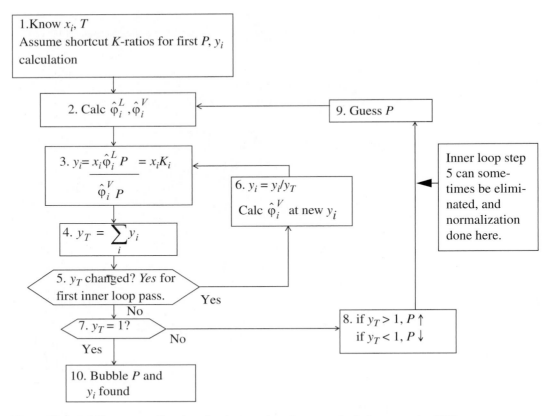

Figure 10.2 *Bubble pressure flowsheet for the equation of state method of representing VLE.
Other routines are given in Appendix C. The examples omit the inner loop.*

Example 10.6 Bubble point pressure from the Peng-Robinson equation

Use the Peng-Robinson equation ($k_{ij} = 0$) to determine the bubble-point pressure of an equimolar solution of nitrogen (1) + methane (2) at 100 K.

Solution: The calculations proceed by first calculating the K-values as given by Raoult's law and the shortcut vapor pressure equation as in Example 10.2 on page 328. The ideal-solution (^{is}) bubble pressure was $P \approx \sum_i x_i P_i^{sat} \approx 0.4119$ bars; $y_{N2}{}^{is} = 0.958$. In fact, the K-values for the first iteration have already been determined in that example, in great detail. The values from that example are $K_1 = 1.955$, $K_2 = 0.1106$. The new estimates of vapor mole fractions are obtained by multiplying $x_i \cdot K_i$. In Example 10.2, the sum was found to be $\sum_i y_i = 1.033$. These calculations are summarized in the first column of Table 10.1.

Example 10.6 Bubble point pressure from the Peng-Robinson equation (Continued)

Noting that these sum to a number greater than unity, we must choose a greater value of pressure for the next iteration. Before we can start the next iteration, however, we must develop new estimates of the vapor mole fractions; the ones we have do not make sense because they sum to more than unity. These new estimates can be obtained simply by dividing the given vapor mole fractions by the number to which they sum. This process is known as *normalization* of the mole fractions. For example to start the second iteration, $y_1 = 0.978/1.033 = 0.947$. After repeating the process for the other component, the mole fractions will sum to unity. Since the result for the second iteration is less than one, the pressure guess is too high.

Normalization of mole functions.

The third iteration consists of applying the interpolation rule to obtain the estimate of pressure and use of the normalization procedure to obtain the estimates of vapor mole fractions. $P = 0.4119 + (1 - 1.033)/(0.956 - 1.033) \cdot (0.45 - 0.4119) = 0.428$ MPa. Since the estimated vapor mole fractions after the third iteration sum very nearly to unity, we may conclude the calculations here. This is the bubble pressure. Note how quickly the estimate for y_1 converges to the final estimate of 0.945.

Table 10.1 *Summary of bubble pressure intermediate calculations.*

Comp	x_i	$P = 0.4119$, $y_1 = 0.958$		$P = 0.45$, $y_1 = 0.947$		$P = 0.428$, $y_1 = 0.946$	
		K_i	y_i	K_i	y_i	K_i	y_i
N_2	0.5	1.957	0.978	1.808	0.904	1.890	0.945
CH_4	0.5	0.1106	0.055	0.1031	0.052	0.1073	0.054
			1.033		0.956		0.999

Bubble pressure calculations are available in PRMIX.EXE for the PC and PRMIX for the HP. The spreadsheet PRFUG.xls is useful for visualizing fugacities in a given phase.

Bubble pressure calculations are available in PRMIX.EXE for the PC and PRMIX for the HP. The spreadsheet PRFUG.xls is useful for visualizing fugacities in a given phase.

Example 10.7 Isothermal flash using the Peng-Robinson equation

A distillation column is to produce overhead products having the following compositions:

Component	z_i
Propane	0.23
Isobutane	0.67
n-Butane	0.10

Suppose a partial condenser is operating at 320 K and 8 bars. What fraction of liquid would be condensed according to the Peng-Robinson equation, assuming all binary interaction parameters are zero ($k_{ij} = 0$)?

Example 10.7 Isothermal flash using the Peng-Robinson equation (Continued)

Solution: This is a flash calculation. Refer back to the same problem (Example 9.1 on page 305) for an initial guess based on the shortcut K-ratio equation. $L/F = 0.75 \Rightarrow \{x_i\} = \{0.1829, 0.7053, 0.1117\}$ and $\{y_i\} = \{0.3713, 0.5642, 0.0648\}$. Substituting these composition estimates for the vapor and liquid compositions into the routine for estimating K-values (cf. Example 10.2 on page 328), we can obtain the estimates for K-values given below:

	T_c(K)	P_c(bar)	ω	z_i	K_i
C_3	369.8	42.49	0.152	0.23	1.729
$i\text{-}C_4$	408.1	36.48	0.177	0.67	0.832
$n\text{-}C_4$	425.2	37.97	0.193	0.10	0.640

The computations for the flash calculation are basically analogous to those in Example 9.1, except that K_i values are calculated from Eqn. 10.19. A detailed flowsheet is presented in Appendix C. For this example, the K-values are not modified until the iteration on L/F converges. After convergence on L/F, the vapor and liquid mole fractions are recomputed using Eqns. 9.57 and 9.58, followed by recomputed estimates for the K-values. If the new estimates for K-values are equal to the old estimates for K-values, then the overall iteration has converged. If not, then the new estimates for K-values are substituted for the old values, and the next iteration proceeds just like the last. This method of iteratively solving for the vector of K-values is known in numerical analysis as the "successive substitution" method.

		$L/F = 0.75$	$L/F = 0.90$	$L/F = 0.868$		
z_i	K_i	D_i	D_i	D_i	x_i	y_i
0.23	1.729	−0.142	−0.1563	−0.1529	0.2098	0.3627
0.67	0.832	0.118	0.1145	0.1151	0.6852	0.5701
0.10	0.640	0.040	0.0373	0.0378	0.1050	0.0672
		0.016	−0.004	0.0000	1.000	1.000

Using these x and y for guesses we find $K = 1.7276, 0.8318$, and 0.6407, respectively. These K-values are similar to those estimated at the compositions derived from the ideal-solution approximation, and will yield a similar L/F. Therefore, we conclude that this iteration has converged (a general criterion is that the average % change in the K-values from one iteration to the next is less than 10^{-4}). Comparison to the shortcut K-ratio approximation shows small but significant deviations—$L/F = 0.87$ for Peng-Robinson versus 0.75 for the shortcut K-ratio method.

> ## Example 10.7 Isothermal flash using the Peng-Robinson equation (Continued)
>
> Based on this example, we may conclude that the shortcut K-ratio approximation provides a reasonable first approximation at these conditions. Note, however, that none of the components is supercritical and all the components are saturated hydrocarbons.
>
> | $K_i^{Peng\text{-}Rob}$ | $K_i^{shortcut}$ |
> |---|---|
> | 1.727 | 2.03 |
> | 0.832 | 0.80 |
> | 0.641 | 0.58 |

> ## Example 10.8 Phase diagram for azeotropic methanol + benzene
>
> Methanol and benzene form an azeotrope. For methanol + benzene the azeotrope occurs at 61.4 mole% methanol and 58°C at atmospheric pressure (1.01325 bars). Additional data for this system are available in the *Chemical Engineers' Handbook*.[1] Use the Peng-Robinson equation with $k_{ij} = 0$ (see Eqn. 10.10) to estimate the phase diagram for this system and compare it to the experimental data on a T-x-y diagram. Determine a better estimate for k_{ij} by iterating on the value until the bubble point pressure matches the experimental value (1.013 bar) at the azeotropic composition and temperature. Plot these results on the T-x-y diagram as well. Note that it is impossible to match both the azeotropic composition and pressure with the Peng-Robinson equation because of the limitations of the single parameter, k_{ij}.
>
> The experimental data for this system are as follows:
>
> | x_m | 0.000 | 0.026 | 0.050 | 0.088 | 0.164 | 0.333 | 0.549 | 0.699 | 0.782 | 0.898 | 0.973 | 1.000 |
> |---|---|---|---|---|---|---|---|---|---|---|---|---|
> | y_m | 0.000 | 0.267 | 0.371 | 0.457 | 0.526 | 0.559 | 0.595 | 0.633 | 0.665 | 0.760 | 0.907 | 1.000 |
> | $T(K)$ | 353.25 | 343.82 | 339.59 | 336.02 | 333.35 | 331.79 | 331.17 | 331.25 | 331.62 | 333.05 | 335.85 | 337.85 |
>
> **Solution:** Solving this problem is computationally intensive enough to write a general program for solving for bubble-point pressure. Fortunately, computer and calculator programs are readily available. We will discuss the solution using the PC program PRMIX.EXE. Select the option KI for adjusting the interaction parameter. This routine will perform a bubble calculation for a guessed value of k_{ij}. When prompted, enter the temperature (331.15 K) and liquid composition $x_m = 0.614$. The program will give a calculated pressure and vapor phase composition. The vapor-phase composition will not match the liquid-phase composition because the azeotrope is not perfectly predicted; however, we continue to change k_{ij} until we match the pressure of 1.013 bar. The following values are obtained for the bubble pressure at the experimental azeotropic composition and temperature with various values of k_{ij}.

HP PRMIX offers bubble pressure.

PRMIX offers option KI for iterating on a single point.

---

1. Perry, R.M., Chilton, C.H., *The Chemical Engineers' Handbook,* 6th ed., page p.13–12, McGraw-Hill (1984).

Example 10.8   Phase diagram for azeotropic methanol + benzene (Continued)

k_{ij}	0.0	0.1	0.076	0.084
P(bars)	0.75	1.06	0.9869	1.011

The resulting k_{ij} is used to perform bubble temperature calculations across the composition range resulting in Fig. 10.3. Note that we might find a way to fit the data more accurately than the method given here, but any improvements would be small relative to the improvement obtained by not estimating $k_{ij} = 0$. We see that the fit is not as good as we would like for process design calculations. This solution is so non-ideal that a more flexible model of the thermodynamics is necessary. Note that the binary interaction parameter alters the magnitude of the bubble pressure curve very effectively but hardly affects the skewness at all.

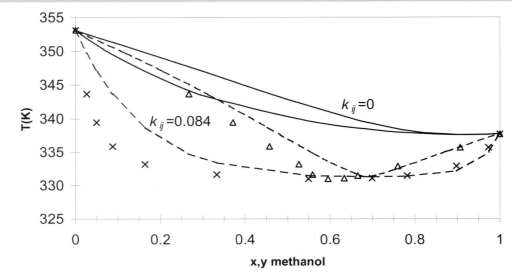

Figure 10.3 *T-x,y diagram for the azeotropic system methanol + benzene. Curves show the predictions of the Peng-Robinson equation ($k_{ij} = 0$) and correlation ($k_{ij} = 0.084$) based on fitting a single data point at the azeotrope. x's and triangles represent liquid and vapor phases, respectively.*

Example 10.9   Phase diagram for nitrogen + methane

HP PRMIX offers bubble pressure.

PRMIX offers other routines as well.

Use the Peng-Robinson equation ($k_{ij} = 0$) to determine the phase diagram of nitrogen + methane at 150 K. Plot P versus x, y and compare the results to the results from the shortcut K-ratio equations.

Example 10.9 Phase diagram for nitrogen + methane (Continued)

Solution: First, the shortcut K-ratio method gives the dotted phase diagram on Fig. 10.4. Applying the bubble pressure option of the program PRMIX on the PC or the HP, we calculate the solid line on Fig. 10.4. For the Peng-Robinson method we assume K-values from the previous solution as the initial guess to get the solutions near $x_{N2} = 0.685$. The program PRMIX assumes this automatically, but we must also be careful to make small changes in the liquid composition as we approach the critical region. The figure below was generated by entering liquid nitrogen compositions of: 0.10, 0.20, 0.40, 0.60, 0.61, 0.62..., 0.68, 0.685. This procedure of starting in a region where a simple approximation is reliable and systematically moving to more difficult regions using previous results is often necessary and should become a familiar trick in your accumulated expertise on phase equilibria in mixtures. We apply a similar approach in estimating the phase diagrams in liquid-liquid mixtures.

The shortcut K-ratio method provides an initial estimate when a supercritical component is at low liquid-phase compositions, but incorrectly predicts VLE at high liquid-phase concentrations of the supercritical component.

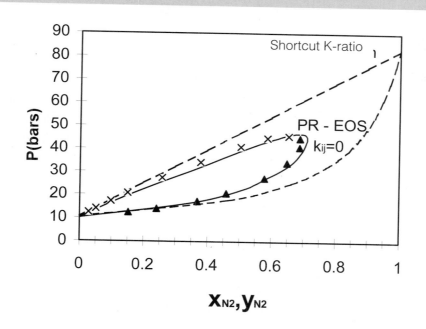

Figure 10.4 *High pressure P-x-y diagram for the nitrogen + methane system comparing the shortcut K-ratio approximation and the Peng-Robinson equation at 150 K. The data points represent experimental results. Both theories are entirely predictive since the Peng-Robinson equation assumes that $k_{ij} = 0$.*

Example 10.9 Phase diagram for nitrogen + methane (Continued)

Comparing the two approximations numerically and graphically, it is clear that the shortcut approximation is significantly less accurate than the Peng-Robinson equation at high concentrations of the supercritical component. This happens because the mixture possesses a critical point, above which separate liquid and vapor roots are impossible, analogous to the situation for pure fluids. Since the mixing rules are in terms of a and b instead of T_c and P_c, the equation of state is generating effective values for A_c and B_c of the mixture. Instead of depending simply on T and P as they did for pure fluids, however, A_c and B_c also depend on composition. The mixture critical point varies from the critical point of one component to the other as the composition changes. Since the shortcut approximation extrapolates the vapor pressure curve to obtain an effective vapor pressure of the supercritical component, that approximation does not reflect the presence of the mixture critical point and this leads to significant errors as the mixture becomes rich in the supercritical component.

The mixture critical point also leads to computational difficulties. If the composition is excessively rich in the supercritical component, the equation of state calculations will obtain the same solution for the vapor root as for the liquid root and, since the fugacities will be equal, the program will terminate. The program may indicate accurate convergence in this case due to some slight inaccuracies that are unavoidable in the critical region. Or the program may diverge. It is often up to the competent engineer to recognize the difference between accurate convergence and a spurious answer. Plotting the phase envelope is an excellent way to stay out of trouble. Note that the mole fraction in the vapor phase is equal to the mole fraction in the liquid phase at P_{max}. What are the similarities and differences between this and an azeotrope?

Example 10.10 Ethane + heptane phase envelopes

 PRMIX offers bubble pressure.

PRMIX offers other routines as well.

 Mixture critical points.

Isopleths.

Use the Peng-Robinson equation ($k_{ij} = 0$) to determine the phase envelope of ethane + n-heptane at compositions of $x_{C7} = [0, 0.1, 0.2, 0.3, 0.5, 0.7, 0.9, 1.0]$. Plot P versus T for each composition by performing bubble-pressure calculations to their terminal point and dew-temperature calculations until the temperature begins to decrease significantly and the pressure approaches its maximum. If necessary, close the phase envelope by starting at the last dew-temperature state and performing dew-pressure calculations until the temperature and pressure approach the terminus of the bubble point curve. For each composition, mark the points where the bubble and dew curves meet with X's. These X's designate the "mixture critical points." Connect the X's with a dashed curve. The dashed curve is known as the "critical locus" of the mixture.

Solution: Note that these phase envelopes are similar to the one from the previous problem, except that we are changing the temperature instead of the composition along each curve. They are more tedious in that both dew and bubble calculations must be performed to generate each curve. The lines of constant composition are sometimes called *isopleths*. The results of the calculations are illustrated in Fig. 10.5. The results at mole fractions of 0 and 1.0 are indicated by dash-dot curves to distinguish them as the vapor pressure curves. Phase equilibria on the P-T plot occurs at the conditions where a bubble line of one composition intersects a dew line of a different composition.

Example 10.10 Ethane + heptane phase envelopes (Continued)

Some practical considerations for high pressure processing can be inferred from the diagram. Consider what happens when starting at 90 bars and ~445 K and dropping the pressure on a 30 mole% C7 mixture at constant temperature. Similar situations could arise with flow of natural gas through a small pipe during natural gas recovery. As the pressure drops, the dew-point curve is crossed and liquid begins to precipitate. Based on intuition developed from experiences at lower pressure, one might expect that dropping the pressure should result in more vapor-like behavior, not precipitation. On the other hand, dropping the pressure reduces the density and solvent power of the ethane-rich mixture. This phenomenon is known as *retrograde condensation*. It occurs near the critical locus when the operating temperature is less than the maximum temperature of the phase envelope. Since this maximum temperature is different from the mixture critical temperature, it needs a distinctive name. The name applied is the "critical condensation temperature" or *cricondentherm*. Similarly, the maximum pressure on the phase envelope is known as the *cricondenbar*. Note that an analogous type of phase transition can occur near the critical locus when the pressure is just above the critical locus and the temperature is changed.

To extend the analysis, imagine what happens in a natural gas stream composed primarily of methane but also containing small amounts of components as heavy as C80. A retrograde condensation region will exist where the heavy components begin to precipitate, as you should now understand. But a different possibility also exists because the melting temperature of the heavy components may often exceed the operating temperature, and the precipitate that forms might be a solid that could stick to the walls of the pipe. This in turn generates a larger constriction which generates a larger pressure drop during flow, right in the vicinity of the deposit. In other words, this deposition process may tend to promote itself until the flow is substantially inhibited. Wax deposition is a significant problem in the oil and natural gas industry and requires considerable engineering expertise because it often occurs away from critical points, as well as in the near-critical regions of this discussion. A wide variety of solubility behavior can occur, although we will delay the discussion until Chapter 13, and concentrate on the development of other calculational skills in the next chapters.

❶ Retrograde condensation.

Example 10.10 Ethane + heptane phase envelopes (Continued)

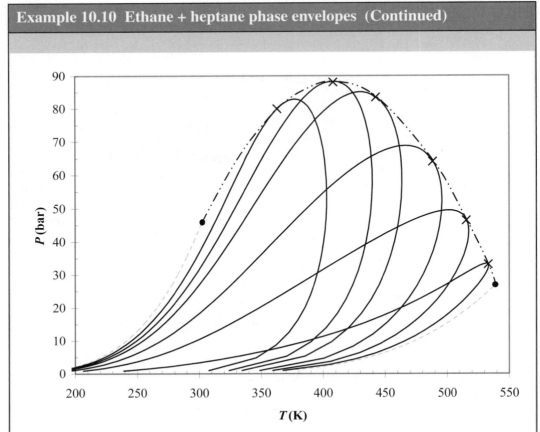

Figure 10.5 *High pressure phase envelopes for the ethane + heptane system comparing the effects of composition according to the Peng-Robinson equation. The theory is entirely predictive because the Peng-Robinson equation has been applied with $k_{ij} = 0$. X's mark the mixture critical points and the dashed line indicates the critical locus. The first and last curves represent the vapor pressures for the pure components.*

10.5 STRATEGIES FOR APPLYING VLE ROUTINES

For some problems, such as generation of a phase diagram, the examples of this chapter can be followed directly. Often, however, it takes some thought to decide which type of VLE routine is appropriate to apply in a given situation. Do not discount this step in the problem solution strategy. Part of the objective of the homework problems is to increase understanding of phase behavior by encouraging thought about which routine to use. Since the equation of state routines are complicated enough to require a computer, they are also solved relatively rapidly, *once the VLE routine has been identified*. Use Table 9.2 on page 302 and the information in Section 9.1 to identify the correct procedure to apply before turning to computational techniques. Use the problem statement to identify whether the liquid, vapor, or overall mole fractions are known. Study the problem statement to see whether the P and T are known. Often this information together with Table 9.2 will determine the routines to apply. Sometimes more than one approach is satisfactory. Sometimes the ideal-solution approximation may be applied, and a review of Section 9.2 may be helpful. Before using software that accompanies the text, be sure to read the appropriate section of Appendix P, and the instructions or readme.txt files that accompany the program.

10.6 SUMMARY AND CONCLUDING REMARKS

The essence of the equation of state approach to mixtures is that the equation of state for mixtures is the same as the equation of state for pure fluids. The expressions for Z, A, and U are exactly the same. The only difference is that the parameters (a and b of the Peng-Robinson equation) are dependent on the composition. That should come as no large surprise when you consider that these parameters must transform from pure component to pure component in some continuous fashion as the composition changes. What may be surprising is the wealth of behaviors that can be inferred from some fairly simple rules for modeling this transformation. Everything from azeotropes to retrograde condensation, and even liquid-liquid separation (to be discussed in Chapter 12) can be represented with qualitative accuracy based on this simple extension of the equation of state.

So, are we done? Unfortunately, the keyword in the preceding paragraph is "qualitative." Equations of state are sufficiently accurate for most applications involving hydrocarbons, gases, and to some extent, ethers, esters, and ketones. But any applications involving strongly hydrogen-bonding species like methanol or water tend to require greater accuracy than can be obtained currently from equations like the Peng-Robinson equation. More adaptable models are treated in the upcoming chapter.

10.7 PRACTICE PROBLEMS

P10.1 Repeat all the practice problems from Chapter 9, this time applying the Peng-Robinson equation.

P10.2 Acrolein (C_3H_4O) + water exhibits an atmospheric (1 bar) azeotrope at 97.4 wt% acrolein and 52.4°C. For acrolein: $T_c = 506$ K; $P_c = 51.6$ bar; and $\omega = 0.330$; $MW = 56$

(a) Determine the value of k_{ij} for the Peng-Robinson equation that matches this bubble pressure at the same liquid composition and temperature. (ANS. 0.015)

(b) Tabulate P, y at 326.55 K and $x = \{0.57, 0.9, 0.95, 0.974\}$ via the Peng-Robinson equation using the k_{ij} determined above.
(ANS. (1.33, 0.575), (1.16, 0.736), (1.06, 0.841), (1.0, 0.860))

P10.3 Laugier and Richon (*J. Chem. Eng. Data*, 40:153, 1995) report the following data for the H_2S + benzene system at 323 K and 2.010 MPa: $x_1 = 0.626$; $y_1 = 0.986$

(a) Quickly estimate the vapor-liquid K-value of H_2S at 298 K and 100 bar. (ANS. 0.21)

(b) Use the data to estimate the k_{ij} value, then estimate the error in the vapor phase mole fraction of H_2S. (ANS. 0.011, 0.1%)

P10.4 The system ethyl acetate + methanol forms an azeotrope at 27.8 mol% EA and 62.1°C. For ethyl acetate: $T_c = 523.2$ K; $P_c = 38.3$ bar; $\omega = 0.362$

(a) What is the estimate of the bubble point pressure from the Peng-Robinson equation of state at this composition and temperature when it is assumed that $k_{ij} = 0$? (ANS. 0.98 bars)

(b) What value of k_{ij} gives a bubble point pressure of 1 bar at this temperature and composition? (ANS. 0.0054)

(c) What is the composition of the azeotrope and value of the bubble point pressure at the azeotrope estimated by the Peng-Robinson equation when the value of k_{ij} from part (b) is used to describe the mixture? (ANS. $x_{EA} = 0.226$)

P10.5 (a) Assuming zero for the binary interaction parameter ($k_{ij} = 0$) of the Peng-Robinson equation, predict whether an azeotrope should be expected in the system CO_2 + ethylene at 222 K. Estimate the bubble point pressure for an equimolar mixture of these components. (ANS. No, 8.7 bar)

(b) Assuming a value for the binary interaction parameter ($k_{ij} = 0.11$) of the Peng-Robinson equation, predict whether an azeotrope should be expected in the system CO_2 + ethylene at 222 K. Estimate the bubble point pressure for an equimolar mixture of these components. (ANS. Yes, 11.3 bar)

P10.6 (a) Assuming zero for the binary interaction parameter ($k_{ij} = 0$) of the Peng-Robinson equation, estimate the bubble pressure and vapor composition of the pentane + acetone system at $x_p = 0.728$, 31.9°C (ANS. 0.78 bars, $y_1 = .83$)

(b) Use the experimental liquid composition and bubble condition of the pentane + acetone system at $x_P = 0.728$, $T = 31.9$°C, $P = 1$ bar to estimate the binary interaction parameter (k_{ij}) of the Peng-Robinson equation, then calculate the bubble pressure of a 13.4 mol% pentane liquid solution at 39.6°C (ANS. 0.117, 1.12 bar)

P10.7 Calculate the dew-point pressure and corresponding liquid composition of a mixture of 30 mol% carbon dioxide, 30% methane, 20% propane, and 20% ethane at 298 K.

(a) using shortcut K-ratios (ANS. 32 bar)

(b) for the Peng-Robinson equation with $k_{ij} = 0$ (ANS. 44 bar)

P10.8 The equation of state below has been suggested for a new equation of state. Derive the expression for the fugacity coefficient of a component.

$$Z = 1 + 4cb\rho/(1 - b\rho)$$

where $b = \sum_i x_i b_i$

$$c = \sum_i \sum_j x_i x_j c_{ij}$$

$$c_{ij} = \sqrt{c_{ii} c_{jj}}$$

(ANS. $\ln \hat{\varphi}_k = 4(c - 2\sum_i x_i c_{ik}) \ln(1 - b\rho) + \frac{b_k}{b}(Z - 1) - \ln Z$)

10.8 HOMEWORK PROBLEMS

10.1 Using Fig. 10.5 on page 344, without performing additional calculations, sketch the P-x-y diagram at 400 K showing the two-phase region. Make the sketch semi-quantitative to show the values where the phase envelope touches the axes of your diagram. Label the bubble and dew lines. Also indicate the approximate value of the maximum pressure.

10.2 Consider two gases that follow the virial equation. Show that an ideal solution of the two gases follows the relation $B = y_1 B_{11} + y_2 B_{22}$.

10.3 Consider phase equilibria modeled with $y_i \hat{\varphi}_i^V P = x_i \hat{\varphi}_i^L P$. When might $\hat{\varphi}_i$ be replaced by φ_i for each phase? When might $\hat{\varphi}_i = 1$ be used for each phase? Discuss the appropriateness of using the virial equation for mixtures to solve phase behavior using the expression $y_i \hat{\varphi}_i^V P = x_i \hat{\varphi}_i^L P$.

10.4 Calculate the molar volume of a binary mixture containing 30 mol% nitrogen(1) and 70 mol% n-butane(2) at 188°C and 6.9 MPa by the following methods:

(a) Assume the mixture to be an ideal gas.
(b) Assume the mixture to be an ideal solution with the volumes of the pure gases given by

$$Z = 1 + \frac{BP}{RT}$$

and the virial coefficients given below.

(c) Use second virial coefficients predicted by the generalized correlation for B.
(d) Use the following values for the second virial coefficients.

Data:

$B_{11} = 14$ $B_{22} = -265$ $B_{12} = -9.5$ (Units are cm³/gmole)

(e) Using the Peng-Robinson equation.

10.5 For the same mixture and experimental conditions as problem 10.4, calculate the fugacity of each component in the mixture, $\hat{f}_i$. Use methods (a) – (e).

10.6 A vapor mixture of CO_2 (1) and i-butane (2) exists at 120°C and 2.5 MPa. Calculate the fugacity of CO_2 in this mixture across the composition range

(a) using the virial equation for mixtures
(b) using the Peng-Robinson equation
(c) assuming the vapor is an ideal mixture

10.7 Use the virial equation to consider a mixture of propane and n-butane at 515 K at pressures between 0.1 and 4.5 MPa. Verify that the virial coefficient method is valid by using Eqn. 6.10.

(a) Prepare a plot of fugacity coefficient for each component as a function of composition at pressures of 0.1 MPa, 2 MPa, and 4.5 MPa.
(b) How would the fugacity coefficient for each component depend on composition if the mixture were assumed to be ideal, and what value(s) would it have for each of the pressures in part (a)? How valid might the ideal-solution model be for each of these conditions?
(c) The excess volume is defined as $V^E = V - \sum_i x_i V_i$, where V is the molar volume of the mixture, and V_i is the pure component molar volume at the same T and P. Plot the prediction of excess volume of the mixture at each of the pressures from part (a). How does the excess volume depend on pressure?
(d) Under which of the pressures in part (a) might the ideal gas law be valid?

10.8 Consider a mixture of nitrogen(1) + n-butane(2) for each of the options: (*i*) 395 K and 2 MPa; (*ii*) 460 K and 3.4 MPa; (*iii*) 360 K and 1 MPa.

(a) Calculate the fugacity coefficients for each of the components in the mixture using the virial coefficient correlation. Make a table for your results at $y_1 = 0.0, 0.2, 0.4, 0.6, 0.8, 1.0$. Plot the results on a graph. On the same graph, plot the curves that would be used for the mixture fugacity coefficients if an ideal mixture model were assumed. Label the curves.

(b) Calculate the fugacity of each component in the mixture as predicted by the virial equation, an ideal-mixture model, and the ideal-gas model. Prepare a table for each component, and list the three predicted fugacities in three columns for easy comparison. Calculate the values at $y_1 = 0.0, 0.2, 0.4, 0.6, 0.8, 1.0$.

10.9 The virial equation $Z = 1 + BP/RT$ may be used to calculate fugacities of components in mixtures. Suppose $B = y_1 B_{11} + y_2 B_{22}$. (This simple form makes calculations easier. Eqn. 10.1 gives the correct form.) Use this simplified expression and the correct form to calculate the respective fugacity coefficient formulas for component 1 in a binary mixture.

10.10 The Lewis-Randall rule is usually valid for components of high concentration in gas mixtures. Consider a mixture of 90% ethane and 10% propane at 125°C and 170 bar. Estimate $\hat{f}_i$ for ethane.

10.11 One of the easiest ways to begin to explore fugacities in non-ideal solutions is to model solubilities of crystalline solids dissolved in high pressure gases. In this case, the crystalline solids remain as a pure phase in equilibrium with a vapor mixture, the fugacity of the "solid" component must be the same in the crystalline phase as in the vapor phase. Consider biphenyl dissolved in carbon dioxide, using $k_{ij} = 0.100$. The molar volume of crystalline biphenyl is 156 cm³/mol.

(a) Calculate the fugacity (in MPa) of pure crystalline biphenyl at 310 K and 330 K and 0.1, 1, 10, 15, and 20 MPa.

(b) Calculate and plot the biphenyl solubility for the isotherm over the pressure range. Compare the solubility to the ideal gas solubility of biphenyl where the Poynting correction is included, but the gas phase non-idealities are ignored.

10.12 Repeat problem 10.11, except consider naphthalene dissolved in carbon dioxide, using $k_{ij} = 0.109$. The molar volume of crystalline naphthalene is 123 cm³/mol.

10.13 A vessel initially containing propane at 30°C is connected to a nitrogen cylinder, and the pressure is isothermally increased to 2.07 MPa. What is the mole fraction of propane in the vapor phase? You may assume that the solubility of N_2 in propane is small enough that the liquid phase may be considered pure propane. Calculate using the following data at 30°C.

	C_3	N_2
V^L (cm³/gmole)	75.6	
P^{sat} (MPa)	1.065	
B (cm³/mole)	−380	−4.0
B_{12} (cm³/mole)		−70

10.14 A 50-mol% mixture of propane(1) + n-butane(2) enters a flash drum at 37°C. If the flash drum is maintained at 0.6 MPa, what fraction of the feed exits as a liquid? What are the compositions of the phases exiting the flash drum? Work the problem the following two ways:

(a) Use Raoult's law.
(b) Assume ideal mixtures of vapor and liquid (K_i is independent of composition).

Data: $B_{11} = -369.5$ cm^3/mol $B_{22} = -665.1$ cm^3/mol $B_{12} = -486.9$ cm^3/mol

$P_1^{sat} = 1.269$ MPa $P_2^{sat} = 0.343$ MPa

10.15 A mixture containing 5 mol% ethane, 57 mol% propane, and 38 mol% n-butane is to be processed in a natural gas plant. Estimate the bubble-point pressure, the liquid composition, and K ratios of the coexisting vapor for this mixture at all pressures above 1 bar at which two phases exist. Set $k_{ij} = 0$. Use the shortcut K-ratio method. Plot ln P versus $1/T$ for your results. What does this plot look like? Plot log K_i versus $1/T$. What values do the K_i approach?

10.16 Vapor-liquid equilibria are usually expressed in terms of K factors in petroleum technology. Use the Peng-Robinson equation to estimate the values for methane and benzene in the benzene + methane system with equimolar feed at 300 K and a total pressure of 30 bar and compare to the estimates based on the shortcut K-ratio method.

10.17 Benzene and ethanol form azeotropic mixtures. Prepare a y-x and a P-x-y diagram for the benzene + ethanol system at 45°C assuming the Peng-Robinson model and using the experimental pressure at $x_E = 0.415$ to estimate k_{12}. Compare the results with the experimental data of Brown and Smith cited in problem 9.2.

10.18 A storage tank is known to contain the following mixture at 45°C and 15 bar on a mole basis: 31% ethane, 34% propane, 21% n-butane, 14% i-butane. What is the composition of the coexisting vapor and liquid phases, and what fraction of the contents of the tank is liquid?

10.19 The *CRC Handbook* lists the atmospheric pressure azeotrope for ethanol + methylethylketone at 74.8°C and 34 wt% ethanol. Estimate the value of the Peng-Robinson k_{12} for this system.

10.20 The *CRC Handbook* lists the atmospheric pressure azeotrope for methanol + toluene at 63.7°C and 72 wt% methanol. Estimate the value of the Peng-Robinson k_{12} for this system.

10.21 Use the Peng-Robinson equation for the ethane/heptane system.

(a) Calculate the P-x-y diagram at 283 K and 373 K. Use $k_{12} = 0$. Plot the results.
(b) Based on a comparison of your diagrams with what would be predicted by Raoult's law at 283 K, does this system have positive or negative deviations from Raoult's law?

10.22 One mol of n-butane and one mol of n-pentane are charged into a container. The container is heated to 90°C where the pressure reads 7 bar. Determine the quantities and compositions of the phases in the container.

10.23 Consider a mixture of 50 mol% n-pentane and 50 mol% n-butane at 15 bar.

(a) What is the dew temperature? What is the composition of the first drop of liquid?
(b) At what temperature is the vapor completely condensed if the pressure is maintained at 15 bar? What is the composition of the last drop of vapor?

10.24 LPG gas is a fuel source used in areas without natural gas lines. Assume that LPG may be modeled as a mixture of propane and *n*-butane. Since the pressure of the LPG tank varies with temperature, there are safety and practical operating conditions that must be met. Suppose the desired maximum pressure is 0.7 MPa, and the lower limit on desired operation is 0.2 MPa. Assume that the maximum summertime tank temperature is 50°C, and that the minimum wintertime temperature is −10°C. [Hint: On a mass basis, the mass of vapor within the tank is negligible relative to the mass of liquid after the tank is filled.]

(a) What is the upper limit (mole fraction) of propane for summertime propane content?
(b) What is the lowest wintertime pressure for this composition from part (a)?
(c) What is the lower limit (mole fraction) of propane for wintertime propane content?
(d) What is the highest summertime pressure for this composition from part (b)?

10.25 The k_{ij} for the pentane + acetone system has been fitted to a single point in problem P10.6 Generate a *P-x-y* diagram at 312.75 K.

10.26 The synthesis of methylamine, dimethylamine, and trimethylamine from methanol and ammonia results in a separation train involving excess ammonia and converted amines. Use the Peng-Robinson equation with $k_{ij} = 0$ to predict whether methylamine + dimethylamine, methylamine + trimethylamine, or dimethylamine + trimethylamine would form an azeotrope at 2 bar. Would the azeotropic behavior identified above be altered by raising the pressure to 20 bar? Locate experimental data relating to these systems in the library. How do your predictions compare to the experimental data?

Compound	T_c (K)	P_c (MPa)	ω
Methylamine	430.0	7.43	0.292
Dimethylamine	437.7	5.31	0.302
Trimethylamine	433.3	4.09	0.205

10.27 For the gas/solvent systems below, we refer to the "gas" as the low molecular weight component. Experimental solubilities of light gases in liquid hydrocarbons are tabulated below. The partial pressure of the light gas is 1.013 bar partial pressure. For each system:

(a) estimate the partial pressure of the liquid hydrocarbon by calculating the pure component vapor pressure via the Peng-Robinson equation, and by subsequently applying Raoult's law for that component
(b) estimate the total pressure and vapor composition using the results of step (a)
(c) use the Peng-Robinson equation with $k_{ij} = 0$ to calculate the vapor and liquid compositions that result in 1.013 bar partial pressure of the light gas and compare the pressure and gas phase composition with steps (a) and (b)
(d) Henry's law asserts that $\hat{f}_i = h_i x_i$, when x_i is near zero, and h_i is the Henry's law constant. Calculate the Henry's law constant from the calculations of part (c)
(e) calculate the solubility expected at 2 bar partial pressure of light gas by using Henry's law as well as by the Peng-Robinson equation and comment on the results.

	gas	liquid	T (°C)	x_{gas}	Source
(i)	methane	cyclohexane	25	2.83 E-3	1
(ii)	methane	carbon tetrachloride	25	2.86 E-3	1
(iii)	methane	benzene	25	2.07 E-3	1
(iv)	methane	n-hexane	25	3.15 E-3	2
	methane	n-hexane	25	4.24 E-3	3

1. J.H. Hildebrand, R.L. Scott, "The Solubility of Nonelectrolytes," 3rd ed., Table 4, pg. 243, Reinhold, New York, 1950
2. A.S. McDaniel, *J. Phys. Chem, 15*, 587 (1911)
3. D. Guerry, Jr., Thesis, Vanderbilt Univ., (1944)

10.28 Estimate the solubility of carbon dioxide in toluene at 25°C and 1 bar of CO_2 partial pressure using the Peng-Robinson equation with zero binary interaction parameter. The techniques of problem 10.27 may be helpful.

10.29 Oxygen dissolved in liquid solvents may present problems during use of the solvents.

(a) Using the Peng-Robinson equation and the techniques introduced in problem 10.27, estimate the solubility of oxygen in n-hexane at an oxygen partial pressure of 0.21 bar.
(b) From the above results, estimate the Henry's law constant.

10.30 Estimate the solubility of ethylene in n-octane at 1-bar partial pressure of ethylene and 25°C. The techniques of problem 10.27 may be helpful. Does the system follow Henry's law up to an ethylene partial pressure of 3 bar at this temperature? Provide the vapor compositions and total pressures for the above states.

10.31 Henry's law asserts that $\hat{f}_i = h_i x_i$, when x_i is near zero, and h_i is the Henry's law constant. Gases at high reduced temperatures can exhibit peculiar trends in their Henry's law constants. Use the Peng-Robinson equation to predict the Henry's law constant for hydrogen in decalin at $T = [300 \text{ K}, 600 \text{ K}]$. Plot the results as a function of temperature and compare to the prediction from the shortcut prediction. Describe in words the behavior that you observe.

10.32 A gas mixture follows the equation of state

$$\frac{PV}{RT} = 1 + \left(b - \frac{a}{T}\right)\frac{P}{RT}$$

where b is the size parameter, $b = \sum_i x_i b_i$ and a is the energetic parameter,

$a = \sum_i \sum_j x_i x_j a_{ij}$. Derive the formula for the partial molar enthalpy for component 1 in a

binary mixture, where the reference state for both components is the ideal gas state of T_R, P_R, and the pure component parameters are temperature-independent.

10.33 The procedure for calculation of the enthalpy departure for a pure gas is shown in Example 7.3 on page 240. Now consider the enthalpy departure for a binary gas mixture. For this calculation, it is necessary to determine da/dT for the mixture.

(a) Write the form of this derivative for a binary mixture in terms of da_1/dT and da_2/dT based on the conventional quadratic mixing rule and geometric mean combining rule with a non-zero k_{ij}.

(b) Provide the expression for the enthalpy departure for a binary mixture that follows the Peng-Robinson equation.

(c) A mixture of 50 mol% CO_2 and 50 mol% N_2 enters a valve at 7 MPa and 40°C. It exits the valve at 0.1013 MPa. Explain how you would determine whether CO_2 precipitates, and if so, whether it would be a liquid or solid.

10.34 A gaseous mixture of 30 mol% CO_2 and 70 mol% CH_4 enters a valve at 70 bar and 40°C and exits at 1.013 bar. Does any CO_2 condense? Assume that the mixture follows the virial equation. Assume that any liquid that forms is pure CO_2. The vapor pressure of CO_2 may be estimated by the shortcut vapor pressure equation. CO_2 sublimes at 0.1013 MPa and −78.8°C, although freezing is less likely.

10.35 The vapor-liquid equilibria for the system acetic acid(1) + acetone(2) needs to be characterized in order to simulate an acetic anhydride production process. Experimental data for this system at 760 mmHg have been reported by Othmer (1943)[1] as summarized below. Use the data at the equimolar composition to determine a value for the binary interaction parameter of the Peng-Robinson equation. Based on the value you determine for the binary interaction parameter, determine the percent errors in the Peng-Robinson prediction for this system at a mole fraction of $x_{(1)} = 0.3$.

T(C)	103.8	93.1	85.8	79.7	74.6	70.2	66.1	62.6	59.2
$x_{(1)}$	0.9	0.8	0.7	0.6	0.5	0.4	0.3	0.2	0.1

10.36 A mixture of methane and ethylene exists as a single gas phase in a spherical tank (10 m³) on the grounds of a refinery. The mixture is at 298 K and 1 MPa. It is a spring day, and the atmospheric temperature is also 298 K. The mole fraction of ethylene is 20 mol%. Your supervisor wants to draw off gas quickly from the bottom of the tank until the pressure is 0.5 MPa. However, being astute, you suggest that depressurization will cause the temperature to fall, and might cause condensation.

(a) Provide a method to calculate the change in temperature with respect to moles removed or tank pressure valid up until condensation starts. Assume the depressurization is adiabatic and reversible. Provide relations to find answers, and assure that enough equations are provided to calculate numerical values for all variables, but you do not need to calculate a final number.

(b) Would the answer in part (a) provide an upper or lower limit to the expected temperature?

(c) Outline how you could find the P, T, n of the tank where condensation starts. Provide relations to find answers, and assure that enough equations are provided to calculate numerical values for all variables, but you do not need to calculate a final number.

1. Gmehling, J., Onken, U., Arlt, W. Vapor-Liquid Equilibrium Data Collection, DECHEMA, Frankfurt/Main, 1977-.

10.37 (a) At 298 K, butane follows the equation of state: $P(V - b) = RT$ at moderate pressures, where b is a function of temperature. Calculate the fugacity for butane at a temperature of 298 K and a pressure of 1 MPa. At this temperature, $b = -732$ cm^3/mol.

(b) Pentane follows the same equation of state with $b = -1195$ cm^3/mol at 298 K. In a mixture, b follows the mixing rule: $b = x_1^2 b_1 + 2x_1 x_2 b_{12} + x_2^2 b_2$ where $b_{12} = -928$ cm^3/mol. Calculate the fugacity of butane in a 20 mol% concentration in pentane at 298 K and 1 MPa, assuming the mixture is in ideal solution.

10.38 The Soave equation of state is:

$$Z = \frac{1}{1 - b\rho} - \frac{a\rho}{RT} \frac{1}{(1 + b\rho)}$$

where the mixing and combining rules are given by Eqns. 10.9 and 10.10. Develop an expression for the fugacity coefficient and compare it to the expression given by Soave (*Chem. Eng. Sci.*, 1972, 27:1197).

10.39 The following equation of state has been proposed for hard sphere mixtures:

$$\frac{A - A^{ig}}{RT} = \frac{2}{(1 - \eta)^2}$$

where $\eta = \displaystyle\sum_i x_i b_i / V$

Derive an expression for the fugacity coefficient.

10.40 The equation of state below has been suggested. Derive the expression for the fugacity coefficient.

$Z = 1 + 4c\rho/(1 - b\rho)$

where $b = \displaystyle\sum_i x_i b_i$

$$c = \sum_i \sum_j x_i x_j c_{ij}; \quad c_{ij} = \text{constant}$$

10.41 The following free energy model has been suggested as part of a new equation of state for mixtures. Derive the expression for the fugacity coefficient of component 1.

$$\frac{A(T, V) - A^{ig}(T, V)}{RT} = -\frac{B^2}{C} \ln\left(1 + \frac{C\rho}{B}\right)$$

where

$$C = \sum_i \sum_j \sum_k x_i x_j x_k C_{ijk}$$

$$B = \sum_i \sum_j x_i x_j B_{ij}$$

CHAPTER

11

ACTIVITY MODELS

I have constructed three thousand different theories in connection with the electric light . . . Yet, in only two cases did my experiments prove the truth of my theory.

Thomas A. Edison

The subject of non-ideal solutions includes just about everything from aqueous acids to polymers to semiconductors. Not surprisingly, *there is no completely general model* for non-ideal solutions. But there are some popular approaches for specific situations like VLE of alcohols or LLE of solvents with water. We will discuss some of these models and provide rationalizations for their forms. Moreover, the challenge of developing accurate descriptions of non-ideal solution behavior means that model development is still an active research area. The presentation here should provide the necessary background to understand the rationales behind new developments as well as the old.

In Chapter 10, we showed the ideal-solution model of Chapter 9 to be a valid first approximation for more accurate calculations using fugacity coefficients that quantify deviations from the ideal-gas model. We demonstrated in Chapter 10 that the Peng-Robinson equation is capable of qualitative representation of systems like methanol + benzene, but the accuracy is not sufficient for process design. This is a general shortcoming of cubic equations in treating polar fluids.[1] In this chapter we will demonstrate how to represent polar systems with activity coefficient models which are typically applied to liquid phases to model deviations from ideal-solution behavior. Section 9.2 briefly shows how this approach is related to the other methods.

> *Keep in mind that a primary goal of this chapter is the same as Chapters 9 and 10— we want to model bubble, dew, flash, or other phase equilibria conditions. The iteration procedures will be very similar to the techniques set forth in Section 9.4, except that the K-ratio will be changed to incorporate the activity coefficients. A second goal is to infer activity coefficient parameters from experimental data.*

---

1. The reason for this is that the pure liquids themselves deviate so significantly from ideal-gas behavior that calculation of the deviations due to non-ideal mixing are a small correction superimposed on a large deviation from ideal-gas behavior. Polar fluids are not so easily characterized by random mixing rules used in many equations of state. Chapter 15, which requires significant background, covers advanced techniques for reliable calculation of fugacity coefficients of polar fluids using deviations from ideal-gas behavior.

As a brief overview of this chapter, the theories to be presented in Sections 11.4–11.5 are simply related to reduced forms of the van der Waals' equation. One basic simplification that occurs frequently is that we skip the calculation of molar volume since the experimental excess volume for liquids is usually near zero. The other theories introduced in Section 11.6 are based on the local composition concept. In both approaches, accurate representation of highly non-ideal solutions requires the introduction of at least two adjustable parameters. These adjustable parameters permit us to compensate for our ignorance in a systematic fashion. By determining reasonable values for the parameters from experimental data, we can interpolate between several measurements, and in some cases extrapolate to systems where we have no measurements. Learning how to determine reasonable values for the parameters and apply the final equations is an important part of this chapter. We also introduce the UNIFAC model, which is useful for predicting behavior when no experimental measurements are available.

11.1 EXCESS PROPERTIES

Excess volume.

The deviation of a property from its ideal-solution value is called the *excess* property. For a generic property M, the excess property is given the symbol M^E, and M^E is the value of the property for the mixture relative to the property for an ideal mixture, $M^E = M - M^{is}$. Ideal solutions have previously been discussed in Section 9.3. For example, the molar volume of an ideal solution is just the weighted sum of the molar volumes of the components, $V^{is} = \sum_i x_i V_i$. The excess volume is then, $V^E = V - V^{is} = V - \sum_i x_i V_i$. Although the excess volumes *of liquids* are typically a very small percentage of the volume, the concepts of excess properties are easily grasped by first studying the excess volume and then exploring the more abstract quantities of excess enthalpy, entropy, or Gibbs energy.

The excess volume of the system 3-pentanone (1) + 1-chlorooctane (2) at 298.15 K has been measured by Lorenzana, et al.,[1] and is shown in Fig. 11.1. The molar volumes of the pure components are $V_1 = 106.44$ cm³/mol and $V_2 = 171.15$ cm³/mol. At the equimolar concentration, the excess volume is 0.204 cm³/mol. Therefore the molar volume is $V = V^E + V^{is} = 0.204 + 0.5 \cdot 106.44 + 0.5 \cdot 171.15 = 139.00$ cm³/mol. The excess volume is only 0.15% of the total volume. The partial molar excess volume is calculated in a manner analogous to the partial molar volume,

$$\overline{V}_i^E = \left(\frac{\partial V^E}{\partial n_i} \right)_{T, P, n_{j \neq i}}$$. If an algebraic expression is available for the excess volume, it may be differentiated by this relation to yield formulas for the excess volumes. Graphically, the partial molar volumes at any point may be found by drawing the tangent line to the excess volume curve and reading the intercepts. At the composition shown at the tangent point in Fig. 11.1, the intercepts give $\overline{V}_1^E \approx 0.24$ cm³/mol and $\overline{V}_2^E \approx 0.17$ cm³/mol. The partial molar volumes depend on composition.

In directly analogous fashion, the excess Gibbs energy can be defined as the difference between the Gibbs energy of the mixture and the Gibbs energy of an ideal solution, $G^E = G - G^{is}$. Then, instead of speaking of partial molar volumes, we speak of partial molar Gibbs energies. But you should recognize the partial molar Gibbs energy as the chemical potential as introduced in section 9.1, and the significance of the chemical potential to phase equilibrium calculations should

1. Lorenzana, T., C. Franjo, E. Jiménez, J. Fernández, M.I. Paz-Andrade, *J. Chem. Eng. Data*, 39:172 (1994).

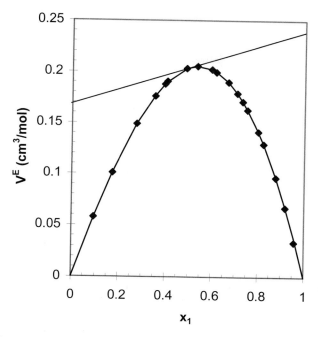

Figure 11.1 *Excess volume for the 3-pentanone (1) + 1-chlorooctane (2) system at 298.15 K.*

resonate strongly after reading Chapters 9 and 10. What remains is to rearrange the mathematics into the final working equations for *K*-ratios. These rearrangements are very similar to the analysis in Chapter 10, with minor differences. The thermodynamic property of interest is the excess Gibbs energy and the derivatives are taken at constant pressure instead of constant volume, so we do not need to interject the Helmholtz energy before deriving the chemical potential. This actually makes the derivations slightly simpler.

11.2   MODIFIED RAOULT'S LAW AND EXCESS GIBBS ENERGY

The relations that we use begin with the criteria $\hat{f}_i^{\,V} = \hat{f}_i^{\,L}$. For the vapor phase, we may use $\hat{f}_i^{\,V} = y_i \hat{\varphi}_i P$. For the liquid phase, we may use an activity coefficient, γ_i, giving $\hat{f}_i^{\,L} = x_i \gamma_i f_i^L$. Typically the Poynting method is used to calculate the pure-component liquid phase fugacities, $f_i^L = \varphi_i^{sat} P_i^{sat} \exp\!\left(\dfrac{V_i^L (P - P_i^{sat})}{RT}\right)$. Combining these expressions,

$$y_i \hat{\varphi}_i P = x_i \gamma_i \varphi_i^{sat} P_i^{sat} \exp\!\left(\frac{V_i^L (P - P_i^{sat})}{RT}\right)$$

This may be written in terms of the *K*-ratio, $K_i = y_i / x_i$,

$$K_i = \frac{\gamma_i^L P_i^{sat}}{P} \left[\frac{\varphi_i^{sat} \exp[V_i^L (P - P_i^{sat})\,/\,RT]}{\hat{\varphi}_i} \right] \qquad\qquad 11.1$$

At the low pressures of many chemical engineering processes the Poynting corrections and the ratios of $\dfrac{\varphi_i^{sat}}{\hat{\varphi}_i}$ for the components approach unity so we get

$$\boxed{K_i = \frac{\gamma_i^L P_i^{sat}}{P}} \quad \text{or} \quad \boxed{y_i P = x_i \gamma_i P_i^{sat}} \qquad 11.2$$

This is the form for the K-ratio encountered most often when dealing with non-ideal solutions. We will refer to this equation as the *Modified Raoult's Law*. Even though the activity coefficient is a kind of correction factor for non-ideal-solution behavior, the equations of thermodynamics assign it a unique mathematical meaning. The activity coefficient is related to the excess Gibbs energy. The excess Gibbs energy is

$$
\begin{aligned}
G^E &\equiv G - G^{is} \\
&= \left(G - \sum_i x_i G_i\right) - \left(G^{is} - \sum_i x_i G_i\right) \qquad 11.3 \\
&= \Delta G_{mix} - \Delta G_{mix}^{is} = \Delta G_{mix} - RT \sum_i x_i \ln(x_i)
\end{aligned}
$$

Recall that $G = \sum_i x_i \mu_i$, (Eqn. 9.8), and letting $\gamma_i^L \equiv \dfrac{\hat{f}_i^L}{x_i f_i^\circ}$, (Eqn. 9.31), where $f_i^\circ \equiv f_i$ for the pure fluid at T and P, we can develop an expression for ΔG_{mix},

$$\Delta G_{mix} = G - \sum_i x_i G_i = \sum_i x_i (\mu_i - G_i) = RT \sum_i x_i \ln\left(\frac{\hat{f}_i}{f_i^\circ}\right) = RT \sum_i x_i \ln(x_i \gamma_i) \qquad 11.4$$

Substituting into Eqn. 11.3

$$G^E \equiv \Delta G_{mix} - RT \sum_i x_i \ln(x_i) = RT \sum_i x_i \ln(x_i \gamma_i) - RT \sum_i x_i \ln(x_i) = RT \sum_i x_i \ln(\gamma_i) \quad 11.5$$

$$\boxed{G^E = RT \sum_i x_i \ln(\gamma_i)} \qquad 11.6$$

Note that the activity coefficients and excess Gibbs energy are coupled—when the activity coefficients of all components are unity, the excess Gibbs energy goes to zero. The excess Gibbs energy is zero for an ideal solution.

Activity coefficients as derivatives

We have seen in the last chapter that fugacity coefficients are related to derivatives of the Gibbs departure function. Similarly, activity coefficients are related to derivatives of the excess Gibbs energy, specifically the partial molar excess Gibbs energy. We have a very simple relation between partial molar quantities and molar quantities developed in Eqn. 9.7,

$$M = \sum_i x_i \overline{M}_i$$

Applying this relation to excess Gibbs energy,

$$G^E = \sum_i x_i \overline{G}_i^E$$

Comparing this with Eqn. 11.6, we see that

$$\boxed{\left(\frac{\partial \underline{G}^E}{\partial n_i}\right)_{T,P,n_{j \neq i}} = \overline{G}_i^E = RT\ln\gamma_i}$$

11.7   ❶ Activity coefficients are related to the partial molar excess Gibbs energy.

So, for any expression of $G^E(T,P,x)$, we can derive γ's as we will show in Example 11.4 on page 365.

Example 11.1   Activity coefficients and the Gibbs-Duhem relation (optional)

This example uses the Gibbs-Duhem Equation that was in optional Section 9.7 on page 313. Provide an alternate derivation of Eqn. 11.7 without directly using the Gibbs energy of a mixture and extend the Gibbs-Duhem relation to the excess properties and activity coefficients.

Solution:

The excess chemical potential will be used for the derivation since $\mu_i = \overline{G}_i$ and thus $\mu_i^E = \overline{G}_i^E$. Starting with Eqn. 9.12 and integrating from the pure state to the mixed state for component i, $\mu_i - \mu_{i,pure} = RT\ln(f_i/f_{i,pure})$. Integrating Eqn. 9.12 from the pure state to an ideal solution state,

$$\mu_i^{is} - \mu_{i,pure} = RT\ln(\hat{f}_i^{is}/f_{i,pure}) = RT\ln(x_i f_{i,pure}/f_{i,pure}) = RT\ln x_i$$

Therefore,

$$\mu_i^E = \mu_i - \mu_i^{is} = (\mu_i - \mu_{i,pure}) - (\mu_i^{is} - \mu_{i,pure})$$
$$= RT\ln(\hat{f}_i/(x_i f_{i,pure})) = RT\ln\gamma_i$$

and

$$\boxed{\mu_i^E = \overline{G}_i^E = RT\ln\gamma_i}$$

11.8

To extend the Gibbs-Duhem equation to excess properties, the excess Gibbs energy can be manipulated in an manner analogous to the derivation in Section 9.7. Therefore,

$$d\underline{G}^E = -\underline{S}^E dT + \underline{V}^E dP + \sum_i \mu_i^E dn_i$$

and performing analogous steps results in $\sum_i n_i d\mu_i^E = 0$ at constant T and P. Differentiating Eqn. 11.8 at constant T and P, and substituting,

$$\boxed{\sum_i n_i d\mu_i^E = RT\sum_i n_i d\ln\gamma_i = 0}$$

11.9

> ## Example 11.1 Activity coefficients and the Gibbs-Duhem relation (optional) (Continued)
>
> This equation means that the activity coefficients of components in an binary system, when plotted versus composition, must have slopes with opposite signs, and the slopes are related in magnitude by Eqn. 11.9. A further deduction is that if one of the activity coefficients in a binary system exhibits a maxima, the other must exhibit a minima at the same composition.

An Initial Glance at UNIFAC—a predictive method

This chapter is extremely long and densely packed with information. In the interest of "telling you what we are going to tell you," it is useful to see how the final equations are applied before being concerned with their derivations. One popular activity coefficient models is the predictive model of UNIFAC. It is the closest thing to a universally applicable predictive model that we currently have, so it makes sense to get right to the point and introduce the rudiments of implementing this model at an early stage. It is a rather complicated model and deriving it must await several other derivations. Nevertheless, the availability of a computer program for applying the method makes it possible to apply it to encourage you to understand it. As an introduction to activity coefficient methods, this is highly worthwhile. We can apply the program as a "black-box" at this stage, and the utility of the model should inspire us to learn more about it. Detailed calculations are illustrated in the UNIFAC spreadsheet in the file ACTCOEFF.xls, discussed in detail in Section 11.6. The UNIFAC model is also available as an HP calculator program.

 Actcoeff.xls, UNIFAC.

 UNIFAC

> ## Example 11.2 VLE prediction using UNIFAC activity coefficients
>
> The 2-propanol (also known as isopropyl alcohol or "*IPA*") + water (*W*) system is known to form an azeotrope at atmospheric pressure and 80.37°C ($x_W = 0.3146$).[1] Use UNIFAC to estimate the conditions of the azeotrope at 760 mmHg. Is UNIFAC accurate for this mixture?
>
> **Solution:** The Antoine coefficients for IPA and water are given in Appendix E. To begin, the VLE can be computed at the experimental conditions and see if it looks like there is an azeotrope nearby. By looking at Figs. 9.6 and 9.7, we can see that on one side of an azeotrope $x > y$ and on the other side of the azeotrope $x < y$. This is the principle that we will use to determine if an azeotrope exists. Also note that the azeotrope condition is at a maximum or minimum on the diagrams.
>
> The pressure is given as 760 mmHg. As a typical application, distillation columns often operate at nearly constant, atmospheric, pressure. Therefore, bubble-temperature calculations make sense for more than this sample calculation. We will develop a detailed description of the UNIFAC model later, but you need to know a little bit about it just to run the program. The UNIFAC model is based on structural and energetic information for the functional groups that comprise the molecules in the mixture. The UNIFAC model estimates the activity coefficients using the groups which comprise the molecules by calculating size, shape, and energy parameters based on the number and types of groups in the molecules.[2] The structures of the molecules for this problem and the UNIFAC groups are:

---

1. Perry, R.H., Chilton, C.H. *Chemical Engineers' Handbook*, 5 ed, Chapter 13, McGraw-Hill (1973).
2. The functional groups for a given molecule are often determined automatically by process design software. Several examples of group assignments are given in Table 11.3 on page 391.

Example 11.2   VLE prediction using UNIFAC activity coefficients (Continued)

	Water	IPA
	H2O - 1	CH3 - 2
		CH - 1
		OH - 1

The UNIFAC model can be operated from either the Excel spreadsheet or the HP calculator. The spreadsheet procedure is described here and the calculator method is described in Appendix P. To operate the spreadsheet program, simply type the temperature of interest (80.37) and the number of functional groups of each type in the appropriate column for each component. Enter the mole fractions, (e.g., [$x_W = 0.3146$, $x_{IPA} = 0.6854$]). The activity coefficients resulting are $\Rightarrow$ $\gamma_w = 2.1108$; $\gamma_{IPA} = 1.0885$. According to Modified Raoult's Law, $P = y_1 P + y_2 P = x_1 \gamma_1 P_1^{sat} + x_2 \gamma_2 P_2^{sat}$. By entering the proper Antoine coefficients, the pressure is computed automatically using this formula in the spreadsheet, and we can keep guessing temperatures (which changes γs and P^{sat}s) until the pressure equals 760 mm Hg (or we can apply the Excel solver tool $\Rightarrow T = 80.47°C$):

> Bubble temperature calculation.

T	P_{IPA}^{sat}	P_W^{sat}	$P = \sum_i x_i \gamma_i P_i^{sat}$	y_W
80.37	695	360	757	0.3158
82.50	760	395	829	0.3164
80.47	697	361	760	0.3158

The vapor phase mole fractions are calculated using $y_1 = (y_1 P)/P = (x_1 \gamma_1 P_1^{sat})/P$ and an analogous expression for the second component (or by using $y_2 = 1 - y_1$). Since the vapor and liquid compositions are not equal, $(x_w = 0.3146 \neq 0.3158 = y_w)$, we did not find the azeotrope. We must try several values of x to find the azeotrope composition.
try $x_w = 0.3177 \Rightarrow \gamma_W = 2.1035$; $\gamma_{IPA} = 1.0904$

T	P_{IPA}^{sat}	P_W^{sat}	$\sum_i x_i \gamma_i P_i^{sat}$	y_W
80.47	697	361	760	0.3177

Since $x_w = 0.3177 = y_w$, this is the composition of the azeotrope estimated by UNIFAC. UNIFAC seems to be fairly accurate for this mixture at the azeotrope. Also note that T versus x is fairly flat near an azeotrope; this is why it was unnecessary to modify the guess for bubble temperature at the new composition. This is generally true, and important in the distillation of azeotropic systems.

Activity Coefficients and Deviations from Raoult's Law

Bubble pressure calculation.

In Section 9.6, the text introduced concepts of positive and negative deviations from Raoult's law. For modified Raoult's law, the bubble pressure is calculated by $P = y_1 P + y_2 P = x_1 \gamma_1 P_1^{sat} + x_2 \gamma_2 P_2^{sat}$. When a system has positive deviations from Raoult's law, the bubble line lies above the Raoult's law bubble line ($P = x_1 P_1^{sat} + x_2 P_2^{sat}$), therefore $\gamma_i > 1$. When a system has negative deviations, $\gamma_i < 1$. Therefore, P-x-y data are related to the deviations of the activity coefficients from unity.

Comparison with Equation of State Methods

In the interest of providing a comparison of the activity (gamma) method with the equation of state method from the previous chapter, we can revisit the diagram illustrating the relationships between the levels of approximation in solution models as shown in Fig. 11.2. The distinctive feature of the activity coefficient approach is that it treats the liquids differently from the way it treats vapors. Liquids are treated in terms of deviations from ideal-solution behavior, and vapors are treated in terms of deviations (often negligible) from ideal-gas behavior. Since the vapor and liquid techniques are fundamentally different, it is awkward to model mixtures by activity coefficient methods at high pressure near the critical point where both phases become identical. This lack of generality detracts from the activity approach. On the other hand, most engineering applications are far from the critical point, so this shortcoming is usually less important than the improved accuracy that the approach provides.

See section 9.2 to review the relationships.

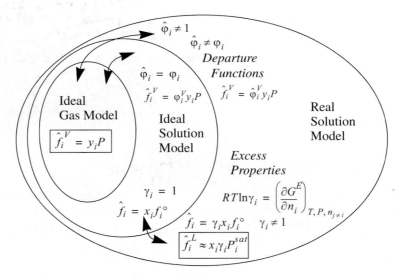

Figure 11.2 *Schematic of the relations between different fluid models. The solution model from an equation of state is applied to the vapor while the liquid is represented by a more flexible Gibbs excess energy model that is often tuned specifically for the mixture of interest. The boxed relations apply to Modified Raoult's Law.*

11.3   DETERMINATION OF G^E FROM EXPERIMENTAL DATA

Suppose that isothermal P-x-y data are available as well as the pure component vapor pressures. According to Modified Raoult's law,

$$\gamma_i = \frac{y_i P}{x_i P_i^{sat}} \qquad\qquad 11.10$$

Numerical values for all variables on the right-hand side are known from a single experiment, which permits the γ_i to be determined for each species. From the values of the activity coefficients, a value for excess Gibbs energy can be calculated:

$$\frac{G^E}{RT} = x_1 \ln\gamma_1 + x_2 \ln\gamma_2 \qquad\qquad 11.11$$

These values may be tabulated to provide G^E/RT vs. x_1 and the resulting curve can be regressed to fit a reasonable analytical expression for convenient interpolation at all compositions. A similar approach can be applied to isobaric data, but then the temperature dependence of the vapor pressures must be taken into account.

Example 11.3   Gibbs excess energy for system 2-propanol + water

Using data from the 2-propanol(1) + water(2) system presented in Fig. 9.6 calculate the excess Gibbs energy at $x_1 = 0.6369$. Data from the original citation provide $T = 30°C$, $P_1^{sat} = 60.7$ mmHg, $P_2^{sat} = 32.1$ mmHg, $y_1 = 0.6462$ when $x_1 = 0.6369$ at $P = 66.9$ mmHg.

Solution: Our approach will be to determine the activity coefficients and then relate them to the excess Gibbs energy,

$$\gamma_1 = \frac{y_1 P}{x_1 P_1^{sat}} = \frac{0.6462 \cdot 66.9}{0.6369 \cdot 60.7} = 1.118$$

$$\gamma_2 = \frac{y_2 P}{x_2 P_2^{sat}} = \frac{0.3538 \cdot 66.9}{0.3631 \cdot 32.1} = 2.031$$

Then,

$$\frac{G^E}{RT} = x_1 \ln\gamma_1 + x_2 \ln\gamma_2 = 0.6369 \ln (1.118) + 0.3631 \ln (2.031) = 0.328 \qquad 11.12$$

We could repeat the calculation for each data point, thus creating a plot of G^E versus x_1 like the one shown in Fig. 11.3. The plot also shows three models to be discussed in upcoming examples.

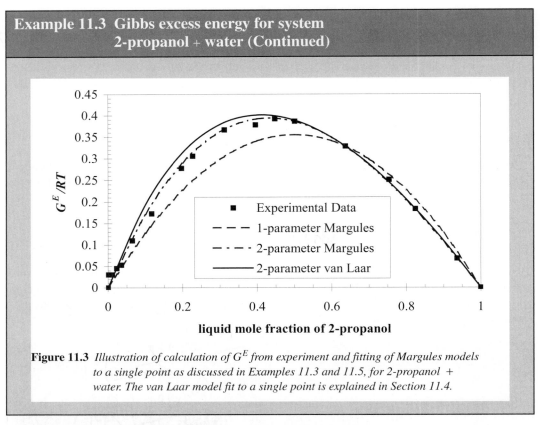

Example 11.3 Gibbs excess energy for system 2-propanol + water (Continued)

Figure 11.3 *Illustration of calculation of G^E from experiment and fitting of Margules models to a single point as discussed in Examples 11.3 and 11.5, for 2-propanol + water. The van Laar model fit to a single point is explained in Section 11.4.*

Fitting the Excess Gibbs Energy—Overview

Excess Gibbs energy models fitted to data like those in Fig. 11.3 provide the capability to take VLE data at a limited number of compositions (as few as one) and extrapolate behavior across the composition range. After developing a model for the composition dependence of G^E, the expression can be differentiated using Eqn. 11.7 to obtain expressions for the activity coefficients as will be shown in upcoming examples. The activity coefficient models resulting from these techniques enable a broad range of engineering analyses. For example, we may wish to design a distillation column that operates at constant pressure and requires *T-x-y* data. However, the available VLE data may exist only as constant temperature *P-x-y* data. We may use the activity coefficient models to convert isothermal *P-x-y* data to isobaric *T-x-y* data, and vice versa. Furthermore, parameters from binary data can be combined and extended to multicomponent systems, even if no multicomponent data are available. Elementary techniques for fitting G^E (or activity coefficient) models are presented in this section and Section 11.4. Advanced techniques for fitting G^E are presented in specific examples and Section 11.7.

The One-Parameter Margules Equation

The one-parameter Margules is the simplest excess Gibbs expression.

Perhaps the simplest expression for the Gibbs excess function is the one-parameter Margules equation (also known as the two-suffix Margules equation).

$$\frac{G^E}{RT} = Ax_1x_2 \qquad\qquad 11.13$$

Note: The parameter A is a constant which is not associated with the other uses of the variable (equation of state parameters, Helmholtz energy, Antoine coefficients) in the text. The parameter A is typically used in discussions of the Margules equation, so we use it here.

The one-parameter Margules equation is symmetrical with composition and will have an extremum at $x_1 = 0.5$ in a binary system.

Example 11.4 Activity coefficients by the one-parameter Margules equation

Derive the expressions for the activity coefficients for the one-parameter Margules equation.

Solution:

$$\frac{G^E}{RT} = An_2\frac{n_1}{n} \qquad 11.14$$

Applying Eqn. 11.7

$$\frac{1}{RT}\left(\frac{\partial G^E}{\partial n_1}\right)_{T,P,n_{j\neq1}} = An_2\left[\frac{1}{n} - \frac{n_1}{n^2}\right] = A\frac{n_2}{n}\left[1 - \frac{n_1}{n}\right] = Ax_2(1 - x_1) \qquad 11.15$$

$$\Rightarrow \ln\gamma_1 = Ax_2^2 \qquad 11.16$$

Similarly,

$$\Rightarrow \ln\gamma_2 = Ax_1^2 \qquad 11.17$$

As we have seen in previous examples, the activity coefficients may be found from experimental data. From a single data point, if we try to determine a single value of A, Eqns. 11.16 and 11.17 will be overspecified. Each can be solved independently for A, and the average taken as one method of fitting. Another option is to use the activity coefficients to calculate excess Gibbs energy, and then fit Eqn. 11.14.

Example 11.5 VLE predictions from the Margules one-parameter equation

The data used for Fig. 9.5 from Udovendo et al. for 2-propanol(1) + water(2) at 30°C show $x_1 = 0.1168$ and $y_1 = 0.5316$ at $P = 60.3$ mmHg. What are the pressure and vapor phase compositions predicted by the one-parameter Margules equation (Example 11.4) at this liquid composition based on the fit of G^E at the composition from Example 11.3?

Solution: At $x_1 = 0.6369$, we found in Example 11.3 that $G^E/RT = 0.328$. Fitting the Margules model,

$$\frac{G^E}{RT} = Ax_1x_2 \Rightarrow A = 0.328/[(0.6369)(0.3631)] = 1.42$$

Example 11.5 VLE predictions from the Margules one-parameter equation (Continued)

Now, at the new composition,

$$x_1 = 0.1168, \text{ we find } \ln \gamma_1 = Ax_2^2 = 1.42(0.8832)^2 = 1.107, \Rightarrow \gamma_1 = 3.03$$

$$\ln \gamma_2 = Ax_1^2 = 1.42(0.1168)^2 = 0.0194 \Rightarrow \gamma_2 = 1.02$$

Note that these activity coefficients differ substantially from those calculated in Example 11.3 because the liquid composition is different.

Therefore, substituting into Modified Raoult's Law to perform a bubble pressure calculation:

$$y_1 P = x_1 \gamma_1 P_1^{sat} = (0.1168)(3.03)(60.7) = 21.48 \text{ mmHg}$$

$$y_2 P = x_2 \gamma_2 P_2^{sat} = (0.8832)(1.02)(32.1) = 28.92 \text{ mmHg}$$

$$P = y_1 P + y_2 P = 50.4 \text{ mmHg}$$

$$y_1 = y_1 P/P = 21.48/50.4 = 0.426$$

Therefore, we underestimate the pressure and the vapor composition of y_1 is too low, but the use of one measurement and one parameter is a great improvement of Raoult's law. The estimation can be compared with the data by repeating the bubble pressure calculation across the composition range; the results are shown in Fig. 11.4. The excess Gibbs energy model using $A = 1.42$ is shown in Fig. 11.3, where we note that the model fails to capture the skewness of the excess Gibbs energy curve.

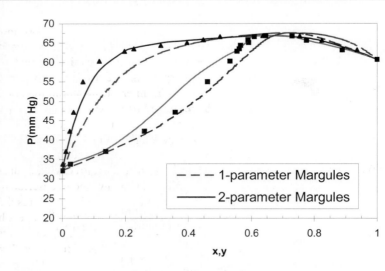

Figure 11.4 *Comparison of the one-parameter and two-parameter Margules equation fitted to a single measurement from Example 11.5 (and following) with the experimental data (points) from Fig. 9.6 on page 311.*

Bubble pressure calculation.

Two-Parameter Margules Equation

Representation of the skewness of experimental excess Gibbs energy is possible with the two-parameter Margules equation

$$\frac{G^E}{RT} = x_1 x_2 (A_{21} x_1 + A_{12} x_2) \qquad 11.18$$

$$\ln\gamma_1 = x_2^2 [A_{12} + 2(A_{21} - A_{12}) x_1] \qquad \ln\gamma_2 = x_1^2 [A_{21} + 2(A_{12} - A_{21}) x_2] \qquad 11.19$$

where A_{21} and A_{12} are fitted to experiment. Note that if $A_{21} = A_{12}$, the expression reduces to the one-parameter model. The two parameters can be fitted to a single VLE measurement using

$$A_{12} = \left(2 - \frac{1}{x_2}\right) \frac{\ln\gamma_1}{x_2} + \frac{2\ln\gamma_2}{x_1} \qquad A_{21} = \left(2 - \frac{1}{x_1}\right) \frac{\ln\gamma_2}{x_1} + \frac{2\ln\gamma_1}{x_2} \qquad 11.20$$

where the activity coefficients are calculated from the VLE data. Extreme care must be used before accepting the values from Eqn. 11.20 applied to a single measurement because experimental errors can occasionally result in questionable parameter values. Eqn. 11.20 applied to the activity coefficients values calculated in Example 11.3 results in $A_{12} = 1.99$, $A_{21} = 1.09$, which provides the representation of G^E shown in Fig. 11.3 and the VLE modeling of Fig. 11.4 designated as the two-parameter model. One disadvantage of the Margules equation, compared to other models to be discussed in upcoming sections, is the lack of theoretical justification for Eqn. 11.18. More fundamental expressions for the excess Gibbs energy are desirable for extending binary data fits to multicomponent applications.

11.4 THE VAN DER WAALS' PERSPECTIVE

Recognizing the significance of the Gibbs excess function, it should not be surprising that many researchers have studied its behavior and developed equations which can represent its various shapes. In essence, these efforts attempt to apply the same reasoning for mixtures that was so successful for pure fluids in the form of the van der Waals equation. The resulting expressions contain parameters which are intended to characterize the molecular interactions within the context of the theory. The values of these parameters must then be regressed from experimental data and the utility of the theory is judged by how accurately the experimental data are correlated. Given that molecules in solution must actually interact according to some single set of laws of nature, one might wonder why there are so many different theories. The problem with mixtures is that there are many different kinds of interactions occurring simultaneously, e.g., disperse attractions, hydrogen bonding, and size asymmetries. As a result, many specific terms must be invoked to describe these many specific interactions. If all the interactions were to be treated in their entirety, the resulting theory would be too unwieldy to be of any practical use. Therefore, many different researchers have made different approximations and developed different theories. Each theory has its proponents, and it is difficult even to describe the various theories without expressing personal prejudices. For practical applications, the perspective we will adopt is that these are empirical equations which can usually fit the data, and that extrapolations beyond the available experimental data must be performed at some risk.

"Regular" Solutions

The Gibbs energy is comprised of two contributions, energetic and entropic. Suppose we were to assume ideal mixing or nearly ideal mixing as a model of the entropic contribution, and concentrate on accurate consideration of the energetic contributions. In further developments, we also make the somewhat related assumption that the solution volume follows ideal mixing rules. These simple suppositions provide all of what is essential in developing the theories of van Laar or Scatchard and Hildebrand. We will call these "regular-solution theories." In this way, we can cover these two theories in short order.

The energetics of mixing are described by the same energy equation for mixtures that we previously developed in discussing the simple basis for mixing rules.

$$U - U^{ig} = -\rho \sum_i \sum_j x_i x_j a_{ij}$$

Noting that $1/\rho = V = \sum_i x_i V_i$ according to "regular-solution theory,"

$$\left(U - U^{ig}\right) = \frac{-\sum_i \sum_j x_i x_j a_{ij}}{\sum_i x_i V_i} \qquad 11.21$$

For the pure fluid, taking the limit as $x_i \to 1$,

$$\left(U - U^{ig}\right)_i = \frac{-a_{ii}}{V_i} \Rightarrow \left(U - U^{ig}\right)^{is} = -\sum_i \left(x_i a_{ii}/V_i\right) \qquad 11.22$$

For a binary mixture, subtracting the ideal solution result to get the excess energy gives,

$$U^E = x_1 \frac{a_{11}}{V_1} + x_2 \frac{a_{22}}{V_2} - \left(\frac{x_1^2 a_{11} + 2x_1 x_2 a_{12} + x_2^2 a_{22}}{x_1 V_1 + x_2 V_2}\right) \qquad 11.23$$

Collecting over a common denominator

$$U^E = \frac{x_1 \frac{a_{11}}{V_1}(x_1 V_1 + x_2 V_2) + x_2 \frac{a_{22}}{V_2}(x_1 V_1 + x_2 V_2) - (x_1^2 a_{11} + 2x_1 x_2 a_{12} + x_2^2 a_{22})}{x_1 V_1 + x_2 V_2} \qquad 11.24$$

$$U^E = \frac{x_1^2 a_{11} + x_1 x_2 a_{11} \frac{V_2}{V_1} + x_2^2 a_{22} + x_1 x_2 a_{22} \frac{V_1}{V_2} - (x_1^2 a_{11} + 2x_1 x_2 a_{12} + x_2^2 a_{22})}{x_1 V_1 + x_2 V_2} \qquad 11.25$$

$$U^E = \frac{x_1 x_2 a_{11} \frac{V_2}{V_1} + x_1 x_2 a_{22} \frac{V_1}{V_2} - 2 x_1 x_2 a_{12} \frac{V_2 V_1}{V_1 V_2}}{x_1 V_1 + x_2 V_2}$$

$$= \frac{x_1 x_2 V_1 V_2}{x_1 V_1 + x_2 V_2} \left(\frac{a_{11}}{V_1^2} + \frac{a_{22}}{V_2^2} - 2 \frac{a_{12}}{V_1 V_2} \right) \qquad \text{11.26}$$

The van Laar Equation

We can simplify the equation for the excess internal energy by arbitrarily defining a single symbol, "Q," to represent the final term in the equation:

$$U^E = \frac{x_1 x_2 V_1 V_2}{x_1 V_1 + x_2 V_2} Q \quad \text{where} \quad Q \equiv \left(\frac{\sqrt{a_{11}}}{V_1} - \frac{\sqrt{a_{22}}}{V_2} \right)^2$$

It would appear that this equation contains three unknown parameters (V_1, V_2, and Q), but van Laar recognized that it could be rearranged such that only two adjustable parameters need to be determined.

$$A_{12} = \frac{Q V_1}{RT} ; \quad A_{21} = \frac{Q V_2}{RT} ; \quad \frac{A_{12}}{A_{21}} = \frac{V_1}{V_2}$$

with the final result:

$$\boxed{\frac{G^E}{RT} = \frac{U^E}{RT} = \frac{A_{12} A_{21} x_1 x_2}{(x_1 A_{12} + x_2 A_{21}}} \qquad \text{11.27}$$

Differentiating using Eqn. 11.7 gives expressions for the activity coefficients,

$$\boxed{\ln \gamma_1 = \frac{A_{12}}{\left[1 + \frac{A_{12} x_1}{A_{21} x_2} \right]^2}} ; \qquad \boxed{\ln \gamma_2 = \frac{A_{21}}{\left[1 + \frac{A_{21} x_2}{A_{12} x_1} \right]^2}} \qquad \text{11.28}$$

🔴 van Laar Model.

Note: the parameter A_{12} and A_{21} for the van Laar and Margules equations have different values for the same data. Do not interchange them.

When applied to binary systems, it is useful to note that these equations can be rearranged to obtain A_{12} and A_{21} from γ_1 and γ_2 given any one VLE point. This is the simple manner of estimating the parameters that we will generally apply in this chapter. Methods of fitting the parameters in optimal fashion for many data are covered in Section 11.7.

$$A_{12} = (\ln \gamma_1) \left[1 + \frac{x_2 \ln \gamma_2}{x_1 \ln \gamma_1} \right]^2 \qquad A_{21} = (\ln \gamma_2) \left[1 + \frac{x_1 \ln \gamma_1}{x_2 \ln \gamma_2} \right]^2 \qquad \text{11.29}$$

Extreme care must be used before accepting the values of Eqn. 11.29 applied to a single measurement because experimental errors can occasionally result in questionable parameter values. Eqns. 11.29 applied to the activity coefficients from Example 11.3 result in $A_{12} = 2.38$, $A_{21} = 1.15$ and G^E is plotted in Fig. 11.3.

A particularly useful data point for VLE is the azeotrope because

1. $x_1 = y_1 \Rightarrow \gamma_1 = P/P_1^{sat}$; $\gamma_2 = P/P_2^{sat}$

The azeotrope is a useful point to fit parameters.

2. Many tables of known azeotropes are commonly available.

3. The location of an azeotrope is very important for distillation design because it represents a point at which further purification in a single distillation column is impossible.

HP γ

Example 11.6 Application of the van Laar equation

Consider the benzene(1) + ethanol(2) system which exhibits an azeotrope at 760 mmHg and 68.24°C containing 44.8 mole% ethanol. Calculate the composition of the vapor in equilibrium with an equimolar liquid solution at 760 mmHg given the Antoine constants,

$$\log P_1^{sat} = 6.87987 - 1196.76/(T + 219.161)$$

$$\log P_2^{sat} = 8.1122 - 1592.86/(T + 226.18).$$

Solution:

at $T = 68.24°C$, $P_1^{sat} = 519.7$ mmHg; $P_2^{sat} = 503.5$ mmHg

$\gamma_1 = 760/519.7 = 1.4624$; $\gamma_2 = 760/503.5 = 1.5094$

$x_1 = 0.552$; $x_2 = 0.448$

Using Eqn. 11.29, $A_{12} = 1.3421$, $A_{21} = 1.8810$

Now consider $x_1 = x_2 = 0.5$. Using Eqn. 11.28, $\gamma_1 = 1.579$; $\gamma_2 = 1.386$

Problem statement $\Rightarrow$ bubble-temperature is required

Bubble temperature calculation.

Guess $T = 60°C \Rightarrow P_1^{sat} = 391.6$ mmHg; $P_2^{sat} = 351.8$ mmHg

$y_i = x_i \gamma_i \ P_i^{sat}/P \Rightarrow y_1 = 0.407$; $y_2 = 0.321$; $\sum_i y_i = 0.728 \Rightarrow T_{guess}$ is too low.

at $T = 68.24°C$, $P_1^{sat} = 519.7$ mmHg; $P_2^{sat} = 503.5$ mmHg

$y_i = x_i \gamma_i \ P_i^{sat}/P \Rightarrow y_1 = 0.540$; $y_2 = 0.459$; $\sum_i y_i = 0.999 \Rightarrow T_{guess}$ is practically $T_{azeotrope}$.

A couple of important points can be inferred from this example:

1. The T,x diagram is fairly flat near an azeotrope. This has an important effect on temperature profiles in distillation columns.

2. The system has a minimum boiling azeotrope. The system has positive deviations from Raoult's Law. Due to the molecular nature of the components, they have a greater fugacity (escaping tendency) from the liquid solution than they would have in an ideal solution. The activity coefficients are greater than one.

Taking the composition limits of Eqn. 11.28, the van Laar parameters can be estimated from the infinite dilution activity coefficients as $A_{12} = \ln\gamma_1^\infty$, $A_{21} = \ln\gamma_2^\infty$. This approach can be useful if no data are available near the composition range of interest, but it should be recalled that extrapolations are less reliable than interpolations. In other words, one might experience significant errors in predictions of bubble-pressures near equimolar compositions when basing van Laar parameters on infinite dilution activity coefficients.

Example 11.7 Infinite dilution activity coefficients from van Laar theory

n-Propyl alcohol (1) forms an azeotrope with toluene (2) at 60 mol% NPA, 92.6°C, and 760 mmHg. Use the van Laar model to estimate the infinite dilution activity coefficients of these two species at this temperature.

Solution: The vapor pressures are:

$$P_1^{sat} = 10\wedge(7.74416 - 1437.69/(92.6 + 198.463)) = 637.86 \text{ mmHg}$$

$$P_2^{sat} = 10\wedge(6.95087 - 1342.31/(92.6 + 219.187)) = 442.24 \text{ mmHg}$$

Applying the azeotropic data gives: $\gamma_1 = P/P_1^{sat} = 760/637.86 = 1.191$; $\gamma_2 = P/P_2^{sat} = 1.719$; Eqn. 11.29 gives: $A_{12} = 1.643$; $A_{21} = 1.193$;

Taking the limits of Eqn. 11.28 as the respective components approach zero composition results in: $\ln\gamma_1^\infty = A_{12}$ and $\gamma_1^\infty = 5.17$; Similarly $\gamma_2^\infty = 3.30$.

Note: $\lim\limits_{x_i \to 1} \gamma_i \approx 1.00$; *the value of* $\lim\limits_{x_i \to 0} \gamma_i$ *is the maximum or minimum value in the majority of homogeneous systems.*

Scatchard-Hildebrand Theory

Returning to Eqn. 11.26, Scatchard and Hildebrand made an assumption which is very similar to assuming $k_{ij} = 0$ in an equation of state. Setting $a_{12} = \sqrt{a_{11}a_{22}}$, and collecting terms,

$$U^E = \frac{x_1 x_2 V_1 V_2}{x_1 V_1 + x_2 V_2}\left(\frac{a_{11}}{V_1^2} + \frac{a_{22}}{V_2^2} - 2\sqrt{\frac{a_{11}}{V_1^2}\frac{a_{22}}{V_2^2}}\right) = \frac{x_1 x_2 V_1 V_2}{x_1 V_1 + x_2 V_2}\left(\frac{\sqrt{a_{11}}}{V_1} - \frac{\sqrt{a_{22}}}{V_2}\right)^2 \qquad 11.30$$

Scatchard and Hildebrand recognized the unknown parameters in terms of volume fractions and disperse attraction energies that could be related to the pure component values. Defining a term called the "solubility parameter"

$$U^E = \Phi_1\Phi_2(\delta_1 - \delta_2)^2(x_1 V_1 + x_2 V_2) \qquad 11.31$$

where $\Phi_i \equiv x_i V_i / \sum\limits_i x_i V_i$ is known as the "volume fraction" $\qquad 11.32$

$\delta_i \equiv \sqrt{a_{ii}} / V_i$ is known as the "solubility parameter" $\qquad 11.33$

To estimate the value of δ_i, Scatchard and Hildebrand suggested that experimental data be used such that

$$\delta_i \equiv \sqrt{\frac{\Delta U^{vap}}{V_i}} = \sqrt{\frac{\Delta H^{vap} - RT}{V_i}} \qquad 11.34$$

(*Note the units on the "a" parameter and the way V_i moves inside the root.*)

In other words, δ_i is assumed to provide a standardized measure of the attractive energy density for each component. As long as a standard reference condition is used, any convenient set of ΔU^{vap} and V_i may be applied. A convenient set of conditions which has become customary is the saturated liquid at 298 K. On this basis, we can tabulate a fair number of solubility parameters for ready reference, as shown in Table 11.1. Note that many similar compounds have similar values for their solubility parameters. Since similar solubility parameters yield small excess energies, solutions of similar components will be nearly ideal, as we would intuitively expect. By scanning the tables for the values of solubility parameters, we can quickly estimate whether the ideal solution model will be accurate or not. This approach gives a quantitative flavor to the old axiom "like dissolves like."

Table 11.1 *Solubility parameters in $(cal/cm^3)^{1/2}$ and molar volumes (cm^3/mol) for various substances as liquids at 298 K.*

1-Olefins	δ	V^L	Naphthenics	δ	V^L	Aromatics	δ	V^L
1-pentene	6.9	109	cyclopentane	8.7	93	benzene	9.2	88
1-hexene	7.4	124	cyclohexane	8.2	107	toluene	8.9	106
1,3 butadiene	7.1	86	Decalin	8.8	156	ethylbenzene	8.8	122
Amines	δ	V^L	**Ketones**	δ	V^L	styrene	9.3	114
ammonia	16.3	28	acetone	9.9	73	*n*-propylbenzene	8.6	139
methyl amine	11.2	46	2-butanone	9.3	89	anthracene	9.9	145
ethyl amine	10.0	65	2-pentanone	8.7	106	phenanthrene	9.8	186
pyridine	14.6	80	2-heptanone	8.5	139	naphthalene	9.9	125
n-Alkanes	δ	V^L	**Alcohols**	δ	V^L	**Ethers**	δ	V^L
n-pentane	7.0	114	water	23.4	18	dimethyl ether	8.8	68
n-hexane	7.3	130	methanol	14.5	40	diethyl ether	7.4	103
n-heptane	7.4	145	ethanol	12.5	58	dipropyl ether	7.8	136
n-octane	7.6	162	*n*-propanol	10.5	74	furan	9.4	72
n-nonane	7.8	177	*n*-butanol	13.6	91	THF	9.1	81
n-decane	7.9	194	*n*-hexanol	10.7	124			
			n-dodecanol	9.9	222			

Turning to the Gibbs energy, with the elimination of excess entropy and excess volume at constant pressure, we have,

$$G^E = U^E = \Phi_1 \Phi_2 (\delta_1 - \delta_2)^2 (x_1 V_1 + x_2 V_2) \qquad 11.35$$

And the resulting activity coefficients are

$$RT \ln \gamma_1 = V_1 \Phi_2^2 (\delta_1 - \delta_2)^2$$

11.36   ❶ Scatchard-Hildebrand Theory.

$$RT \ln \gamma_2 = V_2 \Phi_1^2 (\delta_1 - \delta_2)^2$$

11.37

Example 11.8 VLE predictions using regular-solution theory

🖫 Actcoeff.xls, REGULAR

Benzene and cyclohexane are to be separated by distillation at 1 bar. Use regular solution theory to predict whether an azeotrope should be expected for this mixture.

Solution: Consider y_B where $x_B = 0.01$ and 0.99. If $y_B > x_B$ at $x_B = 0.01$ and $y_B < x_B$ at $x_B = 0.99$, then $y_B = x_B$ (i.e., there is an azeotrope) somewhere in between. If $y_B > x_B$ or $y_B < x_B$ for all x_B, then there is no azeotrope. Given x_B and P, we should perform bubble-temperature calculations.

Using parameters from Table 11.1, at $x_B = 0.99$, guess $T = 350$ K

$$\Rightarrow \quad \Phi_B = 0.99(88)/[0.99(88) + 0.01(107)] = 0.9879$$

Calculating vapor pressures:

$$P_B^{sat} = 10^{\wedge}(6.87982 - 1196.76/(76.85 + 219.161)) = 686.9 \text{ mmHg}$$

$$P_C^{sat} = 10^{\wedge}(7.26475 - 1434.15/(76.85 + 246.721)) = 680 \text{ mmHg}$$

Applying equations 11.36 and 11.37,

$$\ln\gamma_B = 88(1 - 0.9879)^2(9.2 - 8.2)^2/1.987(350) = 1.853\text{E-}5 \Rightarrow \gamma_B = 1.00002$$

$$\ln\gamma_C = 107(0.98789)^2 (9.2 - 8.2)^2/1.987(350) = 0.1502 \Rightarrow \gamma_C = 1.162$$

$$\sum_i y_i P = \sum_i x_i \gamma_i P_i^{sat} = 0.99(686.9)1.00 + 0.01(680.0)1.162 = 687 \text{ mmHg}$$

$$\Rightarrow y_B = 0.99(686.9)1.00/687 = 0.895, \ y_C = 0.01(680.0)1.162/687 = 0.010$$

Since $\sum_i y_i < 1$, we must guess a higher temperature.

Guess $T = 354$ K $\Rightarrow P_B^{sat} = 777.7$; $P_C^{sat} = 770.2$; $\gamma_B = 1.00$; $\gamma_C = 1.160$

$$\sum_i y_i P = \sum_i x_i \gamma_i P_i^{sat} = 0.99(777.7)1.00 + 0.01(770.2)1.160 = 778.9$$

$$\Rightarrow y_B = 1.013, \ y_C = 0.0113$$

$$T \approx 350 + 4(760 - 687)/(778.9 - 687) = 353.2$$

$$P_B^{sat} = 758.8; \ P_C^{sat} = 751.5; \gamma_B = 1.00; \ \gamma_C = 1.1604$$

$$\sum_i y_i P = 0.99(758.8)1.0 + 0.01(751.5)1.160 = 760 \text{ mmHg}$$

$$\Rightarrow y_B = 0.9885 < x_B = 0.99$$

❶ Bubble temperature calculation.

Example 11.8 VLE predictions using regular-solution theory (Continued)

At $x_B = 0.01$, guess $T = 353.5$ K $\Rightarrow \Phi_B = 0.01(88)/[0.01(88) + 0.99(107)] = 0.0082$

$\ln\gamma_C = 107(0.0082)^2(9.2 - 8.2)^2/1.987(353.5) \approx 0 \quad \Rightarrow \gamma_C = 1.00$

$\ln\gamma_B = 88(1 - 0.0082)^2(9.2 - 8.2)^2/1.987(353.5) = 0.1232 \Rightarrow \gamma_B = 1.131$

$$\sum_i y_i P = \sum_i x_i\gamma_i P_i^{sat} = 0.01(765.9)1.131 + 0.99(758.4)1.00 = 759.5 \text{ mmHg}$$

$\Rightarrow y_B = 0.011 > x_B = 0.01$

Therefore, $(y_B - x_B)$ changes sign between 0.01–0.99 so the system has an azeotrope.

Note:

1. *γ is a strong function of composition but a weak function with respect to temperature.*

2. *If $0.95 < \sum_i y_i < 1.05$, then $y_i = x_i\gamma_i P_i^{sat}/(P\sum_i y_i)$ is a good estimate for y_i even though the pressure is inaccurate.*

3. *If $P_B^{sat} \approx P_C^{sat}$, then a small non-ideality can cause an azeotrope.*

When the Scatchard-Hildebrand solution theory is used, the $\{\delta_i\}$ and $\{V_i\}$ are available directly from pure component data, and in principle, there are no adjustable parameters. The theory is entirely predictive. The van Laar theory, on the other hand, generally treats both A_{12} and A_{21} as adjustable parameters. Ignoring the volumetric ratios as well as the energetic parameter permits adjusting the skewness of the excess free energy as well as the magnitude. You should note this behavior in performing the homework. We can also obtain a compromise by assuming

$$a_{12} = \sqrt{a_{11}a_{22}} \, (1 - k_{ij})$$

where k_{ij} is an adjustable parameter called the binary interaction parameter, just as it was for equations of state. The activity coefficient expressions become:

$$RT \ln \gamma_1 = V_1\Phi_2^2\left[(\delta_1 - \delta_2)^2 + 2k_{12}\delta_1\delta_2\right] \quad\quad RT \ln \gamma_2 = V_2\Phi_1^2\left[(\delta_1 - \delta_2)^2 + 2k_{12}\delta_1\delta_2\right] \quad 11.38$$

In mixtures of compounds that deviate moderately from ideal-solution behavior, the Scatchard-Hildebrand solution theory with binary interaction parameters can be extremely helpful. The binary interaction parameter in those cases serves to adjust the magnitude of the excess Gibbs energy without addressing the skewness directly. Large deviations, however, are generally accompanied by non-ideal mixing in the volume and entropy. In those cases, the van Laar equations can often be useful in correlation, but the physical meaning behind the parameters is generally lost.

Example 11.9 Scatchard-Hildebrand versus van Laar theory for methanol + benzene

In Chapter 10 we illustrated the role of the binary interaction parameter in the Peng-Robinson equation. Return to that system to illustrate the roles of the estimated parameters in the Scatchard-Hildebrand and van Laar theories. In all cases, estimate the binary parameters by matching the azeotropic pressure (and the composition in the case of the van Laar two-parameter model). The azeotrope appears at 58.3°C and $x_m = 0.614$. The vapor pressures at 58.3°C are 591.3 mmHg for methanol, 368.7 mmHg for benzene.

Solution: The van Laar parameters and binary interaction parameter are determined by matching the azeotropic pressure (and composition for the van Laar case) as described in previous examples. The resulting calculations are described in the spreadsheet REGULAR in the workbook ACTCOEFF.xls. Fig. 11.5 illustrates the results.

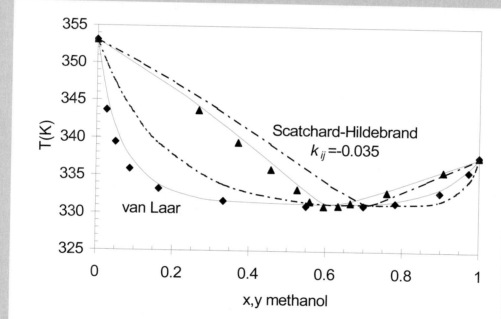

Figure 11.5 *T-x-y diagram for methanol and benzene for Example 11.9. The compositions are plotted in terms of mole fractions of methanol.*

The Flory Model

In deriving the entropy of mixing ideal gases in Example 3.2 on page 94, we applied the notion that ideal gases are point masses and have no volume. We considered the entropy of mixing to be determined by the total volume of the mixture. When we consider the entropy of mixing liquids, however, we realize that the volume occupied by the molecules themselves is a significant part of the total liquid volume. The volume occupied by one molecule is not accessible to the other molecules, and therefore, our assumptions regarding entropy may be inaccurate. One simple way of correcting for this effect is to subtract the volume occupied by the molecules from the total volume and treat the resulting "free volume" in the same way we treated ideal gas volume.

> *Free volume* is the difference between the volume of a liquid and the minimum volume which it would occupy if its molecules were packed firmly in contact with each other.

To use the concept, we assume that there is a global fraction of any liquid which is free volume, λ. This fraction of volume will be the same for all liquids and liquid mixtures. Let us further assume that the entropy change for a component is given by the change in free volume available to that component.

The free volume available to any pure component is

$$\underline{V}_{f,i} = n_i V_i \lambda \qquad 11.39$$

If we assume that there is no volume change on mixing, the resulting free volume in the mixture will be given by the same fraction, λ, of the mixture volume

$$\underline{V}_{f,mixture} = (n_1 V_1 + n_2 V_2)\lambda \qquad 11.40$$

When we mix two components, each component's entropy increases according to how much more space it has by an extension of Eqn. 3.5:

$$\bar{S}_i - S_i = R \ln \frac{\underline{V}_{f,mixture}}{\underline{V}_{f,i}} = R \ln \frac{(n_1 V_1 + n_2 V_2)\lambda}{(n_i V_i)\lambda} = -R \ln \Phi_i \qquad 11.41$$

$$\Delta S_{mix} = S - \sum_i x_i S_i = \sum_i x_i(\bar{S}_i - S_i) = -R \sum_i x_i \ln \Phi_i \qquad 11.42$$

Note that Eqn. 11.42 will reduce to the ideal solution result, Eqn. 9.40, when $V_1 = V_2$. The excess entropy is

$$S^E = -R \sum_i x_i \ln(\Phi_i / x_i) \qquad 11.43$$

This expression provides a simplistic representation of deviations of the entropy from ideal mixing. The entropy of mixing given by Eqn. 11.43 is frequently called the combinatorial entropy of mixing because it derives from the same combinations and permutations that we discussed in the case of particles in boxes. If entropy is the dominant factor in mixing, this formula can be used to find the excess Gibbs energy. When the excess enthalpy is zero, the mixture is called *athermal*.

❶ Flory's equation.

$$\boxed{G^E = H^E - TS^E = RT\left(x_1 \ln \frac{\Phi_1}{x_1} + x_2 \ln \frac{\Phi_2}{x_2}\right)} \qquad 11.44$$

It can also be combined with regular solution theory to derive the predictive theory of Blanks and Prausnitz[1] or the more common "Flory-Huggins" theory. These expressions are particularly important for solutions containing large molecules like polymers.

---

1. Blanks, R.F., Prausnitz, J.M. *Ind. Eng. Chem. Fundam.*, 3:1, (1964).

For a binary solution,

$$G^E = H^E - TS^E = RT\left(x_1 \ln \frac{\Phi_1}{x_1} + x_2 \ln \frac{\Phi_2}{x_2}\right) + \Phi_1\Phi_2(\delta_1 - \delta_2)^2(x_1 V_1 + x_2 V_2) \qquad 11.45$$

$$\ln \gamma_1 = \ln(\Phi_1/x_1) + (1 - \Phi_1/x_1) + \frac{V_1}{RT}\Phi_2^2(\delta_1 - \delta_2)^2 \qquad 11.46$$

$$\ln \gamma_2 = \ln(\Phi_2/x_2) + (1 - \Phi_2/x_2) + \frac{V_2}{RT}\Phi_1^2(\delta_1 - \delta_2)^2 \qquad 11.47$$

Frequently, for mixtures of polymer and solvent, the enthalpic term is fitted empirically to experimental data by adjusting the form of the equation to be:

$$G^E = RT\left(x_1 \ln \frac{\Phi_1}{x_1} + x_2 \ln \frac{\Phi_2}{x_2}\right) + \Phi_1\Phi_2(x_1 + x_2 R)\chi RT \qquad 11.48$$

❶ Flory-Huggins equation.

where component 1 is always the solvent, and component 2 is always the polymer. The variable $R = V_2/V_1$ denotes the ratio of volume of the polymer to the solvent. Similarly, $\chi \equiv V_1(\delta_1 - \delta_2)^2/RT$. A solvent for which $\chi = 0$ is an athermal mixture. Plotting the result for S^E vs. mole fraction for several size ratios, Fig. 11.6 shows that it is always positive, and it becomes larger and more skewed as the size ratio increases. Thus, the size ratio has a large effect on the phase stability when the ratio is large.

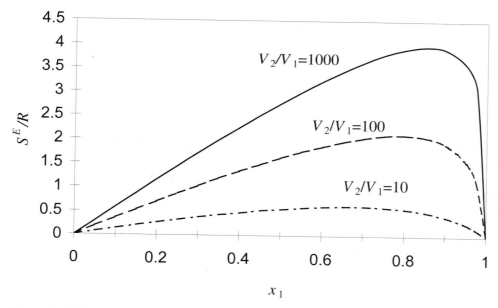

Figure 11.6 *Illustration of excess entropy according to Flory's equation for various pure component volume ratios.*

Example 11.10 Combinatorial contribution to the activity coefficient

The effect of entropy on the Gibbs energy of mixing is derived from considering the combinations possible for particles in boxes. Therefore, it is often referred to as the "combinatorial contribution." The energetic effect is then what is left over, and it is referred to as the "residual contribution." In polymer solutions, it is not uncommon for the solubility parameter of the polymer to nearly equal the solubility parameter of the solvent. To illustrate, consider the case when 1g of benzene is added to 1g of pentastyrene to form a solution. Estimate the activity coefficient of the benzene in the pentastyrene if $\delta_{ps} = \delta_B = 9.2$ and V_{ps} and V_B are estimated using the "R" volume parameters from UNIQUAC/UNIFAC.[1] These group parameters can be used to estimate the relative size of molecules according to the functional groups that comprise them. The group volumes are relative to a $-CH_2-$ group in a long chain alkane, so they are dimensionless, and are always used in ratios for this reason.

Solution:

Since $\delta_{ps} = \delta_B = 9.2$, we can ignore the residual contribution. Therefore,

$$\ln \gamma_B = \ln(\Phi_B / x_B) + (1 - \Phi_B / x_B) \qquad 11.49$$

Benzene is comprised of 6(ACH) groups @ 0.5313 R-units per group $\Rightarrow V_B \propto 3.1878$

Pentastyrene is composed of 25(ACH) + 1(ACCH2) + 4(ACCH) + 4(CH2) + 1(CH3)
$\Rightarrow V_{ps} \propto 21.17$

$$MW_B = 78.114 \text{ and } MW_{ps} = 520.76 \Rightarrow x_B = 0.8696 \qquad 11.50$$

$$\Phi_B = 0.8696(3.1878)/[0.8696(3.1878) + 0.1304(21.17)] = 0.5010 \qquad 11.51$$

Note: The volume fraction is very close to the weight fraction.

$$\ln \gamma_B = \ln(0.5010 / 0.8696) + (1 - 0.5010 / 0.8696) = -0.1275 \Rightarrow \gamma_B = 0.8803 \qquad 11.52$$

Flory's model (no energetic contribution) predicts that the partial pressure of benzene in the vapor would be about 12% *less* than the ideal solution model.

Example 11.11 Polymer mixing

One of the major problems with recycling polymeric products is that different polymers do not form miscible solutions with each other; rather, they form highly non-ideal solutions. To illustrate, suppose 1g each of two different polymers (polymer A and polymer B) is heated to 127°C and mixed as a liquid. Estimate the activity coefficients of A and B using the Flory-Huggins model.

	MW	$V(cm^3/mol)$	$\delta(cal/cm^3)^{1/2}$
A	10,000	1,540,000	9.2
B	2,000	1,680,000	9.3

---

1. The values are available in Table 11.3 on page 391.

> ### Example 11.11 Polymer mixing (Continued)
>
> **Solution:**
>
> $$x_A = (1/10,000)/(1/10,000 + 1/12,000) = 0.546; \quad x_B = 0.454 \tag{11.53}$$
>
> $$\Phi_A = 0.546(1.54)/[0.546(1.54) + 0.454(1.68)] = 0.524; \quad \Phi_B = 0.476 \tag{11.54}$$
>
> $$\begin{aligned} \ln\gamma_A &= \ln(0.5238/0.5455) + (1 - 0.5238/0.5455) \\ &\quad + 1.54E6(9.3 - 9.2)^2(0.4762)^2/1.987(400) \end{aligned} \tag{11.55}$$
>
> $$= -0.0008 + 4.395 \implies \gamma_A = 81 \tag{11.56}$$
>
> $$\begin{aligned} \ln\gamma_B &= \ln(0.4762/0.4545) + (1 - 0.4762/0.4545) \\ &\quad + 1.68E6(9.3 - 9.2)^2(0.5238)^2/1.987(400) \end{aligned} \tag{11.57}$$
>
> $$= +0.0008 + 5.800 \implies \gamma_B = 330 \tag{11.58}$$
>
> *Note:*
>
> 1. *Highly non-ideal mixing is exhibited by this predominately "regular" solution (note that the excess entropy is not what is making this solution non-ideal). This non-ideality arises from the excess internal energy being proportional to $(V) \cdot (\delta_1 - \delta_2)^2$ where δ_i is relatively constant for a polymer of any size as long as it is in the same family. The proportionality to molar volume for the excess property arose when the properties of the pure fluids were subtracted, combined with the observation that δ_i is relatively constant with respect to molecular weight. The energetic differences between the species are greatly magnified while the entropic driving for them to mix has greatly diminished.*
>
> 2. *The large activities computed for these components mean that the fugacities of these components would be greatly enhanced if intermingled at this composition. This means that they show a strong tendency to escape from each other. On the other hand, polymer compounds are too non-volatile to escape to the vapor phase. The only alternative is to escape into separate liquid phases. In other words, the liquids become immiscible. Computations of activity coefficients like those above play a major role in the liquid-liquid phase equilibrium calculations detailed in Chapter 12.*

11.5 FLORY-HUGGINS & VAN DER WAALS' THEORIES (OPTIONAL)

We have shown that the contribution to the excess internal energy in Flory-Huggins theory is identical to that in Scatchard-Hildebrand theory. We derived Scatchard-Hildebrand theory from the excess internal energy function of the van der Waals equation on page 368 and on page 371. Therefore, any potential difference between Flory-Huggins theory and the van der Waals equation must pertain to the excess entropy term. Reviewing briefly, the van der Waals equation of state gives,

$$\frac{G(T, P) - G^{ig}(T, P)}{RT} = -\ln(1 - b\rho) - \frac{a\rho}{RT} + Z - 1 - \ln(Z) \tag{11.59}$$

$$\frac{G^E}{RT} = -\ln(1 - b\rho) - \frac{a}{RTV} + Z - 1 + \sum_i x_i \ln(1 - b_i\rho_i)$$

$$+ \sum_i x_i \frac{a_{ii}}{RTV_i} - \sum_i x_i Z_i + 1 - \sum_i x_i \ln(Z/Z_i) \qquad 11.60$$

Recall that van der Waals' equation gives $U - U^{ig} = \frac{a}{V}$.

Therefore, $U^E = U - \sum_i x_i U_i = -\frac{a}{V} + \sum_i x_i \frac{a_{ii}}{V_i}$. Comparing $\left(-\frac{a}{V} + \sum_i x_i \frac{a_{ii}}{V_i}\right)$ to the result for regular solutions, we see that,

$$U^E = \Phi_1\Phi_2(\delta_1 - \delta_2)^2(x_1 V_1 + x_2 V_2)$$

which is the same. We may also note that $Z - \sum_i x_i Z_i$ is a very small number because: 1) these are liquid compressibility factors so all Z are small numerically; 2) The excess volume is usually a small percentage of the total volume, $V \approx \sum_i x_i V_i$. Thus we may neglect $Z - \sum_i x_i Z_i$.

Turning to the differences between the entropy terms, van der Waals' equation gives

$$TS^E = H^E - G^E = U^E - RT\left(Z - \sum_i x_i Z_i\right) - G^E \qquad 11.61$$

$$\frac{S^E}{R} = \sum_i x_i \ln\left(\frac{1 - b\rho}{1 - b_i\rho_i}\right) + \sum_i x_i \ln\left(\frac{PV / RT}{PV_i / RT}\right) \qquad 11.62$$

Note: $(1 - b_i\rho_i) = (V_i - b_i)/V_i \equiv (V_i\lambda)/V_i = \lambda$

If we assume that λ is a universal constant for all fluids, including the mixture, then

$$\frac{S^E}{R} = \sum_i x_i \ln\left(\frac{PV / RT}{PV_i / RT}\right) = -\sum_i x_i \ln\left(\frac{V_i}{V}\right) = -\sum_i x_i \ln\left(\frac{x_i V_i}{V} \frac{1}{x_i}\right) = -\sum_i x_i \ln\left(\frac{\Phi_i}{x_i}\right) \qquad 11.63$$

This expression is identical to Flory's equation (and note the importance of the $\ln(Z)$ term as the second term on the right-hand side of Eqn. 11.62, which derived from the ideal gas reference state). Therefore, the only difference between van der Waals' and Flory's theories is the assumption that λ is a universal constant. This is equivalent to saying that the packing fraction ($b\rho$) is a constant (the packing fraction is one minus the void fraction). *In other words, Flory-Huggins theory is simply van der Waals theory with the assumption that $b\rho = constant$.* The suggestion that $b\rho = constant$ is actually quite consistent with another observation that should seem more familiar, that is, that the mass density of a polyatomic species is only weakly dependent on its molecular weight. For example, the mass density of decane is 0.73 g/cm³ and the density of *n*-hexadecane is 0.77 g/cm³. Since the molar density decreases inversely as molecular weight increases but the *b*-parameter increases proportionally as molecular weight increases, a constant value for the mass density implies a constant value for $b\rho$. When you consider that the mass density for almost all hydrocarbons, alcohols, amides, and amines lie between 0.7 and 1.3 g/cm³, you begin to get an idea of how broadly applicable this approximation is.

Nevertheless, there are some obvious limitations to the assumption of a constant packing fraction. A little calculation would make it clear that the λ for liquid propane at $T_r = 0.99$ is significantly larger than λ for toluene at $T_r = 0.619$. Thus, a mixture of propane and toluene at 366 K would not be very accurately represented by the Flory-Huggins theory. Note that deviations of λ from each other are related to differences in the compressibilities of the components. Thus, it is common to refer to Flory-Huggins theory as an "incompressible" theory and to develop alternative theories to represent "compressible" polymer mixtures. Not surprisingly, these alternative theories closely resemble the van der Waals' equation (with a slightly modified temperature dependence of the a parameter). This observation lends added significance to Gibbs' quote: "The whole is simpler than the sum of its parts" and to Rayleigh's quote: "I am more than ever an admirer of van der Waals."

11.6   LOCAL COMPOSITION THEORY

One of the major assumptions of regular solution theory was that the mixture interactions were independent of each other such that quadratic mixing rules would provide reasonable approximations as shown in Section 10.1 on page 322. But in some cases, like radically different strengths of attraction, the mixture interaction can be strongly coupled to the mixture composition. That is, for instance, the cross parameter could be a function of composition. $a_{12} = a_{12}(x)$. One way of treating this prospect is to recognize the possibility that the "local compositions" in the mixture might deviate strongly from the bulk compositions. As an example, consider a lattice consisting primarily of type A atoms but with two B atoms right beside each other. Suppose all these atoms were the same size and that the coordination number was 10. Then the local compositions around a B atom are $x_{AB} = 9/10$ and $x_{BB} = 1/10$ (notation of subscripts is $AB \Rightarrow$ "A around B"). Specific interactions such as hydrogen bonding and polarity might lead to such effects, and thus, the basis of the hypothesis is that *energetic differences lead to the nonrandomness that causes the quadratic mixing rules to break down.* Excess Gibbs models based on this hypothesis are termed local composition theories, and were first introduced by Wilson in 1964.[1] To develop the theory, we first introduce nomenclature to identify the local compositions summarized in Table 11.2

Table 11.2 *Nomenclature for local composition variables.*

Composition around a "1" molecule	Composition around a "2" molecule
x_{21} – mole fraction of "2's" around "1"	x_{12} – mole fraction of "1's" around "2"
x_{11} – mole fraction of "1's" around "1"	x_{22} – mole fraction of "2's" around "2"
local mole balance, $x_{11} + x_{21} = 1$	local mole balance, $x_{22} + x_{12} = 1$

We assume that the local compositions are given by some weighting factor, Ω_{ij}, relative to the overall compositions.

$$\frac{x_{21}}{x_{11}} = \frac{x_2}{x_1}\Omega_{21} \qquad\qquad 11.64$$

1. Wilson, G.M., *J. Am. Chem. Soc.* 86:127 (1964).

$$\frac{x_{12}}{x_{22}} = \frac{x_1}{x_2}\Omega_{12}$$

11.65

Therefore, if $\Omega_{12} = \Omega_{21} = 1$, the solution is random. Before introducing the functions that describe the weighting factors, let us discuss how the factors may be used.

Local Compositions around "1" molecules

Let us begin by considering compositions around "1" molecules. We would like to write the local mole fractions x_{21} and x_{11} in terms of the overall mole fractions, x_1 and x_2. Using the local mole balance

$$x_{11} + x_{21} = 1$$

11.66

Rearranging Eqn. 11.64

$$x_{21} = x_{11}\frac{x_2}{x_1}\Omega_{21}$$

11.67

Substituting 11.67 into 11.66

$$x_{11}\left(1 + \frac{x_2}{x_1}\Omega_{21}\right) = 1$$

11.68

Rearranging

$$x_{11} = \frac{x_1}{x_1 + x_2\Omega_{21}}$$

11.69

Substituting 11.69 into 11.67

$$x_{21} = \frac{x_2\Omega_{21}}{x_1 + x_2\Omega_{21}}$$

11.70

Local Compositions around "2" molecules

Similar derivations for molecules of type "2" results in

$$x_{22} = \frac{x_2}{x_1\Omega_{12} + x_2}$$

11.71

$$x_{12} = \frac{x_1\Omega_{12}}{x_1\Omega_{12} + x_2}$$

11.72

Example 11.12 Local compositions in a 2-dimensional lattice

The following lattice contains x's, o's and void spaces. The coordination number of each cell is 8. Estimate the local composition (x_{XO}) and the parameter Ω_{XO} based on rows and columns away from the edges.

		o		o		x		o	
	x		o	x			x		
	x		x	x		o	x	o	
o		x		o			x		
						o			x
	o		x		o			x	
		x		o		o		x	
o	x				x		x		o

Solution:

There are 9 o's and 13 x's that are located away from the edges. The number of x's and o's around each o are:

o#	1	2	3	4	5	6	7	8	9	
#x's	3	3	3	2	1	1	0	2	2	= 17
#o's	2	0	0	0	1	0	3	1	1	= 8
$x_{XO} = 17/25 = 0.68$; $x_O = 9/22$; $\Omega_{XO} = 17/8 \cdot 9/13 = 1.47$										= 1.47

Note: Fluids do not really behave as though their atoms were located on lattice sites, but there are many theories based on the supposition that lattices represent reasonable approximations. In this text, we have elected to omit detailed treatment of lattice theory on the basis that it is too approximate to provide an appreciation for the complete problem and too complicated to justify treating it as a simple theory. This is a judgment call and interested students may wish to learn more about lattice theory. Sandler presents a brief introduction to the theory which may be acceptable for readers at this level.[1]

---

1. Sandler, S.I., *Chemical Engineering Thermodynamics*, 2nd ed, p. 366, Wiley, 1989.

Applying the Local Composition Concept to Obtain the Excess Gibbs Energy

We need to relate the local compositions to the excess Gibbs energy. The perspective of representing all fluids by the square-well potential lends itself naturally to the local composition concept. Then the intermolecular energy is given simply by the local composition times the well-depth for that interaction. In equation form, the energy equation for mixtures can be reformulated in terms of local compositions. The local mole fraction can be related to the bulk mole fraction by defining a quantity Ω_{ij} as follows:

$$\frac{x_{ij}}{x_{jj}} \equiv \frac{x_i}{x_j}\Omega_{ij} \qquad\qquad 11.73$$

The next step in the derivation requires scaling up from the molecular scale, local composition to the macroscopic energy in the mixture. The rigorous procedure for taking this step requires integration of the molecular distributions times the molecular interaction energies, analogous to the procedure for pure fluids as applied in Section 6.8. This rigorous development is presented below in Section 11.9. On the other hand, it is possible to simply present the result of that derivation for the time being. This permits a more rapid exploration of the practical implications of local composition theory. The form of the equation is not so difficult to understand from an intuitive perspective, however. The energy departure is simply a multiplication of the local composition (x_{ij}) by the local interaction energy (ε_{ij}). The departure properties are calculated based on a general model known as the *two-fluid theory*. According to the two-fluid theory, any intensive departure function in a binary is given by

$$(M - M^{ig}) = x_1(M - M^{ig})^{(1)} + x_2(M - M^{ig})^{(2)} \qquad\qquad 11.74$$

where the local composition environment of the type 1 molecules determines $(M - M^{ig})^{(1)}$, and the local composition environment of the type 2 molecules determines $(M - M^{ig})^{(2)}$. Note that $(M - M^{ig})^{(i)}$ is composition-dependent and is equal to the pure component value only when the local composition is pure i.

Noting that $\varepsilon_{12} = \varepsilon_{21}$, and recalling that the local mole fractions must sum to unity, we have for a binary mixture

$$U - U^{ig} = \frac{N_A}{2}\left[x_1 Nc_1(x_{11}\varepsilon_{11} + x_{21}\varepsilon_{21}) + x_2 Nc_2(x_{12}\varepsilon_{12} + x_{22}\varepsilon_{22})\right] \qquad 11.75$$

where Nc_j is the coordination number (total number of atoms in the neighborhood of the jth species), and where we can identify

$$(U - U^{ig})^{(1)} = \frac{N_A}{2}Nc_1(x_{11}\varepsilon_{11} + x_{21}\varepsilon_{21}) \text{ and } (U - U^{ig})^{(2)} = \frac{N_A}{2}Nc_2(x_{12}\varepsilon_{12} + x_{22}\varepsilon_{22}) \quad 11.76$$

When x_1 approaches unity, x_2 goes to zero, and from Eqn. 11.64 x_{21} goes to zero, and x_{11} goes to one. The limit applied to Eqn. 11.75 results in $(U - U^{ig})_{pure1} = (N_A/2)Nc_1\varepsilon_{11}$. Similarly, when x_2 approaches unity, x_1 goes to zero, x_{12} goes to zero, and x_{22} goes to one, resulting in $(U - U^{ig})_{pure2} = (N_A/2)Nc_2\varepsilon_{22}$. For an ideal solution

$$(U - U^{ig})^{is} = x_1(U - U^{ig})_{pure1} + x_2(U - U^{ig})_{pure2} = \frac{N_A}{2}[x_1 Nc_1\varepsilon_{11} + x_2 Nc_2\varepsilon_{22}] \qquad 11.77$$

The excess energy is obtained by subtracting Eqn. 11.77 from 11.75

$$U^E = U - U^{is} = \frac{N_A}{2}[x_1 Nc_1((x_{11}\varepsilon_{11} + x_{21}\varepsilon_{21}) - \varepsilon_{11}) + x_2 Nc_2((x_{12}\varepsilon_{12} + x_{22}\varepsilon_{22}) - \varepsilon_{22})] \qquad 11.78$$

Collecting terms with the same energy variables, and using the local mole balance from Table 11.2 on page 381, $(x_{11}-1)\varepsilon_{11} = -x_{21}\varepsilon_{11}$, and $(x_{22}-1)\varepsilon_{22} = -x_{12}\varepsilon_{22}$, resulting in

$$U^E = \frac{N_A}{2}\left[x_1 x_{21} Nc_1(\varepsilon_{21} - \varepsilon_{11}) + x_2 x_{12} Nc_2(\varepsilon_{12} - \varepsilon_{22})\right] \qquad 11.79$$

If we assume that Nc_j does not change with mixing and that $Nc_j = Nc_i \equiv z$ where z is the coordination number of all the components, then we can formulate the excess energy relative to the ideal solution (note that this assumption would be weak if i and j were very different in size). Substituting Eqn. 11.70 and Eqn. 11.72

$$U^E = \frac{N_A z}{2}\left[\frac{x_1 x_2 \Omega_{21}(\varepsilon_{21} - \varepsilon_{11})}{x_1 + x_2\Omega_{21}} + \frac{x_2 x_1 \Omega_{12}(\varepsilon_{12} - \varepsilon_{22})}{x_1\Omega_{12} + x_2}\right] \qquad 11.80$$

At this point, the traditional local composition theories deviate from regular solution theory in a way that really has nothing to do with local compositions. Instead, the next step focuses on one of the subtleties of classical thermodynamics. Recalling, $A = U - TS \Rightarrow A/RT = U/RT - S/R$

$$T\left(\frac{\partial(A/RT)}{\partial T}\right)_V = \frac{T}{RT}\left(\frac{\partial U}{\partial T}\right)_V - \frac{TU}{RT^2} - \frac{T}{R}\left(\frac{\partial S}{\partial T}\right)_V = \frac{C_V}{R} - \frac{U}{RT} - \frac{T}{R}\frac{C_V}{T} = -\frac{U}{RT} \qquad 11.81$$

Therefore,

$$\frac{A^E}{RT} = \int \frac{-U^E}{RT}\frac{dT}{T} + C \qquad 11.82$$

where C is an integration constant, independent of temperature but possibly dependent on composition or density. In local composition theory, the temperature dependence shows up in Ω_{ij}. Thus, we should truly integrate this expression. But first, we need to have some algebraic expression for the dependence of Ω_{ij} on temperature, which is what distinguishes the local composition theories from each other.

Wilson's Equation

Wilson made a bold assumption regarding the temperature dependence of Ω_{ij}. Wilson's original parameter used in the literature is Λ_{ji}, but it is related to Ω_{ij} in a very direct way. Wilson *assumes*,[1]

$$\Omega_{ij} = \Lambda_{ji} = \frac{V_i}{V_j}\exp\left(\frac{-N_A Nc_j(\varepsilon_{ij} - \varepsilon_{jj})}{2RT}\right) = \frac{V_i}{V_j}\exp\left(\frac{-A_{ji}}{RT}\right) \qquad 11.83$$

(note: $\Lambda_{ii} = \Lambda_{jj} = 1$, and $A_{ij} \neq A_{ji}$ even though $\varepsilon_{ij} = \varepsilon_{ji}$), and integration with respect to T becomes very simple. Assuming $Nc_j = 2$ for all j at all ρ,

$$\frac{A^E}{RT} = -x_1 \ln(x_1 + x_2\Lambda_{12}) - x_2 \ln(x_1\Lambda_{21} + x_2) + C \qquad 11.84$$

Wilson assumed $C = 0$. The customary way of interpreting A^E/RT is to separate it into an energetic part known as the *residual* part, $(A^E/RT)^{RES}$, and a size/shape part known as the *combinatorial* part, $(A^E/RT)^{COMB}$. The residual part should go to zero when $\varepsilon_{12} - \varepsilon_{22} = 0$ and $\varepsilon_{21} - \varepsilon_{11} = 0$. Some rearrangement[2] shows that $(A^E/RT)^{RES} = -x_1\ln(\Phi_1 + \Phi_2\exp(-A_{12}/RT)) - x_2\ln(\Phi_1\exp(-A_{21}/RT) + \Phi_2)$ and combinatorial part is Flory's equation, $(A^E/RT)^{COMB} = x_1\ln(\Phi_1/x_1) + x_2\ln(\Phi_2/x_2)$. It should be noted that the assumption of this temperature dependence has been made largely for convenience, but there is some justification for it, as we show in Section 11.9. Another convenient simplifying assumption is that $G^E = A^E$. This corresponds to neglecting the excess volume of mixing relative to the other contributions and is really quite acceptable for liquids.

One limitation of Wilson's equation is that it is unable to model liquid-liquid equilibria, but it is reasonably accurate for modeling the liquid phase when correlating the vapor-liquid equilibria of many mixtures. Extending Eqn. 11.84 to a multicomponent solution,

$$\boxed{\frac{G^E}{RT} = -\sum_j x_j \ln\left(\sum_i x_i\Lambda_{ji}\right)} \Rightarrow \frac{G^E}{RT} = -\frac{1}{n}\sum_j n_j \ln\left(\sum_i \frac{n_i\Lambda_{ji}}{n}\right) \qquad 11.85$$

$$\frac{G^E}{RT} = -\left[\sum_j n_j \ln\left(\sum_i n_i\Lambda_{ji}\right) - n\ln(n)\right] \qquad 11.86$$

To determine activity coefficients, the excess Gibbs energy is differentiated. Differentiating the last term,

$$\left(\frac{\partial(n\ln n)}{\partial n_k}\right)_{T,P,n_{i,j\neq k}} = \ln n + 1$$

and letting "sum" stand for the summation of Eqn. 11.86

---

1. Advanced readers may note that our definition of local compositions differs slightly from Wilson's. Wilson's original derivation combined the two-fluid theory of local compositions with an *ad hoc* "one-fluid" Flory equation. The same result can be derived more consistently using a two-fluid theory. The difference is that the local compositions are dependent on size as well as energies as defined by Eqns. 11.64, 11.65, and 11.83. This gives $x_{ij}/x_{jj} = (\Phi_i/\Phi_j)\exp(-A_{ji}/(RT))$ where the original was $x_{ij}/x_{jj} = (x_i/x_j)\exp(-A_{ji}/(RT))$.
2. This is one topic of homework problem 11.41.

$$\left(\frac{\partial(\text{sum})}{\partial n_k}\right)_{T, P, n_{i, j \neq k}} = -\ln\left(\sum_i n_i \Lambda_{ki}\right) - \sum_j \left(\frac{n_j \Lambda_{jk}}{\sum_i n_i \Lambda_{ji}}\right)$$

combining,

$$\ln \gamma_k = 1 - \ln\left(\sum_i x_i \Lambda_{ki}\right) - \sum_j \left(\frac{x_j \Lambda_{jk}}{\sum_i x_i \Lambda_{ji}}\right) \qquad 11.87$$

For a binary system, the activity coefficients from the Wilson equation are:

$$\ln \gamma_1 = 1 - \ln(x_1 \Lambda_{11} + x_2 \Lambda_{12}) - \frac{x_1 \Lambda_{11}}{x_1 \Lambda_{11} + x_2 \Lambda_{12}} - \frac{x_2 \Lambda_{21}}{x_1 \Lambda_{21} + x_2 \Lambda_{22}}$$

$$\ln \gamma_2 = 1 - \ln(x_1 \Lambda_{21} + x_2 \Lambda_{22}) - \frac{x_1 \Lambda_{12}}{x_1 \Lambda_{11} + x_2 \Lambda_{12}} - \frac{x_2 \Lambda_{22}}{x_1 \Lambda_{21} + x_2 \Lambda_{22}}$$

Noting that $\Lambda_{11} = \Lambda_{22} = 1$, and looking back at Eqn. 11.69 we can also see that for the first equation $1 - \dfrac{x_1}{x_1 + x_2 \Lambda_{12}} = x_{21} = \dfrac{x_2 \Lambda_{12}}{x_1 + x_2 \Lambda_{12}}$. We can rearrange this expression to obtain the slightly more compact relation:

$$\ln \gamma_1 = -\ln(x_1 + x_2 \Lambda_{12}) + x_2 \left(\frac{\Lambda_{12}}{x_1 + x_2 \Lambda_{12}} - \frac{\Lambda_{21}}{x_1 \Lambda_{21} + x_2}\right) \qquad 11.88$$

⚠ Wilson's Equation.

Similar rearrangement of the second expression gives:

$$\ln \gamma_2 = -\ln(x_1 \Lambda_{21} + x_2) - x_1 \left(\frac{\Lambda_{12}}{x_1 + x_2 \Lambda_{12}} - \frac{\Lambda_{21}}{x_1 \Lambda_{21} + x_2}\right) \qquad 11.89$$

$$\Lambda_{12} = \frac{V_2}{V_1} \exp\left(\frac{-A_{12}}{RT}\right) \quad \text{and} \quad \Lambda_{21} = \frac{V_1}{V_2} \exp\left(\frac{-A_{21}}{RT}\right) \qquad 11.90$$

Example 11.13 Application of Wilson's equation to VLE

For the binary system n-pentanol(1) + n-hexane(2), the Wilson equation constants are $A_{12} = 1718$ cal/mol, $A_{21} = 166.6$ cal/mol. Assuming the vapor phase to be an ideal gas, determine the composition of the vapor in equilibrium with a liquid containing 20 mole% n-pentanol at 30°C. Also calculate the equilibrium pressure.

Given: $P_1^{sat} = 3.23$ mmHg; $P_2^{sat} = 187.1$ mmHg

Solution: From CRC, $\rho_1 = 0.8144$ g/ml (1mole/88g) $\Rightarrow V_1 = 108$ cm³/mole

$$\rho_2 = 0.6603 \text{ g/ml (1mole/86g)} \Rightarrow V_2 = 130 \text{ cm}^3/\text{mole}$$

Note: ρ_1 and ρ_2 are functions of T but $\rho_1/\rho_2 \approx const.$

$V_2/V_1 = 1.205$

utilizing Eqn. 11.90

$\Lambda_{12} = 1.205 \exp(-1718/1.987/303) = 0.070$

$\Lambda_{21} = 1/1.205 \exp(-166.6/1.987/303) = 0.629$

$$\frac{\Lambda_{12}}{x_1 + x_2\Lambda_{12}} - \frac{\Lambda_{21}}{x_1\Lambda_{21} + x_2}$$
$$= 0.070 / (0.2 + 0.8 \cdot 0.070) - 0.629 / (0.8 + 0.2 \cdot 0.629)$$
$$= -0.4062$$

$\ln \gamma_1 = 1.0376 \Rightarrow \gamma_1 = 2.822$

$\ln \gamma_2 = 0.1584 \Rightarrow \gamma_2 = 1.172$

Bubble pressure calculation.

$P = (y_1 + y_2)P = x_1\gamma_1 P_1^{sat} + x_2\gamma_2 P_2^{sat} = 0.2 \cdot 2.822 \cdot 3.23 + 0.8 \cdot 1.172 \cdot 187.1 = 177.2$ mmHg

$y_1 = x_1\gamma_1 P^{sat} / P = 0.2 \cdot 2.822 \cdot 3.23 / 177.2 = 0.0103$

UNIQUAC[1]

UNIQUAC (short for UNIversal QUAsi Chemical model) builds on the work of Wilson by making three primary refinements. First, the temperature dependence of Ω_{ij} is modified to depend on surface areas rather than volumes, based on the hypothesis that the interaction energies that determine local compositions are dependent on the relative surface areas of the molecules. If the parameter q_i is proportional to the surface area of molecule i,

$$\Omega_{ij} = \frac{q_i}{q_j}\exp\left(\frac{-N_A z(\varepsilon_{ij} - \varepsilon_{jj})}{2RT}\right) = \frac{q_i}{q_j}\exp\left(\frac{-a_{ij}}{T}\right) = \frac{q_i}{q_j}\tau_{ij} \qquad 11.91$$

where $z = 10$

---

1. Abrams, E.S., Prausnitz, J.M., *AIChE J. 21:116 (1975).*

The intermediate parameter τ_{ij} is used to write the final expression in compact notation.[1]

$$\tau_{ij} = \exp\left(-\frac{a_{ij}}{T}\right) \qquad 11.92$$

where $\tau_{ii} = \tau_{jj} = 1$. In addition, when the energy equation 11.75 is written with $Nc_j = 10q_j$ for all j at all ρ, the different sizes and shapes of the molecules are implicitly taken into account. Qualitatively, the number of molecules that can contact a central molecule increases as the size of the molecule increases. Using surface fractions attempts to recognize the branching and overlap that can occur between segments in a polyatomic molecule. The inner core of these segments is not accessible, only the surface is accessible for energetic interactions. Therefore, the model of the energy is proposed to be proportional to surface area. Unfortunately, it is not straightforward to construct a more rigorous argument in favor of surface fractions from the energy equation itself. The excess Helmholtz energy for a binary solution becomes

$$\frac{A^E}{RT} = -x_1 q_1 \ln(\theta_1 + \theta_2 \tau_{21}) - x_2 q_2 \ln(\theta_1 \tau_{12} + \theta_2) + C \qquad 11.93$$

where θ_i is the surface area fraction, and $\theta_i = x_i q_i/(x_1 q_1 + x_2 q_2)$ for a binary. The evaluation of the integration constant is the third difference from Wilson's equation. The temperature-independent contribution is attributed to the entropy of mixing hard chains, and an approximate expression for this contribution is applied by Maurer and Prausnitz.[2] This representation of the entropy of mixing traces its roots back to the work of Staverman[3] and Guggenheim[4] and is discussed more recently by Lichtenthaler et al.[5] It is very similar to the Flory term, but it corrects for the fact that large molecules are not always large balls, but sometimes long "strings." By noting that the ratio of surface area to volume for a sphere is different from that for a string, Guggenheim's form (the form actually applied in UNIFAC and UNIQUAC) provides a simple but general correction, giving an indication of the degree of branching and non-sphericity. Nevertheless, the Staverman-Guggenheim term represents a relatively small correction to Flory's term. As shown in the Figure 11.7, the extra correction of including the "surface to volume" parameter serves to decrease the excess entropy to some value between zero and the Flory-Huggins estimate. The combinatorial part of UNIQUAC takes the form

$$\left(\frac{A^E}{RT}\right)^{COMB} = \sum_j x_j \ln(\Phi_j/x_j) - 5\sum_j q_j x_j \ln(\Phi_j/\theta_j) \qquad 11.94$$

1. Note that these assumptions create local compositions of the form $x_{ij}/x_{jj} = (\theta_i/\theta_j)\exp(-a_{ij}/T)$. Compare this with the form of Wilson's equation (footnote page 386). Note that the use of the subscripts for the local composition energetic parameters τ and a are switched for UNIQUAC relative to the Wilson equation.
2. Maurer, G., Prausnitz, J.M., *Fluid Phase Equilibria*, 2:91 (1978).
3. Staverman, A.J., *Recl. Trav. Chem. Pays Bas*, 69:163 (1950).
4. Guggenheim, E.A., *Mixtures,* Oxford University Press, 1952.
5. Lichtenthaler, R.N., Abrams, D.S., Prausnitz, J.M., *Can. J. Chem.* 51:3071 (1973).

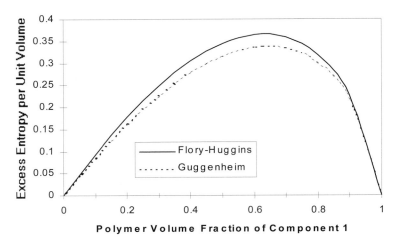

Figure 11.7 *Excess entropy according to the Flory-Huggins equation vs. Guggenheim's equation at $V_2/V_1 = 1695$ for a polymer solvent mixture.*

Instead of using experimental data to calculate volume fractions and surface fractions, they may be calculated from molecular size and surface area parameters that have been tabulated. The size parameters are ratios to the equivalent size for the -CH_2- group in a long chain alkane. These parameters may be calculated in the same manner as the UNIFAC variation discussed in the next section as given in Table 11.3 on page 391. In the table, the uppercase R_k parameter is for the group volume, and the uppercase Q_k parameter is for group surface area. From these values, the molecular size (r_j) and shape (q_j) parameters may be calculated by multiplying the group parameter by the number of times each group appears in the molecule, and summing over all the groups in the molecule.

$$r_j = \sum_k v_k^{(j)} R_k; \qquad q_j = \sum_k v_k^{(j)} Q_k; \qquad\qquad 11.95$$

where $v_k^{(j)}$ is the number of groups of the kth type in the jth molecule. The subdivision of the molecule into groups is sometimes not obvious because there may appear to be more than one way to subdivide, but the conventions have been set forth in examples in the table and these conventions should be followed. The Guggenheim form of the excess entropy is based on the molecular volume fractions, Φ_j, and the surface fractions, θ_j,

$$\Phi_j \equiv \frac{x_j r_j}{\sum_i x_i r_i} \qquad\qquad \theta_j \equiv \frac{x_j q_j}{\sum_i x_i q_i} \qquad\qquad 11.96$$

where R_k and Q_k are group contributions to the molecular parameters r_j and q_j.

Table 11.3 *Group parameters for the UNIFAC and UNIQUAC equations. AC in the table means aromatic carbon. The Main groups serve as categories for similar Sub-groups as explained in the UNIFAC section.*[1]

Main Group	Sub-group	R(rel.vol.)	Q(rel.area)	Example
CH2	CH3	0.9011	0.8480	
	CH2	0.6744	0.5400	*n*-hexane: 4 CH2 + 2 CH3
	CH	0.4469	0.2280	isobutane: 1CH + 3 CH3
	C	0.2195	0	neopentane: 1C + 4 CH3
C=C	CH2=CH	1.3454	1.1760	1-hexene: 1 CH2=CH + 3 CH2 + 1 CH3
	CH=CH	1.1167	0.8670	2-hexene: 1 CH=CH + 2 CH2 + 2 CH3
	CH2=C	1.1173	0.9880	
	CH=C	0.8886	0.6760	
	C=C	0.6605	0.4850	
ACH	ACH	0.5313	0.4000	benzene: 6 ACH
	AC	0.3652	0.1200	benzoic acid: 5 ACH + 1 AC + 1 COOH
ACCH2	ACCH3	1.2663	0.9680	toluene: 5 ACH + 1 ACCH3
	ACCH2	1.0396	0.6600	ethylbenzene: 5 ACH + 1 ACCH2 + 1 CH2
	ACCH	0.8121	0.3480	
OH	OH	1.0000	1.2000	*n*-propanol: 1 OH + 1 CH3 + 2 CH2
CH3OH	CH3OH	1.4311	1.4320	methanol is an independent group
water	H2O	0.9200	1.4000	water is an independent group
furfural	furfural	3.1680	2.484	furfural is an independent group
DOH	(CH2OH)2	2.4088	2.2480	ethylene glycol is an independent group
ACOH	ACOH	0.8952	0.6800	phenol: 1 ACOH + 5 ACH
CH2CO	CH3CO	1.6724	1.4880	dimethylketone: 1 CH3CO + 1 CH3 methylethylketone: 1 CH3CO + 1 CH2 + 1 CH3
	CH2CO	1.4457	1.1800	diethylketone: 1 CH2CO + 2 CH3 + 1 CH2
CHO	CHO	0.9980	0.9480	acetaldehyde: 1 CHO+1 CH3
CCOO	CH3COO	1.9031	1.7280	methyl acetate: 1 CH3COO + 1 CH3
	CH2COO	1.6764	1.4200	methyl propanate: 1 CH2COO + 2 CH3
COOH	COOH	1.3013	1.2240	benzoic acid: 5 ACH + 1 AC + 1 COOH

The parameters to characterize the volume and surface area fractions have already been tabulated, so no more adjustable parameters are really introduced by writing it this way. The only real problem is that including all these group contributions into the formulas makes hand calculations extremely tedious. Fortunately, computers and hand calculators make this task much simpler. As such, we can apply the UNIQUAC method almost as easily as the van Laar method. We should also note, however, that using an equation of state is similarly simplified by using a computer, so the

---

1. Alcohols are usually treated in UNIQUAC without using the group contribution method. Accepted UNIQUAC values for the set of alcohols [MeOH, EtOH, 1-PrOH, 2-PrOH, 1-BuOH] are r = [1.4311, 2.1055, 2.7799, 2.7791, 3.4543], q = [1.4320, 1.9720, 2.5120, 2.5080, 3.0520]. See Gmehling, J., Oken, U., Vapor-Liquid Equilibrium Data Collection, DECHEMA, Frankfort, 1977-.

basic motivation for developing solution models specific to liquids is simultaneously undermined by requiring computers for implementation. From this perspective, what we should be doing is analyzing the mixing rules and models of interaction energies in equations of state if we intend to use a computer anyway. We return to this point in Unit IV, when we discuss hydrogen-bonding equations of state for non-ideal solutions.

Extending Eqn. 11.93 to a multicomponent solution, the UNIQUAC equation becomes

$$\frac{G^E}{RT} = \sum_j x_j \ln(\Phi_j/x_j) - 5\sum_j q_j x_j \ln(\Phi_j/\theta_j) - \sum_j q_j x_j \ln\left(\sum_i \theta_i \tau_{ij}\right) \qquad 11.97$$

Note that the leading term is simply Flory's equation. The second term is the correction for non-sphericity, so these terms are purely geometric. The first two terms are called the *combinatorial* terms to indicate their geometric significance for combining molecules of different shapes and sizes. The energetic parameters are included in the last term, which is called the *residual* term. Since the terms are linearly combined, the combinatorial and residual parts can be individually differentiated to find their contribution to the activity coefficients,

$$\ln \gamma_k = \ln \gamma_k^{COMB} + \ln \gamma_k^{RES} \qquad 11.98$$

 UNIQUAC

$$\ln \gamma_k^{COMB} = \ln(\Phi_k/x_k) + (1 - \Phi_k/x_k) - 5q_k[\ln(\Phi_k/\theta_k) + (1 - \Phi_k/\theta_k)] \qquad 11.99$$

$$\ln \gamma_k^{RES} = q_k\left[1 - \ln\left(\sum_i \theta_i \tau_{ik}\right) - \sum_j \frac{\theta_j \tau_{kj}}{\sum_i \theta_i \tau_{ij}}\right] \qquad 11.100$$

For a binary mixture, the activity equations become

$$\ln\gamma_1 = \ln\frac{\Phi_1}{x_1} + \left(1 - \frac{\Phi_1}{x_1}\right) - 5q_1\left[\ln\frac{\Phi_1}{\theta_1} + \left(1 - \frac{\Phi_1}{\theta_1}\right)\right]$$
$$+ q_1\left[1 - \ln(\theta_1 + \theta_2\tau_{21}) - \frac{\theta_1}{\theta_1 + \theta_2\tau_{21}} - \frac{\theta_2\tau_{12}}{\theta_1\tau_{12} + \theta_2}\right] \qquad 11.101$$

$$\ln\gamma_2 = \ln\frac{\Phi_2}{x_2} + \left(1 - \frac{\Phi_2}{x_2}\right) - 5q_2\left[\ln\frac{\Phi_2}{\theta_2} + \left(1 - \frac{\Phi_2}{\theta_2}\right)\right]$$
$$+ q_2\left[1 - \ln(\theta_1\tau_{12} + \theta_2) - \frac{\theta_1\tau_{21}}{\theta_1 + \theta_2\tau_{21}} - \frac{\theta_2}{\theta_1\tau_{12} + \theta_2}\right] \qquad 11.102$$

Like the Wilson equation, the UNIQUAC equation requires that two adjustable parameters be characterized from experimental data for each binary system. The inclusion of the excess entropy in UNIQUAC by Abrams et al. (1975) is more correct theoretically, but Wilson's equation can be as accurate as the UNIQUAC method for many binary vapor-liquid systems, and much simpler to apply by hand. UNIQUAC supersedes the Wilson equation for describing liquid-liquid systems, however, because the Wilson equation is incapable of representing liquid-liquid equilibria as long as the Λ_{ij} parameters are held positive (as implied by their definition as exponentials, and noting that exponentials cannot take on negative values).[1]

UNIFAC[2]

This is an extension of UNIQUAC with no adjustable parameters for the user to input or fit to experimental data. Instead, all of the adjustable parameters have been characterized by the developers of the model based on group contributions that correlate the data in a very large data base. The assumptions regarding coordination numbers, etc., are similar to the assumptions in UNIQUAC. The same strategy is applied,

$$\ln \gamma_k = \ln \gamma_k^{COMB} + \ln \gamma_k^{RES}$$

The combinatorial term is therefore identical and given by Eqn. 11.99. The major difference between UNIFAC (short for UNIversal Functional Activity Coefficient model) and UNIQUAC is that, *for the residual term,* UNIFAC considers interaction energies between *functional groups* (rather than the whole molecule). Interactions of functional groups are added to predict relative interaction energies of molecules. The large number of possible functional groups are divided into *main groups* and further subdivided into *subgroups*. Examples are shown in Table 11.3. Each of the subgroups has a characteristic size and surface area; however, the energetic interactions are considered to be the same for all subgroups with a particular main group. Thus, representative interaction energies (a_{ij}) are tabulated for only the main functional groups, and it is implied that all subgroups will use the same energetic parameters. An illustrative sample of values for these interactions is given in Table 11.4. Full implementations of the UNIFAC method with large numbers of functional groups are typically available in chemical engineering process design software. A subset of the

Table 11.4 *VLE interaction energies a_{ij} for the UNIFAC equation in units of Kelvin.*

Main Group, i	CH2 $j=1$	ACH $j=3$	ACCH2 $j=4$	OH $j=5$	CH3OH $j=6$	water $j=7$	ACOH $j=8$	CH2CO $j=9$	CHO $j=10$	COOH $j=20$
1,CH2	---	61.13	76.5	986.5	697.2	1318	1333	476.4	677	663.5
3,ACH	−11.12	---	167	636.1	637.3	903.8	1329	25.77	347.3	537.4
4,ACCH2	−69.7	−146.8	---	803.2	603.3	5695	884.9	−52.1	586.8	872.3
5,OH	156.4	89.6	25.82	---	−137.1	353.5	−259.7	84	−203.6	199
6,CH3OH	16.51	−50	−44.5	249.1	---	−181	−101.7	23.39	306.4	−202.0
7,water	300	362.3	377.6	−229.1	289.6	---	324.5	−195.4	−116.0	−14.09
8,ACOH	275.8	25.34	244.2	−451.6	−265.2	−601.8	---	−356.1	−271.1	408.9
9,CH2CO	26.76	140.1	365.8	164.5	108.7	472.5	−133.1	---	−37.36	669.4
10,CHO	505.7	23.39	106.0	529	−340.2	480.8	−155.6	128	---	497.5
20,COOH	315.3	62.32	89.86	−151	339.8	−66.17	−11.00	−297.8	−165.5	---

---

1. Another modification of the Wilson equation that can represent LLE is the NRTL model. See also Appendix D.
2. Fredenslund, Aa., Jones, R.L.; Prausnitz, J.M. *AIChE J.* 21:1086 (1975).

parameters is provided on the UNIFAC spreadsheet in the ACTCOEFF.xls spreadsheet included with the text. Knowing the values of these interaction energies permits estimation of the properties for a really impressive number of chemical solutions. The limitation is that we are *not always entirely sure of the accuracy* of these predictions.

Although UNIFAC is closely related to UNIQUAC, keep in mind that there is no direct extension to a correlative equation like UNIQUAC. If you want to fit experimental data that might be on hand, you cannot do it within the defined framework of UNIFAC; UNIQUAC is the preferred choice when adjustable parameters are desired. Although it is tedious to estimate the a_{ij} parameters of UNIQUAC from UNIFAC, some implementations of chemical engineering process design software have included facilities for estimating UNIQUAC parameters from UNIFAC. This approach can be useful for estimating interactions for a few binary pairs in a multicomponent mixture when most of the binary pairs are known from experimental data specific to those binary interactions.

The basic approach is the generalization of the residual activity coefficient. To understand the development of the predictive technique, imagine the interactions of a CH_3 group in a mixture of isopropanol (1) and component (2). The isopropanol consists of $2(CH_3) + 1(CH) + 1(OH)$. Therefore, in the mixture, a CH_3 will encounter CH_3, CH, OH groups, and the groups of component (2), and the interaction energies depend on the number of each type of group available in the solution. Therefore, the interaction energy of CH_3 groups can be calculated relative to a hypothetical solution of 100% CH_3 groups. The mixture can be approximated as a solution of groups (SOG)[1] (rather than a solution of molecules), and the interaction energies can be integrated with respect to temperature to arrive at chemical potential in a manner similar to the development of Eqns. 11.81 and 11.82.

Therefore, it is possible to calculate $\dfrac{\mu_{CH_3}^{SOG} - \mu_{CH_3}^{o}}{RT} = \ln\Gamma_{CH_3}$ where $\mu_{CH_3}^{o}$ is the chemical potential in a hypothetical solution of 100% CH_3 groups and Γ_{CH_3} is the activity coefficient of CH_3 in the solution of groups. The chemical potential of CH_3 groups in pure isopropanol (1), given by $\mu_{CH_3}^{(1)}$, will differ from $\mu_{CH_3}^{o}$ because even in pure isopropanol CH_3 will encounter a mixture of CH_3, CH, and OH groups in the ratio that they appear in pure isopropanol, and therefore the activity coefficient of CH_3 groups in pure isopropanol, $\Gamma_{CH_3}^{(1)}$, is not unity, where the superscript (1) indicates pure component (1). The difference that is desired is the effect of mixing the CH_3 groups in isopropanol with component (2), relative to pure isopropanol,

$$\frac{\mu_{CH_3}^{SOG} - \mu_{CH_3}^{(1)}}{RT} = \frac{\mu_{CH_3}^{SOG} - \mu_{CH_3}^{o}}{RT} - \frac{\mu_{CH_3}^{(1)} - \mu_{CH_3}^{o}}{RT} = \ln\Gamma_{CH_3} - \ln\Gamma_{CH_3}^{(1)}$$

Fig. 11.8 provides an illustration of the differences that we seek to calculate, with water as a component (2).

---

1. A model exists in the literature that is called ASOG, which is different than the UNIFAC approach, but also uses functional groups.

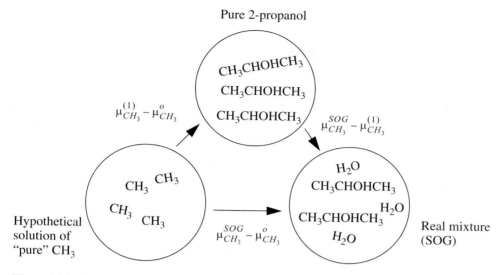

Figure 11.8 *Illustration relating the chemical potential of CH_3 groups in pure 2-propanol, a real solution of groups where water is component (2), and a hypothetical solution of CH_3 groups. The number of groups sketched in each circle is arbitrary and chosen to illustrate the types of groups present. The chemical potential change that we seek is $\mu_{CH_3}^{SOG} - \mu_{CH_3}^{(1)}$. We calculate this difference by taking the difference between the other two paths.*

If the chemical potential of a molecule consists of the sum of interactions of the groups,

$$\mu_1 = 2\mu_{CH_3}^{SOG} + \mu_{CH}^{SOG} + \mu_{OH}^{SOG}$$

$$\mu_1^o = 2\mu_{CH_3}^{(1)} + \mu_{CH}^{(1)} + \mu_{OH}^{(1)}$$

Therefore, we arrive at the important result that is utilized in UNIFAC,

$$\ln\gamma_1^{RES} = \frac{\mu_1 - \mu_1^o}{RT} = \sum_m v_m^{(1)}[\ln\Gamma_m - \ln\Gamma_m^{(1)}] \qquad 11.103$$

where the sum is over all function groups in molecule (1) and $v_m^{(1)}$ is the number of occurrences of group m in the molecule. The activity coefficient formula for any other molecular component can be found by substituting for (1) in Eqn. 11.103. Note that Γ_m is calculated in a solution of groups for all molecules in the mixture, whereas $\Gamma_m^{(1)}$ is calculated in the solution of groups for just component (1). Note that we use uppercase letters to represent the group property analog of the molecular properties, with the following exceptions: uppercase τ looks too much like T, so we substitute Ψ, and the a_{ij} for UNIFAC is understood to be a group property even though the same symbol is represented by a molecular property in UNIQUAC. The relations are shown in Table 11.5.

Table 11.5 *Comparison of group variables and molecular variables for UNIFAC.*

	Group Variable	Molecular Variable
volume	R	r
surface area	Q	q
activity coefficient	Γ	γ
surface fraction	Θ	θ
energy variable	Ψ_{ij}	τ_{ij}
energy parameter	a_{ij}	a_{ij}
mole fraction	X	x

🛑 UNIFAC.

$\ln\Gamma_m$ is calculated by generalizing the UNIQUAC expression for $\ln\gamma_m^{RES}$. Generalizing Eqn. 11.100 and supporting equations

$$\ln\Gamma_m = Q_m\left[1 - \ln\sum_i \Theta_i\Psi_{im} - \sum_j \frac{\Theta_j\Psi_{mj}}{\sum_i \Theta_i\Psi_{ij}}\right] \qquad 11.104$$

$$\Theta_j = (\text{surface area fraction of group } j) \equiv \frac{X_jQ_j}{\sum_i X_iQ_i} \qquad 11.105$$

$$\Psi_{mj} = \exp\left(\frac{-a_{mj}}{T}\right) \qquad 11.106$$

$$X_j = \frac{\sum_{molecules\ i} v_j^{(i)}x_i}{\sum_{molecules\ i}\ \sum_{groups\ k} v_k^{(i)}x_i} \qquad 11.107$$

where $v_k^{(i)}$ is the number of groups of type k in molecule i.

Example 11.14   Calculation of group mole fractions

Calculate the group mole fraction for CH_3 in a mixture of 60 mole% 2-propanol and 40 mole% water.

Solution: The two molecules are illustrated in Example 11.2 on page 360 and the group assignments are tabulated there. On a basis of 10 moles of solution, there are six moles of 2-propanol, and four moles of H_2O. The table below summarizes the totals of the functional groups.

Group	moles	X_j
CH3	12	0.429
CH	6	0.214
OH	6	0.214
H2O	4	0.143
Σ	28	

The mole fraction of CH_3 groups is then $X_{CH_3} = 12/28 = 0.429$. The mole fractions of the other groups are found analogously and also summarized in the table. The results are consistent with Eqn. 11.107 which is more easily programmed,

$$X_{CH_3} = \frac{2(0.6) + 0(0.4)}{(2(0.6) + 1(0.6) + 1(0.6)) + (1(0.4))} = 0.429$$

Example 11.15   Detailed calculations of activity coefficients via UNIFAC

Actcoeff.xls,
UNIFAC

Let's return to the example for the IPA + water system mentioned in Example 11.2. Compute the surface fractions, volume fractions, group interactions and the summations that go into the activity coefficients for this system at its azeotropic conditions. The isopropyl alcohol (*IPA*) + water (*W*) system is known to form an azeotrope at atmospheric pressure and 80.37°C ($x_W = 0.3146$).[1]

Solution: (this calculation can be followed interactively in the UNIFAC spreadsheet):
The molecular size and surface area parameters are found by applying Eqn. 11.95. Isopropanol has 2 CH_3, 1 OH, and 1 CH group. The group parameters are taken from Table 11.3.

For IPA: $r = 2 \cdot 0.9011 + 0.4469 + 1.0 = 3.2491$; $q = 2 \cdot 0.8480 + 0.2280 + 1.2 = 3.124$

For water: $r = 0.920$; $q = 1.40$

At $x_W = 0.3146$, $\Phi_W = 0.1150$, and $\theta_W = 0.1706$

---

1. Perry, R.H., Chilton, C.H., *Chemical Engineers' Handbook, 5 ed.,* Chapter 13, McGraw-Hill (1973).

Example 11.15   Detailed calculations of activity coefficients via UNIFAC (Continued)

$$\ln \gamma_k^{COMB} = \ln(\Phi_k/x_k) + (1 - \Phi_k/x_k) - 5q_k[\ln(\Phi_k/\theta_k) + (1 - \Phi_k/\theta_k)]$$

$$\ln\gamma_W^{COMB} = \ln(\Phi_W/x_W) + (1 - \Phi_W/x_W) - 5q_W[\ln(\Phi_W/\theta_W) + (1 - \Phi_W/\theta_W)] = 0.10724$$

$$\ln\gamma_I^{COMB} = \ln(\Phi_I/x_I) + (1 - \Phi_I/x_I) - 5q_I[\ln(\Phi_I/\theta_I) + (1 - \Phi_I/\theta_I)] = -0.00204$$

Note that these combinatorial contributions are computed on the basis of the total molecule. This is because the space-filling properties are the same whether we consider the functional groups or the whole molecules.

For the residual term, we break the solution into a solution of groups. Then we compute the contribution to the activity coefficients arising from each of those groups. We have four functional groups altogether (2CH3, CH, OH, H2O). We will illustrate the concepts by calculating $\ln\Gamma_{CH_3}^{(1)}$ and simply tabulate the results for the remainder of the calculations since they are analogous. First, let us tabulate the energetic parameters we will need.

We can summarize the calculations in tabular form as follows:

$\Psi_{\text{row-col}}$	CH3	CH	OH	H2O
CH3	1	1	0.061	0.024
CH	1	1	0.061	0.024
OH	0.642	0.642	1	0.368
H2O	0.428	0.428	1.912	1

For pure isopropanol, we tabulate the mole fractions of functional groups, and calculate the surface fractions

j	$\nu_j^{(1)}$	X_j	Q_j	Θ_j
CH_3	2	0.5	0.848	0.543
CH	1	0.25	0.228	0.073
OH	1	0.25	1.2	0.384
			$\sum_j X_j Q_j = 0.781$	

Example 11.15 Detailed calculations of activity coefficients via UNIFAC (Continued)

Applying Eqn. 11.104,

$$\ln \Gamma_{CH_3}^{(1)} = 0.848 \left\{ 1 - \ln[0.543 + 0.073 + 0.384(0.642)] - \frac{0.543}{0.543 + 0.073 + 0.384(0.642)} \right.$$

$$\left. - \frac{0.073}{0.543 + 0.073 + 0.384(0.642)} - \frac{0.384(0.061)}{0.543(0.061) + 0.073(0.061) + 0.384} \right\} = 0.3205$$

The same type of calculations can be repeated for the other functional groups. The calculation of $\ln \Gamma_{H_2O}^{(2)}$ is not necessary, since the whole water molecule is considered a functional group.

Performing the calculations in the mixture, the mole fractions, X_j need to be recalculated to reflect the compositions of groups in the overall mixture. Table 11.6 summarizes the calculations.

Table 11.6 *Summary of calculations for mixture of isopropanol and water at 80.37 °C and $x_w = 0.3146$.*

	CH3 $j = 1$	CH $j = 2$	OH $j = 3$	H2O $j = 4$
Q_j	0.848	0.228	1.200	1.400
Θ_j	0.4503	0.0605	0.3186	0.1706
$\sum_i \Theta_i \Psi_{ij}$	0.7885	0.7885	0.6762	0.3000
$\ln \Gamma_j$	0.4641	0.1248	0.3538	0.6398
$\ln \Gamma_j^{(2)}$	Not Applicable	NA	NA	0
$\ln \Gamma_j^{(1)}$	0.3205	0.0862	0.5927	NA
$\ln \Gamma_j - \ln \Gamma_j^{(i)}$	0.1336	0.0386	−0.2388	0.6398

> **Example 11.15 Detailed calculations of activity coefficients via UNIFAC (Continued)**
>
> The pure component values of $\ln\Gamma_j^{(i)}$ can be easily verified on the spreadsheet after unhiding the columns with the intermediate calculations. Entering values of 0 and 1 for the respective molecular species mole fractions causes the values of $\ln\Gamma_j^{(i)}$ to be calculated. (Note that values will appear on the spreadsheet computed for infinite dilution activity coefficients of the groups which do not exist in the pure component limits, but these are not applicable to our calculation so we can ignore them.) Subtracting the appropriate pure component limits gives the final row in Table 11.6. All that remains is to combine the group contributions together to form the molecules, and to add the residual part to the combinatorial part.
>
> $\ln\gamma_I = \ln\gamma_I^{RES} + \ln\gamma_I^{COMB} = [2(0.1336) + 0.0386 - 0.2388] - 0.00204 = 0.0848; \; \gamma_I = 1.0885$
>
> $\ln\gamma_W = \ln\gamma_W^{RES} + \ln\gamma_W^{COMB} = [0.6398] + 0.1072 = 0.7470; \; \gamma_W = 2.1108$

Fortunately, the spreadsheet and calculator program save us from doing the tedious calculations for UNIFAC, although an understanding of the principles is important.

11.7 FITTING ACTIVITY MODELS TO DATA (OPTIONAL)

Fitting of the Margules and van Laar equations to limited data has been discussed in Examples 11.5, 11.6, and 11.7. Fits to multiple points are preferred, which requires regression of the parameters to optimize the fit.

Fitting of the van Laar Model

Eqn. 11.27 can be rearranged:

$$\frac{x_1 x_2 RT}{G^E} = \left(\frac{1}{A_{21}} - \frac{1}{A_{12}}\right)x_1 + \frac{1}{A_{12}} \qquad 11.108$$

Therefore, if numerical values for the left-hand side are determined using G^E from experimental data as illustrated in Example 11.3 on page 363 and plotted versus x_1, the slope will yield $\left(\frac{1}{A_{21}} - \frac{1}{A_{12}}\right)$, and the intercept will yield $\frac{1}{A_{12}}$. The value of $\frac{1}{A_{21}}$ can also be determined by the value at $x_1 = 1$. Sometimes plots of the data are non-linear when fitting is attempted. This does not necessarily imply that the data are in error. It implies that an alternative model may fit the data better. Another model that is easy to fit is the two-parameter Margules.

🔴 van Laar and Margules models can be linearized for fitting of parameters.

Fitting the Margules equation

Eqn. 11.18 can be linearized:

$$\frac{G^E}{x_1 x_2 RT} = (A_{21} - A_{12})x_1 + A_{12} \qquad 11.109$$

Therefore, plotting $\dfrac{G^E}{x_1 x_2 RT}$ versus x_1 gives a slope of $(A_{21} - A_{12})$ and an intercept of A_{12}. The value of A_{21} can also be determined by the value at $x_1 = 1$.

Alternative fitting techniques

Parameters for excess Gibbs models besides the van Laar and Margules equations are usually non-linearly related to G^E or γ. The Solver tool in EXCEL provides the capability for non-linear fitting techniques. These can be applied to the Wilson equation or the UNIQUAC equation as well as the two equations treated above. The spreadsheet GAMMAFIT.XLS permits non-linear fitting of activity coefficient parameters for the Margules equation by fitting total pressure. It can be easily modified to find parameters for any activity coefficient model.

❶ Non-linear parameter fitting is possible in Excel.

Example 11.16 Using Excel for fitting model parameters

💾 GAMMFIT.xls

Measurements for the 2-propanol + water system at 30°C have been published by Udovenko, et al. (1967).[1] Use the pressure and liquid composition to fit the two-parameter Margules equation. Plot the resulting P-x-y diagram.

Solution: The spreadsheet "P-x-y fit P" in the workbook GAMMAFIT.xls is used to fit the parameters as shown below. Antoine coefficients are entered in the table for the components shown at the top of the spreadsheet. In the experimental data, the researchers report experimental vapor pressures so cells are provided for these values. The flag in the box in the center right determines whether experimental vapor pressures are used in the calculations or values calculated from the Antoine equation. It is best to use experimental values from the same publication to reduce the effect of systematic errors which may be present in the data. Experimental data for x_1 and P_{expt} are entered in columns A and I. Initial guesses for the constants A_{12} and A_{21} are entered in the labeled cells in the top table. Solver is then called to minimize the error in the objective function by adjusting the two parameters. In the spreadsheet provided, the objective function is $\displaystyle\sum_{all\ points} (P_{expt} - P_{calc})^2$. Other objective functions can be used to include error in the calculated values of the vapor compositions if additional columns are added to the table. Explore the spreadsheet to see how the calculations are performed. Calculated pressures are determined by bubble pressure calculations.

---

1. Udovenko, V.V., Mazanko, T.F., *Zh. Fiz. Khim.*, 41:1615 (1967).

GAMMFIT.xls

Example 11.16 Using Excel for fitting model parameters (Continued)

System Components			Parameters to adjust				Antoine Coefficents			Calculated P^{sat}(mm Hg)	Expt P^{sat}(mm Hg)	Selected P^{sat}(mm Hg)
			A_{12}	A_{21}	T(C)		A	B	C			
(1) 2-propanol			2.173055	0.942929	30	1	8.87829	2010.33	252.636	58.277622	60.7	60.7
(2) water						2	8.07131	1730.63	233.426	31.740167	32.1	32.1
				<---optional----->								

x_1	x_2	$\gamma_{1,\,calc}$	$\gamma_{2,\,calc}$	$y_{1,expt}$	$y_{2,expt}$	$y_{1,calc}$	$y_{2,calc}$	P_{expt}	P_{calc}	$(P_{error})^2$	
0	1	8.785079	1	0	1	0	1	32.1	32.1	0	Enter 1 to use Calculated P^{sat}
0.0015	0.9985	8.695982	1.000008	0.0254	0.9746	0.024107	0.975893	33.8	32.84386	0.9141954	Enter 0 to use Expt P^{sat}
0.0111	0.9889	8.152908	1.000416	0.1374	0.8626	0.147468	0.852532	37.1	37.25008	0.0225244	0
0.0231	0.9769	7.535179	1.001787	0.2603	0.7397	0.251681	0.748319	42.3	41.98014	0.102312	
0.0357	0.9643	6.95177	1.004234	0.3577	0.6423	0.326426	0.673574	47.2	46.14952	1.1035166	Objective Function
0.0649	0.9351	5.815504	1.013755	0.4604	0.5396	0.42951	0.57049	55	53.33938	2.7576741	14.26798435
0.1168	0.8832	4.353254	1.043423	0.5316	0.4684	0.510602	0.489398	60.3	60.44532	0.0211192	
0.197	0.803	2.970369	1.119928	0.5547	0.4453	0.551655	0.448345	62.9	64.38697	2.2110888	
0.2271	0.7729	2.62311	1.158008	0.5611	0.4389	0.557245	0.442755	63.5	64.88977	1.9314539	
0.312	0.688	1.945005	1.292474	0.5659	0.4341	0.563409	0.436591	64.4	65.3793	0.9590351	
0.3958	0.6042	1.549309	1.463142	0.5907	0.4093	0.567416	0.432584	65.1	65.59962	0.2496163	
0.4477	0.5523	1.386629	1.586209	0.589	0.411	0.572644	0.427356	65.8	65.80383	1.468E-05	
0.5009	0.4991	1.264065	1.724032	0.6098	0.3902	0.581846	0.418154	66.6	66.05434	0.2977452	
0.6369	0.3631	1.083192	2.106142	0.6462	0.3538	0.630433	0.369567	66.9	66.42417	0.2264146	
0.7542	0.2458	1.01937	2.411709	0.7296	0.2704	0.710348	0.289652	66.8	65.69551	1.2198966	
0.8245	0.1755	1.004463	2.546001	0.7752	0.2248	0.778018	0.221982	65.7	64.61353	1.1804165	
0.9363	0.0637	0.999471	2.622202	0.8892	0.1108	0.913749	0.086251	63.2	62.16513	1.0709608	
1	0	1	2.567492	1	0	1	0	60.7	60.7	0	

The results of the fit are shown by the plot on spreadsheet "P-x-y Plot." This spreadsheet is shown below. Note that the system is the same used in Example 11.5 on page 365, and naturally the fit in this example using the additional data is superior.

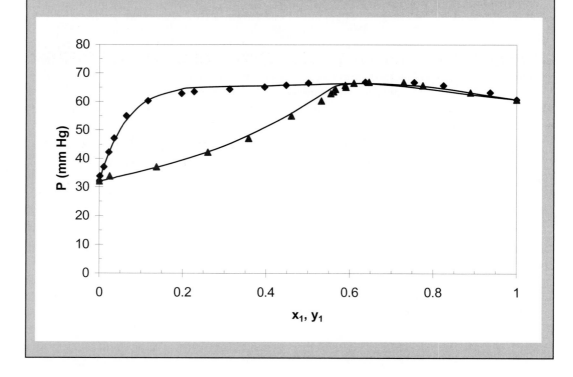

Objective functions

An alternate choice of objective function for a given set of data will usually result in a slightly different set of parameters. In Example 11.16, the total pressure was used, OBJ $= \sum_{points} (P_{expt} - P_{calc})^2$. While the total pressure is often measured accurately, it may be desired to use vapor compositions in the objective function. For example, it is not uncommon for the pure component vapor pressures measured by investigators to differ from literature data which is indicative of a impurity or a systematic error. One method of incorporating additional considerations into the fitting procedure is to use weighted objective functions, where recognition is made of probable errors in measurements. One of the most rigorous methods uses the maximum likelihood principles, which asserts that all measurements are subject to random errors and therefore all measurements have some uncertainty associated with them. Such techniques are discussed by Anderson, et al.,[1] and Prausnitz, et al.[2] The objective function for such an approach takes the form

$$\sum_{points} \left\{ \frac{(P_{true} - P_{expt})_i^2}{\sigma_P^2} + \frac{(T_{true} - T_{expt})_i^2}{\sigma_T^2} + \frac{(x_{1\,true} - x_{1\,expt})_i^2}{\sigma_{1x}^2} + \frac{(y_{1\,true} - y_{1\,expt})_i^2}{\sigma_{1x}^2} \right\} \qquad 11.110$$

where σ represents the variance for each type of measurement. The "true" values are calculated as part of the procedure. Typical values for variances are: $\sigma_P = 2$ mmHg, $\sigma_T = 0.2$ K, $\sigma_x = 0.005$, $\sigma_y = 0.01$, and therefore the weight of a measurement is dependent on the probable experimental error in the value.

11.8 *T* AND *P* DEPENDENCE OF EXCESS GIBBS ENERGY (OPTIONAL)

For the development of accurate process calculations, a thermodynamic model should accurately represent the temperature and pressure dependence of deviations from ideal solution behavior. The excess functions follow the same relations as the total functions, $H^E = U^E + PV^E$, $G^E = H^E - TS^E$, $A^E = U^E - TS^E$. The derivative relations are also followed,

$$\left(\frac{\partial G^E}{\partial T} \right)_{P,x} = -S^E \qquad 11.111$$

$$\left(\frac{\partial G^E}{\partial P} \right)_{T,x} = V^E \qquad 11.112$$

$$\left(\frac{\partial G^E / T}{\partial T} \right)_{P,x} = \frac{-H^E}{T^2} \qquad 11.113$$

1. Anderson, T.F., Abrams, D.S., Grens, E.A., AICHE J., 24:20 (1978).
2. Prausnitz, J., Anderson, T., Grens, E., Eckert, C., Hsieh, R., O'Connell, J., *Computer Calculations for Multicomponent Vapor-Liquid Equilibria,* Prentice-Hall, 1980.

Partial molar quantities may also be used in the equations in this section. Particularly useful is Eqn. 11.113,

$$\left(\frac{\partial \ln \gamma_i}{\partial T}\right)_{P,x} = \frac{-\bar{H}_i^E}{RT^2}$$

11.114

Therefore, excess enthalpy data from calorimetry may be used to check the temperature dependence of the activity coefficient models for thermodynamic consistency. Typically, activity coefficient parameters need to be temperature-dependent for representing data accurately; however, temperature-independent parameters are often suitable for screening and preliminary design. The Gibbs-Duhem equation is also useful for checking thermodynamic consistency of data; however, the applications are subject to uncertainties themselves.[1]

11.9   THE MOLECULAR BASIS OF SOLUTION MODELS (OPTIONAL)

As discussed during the development of quadratic mixing rules, there comes a point at which the assumption of random mixing cannot completely explain the non-idealities of the solution. Local compositions are examples of non-randomness. The popularity of local composition models like Wilson's equation or UNIFAC means that we need to develop some appreciation of the strength of the underlying theory and its limitations. Similar to the situation for the random mixing models, there are limitations to the local composition models. At this time, however, we are not exactly sure what all the limitations are. This is still a question for active research. Nevertheless, we can provide an understanding of the assumptions in these models, because the assumptions are the sources of errors that impose limitations.

Extending the Energy Equation to Mixtures

We begin the discussion with the energy equation, not the pressure equation as we did for pure fluids. This is because we are presently concerned with the Gibbs energy of mixing and its excess change relative to ideal solution behavior. It turns out that the excess Gibbs energy is dominated by the excess internal energy in most cases. In other words, the entropy of mixing is given to reasonable accuracy by the mixing rule on "b" given in Section 10.1. Therefore, we focus our attention on extending the energy equation to mixtures. This requires revisiting our development of the energy equation for pure fluids and applying the same principles to extend it to mixtures. With two small modifications, the energy equation we developed for pure fluids becomes:

$$\frac{U_1 - U_1^{ig}}{N_1 kT} = \frac{N_1}{2V} \int \frac{u_{11}}{kT} g_{11} 4\pi r^2 dr$$

11.115

The small modifications are: (1) we have put the equation on an atomic basis instead of a molar basis by noting $n_i N_A = N_i$ and $Nk = nR$; (2) we recognize that these are the contributions to the energy departure that arise from atoms of type "1." For the pure fluid, it so happens that we only have atoms of type "1."

1. Van Ness, H.C., Byer, S.M., Gibbs, R.E., *AIChE J.*, 19:238 (1973).

In developing the energy equation for pure fluids, we recognized that the average internal energy departure per atom [i.e., $(\underline{U}_1 - \underline{U}_1{}^{ig})/N_1$] was equal to the energy per pair per unit volume times the local density in that volume integrated over the total volume. To make the extension to a mixture, we must simply recognize that there are now atoms of "type 2" around those atoms of "type 1." To illustrate, consider a parking lot full of blue cars and green cars. If one parking lot had only green cars, then the average energy per green car would involve the average number of green cars at each distance around one green car times the energy associated with green cars being that distance from each other. If you pack them too close, you will have to work hard to pack them. Now consider the next parking lot, where blue cars are mixed with green cars. The average energy per *green* car will now involve contributions from green-green interactions and blue-green interactions. In equation form, this becomes,

$$\frac{U_g - U_g^{ig}}{N_g kT} = \frac{N_g}{2V} \int \frac{u_{gg}}{kT} g_{gg} 4\pi r^2 dr + \frac{N_b}{2V} \int \frac{u_{bg}}{kT} g_{bg} 4\pi r^2 dr \qquad 11.116$$

We can check this equation by noting that it approaches the pure green car equation when all the blue cars leave the parking lot (i.e., as $N_b \to 0$). On the other hand, note that the energy per green car does not approach zero as the density of green cars approaches zero. Even if there is only one green car per 10^{23} blue, the energy per green car is not negligible. Its contribution may be negligible relative to the total mixture energy, but when we look at the energy per *green* car, we are looking at the world from the green car's perspective. We may next write the average energy per blue car by symmetry and the total energy by addition.

$$\frac{U_b - U_b^{ig}}{N_b kT} = \frac{N_b}{2V} \int \frac{u_{bb}}{kT} g_{bb} 4\pi r^2 dr + \frac{N_g}{2V} \int \frac{u_{gb}}{kT} g_{gb} 4\pi r^2 dr \qquad 11.117$$

$$\frac{U - U^{ig}}{NkT} = \frac{N_g}{N}\left(\frac{U_g - U_g^{ig}}{N_g kT}\right) + \frac{N_b}{N}\left(\frac{U_b - U_b^{ig}}{N_b kT}\right)$$

$$= \frac{N_g}{N}\left(\frac{N_g}{2V}\int\frac{u_{gg}}{kT}g_{gg}4\pi r^2 dr + \frac{N_b}{2V}\int\frac{u_{bg}}{kT}g_{bg}4\pi r^2 dr\right)$$

$$+ \frac{N_b}{N}\left(\frac{N_b}{2V}\int\frac{u_{bb}}{kT}g_{bb}4\pi r^2 dr + \frac{N_g}{2V}\int\frac{u_{gb}}{kT}g_{gb}4\pi r^2 dr\right)$$

Finally, making the substitutions in terms of the mole fractions and converting back to a molar basis, we see that for multicomponent mixtures,

$$\frac{U - U^{ig}}{RT} = \frac{N_A \rho}{2} \sum_i \sum_j x_i x_j \int \frac{N_A u_{ij}}{RT} g_{ij} 4\pi r^2 dr \qquad 11.118$$ ❗ The energy equation for mixtures.

Comparing to the result for pure fluids

$$a_{ii} = -\tfrac{1}{2}\int N_A^2 u_{ii} g_{ii} 4\pi r^2 dr \qquad 11.119$$

$$\Rightarrow \frac{U - U^{ig}}{RT} = -\frac{\rho}{RT} \sum_i \sum_j x_i x_j a_{ij} \Rightarrow a = \sum_i \sum_j x_i x_j a_{ij} \qquad 11.120$$

where $a_{ij} = -\frac{1}{2} \int N_A^2 u_{ij} g_{ij} 4\pi r^2 dr$ and it is understood that $a_{ij} = a_{ji}$.

In this form we can recognize that the radial distribution functions may be dependent on composition as well as temperature and density. Therefore, assuming the quadratic mixing rule simply neglects the composition dependence of the a parameter, as well as the temperature and density dependence. We found in Unit II that the assumptions about temperature and density in the van der Waals equation were flawed and that is what motivated the Peng-Robinson equation. Similarly, neglecting the composition dependence of the radial distribution functions leads to some limitations that give rise to local composition theory.

Local Compositions in Terms of Radial Distribution Functions

Recalling the energy equation for mixtures,

$$U - U^{ig} = \frac{N_A \rho}{2} \sum_i \sum_j x_i x_j \int N_A u_{ij} \, g_{ij} 4\pi r^2 dr \qquad 11.121$$

We may now define the local compositions in terms of the radial distribution functions.

$$x_{ij} = \frac{N_i \sigma_{ij}^3}{\underline{V} N c_j} \int_0^{R_{ij}} g_{ij} 4\pi r_{ij}^2 dr_{ij} = \frac{x_i n N_A \sigma_{ij}^3}{\underline{V} N c_j} \int_0^{R_{ij}} g_{ij} 4\pi r_{ij}^2 dr_{ij} \qquad 11.122$$

where $r_{ij} = r/\sigma_{ij}$

R_{ij} = "neighborhood"

Nc_i = total # of atoms around sites of type "i," i.e., the coordination number.

Rearrangement gives the molecular definition of the local composition parameter Ω_{ij},

$$\frac{x_{ij}}{x_{jj}} = \frac{Nc_j N_i \sigma_{ij}^3}{Nc_j N_j \sigma_{jj}^3} \frac{\int g_{ij} 4\pi r_{ij}^2 dr_{ij}}{\int g_{jj} 4\pi r_{jj}^2 dr_{jj}} \equiv \frac{x_i}{x_j} \Omega_{ij} \qquad 11.123$$

and we note the similarity between the integral in the energy equation and integral in the definition of local composition.

For a square-well fluid, ε_{ij} = constant so we can factor it out of the integral,

$$\int u_{ij} g_{ij} 4\pi r^2 dr = -\varepsilon_{ij} \int g_{ij} 4\pi r^2 dr \qquad 11.124$$

$$U - U^{ig} = -\frac{1}{2} \sum \sum x_j \frac{x_i N_A \sigma_{ij}^3}{\underline{V}} n N_A \varepsilon_{ij} \int g_{ij} 4\pi r_{ij}^2 dr_{ij} \qquad 11.125$$

Substituting Nc_j, and x_{ij} into the energy equation for mixtures (multiply Eqn. 11.122 by Nc_j and substitute into Eqn. 11.125),

$$(U - U^{ig}) = -\frac{1}{2} \sum_j x_j Nc_j \sum_i x_{ij} N_A \varepsilon_i \qquad \text{11.126}$$

This is the equation previously applied as the starting point for development of the Gibbs excess energy model from a local composition perspective. The rest of the derivation proceeds as before.

Assumptions in Local Composition Models

In the previous section, we discussed some of the currently popular expressions for activity coefficients. We listed the assumptions involved in developing the expressions but we did not take time to discuss those assumptions. Instead, we directly applied the expressions as a practical necessity and moved on. In this section, we recall those assumptions and attempt to put them in perspective. After developing this perspective, we conclude with a word of caution; the reliability of the predictions depends largely on the accuracy of the assumptions.

The local composition theory, upon which UNIQUAC and others are based, has the general intent of correcting regular solution theory for asymmetries in solution behavior due to fluid structure near a central species. Relaxing the assumption that $S^E = 0$ also leads to the necessity of considering the entropy of mixing and differences between the UNIQUAC model and Wilson's model are primarily due to differences in treatment of S^E. In review, four assumptions are shared by the local composition theories when considered with respect to spheres:

1. The average energy of an i-j interaction is independent of temperature, density, and other species present.

2. $(A - A^{is}) = (G - G^{is})$

3. The "coordination number" of a specie in a mixture is the same as that of the pure specie.

4. The temperature-dependent part of the energy of mixing is given by
   $\Omega_{ij} = (\sigma_{ij}/\sigma_{ij})^3 \exp[z(\varepsilon_{ij} - \varepsilon_{jj})/2kT]$ where z is a "coordination number"

Wilson's equation makes the following assumptions:

5w. $Nc_j = z = 2$ for all j at all densities.

6w. $(A^E/RT)^{COMB} = \sum_j x_j \ln(\Phi_j/x_j)$

7w. $(\sigma_{ij}/\sigma_{jj})^3 = V_i/V_j$ for all i, j, and $\dfrac{A_{ji}}{RT} = \dfrac{z(\varepsilon_{ij} - \varepsilon_{jj})}{2kT}$, $\Lambda_{ji} = \Omega_{ij}$.[1]

UNIFAC(QUAC) makes the following assumptions:

5u. $z = 10$, $Nc_j = 10 \cdot \sum_k v_k^{(j)} Q_k$

---

1. Recall that the order of subscripts for the local compositions in Wilson's equation is non-intuitive, see Eqns. 11.64 and 11.65.

6u. $(A^E/RT)^{COMB} = \sum_j [x_j \ln(\Phi_j/x_j) + 1/2 \; Nc_j x_j \ln(\theta_j/\Phi_j)]$

7u. $(\sigma_{ij}/\sigma_{jj})^3 = Nc_i/Nc_j = q_i/q_j$ for all i, j, $\dfrac{a_{ij}}{T} = \dfrac{z(\varepsilon_{ij} - \varepsilon_{jj})}{2kT}$, $\Omega_{ij} = \dfrac{q_i}{q_j}\tau_{ij}$

Assumption 1 involves factoring some average energy out of the energy integral such that the local composition integral is obtained. As noted, this assumption would be correct for a square-well potential so we can probably trust that it would be reasonable for other similar potentials like the Lennard-Jones. The doubt which arises, however, involves the application of this approximation to highly non-ideal mixtures. The square-well and Lennard-Jones potentials rarely give rise to highly non-ideal mixtures when realistic values for their parameters are chosen. There is very little evidence to judge whether the ε_{ij} factored out in this way is really independent of temperature and density for non-ideal mixtures. In fact, in Chapter 1, we showed that dipole interactions *are* temperature dependent.

Assumption 2 has to do with neglecting $\ln(Z/Z^{is})$. For liquids, this may seem dangerous until one realizes that it amounts to neglecting $\ln(1 + \rho^E/\rho) \approx \rho^E/\rho$. Relative to the density of a liquid, the excess density is generally small (but easy to measure with a high degree of accuracy) and this assumption is acceptable.

Assumption 3 has to do with convenience. If the coordination number of each specie was assumed to change with mixing, the theory could become very complicated. That is not a very good physical reason of course. Physically speaking, this assumption could become quite poor if the sizes of the molecules (or segments in the case of UNIFAC) were very different.

Assumption 4 is the primary assumption behind all of the current local composition approaches, but it is not required by the concept of local compositions. It is simply computationally expedient in the equations that develop. The crucial aspect of the assumption is the simple form of the temperature dependence of Ω_{ij}. The main motivation for this assumption appears to be obtaining an expression which can be integrated analytically. But how accurate is this assumption on a physical basis? Moreover, how can we determine the physical behavior for the behavior of Ω_{ij} in an unequivocal manner? Merely fitting experimental data for the Gibbs excess function is equivocal because some set of adjustable parameters will provide a good fit even if the model has no physical basis. An alternative available to us that was not available to van der Waals is to apply computer simulation of square-well mixtures over a specific range of densities and temperatures and test the validity of Wilson's approximation directly through the simulated local compositions. This approach was undertaken by Sandler and Lee.[1]

Sandler and Lee have developed a correlation for what amounts to Ω_{ij} of a square-well potential.

$$\Omega_{ij} = (\sigma_{ij}/\sigma_{jj})^3 \exp[N_A(\varepsilon_{ij} - \varepsilon_{jj})/2RT] \left\{ \frac{\sqrt{2} + \rho\sigma_{jj}^3[\exp(\varepsilon_{jj}/kT) - 1]}{\sqrt{2} + \rho\sigma_{ij}^3[\exp(\varepsilon_{ij}/kT) - 1]} \right\} \qquad 11.127$$

This expression reproduces the local compositions of a substantial set of molecular simulation data which Sandler and coworkers have generated. We can therefore use this expression along with the molecular simulation data as a guide to the accuracy of Wilson's assumption.

---

1. Sandler, S.I., Lee, K.-H., Fluid Phase Equil. 30:135 (1986).

The most important consideration is the temperature dependence of this parameter because it is the integration with respect to temperature that allows us to get from energy to free energy. As for density, it could be argued that the density of all liquids is roughly the same so it is not unreasonable to pick a specific density and just study the temperature effects. Suppose

$$(\sigma_{22}/\sigma_{11})^3 = 1, \quad \rho\sigma^3 = 0.7, \quad \text{and} \quad \varepsilon_{22} = 2\varepsilon_{11} \qquad\qquad 11.128$$

$$\ln \Omega_{ij} = \frac{(\varepsilon_{ij} - \varepsilon_{jj})}{2kT} + \ln\left[\frac{0.714 + 0.7 \exp(\varepsilon_{jj}/kT)}{0.714 + 0.7 \exp(\varepsilon_{ij}/kT)}\right] \qquad\qquad 11.129$$

If the natural logarithm term on the right-hand side in the above equation is small or gives rise to a contribution which is linear in $1/kT$, then Wilson's assumption is basically correct and his definition of ε_{ij} would just be a little different than 11.83. If $\ln \Omega_{ij}$ versus $1/kT$ is not linear, however, then Wilson's assumption is very questionable. Figure 11.9 shows $\ln \Omega_{ij}$ is fairly linear over certain ranges of temperature. This suggests that the primary assumption of Wilson and UNIFAC(QUAC) is not unreasonable. Does this mean that the problem of non-ideal solutions is solved? Maybe, but maybe not. Unfortunately, we must look closely at the range of temperatures that are applicable. This range is limited by the tendency for the mixture to phase separate. That is, dropping the temperature at constant overall density eventually places the conditions inside the binodal curve. For the composition and density listed above, the binodal occurs when $N_A(\varepsilon_{12} - \varepsilon_{22})/RT < -0.4$. This corresponds to a maximum value of A_{12} in Wilson's equation on the order of ~300 cal/mol at temperatures near 400 K. To use larger values for A_{12} would be unsupported, but larger values are often used, as illustrated in Example 11.13.

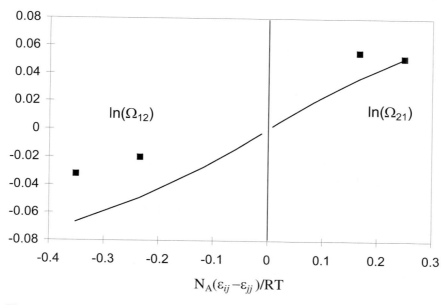

Figure 11.9 *Local composition parameters vs. temperature. Points are molecular simulation data and the curve is the correlation of Sandler and Lee (1986). The approximate linearity of the plot lends support to the assumption applied in integrating the internal energy to obtain the free energy.*

Another indication of the potential for error with the local composition approach is given by experimental data for excess enthalpies of mixing. Relations from classical thermodynamics make it possible to estimate the enthalpy of mixing by taking the derivative of the Gibbs energy of mixing with respect to temperature. Larsen et al., have developed a modified form of UNIFAC to address this problem.[1] Not surprisingly, the modification involves the introduction of a substantial number of additional adjustable parameters. Even though the modified form does improve the accuracy of all the thermodynamic properties for a large number of systems there are many systems for which the predicted heats of mixing are in error by 100–700%. More importantly, there is no way of knowing in advance when the predictions will be in error or when they will be accurate.

So why do these approximations fit the activity coefficient data? Because they have enough adjustable parameters to fit the data. Even the Margules one-parameter equation is good enough for that in many cases, but we suffer few delusions about its physical accuracy. In conclusion, we must say that local composition theory has much to recommend it. It does fit a great wealth of experimental data and there is some justification for its form via the theoretical physics which can be applied. But it is often extrapolated too far and that can lead to miscalculations by unwary users. In the end, we must never underestimate the value of experimental data for non-ideal mixtures and apply the currently available theory with a careful and mildly critical view.

Assumptions 5 through 7 have to do with the entropy of mixing. The inclusion of the Staverman-type modifications to address the differences between surface fraction and volume fraction are generally recognized to have a reasonable physical basis as indicated by polymer lattice computer simulations. This modification and the estimate of molecular volumes by group contributions instead of liquid molar volumes comprise the primary differences between the Wilson and UNIQUAC models.

The UNIFAC theory is distinguished from UNIQUAC in that the solution is assumed to be a mixture of functional groups, not molecules. The UNIQUAC theory is then applied to each type of group interaction. The values for the group interactions are then regressed from a data base that includes phase equilibrium data for many, many systems. In one sense, the UNIFAC method is more like a massive regression than a truly predictive method. Thus it lies somewhere between the purely correlative method of fitting van Laar constants and the purely predictive method of Scatchard-Hildebrand theory (or any equation of state with $k_{ij} = 0$). Like any regressed equation, it can be unreliable if extrapolated far beyond the originally applied data. If you are ever in the position of designing truly novel chemical systems, you should be especially sensitive to the need for specific experimental data.

11.10   SUMMARY

The strategies for problem solving remain much the same as the strategy set forth at the end of Chapter 10, and a review of that strategy is suggested. Like Chapter 10, the objective of a problem may be to fit experimental data, or to solve an applied problem. Use Table 9.2 on page 302 and the information in Section 9.1–9.2 to identify known variables and then the correct routine to use, and the approximations that may be valid. Recognize that the primary difference between Chapters 9, 10, and 11, and is the method used to calculate K_i.

1. Larsen, B., Rasmussen, P.S., Fredenslund, Aa., *Ind. Eng. Chem. Res.*, 26:2274 (1987).

This has been a lengthy and theoretical chapter. We have "constructed three thousand different theories" yet not found any particular arguments which elevate a single theory above all others. What have you learned? What should you have learned?

The big picture that you should retain is that phase equilibrium is controlled by competition between energetic and entropic driving forces. The entropy is driving the molecules to mix, mildly but steadily. The energy of mixing is much less predictable. Sulfuric acid and water may react very favorably towards each other ($G^E \ll 0$), while isopropanol and water have enhanced escaping tendencies because the energy required for forcing them to mix is counteracting the entropic driving force ($G^E \gg 0$). These facts are due in large part to the physical interaction energies characterized by quantities like the solubility parameter. The solubility parameter provides a quantitative interpretation of the familiar guideline that "like dissolves like." The best quantitative models, however, are based on the concept of local compositions, recognizing that the energetic fields around molecules are short-ranged.

Carrying forward the molecular perspective, we can characterize these tendencies in terms of the cross interaction energy (ε_{ij}). If the cross-interaction energy is weaker than the geometric mean ($k_{ij} > 0$), then each component prefers its own company. If the cross-interaction energy is stronger than the geometric mean ($k_{ij} < 0$), then the components are strongly attracted, releasing energy as they fall into the well of their mutual attraction. The energy to break the favorable interactions must be added to separate such a mixture and this may show up as a maximum boiling azeotrope. Conversely, the mixtures with $k_{ij} > 0$ tend to exhibit minimum boiling azeotropes.

At the practical level, it is desirable to clarify when to use a particular solution theory. The detailed considerations of such a selection are presented in Appendix D. An outline is presented here. Briefly, the UNIFAC and UNIQUAC theories stand as the primary options of choice because they can easily describe multicomponent mixtures and liquid-liquid instability (to be discussed in the next chapter). They may appear cumbersome when first encountered, but computers obviate that concern. Occasionally, however, there may not be experimental data to permit the use of UNIQUAC, and the UNIFAC parameters may not be available for a particular mixture of interest. In that case, the Scatchard-Hildebrand theory can provide an approximate guideline. But you really should find a way to get some experimental data if you need to perform a detailed process design, and base your thermodynamic model parameters on that data.

11.11   PRACTICE PROBLEMS

P11.1 Ninov et al. (*J. Chem. Eng. Data*, 40:199, 1995) have shown that the system diethylamine(1) + chloroform(2) forms an azeotrope at 1 bar, 341.55 K and $x_1 = 0.4475$. Determine the temperature and vapor composition at $x_1 = 0.80$. Is this a maximum boiling or minimum boiling azeotrope? (ANS. 331 K, 0.97)

P11.2 The following lattice contains x's, o's, and void spaces. The coordination number of each cell is 8. Estimate the local composition (X_{xo}) and the parameter Ω_{xo} based on rows and columns away from the edges. (ANS. 0.68, 1.47)

		O		O		X		O	
	X		O	X			X		
	X		X	X		O	X	O	
O		X		O			X		
						O			X
	O		X		O			X	
	X			O		O		X	
O	X				X		X		O

P11.3 Acrolein + water exhibits an atmospheric (1 bar) azeotrope at 97.4 wt% acrolein and 52.4°C

(a) Determine the values of A_{ij} for the Van Laar equation that match this bubble point pressure at the same liquid and vapor compositions and temperature. (ANS. 2.97, 2.21)

(You may use the shortcut vapor pressure equation for acrolein: $T_c = 506$ K; $P_c = 51.6$ bar; and $\omega = 0.330$; MW = 56).

(b) Tabulate P at 326.55 K and x = {0.1,0.3,0.5} via the Van Laar equation using the A_{12} and A_{21} determined above. (ANS. P = {1.11, 1.15, 1.03})

P11.4 The system α-epichlorohydrin(1) + n-propanol(2) exhibits an azeotrope at 760 mmHg and 96°C containing 16 mol% epichlorohydrin. Use van Laar theory to estimate the composition of the vapor in equilibrium with a 90 mol% epichlorohydrin liquid solution at 96°C. (α-epichlorohydrin has the formula C_3H_5ClO, and IUPAC name 1-chloro-2,3-epoxypropane. Its vapor pressure can be approximated by: $\log_{10} P^{sat} = 8.0270 - 2007/T$, where P^{sat} is in mmHg and T is in Kelvin. You can use the shortcut vapor pressure equation for n-propanol). (ANS. 0.72, 0.63 bar)

P11.5 The following free energy model has been suggested for a particularly unusual binary liquid-liquid mixture. Derive the expression for the activity coefficient of component 1

$$\frac{\Delta \underline{G}^E}{nRT} = (\Phi_1 \Phi_2)^{0.75} V \frac{(\delta_1 - \delta_2)^2}{RT}$$

where $\Phi_i = x_i V_i \, / \, (\sum_i x_i V_i)$, $V = \sum_i x_i V_i$

(ANS. $\dfrac{V_1 (\delta_1 - \delta_2)^2}{RT} \Phi_2^{0.75} [0.25\Phi_1^{0.75} + 0.75\Phi_1^{0.25}\Phi_2]$)

11.12   HOMEWORK PROBLEMS

11.1 Show that Wilson's equation reduces to Flory's equation when $A_{ij} = A_{ji} = 0$. Further, show that it reduces to an ideal solution if the energy parameters are zero, *and* the molecules are the same size.

11.2 In vapor-liquid equilibria the relative volatility α_{ij} is defined to be the ratio of the K value for species i to that for species j, that is, $\alpha_{ij} = K_i/K_j = (y_i/x_i)/(y_j/x_j)$.

(a) Provide a simple proof that the relative volatility is independent of liquid and vapor composition if a system follows Raoult's Law.

(b) In approximation to a distillation calculation for a non-ideal system, calculate the relative volatility α_{12} and α_{21} as a function of composition for the data provided in problem 11.26.

(c) In approximation to a distillation calculation for a non-ideal system, calculate the relative volatility α_{12} and α_{21} as a function of composition for the data provided in problem 9.2.

(d) Provide conclusions from your analysis.

11.3 The volume change on mixing for the liquid methyl formate(1) + liquid ethanol(2) system at 298.15 K may be approximately represented by J. Polack, B.C.-Y. Lu, *J. Chem Thermodynamics,* 4:469 (1972)

$$\Delta V_{mix} = 0.8 x_1 x_2 \text{ cm}^3/\text{mol}$$

(a) Using this correlation, and the data $V_1 = 67.28$ cm^3/mol, $V_2 = 58.68$ cm^3/mol, determine the molar volume of mixtures at $x_1 = 0, 0.2, 0.4, 0.6, 0.8, 1.0$ in cm^3/mol.

(b) Analytically differentiate the above expression and show that

$$\overline{V}_1^E = 0.8 x_2^2 \text{ and } \overline{V}_2^E = 0.8 x_1^2 \text{ cm}^3/\text{mol}$$

and plot these partial molar excess volumes as a function of x_1.

11.4 Scatchard-Hildebrand, Flory models, UNIFAC, and UNIQUAC all depend on the volume fractions of components. To obtain the activity coefficient expressions, the excess Gibbs energy must be differentiated. Demonstrate that you understand the principles by obtaining

the derivative $\left(\dfrac{\partial \Phi_1}{\partial n_1}\right)_{T, P, n_2}$ in a binary system.

11.5 In the system $A + B$, activity coefficients can be expressed by the one-parameter Margules equation with $A = 0.5$. The vapor pressures of A and B at 80°C are $P_A^{sat} = 900$ mmHg, $P_B^{sat} = 600$ mmHg. Is there an azeotrope in this system at 80°C, and if so, what is the azeotrope pressure and composition?

11.6 The system acetone(1) + methanol(2) is well-represented by the one-term Margules equation using $A = 0.605$ at 50°C.

(a) What is the bubble pressure for an equimolar mixture at 50°C?

(b) What is the dew pressure for an equimolar mixture at 50°C?

(c) What is the bubble temperature for an equimolar mixture at 760 mmHg?

(d) What is the dew temperature for an equimolar mixture at 760 mmHg?

11.7 The excess Gibbs energy for a liquid mixture of n-hexane(1) + benzene(2) at 30°C is represented by $G^E = 1089\ x_1 x_2$ J/mol.

(a) What is the bubble pressure for an equimolar mixture at 30°C?

(b) What is the dew pressure for an equimolar mixture at 30°C?

(c) What is the bubble temperature for an equimolar mixture at 760 mmHg?

(d) What is the dew temperature for an equimolar mixture at 760 mmHg?

11.8 The liquid phase activity coefficients of the ethanol(1) + toluene(2) system at 55°C are given by the two-parameter Margules equation, where $A_{12} = 1.869$ and $A_{21} = 1.654$.

(a) Show that the pure saturation fugacity coefficient is approximately 1 for both components.

(b) Calculate the fugacity for each component in the liquid mixture at $x_1 = 0.0, 0.2, 0.4, 0.6, 0.8, 1.0$. Summarize your results in a table. Plot the fugacities for both components versus x_1. Label your curves. For each curve, indicate the regions that may be approximated by Henry's Law and the ideal solution model.

(c) Using the results of part (b), estimate the total pressure above the liquid mixture at 55°C when a vapor phase coexists. Assume the gas phase is ideal for this calculation. Also estimate the vapor composition.

(d) Comment on the validity of the ideal gas assumption used in part (c).

11.9 The acetone(1) + chloroform(2) system can be represented by the van Laar equation using $A_{12} = -1.0105$, $A_{21} = -0.7560$ at 35.17°C. Use bubble pressure calculations to generate a P-x-y and y-x diagram and compare it with the selected values from the measurements of Zawidzki, *Z. Phys. Chem.*, 35, 129(1900).

P(mmHg)	x_1	y_1
262.6	0.1953	0.1464
248.4	0.4188	0.4368
255.7	0.507	0.5640
272.2	0.6336	0.7271
290.5	0.7296	0.8273
320.1	0.8797	0.9377

11.10 For a particular binary system, data are available:

$$T = 45°C \quad P = 37 \text{ kPa} \quad x_1 = 0.398 \quad y_1 = 0.428$$

In addition, $P_1^{sat} = 27.78$ kPa and $P_2^{sat} = 29.82$ kPa. From these data,

(a) fit the one-parameter Margules equation
(b) fit the two-parameter Margules equation
(c) fit the van Laar Equation

11.11 The compositions of coexisting phases of ethanol(1) + 1-propanol(2) at 55°C are $x_1 = 0.7186$, and $y_1 = 0.7431$ at $P = 307.81$ mmHg, as reported by Kretschmer and Wiebe, *J. Amer. Chem. Soc.*, 71, 1793(1949). Estimate the bubble pressure at 55°C and $x_1 = 0.1$, using:

(a) the one-parameter Margules equation
(b) the two-parameter Margules equation
(c) the van Laar equation

11.12 A vapor/liquid experiment for the carbon disulfide(1) + chloroform(2) system has provided the following data at 298 K: $P_1^{sat} = 46.85$ kPa, $P_2^{sat} = 27.3$ kPa, $x_1 = 0.2$, $y_1 = 0.363$, $P = 34.98$ kPa. Estimate the dew pressure at 298 K and $y_1 = 0.6$, using:

(a) the one-parameter Margules equation
(b) the two-parameter Margules equation
(c) the van Laar equation

11.13 The (1) + (2) system forms an azeotrope at $x_1 = 0.75$ and 80°C. At 80°C, $P_1^{sat} = 600$ mmHg, $P_2^{sat} = 900$ mmHg. The liquid phase can be modeled by the one-parameter Margules equation.

(a) Estimate the activity coefficient of component 1 at $x_1 = 0.75$ and 80°C. [Hint: The relative volatility (given in problem 11.2) is unity at the azeotropic condition].

(b) Qualitatively sketch the P-x-y and T-x-y diagrams that you expect.

11.14 Ethanol(1) + benzene(2) and form azeotropic mixtures. Compare the specified model to the experimental data of Brown and Smith cited in problem 9.2.

(a) Prepare a y-x and P-x-y diagram for the system at 45°C assuming the van Laar model and using the experimental pressure at $x_1 = 0.415$ to estimate A_{12} and A_{21}.

(b) Prepare a y-x and P-x-y diagram for the system at 45°C with the predictions of Scatchard-Hildebrand theory with $k_{12} = 0$.

(c) Prepare a y-x and P-x-y diagram for the system at 45°C assuming the UNIFAC model.

11.15 The CRC Handbook lists the azeotrope for the acetone + chloroform system as 64.7°C and 20 wt% acetone.

(a) Use the van Laar model to estimate the T-x-y diagram at 1 bar

(b) How are the van Laar parameters different for this system relative to the parameters for the ethanol + benzene system of problem 11.14(a).?

11.16 Using the van Laar model and the following data estimate the total pressure and composition of the vapor in equilibrium with a 20 mol% ethanol(1) solution in water(2) at 78.15°C. Data at 78.15°C:

$$P_1^{sat} = 1.006 \text{ bar}$$

$$P_2^{sat} = 0.439 \text{ bar}$$

$$\gamma_1^\infty = 1.6931; \quad \gamma_2^\infty = 1.9523$$

11.17 Using the data from problem 11.16, fit the two parameter Margules equation, and then generate a P-x-y diagram at 78.15°C.

11.18 A liquid mixture of 50 mol% chloroform(1) and 50% 1,4-dioxane(2) at 0.1013 MPa is metered into a flash drum through a valve. The mixture flashes into two phases inside the drum where the pressure and temperature are maintained at 24.95 kPa and 50°C. The compositions of the exiting phases are $x_1 = 0.36$ and $y_1 = 0.62$.

Your supervisor asks you to adjust the flash drum pressure so that the liquid phase is $x_1 = 0.4$ at 50°C. He doesn't provide any *VLE* data, and you are standing in the middle of the plant with only a calculator and pencil and paper, so you must estimate the new flash drum pressure. Fortunately, your supervisor has a phenomenal recall for numbers and tells you that the vapor pressures for the pure components at 50°C are $P_1^{sat} = 69.4$ kPa and $P_2^{sat} = 15.8$ kPa. What is your best estimate of the pressure adjustment that is necessary without using any additional information?

11.19 Suppose a vessel contains an equimolar mixture of chloroform(1) and triethylamine(2) at 25°C. The following data are available at 25°C:

	chloroform(1)	triethylamine(2)
MW	119.4	101.2
V^L (cm³/gmol)	80.19	139.
ΔH^{vap} (kJ/mol)	29.71	31.38
approx $\ln\gamma_i$	$-1.74x_2^2$	$-1.74x_1^2$
P^{sat} (mmHg)	193.4	67.3

(a) If the pressure in the vessel is 90 mmHg, is the mixture a liquid, a vapor, or both liquid and vapor? Justify your answer.
(b) Provide your best estimate of the volume of the vessel under these conditions. State your assumptions.

11.20 The actone(1) + chloroform(2) system has an azeotrope at $x_1 = 0.38$, 248 mmHg, and 35.17°C. Fit the Wilson equation, and predict the P-x-y diagram.

11.21 Model the behavior of ethanol(1) + toluene(2) at 55°C using the UNIQUAC equation and the parameters $r_1 = 2.1055$, $r_2 = 3.9228$, $q_1 = 1.972$, $q_2 = 2.968$, $a_{12} = -76.1573$ K, $a_{21} = 438.005$ K.

11.22 The UNIFAC and UNIQUAC equations use surface fraction and volume fractions. This problem explores the differences.

(a) Calculate the surface area and volume for a cylinder of diameter $d = 1.0$ and length $L = 5$ where the units are arbitrary. Calculate the surface area for a sphere of the same volume. Which object has a higher surface area to volume ratio?
(b) Calculate the volume fractions and surface area fractions for an equimolar mixture of the cylinders and spheres from part (a). Use subscript s to denote spheres and subscript c to denote cylinders.
(c) For this equimolar mixture, calculate the local composition ratios x_{cs}/x_{ss} and x_{sc}/x_{cc} for the UNIQUAC equation if the energy variables τ_{cs} and τ_{sc} are unity. For the equimolar mixture, substitute the values of volume fraction and surface fraction into the expression for UNIQUAC activity coefficients, and simplify as much as possible, leaving the q's as unknowns.
(d) Consider n-pentane and 2,2-dimethyl propane (also known as neopentane). Calculate the UNIQUAC r and q values for each molecule using group contribution methods. Compare the results with part (a). [Hint: You might want to think about the -C-C-C- bond angles.]

11.23 Consider a mixture of isobutene(1) + butane(2). Consider a portion of the calculations that would need to be performed by UNIFAC or UNIQUAC.

$$CH_3$$
$$|$$
$$CH_2 = C - CH_3 \qquad\qquad CH_3CH_2CH_2CH_3$$

$$(1) \qquad\qquad\qquad\qquad (2)$$

(a) Calculate the surface area and volume parameters for each molecule.

(b) Provide reasoning to identify which component has a larger liquid molar volume. Which compound has a larger surface area?

(c) Calculate the volume fractions for a equimolar mixture.

11.24 Solve problem 9.7 using UNIFAC to model the liquid phase.

11.25 The flash point of liquid mixtures is discussed in problem 9.12. For the following mixtures, estimate the flash point temperature of the following components and their equimolar mixtures using UNIFAC:

(a) methanol (LFL = 7.3%) + 2-butanone (LFL = 1.8%)
(b) ethanol (LFL = 4.3%) + 2-butanone (LFL = 1.8%).

11.26 Use the UNIFAC model to predict the VLE behavior of the n-pentane(1) + acetone(2) system at 1 bar and compare to the experimental data of Lo et al. *J. Chem. Eng. Data* 7:327 (1962).

x_1	0.021	0.134	0.292	0.503	0.728	0.953
y_1	0.108	0.475	0.614	0.678	0.739	0.906
T (°C)	49.15	39.58	34.35	33.35	31.93	33.89
P_1^{sat}	1.560	1.146	0.960	0.903	0.880	0.954
P_2^{sat}	0.803	0.551	0.453	0.421	0.410	0.445

11.27 According to Gmehling et al. (1994)[1], the system acetone + water shows azeotropes at: (1) 2793 mmHg, 95.1 mol% acetone, and 100°C; (2) 5155 mmHg, 88.4 mol% acetone and 124°C. What azeotropic pressures and compositions does UNIFAC indicate at 100°C and 124°C? Othmer et al. (1946) (cf. Gmehling[2]) have studied this system at 2570 mmHg. Prepare T-x-y or P-x-y plots comparing the UNIFAC predictions to the experimental data.

11.28 Activity coefficients are an implicit part of the equation of state but they can be determined explicitly by comparing the definitions of the K-ratios. Using the k_{ij} value fit at $x_e = 0.415$, compute the activity coefficients implied by the Peng-Robinson equation for the benzene + ethanol system and compare them with the UNIFAC values and the values determined from the experimental data of Brown and Smith (1954) cited in problem 9.2, using plots of activity coefficient versus composition.

11.29 Flash separations are fundamental to any process separation train. A full steady state process simulation consists largely of many consecutive flash calculations. Use UNIFAC to determine the temperature at which 20 mol% will be vaporized at 760 mmHg of an equimolar mixture liquid feed of n-pentane and acetone.

---

1. Gmehling et al., *Azeotropic Data,* VCH, NY, 1994.
2. Gmehling , J., Onken, V., Arlt, W., *Vapor-Liquid Equilibrium Data Collection,* DECHEMA, Frankfurt, 1977.

11.30 A preliminary evaluation of a new process concept has produced a waste stream of the composition given below. It is desired to reduce the waste stream to 10% of its original mass while recovering essentially pure water from the other stream. Since the solution is very dilute, we can use a simple equation known as Henry's Law to represent the system. According to Henry's Law, $f_i^L = h_i x_i = x_i \gamma_i^\infty P_i^{sat}$. Use UNIFAC to estimate the Henry's Law constants when UNIFAC parameters are available. Use the Scatchard-Hildebrand theory when UNIFAC parameters are not available. Estimate the relative volatilities (relative to water) of each component. Relative volatilities are defined in problem 11.2.

Compositions in mg/Liter are:

Methanol	H2S	Methyl Mercaptan	dimethyl sulfoxide	dimethyl disulfide
5100	30	50	50	100

11.31 Ethanol(1) + benzene(2) form azeotropic mixtures.

(a) From the limited data below at 45 °C, it is desired to estimate the constant A for the one-term Margules equation, $G^E/RT = Ax_1x_2$. Use all of the experimental data to give your best estimate.

x_1	0	0.3141	0.5199	1
y_1	0	0.3625	0.4065	1
$P(bar)$	0.2939	0.4124	0.4100	0.2321

(b) From your value, what are the bubble pressure and vapor compositions for a mixture with $x_1 = 0.8$?

11.32 Fit the data from problem 11.26 to the following model by regression over all points, and compare with the experimental data on the same plot, using:

(a) two-parameter Margules
(b) van Laar
(c) Wilson

11.33 Fit the specified model to the methanol(1) + benzene(2) system P-x-y data at 90°C by minimizing the sum of squares of the pressure residual. Plot the resulting fit together with the original data for both phases (Jost, W., Roek, H, Schroeder, W., Sieg, L., Wagner, H.G., Z. Phys. Chem.(Frankfurt) 10, 133 (1957)), using

(a) two-parameter Margules
(b) van Laar
(c) Wilson

x_1	y_1	P (mmHg)
0.117	0.502	1865
0.257	0.594	2113
0.376	0.618	2218
0.549	0.65	2273
0.707	0.689	2292
0.856	0.765	2208

11.34 Fit the specified model to the methanol(1) + benzene(2) system T-x-y data at 760 mmHg by minimizing the sum of squares of the pressure residual. Plot the resulting fit together with the original data for both phases (Hudson, J.W., Van Winkle, M., *J. Chem. Eng. Data 14*, 310(1969)), using:

(a) two-parameter Margules
(b) van Laar
(c) Wilson

x_1	y_1	$T(°C)$
0.026	0.267	70.67
0.05	0.371	66.44
0.088	0.457	62.87
0.164	0.526	60.2
0.333	0.559	58.64
0.549	0.595	58.02
0.699	0.633	58.1
0.782	0.665	58.47
0.898	0.76	59.9
0.973	0.907	62.71

11.35 VLE data for the system carbon tetrachloride(1) and 1,2-dichloroethane(2) are given below at 760 mmHg, as taken from the literature.[1]

---

1. Yound, H.D., Nelson, O.A., *Ind. Eng. Chem. Anal. Ed.* 4:67 (1932).

x_1	y_1	$T(°C)$
0.040	0.141	81.59
0.091	0.185	80.39
0.097	0.202	80.27
0.185	0.310	78.73
0.370	0.473	76.62
0.506	0.557	75.78
0.880	0.831	75.71
0.900	0.848	75.86
0.923	0.875	75.95
0.960	0.907	76.20

(a) Fit the data to the two-parameter Margules equation.
(b) Fit the data to the van Laar equation.
(c) Fit the data to the Wilson equation.
(d) Plot the P-x-y diagram at 80°C, based on one of the fits from (a) through (c).

11.36 Derive the form of the excess enthalpy predicted by Wilson's equation assuming that A_{ij}'s and ratios of molar volumes are temperature-independent.

11.37 Starting from the excess Gibbs energy formula for Flory's equation, derive the formula for the activity coefficient of component 1 in a binary mixture.

11.38 Orbey and Sandler (*Ind. Eng. Chem. Res.* 34:4351, 1995) have proposed a correction term to be added to the excess Gibbs energy of mixing given by UNIQUAC. To a reasonable degree of accuracy the new term can be written:

$$\frac{G_{HB}^E}{RT} = \sum x_i \left[-2 \ln(1 + a_i F) + \frac{a_i F}{1 + a_i F} - C_i^{pure} \right]$$

where

$$F \equiv \sum \frac{x_j a_j}{1 + a_j F}$$

Derive an expression for the correction to the activity coefficient. [Hint: Do you remember how to differentiate implicitly?]

11.39 The energy equation for mixtures can be written for polymers in the form:

$$\frac{U - U^{ig}}{RT} = \frac{\rho}{2} \sum_i \sum_j x_i x_j Nd_i Nd_j \int \frac{N_A u_{ij}}{RT} g_{ij} 4\pi r^2 dr$$

By analogy to the development of Scatchard-Hildebrand theory, this can be rearranged to:

$$\frac{U - U^{ig}}{RT} = \frac{-\rho}{2RT} \sum_i \sum_j x_i x_j Nd_i Nd_j N_A \varepsilon_{ij} \sigma_{ij}^3 a_{ij}^*$$

where

Nd_i = degree of polymerization for the ith component

ρ = molar density

x_i = mole fraction of the ith component

N_A = Avogadro's number

U = molar internal energy.

$a_{ii}^* = 3 + 2/Nd_i$

$a_{ij}^* = (a_{ii}^* \, a_{jj}^*)^{1/2}$

$\sigma_{ij}^3 = (\sigma_i^3 + \sigma_i^3)/2$

$\varepsilon_{ij} = (\varepsilon_{ii} \, \varepsilon_{jj})^{1/2}$

Derive an expression for $\ln \gamma_1$ for the activity coefficient model presented above.

11.40 Use the UNIFAC model to predict P-x-y data at 90°C and x_1, = {0, 0.1, 0.3, 0.5, 0.7, 0.9, 1.0} for propanoic acid + water. Fit the UNIQUAC model to the predicted P-x-y data and report your UNIQUAC a_{12} and a_{21} parameters in kJ/mole.

11.41 (a) Rearrange Eqn. 11.84 to show $(A^E/RT)^{COMB}$ is Flory's equation when $C = 0$.

(b) Use Eqn. 11.83 in Eqn. 11.80 and perform the integration to obtain Eqn. 11.84.

(c) Use Eqn. 11.91 in Eqn. 11.79 and perform the integration to obtain Eqn. 11.93.

CHAPTER 12

LIQUID-LIQUID PHASE EQUILIBRIA

In the field of observation, chance favors only the prepared mind.

Pasteur

The large magnitudes of the activity coefficients in the polymer mixing example should suggest an interesting possibility. Perhaps the escaping tendency for each of the polymers in the mixture is so high that they would prefer to escape the mixture to something besides the vapor phase. In other words, the components might separate into two distinct liquid phases. This would present quite a problem to blending plastics and recycling them because they would not stay blended. As a matter of fact, this is exactly what happens. The next question is: how can we tell if a liquid mixture is stable as a single liquid phase?

12.1   THE ONSET OF LIQUID-LIQUID INSTABILITY

Our common experience tells us that oil and water do not mix completely, even though both are liquids. If we consider equilibria between the two liquid phases, we can label one phase α and the other β. For such a system we can quickly show that the equilibrium compositions are given by:

$$\hat{f}_i^{\alpha} = \gamma_i^{\alpha} x_i^{\alpha} P_i^{sat} = \gamma_i^{\beta} x_i^{\beta} P_i^{sat} = \hat{f}_i^{\beta}$$

$$\gamma_i^{\alpha} x_i^{\alpha} = \gamma_i^{\beta} x_i^{\beta}$$

12.1   ❶ Equations for LLE.

where superscripts identify the liquid phase. Note that we have assumed an activity coefficient approach here even though we could formulate an entirely analogous treatment by an equation of state approach. There is also the possibility that three phases can coexist, two liquids and a vapor, which is illustrated below. In this case we have an additional fugacity relation for the gas phase, where we assume in the figure that the ideal gas law is valid for the vapor phase. An example of how such a system could be solved is given below. The phase equilibria can be solved by starting with whichever two phases we know the most about, and filling in the details for the third phase.

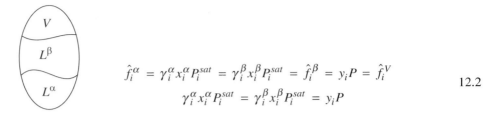

$$\hat{f}_i^{\alpha} = \gamma_i^{\alpha} x_i^{\alpha} P_i^{sat} = \gamma_i^{\beta} x_i^{\beta} P_i^{sat} = \hat{f}_i^{\beta} = y_i P = \hat{f}_i^{V}$$

$$\gamma_i^{\alpha} x_i^{\alpha} P_i^{sat} = \gamma_i^{\beta} x_i^{\beta} P_i^{sat} = y_i P$$

12.2

Example 12.1 Simple liquid-liquid-vapor equilibrium (LLVE) calculations

At 25°C, a binary system containing components 1 and 2 is in a state of three-phase LLVE. Analysis of the two equilibrium liquid phases (α and β) yields the following compositions:

$$x_2^{\alpha} = 0.05; \qquad x_1^{\beta} = 0.01$$

Vapor pressures for the two pure components at 25°C are:

$$P_1^{sat} = 0.7 \text{ bar}; \quad P_2^{sat} = 0.1 \text{ bar}$$

Making reasonable assumptions, determine good estimates for:

(a) the activity coefficients γ_1 and γ_2 (pure components at 25°C and 1 bar standard states)
(b) the equilibrium pressure
(c) the equilibrium vapor composition

Solution:

$$y_i P = x_i \gamma_i P_i^{sat}$$

Assume $\gamma_1^{\alpha} \approx 1$, $\gamma_2^{\beta} \approx 1$ because these are practically pure in the specified phases.

$$y_1 P = \gamma_1^{\alpha} x_1^{\alpha} P_1^{sat} = 0.95(0.7); \quad y_2 P = \gamma_2^{\beta} x_2^{\beta} P_2^{sat} = 0.99(0.1)$$

$$P = \sum_i y_i P = 0.764 \text{ bar}, \qquad y_1 = 0.95(0.7)/0.764 = 0.8704, \qquad y_2 = 0.1296$$

$$\gamma_2^{\alpha} = y_2 P/0.05(0.1) = 19.8; \quad \gamma_1^{\beta} = y_1 P/0.01(0.7) = 95$$

12.2 STABILITY AND EXCESS GIBBS ENERGY

Consider the free energies that would be predicted by a modified Margules one-parameter equation

$$\frac{G^E}{RT} = \frac{A'}{RT} x_1 x_2$$

12.3

Note that this modification will result in temperature-dependent activity coefficients

$$\ln\gamma_1 = \frac{A'}{RT} x_2^2 \quad \text{and} \quad \ln\gamma_2 = \frac{A'}{RT} x_1^2$$

12.4

Keep in mind that nature will dictate phase stability by minimizing the Gibbs energy. Plots for several values of A' are shown in Figure 12.1. In this plot, the important quantity is the Gibbs energy of the mixture G. Since $\Delta G_{mix} = G - \sum_i x_i G_i$, then rearranging,

$$G = \Delta G_{mix} + \sum_i x_i G_i = G^E + \Delta G_{mix}^{is} + \sum_i x_i G_i.$$

In Figure 12.1, the endpoints represent the values of G_i for the pure components, where the references states have been arbitrarily chosen, and only component 2 is at its reference state. The entropy of mixing is the same for each curve, but the increasing excess Gibbs energy of mixing (larger A'/RT) ultimately causes a "w-shaped" curve to form.

At this point we must digress to consider the meaning of the straight line connecting the endpoints of the "w-shaped" curve. If we imagine computing the Gibbs energy of two separate beakers containing the two separate components on the basis of one mole of total fluid, we see that this "average" Gibbs energy is simply a molar average of the two component Gibbs energies. There is no contribution from the entropy of mixing, because they are not mixing; they are in separate beakers. Substituting $(1-x_1)$ for x_2 in this molar average formula shows that this formula is simply a linear relation in terms of x_1. A similar consideration shows that a straight line is also obtained when the components in each beaker are not entirely pure but the amounts in each beaker are varied as described above. When such a straight line connects the compositions of two separate phases, it is called a tie line. The tie line can give the phase amount once the feed composition is known (as illustrated in homework problem 12.1). The question is whether that phase-split tie line gives a lower Gibbs energy than the value obtained through the combination of the energetic and entropic terms at that overall composition, because nature will minimize the Gibbs energy.

As shown in Fig. 12.1, $A'/RT = 1$ leads to a situation where connecting any two compositions by a straight line gives a value of G at the overall composition that is higher than the G along the curve. Try it for yourself for a feed of $z_1 = 0.5$ by trying to find two compositions along $A'/RT = 1$ that can be combined at $z_1 = 0.5$ to give $G/RT < \sim -0.2$. Since the curve is concave up, a straight line between two points always gives a higher value. This situation is quite different when we consider the case where $A'/RT = 3$. Along this curve, the mixture at $z_1 = 0.5$ would have $G \approx 0.31$ if it remained as a single phase. If the solution splits into two phases, a line can be drawn between the compositions of the phases (one point on either side of a) as shown by points c ($x_1^\alpha = 0.07$) and d ($x_1^\beta = 0.93$), and the overall energy is given by the intersection of this line with the overall composition $z_1 = 0.5$ as shown by point b (remember that the overall system is constrained to stay along an imaginary vertical line at $z_1 = 0.5$). The lowest energy is obtained when the line is tangent to the humps as shown in the figure, where $G \approx 0.2$. Then by splitting into two phases, the system clearly has a lower value for G/RT. Any other line that is drawn would force point b to have a higher Gibbs energy than this point. (Try it.) Considering these points at different values of A'/RT indicates that there is no phase split unless there is a hump in G/RT that makes it concave down. Note that A'/RT must be positive to create this curvature that results in a phase split. This means that the activity coefficients will be greater than one, and the system will also have positive deviations from Raoult's Law for VLE.

> ❗ A system will split into two phases if it results in a lower Gibbs energy.

One more point needs to be made before working some examples. Note that the line construction seems similar to what was done for VLE in a flash calculation at the beginning of Chapter 9. In fact, this is completely analogous mathematically to the flash in those diagrams, and the lever rule applies. The ratios of the phases can be found in a similar fashion. The only difference is that two liquid phases are formed upon a liquid-liquid flash rather than a vapor and liquid flash. For the

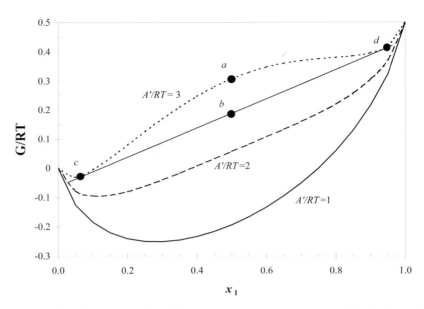

Figure 12.1 *Illustration of the Gibbs energy of a mixture represented by the Margules one-parameter equation.*

example in the figure above, the fraction of the overall mixture that is the α phase is given by

$$\frac{\overline{bd}}{\overline{cd}} = \frac{0.93 - 0.5}{0.93 - 0.07} = 0.5 \text{, so the mixture of this example splits into equal portions of the two phases.}$$

Example 12.2 LLE predictions using Flory-Huggins theory: polymer mixing

One of the major problems with recycling polymeric products is that different polymers do not form miscible solutions with each other, but, rather, they form highly non-ideal solutions. To illustrate, suppose 1g each of two different polymers (polymer *A* and polymer *B*) is heated to 127°C and mixed as a liquid. Estimate the mutual solubilities of *A* and *B* using the Flory-Huggins equation. Predict the energy of mixing using Scatchard-Hildebrand theory. Polymer data:

	MW	V (cm³/mol)	δ (cal/cm³)$^{1/2}$
A	10,000	1,540,000	9.2
B	12,000	1,680,000	9.3

Solution: This is the same mixture that we considered as an equal-weight-fraction mixture in Example 11.11. Based on that calculation, we know that the solution is highly non-ideal. We must now iterate on the guessed solubilities until the implied activity coefficients are consistent. Let's start by guessing that the solutions are infinitely dilute.

Example 12.2 LLE predictions using Flory-Huggins theory: polymer mixing (Continued)

$$\lim_{x_A^\alpha \to 0} \Phi_A^\alpha = 0 \text{ and } \lim_{x_B^\beta \to 0} \Phi_B^\beta = 0$$

$$\lim_{x_i \to 0} \frac{\Phi_i}{x_i} = \lim_{x_i \to 0} \frac{x_i V_i}{x_i V_i + x_j V_j} \cdot \frac{1}{x_i} = \frac{V_i}{V_j}$$

Using Eqns. 11.46 and 11.47,

$$\ln\gamma_A^\alpha = \ln(0.91) + (1 - 0.91) + 1.54E6(9.3 - 9.2)^2 (1.0)^2 /[1.987(400)] = 19.4$$

$$\Rightarrow \gamma_A^\alpha = 10^{8.4} \Rightarrow x_A^\alpha = 10^{-8.4}$$

$$\ln\gamma_B^\beta = \ln(1.09) + (1 - 1.09) + 1.68E6(9.3 - 9.2)^2 (1.0)^2 /[1.987(400)] = 21.2$$

$$\Rightarrow \gamma_B^\beta = 10^{9.2} \Rightarrow x_B^\beta = 10^{-9.2}$$

Good guess. The polymers are immiscible.

Example 12.3 LLE predictions using UNIFAC

In principle, all that is required to make predictions of LLE partitioning is some method of predicting activity coefficients. Clearly, UNIFAC is a convenient means of providing these, but there is a certain danger in applying too much confidence in such predictions. LLE tends to be more sensitive to the accuracy of the activity coefficients than VLE. Furthermore, the empirical nature of UNIFAC means that the same parameter set, $\{a_{mn}\}$, is not generally accurate for both VLE and LLE. This latter problem has been solved and many chemical process design programs are available which incorporate specific parameter sets for specific applications. As for the sensitivity problem, the best advice is not to take any UNIFAC predictions too seriously. They can be used as a guide to assess miscibility in a way that is slightly better than looking at solubility parameter tables, but should never be considered as a substitute for experimental data. With these cautions in mind, it is useful to show how LLE can be predicted using UNIFAC. *We have provided the LLE parameters on the spreadsheet UNIFAC(LLE) within ACTCOEFF.xls.*

❶ UNIFAC parameters for LLE differ from those for VLE.

Approximate miscibility predictions for water + methylethylketone

Arce et al.[1] give the compositions for the tie lines in the system water(1) + propionic acid(2) + methylethylketone (MEK)(3) at 298 K and 1 bar. As limiting conditions the mutual solubilities of water and MEK ($1CH_3 + 1CH_3CO + 1CH_2$) are also listed. Use UNIFAC to roughly estimate these mutual solubilities to within ± 5 mole%.

---

1. *J. Chem. Eng. Data*, 40:225 (1995).

Example 12.3 LLE Predictions Using UNIFAC (Continued)

Solution: Selecting the appropriate groups from the UNIFAC menu, then copying the values of the activity coefficient to the LLE worksheet, we can develop the Figs. 12.2 and 12.3 using increments of $x_w = 0.05$.

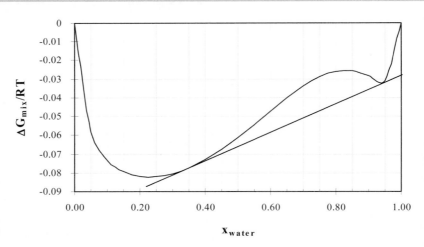

Figure 12.2 *Gibbs energy of mixing in the water + MEK system as predicted by UNIFAC.*

Using the drawing tool to construct a straight line, we obtain tangents at the following compositions:

$$x_1^\alpha = 0.94 \text{ and } x_1^\beta = 0.35$$

These are sufficiently accurate for the problem statement as given above. The humps in the activity plot are characteristic of LLE. Fig. 12.3 exhibits extrema in the plots of activity, which are indicative of LLE. The vertical lines indicate the compositions where the activities are equal in each phase. The horizontal lines indicate the activity values. This analysis is a graphical solution to Eqn. 12.1. Fig 12.2 is typically easier to implement for determining equilibria in a binary system.

Precise miscibility computations for water + MEK by LLE flash calculation

In some cases we may require higher accuracy than obtained by the method described above. Furthermore, we may encounter multicomponent mixtures, for which the extension of the above method is not straightforward. On the other hand, we have previously described an entirely general method for computing the phase partitioning given relative activities. This is simply a LLE flash calculation. We assume a feed composition which is between our expected upper and lower phase compositions, then we compute the estimated K-values (upper phase composition/lower phase composition) and perform a flash calculation to determine the relative phase amounts. The flash calculation develops new estimates for the upper and lower phase compositions, which yield new estimates for the K-values, and so forth. The formula for the K-values is analogous to the VLE formula for an equation of state, where the activity coefficients replace the fugacity coefficients,

$$K_i = \frac{x_i^\beta}{x_i^\alpha} = \frac{\gamma_i^\alpha}{\gamma_i^\beta}$$

12.5

Example 12.3 LLE Predictions Using UNIFAC (Continued)

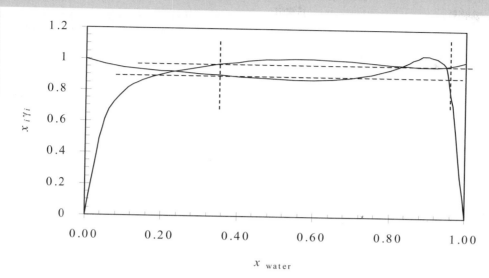

Figure 12.3 *Activities of water and MEK as a function of mole fraction water as predicted by UNIFAC. The activity vs. mole fraction plots will have a maximum when LLE exists. The dashed lines show the compositions where the activities of components are equal in both phases.*

To illustrate the flash calculation, start with a guess derived from the infinite dilution activity coefficients. $K_W = 1/23.55 = 0.0425$; $K_{MEK} = 45.758$. For a binary mixture, we can obtain a first guess for the phase compositions by using our estimates of the K values and by noting that the upper phase β compositions must sum to unity to obtain

$$x_1^\alpha K_1 + (1 - x_1^\alpha)K_2 = 1 \Rightarrow x_1^\alpha = \frac{1 - K_2}{K_1 - K_2}; \text{ and } x_1^\beta = x_1^\alpha K_1 \qquad 12.6$$

Setting water as (1) and MEK as (2), these initial guesses result in $x_{1new}^\alpha = 0.979$; $x_{1new}^\beta = 0.0416$ for the first iteration. Since we have initialized the iterations, we skip the first two steps in the iteration scheme the first time through:

1. Calculating $x_{1,new}^\alpha = \dfrac{1 - K_{2,old}}{K_{1,old} - K_{2,old}}$, $x_{2new}^\alpha = 1 - x_{1new}^\alpha$;

2. Calculating $x_{i,new}^\beta = K_{i,old} x_{i,new}^\alpha$;

3. Determining $\gamma_{i,new}$ values for each liquid phase from the $x_{i,new}$ values;

4. Calculating $K_{i,new} = \dfrac{\gamma_i^\alpha}{\gamma_i^\beta}$;

5. Replacing all $x_{i,old}$ and $K_{i,old}$ values with the corresponding new values;

6. Repeating the iterations until calculations converge.

Iterative flash procedure for binary LLE.

Example 12.3 LLE Predictions Using UNIFAC (Continued)

Unfortunately, the calculations converge much more slowly than VLE flash calculations. The calculations may drift a couple mole percent in compositions after they are changing at step sizes in the tenths of mole percents, so patience is required in converging the calculations. The table below summarizes the iterations. The procedure can be written into a spreadsheet macro to assist in the iterations as outlined in Section 12.6. Proceeding by successive substitution finally results in $x_1^\alpha = 0.96$; $x_1^\beta = 0.37$.

iteration	x_1^α	x_1^β	$K_{1,new}$	$K_{2,new}$
initialize	1	0	0.0425	45.758
1	0.979	0.0416	0.0859	32.202
2	0.972	0.083	0.1129	25.93
3	0.966	0.109	0.1359	22.99
4	0.962	0.131	0.1566	21.28
5	0.9601	0.1504	0.1758	20.199
30	0.958	0.342	0.276	15.716
40	0.958	0.349	0.365	15.555
50	0.958	0.354	0.369	15.47

A similar approach could be applied to solve for the partitioning of the propionic acid between water and MEK starting with the above result and infinite dilution of the propionic acid. By steadily increasing the propionic acid and performing flash calculations each time, the impact of the propionic acid on the water + MEK partitioning could be studied. Can you guess whether the propionic acid will cause relatively more MEK to dissolve into the water phase or vice versa? The answer is explored later in a homework problem.

12.3 PLOTTING TERNARY LLE DATA

Graphical representation of LLE data is important for design of separation processes. For ternary systems, triangular coordinates simultaneously represent all three mole fractions, or alternatively, all three weight fractions. Triangular coordinates are shown in Figure 12.4a, with a few grid lines displayed. The fraction of component A is represented by lines parallel to the BC axis: along $\overline{de}$, the composition is $x_A = 0.25$; along $\overline{ab}$, the composition is $x_A = 0.5$. The fraction of B is represented by lines parallel to the AC axis; along $\overline{ac}$, the composition is $x_B = 0.5$; along $\overline{fg}$, the composition is $x_B = 0.25$. The fraction of C is along lines parallel to the AB axis; along $\overline{bc}$, $x_C = 0.5$. Combining these conventions, at point h, $x_A = 0.25$, $x_B = 0.25$, $x_C = 0.5$. An example of plotted LLE phase behavior is shown on triangular coordinates in Figure 12.4b. The compositions of coexisting α and β phases are plotted and connected with tie lines. The lever rule can be applied. For example, in Fig. 12.4b, a feed of overall composition d will split into two phases: (moles β)/(moles α) = $\overline{de}/\overline{cd}$, and (moles β)/(moles feed) = $\overline{de}/\overline{ce}$.

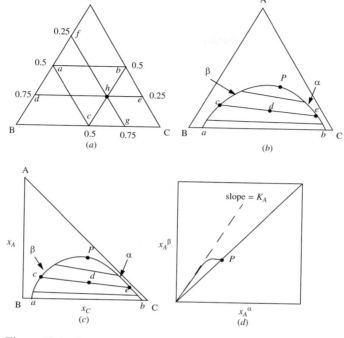

Figure 12.4 *Illustrations of graphical representation of ternary data on triangular diagrams; a) illustration of grid lines on an equilateral triangle; b) illustration of LLE on an equilateral triangle; c) illustration of LLE on a right triangle; d) illustration of tie-line data on a right triangle.*

To specify the composition of an arbitrary phase α in a ternary system, only two mole fractions are required. Since the mole fractions must sum to unity, if x_A^α and x_C^α are known, then the third mole fraction can be found, $x_B^\alpha = 1 - x_A^\alpha - x_C^\alpha$. The subscripts may be interchanged, and the same constraints hold for phase β. Therefore, cartesian coordinates can be used to plot two mole (or weight) fractions of each phase. Fig. 12.4c presents the data from Fig. 12.4b on cartesian coordinates. The tie line and lever rule concepts also apply on this diagram. Tie line data can be plotted on cartesian coordinates as in Fig. 12.4d. This representation of tie line data permits improved accuracy when interpolating between experimental tie lines. The slope of the tie line data, as presented in Fig. 12.4d, is frequently linear near the origin; the slope of the line in this region is x_A^β / x_A^α, which is the distribution coefficient, K_i. For the LLE phase behavior shown in Fig. 12.4, the $B + C$ miscibility is increased by the addition of A. Along the phase boundary, the tie lines become shorter upon approach to point P, the *Plait point*. Note that at the Plait point, all distribution coefficients are one.

Other Examples of LLE Behavior

The ternary LLE of Fig. 12.4 is one of several common types. In this example, $A + C$ is totally miscible (there is no immiscibility on the AC axis), as is $A + B$. When two of the three pairs of components are immiscible, the type of Fig. 12.5a can result, and when all three pairs have immiscibility regions, then type of Fig. 12.5b can result. In Fig. 12.5b, the center region is LLL; since the T and P are fixed, any overall composition in the center triangle will split into the three phases of compositions a, b, and c, because $F = C - P + 2 = 3 - 3 + 2 = 2$.

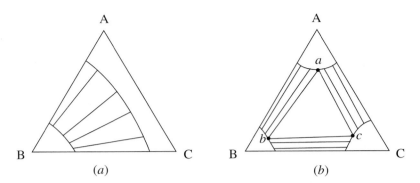

Figure 12.5 *Illustration of other types of LLE behavior.*

12.4 VLLE WITH IMMISCIBLE COMPONENTS

A special case of VLLE is obtained when one of the liquid-phase components is almost entirely insoluble in other components, and all other components are essentially insoluble in it, as occurs with many nonfunctional hydrocarbons with water. When a mixture forms two liquid phases, the mole fractions sum to unity in each of the phases. When a vapor phase coexists, it is a mixture of all components. The bubble pressure can be calculated using Eqn. 12.2, where the liquid phase fugacities are used to calculate the vapor phase partial pressures. For example, water and *n*-pentane are extremely immiscible. Applying the strategy used in Example 12.1, each liquid phase is essentially pure, resulting in the bubble pressure $P = P_{water}^{sat} + P_{pentane}^{sat}$ which is greater than the bubble point of either component. Rather than considering the bubble pressure, think about the bubble temperature. Since each component contributes to the partial pressure, the bubble temperature of a mixture of two immiscible liquids will be lower than the bubble temperature of either pure component at a specified pressure. This phenomena can be use to permit boiling at a lower temperature as shown in the next example.

Example 12.4 Steam distillation

Consider a steam distillation with the vapor leaving the top of the fractionating column and entering the condenser at 0.1 MPa with the following analysis:

	y_i	T_c	P_c	ω
n-C8	0.20	568.8	2.486	0.396
C12 fraction	0.40	660.0	2.000	0.540
H_2O	0.40	(use steam tables)		

Assuming no pressure drop in the condenser and that the water and hydrocarbons are completely immiscible, calculate the maximum temperature which assures complete condensation at 0.1 MPa. Use the shortcut *K*-ratio method for the hydrocarbons.

Example 12.4 Steam distillation (Continued)

Solution:

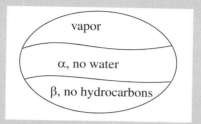

The temperature that we seek is a bubble temperature of the liquid phases. The hydrocarbons and the water are essentially immiscible. We may approximate the hydrocarbon liquid phase, α, as an ideal solution of C8 and C12 with no water present. Therefore, two liquid phases will form: one of pure H_2O and the other a mixture of 1/3 n-C8 + 2/3 C12 fraction. We may apply Raoult's Law with $x_w = 1$ for water in the β phase. The vapor mixture is a single phase, however, and must conform to: $1 = \sum_i y_i$

	phase	x_i	VLE K_i (353 K)	y_i	VLE K_i (368 K)	y_i
n-C8	α	0.333	0.254	0.084	0.415	0.138
C12 fraction	α	0.667	0.015	0.010	0.028	0.019
H_2O	β	1.0	0.474	0.474	0.846	0.846
				0.568		1.0027

So the bubble temperature with water present is ~95°C. What would it be without water?

	x_i	VLE K_i (400K)	y_i	VLE K_i (440K)	y_i
n-C8	0.333	1.049	0.349	2.727	0.908
C12 fraction	0.667	0.092	0.061	0.319	0.213
			0.410		1.121

Then $T \approx 400 + (1 - 0.41)/(1.121 - 0.41)\cdot 40 = 433$ K. Thus we reduced the bubble temperature by 65°C in the steam distillation.

12.5   CRITICAL POINTS IN BINARY LIQUID MIXTURES (OPTIONAL)

Referring back to Figure 12.1, if we wish to find the combination of x_1 and T where the system just begins to phase split (i.e., the liquid-liquid critical point), then we must seek x_1 where the concavity is equal to zero but the concavity at all other x_1 is greater than zero (if it was less than zero anywhere, then we would just have a regular phase split, not a critical point). But if the concavity is equal to zero at x_1 and greater everywhere else, then this must represent a minimum in the concavity as well a point where it equals zero. These two conditions provide two equations for the two unknowns, x_1 and T, involved in determining the critical point. Recalling that concavity is given by the second derivative:

$$\frac{\partial^2 G}{\partial x_1^2} = 0; \qquad \frac{\partial^3 G}{\partial x_1^3} = 0 \qquad\qquad 12.7$$

A generalization of these concepts to multicomponent mixtures gives:

$$\frac{\partial^2 G/RT}{\partial n_i^2} = 0; \qquad \frac{\partial^3 G/RT}{\partial n_i^3} = 0 \qquad\qquad 12.8$$

By analogy to the phase diagram for pure fluids, the locus where the second derivative equals zero represents the boundary of instability and is called the liquid-liquid spinodal.

Example 12.5 Liquid-liquid critical point of the Margules one-parameter model

Based on Fig. 12.1 and the discussion of concavity above, it looks like the value of $A'/RT = 2$ may be close to the critical point. Use Eqns. 12.7 and 12.8 to determine the exact value of the critical parameter.

Solution: To make the notation simpler, define a quantity χ' by $\chi' = A'/RT$. Multiplying through by n and recognizing that we have previously performed the initial part of this derivation (Example 11.4) gives

$$\frac{G}{RT} = n_1 \ln(x_1) + n_2 \ln(1 - x_1) + \chi' n_2 x_1$$

$$\frac{\partial \frac{G}{RT}}{\partial n_1} = \frac{\partial \frac{G^{is}}{RT}}{\partial n_1} + \frac{\partial \frac{G^E}{RT}}{\partial n_1} = \ln(x_i) + \ln(\gamma_i) = \ln(x_i \gamma_i) = \ln(x_i) + \chi' x_2^2 \qquad\qquad 12.9$$

Since this is a binary solution, there is a simple finite relationship between the derivative with respect to mole number and the derivative with respect to mole fraction, leading expeditiously to the expected conclusion

$$\frac{\partial^2 \frac{G}{RT}}{\partial n_1^2} = \frac{\partial^2 \frac{G}{RT}}{\partial x_1^2} \left(\frac{\partial x_1}{\partial n_1}\right)^2 = 0 \Rightarrow \frac{\partial^2 \frac{G}{RT}}{\partial x_1^2} = 0 = \frac{1}{x_1} - 2(1 - x_1)\chi'$$

$$\frac{\partial^3 \frac{G}{RT}}{\partial n_1^3} = \frac{\partial^3 \frac{G}{RT}}{\partial x_1^3} \left(\frac{\partial x_1}{\partial n_1}\right)^3 = 0 \Rightarrow \frac{\partial^3 \frac{G}{RT}}{\partial x_1^3} = 0 = \frac{-1}{x_1^2} + 2\chi' \qquad\qquad 12.10$$

$$2\chi' = \frac{1}{x_1^2} \Rightarrow 0 = 1 - \frac{1 - x_1}{x_1} \Rightarrow x_{1,c} = \frac{1}{2}; \; \chi'_c = 2; T_c = \frac{A'}{2R}$$

Example 12.6 Liquid-liquid critical point of the Flory-Huggins model

Based on the preceding example, it is possible to convey some background on one of the most significant scientific challenges currently occupying modern thermodynamicists. This is the problem of phase separation in polymer solutions and polymer blends. The same forces driving these phase separations lead to the extremely important problem of the collapse transition of a single polymer chain in a solvent or mixture. A special kind of collapse describes the protein folding problem. Imagine a bowl of spaghetti formed from a single noodle. After stretching the noodle until it is completely straight, release and watch it collapse into the bowl again. If the noodle was really a protein, it would collapse into exactly the same hooks and crooks as it was before being stretched. A complete understanding of that phenomenon would greatly facilitate drug design.

Limiting this introductory presentation to liquid-liquid equilibria, the phase partitioning of polymer mixtures is somewhat simpler in that we care only about collections of chains rather than the details of individual chains. Polymer solutions are classified differently from polymer blends in a manner that is superfluous to our mathematical analysis: polymer solutions are blends where the degree of polymerization of one of the components is unity (the small one is known as the solvent). With this minimal background, we can phrase the following problem: determine the critical value of the Flory-Huggins χ parameter considering the degrees of polymerization of each component.

Solution: Note that we have already solved this problem for the special case where the two components are identical in size. Then the excess entropy is zero, the volume fractions are equal to the mole fractions, and the Margules one-parameter model is recovered with χ' having nearly the same meaning as the Flory-Huggins χ parameter. To consider the problem of including the degree of polymerization, Nd, we must define the χ' parameter with respect to a standard unit of volume. Nd is the number of monomer repeat units in the polymer. In the presentation below (and most other presentations of the same material), the volume of a *monomer* of component 1 is assigned as this standard volume ($\chi' = V_{std} \cdot [\delta_1 - \delta_2]^2/RT$; $V_{std} = V_1/Nd_1$). Recalling the formula for the activity coefficient with this notational adaptation, the starting point (Eqn. 11.46) for this derivation becomes:

$$\frac{\partial \frac{G}{RT}}{\partial n_1} = \frac{\partial \frac{G^{is}}{RT}}{\partial n_1} + \frac{\partial \frac{G^E}{RT}}{\partial n_1} = \ln(x_1) + \ln(\gamma_1) = \ln(x_1\gamma_1) \qquad 12.11$$

$$= \ln(\Phi_1) + \left(1 - \frac{\Phi_1}{x_1}\right) + Nd_1\chi' \, \Phi_2^2$$

The next step is greatly simplified if we recognize a simple relationship that is very similar to the formula for computing the number average molecular weight from the weight fractions of each component. The analogous formula for the volume can be rearranged in terms of the volume ratio $R = V_2/V_1$ as follows:

$$V = \left[\frac{\Phi_1}{V_1} + \frac{\Phi_2}{V_2}\right]^{-1} \Rightarrow \frac{\Phi_1}{x_1 V_1} = \frac{1}{V} = \frac{\Phi_1}{V_1} + \frac{\Phi_2}{V_2}$$

$$\Rightarrow \frac{\Phi_1}{x_1} = \Phi_1 + \Phi_2 1/R \Rightarrow 1 - \frac{\Phi_1}{x_1} = \Phi_2(1 - 1/R) \qquad 12.12$$

Example 12.6 Liquid-liquid critical point of the Flory-Huggins model (Continued)

Since this is a binary solution, there is a simple finite relationship between the derivative with respect to mole number and the derivative with respect to *volume* fraction, leading expeditiously to the *general* conclusion (note $d\Phi_1 = -d\Phi_2$)

$$\frac{\partial^2 \frac{G}{RT}}{\partial n_1^2} = \frac{\partial^2 \frac{G}{RT}}{\partial \Phi_1^2}\left(\frac{\partial \Phi_1}{\partial n_1}\right)^2 = 0 \Rightarrow \frac{\partial^2 \frac{G}{RT}}{\partial \Phi_1^2} = 0 = \frac{1}{\Phi_1} - (1 - 1/R) - 2(1 - \Phi_1)Nd_1\chi'$$

$$\frac{\partial^3 \frac{G}{RT}}{\partial n_1^3} = \frac{\partial^3 \frac{G}{RT}}{\partial \Phi_1^3}\left(\frac{\partial \Phi_1}{\partial n_1}\right)^3 = 0 \Rightarrow \frac{\partial^3 \frac{G}{RT}}{\partial \Phi_1^3} = 0 = \frac{-1}{\Phi_1^2} + 2Nd_1\chi'$$

12.13

which leads to two important results:

$$2Nd_1\chi' = \frac{1}{\Phi_1^2} \Rightarrow 0 = \frac{1}{\Phi_1} - (1 - 1/R) - \frac{(1 - \Phi_1)}{\Phi_1^2} \Rightarrow \Phi_{1,c} = \frac{1}{1 + \sqrt{1/R}} = \frac{\sqrt{V_2}}{\sqrt{V_1} + \sqrt{V_2}}$$

$$\chi_c = \frac{1}{2Nd_1}\left(1 + \sqrt{1/R}\right)^2 = \frac{V_1}{2Nd_1}\left(\frac{1}{\sqrt{V_1}} + \frac{1}{\sqrt{V_2}}\right)^2 \Rightarrow \frac{(\delta_1 - \delta_2)^2}{RT_c} = \frac{1}{2}\left(\frac{1}{\sqrt{V_1}} + \frac{1}{\sqrt{V_2}}\right)^2$$

12.14

There are a number of significant conclusions that may be drawn from the above equations. First, for polymer solutions where (1) is the solvent (i.e., $V_1 \ll V_2$), the critical composition of polymer (2) approaches zero, and the critical temperature is a finite value that depends on the solvent and polymer molar volumes and solubility parameters. Furthermore, the critical temperatures for all polymers in a given solvent should be given by a universal curve with respect to molecular weight when reduced by the solubility parameter difference although these predictions are only semiquantitative due to the approximate nature of Scatchard-Hildebrand theory. For blends where $V_1 \sim V_2$, the critical temperature should be proportional to the molecular weight. We have applied several approximations in deriving these formulas, however. Therefore, significant efforts are currently underway to determine whether the formulas presented above are sufficiently accurate to describe the complex phase behavior often observed in polymer solutions and blends. Hanging in the balance is the ability to tailor-make polymer solutions and blends with many commercial advantages, because the ability to manipulate phase behavior successfully often relies on operating within the very sensitive critical region and knowing how to maneuver appropriately.

12.6 EXCEL PROCEDURE FOR BINARY, TERNARY LLE (OPTIONAL)

For a binary system, the LLE iteration procedure has been outlined in Example 12.3. This section provides the detailed steps to apply that method using a spreadsheet to determine the mutual solubilities predicted by an activity coefficient expression. Two activity coefficient tables are required, one for phase α and one for phase β. The procedure outlined in the earlier example was based on successively updating the K_i values. This will be done by using cells to hold the old and new values and copying and pasting values for updating. Two variations have been provided on the spreadsheets, UNIQUAC and UNIFAC(LLE). In both cases, an iteration table has been provided. For UNIFAC(LLE) the activity coefficient calculations are complex enough to require two spreadsheets, but cells can be referenced on the second spreadsheet by including the worksheet name in the formula.

Step 1. Select the activity method and open the spreadsheet, and locate the iteration table with rows for $K_{i,old}$, $K_{i,new}$, $x_{i,old}^\alpha$, $x_{i,new}^\alpha$, $x_{i,old}^\beta$, and $x_{i,new}^\beta$. The $x_{i,old}^\alpha$ and $x_{i,old}^\beta$ rows in the iteration tables will duplicate values in the activity coefficient tables, and they are for convenience to see how the values are changing since the screen is pretty crowded, and therefore during iteration only the iteration table will need to be monitored. Program the *iteration table* $x_{i,old}^\alpha$ and $x_{i,old}^\beta$ rows to duplicate the values used in the *activity coefficient table* for the specified phase by entering the cell name with the desired value. (e.g. "=A1" to duplicate cell A1). (If you are using UNIFAC with two separate spreadsheets, you can still reference values from the sheets that are not on top, just use the mouse to click on the correct cell on the correct sheet when entering the formula.)

Step 2. Initialize the compositions using the following steps. Before you proceed, decide which components will be rich in the α and β phases, or follow the problem statement. It is usually easiest if component numbers are chosen so that component three is completely miscible in the other components. Follow the steps carefully, saving the spreadsheet often, and if the initialization goes awry, repeat the procedure.

(a) For a binary system, enter the compositions in the *activity coefficient tables* for each phase assuming infinite dilution for component 1 in one phase and infinite dilution for component 2 in the other phase. Enter these compositions for each phase before proceeding to assure proper initialization.

(b) For a ternary system, the procedure that we develop here will only work for component 3 appearing at a fixed mole fraction. The iterations are best started from a previously converged binary result, with the third component at infinite dilution in each phase, or from a previously converged ternary result. Enter these compositions for each phase before proceeding to assure proper initialization. (You may use a small number like 1E-11 for infinite dilution.)

Step 3. Initialize the K-ratios and perform the first iteration with the following steps. Enter formulae in the *iteration table* to calculate the $K_{i,new} = \gamma_i^\alpha / \gamma_i^\beta$ (Eqn. 12.5) for each component. This creates a first guess for all K_i. Copy the *values* from $K_{i,new}$ to $K_{i,old}$ (To do this, right-click before pasting, and choose **Paste Special...** then choose **Values** from the dialog box. Below, this procedure will be referred to as an instruction to paste *values*). For the iteration table row for $x_{i,new}^\alpha$ values, enter in the following order (using cell references in the formulas): Binary calculation—enter

$$x_{1,new}^\alpha = (1 - K_{2,old})/(K_{1,old} - K_{2,old}) \quad \text{and} \quad x_{2,new}^\alpha = 1 - x_{1,new}^\alpha, \quad \text{and for the } x_{i,new}^\beta \text{ row,}$$

enter the formulas to calculate $x_{i,new}^\beta = K_{i,old} x_{i,new}^\alpha$ for *both* components; Ternary calculation—first enter the desired value at which $x_{3,new}^\alpha$ is to be constrained (use a small number like 1E-11 for zero), then enter $x_{1,new}^\alpha = (1 - x_{3,old}^\alpha(K_{3,old} - K_{2,old}) - K_{2,old})/(K_{1,old} - K_{2,old})$ (which is derived in a homework problem), and $x_{2,new}^\alpha = 1 - x_{1,new}^\alpha - x_{3,new}^\alpha$, and then for the $x_{i,new}^\beta$ row, enter the formulas to calculate $x_{i,new}^\beta = K_{i,old} x_{i,new}^\alpha$ for *all three* components. Note that the ternary formula reduces to the binary formula when $x_3^\alpha = 0$, so a binary calculation can be performed from the ternary spreadsheet.

Step 4. Iterate. The strategy will be to update *values* for x_i^α and x_i^β in the activity calculation tables and K_i in the iteration table. Because some of the values are immediately recalculated, the order of the substitutions is important:

(a) Update the x_i^α values in the activity coefficient table for α by copying and pasting *values* from the $x_{i,new}^\alpha$ row of the iteration table.

(b) Update the x_i^β values in the activity coefficient table for β by copying and pasting *values* from the $x_{i,new}^\beta$ row of the iteration table.

(c) Update the $K_{i,old}$ values by copying and pasting the $K_{i,new}$ *values* into the $K_{i,old}$ row. The calculations should converge by repeating these steps over and over. When performing iterations by hand, note that *either* step a or b can be skipped four or five times if the compositions in the given phase have not change significantly. Continue iterating to determine the compositions of the coexisting phases. (The convergence of LLE can be slow. Compositions may creep a couple mole percent after they are changing on the tenths of mole percents. You may wish to write a macro to perform the iterations of step 4. See Appendix B for macro instructions.)

> *Note: For ternary systems, the value of x_3^α can be systematically changed to generate tie lines. Each time the value is changed the calculations need to be converged. The ternary method that is presented here is not completely general since the composition is constrained in one of the phases. A general technique can be written, but it involves a generalized flash routine which is an extension of the VLE flash routine involving nested iterations.*

12.7 SUMMARY

The presentation here has been brief, but has opened a broad new frontier of phase equilibrium analysis: multiphase equilibrium. You might find it incredible to see how many phases and peculiar behaviors can be observed with just "oil," water, and special third components known as surfactants. (Common hand soap is an example of a surfactant.) The short introduction here is a branching point that barely scratches the surface of such oleic and aqueous systems.[1] Far down this road you may begin to gain some understanding of the forces which hold cell membranes and living organisms together. At the more practical engineering level, you should be able to perform preliminary designs of liquid extraction equipment with little more thermodynamical background than has been presented here.

This kind of breadth is possible in such a short chapter because the fundamentals have been laid out previously.   The liquid phases split because the Gibbs excess energy becomes so large that the stability limit is exceeded. But the procedure for calculating Gibbs energy is the same as for VLE. The calculation of the phase distributions requires a flash calculation in terms of liquid-liquid K-ratios. But the flash algorithm has been discussed before and liquid-liquid flash calculations are hardly different from vapor-liquid flash calculations. The only trick here is to know when to search for additional phases. We have treated that problem for binary and ternary mixtures but have not generalized to multicomponent systems. The generalization of critical solution behavior requires a fair amount of calculus and matrix manipulation,[2] but leads to a much deeper understanding of critical behavior than we have provided here. As a practical matter, the process flowsheet usually makes it clear when a liquid-liquid separation might be suspected, and performing a flash calculation

1. Hiemenz, P.C., *Principles of Colloid and Surface Chemistry,* 2nd ed. Marcel Dekker, NY, 1986.
2. Tester, J.W. and Modell, M., *Thermodynamics and Its Applications*, 3rd ed. Prentice-Hall, 1997.

to make certain is not too difficult. The activity coefficient models must be used carefully because the accuracy of the models is usually inferior to VLE for the same systems. In particular, the temperature dependence of LLE is usually not predicted well. Frequently, the miscibility increases with temperature more quickly than predicted by the activity coefficients even with UNIQUAC or UNIFAC. Temperature-dependent parameters are usually required.

Altogether, you should retain the idea that high escaping tendencies may lead to peculiar phase behavior beyond the azeotropes and vapor-liquid phase envelopes of the previous chapters. If the vapor pressures are too low, the components may escape each other by simply forming separate condensed phases. An entropic penalty must be paid, but a highly unfavorable energy of interaction may more than compensate.

12.8   PRACTICE PROBLEMS

P12.1 It has been suggested that the phase diagram of the hexane + furfural system can be adequately represented by the Margules 1-parameter equation, where $\ln \gamma_i = x_j^2 \cdot 800/T\,(K)$. Estimate the liquid-liquid mutual solubilities of each component in each liquid phase at 298 K. (ANS. ~10% each, by symmetry)

P12.2 Suppose the solubility of water in ethyl benzene was measured by Karl-Fisher analysis to be 1 mol%. Use UNIFAC to estimate the solubility of ethylbenzene in the water phase. (ANS. 0.003wt% or 30ppmw).

P12.3 According to Perry's Handbook, the system water + isobutanol forms an atmospheric pressure azeotrope at 67.14 mol% water and 89.92°C. Based on these data, we can estimate the van Laar coefficients to be $A_{12} = 1.566$; $A_{21} = 3.833$ at 273 K. Estimate the liquid-liquid mutual solubilities of each component in each liquid phase at 273 K. (ANS. 0.33,0.97)

12.9   HOMEWORK PROBLEMS

12.1 Suppose the (1) + (2) system exhibits liquid-liquid immiscibility. Suppose we are at a state where $G_1/RT = 0.1$ and $G_2/RT = 0.3$. The Gibbs energy of mixing quantifies the Gibbs energy of the mixture relative to the Gibbs energies of the pure components. Suppose the excess Gibbs energy for the (1) + (2) mixture is given by:

$$G^E/RT = 2.5\, x_1\, x_2$$

(a) Combine this with the Gibbs energy for ideal mixing to calculate the Gibbs energy of mixing across the composition range and plot the results against x_1 to illustrate that the system exhibits immiscibility.

(b) Draw a tangent to the humps to illustrate that the system will be one phase at compositions greater than $z_1 = 0.854$ and less than $z_1 = 0.145$, but will split into two phases with these compositions at any intermediate overall composition. Most systems with liquid-liquid immiscibility must be modeled with a more complex formula for excess Gibbs energy. The humps on the diagram are usually off center, as in Fig. 12.2 on page 428 in the text. The simple model used for the calculations here results in the symmetrical diagram.

(c) When a mixture splits into two phases, the over-all fractions (of total moles) of the two phases are found by the lever rule along the composition coordinate. Suppose 0.6 mol of (1) and 0.4 mol of (2) are mixed. Use the lever rule to calculate the total number of moles which would be found in each phase of the actual system. Specify the phases as the (1)-rich phase and the (2)-rich phase.

(d) What is the value of the hypothetical Gibbs energy, (expressed as G/RT), of a mixture of 0.6 mol of (1) and 0.4 mol of (2) if the mixture were to remain as one phase? Calculate the Gibbs energy of the total system considering the phase split into two phases, and show that the Gibbs energy is less than the Gibbs energy of the single-phase system.

12.2 Assume solvents A and B are virtually insoluble in each other. Component C is soluble in both.

(a) Use regular solution theory to estimate the distribution coefficient at low concentrations of C given as (mole fraction C in A)/(mole fraction C in B).
(b) If the phase containing A is 0.1 mol% C, estimate the composition of the phase containing B.
(c) If an extractor was designed and constructed, is the distribution coefficient favorable for extraction from B into A? Data:

	Volume (cc/mole)	δ (cal/cc)$^{1/2}$
A	50	5.8
B	250	10.4
C	100	9.8

12.3 A new drug is to have the formula para-CH_3CH_2-(C_6H_4)-CH_2CH_2COOH, where (C_6H_4) designates a phenyl ring. A useful method for assessing the extent of partitioning between the bloodstream and body fat is to determine the infinite dilution partitioning coefficient for the drug between water and n-octanol. Use UNIFAC to make this determination. Will the new drug stay in the bloodstream or move into fatty body parts?

12.4 Use the Scratchard-Hildebrand theory to generate figures of activity as a function of composition and ΔG_{mix} as a function of composition for neopentane and dichloromethane at 0°C. Determine the compositions of the two phases in equilibrium. Data:

	V (cm^3/mole)	δ (cal/cm^3)$^{1/2}$
Neopentane	122	6.2
Dichloromethane	64	9.7

12.5 The bubble point of a liquid mixture of n-butanol and water containing 4 mol% butanol is 92.7°C at 1.013 bar. At 92.7°C the vapor pressure of pure water is 0.776 bar and that of pure n-butanol is 0.383 bar. Assuming the activity coefficient of water in the 4% butanol solution is near unity, estimate the composition of the vapor and activity coefficient of butanol that gives the correct bubble pressure and compare to the values estimated by UNIFAC.

12.6 Schulte et al.[1] discuss a linear solvation energy relationship (LSER) method for the partitioning of 41 environmentally important compounds between hexane + water phases at 25°C. The LSER method is based on the idea that contributions to the Gibbs excess energy

1. *J. Chem. Eng. Data*, 43:72 (1998)

(and to the logarithm of the partition coefficient) from effects like van der Waals forces and hydrogen bonding are independent of each other. Therefore, these contributions can be added up as separate linear contributions. We can test this hypothesis by plotting partition data for several compounds based on experimental data, LSER, and UNIFAC. The following table presents the required parameters for the LSER method for several compounds. These parameters are to be substituted into the equation:

$$\log K_{H/W} = 0.404 + 5.382v_i - 1.786\pi_i + 0.856\delta_i - 4.644\beta_i - 3.078\alpha_i$$

where v = volume parameter, π = polarity parameter, δ = polarizability parameter, β = hydrogen bond acceptor parameter, α = hydrogen bond donor parameter. Compute the log partition coefficients for the following compounds by the LSER method and by UNIFAC and plot them against the experimental values listed in the table below.

Table 12.1 *LSER parameters and experimental hexane + water partition coefficients for several compounds.*

Compd	logK	v	π	δ	β	α
Phenol	−0.96	0.536	0.72	0.52	0.33	0.61
o-cresol	−0.12	0.637	0.68	1.0	0.41	0.54
2,4-dimethyl phenol	0.36	0.738	0.64	1.0	0.42	0.54
benzaldehyde	0.36	0.606	0.92	1.39	0.44	0
p-chlorobenzaldehyde	1.60	0.698	0.92	1.39	0.40	0
benzene	2.06	0.491	0.59	0.68	0.10	0
toluene	2.75	0.592	0.55	1.0	0.11	0

12.7 Use UNIFAC to predict the compositions of the coexisting liquid phases for the system methanol (1) + cyclohexane (2) at 298 K. Let α be the methanol-rich phase.

12.8 Use UNIFAC to predict the compositions of the coexisting liquid phases for the system methanol (1) + cyclohexane (2) at 285.15 K and 310.15 K. Let α be the methanol-rich phase. Compare quickly with the data from Fig. Fig. 15.9 on page 572 and comment on the accuracy of the results. (Include a printout of your results including converged compositions and activity coefficients of both phases at one of the temperatures.)

12.9 Benzene and water are virtually immiscible. What is the bubble pressure of an overall mixture that is 50 mol% of each at 75°C?

12.10 Water + hexane and water + benzene are immiscible pairs.

(a) The binary system water + benzene boils at 69.4°C and 760 mmHg. What is the activity coefficient of benzene in water if the solubility at this point is $x_B = 1.6E-4$, using only this information and the Antoine coefficients?

(b) What is the vapor composition at the bubble pressure at room temperature (292 K) for a ternary mixture consisting of 1 mole overall of each component if the organic layer is assumed to be an ideal solution?

(c) What is the vapor composition at the bubble pressure at room temperature (292 K) for a ternary mixture consisting of 1 mole overall of each component if the activity coefficients of the organic layer are predicted by UNIFAC?

The following problems concern use of UNIFAC and UNIQUAC to find LLE in ternary systems. Experimental data for the systems are summarized below.

Table 12.2 *Water(1) + methylethylketone(MEK)(2) + acetic acid(3) system at T = 299.85 K, reported by Skrzec, A.E., Murphy, N.F., Ind. Eng. Chem., 46:2245(1954). Compositions in mole percents.*

α phase			β phase		
1	2	3	1	2	3
92.689	7.311	0.	36.383	63.617	0.
91.644	8.049	0.307	38.601	60.547	0.851
90.839	8.623	0.538	40.531	57.986	1.482
90.681	8.733	0.586	41.835	56.287	1.878
89.325	9.717	0.958	44.866	52.640	2.494
88.631	10.228	1.140	49.069	47.994	2.937
88.084	10.669	1.247	52.180	44.769	3.051

Table 12.3 *1-Butanol(1) + water(2) + methanol(3) at 288.15 K as reported by Mueller, A.J., Pugsley, L.I., Ferguson, J.B., J. Phys. Chem., 35:1314(1931). Compositions in mole percents.*

α phase			β phase		
1	2	3	1	2	3
2.115	95.071	2.813	45.254	51.598	3.148
2.319	92.876	4.804	41.183	53.459	5.358
2.548	91.304	6.148	35.997	56.276	7.727
2.966	89.460	7.574	30.372	59.851	9.777
4.043	86.874	9.083	23.296	65.170	11.534
5.171	85.094	9.736	18.482	69.704	11.814

12.11 Consider the system water(1) + MEK(2) at 299.85 K. The solubilities measured by Skrzec, A.E., Murphy, N.F., *Ind. Eng. Chem.*, 46(1954), 2245, are $x_1^\alpha = 0.927$ and $x_1^\beta = 0.364$. For a binary system, the LLE iteration procedure has been outlined in Example 12.3. Apply the procedure to determine the mutual solubilities predicted by UNIQUAC. The mixture parameters are $r = [0.92, 3.2479]$, $q = [1.40, 2.876]$, $a_{12} = -2.0882$ K, $a_{21} = 345.53$ K. Let phase α be the water-rich phase.

12.12 For a binary system, iterations can be performed by finding a new value of $x_{1,new}^\alpha$ from only the K-ratios as shown in Eqn. 12.6. For a ternary system, we need at least one composition. Derive the iteration formula, using $x_{3,old}^\alpha$ as the specified composition

$$x_{1,new}^\alpha = \frac{1 - x_{3,old}^\alpha(K_{3,old} - K_{2,old}) - K_{2,old}}{K_{1,old} - K_{2,old}}.$$

12.13 Consider the system water(1) + methylethylketone(MEK)(2) + acetic acid(AA)(3) at 299.85 K. For a ternary LLE system, estimate tie lines at $x_3^\alpha = 0.005, 0.01, 0.02$, using UNIQUAC, where the parameter values are $r = [0.92, 3.2479, 2.2024]$, $q = [1.40, 2.876, 2.072]$, and the a values (in K) are $a_{12} = -2.0882$, $a_{21} = 345.53$, $a_{13} = 254.15$, $a_{31} = -301.02$, $a_{23} = -254.13$, $a_{32} = -4.5537$. Let α be the water-rich phase. Plot the results on rectangular coordinates, using x_1 as the abscissa and x_3 as the ordinate. Connect the tie lines on the plot. Add the experimental tie lines (Table 12.2) to the same plot using different symbols.

12.14 One mole of a stream containing pentane, acetone, methanol, and water in proportions $z = 0.75, 0.13, 0.11, 0.01$ respectively is to be mixed and decanted with 1 mol of pure water at 25°C. Use the UNIFAC model to estimate the partition coefficients for each of the components and the proportion of lower-phase/Feed where Feed, includes both the pentane-rich and pure water streams.

12.15 Calculate the LLE in the system 1-butanol(1) + water(2) + methanol(3) at 288.15 K, using UNIQUAC with the following parameters: $r = [3.4543, 0.92, 1.4311]$; $q = [3.052, 1.4, 1.432]$; and the a values (in K) are $a_{13} = 355.54$; $a_{31} = -164.09$; $a_{12} = -82.688$; $a_{21} = 443.56$; $a_{32} = -85.451$; $a_{23} = -321.92$; and compare graphically with the data from Table 12.3.

12.16 Solve problem 12.13, except use UNIFAC.

12.17 Consider the system water(1) + methylethylketone(MEK)(2) + propionic acid(PA)(3). Use UNIFAC to predict the compositions for the coexisting phases at $x_3^\alpha = 0.01, 0.05$, and 0.10 at 298.15 K. Let α be the water-rich phase. Plot the results on rectangular coordinates by using x_1 as the abscissa and x_3 as the ordinate. Connect the tie lines on the plot.

12.18 Solve problem 12.15, except use UNIFAC.

12.19 Derive the formulas for the spinodal curves of the Flory-Huggins model and plot the spinodals (T versus Φ^α, Φ^β) of several polystyrenes in cyclohexane using UNIFAC parameters to estimate the volume of polystyrene relative to ethylbenzene and taking the experimental values of solubility parameters and molar volumes for the species when $Nd = 1$. Take the degrees of polymerization of polystyrene to be 100, 200, 500, 1000. Plot the estimated reciprocal critical temperatures vs. $Nd^{-1/2}$. Mark the infinite molecular weight critical temperature on both plots with a big **X**.

CHAPTER

13

SPECIAL TOPICS

Whenever new fields of technology are developed, they will involve atoms and molecules. Those will have to be manipulated on a large scale, and that will mean that chemical engineering will be involved — inevitably.

Isaac Asimov (1988)

This chapter discusses three topics that can each be studied independently: phase behavior; solid-liquid equilibria; and residue curve analysis.

13.1 PHASE BEHAVIOR

Several types of phase behavior may occur in binary systems.[1] In earlier chapters, we explored phase behavior by examining *P-x-y* or *T-x-y* diagrams. In this section we demonstrate how these phase diagrams are related to the three-dimensional *P-T-x-y* diagrams. The *P-x-y* and *T-x-y* diagrams are two-dimensional cross sections of the three-dimensional phase envelope, and by studying the phase envelope, the progressions of changing shapes of the two-dimensional cross sections can be more easily grasped. The relations between the three-dimensional phase envelope and the cross sections are shown in Fig. 13.1 for three different systems that all fall under the classification as Type I systems. Type I is the simplest class of phase behavior because there is no LLE behavior. Note that non-azeotrope as well as minimum boiling and maximum boiling homogeneous azeotropes are in this class. In each of the three-dimensional diagrams, short dashes are used to denote the pure component vapor pressures which terminate at the critical points denoted by *A* and *B*. Horizontal cross sections of the three-dimensional phase envelopes are shown with long dashes and are *T-x-y* diagrams. Vertical cross sections of the three-dimensional phase envelopes are shown with solid lines, and are *P-x-y* diagrams. (A summary of special notation used in this section appears at the end of the section.) There is a one-to-one correspondence of the cross sections in the three-dimensional plots to the phase envelopes plotted on the *P-x-y* and *T-x-y* cross sections. The solid line running

> ❶ A summary of special notation appears at the end of this section.

---

1. Much of this section has been published in C.T. Lira, "Thermodynamics of Supercritical Fluids with Respect to Lipid-Containing Systems," in *Supercritical Fluid Technology in Oil and Lipid Chemistry,* J.W. King, G.R. List, eds., AOCS Press, Champaign, IL, 1996. Reproduced with permission.

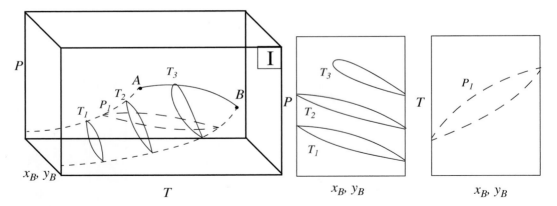

Figure 13.1a *Illustration of a system which does not form an azeotrope. The two-dimensional envelopes correspond to cross-sections shown on the three-dimensional diagram. The P-x-y diagrams are vertical cross-sections, and the T-x-y is a horizontal cross section.*

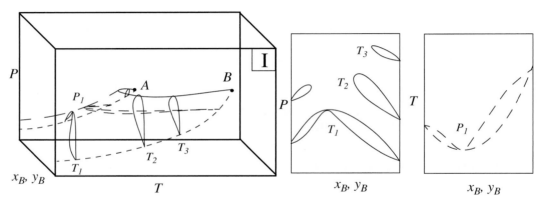

Figure 13.1b *Illustration of a system which forms a minimum boiling azeotrope due to positive deviations from Raoult's Law. The two-dimensional envelopes correspond to cross-sections shown on the three-dimensional diagram.*

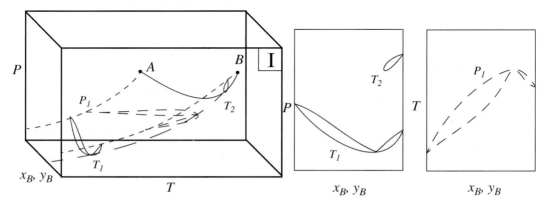

Figure 13.1c *Illustration of a system which forms a maximum boiling azeotrope due to negative deviations from Raoult's Law. The two-dimensional envelopes correspond to cross-sections shown on the three-dimensional diagram.*

from the critical point of A to the critical point of B is the locus of critical points of the mixture where the vapor and liquid become identical. Note the branches of the two dimensional cross sections do not span the composition range when the critical locus is intersected. Each lobe on the cross sections will have a critical point. As discussed in Example 10.10 on page 342, the critical points are frequently not at the maximum temperature or pressure of the phase envelope. Refer back to phase diagrams from Chapters 9–12 to see how the phase diagrams in those chapters relate to the diagrams shown here.

> *P-x-y* and *T-x-y* diagrams are cross sections of 3-*D* phase envelopes.

> *P-x-y* and *T-x-y* diagrams for systems with LLE are shown in Fig. 13.2a. The liquid-liquid behavior occurs in the region to the left of the critical endpoint U, in the U-shaped region above the vapor-liquid envelope. Three-phase *llv* occurs on the surface marked with tie lines. The intersection of this surface with *P-x-y* and *T-x-y* diagrams results in the *llv* tie lines on the cross section diagrams. At the upper critical endpoint, U, a vapor and liquid phase become identical, denoted with the notation, $l - l = v$. When the vapor pressures of the components are significantly different from each other, the system may not have an azeotrope as shown in Fig. 13.2a*a-c*. When the vapor

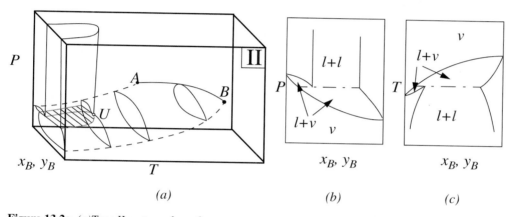

Figure 13.2a *(a)Type II system where the vapor pressures are significantly different; (b) P-x-y at a temperature below T_U; (c) T-x-y at pressure below P_U.*

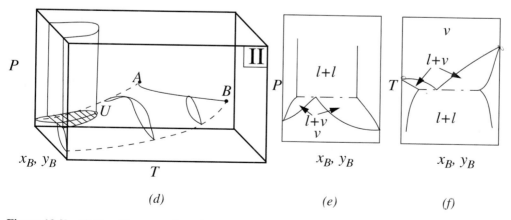

Figure 13.2b *(d) Type II system where the vapor pressures are relatively similar; (e) P-x-y at a temperature below T_U; (f) T-x-y at pressure below P_U. It is also possible for an azeotrope to lie to the right or left of the liquid-liquid region, rather than the heteroazeotropic behavior that is shown.*

pressures are closer to each other, azeotropes and heteroazeotropes will form. Fig. 13.2a*d-f* shows a system with heteroazeotropic behavior below the temperature of the upper critical endpoint, T_U, and azeotropic behavior above T_U.

Perspective

Before advancing further into the phase behavior classifications, some justification for such study is offered. High pressure can be used to create dense fluids that are useful for processing. For example, high pressure gases are employed industrially for petroleum fractionations in the oil industry and for hops and spice extractions and coffee decaffeination in the food industry. Dense gases are also under study for fractionation of specialty vegetable and fish oil components, as reaction media for polymerizations and other chemical synthesis and separations. High pressure processing and supercritical fluid extraction rely on control of solubility through manipulation of temperature and pressure. Solubility behaviors follow clear patterns which depend on similarities/differences in the thermodynamic and structural properties of the solute and the solvent. This section serves as an overview of phase behavior and systematic trends in phase behavior.

Natural materials such as foods and oils are multicomponent mixtures. Polymers typically contain a molecular weight distribution. Frequently, these types of mixtures are not well identified or characterized. Solubilities and extractabilities for these mixtures are currently difficult to predict quantitatively; however, significant knowledge regarding solubility trends may be obtained by studying simpler binary and ternary systems. Solubility represents a saturation condition; therefore, solubility is represented as a boundary on a phase diagram. Systematic study of binary and ternary systems shows that the phase boundaries of ternary systems are intermediate to the constituent binary systems, and many of the same trends continue in multicomponent systems although fundamental exploration of the trends is an ongoing research topic.[2] Process simulators and computers continue to simplify the calculation of phase equilibria, however the interpretation of the resulting phase behavior is aided by a general understanding of the classes of phase behavior presented in this section.

Classes of Binary Phase Behavior

 There are six major types of phase behavior.

Since 1970 there have been several reviews and classifications of phase behavior[2,3,4,5,6,7,8,9,10,11]. The types are usually summarized by the projection of their phase boundaries onto two-dimensional pressure-temperature diagrams. Type I and II phase behavior have already been discussed, and they are shown by the upper two plots in Fig. 13.3. Note that azeotrope behavior is a subset of the major classes of behavior and it is not shown explicitly on the projections in Fig. 13.3. For this discussion, the convention of van Konynenburg and Scott[10] will be followed for classification of phase behavior. There are six major types of phase behavior shown in the plots which use

2. Luks, K.D. (1986) *Fluid Phase Equil. 29*, 209–24.
3. Jangkamolkulchai, A. D.H. Lam, K.D. Luks (1989), *Fluid Phase Equil. 50*, 175–187.
4. Estrera, S.S., M.M. Arbuckle, K.D. Luks, (1987) *Fluid Phase Equil. 35*, 291–307.
5. McHugh, M.A. and V.J. Krukonis (1986) *Supercritical Fluid Extraction: principles and practice*, Butterworths, Stoneham, MA.
6. Miller, M.M., K.D. Luks (1989), *Fluid Phase Equil. 44*, 295–304.
7. Schneider, G.M. (1991) *Pure and Applied Chem. 63*, 1313–1326. Also published as Schneider, G.M., (1991) *J. Chem. Therm. 23*, 301–26, 15. Schneider, G.M. (1970) *Adv. Chem. Phys. 1*, 1–42.
8. White, G.L., C.T. Lira, (1992), *Fluid Phase Equil.*, *78*, 269; White, G.L., C.T. Lira (1989) in *Supercritical Fluid Science and Technology*, K.P. Johnston and J.M.L. Penninger, eds., American Chemical Society, Washington, D.C. pp 111–120
9. Schneider, G.M. (1978) *Angew. Chem. Int. Ed. Engl. 17*, 716–27.
10. van Konynenburg, P.H., Scott, R.L. (1980) *Phil. Trans. Roy. Soc. London, Ser. A 298:1442*, 495–540.

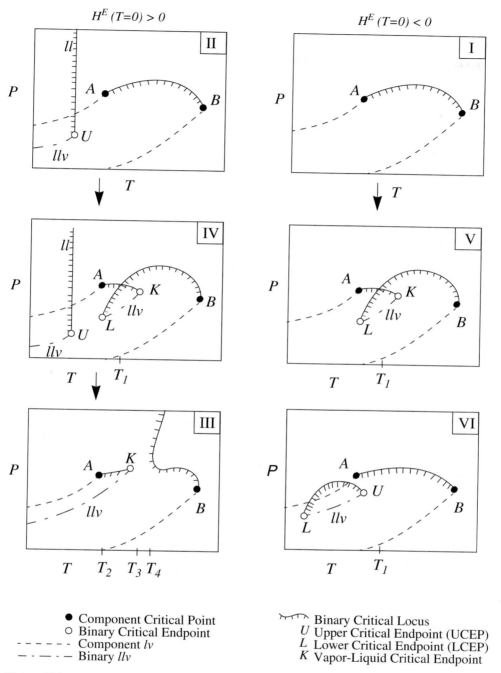

Figure 13.3 *Progression of binary phase behavior with increasing molecular asymmetry according to van Konynenburg and Scott[10]. Arrows denote progressions of phase behavior expected by theory. Experimental progressions frequently differ.*

special notation denoted in the figures and at the end of the section. The types of phase behavior are indicated with the Roman numerals in the upper left of each plot, and the distinctions between the types are due to the location of critical points and critical loci. Type VI phase behavior appears to be exclusive to aqueous systems.[11]

❶ The Gibbs phase rule is helpful in interpreting the *P-T* projections.

The Gibbs phase rule is helpful in interpreting the *P-T* projections. In a pure system, if two phases coexist, one degree of freedom is available; therefore, two-phase coexistence appears as a line on a *P-T* projection. At the triple point, solid, liquid, and gas coexist, and no degrees of freedom are available; therefore the condition is a point on the *P-T* projection. At critical points, all intensive properties of two phases become identical, so the number of degrees of freedom are reduced, and this also appears as a point for pure systems. To avoid misuse of the phase rule, only intensive variables are used for the degrees of freedom. Also, the intensive variables must be varied over a finite range, e.g., two fluid phases are impossible above the critical temperature of a pure substance. More detailed discussions of the correct use of the degrees of freedom are available.[12]

In a binary system, liquid-vapor phase behavior may occur with two degrees of freedom. Liquid-vapor critical behavior occurs with one degree of freedom. Therefore, liquid-vapor behavior appears within a region on the *P-T* projection, and critical behavior occurs along a line known as the critical locus. In Fig. 13.3, pure component *l-v* lines are indicated by the dashed lines. Pure component critical points are indicated by solid circles. Invariant critical points in the binary are indicated by open circles and critical lines are indicated by solid lines. Critical lines are hashed to indicate the side on which two phases coexist. Since the critical locus is a projection on the *P-T* diagram, the diagram and phase rule indicate that two phases coexist on one side of the curve *over a finite composition range*. The diagram does not imply that two phases will coexist at *all* compositions below the critical locus. Also, in some areas, two critical curves are superimposed over the same temperature-pressure ranges, e.g., Types IV and V; however, the overlaps are due to the projection onto the two-dimensional diagram—the two overlapping regions of critical behavior occur over different composition ranges. The critical locus represents conditions where two phases become identical but this does not necessarily imply that only one phase exists above the critical locus because: (1) the two phases which become identical at the critical locus may coexist at temperatures and/or pressures slightly above the critical locus as non-identical phases (see Example 10.10); and (2) if three phases exist below the critical line, two phases will coexist above the critical line. Also, the absence of a critical line on a region of a *P-T* trace does not imply that only one phase is present; it simply means that no phases become identical within the range of the diagram.

❶ The type of phase behavior depends on the molecular asymmetry of the mixture.

The type of phase behavior depends on the molecular asymmetry of the mixture. *Molecular asymmetry* is a term used to describe size differences (molecular weight) for functionally similar molecules or polarity or functional differences for molecules of similar molecular weight. As the molecular asymmetry of the system increases, the critical points of the species generally move farther from each other on the *P-T* traces. With increasing disparity of the critical points, all phase behavior spans increasingly larger areas of *P-T* space.

Type I and Type II phase behavior occurs in systems where the molecular asymmetry is relatively small. When the molecular asymmetry is greater, the immiscibility regions will be larger, as illustrated by the three-dimensional diagrams for Types III and IV in Fig. 13.4. Note that the region

11. Streett, W.B. (1983) in *Chemical Engineering at Supercritical Fluid Conditions,* M.E. Paulaitis, J.M.L. Penninger, R.D. Gray, P. Davidson, eds., Ann Arbor Science, Ann Arbor, MI, pp. 3–30.

12. Modell, M. and R. Reid (1983) *Thermodynamics and Its Applications,* 2ed., Prentice-Hall, Englewood Cliffs, N.J., 259–264; Denbigh, K. (1981) *The Principles of Chemical Equilibria,* 4ed., Cambridge University Press, New York, pp. 188–190.

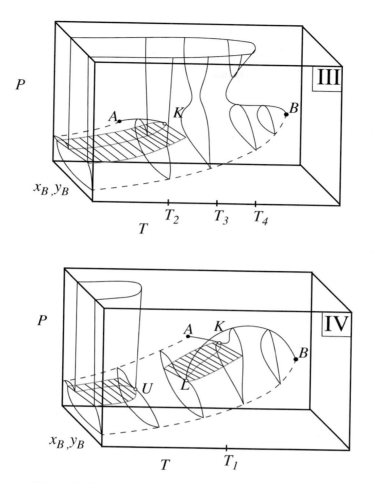

Figure 13.4 *Type III and IV phase behavior illustrated on three-dimensional diagrams. Symbols are the same as Fig. 13.3. The labeled temperatures are the same as Fig. 13.3 and are used for plotting cross-sections in other figures.*

below U in Type IV can look the same as the region below U in Type II. Fig. 13.5 shows some isothermal sections where the temperatures used for the plots are denoted in Figs. 13.3 and 13.4. Note that the phase behavior in Fig. 13.5a and 13.5b differ only in the existence of a *ll* critical point of the first diagram. The effect of temperature on the phase behavior of a Type III system is interesting because the narrow neck region can pinch off as the temperature is increased resulting in the cross sections of Fig. 13.5c and d.

Phase Behavior in the Presence of Solids

Figs. 13.1–13.5 illustrate only fluid phase behavior. Solid-liquid-vapor coexistence can interfere with the fluid phase behavior when the triple point temperature of the higher molecular weight *(heavier)* component approaches the range of experiments. (The general trend is for the melting point to increase with molecular weight.) The superposition of solidus lines on these diagrams will be discussed later. For all the phase diagrams sketched here, the light component A, has the lower critical temperature and appears at the left in Fig. 13.3 and in the rear of Fig. 13.4.

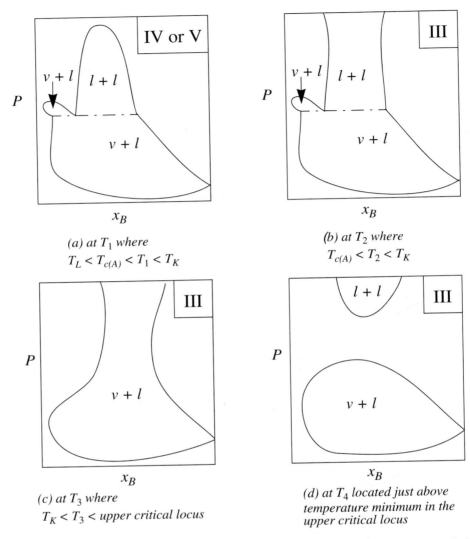

(a) at T_1 where
$T_L < T_{c(A)} < T_1 < T_K$

(b) at T_2 where
$T_{c(A)} < T_2 < T_K$

(c) at T_3 where
$T_K < T_3 <$ upper critical locus

(d) at T_4 located just above
temperature minimum in the
upper critical locus

Figure 13.5 *Illustration of selected isothermal P-x sections for Types IV, V, and III. Symbols and labeled temperatures are the same as Fig. 13.3 and 13.4.*

Experimental Studies of Homologous Series

One way to understand the progression of phase behavior is to review literature measurements which classify the phase behaviors. Classifications are based on experimental studies and may require revision if additional phase transitions are found to occur.[13,14] Types I or II occur when the molecules are fairly similar in structure or critical properties. Van Konynenburg and Scott found the van der Waals equation predicted the progression II $\Rightarrow$ V $\Rightarrow$ III for increasing molecular asymmetry

13. Peters, C.J., R.N. Lichtenthaler, J. de Swaan Arons, (1986) *Fluid Phase Equil. 29*, 495–504.
14. Enick, R., G.D. Holder, B.I. Morsi (1985) *Fluid Phase Equil. 22*, 209–24.

in systems with an endothermic low-temperature heat of mixing, and they found the progression I $\Rightarrow$ V for increasing asymmetry in systems with an exothermic low temperature heat of mixing. Most nonpolar systems are expected to have endothermic heats of mixing, e.g., ethane-butane, benzene-hexane, and might be expected to be Type II by theory; however, they are classified as based on experimental measurements, and most are Type I. Type II phase behavior includes a liquid-liquid immiscibility below the critical temperature of both components. The liquid-liquid behavior is often relatively insensitive to pressure when the liquids are incompressible far below the critical temperature. In the liquid-liquid region, as the temperature is increased, the liquids become increasingly miscible in each other until they become identical at the UCEP. The UCEP occurs at extremely low temperatures for small endothermic heats of mixing, and is frequently not found experimentally due to lack of experimental exploration at low temperatures or due to freezing of the liquids before they become immiscible. While a significant number of liquid-liquid experiments have been performed to locate liquid-liquid UCEPs at atmospheric pressure,[15] phase behavior characterization near the critical point of the lighter component is necessary to permit classification as Type II or IV. A summary of some homologous series are provided below to explore the progressions of phase behavior. To study the series, the asymmetry of the systems is systematically increased by varying the molecular weight of the heavier component by one functional group at a time, and observing the trends in the location of critical points and the class of phase behavior.

Ethane/*n*-alkanes: Studies of the ethane family are summarized by Peters, et al.,[1] and Miller and Luks.[6,13] Type I behavior exists up through *n*-heptadecane because an upper critical end point (UCEP) has not been reported, possibly because the components freeze before they become immiscible. Beginning with *n*-octadecane, a liquid-liquid-vapor region develops near the ethane critical point characteristic of Type V. Once again, an UCEP is not found experimentally. This phase behavior continues through *n*-tricosane. With *n*-tetracosane and *n*-pentacosane a modification of Type V or III occurs where the $s_B l_2 v$ line interferes with the fluid behavior as shown in Fig. 13.6*a*. The three phase $s_B l_2 v$ line extends from the triple point of the heavier component and intersects the *llv* line at a quadruple point, *Q* where four phases coexist. Beginning with *n*-hexacosane, the $s_B l_2 v$ line moves to temperatures above the *K* point and the Fig. 13.6*d* phase behavior begins.[13] In Fig. 13.6, the line labeled $s_B l_2 v$ denotes the conditions where the solid melts over some composition ranges[8]. Schneider reports that squalane, a branched C_{30}, exhibits Type III,[9] which does not fit the pattern of *n*-alkanes at the same carbon number.

CO_2/*n*-alkanes: Studies of the CO_2 family are summarized by Schneider,[7,9] Miller and Luks[6] and Enick et al.[14] Pure CO_2 freezes at a relatively high reduced temperature ($T_r = 0.712$) while ethane freezes at a comparatively low reduced temperature ($T_r = 0.295$). Therefore it may initially seem less likely to observe liquid-liquid immiscibility in CO_2 systems because the systems might freeze before liquid-liquid critical point, *U,* is reached, however the opposite is found and although the liquid-liquid region is not seen for the lightest alkanes, *n*-heptane clearly shows Type II behavior[6]. Type II behavior continues through *n*-dodecane. Enick, et al.,[14] show that *n*-tridecane is Type IV, and that beginning with *n*-tetradecane, the type of Fig. 13.6*a* emerges. Schneider[9] reports the type of Fig. 13.6*a* for *n*-hexadecane. The type of Fig. 13.6*a* is exhibited at *n*-heneicosane[16] and the type of Fig. 13.6*d* appears with *n*-docosane[6,17]. Schneider[9] reports that squalane, a branched C_{30} is Type III, and as with ethane, varies from the pattern with *n*-alkanes.

15. Hildebrand, J.H., J.M. Prausnitz; R.L. Scott (1970) *Regular and Related Solutions,* Van Nostrand.
16. Fall, D.J., J.L. Fall, K.D. Luks (1985), *J. Chem. Eng. Data 30*, 82–88.
17. Fall D.J., K.D. Luks (1984), *J. Chem. Eng. Data 29*, 413–417.

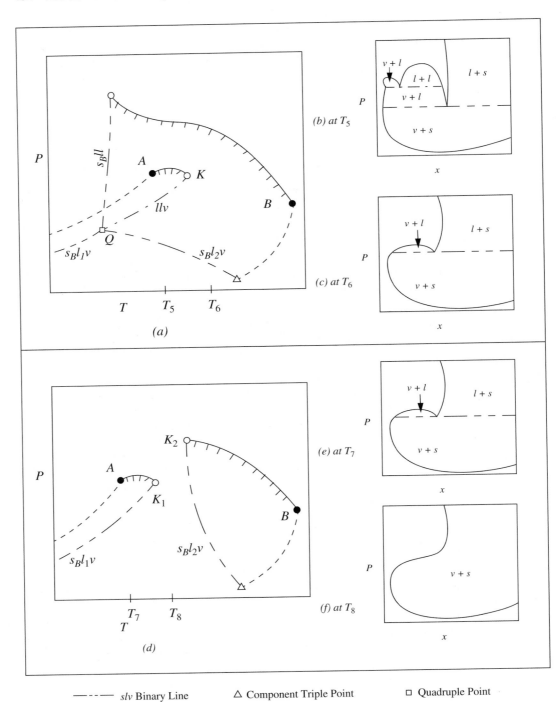

$\cdots\cdots$ *slv* Binary Line $\quad$ △ Component Triple Point $\quad$ □ Quadruple Point

Figure 13.6 *Pressure-temperature projections of phase behavior when the triple point of B is in the vicinity of the critical point of A. Symbols are the same as Fig. 13.3, except as noted.*

The homologous series of n-alkanes with CO_2 exhibits liquid-liquid behavior at much higher reduced temperatures compared to ethane.[18] This may be understood by considering the heat of mixing using the propane system as an example. The low temperature heat of mixing is estimated from the van der Waals equation by:

$$\lim_{T \to 0} H^E = \left(\frac{a_1}{b_1^2} - \frac{2a_{12}}{b_1 b_2} + \frac{a_2}{b_2^2} \right) \frac{x_1 x_2 b_1 b_2}{x_1 b_1 + x_2 b_2} \qquad 13.1$$

where the a's and b's are the van der Waals' parameters. Using Eqn. 13.1 to approximate the low temperature heat of mixing and the normal van der Waals' geometric mean combining rule for the cross parameter a_{12}, both the ethane/propane and CO_2/propane systems are endothermic; however, the heat of mixing in the CO_2/propane system is roughly 100 times the heat of mixing for ethane/propane. Therefore, the excess Gibbs energy of mixing will also be considerably larger and liquid-liquid behavior is expected to occur to higher temperatures. In general, the n-alkane series with CO_2 shows greater asymmetry than the same ethane series, as might be expected. Therefore, the solubilities are lower at a given temperature and pressure, and the progression of behavior from Type II to the type of Fig. 13.6d occurs at lower carbon numbers, even though the critical temperatures of ethane and CO_2 are approximately the same.

CO2/alcohols: Schneider[9] reports 2-hexanol and 2-octanol as Type II and 2,5-hexanediol and 1-dodecanol[7] as Type III. As expected, the addition of hydroxyls on component B increases the asymmetry of the system and the same carbon number by raising the critical point of B, and lowers the carbon numbers for transitions between the phase behavior types.

CO2/aromatics and ethane/aromatics: Both CO_2 and ethane have been studied with n-alkylbenzenes through C_{21} and C_{20} respectively.[6] The over-all trends in behavior are very similar to the comparisons made in the n-alkane series. Polycyclic aromatics have also been studied, and are typically the type of Fig. 13.6d.[8]

N2O/n-alkanes: Nitrous oxide has been studied with the n-alkanes, and the phase behavior is very similar to the ethane series[3] in both the carbon numbers, where phase behavior changes and the critical endpoint temperatures.

CO2/triglycerides/fatty acids: Many common triglycerides and fatty acids are solids at 30°C, and thus exhibit the type of Fig. 13.6a or Fig. 13.6d behavior in binaries with CO_2. Compounds which are normally liquids will probably be Types III, IV, or V. Relatively few studies of model systems are available and in all studies only solubilities are reported.[19,20,21,22] Bamberger, et al.,[22] did determine that lauric and myristic acid, trilaurin and trimyristin melted at the average pressure of their experiments, but the experiments were insufficient to characterize the phase behavior types as Fig. 13.6a or d. Tripalmitin is reported to remain a solid although the triple point is only 5°C greater than trimyristin. Also, most of the mixtures involving tripalmitin were reported to remain solid, which would not be expected if the solids formed an eutectic mixture.[8] The other publications have

---

18. For comparative purposes the reduced temperature is calculated based on the critical temperature of ethane or CO_2 for the respective systems.

19. King, M.B., T.R. Bott, M.J. Barr, R.S. Mahmud (1987) *Sep. Sci. Techn. 22*, 1103–20; Brunetti, L., A. Daghetta, E. Fedeli, I. Kikic, L. Zanderighi (1989) *J. Am. Oil Chem. Soc. 66*, 209–17; Chrastil, J. (1982) *J. Phys. Chem. 86*, 3016–21.

20. Goncalves, M; A.M.P. Vasconcelos; E.J.S. Gomes de Azevedo; H.J. Chaves das Neves; M. Nunes da Ponte (1991) *J. Am. Oil Chem. Soc. 68*, 474–80.

21. Nilsson, W.B., E.J. Gauglitz, J.K. Hudson (1991) *J. Am. Oil Chem. Soc. 68*, 87–91.

22. Bamberger, T., J.C. Erickson, C.L. Cooney, S.K. Kumar (1988) *J. Chem. Eng. Data 33*, 23.

not provided any characterization of the melting behavior of the solids. Unfortunately, disagreements of up to an order of magnitude exist in a few of the solubility measurements. The disagreements have been attributed in part to purity of materials.[20,21] Nilsson, et al.[21] report solubility data for mono- and di-glycerides. One interesting conclusion from the analysis of Czubryt, et al.[23] is that stearic acid appears to be dimerized in the CO_2 phase. Phase behavior of oil mixtures will be discussed in the section regarding solubility behavior.

Propane/triglycerides/fatty acids: Hixson, et al., published most of the available information on these systems in the 1940's,[24,25,26] and recently, more complete information has been published on the propane/tripalmitin system.[27] Propane refining of oils was practiced industrially as the Solexol process and descriptions are available.[28] Similar to the case of stearic acid in CO_2 discussed above, the acids appear to dimerize in the fluid phase. This hypothesis is supported by the correlation of phase behavior with the effective molecular weight, which is double the molecular weight of an acid and equal to the molecular weight of an ester or triglyceride. Binary mixtures of propane with lighter molecular weight acids and esters is Type I (a UCEP has not been located), while the triglycerides and heavier acids and esters exhibit Type IV. For lauric acid (effective M.W. = 400.6) and for myristic acid (effective M.W. = 456.7) a lower critical end point (LCEP) has not been located at propane concentrations between 25–95 wt%.[26] Tricaprylin (M.W. = 470.7) and higher molecular weight compounds exhibit Type IV. The correlation of the LCEP with temperature is shown in Fig. 13.7. The correlation is based on only saturated and mono-unsaturated compounds. Conjugation of unsaturation may have a dramatic effect as exhibited by abietic acid which does not show a LCEP above 30°C.[25]

The Solexol process is based on refining of the oils using the *llv* region in Type IV or V systems between the LCEP and the vapor-liquid critical endpoint, *K*. The phases of interest are the two liquid phases, and Hixson and coworkers did not measure the vapor phase solubilities. As might be expected, the compounds with the lowest effective molecular weight have the greatest solubility in the propane-rich liquid, and for a given chain length solubility increases with decreasing *effective molecular weight: triglyceride - fatty acid - ester.* Hixson's studies concentrate on the measurement of *l-l* solubilities on the three-phase *llv* line in binaries and extension of the measurements to model ternary systems. Hixson's publications address the incorporation of staged countercurrent flow, the importance of density differences for counter-current flow, and internal column recycle using temperature gradients which have been discussed recently in applications with CO_2 extractions.[29,30] These engineering concepts remain important in industrial applications of high pressure technology. In the Solexol process, virtually all the feed oil leaves the top of the first tower dissolved in propane, and the stream is then fractionated by molecular weight and saturation. The majority of color bodies leave the bottom of the first column. The conjugated unsaturated components are in

❶ Critical points vary systematically as molecular asymmetry is changed.

23. Czubryt, J.J., M.N. Myers, J.C. Giddings (1970) *J. Phys. Chem.* **74**, 4260–66.
24. Hixson, A.W., J.B. Bockelmann (1942) *Trans. Am. Inst. Chem. Eng.* **38**, 891–930; Drew, D.A., A.N. Hixson (1944) *Trans. Am. Inst. Chem. Eng.* **40**, 675–694; Hixson, A.N., R. Miller (1940), U.S. Pat. 2,219,652; (1944), U.S. Pat. 2,344,089; (1945), U.S. Pat. 2,388,412.
25. Hixson, A.W., A.N. Hixson (1941) *Trans. Am. Inst. Chem. Eng.* **37**, 927–957.
26. Bogash, R., A.N. Hixson (1949) *Chem. Eng. Progress* **45**, 597–601.
27. Coorens, H.G.A., Peters, C.J., de Swaan Arons, J. (1988) *Fluid Phase Equil.* **40**, 135–151.
28. Passino, H.J. (1949) Ind. Eng. Chem. 41, 280–287, Dickinson, N.L., J.M. Meyers (1952) J. Am. Oil Chem. Soc. 29, 235–39.
29. Stahl, E., K.-W. Quirin, D. Gerard (1988) *Dense Gases for Extraction and Refining,* Springer-Verlag, New York.
30. Eisenbach, W. (1984) *Ber. Bunsenges. Phys. Chem.* **88**, 882; Nilsson, W.B., E.J. Gauglitz, Jr., J.K. Hudson, V.F. Stout, J. Spinelli (1988) *J. Am. Oil Chem. Soc.* **65**, 109–117; Nilsson, W.B., E.J. Gauglitz, J.K. Hudson (1989) *J. Am. Oil Chem. Soc.* **66**, 1596–1600.

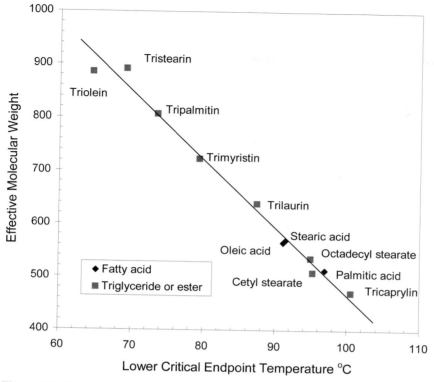

Figure 13.7 *Trend in the lower critical endpoint (LCEP) for components in binary
mixtures with propane (Type IV systems).*

general less soluble than the saturated or mono-unsaturated components. One of the distinct capabilities of the Solexol process was the concentration of vitamin A but, as alternative methods for obtaining vitamin A became available, the process was abandoned.

Assimilation of experimental data on homologous series: Schneider[7] suggests that Types II, IV, and III follow a trend which can be understood if the phase behaviors are considered as a result of effects which are superimposed on each other and can be best visualized on Fig. 13.3. When a homologous series of systems is studied, with only the heavier compound varying, the first system to exhibit Type III behavior has a strong maximum and minimum in the critical line extending from the heavier component critical point. If this strong minimum exists for less dissimilar systems, it is expected to intersect the *llv* line in two places, giving Type IV behavior. These trends also appear consistent with the calculations of Chai[31] using the Peng-Robinson equation. Another trend is obvious which supports this theory. As the dissimilarity of components increases, the UCEP temperature at *U* increases which tends to decrease the difference in the temperatures of *U* and *L* of Type IV systems until they merge and Type III results (Fig. 13.8). It is also possible for the phase behavior to progress directly from Type II to Type III.[10]

There is a systematic progression from Type II → IV → III.

31. Chai, C.-P., (1981) "Phase Equilibrium Behavior for Carbon-Dioxide and Heavy Hydrocarbons," Ph.D. dissertation, University of Delaware.

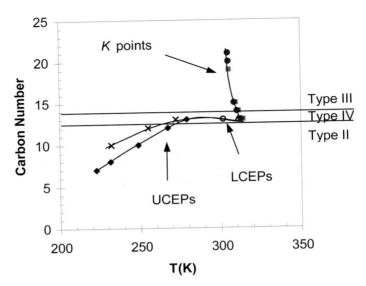

Figure 13.8 *Illustration of trends in critical endpoint temperatures as the molecular asymmetry of systems increases for carbon dioxide + n-alkyl benzenes (diamonds, triangle, filled circles) and carbon dioxide + n-paraffins (x's, o, filled squares) as compiled by Miller and Luks.*[11] *In both series, the systems are Type II below C13, Type IV for C13, then Type III above C13.*

Phase Behavior Summary

Phase behavior is determined by the degree of molecular asymmetry. A variety of interesting phase behavior exists, and a natural progression of phase behaviors occurs as the asymmetry of the system is increased. Trends in critical behavior suggest that organic acids dimerize in non-polar solvents.

List of Symbols for This Section

1–2 = 3	phase 1,2,3 coexist, and phases 2 and 3 are identical at a critical state.
1 = 2–3	phase 1,2,3 coexist, and phases 1 and 2 are identical at a critical state.
A	component with lower critical temperature.
B	component with higher critical temperature.
K	a *K* point—a critical endpoint where the phases are $l - l = v$.
l	liquid phase.
LCEP	a lower critical endpoint where the phases are $l = l - v$.
Q	a *Q* point, or quadruple point, which has four phases in equilibria; herein, the four phases are *sllv*.
s	solid phase.
UCEP	an upper critical endpoint where the phases are $l - l = v$.
v	vapor phase.

13.2   SOLID-LIQUID EQUILIBRIA

Solid-liquid equilibria (SLE) calculations differ slightly from the vapor-liquid calculations we have performed in earlier chapters. Although the equations look different, the principles are the same; the chemical potentials (or fugacities) of the components in the solid phase must be the same as the component fugacities in the liquid phase.

Pressure Effects

For vapor-liquid calculations, pressure changes have large effects on the results of the calculations. This is because the fugacity of a component in the vapor phase is highly dependent on the pressure. For solid-liquid equilibria, pressure changes usually have very small effects on the equilibria unless the pressure changes are large (10 to 100 MPa), because the enthalpies and entropies of *condensed* phases are only weakly pressure dependent. Since $dG = RT\, d \ln f = dH - T\, dS = V\, dP$ for a pressure change at constant T, the Gibbs energy and fugacity exhibit only small changes with pressure when the enthalpy and entropy exhibit small changes. (Recall that the Poynting correction term is usually very near one.) In a mixture of liquids, the analysis must be done with partial molar enthalpies and partial molar entropies; however, these properties are also only weakly dependent on pressure. In the following subsections, we calculate properties at the triple-point, or other low pressures, and use the results at atmospheric pressure without pressure correction due to the weak pressure dependence.

❶ Pressure effects on SLE will be neglected.

SLE in a Single Component System

To begin our discussion, we will consider a single component. At 1 bar, water freezes at 0°C. Ice exists below this temperature and liquid exists above. From the principles of thermodynamics, the Gibbs energy is minimized at constant pressure and temperature; therefore, above 0°C, the Gibbs energy of liquid water must be lower than the Gibbs energy of solid water. In order to express this concept quantitatively, we must consider how the Gibbs energies of each of these phases changes with temperature.

The effect of temperature on the Gibbs energy of any phase may be determined most easily at constant pressure. We may write

$$dG = -S\, dT + V\, dP$$

but at constant pressure:

$$dG = -S\, dT \text{ (constant } P)$$

and thus:

$$(\partial G/\partial T)_P = -S$$

The temperature dependence of Gibbs energy is then dependent on the entropy of the phase. Entropy is a positive quantity; therefore, the Gibbs energy of any phase must decrease with increasing temperature. However, the entropy of a liquid phase is greater than the entropy of a solid phase; thus the Gibbs energy of a liquid phase decreases more rapidly as the temperature increases.[32] Since the Gibbs energies of the phases are equal at the freezing temperature, the Gibbs energy of the liquid will lie below the Gibbs energy of the solid above the freezing temperature (see Fig.

---

32. To begin the calculations, we must specify a reference state. In any thermodynamic analysis, we must have only one reference state for each chemical species; for our example here, the reference state for water must be the same whether the water is solid or liquid. See Section 8.12 on page 278 and footnote therein.

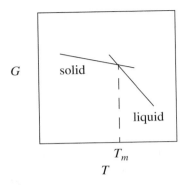

Figure 13.9 *Illustration of Gibbs energies for pure SLE.*

13.9). The portion of the solid Gibbs energy curve above the melting temperature represents the Gibbs energies of a hypothetical solid, and a melting process will occur spontaneously at constant temperature and pressure because the ΔG for the process is negative. Alternatively, an equilibrium solid will not be formed above the melting temperature because ΔG is positive. A discussion for behavior below the melting temperature is not presented although the ideas are similar. Melting of a solid below the freezing temperature requires a positive change in Gibbs energy at constant temperature and pressure.

⚠ $\Delta G^{fus} > 0$ below the normal melting temperature.

The Calculation Pathway

We know that solids dissolve in liquid mixtures well below their normal melting temperatures. Sugar and salt both dissolve in water at room temperature although the pure solid melting temperatures are far above room temperature. Also, salt is spread on the highways in the winter to lower the temperature at which solid ice forms. We may choose to address several problems such as: 1) How much sugar may be dissolved in water before the solubility limit is reached? 2) In a water/salt solution, at what temperature will a solid form, and will the crystals be water or salt? (Salt is introduced as a practical example, although rigorous treatment of the problem involves electrolyte thermodynamics.) 3) How may a solvent or solvent mixture be modified to regulate crystallization? In order to deal with these concepts mathematically, we will use state properties to calculate thermodynamic changes along convenient pathways that involve hypothetical steps.

Let us consider a practical example of dissolving naphthalene (2) in *n*-hexane (1) at 298 K. Since the normal melting temperature for pure naphthalene is 353.3 K, how can we explain the phenomena that the naphthalene dissolves in hexane? First, recall that the naphthalene will dissolve in the *n*-hexane if the total Gibbs energy of the system (*n*-hexane and naphthalene) decreases upon dissolution. Thus, more and more solid may be added to the liquid solution until any further addition causes the total Gibbs energy to increase rather than decrease. This method of calculating equilibrium is fairly tedious to apply, and a preferred method is used that gives identical results: the solubility limit will be reached when the chemical potential of the naphthalene in the liquid is the same as the chemical potential of the pure solid naphthalene. Therefore, we will solve the problem by equating the chemical potentials for naphthalene in the liquid and solid phases.

The equilibrium can be written as $\mu_2 = \mu_2^S$ or, recognizing the notational definitions $\mu_2 = \overline{G}_2$ and $\mu_2^S = G_2^S$, therefore,

$$\overline{G}_2 - G_2^S = 0 \qquad\qquad 13.2$$

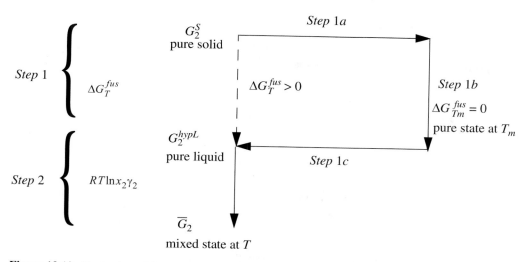

Figure 13.10 *Illustration of the two-step process for calculating solubility of solids in liquids.*

Next, envision the hypothetical pathway shown in Fig. 13.10 to calculate the chemical potential difference given by Eqn. 13.2, which consists of two primary steps.

Step 1. Naphthalene is melted to form a hypothetical liquid at 298 K. The Gibbs energy change for this step is positive as discussed above. The Gibbs energy change will be:

$$\Delta G_T^{fus} = G_2^{hypL} - G_2^{S}$$
  13.3

where the superscript *hypL* indicates a hypothetical liquid.

Step 2. The hypothetical liquid naphthalene is mixed with liquid *n*-hexane. If the solution is non-ideal, the Gibbs energy change for component 2 is

$$\overline{G}_2 - G_2^{hypL} = RT\ln(x_2\gamma_2)$$
  13.4

The Gibbs energy change for this step is always negative if mixing occurs spontaneously, and must be large enough to cancel the Gibbs energy change from Step 1.

> Solubility is determined by a balance between the positive ΔG^{fus} and the negative Gibbs energy effect of mixing.

Then clearly, from Fig. 13.10 and Eqn. 13.2 through 13.4,

$$\overline{G}_2 - G_2^{S} = (\overline{G}_2 - G_2^{hypL}) + (G_2^{hypL} - G_2^{S}) = RT\ln(x_2\gamma_2) + \Delta G_T^{fus}$$

or

$$\boxed{\Delta G_T^{fus} = -RT\ln(x_2\gamma_2)}$$
  13.5

where T is 298 K for our example. Relations for the activity coefficients in the right-hand side of the equation have been developed in previous chapters. In the next subsection, the calculation of ΔG_T^{fus} is explained.

Formation of a Hypothetical Liquid

The Gibbs Energy change for step one is most easily calculated using the entropies and enthalpies. For an isothermal process, we can write:

$$dG = dH - T\,dS$$

Continuing with the example of dissolving naphthalene at 298 K, the Gibbs energy change for melting naphthalene at 298 K can be calculated from the enthalpy and entropy of melting,

$$\Delta G_T^{fus} = \Delta H_T^{fus} - T\Delta S_T^{fus} \tag{13.6}$$

where $T = 298$ K. ΔH_{298}^{fus} can be calculated by determining the enthalpies of the liquid and solid phases relative to the normal melting temperature, where the heat of fusion, $\Delta H_{353.3}^{fus} = 18.8$ kJ/mol. The enthalpy of solid at 353.3 K relative to solid at 298 K (Step 1a of Fig. 13.10) is:

$$H_{353.3}^{S} - H_{298}^{S} = \int_{298}^{353.3} C_P^S dT$$

The enthalpy of the liquid at 298 K relative to liquid at 353.3 K (Step 1c of Fig. 13.10) is:

$$H_{298}^{hypL} - H_{353.3}^{L} = \int_{353.3}^{298} C_P^L dT$$

Thus the ΔH_{298}^{fus} for melting:

$$\Delta H_{298}^{fus} = H_{298}^{hypL} - H_{298}^{S} = (H_{298}^{hypL} - H_{353.3}^{L}) + (H_{353.3}^{L} - H_{353.3}^{S}) + (H_{353.3}^{S} - H_{298}^{S})$$

$$= \Delta H_{353.3}^{fus} + \int_{353.3}^{298} (C_P^L - C_P^S) dT$$

which we can generalize to:

$$\boxed{\Delta H_T^{fus} = H_T^{hypL} - H_T^{S} = \Delta H_{T_m}^{fus} + \int_{T_m}^{T} (C_P^L - C_P^S) dT} \tag{13.7}$$

A similar derivation for the entropy gives:

$$\Delta S_{298}^{fus} = \Delta S_{353.3}^{fus} + \int_{353.3}^{298} \frac{(C_P^L - C_P^S)}{T} dT$$

which we can generalize to:

$$\boxed{\Delta S_T^{fus} = \Delta S_{T_m}^{fus} + \int_{T_m}^{T} \frac{(C_P^L - C_P^S)}{T} dT} \tag{13.8}$$

In addition to these relationships, at the normal melting temperature, since $\Delta G_{T_m}^{fus} = 0$,

$$\Delta S_{T_m}^{fus} = \frac{\Delta H_{T_m}^{fus}}{T_m} \tag{13.9}$$

where $T_m = 353.3$ K. Combining the results and neglecting the integrals (which are nonzero, but essentially cancel each other), we have

$$\boxed{\Delta G_T^{fus} = \Delta H_{T_{m,2}}^{fus}\left(1 - \frac{T}{T_{m,2}}\right)} \tag{13.10}$$

where for our example, $T = 298$ and $T_m = 353.3$. This is the ΔG_T^{fus} that we desire for Eqn. 13.5.

Criteria for Equilibrium

In general then, combining Eqn. 13.10 with 13.5, we arrive at the equation for the solubility of component 2:

$$\boxed{\ln x_2 \gamma_2 = -\frac{\Delta H_2^{fus}}{RT_{m,2}}\left(\frac{T_{m,2}}{T} - 1\right)} \tag{13.11}$$

❶ Solubility equation for crystalline solids.

where the heat of fusion is for component 2 at the normal melting temperature of 2, $T_{m,2}$, and the heat capacity integrals are neglected.

Hexane will also dissolve in a hexane–naphthalene solution below the hexane melting temperature. The relationship for this behavior is

$$\boxed{\ln x_1 \gamma_1 = -\frac{\Delta H_1^{fus}}{RT_{m,1}}\left(\frac{T_{m,1}}{T} - 1\right)} \tag{13.12}$$

where the heat of fusion is for pure component 1 at the normal melting temperature of component 1, $T_{m,1}$. Note that Eqns. 13.11 and 13.12 may also be use to determine crystallization temperatures at specified compositions.

Example 13.1 Eutectic behavior of chloronitrobenzenes

Fig. 13.11 illustrates application to the system o-chloronitrobenzene (1)+p-chlorornitrobenzene (2). The compounds are chemically similar; thus, the liquid phase may be assumed to be ideal, and the activity coefficients may be set to 1. The two branches represent calculations performed from Eqns. 13.11 and 13.12, each giving one-half the diagram. The curves are hypothetical below the point of intersection. This temperature at the point of intersection of the two curves is called the *eutectic temperature*, and the composition is the *eutectic composition*.

❶ Eutectic.

Example 13.1 Eutectic behavior of chloronitrobenzenes (Continued)

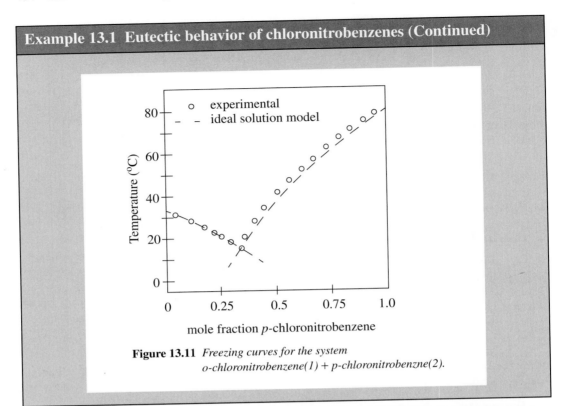

Figure 13.11 *Freezing curves for the system o-chloronitrobenzene(1) + p-chloronitrobenzne(2).*

Example 13.2 Eutectic behavior of benzene + phenol

In most systems, an activity coefficient model must be included. Fig. 13.12 shows an example where the ideal solution model is not a good approximation, and the activity coefficients are modelled with the UNIFAC activity coefficient model. To solve for solubility given a temperature, the following procedure may be used (taking component 2 for example):

1. Assume the $\gamma_2 = 1$
2. Solve Eqn. 13.11 for x_2
3. At this value of x_2, determine γ_2 from the activity model.
4. Return to step 2, this time including the value of γ_2 in Eqn. 13.11, until subsequent iterations converge.

This is a relatively stable iteration, and Excel Solver can iterate on activity by adjusting composition, bypassing the successive substitution method above.

Example 13.2 Eutectic behavior of benzene + phenol (Continued)

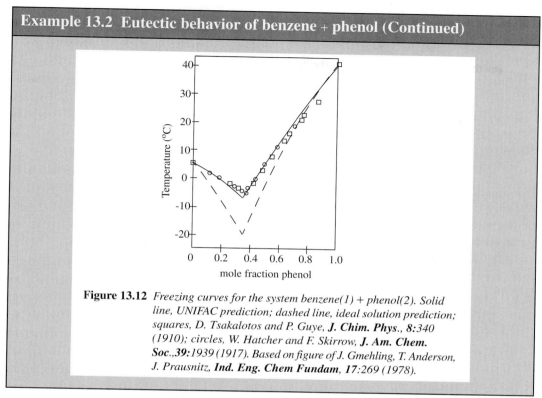

Figure 13.12 *Freezing curves for the system benzene(1) + phenol(2). Solid line, UNIFAC prediction; dashed line, ideal solution prediction; squares, D. Tsakalotos and P. Guye,* **J. Chim. Phys.**, *8:340 (1910); circles, W. Hatcher and F. Skirrow,* **J. Am. Chem. Soc.**, *39:1939 (1917). Based on figure of J. Gmehling, T. Anderson, J. Prausnitz,* **Ind. Eng. Chem Fundam**, *17:269 (1978).*

SLE with Solid Mixtures

So far, we have only covered phase behavior in systems where the solids are completely immiscible in each other. Fig. 13.13 illustrates a case where the solids form solid solutions, and Fig. 13.14 illustrates behavior where compounds are formed in the solid complexes. Also, the case of wax precipitation from petroleum results in a range of n-C_{20} to n-C_{35} straight chain alkanes being mixed in the solid phase. Paraffin wax that you can purchase in the grocery store is primarily composed of n-C_{20} to n-C_{35} straight chain alkanes. For the case of liquid and solid mixtures in equilibrium, the derivation of the equilibrium relationship can be modified by adding a step for "unmixing" of solid solutions to the schematic of Fig. 13.10 on page 461. This step is analogous to a reversal of Step 2 of the diagram except involving solid solutions and pure solids rather than liquid solutions and pure liquids. For each component in the mixtures:

$$-RT \ln(x_i^S \gamma_i^S) + \Delta G_i^{fus} + RT \ln(x_i^L \gamma_i^L) = 0 \qquad 13.13$$

Example 13.3 Wax precipitation

An especially difficult problem in the recovery of natural gas is the clogging of pipes caused by small amounts of wax that accumulate over time. In the Gulf of Mexico, natural gas at the bottom of the well can be 250 bars and 100°C, but it must be reduced to 100 bars to be permitted in the pipeline, and the sea floor can drop to 5°C. The reduction in pressure and temperature results in a loss of carrying power and the small amounts of heavy liquid hydrocarbons can precipitate, eventually coating the walls with viscous liquid. After the liquid has formed, further cooling can cause solid wax to deposit on the walls of the pipe. These deposits cause constrictions and larger pressure drops that lead to more deposits, and so forth.

Example 13.3 Wax precipitation (Continued)

You should know that naturally occurring petroleum is a complex mixture of hundreds of individual components. Rather than attempt to specify the identity and composition of every component, it is conventional to collect several fractions of the original according to the ranges of their molecular weights. Hansen et al.[33] provide the data in the first four columns of Table 13.1 for the composition, mass density, and molecular weight of several fractions from a typical petroleum stream. This kind of data is typically collected by distilling the initial sample and collecting fractions over time. As the lower molecular weight species are removed, the boiling temperature rises and the distillate collected over each particular temperature range is stored in a separate container. The weight of each fraction relative to the weight of the initial sample gives the composition of that species fraction. The mass of each fraction divided by its volume gives the density. And the average molecular weight can be characterized by gas chromatography or through correlations with viscosity. Note that the molecular weight for any particular species is not necessarily equal to the molecular weight for the corresponding saturated hydrocarbon. This is an indication that olefins, naphthenics, and aromatics are present in significant compositions. The objective of this example problem is to treat the data of Hansen et al. as characteristic of a gas condensate and compute the fractions of the stream that form solids at each temperature.

The fusion (melting) temperatures and heats of fusion for n-paraffins can be calculated according to the correlations of Won.[34]

$$T_i^{fus} = 374.5 + 0.02617 M_i - \frac{20172}{M_i} \qquad 13.14$$

$$\Delta H_i^{fus} = 0.1426 M_i T_i^{fus} \qquad 13.15$$

Noting that each species fraction can contain many species besides the n-paraffins that are responsible for practically all wax formation, it is necessary to estimate the portion of each species fraction that can form a wax. Since the densities of n-paraffins are well-known, it is convenient to use the difference between the observed density and the density of an n-paraffin of the same molecular weight as a measure of the n-paraffin content. The correlations for estimating the percentage of wax-forming components in the feed (z_i^W) are taken from Pedersen.[35]

$$z_i^W = z_i^{tot}\left[1 - \left(0.8824 + \frac{0.5353 M_i}{1000}\right)\left(\frac{\rho_i - \rho_i^P}{\rho_i^P}\right)^{0.1144}\right] \qquad 13.16$$

$$\rho_i^P = 0.3915 + 0.0675 \ln(M_i) \qquad 13.17$$

where z_i^{tot} is the species mole fraction of all in the initial sample
z_i^W is the portion of that fraction which is wax forming (i.e. n-paraffin).

Use these data and correlations to estimate the solid wax phase amount and the composition of the solid as a function of temperature. Use your estimates to predict the temperature at which wax begins to precipitate. Hansen et al. give the experimental value as 304 K.

---

33. Hansen, J.H., Fredenslund, Aa., Pedersen, K.S., Ronningsen, H.P. *AIChE J.* 34:1937 (1988).
34. Won, K.W. *Fluid Phase Equil.*, 53:377 (1989).
35. Pedersen, K.S. *SPE Prod. and Fac.*, Feb:46 (1995).

Example 13.3 Wax precipitation (Continued)

Solution: This problem is basically a multicomponent variation of the binary solid-liquid equilibrium problems discussed above. The main difference is that the solid phase is not pure. We can adapt the algorithm as follows.

Assuming ideal solution behavior for both the solid and liquid phases, we define $K_i = x_i^L/x_i^S$, and as before, we assume the difference in heat capacities between liquid and solid is negligible relative to the heat of fusion,

$$K_i = \exp\left[\frac{\Delta H_i^{fus}}{RT}\left(1 - \frac{T}{T_i^{fus}}\right)\right] \qquad 13.18$$

which is independent of the compositions of the liquid and solid phases because of the ideal solution assumptions. The solid solution mole fraction is given by $x_i^S = x_i^L/K_i$. Compare this method to the vapor-liquid calculations using the shortcut K-ratio in Chapter 9. This is a liquid-solid freezing temperature analog to the vapor-liquid dew temperature procedure. The liquid mole fractions are given by the z_i^W values in the table below. All that remains is to guess values of T, which changes all K_i until $\sum_i x_i^S = 1.0$. Hand calculations would be easy with a couple of components, but spreadsheets are recommended for a multicomponent mixture. Using Solver for spreadsheet WAX.xls gives $T = 320.7$ K. This is slightly higher than the experimental value, but reasonably accurate considering the complex nature of the petroleum fractions and their variabilities from one geographic location to another.

Example 13.3 Wax precipitation (Continued)

Table 13.1 *Petroleum characterization for wax point (Hansen, et al., 1988)*

Species	Wt%	M_i	$\rho(g/cm^3)$	$\rho^P(g/cm^3)$	z_i^{tot}	z_i^W	T^{fus}	ΔH^{fus} (J/mol)	K_i^{320}	x_i^S
<c4	0.031	29	0.416	0.619	0.003	0.000	---	---		
c5	0.855	71	0.632	0.679	0.031	0.000	92	3904		
c6	0.377	82	0.695	0.689	0.012	0.000	131	6386		
c7	2.371	91	0.751	0.696	0.068	0.021	155	8419	28.98	0.0007
c8	2.285	103	0.778	0.704	0.058	0.016	181	11134	24.75	0.0006
c9	2.539	116	0.793	0.712	0.057	0.015	204	14080	20.81	0.0007
c10	2.479	132	0.798	0.721	0.049	0.013	225	17714	16.78	0.0008
c11	1.916	147	0.803	0.728	0.034	0.009	241	21128	13.67	0.0006
c12	2.352	163	0.817	0.735	0.038	0.009	255	24777	10.95	0.0008
c13	2.091	175	0.836	0.740	0.031	0.007	264	27519	9.26	0.0008
c14	3.677	190	0.843	0.746	0.051	0.011	273	30952	7.49	0.0015
c15	3.722	205	0.849	0.751	0.047	0.010	281	34393	6.04	0.0017
c16	2.034	215	0.853	0.754	0.025	0.005	286	36691	5.22	0.0010
c17	4.135	237	0.844	0.761	0.046	0.010	296	41757	3.78	0.0026
c18	3.772	251	0.846	0.764	0.039	0.008	301	44989	3.07	0.0027
c19	3.407	262	0.857	0.767	0.034	0.007	304	47532	2.60	0.0026
c20	2.781	268	0.868	0.769	0.027	0.005	306	48921	2.38	0.0021
c21	3.292	284	0.862	0.773	0.030	0.006	311	52631	1.86	0.0031
c22	3.14	299	0.863	0.776	0.027	0.005	315	56116	1.48	0.0035
c23	3.445	315	0.963	0.780	0.029	0.003	319	59841	1.15	0.0027
c24	3.254	330	0.865	0.783	0.026	0.005	322	63340	0.91	0.0052
c25	2.975	342	0.867	0.785	0.023	0.004	324	66144	0.75	0.0054
c26	3.038	352	0.869	0.787	0.023	0.004	326	68485	0.64	0.0061
c27	2.085	371	0.873	0.791	0.015	0.002	330	72941	0.47	0.0052
c28	2.74	385	0.877	0.793	0.019	0.003	332	76231	0.37	0.0079
c29	3.178	399	0.881	0.796	0.021	0.003	334	79527	0.29	0.0107
>c30	31.12	578	0.905	0.821	0.141	0.011	355	122213	0.01	0.9308
SUM					1.00	0.193				1.000

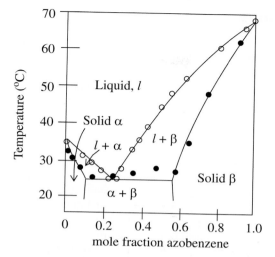

Figure 13.13 *Freezing curves for the Azoxybenzene(1) + azobenzene(2) system illustrating a system with solid-solid solubility. Based on Hildebrand and Scott, Solubility of Non-electrolytes, Dover, 1964.*

Lever rule

$$f = \frac{\alpha - p}{\alpha - l}$$

$$\alpha = \frac{p - l}{\alpha - l}$$

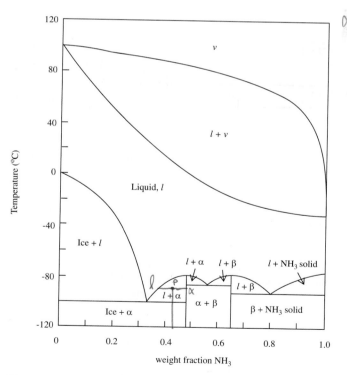

Figure 13.14 *Solid-liquid and vapor-liquid behavior for the ammonia(1) + water(2) system at 1.013 bar. NH_3 and H_2O form two crystals in the stoichiometries:* α) $NH_3 \cdot H_2O$; β) $2NH_3 \cdot H_2O$. *(Based on Landolt-Börnstein, II/2a:377, 1960)*

Solid-Liquid Equilibria Summary

Phase equilibrium involving solids is an extension of previous modeling concepts. Like liquid-liquid equilibria, the condensed phase fugacities are quite insensitive to pressure, so the partition coefficients are simply functions of temperature. The main difference is that the heats of fusion are used to relate the component fugacities in their various states of matter. In multicomponent mixtures, the solid-liquid procedures for calculation are analogs of the vapor/liquid procedures, where the partition coefficients are calculated in a different manner.

13.3 RESIDUE CURVES

Distillation is among the most highly developed and reliable separation techniques for the chemical industry, so it is usually among the first separation techniques considered during process design. Design of multicomponent distillation columns can be complicated by azeotrope and heteroazeotropes. Shortcut distillation techniques are useful for distillation column screening, but azeotrope systems require modified shortcut equations and relative volatilities[36]. Design of multicomponent steady-state distillation columns is usually performed by process simulators, nevertheless many design hours are saved if the phase behavior is explored using residue curves before working with the detailed column calculations. In this text, residue curves for ternary systems with a single feed and one distillate product and one bottoms product will be covered.[37] Residue curve analysis for ternary systems, under these conditions, involves the following concepts that will be described below:

> ❶ Using residue curves can save many design hours.

1. The steady-state column feed, distillate, and bottoms compositions are co-linear on ternary diagrams in accordance with the lever rule.

2. Two combined streams may be represented by a mixing point co-linear with the feed streams at a point between the streams given by the lever rule.

3. Binary and ternary azeotropes are connected on ternary diagrams by boundary lines called separatrices.

4. Separatrices divide the ternary residue curve map diagram into regions.

5. As a reliable rule, the distillate, feed, and bottoms compositions will all be in the same residue curve map region.[38]

6. Residue curves begin and end in a single region—they do not cross separatrices.

> ❶ Residue curves can provide a guide to attainable compositions.

Therefore, only certain composition regions in a ternary system describe possible products for a given feed, and the residue curves provide the guide to attainable compositions.

Residue curves represent the trace of liquid-phase (the residue) composition during a single-stage constant-pressure batch distillation. Single-stage batch distillation calculations are not complex, and will be introduced below using the VLE K-ratios that have been the topics of earlier chapters. As the batch is distilled, the boiling point of the residue will increase until the composition reaches either a maximum boiling azeotrope or the pure composition of the highest boiling component. The residue curves will *terminate* at these compositions. The residue compositions will move away

---

36. Frank, T.C., Chem. Eng. Prog., 93(No. 4):52–63 (1997).

37. Interested readers may find a review article helpful. See Fien, G.-J.A.F., Liu, Y.A., *Ind. Eng. Chem. Res.*, 33:2505 (1994). The topic is also becoming available in chemical engineering textbooks intended for separations courses.

38. This guideline may be overcome when the region boundary is highly curved and the feed is near the boundary on the concave side, however, such application is rarely economical.

from any minimum boiling azeotropes and/or the pure composition of the lowest boiling component. The residue curves *originate* at these compositions. In the diagrams in this section, we use the letter *o* to represent the low boiling point residue curves' origin and the letter *t* to represent the high boiling point residue curves' terminus.

The Bow-Tie Approximation

Let us begin analysis by considering residue curves for a ternary system without azeotropes, propane + butane + pentane. As this ternary mixture boils, bubble point calculations show that propane and butane are the most prevalent vapor species. As the vapor is removed, the residue becomes increasingly rich in pentane. Residue curves calculated with the Peng-Robinson equation are shown in Fig. 13.15a. Starting from any initial composition, the curves on the diagram can be used to follow the trace of the residue composition. The residue from any ternary composition in this system will become increasingly rich in pentane. The residue curve map has a single region, and the residue curves originate at pure propane (the lowest boiling composition in the region) and terminate at pentane (the highest boiling point in the region). Residue curves are helpful in the design of steady-state continuous flow distillation columns because the accessible compositions of distillate and bottoms products from a steady-state distillation column can be found by studying the region of the residue curve map that contains the feed. *The distillate of a one-feed steady-state column will tend to approach the origin of the residue curves, and bottoms of the column will tend to approach the terminus of the residue curves within the restrictions summarized in points (1) through (6) above.*

Recognize the importance of point 1, that the feed, distillate and bottoms for a ternary system must be colinear when plotted. Consider the implications for the propane + butane + pentane system shown in Fig. 13.15a. When a column is designed for a specified feed, F, the distillate or bottoms composition may be specified also, but not both. For a feed, F, if the bottoms product of a distillation column is B_1, the distillate (top product), D_1, must also be colinear on the plot, and F must fall between B_1 and D_1 to satisfy the lever rule material balance, $D_1/F + B_1/F = 1$, and $D_1/B_1 = (\overline{FB_1})/(\overline{FD_1})$. Another option for operating the column is to achieve pure propane distillate, D_2. In this case, the feed must lie on a line between D_2 and B_2. *The first approximation of reachable compositions is given by the bow-tie shaped region D_1-D_2-B_2-B_1.*[39] The distillate and bottoms compositions are not required to be on the boundary of the region as we have chosen to show in the figures, but the colinearity and lever rule must be followed. Note that it is not possible to simultaneously obtain pure propane and pure pentane in a single-feed, two-product (distillate and bottoms) column from feed F. Note that products near pure butane are unattainable, so design can focus on other alternatives. The bow-tie shaped region is an approximation that will be adequate for the screening introduction intended here.

❶ The bow-tie method provides a first approximation for attainable compositions.

Azeotropes

When one azeotrope exists in a ternary system, the residue curve map will still be a single region, but the azeotrope will be an origin or a terminus for the residue curves. Maximum boiling azeotropes may become the terminus of some residue curves, and minimum boiling azeotropes may become the origin of some residue curves. For example, in the methanol + acetone + water system shown in Fig. 13.15b, the methanol + acetone system forms a minimum boiling azeotrope that becomes an origin for the residue curves because it is the lowest boiling point in the region. The distillate product for any steady-state column with a feed in the region will tend toward this

---

39. The products of a two-feed column may lie outside the bow-tie region drawn from the overall feed, a principle exploited in extractive distillation. See Wahnschaft, O.M., Westerberg, A.W., *Ind. Eng. Chem. Res.*, 32:1108 (1993).

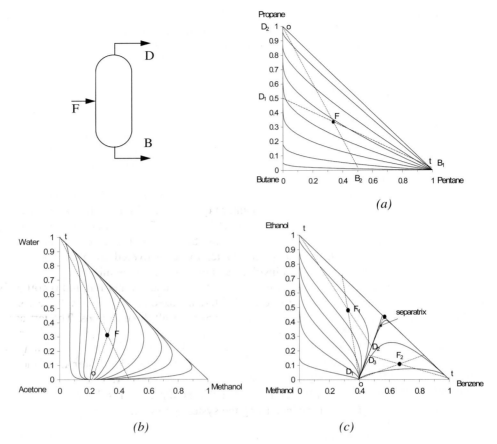

Figure 13.15 *(a) Residue curves for the system propane + n-butane + n-pentane as calculated with the Peng-Robinson equation at 1 bar; (b) Methanol + acetone + water as modeled with the UNIQUAC equation at 1 bar; (c) Methanol + ethanol + benzene as modeled with the UNIQUAC equation at 1 bar. Calculations performed using Aspen Plus® ver. 9.2.*

composition. The highest boiling point in the region is the terminus of the residue curves at the composition of pure water, and the steady-state column bottoms will tend toward this composition. The bow-tie approximation of attainable compositions is shown for a feed *F*, and note that a distillate composition of pure methanol or pure acetone is not attainable because the residue curves' origin is the azeotrope. The lever rule concepts apply as in non-azeotropic systems.

When two or more azeotropes exist, the residue curve map will often be divided into two or more regions of residue curves. The residue curve maps show the distillation regions of attainable compositions, and the location of the separatrices. For example, residue curves for methanol + ethanol + benzene are shown in Fig. 13.15c. In the left region, ethanol is the high boiling point (residue curve terminus) and the methanol + benzene azeotrope is the low boiling point (residue curve origin). In the right region, the methanol + benzene azeotrope is also the residue curve origin (low boiling point) and benzene is the residue curve terminus (high boiling point). The ethanol + benzene azeotrope is at an intermediate temperature between the origin and terminus of each region, and therefore, is neither an origin or terminus, and is called a saddle point. Two different arbitrary feed compositions are plotted on the diagram, and the corresponding bow-tie shaped regions of approximate accessible compositions are shown. Since the design rule is that a single column cannot cross a separatrix, the practical boundary for distillate compositions from the feed F_1 is approximated by the region F_1-D_1-D_2. For a feed of composition F_2, the practical region of attainable distillate compositions is approximated by the region F_1-D_1-D_3. Although an example with a maximum boiling azeotrope is not shown, systems with these azeotropes can be screened using the same lever rule techniques; however, the bottoms of the column will be affected by the azeotropes.

Heteroazeotropes—Systems with LLE

Ternary systems exhibiting LLE form minimum-boiling heteroazeotropes and can often be separated in a system of two columns as shown in Fig. 13.16. The LLE behavior often spans a separatrix on the residue curve map. This means that an overhead stream, D_M, can be condensed in a decanter, and one of the liquid phases will be on the same side of the separatrix as D_1 and can be returned into that column (left in the figure). The other decanter liquid phase will be on the other side of the separatrix and can be used as a feed to another column (right in the figure). An example of this procedure is given by the separation of ethanol + water using benzene. In this case, benzene is intentionally added to break the azeotrope and permit water to be recovered from one column and ethanol from the other column. The system involves an ethanol + benzene minimum boiling azeotrope, an ethanol + water minimum boiling azeotrope, and a benzene + water minimum boiling heteroazeotrope. The system has three separatrices. The ternary azeotrope, o, is the lowest boiling point in the system and it is the origin of the residue curves for all three regions. The left column operates in the right region of the residue curve map, and the right column operates in the upper left region. Care is taken to avoid having F_2 fall in the lower left region of the residue curve map; such a feed would result in a bottoms of benzene rather than ethanol. Illustrative material balance lines are provided on the diagram, and the LLE curve is superimposed on the residue curve map to clearly show how the tie lines span the separatrices. For this example, the residue curves, origin can be moved farther into the LLE region by increasing the system pressure.

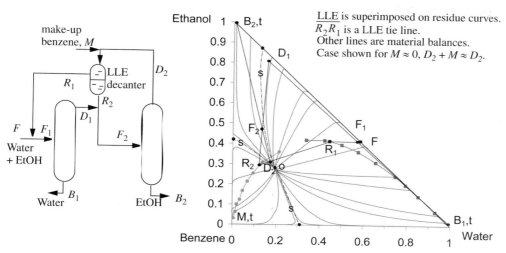

Figure 13.16 *Illustration of the column configuration, residue curves and LLE behavior for the separation of ethanol and water using benzene, residue curves calculated using Aspen Plus® ver. 9.2 with UNIQUAC at 760 mmHg, LLE data at 25°C from Chang, Y.-I., Moultron, R.W., Ind. Eng. Chem., 45(1953)2350. LLE tie lines are not plotted to avoid clutter, but tie-line data appear in pairs of points along the binodal line.*

Generating Residue Curves

Residue curves are generated by a bubble temperature algorithm for a simple single-stage batch distillation without reflux. If the total moles vaporized from a multicomponent mixture is dn^L, the moles of component i leaving is calculated by the composition of the vapor phase, $dn_i = y_i dn^L$. For component i in the liquid phase, $dn_i = d(x_i n^L) = n^L dx_i + x_i dn^L$. Equating the vapor and liquid expressions for dn_i,

$$y_i \, dn^L = x_i \, dn^L + n^L dx_i \qquad\qquad 13.19$$

rearranging

$$(y_i - x_i) dn^L / n^L = dx_i \qquad\qquad 13.20$$

which may be written

$$(K_i - 1) x_i d(\ln[n^L]) = dx_i \qquad\qquad 13.21$$

where K_i values can be calculated by any appropriate bubble temperature method from Chapter 9–12 for the system at composition x_i. To generate residue curves in an N-component mixture, differential values of $d(\ln[n^L])$ may be chosen arbitrarily to generate differential values of dx_i. Only $N - 1$ values of dx_i are required since

$$\sum_{i=1}^{N-1} dx_i + dx_N = 0 \qquad\qquad 13.22$$

From an arbitrarily selected initial composition, the trace of x_i values yields the residue curve. Special care should be taken to include liquid-liquid behavior, if present in the system, requiring a three-phase bubble temperature calculation as illustrated in Chapter 12. Table 13.2 presents the first few residue curve calculations for the n-pentane(1) + n-hexane(2) + n-heptane(3) system at 0.1013 MPa via Eqn. 13.21 using the shortcut K-ratio method to calculate the K_i values. The initial composition for the residue curve is arbitrarily selected as $x_1 = 0.1$, $x_2 = 0.6$ and $x_3 = 0.3$ denoted in cells with double ruling in Table 13.2. The step size for Eqn. 13.21 is arbitrarily selected as $d(\ln[n^L]) = -0.15$ to generate the dx_i values listed in the table. Eqn. 13.21 is used directly as a finite difference formula to provide a quick estimate as $x_i^{new} = x_i^{old} + dx_i^{old}$ to move down the table from the initial point. The residue curve is generated towards the increasing n-pentane liquid compositions by moving

Table 13.2 *Example calculation of residue curves for n-butane(1) + n-hexane(2) + n-heptane(3) as predicted by the shortcut K-ratio method.*

x_1	x_2	x_3	$T(°C)$	P_1^{sat} (MPa)	P_2^{sat} (MPa)	P_3^{sat} (MPa)	K_1	K_2	dx_1	dx_2
0.156	0.596	0.248	65.02	0.2477	0.0902	0.0360	2.446	0.890	−0.034	9.8e−3
0.126	0.601	0.273	67.09	0.2622	0.0961	0.0386	2.588	0.949	−0.030	4.6e−3
0.1	0.6	0.3	69.07	0.2766	0.1021	0.0412	2.731	1.008	−0.026	−7.3e−4
0.074	0.600	0.327	71.17	0.2926	0.1088	0.0442	2.888	1.074	−0.021	−6.7e−3
0.053	0.593	0.354	73.09	0.3078	0.1152	0.0471	3.039	1.137	−0.016	−0.012

up the table from the initial point and using $x_i^{new} = x_i^{old} - dx_i^{old}$. More accurate finite difference methods can be employed for important applications. Residue curves are calculated in other composition ranges by selecting other initial compositions. The location of the separatrices is obvious after generating enough residue curves. Residue curve calculations are offered by some process simulation software due to the importance of screening separations in the chemical industry. More detailed discussions are available.[40]

Residue Curve Summary

Residue curves provide useful tools for designing separation schemes. Although a single feed column with one overhead and one bottoms product cannot cross a separatrix, a separation scheme can be constructed using multiple columns to achieve most separations. The residue curve maps are useful in screening additives for selection of the most promising systems. Residue curve calculations for homogeneous liquids are straightforward, and an algorithm has been presented.

13.4 HOMEWORK PROBLEMS

Phase Behavior

13.1 A binary mixture obeys a simple one-term equation for excess Gibbs energy, $G^E = Ax_1x_2$, where A is a function of temperature: $A = 2930 + 5.02E5/T(K)$ J/mol.

(a) Does this system exhibit partial immiscibility? If so, over what temperature range?

(b) Suppose component 1 has a normal boiling temperature of 310 K, and component 2 has a normal boiling temperature of 345 K. The enthalpies of vaporization are equal, both being 4475 cal/mol. Sketch a qualitatively correct T-x-y diagram including all LLE, VLE behavior at 1 bar.

13.2 As the research scientist of a company where phase equilibria is under study, you are approached by a laboratory technician with the most recent (and incomplete) high pressure phase equilibria results. The technician expresses concern that the data have not been collected correctly, or that some of the samples may have been mislabeled for some of the runs, or there is a problem with the automated equipment, because the data appear different from what he has seen before. The most recent results are summarized on the two P-x-y diagrams over the same P range shown below at T_1 and T_2, where $T_1 < T_2$. The circles and triangles give coexisting phase compositions at each pressure. What is your response?

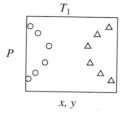

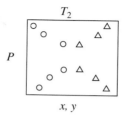

40. Barnicki, S.D., Siirola, J.J., "Separations Process Synthesis," and Doherty, M.F., Knapp, J.P., "*Distillation, Azeotropic and Extractive,*" in *Kirk-Othmer Encyclopedia of Chemical Technology,* 4th ed., Wiley, 1997; also Van Dongen, D.B., Doherty, M.F., *Ind. Eng. Chem. Fundam.,* 24, 454(1985).

13.3 Using the value $k_{ij} = 0.084$ fitted to the system methanol(1) + benzene(2), use the Peng-Robinson equation to plot P-x-y behavior of the system at the following temperatures: 350 K, 500 K, 520 K. Plot the T-x-y behavior at the following pressures: 0.15 MPa, 2 MPa, 5.5 MPa. Sketch the P-T projection of the system.

13.4 Use UNIFAC to model the methanol(1) + methylcyclohexane(2) system. Provide a P-x-y diagram at 60°C and 90°C, and a T-x-y at 0.1013 MPa.

13.5 Use UNIFAC to model the methanol(1) + hexane(2) system. Provide a P-x-y diagram at 50°C and 25°C, and a T-x-y at 0.1013 MPa.

13.6 Model the system CO_2(1) + tetralin(2) using $k_{ij} = 0.10$. Generate a P-x-y diagram at 20°C, 45°C, and a T-x-y at 12 MPa and 22 MPa. What type of phase behavior does the system exhibit?[41]

13.7 The system ethanol(1) + bromoethane(2) forms an azeotrope containing 93.2 mol% bromoethane and having a minimum boiling point of 37.0°C at 760 mmHg. The following vapor pressure data are available:

| | T^{sat}(°C) | |
P (mmHg)	ethanol	bromoethane
100	34.9	---
400	63.5	21.0
760	78.4	38.4
1520	97.5	60.2
3800	---	95.0

(a) Use all of the available data to determine coefficients for equation $\log_{10} P^{sat} = A + B/T$ by performing a linear regression to find the coefficients A and B. Be sure to use temperature in Kelvin.
(b) Use the azeotropic point to determine the parameters for the van Laar equation.
(c) Generate a P-x-y diagram at 100°C. Tabulate the data used to plot the diagram.
(d) Generate a T-x-y diagram at 760 mmHg. Assume that the activity coefficients are independent of temperature over the range of temperature involved. (This is the type of calculation necessary for generating an x-y diagram for a distillation at 760 mmHg).
(e) Experimentally, the system is found to have a homogeneous maximum pressure azeotrope (Smyth, C.P., Engel, E.W. *J. Am. Chem. Soc.,* 51, (1929), 2660. How does this compare with your findings?

13.8 Benzaldehyde is known in the flavor industry as bitter almond oil. It has a cherry or almond essence. It may be possible to recover it using CO_2 for a portion of the processing. Explore the phase behavior of the CO_2(1) + benzaldehyde(2) system, using the Peng-Robinson equation to categorize the system among the types discussed in this chapter. Experimental data are not available, so use an interaction coefficient $k_{ij} = 0.1$, which has been fitted to the

41. Experimental data are available in Leder, F., Irani, C.A., *J. Chem. Eng. Data,* 20:323 (1975).

CO_2-benzene system. Calculate P-x-y diagrams at 295 K and 323 K. Also determine T-x-y diagrams at 34.5 bar and 206.9 bar. Note that 34.5 bar is below the critical pressure of both substances. If a bubble calculation fails, try a dew, and vice-versa.

Solid-Liquid Behavior

13.9 In the treatment of solid-liquid equilibria, the effects of pressure on melting points are neglected.

    (a) Draw a schematic of the Gibbs energy of liquid and solid phases versus pressure at constant temperature for a compound for which the molar volume of the solid is less than the molar volume of the liquid. Plot both curves on the same figure, and indicate the melting pressure. Most chemicals follow this type of behavior.

    (b) For water, the molar volume of the solid is greater than the molar volume of the liquid. Sketch the Gibbs energy of liquid and solid phases as a function of pressure at constant temperature for this type of behavior. Plot both curves on the same figure, and indicate the melting pressure.

    (c) Calculate the hypothetical Gibbs energy for melting solid naphthalene at 5 bar and the normal melting temperature, 80.2°C. You may assume that the liquid and solid are incompressible. Be sure to clearly specify the path you use for your calculation. $V^S = 124.8$ cm^3/gmole, $V^L = 133$ cm^3/gmole.

    (d) Calculate the hypothetical Gibbs energy for melting solid naphthalene at 1 atm and 78°C. Compare the magnitude with the results of part (c) to verify that the pressure effects are small relative to temperature effects.

13.10 Generate a solid-liquid equilibrium T-x diagram for naphthalene(1) + biphenyl(2) assuming ideal solutions. What are the predicted eutectic temperature and composition? The experimental eutectic point is 39.4°C and x biphenyl = 0.555 (Lee, H.H., Warner, J.C., *J. Amer. Chem. Soc.*, **57**:318,1935).

13.11 At 25°C, the solubility of naphthalene in *n*-hexane is 11.9 mol%. The liquid phase is non-ideal. Use the simple solution model $G^E/RT = Ax_1x_2$ to predict the solubility at 10°C. (The experimental solubility at 10°C is 6.5mol%, Sunier, A. *J. Phys. Chem*, 34, 2582,1930).

13.12 Phenanthrene and anthracene are structurally very similar. Would you expect them to have similar solubilities in benzene at 25°C? Provide a quantitative answer, and an explanation.

13.13 Use UNIFAC to determine the solubility (in mole fraction) of phenol at the cited conditions.

    (a) Solubility in *n*-heptane at 25°C

    (b) Solubility in ethanol at 25°C

    (c) Solubility in a 50/50 mole ratio of heptane and ethanol at 25°C

13.14 A 50 wt% (22.5 mol%) solution of ethylene glycol + water freezes at about 240 K.

    (a) What freezing temperature would be predicted by assuming that ethylene glycol and water form an ideal solution? The freezing occurs by formation of water crystals.

    (b) Does your calculation indicate that the system has positive or negative deviations from Raoult's law? Why?

13.15 Determine the ideal solubility of naphthalene in any solvent at 40°C. Then calculate the solubility predicted by UNIFAC and compare with the experimental mol% solubility (shown in parentheses) for the specified solvent:

(a) methanol (4.4)
(b) ethanol (7.3)
(c) 1-propanol (9.4)
(d) 2-propanol (7.6)
(e) 1-butanol (11.6)
(f) *n*-hexane (22.2)
(g) cyclohexanol (22.5)
(h) acetic acid (11.7)
(i) acetone (37.8)
(j) chloroform (47.3)

13.16 Determine the ideal solubility of anthracene in any solvent at 20°C. Then calculate the solubility predicted by UNIFAC and compare with the experimental mol% solubility (shown in parentheses) for the specified solvent:

(a) acetone (0.31)
(b) chloroform (0.94)
(c) ethanol (0.05)
(d) methanol (0.02)

13.17 Determine the ideal solubility of phenanthrene in any solvent at 20°C. Then calculate the solubility predicted by UNIFAC and compare with the experimental solubility (shown in parentheses) for the specified solvent:

(a) acetone (14.5)
(b) chloroform (23.8)
(c) ethanol (1.23)
(d) acetic acid (1.92)
(e) methanol (0.64)

13.18 Determine the solubility curve for naphthalene in the specified solvent, and compare with the literature data:

(a) acetic acid (H. Ward, *J. Phys. Chem.*, 30: 1316(1926))
(b) *n*-hexane (H. Ward, *J. Phys. Chem.*, 30: 1316(1926))
(c) cyclohexanol (G. Weissenberger, *Z. Agnew. Chem.*, 40: 776(1927))
(d) acetone (H. Ward, *J. Phys. Chem.*, 30: 1316(1926))
(e) chloroform (J. Hildebrand, *J. Am. Chem. Soc.*, 42: 2180(1920))
(f) methanol (H. Ward, *J. Phys. Chem.*, 30: 1316(1926))
(g) *n*-butanol (H. Ward, *J. Phys. Chem.*, 30: 1316(1926))
(h) ethanol (A. Sunier, *J. Phys. Chem.*, 34: 2582,1930)
(i) *n*-propanol (A. Sunier, *J. Phys. Chem.*, 34: 2582,1930)
(j) 2-propanol (A. Sunier, *J. Phys. Chem.*, 34: 2582,1930)

13.19 The gas condensate from a new gas well in Prudhoe Bay Alaska has the following weight% of C5, C10, C15, C20, C25, C30, C35, C40, C45, C50, and >C50, respectively: 1, 4, 7, 10, 12, 12, 12, 12, 8, 8, 14. Estimate the temperature at which wax may begin to precipitate from this liquid.

13.20 Generate a SLE phase diagram for phenol(1) + cyclohexane(2).

(a) Assume an ideal solution.
(b) Use UNIFAC to model liquid phase non-idealities.
(c) Make a comment about how the solubility of phenol in cyclohexane differs from the solubility in benzene at the same temperature. (See Fig. 13.12 on page 465.)

13.21 Create a flowsheet analog to VLE or LLE calculations to find the melting temperature and liquid phase composition for a given solid mixture composition

(a) for ideal solutions of solid and liquid
(b) for non-ideal solutions of solid and liquid

13.22 Create a flowsheet analog to VLE or LLE calculations to find the freezing temperature and solid composition for a given liquid composition when the liquids and solids form a non-ideal solution.

13.23 Create a flowsheet analog to VLE or LLE flash calculations to find the coexisting liquid and solid compositions that exist for a liquid-solid mixture of specified overall composition that is between the conditions of first freezing and first melting.

Residue Curves

13.24 For the systems specified below, obtain residue curve maps and plot the range of possible distillate and bottoms compositions for the given feeds. UNIQUAC parameters are provided.

(a) methanol(1) + ethanol(2) + ethyl acetate(3), $z_1 = 0.25$, $z_2 = 0.25$, $z_3 = 0.5$,
$r = [1.4311, 2.1055, 3.4786]$, $q = [1.432, 1.972, 3.116]$,
a parameters (in K); $a_{12} = -91.226$, $a_{21} = 124.485$, $a_{13} = -74.2746$, $a_{31} = 389.285$,
$a_{23} = -31.5531$, $a_{32} = 183.154$.
(b) Solve part (a), but use $z_1 = 0.4$, $z_2 = 0.4$, $z_3 = 0.2$.
(c) methanol(1) + 2-propanol(2) + water(3), $z_1 = 0.2$, $z_2 = 0.6$, $z_3 = 0.2$,
$r = [1.4311, 2.7791, 0.92]$, $q = [1.432, 2.508, 1.4]$,
a parameters (in K); $a_{12} = 39.717$, $a_{21} = -26.6935$, $a_{13} = -112.697$, $a_{31} = 176.077$,
$a_{23} = 151.061$, $a_{32} = 55.1272$.
(d) Solve part (c), but use $z_1 = 0.6$, $z_2 = 0.2$, $z_3 = 0.2$.
(e) methanol(1) + ethanol(2) + benzene(3), $z_1 = 0.25$, $z_2 = 0.25$, $z_3 = 0.5$,
$r = [1.4311, 2.1055, 3.1878]$, $q = [1.432, 1.972, 2.4]$,
a parameters (in K); $a_{12} = -91.2264$, $a_{21} = 124.485$, $a_{13} = -27.2253$, $a_{31} = 559.486$,
$a_{23} = -42.6567$, $a_{32} = 337.739$.
(f) Solve part (e), but use $z_1 = 0.4$, $z_2 = 0.4$, $z_3 = 0.2$.

13.25 Consider an equimolar mixture of methanol(1) + water(2) + acetone(3). Demonstrate that it is not possible to obtain pure acetone as distillate from a single-feed column by generating a residue curve map and applying the bow-tie approximation.[42] What compositions are attainable from this feed? Data: $r = [1.4311, 0.92, 2.5735]$, $q = [1.432, 1.4, 2.3360]$,
a parameters (in K); $a_{12} = -112.697$, $a_{13} = -50.9396$, $a_{21} = 176.077$, $a_{23} = 29.1681$,
$a_{31} = 218.872$, $a_{32} = 228.990$.

---

42. It is possible to obtain acetone distillate from a two-feed column if an equimolar acetone(1) + methanol(2) is fed near the center of the column and water is fed near the top of the column, however such separations are not predicted by the bow-tie approximation. A separation of this type is called an extractive distillation.

UNIT

REACTING SYSTEMS

Problems involving reactions are affected by equilibrium limitations as well as their kinetics. Briefly, the Gibbs energy related to the "partitioning" of atoms between species must be minimized in a fashion analogous to the way that it was minimized for phase partitioning. For example, the formation of methanol from carbon monoxide and hydrogen is favorable based on the energy from the bond formation, but limited by the reduction in entropy as the carbon monoxide and hydrogen are squeezed into a single molecule. Thus, even though the heat of reaction may be favorable, a balance is struck between the loss in entropy and the favorable energy as the methanol is formed. At 100 bars and 510 K, this balance occurs at about 60% conversion of stoichiometric $CO + H_2$ to CH_3OH. This unit introduces the principles for determining the effect of temperature and pressure on equilibrium conversion. A few overly zealous practitioners of the past have erroneously assumed that conversion of reactants is determined solely by the rate of reaction, and have attempted to increase conversion beyond the equilibrium limit by increasing reactor size. For example, LeChatelier (1888) wrote:

> "It is known that in the blast furnace the reduction of iron oxide is produced by carbon monoxide, according to the reaction: $Fe_2O_3 + 3CO = 2Fe + 3CO_2$, but the gas leaving the chimney contains a considerable proportion of carbon monoxide, which thus carries away an important quantity of unutilized heat. Because this incomplete reaction was thought to be due to an insufficiently prolonged contact between carbon monoxide and the iron ore, the dimensions of the furnaces have been increased. In England they have been made as high as thirty meters. But the proportion of carbon monoxide escaping has not diminished, thus demonstrating, by an experiment costing several hundred thousand francs, that the reduction of iron oxide by carbon monoxide is a limited reaction. Acquaintance with the laws of chemical equilibrium would have permitted the same conclusion to be reached more rapidly and far more economically."

Let us hope that this one piece of history which you will not repeat.

CHAPTER 14

REACTING SYSTEMS

We must first speak a little concerning contact or mutual touching, action, passion and reaction.

Daniel Sennert (1660)

Another important aspect of the thermodynamics of multicomponent systems is the rearrangement of atoms within and between molecules, known as chemical reaction. Equilibrium thermodynamic considerations tell us the direction and extent to which a reaction will go. As always, the constraint of minimum Gibbs energy dictates the equilibrium results at a fixed T and P.

14.1 REACTION COORDINATE

It is convenient to adopt the conventions of stoichiometry

$$v_1 C_1 + v_2 C_2 + v_3 C_3 + v_4 C_4 = 0$$

where the C's represent the species, and reactants have negative v's and products have positive v's[1] (e.g., $CH_4 + H_2O = CO + 3H_2$, numbering from left to right),

$$\Rightarrow \quad v_1 = -1; \quad v_2 = -1; \quad v_3 = +1; \quad v_4 = +3.$$

Consider what would happen if a certain amount of component 1 were to react with component 2 to form products 3 and 4. We see $dn_1 = dn_2 (v_1/v_2)$ because v_1 moles of component 1 requires v_2 moles of component 2 in order to react. Rearranging:

$$\frac{dn_1}{v_1} = \frac{dn_2}{v_2} \text{ and similarly: } \frac{dn_1}{v_1} = \frac{dn_3}{v_3} \qquad\qquad 14.1$$

---

1. The v's are called the *stoichiometric numbers*. The absolute values of the v's are the *stoichiometric coefficients*.

Since all these quantities are equal, it is convenient to define a variable which represents this quantity.

$$d\xi = dn_i / \nu_i \qquad 14.2$$

ξ is called the *reaction coordinate*.[1]

Integrating:

$$\int_0^\xi d\xi = \frac{1}{\nu} \int_{n^i}^{n^f} dn$$

$$\xi = \frac{1}{\nu}(n^f - n^i)$$

or in a more useful form for any component i:

$$\boxed{n_i^f = n_i^i + \nu_i \xi} \qquad 14.3$$

In a flowing system, n_i^f represents the outlet, n_i^i represents the inlet and thus for component i:

$$\boxed{\dot{n}_i^{out} = \dot{n}_i^{in} + \nu_i \dot{\xi}} \qquad 14.4$$

where $\dot{\xi}$ represents the overall rate of species interconversion. *Moles are not conserved in a chemical reaction*, which can be quantified by summing Eqn. 14.3 or 14.4 over all species—for a flowing system, $\dot{n}^{in} = \sum_i \dot{n}_i^{in}$ and $\dot{n}^{out} = \sum_i \dot{n}_i^{out}$, thus

$$\boxed{\dot{n}^{out} = \dot{n}^{in} + \dot{\xi} \sum_i \nu_i} \qquad 14.5$$

In closed systems, the value of ξ is determined by chemical equilibria calculations; ξ may be positive or negative. The only limit on ξ is that $n_i^f \geq 0$ for all i. The boundary values of ξ may be determined in this manner before beginning an equilibrium calculation. Naturally, in a flowing system, the same arguments apply to $\dot{\xi}$ and $\dot{n}_i^{out} \geq 0$.

---

1. Another common measure of reaction progression is conversion. In reaction engineering, it is common to follow the conversion of a particular reactant species, say species A. If X_A is the conversion of A, then $n_A = n_A{}^{in}(1 - X_A)$, and $X_A = a\,\xi/n_A{}^{in}$, where a is the stoichiometric coefficient for A as written in the reaction.

Example 14.1 Stoichiometry and the reaction coordinate

Five moles of hydrogen, two moles of CO, and 1.5 moles of CH_3OH vapor are combined in a closed system methanol synthesis reactor at 500 K and 1 MPa. Develop expressions for the mole fractions of the species in terms of the reaction coordinate. The components are known to react with the following stoichiometry:

$$2H_{2(g)} + CO_{(g)} = CH_3OH_{(g)}$$

Solution: Although the reaction is written as though it will proceed from left to right, the direction of the actual reaction does not need to be known. If the reverse reaction occurs, this will be obvious in the solution because a negative value of ξ will be found. The task at hand is to develop the mole balances that would be used toward determining the value of ξ.

	n^i	n^f
H_2	5	$5 - 2\xi$
CO	2	$2 - \xi$
CH_3OH	1.5	$1.5 + \xi$
n_T		$8.5 - 2\xi$

The total number of moles at any time is $8.5 - 2\xi$. The mole fractions will be

$$y_{H_2} = \frac{n^f_{H_2}}{n_T} = \frac{5 - 2\xi}{8.5 - 2\xi}$$

$$y_{CO} = \frac{2 - \xi}{8.5 - 2\xi}$$

$$y_{CH_3OH} = \frac{1.5 + \xi}{8.5 - 2\xi}$$

To assure that all $n^f_i \geq 0$, the acceptable upper limit of ξ is determined by CO, and the acceptable lower limit is determined by CH_3OH.

$$-1.5 \leq \xi \leq 2$$

14.2 EQUILIBRIUM CONSTRAINT

If the composition of a system is changing, the change in the Gibbs energy is given by:

$$dG = -SdT + VdP + \sum_i \mu_i dn_i \quad (\text{Recall } \mu_i \equiv (\partial G/\partial n_i)_{T,P,n_{i \neq j}}) \qquad 14.6$$

The fact that species are being created or consumed by a reaction does not alter this equation. At constant temperature and pressure,

$$dG = \sum_i \mu_i dn_i \qquad 14.7$$

Substituting the definition of reaction coordinate from Eqn. 14.2,

$$dG = \sum_i \mu_i \nu_i d\xi \qquad 14.8$$

But G is minimized at equilibrium at fixed T and P, therefore

$$dG / d\xi = \sum_i \mu_i \nu_i = 0 \qquad 14.9$$

Now we have one unknown, ξ, in terms of which we can determine the changes in moles for all of the components. We make a further manipulation before we apply the equilibrium constraint. In phase equilibria, we found fugacity to be a convenient property to use because it reduced to the partial pressure for a component in an ideal gas mixture. We can rewrite Eqn. 14.9 in terms of fugacities. We recall our definition of fugacity $dG = RT\, d\ln f$. Integrating from the standard state to the mixture state of interest:

$$\int_{G^o}^{\mu} dG = RT \int_{f^o}^{\hat{f}} d\ln f_i \Rightarrow \mu - G^o = RT\ln\left[\frac{\hat{f}_i}{f_i^o}\right] \Rightarrow \mu = G^o + RT\ln\left[\frac{\hat{f}_i}{f_i^o}\right] \qquad 14.10$$

where G_i^0 is the standard state Gibbs energy of species i and f_i^0 is the standard state fugacity. Substitution into Eqn. 14.9

$$0 = \sum_i \nu_i \left\{ G_i^o + RT \ln\left[\frac{\hat{f}_i}{f_i^o}\right] \right\}$$

or $\qquad\qquad\qquad$ 14.11

$$0 = \sum_i \nu_i G_i^o + RT \sum_i \nu_i \ln\left[\frac{\hat{f}_i}{f_i^o}\right]$$

We will now consider separately the terms appearing in Eqn. 14.11, and then combine the results. First, note that $\sum_i \nu_i G_i^o$ is called the *standard state Gibbs energy of reaction* at the temperature of interest, which we will denote ΔG_T^o. The standard state Gibbs energy for reaction can be calculated using Gibbs energies of formation.

$$\Delta G_T^o \equiv \sum_i v_i G_i^o = \sum_{products} |v_i| \Delta G_{f,i}^o - \sum_{reactants} |v_i| \Delta G_{f,i}^o \qquad 14.12$$

For $CH_4 + H_2O \rightarrow CO + 3H_2$

$$\Delta G_T = \sum_i v_i G_i^o = \Delta G_{f(CO_{(g)})}^o + 3\Delta G_{f(H_{2(g)})}^o - \Delta G_{f(CH_{4(g)})}^o - \Delta G_{f(H_2O_{(g)})}^o$$

Thus,

$$\Delta G_T^o = \sum_i v_i \Delta G_{f,i}^o \qquad 14.13$$

❗ Standard state Gibbs energy of reaction.

The Gibbs energies of formation are typically tabulated at 298.15 K and 1 bar, and special calculations must be performed to calculate ΔG_T^o at other temperatures—the calculations will be covered in Section 14.4. The Gibbs energy of formation is taken as zero for elements that naturally exist as molecules at 298.15 K and 1 bar. The state of aggregation for the zero value is taken as the naturally occurring state at 298.15 K and 1 bar. For example, H exists as $H_{2(g)}$, so the $\Delta G_{f\,298}^o$ is zero for $H_{2(g)}$. Carbon is naturally a solid, so the value of $\Delta G_{f\,298}^o$ is zero for $C_{(s)}$. The zero values for elements in the naturally occurring state are not listed in the tables in reference books. Gibbs energies of formation are tabulated for many compounds in Appendix E at 298.15 K and 1 bar.

One special note about the state of aggregation is important. Some molecules, like water, commonly exist as vapor (g), or liquid (l). The state of aggregation must be specified in the reaction. Care should be used to obtain the correct value from the tables. Note that for water, the difference between $\Delta G_{f\,298(g)}^o$ and $\Delta G_{f\,298(l)}^o$ is essentially the Gibbs energy of vaporization at 298 K. The difference is non-zero because liquid is more stable. (Which phase will have a lower Gibbs energy of formation at 298.15 K and 1 bar?)

Example 14.2 Calculation of standard state Gibbs energy of reaction

Butadiene is prepared by the gas phase catalytic dehydrogenation of 1-butene:

$$C_4H_{8(g)} = C_4H_{6(g)} + H_{2(g)}$$

Calculate the standard state Gibbs energy of reaction at 298.15 K.

Solution: We find values tabulated for the standard state enthalpies of formation and standard state Gibbs energy of formation at 298.15 K.

Compound	$\Delta H_{f,298}^o$(J/mole)	$\Delta G_{f,298}^o$(J/mole)
1-butene (g)	−540	70,240
1,3-butadiene (g)	109,240	149,730
Hydrogen (g)	0	0

$$\Delta G_{298}^o = 149,730 + 0 - 70,240 = 79,490 \text{ J/mole}$$

Now, let us explore how the numerical value of the standard state Gibbs energy of reaction is used to determine the reaction coordinate at equilibrium. Rearranging Eqn. 14.11,

$$\frac{\Delta G_T^o}{RT} = -\sum_i v_i \ln\left[\frac{\hat{f}_i}{f_i^o}\right]$$

14.14

Note that the sum of the logs equals the log of the products, and the right-hand side may be written,

$$\sum_{i=1}^{NC} v_i \ln\left[\frac{\hat{f}}{f_i^o}\right] = \ln\left[\prod_{i=1}^{NC}\left[\frac{\hat{f}}{f_i^o}\right]^{v_i}\right]$$

where $\Pi \equiv$ product from $i = 1$, NC and where NC is the number of components. *(Note: if $v_i < 0$,*

then $\left(\dfrac{\hat{f}}{f_i^o}\right)^{v_i} = \left(\dfrac{f_i^o}{\hat{f}}\right)^{|v_i|}$ *).*

We define the product term at equilibrium as the equilibrium constant K_a

General equilibrium constraint.

$$\boxed{K_a = \prod\left[\frac{\hat{f}_i}{f_i^o}\right]^{v_i}}$$

14.15

Therefore, at equilibrium, we may write:

$$\boxed{K_a = \exp\left(\frac{-\Delta G_T^o}{RT}\right)}$$

14.16

Suppose for *gas-phase reactants and products*, we choose our standard state to be the ideal gas at 1 bar pressure, where $f^o = P = 1$ bar.

Substitution yields,

Gas-phase equilibrium constraint.

$$\boxed{\prod \hat{f}_i^{v_i} = \exp(-\Delta G_T^o / RT)}$$

14.17

because $f_i^o = 1$ bar.

> Note that f_i [$\equiv$] bar in order for this simplification to be valid, and most importantly, ΔG_T^o must be evaluated at exactly 1 bar (because the pressure dependence is included in the fugacity term we are using). Fortunately, this is the pressure at which our compilations of free energies of formation are tabulated. But we must carefully convert all pressures to bar before solving any specific problems.

14.3   REACTION EQUILIBRIA FOR IDEAL SOLUTIONS

To provide the activation energy necessary to initiate a reaction, reaction temperatures are generally quite high. This means that the reactions often involve gases. For simplification, gases can be modeled as ideal solutions, i.e. $\hat{f}_i^{is} = y_i\varphi_i P$, or an ideal gas, $\hat{f}_i^{ig} = y_i P$.

For a *gas phase* ideal solution, substitution yields

$$\prod \hat{f}_i^{\nu_i} = \exp\left(-\frac{\Delta G_T^o}{RT}\right) = K_a \qquad\qquad 14.18$$

$$\left(\prod \varphi_i^{\nu_i}\right)\left(\prod y_i^{\nu_i}\right)P^{\Sigma\nu_i} = \exp\left(-\frac{\Delta G_T^o}{RT}\right) = K_a \qquad\qquad 14.19$$

Note that if $\varphi_i = 1$ for all i, then

$$\boxed{\left(\prod y_i^{\nu_i}\right)P^{\Sigma\nu_i} = \exp\left(-\frac{\Delta G_T^o}{RT}\right) = K_a} \qquad\qquad \text{(ig) } 14.20$$

Ideal-gas-phase equilibrium constraint.

This is the form that should be familiar from freshman chemistry. It has become convention to refer to $\exp(-\Delta G_T^o/RT)$ as the reaction equilibrium constant, "K_a." Another convention is to refer to the different contributions on the left-hand side as K_φ and K_y, where

$$\left(\prod y_i^{\nu_i}\right) = K_y, \quad\text{and}\quad \left(\prod \varphi_i^{\nu_i}\right) = K_\varphi. \qquad\qquad 14.21$$

Note that $\prod\left(\hat{f}_i / f_i^o\right)^{\nu_i} = K_a$ and that $a_i = \hat{f}_i / f_i^o$ so this convention for defining K_a is consistent with the other definitions. Nevertheless, these semantics can often be the most confusing aspects of reaction equilibrium problems.

Example 14.3   Computing the reaction coordinate

Stoichiometric ratios of CO and H_2 are fed to a reactor.

$$CO_{(g)} + 2H_{2(g)} = CH_3OH_{(g)}$$

Suppose this reaction is carried out at 500 K and 1 bar.

We can determine that $\exp(-\Delta G_{500}^o/RT) = 0.00581$ at 500 K. (We illustrate how to perform this determination in Section 14.4.) Compute the equilibrium conversion of CO.

Solution: Since this reaction is being carried out at 1 bar, the reaction medium is approximately an ideal gas) $\Rightarrow \hat{f}_i = y_i P$.

$$\frac{P}{P^3}\frac{y_{CH_3OH}}{y_{CO}y_{H_2}^2} = 0.00581 \text{ but } P = 1 \Rightarrow \frac{y_{CH_3OH}}{y_{CO}y_{H_2}^2} = 0.00581$$

Example 14.3 Computing the reaction coordinate (Continued)

Basis: 1 mole CO fed

$$
\begin{array}{rcl}
n_{CO} & = & 1 - \xi \\
n_{H_2} & = & 2 - 2\xi \\
n_{CH_3OH} & = & \xi \\
\hline
n_T & = & 3 - 2\xi
\end{array}
$$

$$
y_{co} = \frac{1 - \xi}{3 - 2\xi}; \quad y_{H_2} = \frac{2(1 - \xi)}{3 - 2\xi}; \quad y_{CH_3OH} = \frac{\xi}{3 - 2\xi}
\qquad 14.22
$$

$$
\text{Substitution} \Rightarrow \frac{\dfrac{\xi}{(3 - 2\xi)}}{\dfrac{4(1 - \xi)^3}{(3 - 2\xi)^3}} = 0.00581 \quad \Rightarrow \quad \frac{\xi(3 - 2\xi)^2}{4(1 - \xi)^3} = 0.00581
$$

$$
\Rightarrow F(\xi) \equiv 0.00581 \cdot 4(1 - \xi)^3 - \xi(3 - 2\xi)^2
$$

One equation and one unknown $\Rightarrow$ solve by trial and error and substitute to get ξ.

ξ	$F(\xi)$
0.10	−0.767
0.01	−0.0662
0.001	0.0142
0.00257	1E-5

Now Eqn. 14.22 may be used to find the y's. Note the small conversion at these conditions. A more likely way to operate this process is discussed in Section 14.7 on page 502.

Example 14.4 Butadiene revisited

Consider again the Butadiene reaction of Example 14.2 on page 487. Butadiene is prepared by the gas phase catalytic dehydrogenation of 1-butene, at 900 K and 1 bar.

$$
C_4H_{8(g)} = C_4H_{6(g)} + H_{2(g)}
$$

(a) In order to suppress side reactions, the butene is diluted with steam before it passes into the reactor. Estimate the equilibrium composition for a feed consisting of 10 moles of steam per mole of 1-butene.

(b) Find the equilibrium composition of the mixture in the reactor if the inerts were absent.

(c) Find the total pressure that would be required to obtain the same conversion as in (a) if no inerts were present.

Example 14.4 Butadiene revisited (Continued)

In our earlier example, we determined the value at 298.15 K for ΔG_f^o. Now we need a value at 900 K. The next section will explain how the value at 900 K may be obtained. For our purposes, let us use the following data for ΔG_f^o (kJ/mole) at 900 K and 1 bar:

Component	ΔG_f^o (kJ/mole)
1,3 Butadiene	243.474
1-Butene	232.854
Hydrogen	0

$$\Delta G_{900}^o = 243.474 - 232.854 = 10.62 \text{ kJ/mole}$$

$$K_a = \exp(-\Delta G_{900}^o / RT) = 0.242$$

(a) Set up reaction coordinate, using I to indicate inerts,

$n_{C_4H_8}$	$1 - \xi$
$n_{C_4H_6}$	ξ
n_{H_2}	ξ
n_I	10
n_T	$11 + \xi$

$$0.242 P^{-1} \left(\frac{1 - \xi}{11 + \xi} \right) = \left(\frac{\xi}{11 + \xi} \right)^2$$

$P = 1$ bar $\Rightarrow 1.242\xi^2 + 2.42\,\xi - 2.662 = 0 \Rightarrow \xi = 0.784$

(b) $n_I = 0$

$n_T = 1 + \xi$; $1.242\xi^2 - 0.242 = 0$; $\xi = 0.44$, so the conversion decreases without the inert.

(c) $P^{-1} = \xi^2 / [0.242 \cdot (1 - \xi) \cdot (1 + \xi)]$ and $P = 0.152$ bar. So we would need to run at a much lower pressure without the inerts to achieve the same conversion. In other words, the inerts serve to dilute the fugacities of the products and suppress the reverse reaction since there are more moles of product than reactant.

Generalized Evaluation of ΔG_T^o

Always remember that ΔG_T^o depends on the standard state, which changes with temperature. Comparing Examples 14.2 and 14.4, $\Delta G_T^o = 79,490$ J/mole at 298 K, but decreases to $\Delta G_T^o = 10,620$ J/mol at 900 K. In order to calculate ΔG_T^o, it may seem that we need to know ΔG_f^o for each compound at all temperatures. Fortunately this is not necessary because the ΔG_T^o can be determined from the Gibbs energy for the reaction at a certain *reference* temperature (usually 298.15 K) together with the enthalpy for the reaction and the heat capacities of the species. The next section explains these important calculations.

14.4 TEMPERATURE EFFECTS

Suppose we have a table of standard energies of formation at 298.15 K but we would like the value for ΔG_T^o at some other temperature. We can account for temperature effects by applying classical thermodynamics to the change in Gibbs energy with respect to temperature.

$$\frac{\partial(\Delta G / RT)}{\partial T} = \frac{1}{RT}\left(\frac{\partial \Delta G}{\partial T}\right)_P - \frac{\Delta G}{RT^2} = -\frac{\Delta S}{RT} - \left(\frac{\Delta H}{RT^2} - \frac{\Delta S}{RT}\right) \qquad 14.23$$

which results in the van't Hoff equation

❶ van't Hoff equation.

$$\boxed{\frac{\partial(\Delta G_T^o / RT)}{\partial T} = -\frac{\Delta H_T^o}{RT^2}} \qquad 14.24$$

$$\boxed{\frac{\Delta G_T^o}{RT} = -\int_{T_R}^{T} \frac{\Delta H_T^o}{RT^2} \, dT + \frac{\Delta G_R^o}{RT_R}} \qquad 14.25$$

$$\Delta H_T^o = \Delta H_R^o + \int_{T_R}^{T} \Delta C_P \, dT \qquad 14.26$$

A reaction with a negative value of ΔH_T^o is called an exothermic reaction. A reaction with a positive value of ΔH_T^o is called an endothermic reaction. In this equation, ΔH_T^o is easily determined if the *standard heat of reaction* ΔH_R^o is known at a single reference temperature and 1 bar.

❶ Standard heat of reaction.

$$\boxed{\Delta H_R^o = \sum v_i H_{R,i}^o = \sum v_i \Delta H_{fR,i}^o = \sum_{products} |v_i| \Delta H_{fR,i}^o - \sum_{reactants} |v_i| \Delta H_{fR,i}^o} \qquad 14.27$$

Almost always, the best reference state to use is $T_R = 298.15$ K and 1 bar, because this is the temperature where the standard state enthalpies of formation are tabulated. The heat of formation is taken as zero for elements that naturally exist as molecules at 298.15 K and 1 bar, in an analogous way that the Gibbs energies of formation were handled in Section 14.2. The state of aggregation for the zero value is taken as the naturally occurring state at 298.15 K and 1 bar. For example, H exists as $H_{2(g)}$, so $\Delta H_{f\,298}^o$ is zero for $H_{2(g)}$. Carbon is a solid, so the value of $\Delta H_{f\,298}^o$ is zero for $C_{(s)}$. The zero values for elements in the naturally occurring state are not listed in the tables in reference books. Enthalpies of formation are tabulated for many compounds in Appendix E at 298.15 K and 1 bar.

The caution mentioned in Section 14.2 regarding the state of aggregation is important here also. Some molecules, like water, commonly exist as vapor (*g*), or liquid (*l*). The state of aggregation must be specified in the reaction. Care should be used to obtain the correct value from the tables. Note that for water, the difference between $\Delta H^o_{f\,298(l)}$ and $\Delta H^o_{f\,298(g)}$ is essentially the heat of vaporization at 298 K that can be obtained from the steam tables.

The full form of the integral of Eqn. 14.26 is tedious to calculate, e.g., if $C_{P,\,i} = a_i + b_i T + c_i T^2 + d_i T^3$, Eqn. 14.26 becomes

$$\Delta H^o_T = \Delta H^o_R + \Delta a(T - T_R) + \frac{\Delta b}{2}(T^2 - T_R^2) + \frac{\Delta c}{3}(T^3 - T_R^3) + \frac{\Delta d}{4}(T^4 - T_R^4)$$

$$= J + \Delta a T + \frac{\Delta b}{2}T^2 + \frac{\Delta c}{3}T^3 + \frac{\Delta d}{4}T^4 \qquad \text{14.28}$$

where $\Delta a = \sum_i v_i a_i$, and heat capacity constants *b*, *c*, and *d* are handled analogously. The value of the constant *J* is found by using a known numerical value of ΔH_R^o on the left-hand side and setting the right-hand side temperature to T_R. Substituting into the van't Hoff equation (Eqn. 14.24) and integrating again,

$$\boxed{\frac{\Delta G^o}{RT} = \frac{J}{R}\left(\frac{1}{T} - \frac{1}{T_R}\right) - \frac{\Delta a}{R}\ln\frac{T}{T_R} - \frac{\Delta b}{2R}(T - T_R) - \frac{\Delta c}{6R}(T^2 - T_R^2) - \frac{\Delta d}{12R}(T^3 - T_R^3) + \frac{\Delta G_R^o}{RT_R}} \qquad \text{14.29}$$

If desired, all values at T_R can be lumped together in a constant, *I*.

$$\boxed{\frac{\Delta G^o}{RT} = \frac{J}{RT} - \frac{\Delta a}{R}\ln T - \frac{\Delta b T}{2R} - \frac{\Delta c T^2}{6R} - \frac{\Delta d T^3}{12R} + I} \qquad \text{14.30}$$

The constant *I* may be evaluated from a knowledge of ΔG^o_{298} by plugging in $T = 298.15$ on the right-hand side as illustrated below.

Example 14.5 Equilibrium constant as a function of temperature

The heat capacities of ethanol, ethylene and water can be expressed as $C_P = a + bT + cT^2 + dT^3$ where values for *a*, *b*, *c*, *d* are given below along with standard energies of formation. Calculate the equilibrium constant [$\exp(-\Delta G^o_T/RT)$] for the vapor phase hydration of ethylene at 145 and 320°C.

$$C_2H_{4(g)} + H_2O_{(g)} = C_2H_5OH_{(g)}$$

	$\Delta H^o_{f,\,298}$ kJ/mol	$\Delta G^o_{f,\,298}$ kJ/mol	*a*	*b*	*c*	*d*
Ethylene	52.51	68.43	3.806	1.566E-01	−8.348E-05	1.755E-08
Water	−241.835	−228.614	32.24	1.924E-03	1.055E-05	−3.596E-09
Ethanol	−234.95	−167.73	9.014	2.141E-01	−8.390E-05	1.373E-09

The spread-sheet KCALC.xls is helpful in doing these calculations.

Example 14.5 Equilibrium constant as a function of temperature (Continued)

Solution:

$$C_2H_4 \;+\; H_2O \;=\; C_2H_5OH$$

$$\nu_i \qquad\quad -1 \qquad -1 \qquad\quad +1$$

Taking 298.15 K as the reference temperature,

$$\Delta H_R^o \;=\; \Delta H_{298}^o = -234.95 - [52.51 + (-241.835)] = -45{,}625 \text{ J/mole}$$

$$\Delta G_R^o \;=\; \Delta G_{298}^o = -167.73 - [68.43 + (-228.614)] = -7546 \text{ J/mole}$$

The variable J may be found with Eqn. 14.28 at 298.15 K

$$\Delta H_R^o = -45{,}625 = J + \Delta a T + (\Delta b/2)\cdot T^2 + (\Delta c/3)\cdot T^3 + (\Delta d/4)T^4 =$$

$$J + (9.014 - 3.806 - 32.24)\,T + [(0.2141 - 0.1566 - 0.0019)/2]\,T^2 +$$

$$[(-8.39 + 8.348 - 1.055)(1\text{E-}5)/3]\,T^3 + [(1.373 - 17.55 + 3.596)(1\text{E-}9)/4]\,T^4$$

$$= J - 27.032\,T + 0.02779\,T^2 - (3.657\text{E-}6)T^3 - (3.145\text{E-}9)T^4$$

Plugging in $T = 298.15$ K, and solving for J, $J = -39.914$ kJ/mole. Using this result in Eqn. 14.30 at 298.15 K will yield the variable I.

$$\Delta G^o/RT = -39{,}914/(8.314\cdot T) + 27.032/8.314 \ln T - [(5.558\text{E-}2)/(2\cdot 8.314)]\,T$$

$$+ [(1.097\text{E-}5)/(6\cdot 8.314)]\,T^2 + [(1.258\text{E-}8)/(12\cdot 8.314)]\,T^3 + I$$

Plugging in ΔG^o at 298.15K, $\Delta G^o/RT = -7546/8.314/298.15 = 3.0442$. Plugging in for T on the right-hand side results in $I = -4.494$.

The resulting formula to calculate ΔG_T^o at any temperature:

$$\Delta G_T^o = -39{,}914 + 27.032\,T \ln T - 0.0278\,T^2 + (1.828\text{E-}6)T^3 + (1.048\text{E-}9)T^4 - 37.363\,T$$

$$\text{at } 145°C \quad \Delta G_T^o = 7997 \text{ J/mol} \Rightarrow K_a = 0.1002;$$

$$\text{at } 320°C \quad \Delta G_T^o = 31{,}045 \text{ J/mol} \Rightarrow K_a = 0.00185$$

14.5 SHORTCUT ESTIMATION OF TEMPERATURE EFFECTS

Recall Eqn. 14.24, which we refer to as the *general* van't Hoff equation:

$$\frac{\partial(\Delta G_T^o / RT)}{\partial T} = -\frac{\Delta H_T^o}{RT^2} = -\frac{\partial \ln(K_a)}{\partial T}$$

Suppose we make the approximation that ΔH_T^o is independent of temperature. That is, suppose $\Delta C_P = \Delta a = \Delta b = \Delta c = \Delta d = 0$. Then in the equations above, $J = \Delta H_R^o$ in Eqn. 14.29, or we can integrate Eqn. 14.24 directly to obtain what we refer to as the shortcut van't Hoff equation:

$$\ln\left(\frac{K_a}{K_{aR}}\right) = \frac{\Delta G_R^o}{RT_R} - \frac{\Delta G_R^o}{RT} = \frac{-\Delta H_R^o}{R}\left(\frac{1}{T} - \frac{1}{T_R}\right)$$

T not R (handwritten annotation pointing to RT term)

14.31   ❗ Shortcut van't Hoff equation.

This equation enables us to obtain rapid insight about the effects of temperature and we can often fill in the details later. It is such a popular equation that people often forget that it is only a shortcut, and not generally applicable to temperature ranges broader than about +/– 100 K for high accuracy. As a particular observation, we take special note from the above equation that exothermic reactions ($\Delta H_T < 0$) lead to K_a decreasing as temperature increases, and endothermic reactions ($\Delta H_T > 0$) lead to K_a increasing as temperature increases. This means *equilibrium* conversion (for a specified feed) will increase with increasing temperature for endothermic reactions, and decrease with increasing temperature for exothermic reactions. This effect is illustrated in Fig.14.1.

Example 14.6   Application of the shortcut van't Hoff equation

Apply the shortcut approximation to the vapor phase hydration of ethylene. This reaction has already been studied in the previous example, and the Gibbs energy of reaction and heat of reaction can be obtained from that example.

$$K_{a298} = \exp(+7546/8.314/298) = 21.03$$

$$\ln\left(\frac{K_a}{21.03}\right) = \frac{+45,625}{8.314}\left(\frac{1}{T} - \frac{1}{298.15}\right)$$

$$K_a = 0.106 \text{ at } 145°C; K_a = 0.0022 \text{ at } 320°C$$

The results are very similar to the answer obtained by the general van't Hoff equation in Example 14.5.

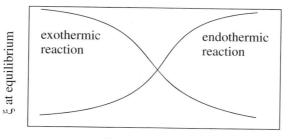

Figure 14.1 *Qualitative behavior of equilibrium conversion for exothermic and endothermic reactions.*

14.6 ENERGY BALANCES FOR REACTIONS

To calculate heat transfer to or from a reactor system, we must use the energy balance. The energy balance used in earlier chapters, however, requires further consideration. To simplify the derivation of the energy balance for reactive systems, consider a single inlet stream and single outlet stream flowing at steady-state.

$$0 = H^{in}\dot{n}^{in} - H^{out}\dot{n}^{out} + \dot{Q} + \underline{\dot{W}}_S \qquad 14.32$$

For either the inlet stream or the outlet stream, the total enthalpy can be calculated by summing the enthalpy of the components plus the enthalpy of mixing at the stream temperature and pressure. For a system without phase transitions between T_R and T,

! Enthalpy of a mixed stream where there are no phase changes between T_R and T.

$$H\dot{n} = \underline{\dot{H}} = \sum_{components} \dot{n}_i\left(H^o_{i,R} + \int_{T_R}^T \left(\frac{\partial H}{\partial T}\right)_{P^o,i} dT + \overset{\text{small}}{\int_{P^o}^P \left(\frac{\partial H}{\partial P}\right)_{T,i} dP} \right) + \overset{\text{small}}{\dot{n}\Delta H_{mix}}$$

$$\approx \sum_{components} \dot{n}_i\left(H^o_{i,R} + \int_{T_R}^T C_P dT\right) \qquad 14.33$$

where the pressure dependence of the enthalpy and the heat of mixing have been assumed to be small relative to the heat of reaction. Both assumptions are often valid. It also should be stressed that Eqn. 14.33 does not include phase changes. It might not be immediately obvious that Eqn. 14.32 includes the heat of reaction. Considering *just the flow terms of the energy balance*, by plugging Eqn. 14.33 into 14.32 the flow terms become

$$H^{in}\dot{n}^{in} - H^{out}\dot{n}^{out} = \sum_{components} (\dot{n}_i^{in} - \dot{n}_i^{out})H^o_{i,R} +$$

$$\sum_{components} \dot{n}_i^{in}\int_{T_R}^{T^{in}} C_{P,i} dT - \sum_{components} \dot{n}_i^{out}\int_{T_R}^{T^{out}} C_{P,i} dT \qquad 14.34$$

where the inlet temperature of all reactants is the same. The first term on the right of Eqn. 14.34 can be related to the heat of reaction using Eqn. 14.4 to introduce ξ and Eqn. 14.27 to insert the heat of reaction:

$$\sum_{components} (\dot{n}_i^{in} - \dot{n}_i^{out})H^o_{i,R} = -\dot{\xi}\sum_{components} \nu_i H^o_{i,R} = -\dot{\xi}\Delta H^o_R \qquad 14.35$$

Therefore the steady-state energy balance can be calculated using the result that we will call Method 1.

Method 1

$$\boxed{0 = \sum_{components} \dot{n}_i^{in}\int_{T_R}^{T^{in}} C_{P,i} dT - \sum_{components} \dot{n}_i^{out}\int_{T_R}^{T^{out}} C_{P,i} dT + \dot{Q} + \underline{\dot{W}}_S - \dot{\xi}\Delta H^o_R} \qquad 14.36$$

where the integrals represent the enthalpies of the components relative to the reference temperature. To use this expression correctly, *the enthalpies of the inlet and outlet streams must be calculated relative to the same reference temperature where* ΔH^o_R *is known* and any phase transitions at temperatures between the reference state and the inlet or outlet states must be added. The temperature of 298.15 K is almost always the reference temperature for balances involving chemical reactions. There is less flexibility in choosing the reference temperature than for non-reactive systems. This method is easiest to apply with one or two reactions where the stoichiometry is known.

Method 2

Note in Eqn. 14.35 that $\sum\limits_{components} v_i H_{i,R}^o$ can be calculated alternatively using the enthalpies of

formation using Eqn. 14.27, $\sum\limits_i v_i \Delta H_{R,i}^o = \sum\limits_i v_i \Delta H_{fR,i}^o$. Carrying this substitution of $\Delta H_{fR,i}^o$ for

$\Delta H_{R,i}^o$ back to Eqn. 14.33, note that the energy balance can be used as given in Eqn. 14.32 provided that the enthalpies are calculated relative to the elements in the standard state, in the form we call Method 2.

$$0 = H^{in} \dot{n}^{in} - H^{out} \dot{n}^{out} + \underline{\dot{Q}} + \underline{\dot{W}}_S \qquad 14.37$$

where

$$H^{in} \dot{n}^{in} = \sum_{components} \dot{n}_i^{in} \left(\Delta H_{fR,i}^o + \int_{T_R}^{T^{in}} C_P dT \right) \qquad 14.38$$

and

$$H^{out} \dot{n}^{out} = \sum_{components} \dot{n}_i^{out} \left(\Delta H_{fR,i}^o + \int_{T_R}^{T^{out}} C_P dT \right) \qquad 14.39$$

subject to the approximations included in Eqn. 14.33. Eqns. 14.38 and 14.39 do not include phase transitions occurring between T_R and T^{in} or T^{out}. Method 2 is easiest to apply for multireaction equilibria, or when stoichiometry is unknown.

Work effects

Usually shaft work and expansion/contraction work are negligible relative to other terms in the energy balance. They may usually be neglected without significant error.

Adiabatic Reactors

Suppose that a reactor is adiabatic ($\underline{\dot{Q}} = 0$). The energy balance becomes

$$0 = \sum_{components} \dot{n}_i^{in} \int_{T_R}^{T^{in}} C_{P,i} dT - \sum_{components} \dot{n}_i^{out} \int_{T_R}^{T^{out}} C_{P,i} dT - \dot{\xi} \Delta H_R^o \qquad 14.40$$

An exothermic heat of reaction will raise the outlet temperature above the inlet temperature. For an endothermic heat of reaction, the outlet temperature will be below the inlet temperature. At steady-state, the system finds a temperature where the heat of reaction is just absorbed by the enthalpies of the process streams. This temperature is known as the adiabatic reaction temperature, and the maximum reactor temperature change is dependent on the kinetics and reaction time, or on equilibrium. For fixed quantities and temperature of feed, Eqn. 14.40 involves two unknowns, T^{out} and ξ, and, *if the reaction is not limited by equilibrium*, the kinetic model and reaction time determine these variables. If a reaction time is sufficiently large, equilibrium may be approached. The variables T^{out} and ξ also appear in the equilibrium constraint that will govern maximum conversion, so the energy balance together with the equilibrium constraint determine the maximum conversion and adiabatic outlet temperature for an equilibrium reactor. To determine the adiabatic reaction temperature for a reaction where equilibrium limits conversion using Method 1:

1. Write the energy balance. Calculate the enthalpy of the inlet components at T^{in}.
2. Guess the outlet temperature, T^{out}. Calculate the enthalpy of the outlet components at T^{out}.
3. Determine $\dot{\xi}$ at T^{out} using the chemical equilibrium constant constraint.
4. Calculate $\dot{\xi}\Delta H_R^o$ for this conversion.
5. Check the energy balance for closure.
6. If the energy balance does not close, go to step 2.

As you might expect, this type of calculation lends itself to numerical solution techniques such as Solver in Excel.

Example 14.7 Adiabatic reaction in an ammonia reactor

Roughly (and quickly) estimate the outlet temperature and equilibrium mole fraction of ammonia synthesized from a stoichiometric ratio of N_2 and H_2 fed at 25°C and reacted at 100 bar. How would these change if the pressure was 200 bar?

$$1/2\ N_{2(g)} + 3/2\ H_{2(g)} = NH_{3(g)}$$

Solution: For a rough estimate we will use the shortcut approximation of temperature effects. Furthermore, we will assume $K_\varphi \approx 1$. (Is this a good approximation or not?[1]) Therefore we obtain,

$$P\,K_a = y_{NH3} / [(y_{N2})\,(y_{H2})^3]^{1/2}$$

Basis: Stoichiometric ratio in feed.

$$\dot{n}_{N_2}^{out} = \frac{1}{2} - \frac{1}{2}\dot{\xi}$$

$$\dot{n}_{H_2}^{out} = \frac{3}{2} - \frac{3}{2}\dot{\xi}$$

$$\dot{n}_{NH_3}^{out} = \dot{\xi}$$

$$\dot{n}_T^{out} = 2 - \dot{\xi}$$

---

1. We can evaluate this assumption by calculating the reduced temperatures at the end of our calculation and estimating the virial coefficients.

Example 14.7 Adiabatic reaction in an ammonia reactor (Continued)

Component	ΔH_f° (298)J/mole	ΔG_f° (298)J/mole
N_2	0	0
H_2	0	0
$NH_3\,(g)$	−45,940	−16,401.3

Calculating the equilibrium constant as a function of T:

$$K_{a298} = \exp\,(16{,}401.3/8.314/298.15) = 742.91$$

Using the shortcut van't Hoff,

$$\ln\!\left(\frac{K_a}{742.91}\right) = \frac{+\,45{,}940}{8.314}\left(\frac{1}{T} - \frac{1}{298.15}\right) \Rightarrow K_a = 742.91\,\exp\!\left[5525.6\!\left(\frac{1}{T} - \frac{1}{298.15}\right)\right] \qquad 14.41$$

Plugging the mole fraction expressions into Eqn. 14.20, and collecting the fractions 1/2 and 3/2,

$$PK_a\!\left(\frac{3^3}{2^4}\right)^{\!1/2} = \frac{\dot\xi(2 - \dot\xi)}{(1 - \dot\xi)^2} \Rightarrow 2\dot\xi - \dot\xi^2 = \frac{\sqrt{27}}{4}\,PK_a(1 - \dot\xi)^2$$

let $M = \dfrac{\sqrt{27}}{4}\,PK_a \;\Rightarrow\; (M + 1)\xi^2 - 2(M + 1)\xi + M = 0 \Rightarrow \xi^2 - 2\xi + \left(\dfrac{M}{M + 1}\right) = 0$

Applying the quadratic formula

$$\xi = \frac{2 \pm \sqrt{4 - 4M\,/\,(1 + M)}}{2} = 1 - \sqrt{1 - M\,/\,(1 + M)} \qquad 14.42$$

Heat capacity information and the enthalpy balance have been entered in the spreadsheet RXNADIA.xls. The objective is to find the temperature where the energy balance closes,

RXNADIA.xls.

$$F(T) = \sum_{components} \dot n_i^{\,in}\!\int_{T_R}^{T^{in}}\!\! C_{P,\,i}\,dT - \sum_{components} \dot n_i^{\,out}\!\int_{T_R}^{T^{out}}\!\! C_{P,\,i}\,dT - \dot\xi\Delta H_R^o = 0 \qquad 14.43$$

At 100 bar, guessing the outlet temperature of 500 K gives the following results:

T (K)	K_a	$\dot\xi$	Energy Balance (J)
500	0.4185	0.866	31,399

Example 14.7 Adiabatic reaction in an ammonia reactor (Continued)

As temperature is increased, the $F(T)$ of Eqn. 14.43 will become increasingly negative. This is because the positive term, $-\dot{\xi}\Delta H_R^o$, becomes smaller as T is increased (because $\dot{\xi}$ decreases for an exothermic reaction, see Fig. 14.1), and the negative term $-\sum\limits_{components} \dot{n}_i^{out} \int_{T_R}^{T^{out}} C_{P,i} dT$

becomes larger. The initial temperature guess and $\dot{\xi}$ are too low. We can use Solver to find the answer quickly. For the purposes of the example here, we demonstrate the results using trial and error. Raising our guess, then converging after a few guesses, gives the following results.

Table 14.1 *Iterations to determine the adiabatic reaction temperature at 100 bar.*

T (K)	K_a	$\dot{\xi}$	Energy Balance (J)
600	0.06633	0.678	17,163
700	0.0178	0.451	121.58
700.7	0.01766	0.449	3.72

We are very close to the answer with 700.7 K.

At 200 bar, let us start with our earlier result of 700.7 K. The results are given in Table 14.2.

Table 14.2 *Iterations to determine the adiabatic reaction temperature at 200 bar.*

T (K)	K_a	$\dot{\xi}$	Energy Balance (J)
700.7	0.01766	0.577	6745.2
750	0.01052	0.482	−930.5
744.0	0.0112	0.494	−4.2

Note: Conversion goes up but not so much. Reassure yourself by comparing Tables 14.1 and 14.2 at 700.7 K that K_a is not a function of pressure, only temperature.

Example 14.7 Adiabatic reaction in an ammonia reactor (Continued)

Adiabatic Synthesis of Ammonia   Protected without a password

Feed Temperature (K)	298
Outlet Temperature(K)	700.7221
P(bar)	100
Standard State Heat of Reaction at 298K	-45940 J/mol
K_a (298.15)	742.91

Heat Capacity Constants

	a	b	c	d
H2	2.71E+01	9.27E-03	-1.38E-05	7.65E-09
N2	3.12E+01	-1.36E-02	2.68E-05	-1.17E-08
NH3	2.73E+01	2.38E-02	1.71E-05	-1.19E-08

| K_a at reaction T | 0.017653 | M | 2.293223 | ξ | 0.448952 |

	Inlet			Outlet		
	moles	H(J/mol)	totals	moles	H(J/mol)	totals
H2	1.5	0	0	0.826572	11778.61	9735.873
N2	0.5	0	0	0.275524	11972.07	3298.593
NH3	0	0	0	0.448952	16906.91	7590.39
Total			0			20624.86

| Balance($\Sigma H^{in}n^{in} - \Sigma H^{out}n^{out} - \xi\Delta H$)= | 1.21E-07 J |

NOTE: The inlet moles cannot be changed without recalculating a formula for ξ

Use solver to set value of Balance to zero by adjusting Feed Temperature, Outlet Temperature, or P.

Figure 14.2 *Display from RXNADIA.XLS showing a converged answer.*

Graphical Visualization of the Energy Balance

The energy balance as presented by Method I (Eqn. 14.36) can be easily plotted for an adiabatic reaction. Let us use an approximate form of the adiabatic reaction energy balance of Eqn. 14.40. Frequently the overall mass heat capacities of the inlet and outlet streams are approximately equal. Most hydrocarbons have a mass heat capacity near the range of 1–1.5 J/g, even though the molar heat capacities vary significantly. Hydrogen has a much higher high mass heat capacity, but inlet and outlet stream heat capacities on a mass basis are still often approximately the same. For example, in the reaction of nitrogen and hydrogen to form ammonia, a stoichiometric feed has a heat capacity of 3.4 J/g at 298 K, and pure ammonia product has a heat capacity of 2.1 J/g at the same temperature. An average heat capacity can be used. This technique of assuming equal heat capacities provides a screening technique that can be refined by more detailed calculations, such as those given above. Since a steady-state process requires the mass flows on the inlet and outlet to be equal, and the heat capacities are about the same, then the balance of Eqn. 14.40 can be approximated as

$$T^{out} \approx T^{in} - \frac{\dot{\xi}\Delta H_R^o}{\dot{m}C_{Pm}}$$

14.44

where C_{Pm} is the average mass heat capacity of the stream. Consider the case of an exothermic reaction, where conversion is limited by equilibrium. A schematic of the equilibrium curve and the energy balance are shown in Fig. 14.3. For the case of the ammonia reaction, the equilibrium constraint curve could be generated by inserting various temperatures in Eqn. 14.41 and then determining the reaction coordinate from Eqn. 14.42. The energy balance is plotted using Eqn. 14.44. The

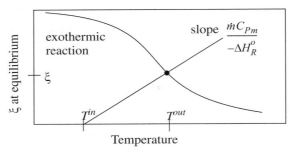

Figure 14.3 *Equilibrium curve and approximate energy*
balance for an exothermic reaction. The dot
represents the maximum reaction coordinate
value and the adiabatic temperature rise if
the reaction goes to equilibrium. The plot
for an endothermic reaction will be a mirror
image of this figure.

dot in the figure represents the point where the energy balance and equilibrium constraint are both satisfied. It is possible for the conversion to be less than the equilibrium conversion, if the conversion is limited by kinetics or reaction time. Note that an endothermic reaction will have an energy balance with a negative slope, and the equilibrium line will change shape as shown in Fig. 14.1, making the plot for an endothermic reaction a mirror image of Fig. 14.3.

14.7   GENERAL OBSERVATIONS ABOUT PRESSURE EFFECTS

We have obtained the following expression for the equilibrium constraint for vapor phase reactions:

$$\prod \hat{f}_i^{\nu_i} = (\prod \hat{\varphi}_i^{\nu_i})(\prod y_i^{\nu_i})P^{\Sigma \nu_i} = \exp(-\Delta G_T^o / RT)$$

where the standard pressure is 1 bar. Furthermore, the ideal solution approximation (as opposed to the ideal gas approximation) is often quite reasonable for vapor phases, resulting in

$$(\prod \varphi_i^{\nu_i})(\prod y_i^{\nu_i})P^{\Sigma \nu_i} \equiv K_\varphi K_y P^{\Sigma \nu_i} \qquad 14.45$$

or in terms of the partial pressures ($p_i = y_i P$)

$$(\prod \varphi_i^{\nu_i})(\prod p_i^{\nu_i}) \equiv K_\varphi K_p \qquad 14.46$$

1. Between vapor phase non-idealities and pressure, the strongest effect on equilibrium coordinate is $P^{\Sigma \nu_i}$. $\Sigma \nu_i$ will be > 0 when the number of moles of vapor phase products is greater than the number of the moles of vapor phase reactants. Since $\exp(-\Delta G_T^o / RT)$ is pressure-independent, K_y decreases to compensate for an increase in $P^{\Sigma \nu_i}$. Thus, increasing the pressure in this case would cause less formation of products. This effect is often referred to as one expression of "LeChatelier's Principle." For example, suppose the methanol synthesis reaction of Example 14.3 was carried out at 50 bar and 500 K. What would be the equilibrium conversion then?

$$0.00581 = K_y \, P^{\Sigma v_i} = K_y \, P^{-2}$$
$$0.00581 \cdot 50^2 \cdot 4(1 - \xi)^3 - \xi(3 - 2\xi)^2 \quad = 0 \Rightarrow \xi = 0.685 \Rightarrow y_{MeOH} = 0.42$$

Increasing the pressure has substantially enhanced the value of ξ. These are typical operating conditions for methanol synthesis.

2. Adding inerts to vapor phase reactions decreases the effect of $P^{\Sigma v_i}$. This effect cannot be seen directly from Eqns. 14.45 and 14.46, but becomes apparent when the mole fractions are written in terms of reaction coordinate.

3. The effect of $(\prod \varphi_i^{\,v_i}) = K_\varphi$ is much less pronounced than the other effects and less susceptible to broad generalization. The direction and magnitude of the effect depend on the nature of the products themselves. If the products are at significantly higher reduced pressures and lower reduced temperatures than the reactants, then $K_\varphi < 1$ and the formation of products would be favored.

14.8   MULTIREACTION EQUILIBRIA

When the equilibrium state in a reacting system depends on two or more simultaneous chemical reactions, the equilibrium composition can be found by a direct extension of the methods developed for single reactions. Each reaction will have its own reaction coordinate in which the compositions can be expressed. Some of the products of one reaction may act as reactants in another reaction, but the amount of that substance can still be written in terms of the extents of the reactions. Eventually, the material balances lead us to a system of N nonlinear equations in terms of N unknowns. We know in principle that these systems can be solved. Fortunately, there are standard programs available which are formulated to solve the general problem of multiple non-linear systems of equations, so we can concentrate on applying the program to thermodynamics instead of developing the numerical analysis. Many software packages like MATHEMATICA®, MATHCAD®, and even Excel offer the capability to solve non-linear systems of equations. Excel provides an especially convenient basis for illustrating the methods presented here. We will summarize our introduction to multireaction equilibria and simultaneous phase and reaction equilibria by means of three Excel spreadsheets: one for direct minimization of the Gibbs energy, one for solving a system of two simultaneous reactions, and one for two simultaneous reactions with phase partitioning. *All of the worksheets referenced in this section are available in the workbook RXNS.xls.*

Example 14.8 Simultaneous reactions that can be solved by hand

We can occasionally come across multiple reactions which can be solved without a computer. These are generally limited to textbook problems, but provide a starting point and test case for applying the general approach. Consider the two gas phase reactions:

$$A + B = C + D \qquad K_{a1} = 2.667$$
$$A + C = 2E \qquad K_{a2} = 3.200$$

Example 14.8 Simultaneous reactions that can be solved by hand (Continued)

The pressure in the reactor is 10 bar, and the feed consists of 2 moles of A and 1 mole of B. Calculate the composition of the reaction mixture if equilibrium is reached with respect to both reactions.

$$n_A = 2 - \xi_1 - \xi_2$$
$$n_B = 1 - \xi_1$$
$$n_C = \xi_1 - \xi_2$$
$$n_D = \xi_1$$
$$n_E = 2\xi_2$$
$$\overline{n_T = 3}$$

$$\frac{y_C y_D}{y_A y_B} = 2.667 ; \quad \frac{y_E^2}{y_A y_C} = 3.200$$

$$\frac{(\xi_1 - \xi_2)\xi_1}{(2 - \xi_1 - \xi_2)(1 - \xi_1)} = 2.667 ; \quad \frac{4\xi_2^2}{(2 - \xi_1 - \xi_2)(\xi_1 - \xi_2)} = 3.2$$

Solving the first equation for ξ_1 by quadratic equation,

$$\xi_1 = 2.4 - 1.1\xi_2 - \sqrt{(2.4 - 1.1\xi_2)^2 - 1.6(2 - \xi_2)} \qquad 14.47$$

Similarly, for the second reaction,

$$\xi_2 = -4 + \sqrt{16 + 4\xi_1(2 - \xi_1)} \qquad 14.48$$

We may now solve by trial and error. The procedure is: 1) Guess ξ_1; 2) Solve Eqn. 14.48 for ξ_2; 3) Solve Eqn. 14.47 for ξ_1^{new}; 4) If $\xi_1^{new} \neq \xi_1$, go to step 1. The iterations are summarized below.

ξ_1	ξ_2	ξ_1^{new}
1.0000	0.4721	0.8355
0.8355	0.4600	0.8342
0.8342	0.4598	0.83415

Example 14.9   Solving multireaction equilibrium equations by EXCEL

Methanol has a lower vapor pressure than gasoline. That can make it difficult to start a car fueled by pure methanol. One solution is to convert some of the methanol to methyl ether *in situ* during the start-up phase of the process (i.e., automobile). At a given temperature, 1 mole of MeOH is fed to a reactor at atmospheric pressure. It is assumed that only the two reactions given below take place. Compute the extents of the two simultaneous reactions over a range of temperatures from 200°C to 300°C. Also include the equilibrium mole fractions of the various species.

$$CH_3OH_{(g)} = CO_{(g)} + 2H_{2\,(g)} \qquad\qquad (1)$$

$$2CH_3OH_{(g)} = CH_3OCH_{3(g)} + H_2O_{(g)} \qquad\qquad (2)$$

Solution: Data for reaction (1) have been tabulated by Reactions Ltd.[1]—at 473.15 K, $\Delta H_T = 96{,}865$ J/mol and $\ln K_{a1,473} = 3.8205$. Over the temperature range of interest we can apply the shortcut van't Hoff equation assuming constant heat of reaction using the data at 200°C as a reference.

$$\ln K_{a1} = \frac{-96865}{8.314}\left(\frac{1}{T} - \frac{1}{473.15}\right) + 3.8205$$

Data for reaction (2) can be obtained from Appendix E for MeOH and water. For DME, the values are from Reid et al. (1987).[2]

Component		ΔH_f^o kJ/mole	ΔG_f^o kJ/mole
MeOH	−2	−200.94	−162.24
H$_2$O	1	−241.835	−228.614
MeOMe	1	−184.2	−113.0
		−24.155	−17.134

The shortcut van't Hoff equation for this reaction gives:

$$\text{at 298 K,}\quad \ln K_{a2,\,298} = \frac{17134}{8.314(298)} = 6.9156$$

$$\ln K_{a2} = \frac{24,155}{8.314}\left(\frac{1}{T} - \frac{1}{298.15}\right) + 6.9156$$

---

1. These data are slightly different from values calculated using tabulated properties from Appendix E, but such variations are common in thermochemical data. In this case the equilibrium compositions are about the same if the example is reworked using data from Appendix E.

2. Reid, R., Prausnitz, J.M., Poling, B. *The Properties of Gases and Liquids,* 4th ed. McGraw-Hill, NY, 1987.

Example 14.9 Solving multireaction equilibrium equations by EXCEL (Continued)

Material Balances:

Specie	initial	final
1 MeOH	1	$1 - \xi_1 - 2\xi_2$
2 CO	0	ξ_1
3 H$_2$	0	$2\xi_1$
4 MeOMe	0	ξ_2
5 H$_2$O	0	ξ_2
Total	1	$1 + 2\xi_1$

Writing equations for reaction coordinates for reaction 1:

$$\frac{4\xi_1^3}{(1 - \xi_1 - 2\xi_2)(1 + 2\xi_1)^2} = K_{a1} \Rightarrow \quad 4\xi_1^3 - K_{a1} \cdot (1 - \xi_1 - 2\xi_2) \cdot (1 + 2\xi_1)^2 = 0 = \text{``err1''}$$

and for reaction 2:

$$\frac{\xi_2^2}{(1 - \xi_1 - 2\xi_2)^2} = K_{a2} \quad \Rightarrow \quad \xi_2^2 - K_{a2} \cdot (1 - \xi_1 - 2\xi_2)^2 = 0 = \text{``err2''}$$

These two equations could be solved simultaneously for ξ_1 and ξ_2. Alternatively, we could guess ξ_1 and ξ_2, calculate the mole fractions, and then set up objective functions based on the five mole fractions determined from the two unknowns (using specie numbering from the above table):

$$K_{a1} \cdot y_1 - y_2 \cdot y_3 \cdot P_2 = 0$$

$$\text{and } K_{a2} \cdot y_1 - y_4 \cdot y_5 = 0$$

We have chosen the former method since it approaches a polynomial form which should be easier to solve numerically. The solution is implemented in the spreadsheet DUALRXN in RXNS.xls. In the example here (see Fig. 14.4), the ΔC_P for both reactions is neglected. The equations derived above are entered directly into the cells, and the Solver tool is called.[1] You will need to designate one of the reaction equations as the target cell, the value of which is set to zero. The other reaction equation should be designated as a constraint (also set to zero). The cells with the reaction coordinates are the variables to be changed to obtain a solution. Under "options," you may want to specify the "conjugate" method, since that generally seems to converge more robustly for the reacting systems typically encountered. Generally, the Solver tool will require a reasonably accurate initial guess to keep it from converging on absurd results (e.g., $y_i < 0$). The initial guess can be easily developed by varying the values in the reaction-extent cells until the target cells move in the right direction. It sounds difficult, but the given spreadsheet will get you started, then you can experiment with initial guesses and experience how good your initial guesses need to be.

RXNS.xls, DUALRXN.xls.

---

1. See Appendix P for an introduction to Solver.

Example 14.9   Solving multireaction equilibrium equations by EXCEL (Continued)

Sample solution of two simultaneous reactions:

$$CH_3OH = CO + 2H_2$$

$$2CH_3OH = CH_3OCH_3 + H_2O$$

(Details of input equations described in text by Elliott and Lira)

T(K)	473	493	513	533	553	573
K_{a1}	45.272	122.971	308.986	724.512	1597.281	3332.341
K_{a2}	27.4786	21.4179	17.0214	13.7627	11.3002	9.4069
ξ_1	0.9048	0.9651	0.9870	0.9951	0.9979	0.9991
ξ_2	0.0435	0.0158	0.0058	0.0022	0.0009	0.0004
y_1	0.0030	0.0012	0.0005	0.0002	0.0001	0.0000
y_2	0.3220	0.3294	0.3319	0.3328	0.3331	0.3332
y_3	0.6441	0.6587	0.6637	0.6656	0.6662	0.6665
y_4	0.0155	0.0054	0.0020	0.0007	0.0003	0.0001
y_5	0.0155	0.0054	0.0020	0.0007	0.0003	0.0001

Objective Functions

err1	0.0000	0.0000	0.0000	0.0000	0.0000	0.0000
err2	0.0000	0.0000	0.0000	0.0000	0.0000	0.0000

Figure 14.4   *Spreadsheet DUALRXN from workbook RXNS.xls for Example 14.9 showing converged answers at several temperatures.*

Example 14.10   Direct minimization of the Gibbs energy with EXCEL

A remarkably simple technique can be applied to solve for the equilibrium compositions of species when only gas phases are present. This technique recognizes the simplicity of the fundamental problem of minimizing the Gibbs energy at equilibrium. By expressing the total Gibbs energy of the mixture in terms of its ideal solution components, we can simply request that the value of the Gibbs energy be minimized. The target cell becomes the value of the Gibbs energy of the mixture:

$$\underline{G} = \sum n_i \overline{G}_i \quad \text{and} \quad \frac{\overline{G}_i}{RT} = \frac{G_i}{RT} + \left(\frac{\overline{G}_i - G_i}{RT}\right) = \frac{G_i}{RT} + \ln y_i \tag{ig}$$

where in the last equality we have assumed all components are ideal gas vapors. Taking reference state to be the elements in their naturally occurring molecular form at the standard state, then, at the standard state pressure, $\dfrac{G_i}{RT} = \dfrac{\Delta G_f^o}{RT}$, resulting in $\underline{G}/RT = \sum n_i \Delta G_{f,i}^o/RT + \sum n_i \ln y_i$

Example 14.10 Direct minimization of the Gibbs energy with EXCEL (Continued)

What could be simpler? We do not even need to explicitly express what the reactions are. This method assumes that equilibrium is reached by whatever system of reactions is necessary. Apply this approach to the problem of steam cracking of ethane at 1000 K where the ratio of steam to ethane in the feed is 4:1. Determine the distribution of C_1 and C_2 products, neglecting the possible formation of aldehydes, carboxylic acids, and higher hydrocarbons.

Solution: One fundamental problem and one practical problem remain to be faced. The fundamental problem is that there are several constraints that must be respected during the minimization process. These are the atom balances. We must keep in mind that we are not destroying matter, only rearranging it. So the number of carbon atoms, say, must be the same at the beginning and end of the process. Atom balance constraints must be written for every atom present. The atom balances are given straightforwardly by the stoichiometry of the species. For this example the balances are:

O-balance: $n_O^{feed} - 2n_{CO_2} - n_{CO} - 2n_{O_2} - n_{H_2O} = 0$

H-balance: $n_H^{feed} - 4n_{CH_4} - 4n_{C_2H_4} - 2n_{C_2H_2} - 6n_{C_2H_6} - 2n_{H_2} - 2n_{H_2O} = 0$

C-balance: $n_C^{feed} - n_{CH_4} - 2n_{C_2H_4} - 2n_{C_2H_2} - n_{CO_2} - n_{CO} - 2n_{C_2H_6} = 0$

The practical problem that remains is that the numerical solver often attempts to substitute negative values for the prospective species. This problem is easily treated by solving for the $\log(n_i)$ during the iterations and determining the values of n_i after the solution is obtained. Large negative values for the $\log(n_i)$s cause no difficulty. They simply mean that the concentrations of those species are small.

RXNS.xls, GIBBSMIN.

The solution is obtained using the spreadsheet GIBBSMIN contained in the workbook RXNS.xls (see Fig. 14.5). In order to apply the Gibbs minimization, the Gibbs energy of formation is required for each component at the reaction temperature. This preliminary calculation is the same type of calculation as performed in Example 14.5 on page 493. For example, the Gibbs energy of methane is simply the Gibbs energy of the formation reaction $C_{(s)} + 2H_{2(g)} \Rightarrow CH_{4(g)}$ at 1000 K.

The primary product of this particular process is hydrogen. Fracturing hydrocarbons is a common problem in the petrochemical industry. This kind of process provides the raw materials for many downstream processes. The extension of this method to other reactions is straightforward. Some examples of interest would include several systems with environmental applications: carbon monoxide and NO_x from a catalytic converter, or by-products from catalytic destruction of chlorinated hydrocarbons. Evaluating the equilibrium possibilities by this method is so easy that it should be a required preliminary to any gas phase process reaction study.

Example 14.10 Direct minimization of the Gibbs energy with EXCEL (Continued)

Values for $\Delta G_f^o/RT$ are determined by using Kcalc.xls at 1000 K prior to using this spreadsheet.

	G_i/RT 1000 K	moles fed	log(n_i)	n_i	y_i	$n_i(G_i/RT + \ln y_i)$
CH_4	2.4622		-1.210455673	0.061594839	0.007	-0.15451
C_2H_4	14.27		-5.467702147	3.40642E-06	0.000	0.00000
C_2H_2	20.427		-5.417178435	3.82667E-06	0.000	0.00002
CO_2	-47.612		-0.259817523	0.549771822	0.062	-27.70503
CO	-24.089		0.142583689	1.388620872	0.156	-36.02656
O_2	0		-13.59340412	2.55033E-14	0.000	0.00000
H_2	0		0.729567143	5.364968093	0.604	-2.70153
H_2O	-23.178	4	0.179504587	1.511835668	0.170	-37.71745
C_2H_6	13.33	1	-18.77525211	1.67783E-19	0.000	0.00000
Total				8.876798528		-104.3050627

Balances	O-bal	4
	H-bal	14
	C-bal	2

Figure 14.5 *Spreadsheet GIBBSMIN from the workbook RXNS.xls for Example 14.10.*

Example 14.11 Pressure effects for Gibbs energy minimization

One complication with the Gibbs energy minimization approach is that the manner of accounting for pressure effects is not immediately obvious. Keeping the same reference state as the previous problem, the effect of pressure on Gibbs energy of an ideal gas is given by Eqn. 8.17, therefore,

$$\frac{G_i}{RT} = \frac{\Delta G_{f,i}^o}{RT} + \ln\frac{P}{P^o} \qquad \text{(ig) 14.49}$$

Demonstrate the viability of this approach by adapting the Gibbs energy minimization method to the methanol synthesis reaction using stoichiometric feed at 50 bar and 500 K. The reaction has been discussed in Example 14.3 on page 489, and Section 14.7 on page 502.

Solution: It is convenient to first find $\Delta G_{f,1\,bar}^o$ and then find G_i/RT for each species, and then apply these values in the Gibbs minimization.

Example 14.11 Pressure effects for Gibbs energy minimization (Continued)

Compound	$\Delta G_{f,1bar}$(kJ/mole)	G_i/RT
Methanol	−134.04	$-134.04/8.3143E\text{-}3/500 + \ln(50) = -28.332$
CO	−155.38	$-155.38/8.3143E\text{-}3/500 + \ln(50) = -33.466$
H_2	0	$\ln(50)$ $= 3.9120$

for a basis of 1 mole CO: $n_{C,feed} = 1 = n_{CO} + n_{MeOH}$; $n_{H,feed} = 4 = 2n_{H2} + 4n_{MeOH}$; $n_{O,feed} = n_{CO} + n_{MeOH}$; the C-balance is redundant with the O-balance so only one of these should be included as a constraint to improve convergence. Minimizing the Gibbs energy gives $y_{MeOH} = 0.42$ in agreement with the other method in Section 14.7 on page 502.

Note: The objective function changes weakly with mole numbers near the minimum, so tighten the convergence criteria or re-run the solver after the first convergence. Convergence is sensitive to the initial guess. An initial guess which works is $\log(n_i) = -0.1$ for all i.

14.9 SIMULTANEOUS REACTION AND PHASE EQUILIBRIUM

It is not difficult to imagine situations in which reactions take place in the presence of multiple phases. Absorption of CO_2 into a NaOH solution involves a reaction of the CO_2 as it dissolves to form sodium bicarbonate and sodium carbonate. Hydrogen bonding in "pure" fluids implies reaction and phase equilibrium at saturation conditions. The production of methyl t-butyl ether (MTBE) as an oxygenated fuel additive is an interesting process in which catalyst is placed on the trays of a distillation column. As catalysts are developed which are active at progressively lower temperatures, multiphase reactions should become even more common. In low-temperature methanol synthesis (~240°C), it can be advantageous to add a liquid phase to absorb the heat of reaction, as described in the example below. Biological pathways are also in development for many products, which frequently involve multiple phases.

The thermodynamic analysis of this seemingly complex kind of process is actually very similar to the analysis of multireaction equilibria. The extent of formation of a second phase is analogous to a reaction coordinate. The easiest way to illustrate the formulation of the problem is to consider an example. Suppose the methanol synthesis reaction was carried out at 75 bars and 240°C such that the gas phase mole fractions were 0.25 CO, 0.25 MeOH, and 0.50 H_2. Based on a stoichiometric feed composition, the conversion would then be 50%. Now suppose this gas phase was placed in contact with a liquid phase with a nonvolatile solvent. The K-ratios at these conditions are about 10 for CO and H_2, and about 1.0 for MeOH. What would be the composition in the liquid phase and the extent of conversion if only liquid was removed? The composition of the liquid would be 0.025 CO and 0.25 MeOH. Therefore, the extent of conversion would be 0.25/0.275 = 91%. Thus, the addition of a liquid phase greatly enhances conversion of this process. The example below elaborates on these findings in a much more formal manner.

Example 14.12   The solvent methanol process

In a process being considered for methanol synthesis, a heavy liquid phase is added directly to the reactor to absorb the heat of reaction. The liquid is then circulated through an external heat exchanger. Usually, the catalyst is slurried in the liquid phase. An alternative to be considered is putting the catalyst in a fixed bed and adding just enough liquid so that a fairly small amount of vapor is left at the end of the reaction. Supposing naphthalene was used as the heavy liquid phase, use the Peng-Robinson equation to obtain approximate vapor-liquid K-value expressions of the form

$$K_i = \frac{a_i 10^{[b_i(1 - 1/Tr)]}}{P}$$

for each component at a temperature of 200–250°C and pressures from 50–100 bars[1].

Solution: Computing the K-value would normally require calling the Peng-Robinson equation during every flash and reaction iteration. This approximate correlation enables you to use Excel to perform the calculations since it is independent of any external programming requirements. The correlation should be suitably accurate if you "guess" compositions for developing the correlation that are reasonably close to the compositions at the outlet of the reactor. We suggest a guess for feed composition of {0.02, 0.10, 0.02, 0.035, 0.005, 0.82} for {CO, H_2, CO_2, methanol, water, and naphthalene}. As an example of a way to develop a synthetic data base, perform flash calculations at 75 bars and temperatures of {200, 210, 220, 230, 240, 250}. Tabulate the K-values for each component and plot them logarithmically with reciprocal temperature on the abscissa. Select a set of points, then select "add trendline" from the Chart menu. Select the options for a logarithmic fit, and displaying the equation on the chart. The coefficients of the equation give the a and b for the local "shortcut" correlation. For greater reliability, you can generate a separate correlation at each pressure of interest, but a better approach for general calculations is to simply gain access to a chemical process simulator. After all, we are only trying to demonstrate the thermodynamic principles here, not supplant commercial products for process simulation.

In the spreadsheet computations, you may assume the K-value of naphthalene to be negligible. Solve for the simultaneous reaction and phase equilibria at 240°C and $P = 100$ bars considering the following two reactions:

$$CO + 2H_2 = CH_3OH \qquad (1)$$

$$CO_2 + 3H_2 = CH_3OH + H_2O \qquad (2)$$

Add moles of the heavy liquid until 9 moles of liquid is obtained for every mole of vapor *output*. The gases are fed in proportions 2:7:1 CO:H_2:CO_2.

Applying the shortcut van't Hoff equation: (calculated at 503 K using $\ln K_a$ and $\Delta H°_R$, all reacting species gases),

$$\ln K_{a1} = 11746/T - 28.951$$

$$\ln K_{a2} = 6940/T - 24.206$$

---

1. Note: the symbols a and b are simply regression coefficients, not the equation of state parameters a and b.

Example 14.12 The solvent methanol process (Continued)

Stoichiometry

	n_i^i	ν_{1i}	ν_{2i}	$n_i^f{}_2$
CO	2	−1	0	$2 - \xi_1$
H_2	7	−2	−3	$7 - 2\xi_1 - 3\xi_2$
CO_2	1	0	−1	$1 - \xi_2$
MeOH	0	1	1	$\xi_1 + \xi_2$
H_2O	0	0	1	ξ_2
				$10 - 2\xi_1 - 2\xi_2$

Numbering components

CO	H_2	CO_2	MeOH	H_2O	SOLVENT
1	2	3	4	5	6

Vapor-liquid K values:

$$K_1 = 9.9/P \ 10^{(2.49(1-133/T))}$$

$$K_2 = 970/P$$

$$K_3 = 23/P \ 10^{(2.87(1-304/T))}$$

$$K_4 = 61/P \ 10^{(3.64(1-513/T))}$$

$$K_5 = 410/P \ 10^{(3.14(1-647/T))}$$

$$K_6 \approx 0$$

Imagine performing a flash at each new extent of conversion:

$$n_T = n_{T0} - 2\xi_1 - 2\xi_2$$

$$z_i = (n_{0i} + \nu_{1i}\xi_1 + \nu_{2i}\xi_2)/n_T$$

$$y_i = \frac{z_i K_i}{K_i + \dfrac{L}{F}(1 - K_i)}$$

Writing objective functions:

$$F(1) = err1 = (P^2 K_{a1} y_1 y_2^2 - y_4)$$

$$F(2) = err2 = (P^2 K_{a2} y_3 y_2^3 - y_4 y_5)$$

$$F(3) = 1 - \sum_i y_i$$

Example 14.12   The solvent methanol process (Continued)

This spreadsheet is called SMPRXN. An example of the output from a feed of 2,7,1,0,0 mole each of CO, H_2, CO_2, CH_3OH, H_2O is shown in Fig. 14.6.

RXNS.xls, SMPRXN.

P(bar)	50					
T(K)	513	533	553	573	593	613
Ka_1	0.002347712	0.000994295	0.0004481	0.0002135	0.00010694	5.6037E-05
Ka_2	2.30525E-05	1.38758E-05	8.66454E-06	5.59125E-06	3.71625E-06	2.53674E-06
K_1	13.839	14.633	15.410	16.170	16.912	17.637
K_2	19.400	19.400	19.400	19.400	19.400	19.400
K_3	6.792	7.867	9.016	10.235	11.520	12.867
K_4	1.220	1.671	2.237	2.934	3.779	4.788
K_5	1.241	1.747	2.399	3.223	4.245	5.491
K_6	0.000	0.000	0.000	0.000	0.000	0.000
ξ_1	1.845	1.604	1.185	0.694	0.302	0.058
ξ_2	0.361	0.253	0.198	0.179	0.180	0.191
L/F	0.900	0.900	0.900	0.900	0.900	0.900
moles solv	16.368	23.680	32.671	41.780	48.658	52.826
y_1	0.043	0.082	0.129	0.168	0.192	0.206
y_2	0.692	0.691	0.691	0.693	0.693	0.692
y_3	0.125	0.116	0.101	0.087	0.080	0.076
y_4	0.120	0.097	0.069	0.043	0.025	0.014
y_5	0.020	0.014	0.010	0.009	0.010	0.012
err1	0.000	0.000	0.000	0.000	0.000	0.000
err2	0.000	0.000	0.000	0.000	0.000	0.000
$\Sigma(y_i)$	1.000	1.000	1.000	1.000	1.000	1.000
z_1	0.007	0.013	0.020	0.026	0.029	0.031
z_2	0.101	0.101	0.101	0.101	0.102	0.101
z_3	0.029	0.025	0.020	0.016	0.014	0.013
z_4	0.101	0.062	0.035	0.017	0.008	0.004
z_5	0.016	0.008	0.005	0.004	0.003	0.003

Figure 14.6 *Worksheet SMPRXN from workbook RXNS.xls for Example 14.12 at several temperatures.*

The method of solving this problem is extremely similar to the DUALRXN problem. The only significant addition is an extra constraint equation which specifies that the vapor mole fractions must sum to unity. Note that ξ_1 is greater than unity. This is because we have 2 moles of CO in the feed, so 1.3 moles converted is about 65%.

Example 14.13 NO$_2$ absorption[1]

The strength of concentrated acid which can be produced is limited by the back pressure of NO$_2$ over the acid leaving the absorbers. The overall reaction, obtained by adding reactions (*a*) and (*b*) is shown as (*c*). Here we assume that N$_2$O$_4$ is equivalent to 2NO$_2$.

$$2NO_2 + H_2O = HNO_3 + HNO_2 \quad (a)$$

$$3HNO_2 = HNO_3 + 2NO + H_2O \quad (b)$$

$$3NO_{2(g)} + H_2O_{(l)} = 2HNO_{3(l)} + NO_{(g)} \quad (c)$$

The gas entering the bottom plate of a nitric acid absorber contains 0.1 mole of NO per mole of mixture and 0.25 mole of NO$_2$ per mole mixture. The entering gas also contains 0.3 bar partial pressure of oxygen, in addition to inert gas. The total pressure is 1 bar. The acid made by the absorption operation contains 50 percent by weight of HNO$_3$, and the operation is isothermal at 86°F. Estimate the composition of the gas entering the second plate and the strength of the gas leaving the second plate.

Solution: (Basis: 1 mole gaseous feed)

Assume $y_w = y_{HNO_3} = 0$ and $x_{NO2} = x_{NO} = 0$.

For liquid:

	In	Δ	Out
H$_2$O	W	$-\xi$	$W - \xi$
HNO$_3$	0	2ξ	2ξ
Total			$W + \xi$

For vapor:

	In	Δ	Out
NO$_2$	0.25	-3ξ	$0.25 - 3\xi$
NO	0.10	ξ	$0.10 + \xi$
O$_2$	0.30	0	0.30
I	0.35	0	0.35
Total			$1 - 2\xi$

1. S. Lee, Personal Communication, 1993.

Example 14.13 NO_2 absorption (Continued)

$$K_a = \frac{\left(\dfrac{\hat{f}_{NO}}{f_{NO}^o}\right)\left(\dfrac{\hat{f}_{HNO_3}}{f_{HNO_3}^o}\right)^2}{\left(\dfrac{\hat{f}_{H_2O}}{f_{H_2O}^o}\right)\left(\dfrac{\hat{f}_{NO_2}}{f_{NO_2}^o}\right)^3} = \frac{y_{NO}\left(x_{HNO_3}\gamma_{HNO_3}\right)^2}{y_{NO_2}^3\left(x_{H_2O}\gamma_{H_2O}\right)} \frac{1}{P^2}$$

We can determine the mole fractions from the weight fractions:

$$x_{HNO_3} = (0.5/63)/[(0.5/63) + (0.5/18)] = 0.222$$

Noting from the CRC Handbook[1] the vapor pressure of HNO_3 is 64.6 mmHg, we can estimate the activity coefficients of HNO_3 and water from the x-y data in *The Chemical Engineers' Handbook.*[2]

$$\gamma_{HNO_3} = \frac{y_{HNO_3}P}{x_{HNO_3}P_{HNO_3}^{sat}} = \frac{0.39 \text{ mmHg}}{0.222 \cdot 64.6 \text{ mmHg}} = 0.027; \quad \gamma_{H_2O} = \frac{10.7 \text{ mmHg}}{0.778 \cdot 23.8 \text{ mmHg}} = 0.5785$$

Gibbs energies of formation are available in Appendix E for all but nitrogen dioxide, and Reid et al.[3] give the standard Gibbs energy of formation as 52 kJ/mol and the standard heat of formation as 33.87 kJ/mol. Performing a shortcut calculation using KCALC.xls, the equilibrium constant at 303.15 K is

$$K_a = 0.0054$$

at $P = 1$bar

$$K_a = \frac{y_{NO}(0.222 \cdot 0.027)^2}{y_{NO_2}^3(0.778 \cdot 0.5785)} = 0.0054 = \frac{y_{NO}}{y_{NO_2}^3} \cdot 0.00007983$$

Substituting for the reaction coordinate:

$$\frac{y_{NO}}{y_{NO_2}^3} = \frac{0.054}{0.00007983} = 67.65 = \frac{\left(\dfrac{0.1 + \xi}{1 - 2\xi}\right)}{\left(\dfrac{0.25 - 3\xi}{1 - 2\xi}\right)^3}$$

Solving the cubic equation, $\xi = 0.0431$.

$$x_{HNO_3} = (2\xi/(W + \xi)) = 2 \cdot 0.0431/(W + 0.0431) = 0.222 \text{ tells us that}$$

$$W = -0.0431 + 2 \cdot 0.0431/0.222 = 0.345 \text{moles}$$

So the composition of gas entering the second stage is: $y_{NO} = 0.157$; $y_{NO2} = 0.132$; $y_{O2} = 0.328$; $y_I = 0.383$. Computations for further stages would be similar.

1. Weast, R.C. *CRC Handbook of Chemistry and Physics,* 60th ed, page D-224, CRC Press, 1979.
2. Perry, R.H., Chilton, C.H., *The Chemical Engineers' Handbook*, 6th ed, p. 3–70, McGraw-Hill, 1986.
3. Reid, R.C., Prausnitz, J.M., Poling, B. *The Properties of Gases & Liquids,* 4th ed, McGraw-Hill, 1987.

14.10 ELECTROLYTE THERMODYNAMICS

Electrolyte thermodynamics involves modeling the associations and dissociations of electrolytes, all of which are balanced chemical reactions. Additionally, gas phases may be present where volatile species remain undissociated. Analyzing this reaction + phase equilibrium problem requires learning the terms and conventions of electrolyte thermodynamics. Not surprisingly, a new standard state condition is traditionally defined. The prevalence of electrolytes in practical applications prompts the need for at least a cursory introduction to the terminology and methods. Many texts are available to facilitate more advanced study. The following description has been adapted from Zemaitis et al.[1]

One difference in terminology from our previous calculations is the use of molality as a concentration measure. In the case of electrolytes, it is often easiest to perform calculations based on the number of solute (i.e. electrolyte) molecules relative to the number of solvent (e.g. water) molecules. Molality is a convenient measure of this ratio. "Molality" is the moles of species existing in one kg of solvent (e.g. water) molecules. If water is the solvent, the number of moles per kg of solvent is 55.509. For example, a one *molal* aqueous solution of sodium chloride is prepared by adding one mole of NaCl (58.44g) to one kg of water. The *molarity* is the moles per liter of solution; it will be slightly less than 1M since the total volume after addition will be greater than one liter. Molality and molarity are subtly but distinctly different. Note that the ratio of solute to solvent molecules in 1 molar solutions is dependent on liquid density and therefore dependent on the solute. The ratio of solute to solvent molecules in all 1 molal solutions will be the same for all solutes in a given solvent. A review of an introductory chemistry textbook can be helpful in further understanding the differences.

Dissociations of electrolytes are modeled as chemical reactions. For example, sodium chloride dissociates as

$$NaCl(aq) = Na^+ + Cl^-$$

This is simply a chemical reaction, and the extent of reaction can be modeled by applying Eqns. 14.15 and 14.24. In order to accurately model dissociation reactions like this, it is necessary to represent the activity coefficients in the aqueous phase, which requires the use of a standard state. One nuance of electrolyte thermodynamics is that the standard state is usually selected to be a "hypothetical ideal solution at 1 molal concentration at the system temperature and pressure."[2] This means that the reaction equilibrium constants have been tabulated relative to 1 molal concentrations, so calculating, say, ΔG°_{rxn} is based on these 1 molal concentrations. The standard state is chosen so that the molal activity coefficient ($\gamma_i \equiv a_i/M_i$) tends to unity as the molality M_i approaches zero. This may seem like a strange way to define the standard state until one considers that the electrolyte ions do not exist in a pure state at typical temperatures and pressures, so the standard state must be a selected concentration in solution.

The only remaining clarification is the model for calculating activity coefficients of electrolyte solutions. Fortunately, the relatively simple "extended Debye-Huckel" model is generally accurate to 0.1 molal, and that is sufficient for the present example. Physically, the Debye-Huckel approximation derives from assuming that the dilute ionic species distribute themselves according to:

$$g \sim \exp(-u_{Coul}/kT)$$

1. Zemaitis, J.F., Clark, D.M., Rafal, M., Scrivner, N.C., *Handbook of Aqueous Electrolytes,* AIChE-DIPPR, NY, p. 595, 1986.
2. Zemaitis et al., page 18.

The extended Debye-Huckel model extrapolates to slightly higher concentrations by long division with the following results:

$$\ln K_a = \frac{nFE}{RT}$$

$$K_a \equiv \frac{\gamma_\pm^v m_+^{v+} m_{m-}^{v-}}{\gamma_u m_u} \qquad \log \gamma_\pm = \frac{-a|z_- z_+|\sqrt{I}}{1+\sqrt{I}} \tag{14.50}$$

$$-nFE = RT \ln \frac{[M_{Cu^{2+}} M_{SO_4^{2-}} (C_\pm^2 \gamma_{SO_4^{2-}})]_I}{[M_{Cu^{2+}} M_{SO_4} (C_\pm^2 \gamma_{SO_4^{2-}})]_{II}}$$

$$\log_{10} \gamma_\pm = \frac{-A|z_+ z_-|\sqrt{I}}{1+\sqrt{I}} \qquad M_{Cu^{2+}} = M_{SO_4^{2-}} = M_{CuSO_4} \tag{14.51}$$

$$E = \frac{-RT}{nF} 2\ln \frac{[M_{CuSO_4}]_I}{[M_{CuSO_4}]_{II}} + \frac{2RT}{nf}\ln \frac{(\gamma_\pm)_I}{(\gamma_\pm)_{II}} \qquad I = 0.5\sum m_i z_i^2 \xleftarrow{molality} \qquad F = \frac{96485\,c\,b}{mol} \tag{14.52}$$

$$A = 0.3911\left(\frac{298}{T}\right)^{1.5} \qquad W = nFE \tag{14.53}$$

z_i is the valency of the coulombic charge, v_i is the stoichiometric coefficient in the reaction $U = v_+ C_+ + v\_C\_$. The term $\gamma_\pm$ is the mean activity coefficient for the salt ions in solution. It is necessary to define a mean activity for the salt because the positive ions cannot exist in solution without the negative ions. $v = v_+ + v_-$, U denotes the undissociated specie, I is the ionic strength.

The activity of the water is altered primarily by the ionic strength, reflecting the contributions from all ionic species in the mixture. Specific expressions applicable to the examples considered here are given in the example below. The derivation of the Debye-Huckel thermodynamics from the assumed model for the radial distribution function is best understood in Fourier transform notation.[1] At present, extensions to higher concentrations are largely empirical adaptations of the extended Debye-Huckel formulas. A substantial number of these empirical activity coefficient models are described in detail by Zemaitis, et al.

Example 14.14 Chlorine + water electrolyte solutions

The partial pressure of Cl_2 above aqueous solution is lowered by the reaction of chlorine with water to form hypochlorous acid (HClO) and chlorine ion in addition to the molecular chlorine dissolved in the water. There are three reactions to be taken into account:

$H_2O = H^+ + OH^-$

$Cl_{2(aq)} + H_2O = H^+ + Cl^- + HClO_{(aq)}$

$HClO_{(aq)} = H^+ + ClO^-$

We can also write the liquid-vapor equilibria as "reactions"

$H_2O = H_2O_{(v)}$

$Cl_{2(aq)} = Cl_{2(v)}$

Note that H_2O without any special designation refers to the liquid water. The activity of water in the presence of these ionic species is dictated by the Gibbs-Duhem equation in conjunction with the model for $\ln \gamma_\pm$.[2] For the extended Debye-Huckel model[3]

---

1. Hansen, J.P. and McDonald, I.R. *Theory of Simple Liquids,* 2nd ed, Academic, NY, 1986.
2. Zemaitis et al., p. 20.
3. Zemaitis et al., p. 602.

Example 14.14 Chlorine + water electrolyte solutions (Continued)

$$\ln(a_w)_{mix} = \frac{m_{H+} \cdot m_{Cl-}}{I^2} \ln(a_w)_{H-Cl} + \frac{m_{H+} \cdot m_{ClO-}}{I^2} \ln(a_w)_{H-ClO} \qquad 14.54$$

$$\ln(a_w)_{i-j} = -\frac{18(m_i + m_j)}{1000} \phi_{ij} \qquad 14.55$$

$$\phi_{ij} = 1 - 2.303\left[\frac{A|z_+z_-|}{I}\left(1 + \sqrt{I} - \frac{1}{1+\sqrt{I}} - 2\ln(1+\sqrt{I})\right)\right] \qquad 14.56$$

Develop a spreadsheet that can compute the partial pressure of Cl_2 above an aqueous solution at 298 K and 0.8 atm.

Solution: The equilibrium constants for the five reactions can be obtained from Zemaitis et al. The relevant values are tabulated below and in the spreadsheet: CL2H2O.XLS

Table 14.3 *Thermochemical data for the species.*

Species	ΔG^o_f(kJ/mol)	ΔH^o_f(kJ/mol)
H_2O	−237.18	−285.83
$H_2O_{(v)}$	−228.59	−241.82
H^+	0	0
OH^-	−157.29	−229.99
$Cl_{2(aq)}$	6.90	−23.43
$Cl_{2(v)}$	0	0
$HClO$	−79.71	−120.92
Cl^-	−131.26	−167.16
ClO^-	−36.82	−107.11

First of all, the equilibrium constants for each of the five "reactions" are calculated in the fourth column of Table 14.4. In addition, three atom balances must be satisfied: H,O,Cl. Finally, an additional constraint must be incorporated especially for electrolyte solutions: the electroneutrality constraint. The electroneutrality constraint specifies that the net charge of the solution at equilibrium must be zero (i.e. $\sum_i m_i z_i = 0$). The atom balances and electroneutrality constraint are shown in Table 14.5 for a basis of 1 liter of liquid water and 0.9 moles of Cl_2. To solve these nine equations (five equilibria, three atom balances, electroneutrality constraint) simultaneously, we must identify nine unknowns. The nine unknowns selected here are the species listed in Table 14.3: the liquid moles of H_2O, Cl_2, $HClO$, H^+, Cl^-, ClO^-, and OH^-, and the vapor moles of $H_2O_{(v)}$ and $Cl_{2(v)}$. All that remains is to solve these equations, done in this case by using the

CL2H2O.xls.

Example 14.14 Chlorine + water electrolyte solutions (Continued)

Excel Solver by allowing Solver to adjust the moles of each specie (in the second and sixth columns of Table 14.6) until all equations are simultaneously satisfied. One practical hint for solving problems of this type is that the mole numbers must always be greater than or equal to zero, which can be satisfied by iterating on the log of the mole numbers of ionic species in order to avoid negative mole numbers during the iterations.

Note from Table 14.6 that the $\gamma_\pm$ for all the ionic species is the same. This occurs because the Debye-Huckel model is too simple to make distinctions as long as all species have the same valence. The activity coefficients for Cl_2 and $HClO$ are assumed to be unity because the solution is dilute in all species.

The result of the calculation summarized in Table 14.6 shows that the chlorine solubility is enhanced from the formation of $HClO$ and Cl^- in solution. The Cl in $HClO$ and Cl^- together is about two-thirds of the Cl atoms in $Cl_{2(aq)}$ at $P = 0.8$ atm. This means that the amount of chlorine-containing species in solution is significantly more than would have been estimated from, say, the Peng-Robinson equation. Now that the problem has been solved at 298 K and $P = 0.8$ atm, different temperatures and pressures could be used to explore the changes in ionic species, and partial pressure of Cl_2. Also, it is possible to generate T-x-y or P-x-y diagrams.

Table 14.4 *Electrolyte Reaction Equilibrium Constraints—K_a in the fourth column is calculated from the standard state Gibbs energies and enthalpies of formation. The fifth column is calculated using the activities from the converged answer summarized in Table 14.6, and the two columns are equal at equilibrium as shown.*

$\Delta G^o_{298}/(R298)$	$\Delta H^o_{298}/(R298)$	$\Delta C_P/R$	K_{a298}	$\prod_i [a_i]^{\nu_i}$	
32.2448	22.5374	−0.09034395	9.914E-15	9.92E-15	$H_2O = H^+ + OH^-$
7.70951	8.55054	−0.08544634	0.00044854	0.000448	$Cl_2 + H_2O = H^+ + Cl^- + HClO$
17.3949	5.57314	0	2.7891E-08	2.79E-08	$HClO = H^+ + ClO^-$
−3.4672	−17.765	0.01683763	32.0458427	32.04584	$H_2O_{(v)} = H_2O$
2.78657	−9.4574	−0.01368628	0.06163231	0.061632	$Cl_{2(v)} = Cl_{2(aq)}$

Table 14.5 *Atom Balance and Electroneutrality Constraints, $n^i_{H2O} = 55.51$ and $n^i_{Cl2} = 0.9$ moles in feed. The third column is calculated using the mole numbers from the converged answer summarized in Table 14.6, and the right column shows the balance at equilibrium.*

	initial	final	final-initial
H-bal	111.02	111.02	2.14E-11
Cl-bal	1.8	1.8	−6.29E-13
O-bal	55.51	55.51	−1.85E-12
e-bal	0	−2.3155E-14	−2.32E-14

Example 14.14 Chlorine + water electrolyte solutions (Continued)

Table 14.6 *Electrolyte component mole numbers and activities at the converged composition.*

	LIQUID				VAPOR		
species	moles	molality	γ	species	moles	y
H_2O	55.4464	----	0.99896	H_2O	0.03333	3.90E-02
$Cl_{2(aq)}$	4.73E-02	0.0473846	1.00000	Cl_2	0.82244	9.61E-01
HClO	3.02E-02	0.0302603	1.00000	n^V	0.85577	
H^+	3.02E-02	0.03026038	0.87508			
OH^-	4.27E-13	4.2746E-13	0.87508			
ClO^-	3.64E-08	3.6422E-08	0.87508			
Cl^-	3.02E-02	0.03026034	0.87508			
tot moles	55.5844					
Intermediate Calculations		$I = 0.03026038$		$\ln(a_W)_{HCl} = -0.0010447$		
		$\phi = 0.95902$		$\ln(a_W)_{HClO} = -0.0005224$		
		$\ln(a_W) = -0.001$		$a_W = 0.99895582$		

14.11 SOLID COMPONENTS IN REACTIONS

Consider the reaction:

$$CO_{(g)} + H_{2(g)} = C_{(s)} + H_2O_{(g)}$$

The carbon formed in this reaction comes out as coke, a solid which is virtually pure carbon and separate from the gas phase. What is the fugacity of this carbon? Would its presence in excess ever tend to push the reaction in the reverse direction? The answer to the first question is that the fugacity would roughly equal the partial pressure which would roughly equal the vapor pressure which would be extremely negligible. So the answer to the second question is that the amount of solid carbon cannot influence the extent of this reaction. The next question is, how can we express these observations quantitatively?

The crux of the quantitative answer hinges on treatment of the reference condition. By defining the reference condition of the solid as the pure solid at 1 bar and the temperature of the reaction, and the reference for the gaseous species as the pure gases at 1 bar and the temperature of the reaction, we find

$$K_a = \left[\prod \left[\frac{\hat{f}_i}{f_i^o} \right]^{\nu_i} \right]$$

$$K_a = \frac{\left(\hat{f}_{H2O} / f_{H2O}^o \right) \cdot 1.0}{\left(\hat{f}_{H2} / f_{H2}^o \right)\left(\hat{f}_{CO} / f_{CO}^o \right)} = \frac{y_{H2O}}{y_{H2} y_{CO}} \frac{1}{P} = \exp\left[\frac{-\Delta G_T^o}{RT} \right]$$

To compute ΔG^o_T as a function of temperature, we apply the usual van't Hoff procedure. This means that $C_{P,c}$ can be treated just like C_P of the gaseous species.

Example 14.15   Thermal decomposition of methane

A 2-liter constant-volume bomb is evacuated and then filled with 0.10 moles of methane, after which the temperature of the bomb and its contents is raised to 1273 K. At this temperature the equilibrium pressure is measured to be 7.02 bar. Assuming that methane dissociates according the reaction $CH_{4(g)} = C_{(s)} + 2H_{2(g)}$, compute K_a for this reaction at 1273 K from the experimental data.

Solution:

$$K_a = \frac{\left(\hat{f}_{H2}\Big/ f^o_{H2}\right)^2 \cdot 1.0}{\left(\hat{f}_{CH4}\Big/ f^o_{CH4}\right)} = \frac{y^2_{H2}}{y_{CH4}}P$$

We can calculate the mole fractions of H_2 and CH_4 as follows. Since the temperature is high, the total number of moles finally in the bomb can be determined from the ideal gas law (assuming that the solid carbon has negligible volume): $n = P\underline{V}/RT = 0.702\cdot2000/(8.314)(1273) = 0.1327$. Now assume that ξ moles of CH_4 reacted. Then we have the following total mass balance: $n_T = 0.10 + \xi$. Therefore $\xi = 0.0327$ and

$$K_a = \frac{\left(\dfrac{2\xi}{0.1327}\right)^2 P}{\left(\dfrac{(0.1-\xi)}{0.1327}\right)} = \frac{0.493^2}{0.507}7.02 = 3.37$$

14.12   SUMMARY AND CONCLUDING REMARKS

We have greatly enlarged the scope of our coverage of engineering thermodynamics with very little extension of the conceptual machinery. All we really did conceptually was recall that the Gibbs energy should be minimized. The provision that atoms can be moved from one chemical species to another with commensurate changes in energy and entropy simply means that reference states must be assigned to elemental standard states instead of standard states based on pure components, such that the free energies of all the components can be compared. In this sense we begin to comprehend in a new light the broad range of applications mentioned in Einstein's quote at the end of Chater 1. Instead of conceptual challenges, reaction equilibria focus primarily on the computational aspects of setting up and solving the problems. Notably, equation solvers can provide multidimensional capability. These are tools that can be adapted to many problems, even those beyond the scope of thermodynamics. You should familiarize yourself with such tools and build the expertise that will permit you to enhance your productivity.

Cathode $\Delta G_I = -nFE^\circ + RT \ln \dfrac{a^2_{o^-}}{a_{o_2}}$

$\Delta G_{II} = -nFE^\circ + RT \ln \dfrac{a^2_{o^-}}{a_{o_2}}$

$\Delta G = RT \ln \left(\dfrac{a^I_{o_2}}{a^{II}_{o_2}}\right) = RT \ln \left(\dfrac{P^I_{o_2}/P^\circ_{o_2}}{P^{II}_{o_2}/P^\circ_{o_2}}\right)$

$2O^{2-} + 4e^- \longrightarrow O_2$

$a^I_{o_2} = \dfrac{P^I_{o_2}}{P^\circ_{o_2}}$

$a^{II}_{o_2} = \dfrac{P^{II}_{o_2}}{P^\circ_{o_2}}$

14.13 PRACTICE PROBLEMS

P14.1 An equimolar mixture of H_2 and CO obtained by the reaction of steam with coal. The product mixture is known as "water-gas." To enhance the H_2 content, steam is mixed with water-gas and passed over a catalyst at 550°C and 1 bar so as to convert CO to CO_2 by the reaction:

$$H_2O + CO = H_2 + CO_2$$

Any unreacted H_2O is subsequently condensed and the CO_2 is subsequently absorbed to give a final product that is mostly H_2. This operation is called the water-gas shift reaction. Compute the equilibrium compositions at 550°C based on an equimolar feed of H_2, CO, and H_2O.

Data for 550°C:

	ΔG_f^o (kJ/gmol)	ΔH_f^o (kJ/gmol)
H_2O	−202.25	−246.60
CO	−184.47	−110.83
CO_2	−395.56	−294.26

P14.2 One method for the production of hydrogen cyanide is by the gas-phase nitrogenation of acetylene according to the reaction: $N_2 + C_2H_2 = 2HCN$. The feed to a reactor in which the above reaction takes place contains gaseous N_2 and C_2H_2 in their stoichiometric proportions. The reaction temperature is controlled at 300°C. Estimate the product composition if the reactor pressure is: (a) 1 bar; (b) 200 bar. At 300°C, $\Delta G_T^o = 30.08$ kJ/mole.

P14.3 Butadiene can be prepared by the gas-phase catalytic dehydrogenation of 1-Butene: $C_4H_8 = C_4H_6 + H_2$. In order to suppress side reactions, the butene is diluted with steam before it passes into the reactor.

(a) Estimate the temperature at which the reactor must be operated in order to convert 30% of the 1-butene to 1,3-butadiene at a reactor pressure of 2 bar from a feed consisting of 12 mol of steam per mole of 1-butene.

(b) If the initial mixture consists of 50 mol% steam and 50 mol% 1-butene, how will the required temperature be affected?

ΔG_f^o (kJ/gmol)	600 K	700 K	800 K	900 K
1,3-Butadiene	195.73	211.71	227.94	244.35
1-Butene	150.92	178.78	206.89	235.35
ΔG_T^o	44.81	32.93	21.05	9.00

P14.4 Ethylene oxide is an important organic intermediate in the chemical industry. The standard Gibbs energy change at 298 K for the reaction $C_2H_4 + \frac{1}{2}O_2 = C_2H_4O$ is −79.79 kJ/mole. This large negative value of ΔG_T^o indicates that equilibrium is far to the right at 298 K. However, the direct oxidation of ethylene must be promoted by a catalyst selective to this reaction to prevent the complete combustion of ethylene to carbon dioxide and water. Even with such a catalyst, it is thought that the reaction will have to be carried out at a temperature of about 550 K in order to obtain a reasonable reaction rate. Since the reaction is exothermic, an increase in temperature will have an adverse effect on the equilibrium. Is the reaction feasible (from an equilibrium standpoint) at 550 K, assuming that a suitable catalyst selective for this reaction is available? Heat capacity equations (in J/mole-K) for the temperature range involved may be approximated by:

$C_{P,C2H4O} = 6.57 + 0.1389\ T(K)$; $C_{P,C2H4} = 15.40 + 0.0937\ T(K)$; $C_{P,O2} = 26.65 + 0.0084\ T(K)$; for ethylene oxide: $\Delta H^\circ_{f,298} = -52.63\ kJ/mol$

P14.5 The "water gas" shift reaction is to be carried out at a specified temperature and pressure employing a feed containing only carbon monoxide and steam. Show that the maximum equilibrium mole fraction of hydrogen in the product stream results when the feed contains CO and H_2O in their stoichiometric proportions. Assume ideal gas behavior.

P14.6 Assuming ideal gas behavior, estimate the equilibrium composition at 400 K and 1 bar of a reactive gaseous mixture containing the three isomers of pentane. Standard formation data at 400 K are:

	ΔG°_f (kJ/mol) (400 K)
n-pentane	40.17
isopentane	34.31
neopentane	37.61

P14.7 One method for the manufacture of synthesis gas depends on the vapor-phase catalytic reaction of methane with steam according to the equation $CH_4 + H_2O = CO + H_2$. The only other reaction which ordinarily occurs to an appreciable extent is the water-gas shift reaction. Gibbs energies and enthalpies tabulated in kJ/mol.

	$\Delta H^\circ_f(600\ K)$	$\Delta H^\circ_f\ (1300\ K)$	$\Delta G^\circ_f\ (600\ K)$	$\Delta G^\circ_f\ (1300\ K)$
CH_4	−83.22	−91.71	−22.97	52.30
H_2O	−244.72	−249.45	−214.01	−175.81
CO	−110.16	−113.85	−164.68	−226.94
CO_2	−393.80	−395.22	−395.14	−396.14

Compute the equilibrium compositions based on a 1:1 feed ratio at 600 K and 1300 K and 1 bar and 100 bars.

P14.8 Is there any danger that solid carbon will form at 550°C and 1 bar by the reaction $2CO = C_s + CO_2$? (ANS. Yes)

P14.9 Calculate the equilibrium percent conversion of ethylene oxide to ethylene glycol at 298 K and 1 bar if the initial molar ratio of ethylene oxide to water is 3.0.

$C_2H_4O_{(g)} + H_2O_{(l)} = (CH_2OH)_2$ (1 M aq solution) $\Delta G^o_T = -7824\ J/mole$

To simplify the calculations, assume that the gas phase is an ideal gas mixture, that $\gamma_w = 1.0$, and that the shortcut K value is applicable for ethylene oxide and ethylene glycol.

P14.10 Acetic acid vapor dimerizes according to $2A_1 = A_2$. Assume that no higher-order associations occur. Supposing that a value for Ka is available, and that the monomers and dimers behave as an ideal gas, derive an expression for y_{A1} in terms of P and Ka. Then develop an expression for $P\underline{V}/n_oRT$ in terms of y_{A1}, where n_o is the superficial number of moles neglecting dimerization. Hint: write n_o/n_T in terms of y_{A1} where $n_T = n_1 + n_2$.

14.14   HOMEWORK PROBLEMS

14.1 For their homework assignment three students, Julie, John, and Jacob, were working on the formation of ammonia. The feed is in stoichiometric ratio of nitrogen and hydrogen at a particular T and P.

Julie, who thought in round numbers of product, wrote:

$$\frac{1}{2}N_2 + \frac{3}{2}H_2 = NH_3$$

John, who thought in round numbers of nitrogen, wrote:

$$N_2 + 3H_2 = 2NH_3$$

Jacob, who thought in round numbers of hydrogen, wrote:

$$\frac{1}{3}N_2 + H_2 = \frac{2}{3}NH_3$$

(a) How will John's and Jacob's standard state Gibbs energy of reactions compare to Julie's?
(b) How will John's and Jacob's equilibrium constants be related to Julie's?
(c) How will John's and Jacob's equilibrium compositions be related to Julie's?
(d) How will John's and Jacob's reaction coordinate values be related to Julie's?

14.2 The simple statement of the LeChatelier principle leads one to expect that if the concentration of a reactant were increased, the reaction would proceed so as to consume the added reactant. Nevertheless, consider the gas-phase reaction, $N_2 + 3H_2 = 2NH_3$ equilibrated with excess N_2 such that N_2's equilibrium mole fraction is 0.55. Does adding more N_2 to the equilibrated mixture result in more NH_3? Why?

14.3 The production of NO by the direct oxidation of nitrogen occurs naturally in internal combustion engines. This reaction is also used to produce nitric oxide commercially in electric arcs in the Berkeland-Eyde process. If air is used as the feed, compute the equilibrium conversion of oxygen at 1 bar total pressure over the temperature range of 1300 to 1500°C. Air contains 21 mol% oxygen and 79% N_2.

14.4 The following reaction reaches equilibrium at the specified conditions.

$$C_6H_5CH = CH_{2(g)} + H_{2(g)} = C_6H_5C_2H_{5(g)}$$

The system initially contains 3 mol H_2 for each mole of styrene. Assume ideal gases. For styrene, $\Delta G^\circ_{f,298} = 213.18$ kJ/mol, $\Delta H^\circ_{f,298} = 147.36$ kJ/mol

(a) What is K_a at 600°C?
(b) What are the equilibrium mole fractions at 600°C and 1 bar?
(c) What are the equilibrium mole fractions at 600°C and 2 bar?

14.5 For the cracking reaction,

$$C_3H_{8(g)} = C_2H_{4(g)} + CH_{4(g)}$$

the equilibrium conversion is negligible at room temperature but becomes appreciable at temperatures above 500 K. For a pressure of 1 bar, neglecting any side reactions, determine:

(a) the temperature where the conversion is 75%. [Hint: conversion = amount reacted/amount fed. Relate ξ to the conversion.]

(b) the fractional conversion which would be obtained at 600 K if the feed to a reactor is 50 mol% propane and 50 mol% nitrogen (inert). (Consider the reaction to proceed to equilibrium.)

14.6 Ethanol can be manufactured by the vapor phase hydration of ethylene according to the reaction: $C_2H_4 + H_2O = C_2H_5OH$. The feed to a reactor in which the above reaction takes place is a gas mixture containing 25 mol% ethylene and 75 mol% steam.

(a) What is the value of the equilibrium constant, K_a, at 125°C and 1 bar?

(b) Provide an expression to relate K_a to ξ. Solve for ξ.

14.7 Ethylene is a valuable feedstock for many chemical processes. In future years, when petroleum is not as readily available, ethylene may be produced by dehydration of ethanol. Ethanol may be readily obtained by fermentation of biomass.

(a) What percentage of a pure ethanol feed stream will react at 150°C and 1 bar if equilibrium conversion is achieved?

(b) If the feed stream is 50 mol% ethanol and 50 mol% N_2, what is the maximum conversion of ethanol at 150°C and 1 bar?

14.8 The catalyzed methanol synthesis reaction, $CO_{(g)} + 2H_{2(g)} = CH_3OH_{(g)}$ is to be conducted by introducing equimolar feed at 200°C. What are the mole fractions and the temperature at the outlet if the system is adiabatic at 10 bar and the catalyst provides equilibrium conversion without any competing reactions?

14.9 A gas stream composed of 15 mol% SO_2, 20 mol% O_2, and 65 mol% N_2 enters a catalytic reactor operating at 480°C and 2 bar.

(a) Determine the equilibrium conversion of SO_2.

(b) Determine the heat transfer required per mole of reactor feed entering at 295 K and 1 bar.

14.10 Consider problem 4.16a. Determine the amount of fuel required per mole of air if the fuel is modeled as isooctane and combustion is complete.

14.11 Consider problem 4.18a. Determine the amount of fuel required per mole of air if the fuel is modeled as isooctane and combustion is complete.

14.12 In the event of an explosive combustion of vapor at atmospheric pressure, the vapor cloud can be modeled as adiabatic because the combustion occurs so rapidly. The vapor cloud expands rapidly due to the increase in moles due to combustion, but also due to the adiabatic temperature rise. Estimate the volume increase of a 22°C, 1 m^3 mixture of propane and a stoichiometric quantity of air that burns explosively to completion. Estimate the temperature rise.

14.13 (a) Derive the energy balance for a closed, constant-volume, adiabatic-system vapor phase chemical reaction, neglecting the energy of mixing for reactants and products, and assuming the ideal gas law.

(b) Suppose that a 200 L propane tank is at 0.09 MPa pressure and, due to an air leak, contains the propane with a stoichiometric quantity of air. If a source of spark is present, the system will burn so rapidly that it may be considered adiabatic, and there will not be time for any flow out of the vessel. If ignited at 20°C, what pressure and temperature are generated assuming this is a constant volume system?

14.14 The feed gas to a methanol synthesis reactor is composed of 75 mol% H_2, 12 mol% CO, 8 mol% CO_2, and 5 mol% N_2. The system comes to equilibrium at 550 K and 100 bar with respect to the following reactions:

$$2H_{2(g)} + CO_{(g)} = CH_3OH_{(g)}$$

$$H_{2(g)} + CO_{2(g)} = CO_{(g)} + H_2O_{(g)}$$

Assuming ideal gases, derive the equations that would be solved simultaneously for ξ_1, ξ_2 where 1 refers to the first reaction listed. Provide numerical values for the equilibrium constants. Determine ξ_1 and ξ_2 ignoring any other reactions.

14.15 The 10/25/93 issue of *Chemical and Engineering News* suggests that the thermodynamic equilibrium in the isomerization of *n*-butene (CH3CH = CHCH3, a mix of cis and trans isomers) is reached at a temperature of 350°C using a zeolite catalyst. The products are isobutene and 1-butene ($CH_2 = CHCH_2CH_3$). The isobutene is the desired product, for further reaction to MTBE. Determine the equilibrium composition of this product stream at 1 bar.

14.16 Acrylic acid is produced from propylene by the following gas phase reaction:

$$C_3H_6 + \frac{3}{2}O_2 = CH_2CHCOOH + H_2O$$

A significant side reaction is the formation of acetic acid:

$$C_3H_6 + \frac{5}{2}O_2 = CH_3COOH + CO_2 + H_2O$$

The reactions are carried out at 310°C and 4 bar pressure using a catalyst and air as an oxidant. Steam is added in the ratio 8:1 steam to propylene to stabilize the heat of reaction. If 50% excess air is used (sufficient air so that 50% more oxygen is present than is needed for all the propylene to react by the first reaction), calculate the equilibrium composition of the reactor effluent.

14.17 (a) As part of a methanol synthesis process like problem 14.14, one side reaction that can have an especially unfavorable impact on the catalyst is coke formation. As a first approximation of whether coke (carbon) formation would be significant, estimate the equilibrium extent of coke formation based solely on the reaction: $CO + H_2 = Coke + H_2O$. Conditions for the reaction are 600 K and 100 bar.

(b) Is coke formation by the reaction from part (a) expected at the conditions cited in problem 14.14?

14.18 Hydrogen gas can be produced by the following reactions between propane and steam in the presence of a nickel catalyst:

$$C_3H_8 + 3H_2O = 3CO + 7H_2$$

$$CO + H_2O = CO_2 + H_2$$

Neglecting any other competing reactions:

(a) Compute the equilibrium constants at 700 K and 750 K.
(b) What is the equilibrium composition of the product gas if the inlet to the catalytic reactor is propane and steam in a 1:5 molar ratio at each of the temperatures?

14.19 Write and balance the chemical reaction of carbon monoxide forming solid carbon and carbon dioxide vapor. Determine the equilibrium constant at 700 and 750 K. Will solid carbon form at the conditions of problem 14.18?

14.20 Catalytic converters on automobiles are designed to minimize the NO and CO emissions derived from the engine exhaust. They generally operate between 400–600°C and 1 bar of pressure. K.C. Taylor (*Cat. Rev. Sci. Eng.*, 35:457, 1993) gives the following compositions (in ppm, molar basis) for typical exhaust from the engine:

NO	CO	O_2	CO_2	N_2	H_2	H_2O	hydrocarbons(~propane)
1050	6800	5100	135000	724000	2300	125000	750

The additional products of the effluent stream include NO_2, N_2O, N_2O_4, and NH_3. Estimate the compositions of all species at each temperature {400°C, 500°C, 600°C} and plot the ratio of NH_3/CO as a function of temperature. (Note: Use the options of the Solver software to set the precision of the results as high as possible.)

14.21 Styrene can be hydrogenated to ethyl benzene at moderate conditions in both the liquid and the gas phases. Calculate the equilibrium compositions in the vapor and liquid phases of hydrogen, styrene, and ethyl benzene at each of the following conditions:

(a) 3 bar pressure and 298 K, with a starting mole ratio of hydrogen to styrene of 2:1
(b) 3 bar pressure and 423 K, with a starting mole ratio of hydrogen to styrene of 2:1
(c) 3 bar pressure and 600 K, with a starting mole ratio of hydrogen to styrene of 2:1

14.22 Habenicht et al. (*Ind. Eng. Chem. Res.*, 34:3784, 1995) report on the reaction of t-butyl alcohol (TBA) and ethanol (EtOH) to form ethyltertiary-butyl ether (ETBE). The reaction is conducted at 170°C. A typical feed stream composition (in mole fraction) is:

TBA	EtOH	H_2O
0.027	0.832	0.141

Isobutene is the only significant by-product. Assuming that equilibrium is reached in the outlet stream, estimate the minimum pressure at which the reaction must be conducted in order to maintain everything in the liquid phase. Do isobutene or ETBE exceed their liquid solubility limits at the outlet conditions?

14.23 Limestone ($CaCO_3$) decomposes upon heating to yield quicklime (CaO) and carbon dioxide. At what temperature does limestone exert a decomposition pressure of 1 bar?

14.24 Two-tenths of a gram of $CaCO_{3(s)}$ is placed in a $100 = cm^3$ pressure vessel. The vessel is evacuated of all vapor at 298 K and sealed. The reaction $CaCO_{3(s)} = CaO_{(s)} + CO_{2(g)}$ occurs as the temperature is raised. At what temperature will the conversion of $CaCO_3$ be 50%, and what will the pressure be?

14.25 Plot the "apparent molality" of Cl_2 in solution against the partial pressure of Cl_2. The apparent molality is the sum of all Cl species in solution (Cl_2 counts twice) divided by 2 (to put it on a Cl_2 basis). Compare your plot to the experimental data of Whitney and Vivian (1941).[1]

14.26 Suppose 0.1 mol of CO_2 were mixed with 0.9 mol of Cl_2 and 1 liter of water. What would be the concentrations of the aqueous species and the mole fractions in the vapor phase at 0.8 atm in that case?

1. *Ind. Eng. Chem.* 33:741, (1941).

MOLECULAR ASSOCIATION AND SOLVATION

The satisfaction and good fortune of the scientist lie not in the peace of possessing knowledge, but in the toil of continually increasing it.

Max Planck

When specific chemical forces act between molecules, there is a possibility of complex formation. Chapter 14 dealt with systems where the interactions were strong enough to create new molecules. There are actually chemical interactions that are much weaker that affect solution thermodynamics where chemical bonds are not broken, but complexes may form. The complexes usually cannot be isolated, but their existence is certain from measurements such as spectroscopic studies. Hydrogen bonding is an example of this type of behavior, as well as Lewis acid/base interactions. When complexation occurs between molecules that are all from the same component, the phenomenon is called *association*. For example, acetic acid dimerizes in pure solutions. When complexation occurs between molecules that are from different components, the phenomenon is called *solvation*.

15.1 ASSOCIATION AND SOLVATION

Hydrogen Bonding

Hydrogen bonding is the most common phenomenon leading to association or solvation. As an example, consider the vapor phase compressibility factor of acetic acid. Table 15.1 shows that, even though the pressure is very low for the saturated vapor at low temperature, the vapor compressibility factor of acetic acid deviates considerably from unity, when in fact, we would expect the ideal gas law to hold and $Z = 1$. For comparison, benzene under the same conditions is very nearly an ideal gas.

Table 15.1 *P-V-T evidence of association from the compressibility factors of saturated vapors.*

Acetic Acid			Benzene		
T	P^{sat}(mmHg)	Z	T	P^{sat}(mmHg)	Z
293	11.8	0.507	290	65.78	0.994
323	56.20	0.540	320	241.52	0.989
373	416.50	0.586	370	1240.60	0.962

In the case of acetic acid, the small compressibility factor is due to the dimerization of the carboxylic acid. Even though the pressure is very low, the compressibility factor approaches 0.5 because the number of molecules in the vapor is actually half the superficial amount, owing to the dimerization.[1] Note that the compressibility factor approaches unity as the temperature is increased. Why should the conversion to dimer be more complete at low temperature? Because association is an exothermic reaction and the van't Hoff relation clearly dictates that conversion of exothermic reactions decreases with increasing temperature. Another implication of associating fluids relative to non-associating fluids is that higher temperatures are needed to make the associating network break into a gas, and this means that T_c for an associating component will be significantly higher than a non-associating component of similar structure (e.g., H_2O vs. CH_4). The chemical association of acetic acid during dimerization is illustrated in Fig. 15.1. Note that the structure forms two hydrogen bonds simultaneously which makes the dimerization quite strong. Note also that a property of hydrogen bonding is that the O-H-O bond angle is nearly linear. This means that the carboxylic acid structure is close to a six-sided ring, not an eight-sided ring.

Dimerization of carboxylic acids is very common. In fact, the trend in LCEP discussed in Section 13 shows that dimerization is present even in long-chain fatty acids. Another commonly cited example of association is HF vapor, important in the manufacture of alternative refrigerants, which is modeled as $(HF)_n$ with n predominately 6. Water and alcohols are extremely common substances in the chemical process industry and both exhibit association effects.

Hydrogen bonding is also common between different species. The hydrogen on a highly chlorinated molecule is easily hydrogen-bonded to a Lewis base, such as a carbonyl, ether, or amine group. Well-known examples are chloroform + acetone, chloroform + diethyl ether, and acetylene + acetone, as depicted in Fig. 15.2.

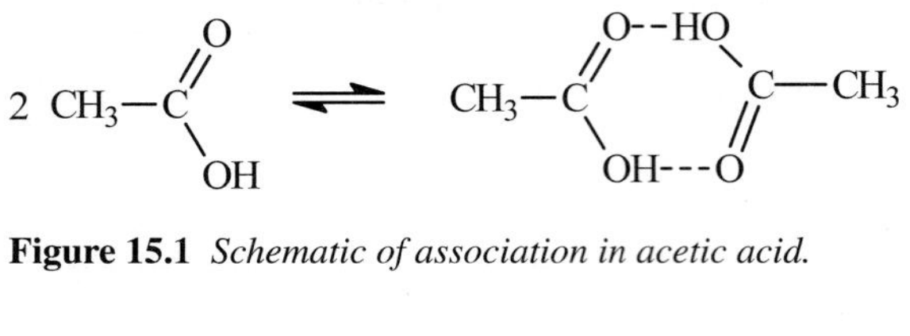

Figure 15.1 *Schematic of association in acetic acid.*

---

1. The superficial number of moles is given by the species mass divided by the molecular weight of the monomer. On a superficial basis, we ignore the effect of association to dimers, timers, etc.

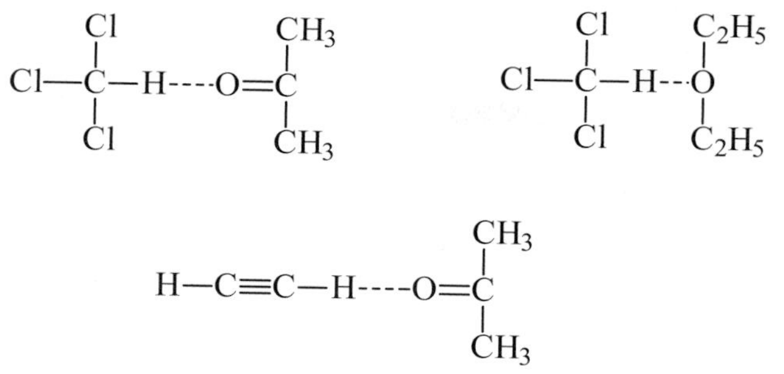

Figure 15.2 *Schematic of solvation in several systems where association is negligible.*

Hydrogen bonding in solvating systems can enhance miscibility. For example, water and triethylamine are miscible in all proportions below 18°C (and above 0°C where water freezes). However, above 18-19°C, the increased thermal energy breaks the hydrogen bonds resulting in an immiscibility region which increases with increasing temperature[1] until the hydrogen bonding energy becomes too weak to influence the phase behavior.

A measure of the relative acidity/basicity is available from spectroscopic measurements tabulated in the Kamlet-Taft acidity/basicity parameters. The Kamlet-Taft parameters are determined by comparing spectroscopic behavior of probe molecules in the solvents of interest. A small table of parameters appears in Table 15.2.[2]

Table 15.2 *Kamlet-Taft acidity/basicity parameters. The α parameter describes the molecule's ability to donate a proton. The β scale provides a measure of the molecule's ability to donate an electron pair. Values in parenthesis are known with less certainty.*

Compound	β	α	Compound	β	α
n-hexane, *n*-heptane	0.00	0.00	chlorobenzene	0.07	0.00
diisopropyl ether	0.49	0.00	acetone	0.48	0.06
diethyl ether	0.47	0.00	ethyl acetate	0.45	0.00
tetrahydrofuran	0.55	0.00	chloroform	0.00	(0.44)
acetic acid		1.12	methylene chloride	0.00	(0.30)
triethylamine	0.71	0.00	2-propanol	(0.95)	0.76
nitrobenzene	0.39	0.00	ethanol	(0.77)	0.83
nitromethane		0.22	methanol	(0.62)	0.93
pyridine	0.64	0.00	*n*-butanol	(0.88)	0.79
toluene	0.11	0.00	water	(0.18)	1.17

1. B. J. Hales, G. L. Bertrand, L. G. Hepler, *J. Phys. Chem.*, 70:3970(1966).
2. M. J. Kamlet, J.-L. M. Abboud, M. H. Abraham, R. W. Taft, *J. Org. Chem.*, 48:2877(1983). For solutes in liquids see Abraham, M. H.; Andonian-Haftvan, J.; Whiting, G. S.; Leo, A.; Taft, R. W., *J. Chem. Soc. Perkin Trans.* 2:1777(1994).

Charge-Transfer Complexes

Complexes can also form between one component that is electron-rich (loosely bound electron) and one that is strongly electron-attracting (low energy vacant orbital). For example, common donors are benzene rings with electron-donating groups like -OH, -OCH$_3$, -N(CH$_3$)$_2$. Common acceptors are compounds with several nitro groups (1,3,5-trinitro benzene), quinones, and compounds with several CN groups (tetracyanoethane, 2,3-dicyano-1,4-benzoquinone). For example, nitrobenzene and mesitylene form a complex when they are mixed as shown in Fig. 15.3. Tetrahydrofuran (THF) and toluene form a charge transfer complex when mixed together. When a charge-transfer complex forms, the energy typically corresponds to about 5 kJ/mole, and the strength is the same order of magnitude as a hydrogen bond.

Perspective

As might be expected, nature provides a spectrum of complex strengths. There is no clear distinction of acidity/basicity values that are "required" for complex formation in solvated systems. Associating systems also cannot be quantitatively characterized for complex strength based solely on the chemical structure or component properties. Only qualitative or semi-quantitative comparisons can be made with other systems at the current time. Weak complex formation can often be completely ignored in thermodynamic models without serious consequences, however systems like carboxylic acids always require representation of complexation. The process design engineer must decide the importance of including the complexation in the particular application.

Spectroscopic data are the best source for characterizing complex formation and further information is available.[1] Solid-liquid phase boundaries are indicative of strong complex formation, and may be used to infer complex stoichiometries. For example, the phase diagram for NH$_3$ + water in Fig. 13.14 shows two complexes that form in the solid phase, and might be expected to appear in fluid phases in addition to ionic species. Acetone and chloroform show a 1:1 compound in the solid phase.[2]

In a generalized binary solution,[3] the ith complex can be represented by the general form $A_{a_i}B_{b_i}$, where the values of a_i and b_i are integers which will depend on the particular system (note

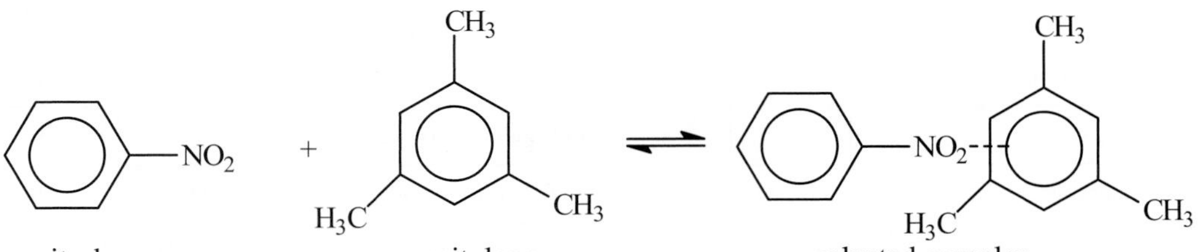

| nitrobenzene | mesitylene | solvated complex |

Figure 15.3 *Schematic of charge-transfer complexation.*

1. J. M. Prausnitz, R. N. Lichtenthaler, E. G. Azevedo, *Molecular Thermodynamics of Fluid-Phase Equilibria,* 2nd ed., Chapters 4 and 7, Prentice-Hall, 1986.
2. A. N. Campbell, E. M. Kartzmark, *Can. J. Chem.,* 38:652(1960)
3. We restrict the discussion to binaries simply because the notation becomes unwieldy in multicomponent solutions, and the reader should recognize that the concepts and proofs extend to additional components.

that for an associated specie, either a_i or b_i is zero, but that won't affect the proof). The integers a_i and b_i are the stoichiometric coefficients,

$$a_i A + b_i B = A_{a_i} B_{b_i}$$

15.1

For example, a hypothetical system is shown in Fig. 15.4 that exhibits both association and solvation where the components A and B are added to the solution in quantities n_A and n_B.

The concentrations $x_A = \dfrac{n_A}{n_A + n_B}$ and $x_B = \dfrac{n_B}{n_A + n_B}$ are the mole fractions that are experimentally important for macroscopic characterization and are the conventional mole fractions. Since these mole fractions do not represent the true species in solution, they are also called the *superficial* mole fractions. The mole fractions in the actual solution are called the *true* mole fractions, and also

denoted by x's, e.g., $x_{A_1} = \dfrac{n_{A_1}}{n_{A_1} + n_{A_2} + n_{B_1} + n_{AB_2}}$, $x_{B_1} = \dfrac{n_{B_1}}{n_{A_1} + n_{A_2} + n_{B_1} + n_{AB_2}}$. Note that x_A

(the superficial mole fraction) is not the same as x_{A_1} (the A monomer mole fraction). Because the

monomer mole fraction will end up being so important in later proofs, we give the special subscript M to help distinguish this as the monomer in the true mixture. Therefore $x_{A_1} \equiv x_{AM}$, and

$x_{B_1} \equiv x_{BM}$. The true species that are present in solution will usually be inferred from fitting experimental data, and the true mole fractions are usually *modeled* quantities rather than experimental quantities since they are subject to the assumptions of the model. The nomenclature is nevertheless the established convention in the literature. Note that the implication that the mixture of this discussion is a *liquid* mixture is not restrictive. The same balances and notation will be used to refer to vapor mixtures, but y's will be substituted for the x's.

❗ The subscript M is used for monomer.

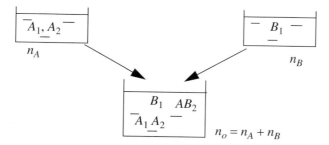

Figure 15.4 *Illustration of mixing of species A which associates with species B to form a solution with both association and solvation. The labels in the containers indicate the true species present, not the relative concentrations. The species have been chosen for illustrative purposes.*

15.2   EQUILIBRIUM CRITERIA

Mole Balance

One may wonder how quantification of the phenomena can be approached in a generalized fashion, but the criteria are presented clearly by Prigogine and Defay (1954) whose proof we reproduce here with modified notation. The first balances that must be satisfied are the material balances. For a solution created from n_A moles of A, and n_B moles of B,

$$n_A = \sum_i a_i n_{A_{a_i} B_{b_i}} \qquad n_B = \sum_i b_i n_{A_{a_i} B_{b_i}}$$

15.2

where the summations are over all i true species found in the solution. For example, in the binary solution of Fig. 15.4, Eqn. 15.2 becomes

$$n_A = 1 n_{AM} + 0 n_{BM} + 2 n_{A_2} + 1 n_{AB_2}$$

$$n_B = 0 n_{AM} + 1 n_{BM} + 0 n_{A_2} + 2 n_{AB_2}$$

Chemical Potential Criteria

The chemical potentials of the true species are designated by μ_{AM}, μ_{BM}, $\mu_{A_{a_i} B_{b_i}}$ etc., and the superficial chemical potentials by μ_A and μ_B. Applying the principles of chemical equilibria to Eqn. 15.1, we find

$$\mu_{A_{a_i} B_{b_i}} = a_i \mu_{AM} + b_i \mu_{BM}$$

15.3

If the total differential of G (Eqn. 14.7) is evaluated at constant T and P, allowing the species to come to equilibria,

$$dG = 0 = \sum_i \mu_{A_{a_i} B_{b_i}} dn_{A_{a_i} B_{b_i}}$$

we find by incorporating Eqn. 15.3 and the differential of Eqn. 15.2

$$0 = \sum_i a_i \mu_{AM} dn_{A_{a_i} B_{b_i}} + \sum_i b_i \mu_{BM} dn_{A_{a_i} B_{b_i}}$$

$$0 = \mu_{AM} dn_A + \mu_{BM} dn_B$$

15.4

On the other hand, for any binary solution at constant P, T, according to the superficial components

$$dG = 0 = \mu_A dn_A + \mu_B dn_B$$

15.5

By comparing 15.4 and 15.5 we conclude

$$\boxed{\mu_A = \mu_{AM} \qquad \mu_B = \mu_{BM}}$$

15.6

Therefore, the superficial (conventional) chemical potential is quantified by a model that calculates the chemical potential of the true monomer species. It should be noted that this proof is independent of the number or stoichiometry of species that are formed in solution.

Fugacity Criteria

The chemical potential criteria may be extended to fugacity. For the superficial chemical potential,

$$\mu_A = \mu_A^o + RT\ln\frac{\hat{f}_A}{f_A^o} \tag{15.7}$$

and for the monomer

$$\mu_{AM} = \mu_{AM}^o + RT\ln\frac{\hat{f}_{AM}}{\hat{f}_{AM}^o} \tag{15.8}$$

Note that for a species that associates the standard state for the monomer is a mixture state since, even when A is pure, there is a mixture of true associated species. Applying Eqn. 15.6

$$\mu_A^o + RT\ln\frac{\hat{f}_A}{f_A^o} = \mu_{AM}^o + RT\ln\frac{\hat{f}_{AM}}{\hat{f}_{AM}^o} \tag{15.9}$$

where the standard state is pure component A.

In the event that component A does not associate, the true solution is completely a monomer when A is pure, and

$$\mu_A^o = \mu_{AM}^o \qquad f_A^o = \hat{f}_{AM}^o = \hat{f}_{AM} \tag{15.10}$$

which leads to

$$\boxed{\hat{f}_A = \hat{f}_{AM}} \tag{15.11}$$

The situation when A associates is slightly more complex. Now, recognizing that the superficial state is one that neglects complexation, the chemical potential of monomer μ_{AM} in Eqn. 15.8 can be calculated

$$\int_{\text{pure monomer}}^{\text{pure component } A} d\mu_{AM} = RT\int_{\text{pure monomer}}^{\text{pure component } A} d\ln\hat{f}_{AM}$$

where the lower limit is the hypothetical state of pure monomer, and the upper limit is the monomer state that actually exists in a pure associating solution of A. Integrating both sides, and recognizing the lower limit of each integral as the superficial standard state

$$\mu_{AM}^o - \mu_A^o = RT\ln\frac{\hat{f}_{AM}^o}{f_A^o}$$

Combining with Eqn. 15.9 again results in Eqn. 15.11. Note that a parallel proof would show

$$\boxed{\hat{f}_B = \hat{f}_{BM}}$$

15.12

Activity Criteria

Returning to Eqn. 15.7, it can be rewritten in terms of the superficial activity and activity coefficient.

$$\mu_A = \mu^\circ_A + RT\ln\frac{x_A\gamma_A f^\circ_A}{f^\circ_A} = \mu^\circ_A + RT\ln x_A\gamma_A$$

15.13

Defining an activity coefficient, α_{AM}, of the true monomer species, the chemical potential is

$$\mu_{AM} = \mu^\circ_{AM} + RT\ln\frac{x_{AM}\alpha_{AM}\hat{f}^\circ_{AM}}{\hat{f}^\circ_{AM}} = \mu^\circ_{AM} + RT\ln x_{AM}\alpha_{AM}$$

15.14

Using Eqn. 15.6 to equate 15.13 and 15.14

$$x_A\gamma_A = x_{AM}\alpha_{AM}\exp\left(\frac{\mu^\circ_{AM} - \mu^\circ_A}{RT}\right)$$

15.15

where the exponential term is a constant at a given temperature. The symmetrical convention of superficial activity requires $\lim_{x_A \to 1} x_A\gamma_A = 1$. For a non-associating species, the exponential term of Eqn. 15.15 is unity by Eqn. 15.10, and thus $\lim_{x_A \to 1} \alpha_{AM} = 1$, $\lim_{x_A \to 1} x_{AM} = 1$. For an associating species $\lim_{x_A \to 1} x_{AM} = x^\circ_{AM}$ which is the true mole fraction of monomer in pure A, which is not unity. Therefore the exponential term is simply the reciprocal of the limiting value of the monomer activity $\exp\left(-\frac{(\mu^\circ_{A1} - \mu^\circ_A)}{RT}\right) = \lim_{x_A \to 1} x_{AM}\alpha_{AM} = (x^\circ_{AM}\alpha^\circ_{AM})$. Therefore we write

$$\boxed{x_A\gamma_A = \frac{x_{AM}\alpha_{AM}}{(x^\circ_{AM}\alpha^\circ_{AM})}}$$

15.16

where $x^\circ_{AM}\alpha^\circ_{AM} = 1$ for a non-associating component. Note that a parallel proof would show that

$$\boxed{x_B\gamma_B = \frac{x_{BM}\alpha_{BM}}{(x^\circ_{BM}\alpha^\circ_{BM})}}$$

15.17

15.3 BALANCE EQUATIONS

The balance equations to be solved take the same form for both vapors and liquids. The liquid equations will be shown, and the reader should recognize the vapor equations by analogy. First, the true mole fractions must sum to unity,

$$\boxed{\sum_i x_i = 1}$$ 15.18

In a binary system, a balance equation can be written for either component to match the superficial mole fraction,

$$x_A = \frac{\text{moles of all A in compounds}}{\text{moles of all A and B in compounds}} = \frac{\sum_i a_i n_i}{\sum_i (a_i + b_i) n_i}$$

Dividing numerator and denominator by the true number of moles, n_T,

$$x_A = \frac{\sum_i a_i x_i}{\sum_i (a_i + b_i) x_i}$$ 15.19

Rearranging Eqn. 15.19,

$$\sum_i x_A (a_i + b_i) x_i - \sum_i a_i x_i = 0$$

or

$$\boxed{\sum_i (b_i x_A - a_i x_B) x_i = 0}$$ 15.20

15.4   IDEAL CHEMICAL THEORY

The simplest method of modeling complex behavior is to neglect the non-idealities by modeling a vapor phase as an ideal gas mixture including the complexes, and model a liquid phase as an ideal solution containing complexes. Using these methods, the true fugacity coefficients become unity for all vapor species. When modeling liquid phases, the true activity coefficients, α_i's, become unity for all liquid species. The approach is called Ideal Chemical Theory. Modeling complex formation with ideal chemical theory, Eqn. 15.1 can be expressed in terms of an equilibrium constant, $x_i = K_{a,i} x_{AM}^{a_i} x_{BM}^{b_i}$. Plugging into Eqns. 15.18 and 15.20, the equations to be solved are obtained:

$$\boxed{\sum_i K_{a,i} x_{AM}^{a_i} x_{BM}^{b_i} - 1 = 0}$$ 15.21

$$\boxed{\sum_i (b_i x_A - a_i x_B) K_{a,i} x_{AM}^{a_i} x_{BM}^{b_i} = 0}$$ 15.22

where for the monomers, $K_{a,i} = 1$. Once the $K_{a,i}$ are known, then x_{AM} and x_{BM} can be determined by solving these two equations. Subsequently, all other x_i and γ_A, γ_B can be calculated. If γ_A, γ_B are known from experiment, and the complex stoichiometry is known, $K_{a,i}$ values can be adjusted to fit the data using optimization methods. FORTRAN code is provided for solving for the true species for given values of $K_{a,i}$ in the program IChemT.exe.

IChemT.exe

For ideal chemical theory applied to the vapor phase, the x_i are replaced with y_i and Eqn. 15.1 is expressed as $y_i = K_{a,i} P^{(a_i + b_i - 1)} y_{AM}^{a_i} y_{BM}^{b_i}$. Eqns. 15.18 and 15.20 then become

$$\sum_i K_{a,i} P^{(a_i + b_i - 1)} y_{AM}^{a_i} y_{BM}^{b_i} - 1 = 0 \qquad \text{(ig) 15.23}$$

$$\sum_i (b_i x_A - a_i x_B) K_{a,i} P^{(a_i + b_i - 1)} y_{AM}^{a_i} y_{BM}^{b_i} = 0 \qquad \text{(ig) 15.24}$$

These equations are marked as ideal gas equations since they are ideal gas equations from the perspective of the true solution. As with the liquid-phase calculation, if the $K_{a,i}$ values are known, y_{AM} and y_{BM} can be determined. The code IChemT.exe can also be used if values for $K_{a,i} P^{(a_i + b_i - 1)}$ are used when prompted for $K_{a,i}$.

Example 15.1 Compressibility factors in associating/solvating systems

Derive a formula to relate the true mole fractions to the compressibility factor of a vapor phase where the true species follow the ideal gas law.

Solution: A vessel of volume $\underline{V}$ holds n_o superficial moles. However, experimentally, in the same total volume, there would be a smaller number of true moles n_T. Applying the ideal gas law,

$$P\underline{V} = n_T RT \qquad \frac{P\underline{V}}{n_T RT} = 1 \qquad \text{(ig) 15.25}$$

Experimentally, we wish to work in terms of the superficial number of moles,

$$Z = \frac{P\underline{V}}{n_o RT} = \frac{P\underline{V}}{n_T RT} \cdot \frac{n_T}{n_o} = \frac{n_T}{n_o} \qquad \text{(ig)}$$

Example 15.1 Compressibility factors in associating/solvating systems (Continued)

Note that this equation is labelled as an ideal gas equation because the true species follow the ideal gas law, even though from the perspective of the superficial species, the ideal gas law will not be followed. From the total mole balances, $n_T = \sum_i n_i$, and $n_o = \sum_i (a_i + b_i)n_i$, therefore

$$Z = \frac{\sum_i n_i}{\sum_i (a_i + b_i)n_i}$$. Dividing numerator and denominator by n_T.

$$Z = \frac{\sum_i y_i}{\sum_i (a_i + b_i)y_i} = \frac{1}{\sum_i (a_i + b_i)y_i}$$ (ig) 15.26

Therefore, once the true mole fractions have been determined, the compressibility factor can be calculated.

Example 15.2 Dimerization of carboxylic acids

P-V-T measurements of acetic and propionic acid vapors are available.[1] The equilibrium constants for acetic and propionic acids at 40°C are 375 bar^{-1} and 600 bar^{-1} respectively. At a pressure of 0.016 bar, determine the true mole fractions, the compressibility factor, and the fugacity coefficients.

Solution: Beginning with Eqn. 15.23, letting A be the acid of interest,

$$y_{AM} + K_{a,A_2}Py_{AM}^2 - 1 = 0$$

Eqn. 15.24 is not required since the system is a single component. This simple equation can be solved with the quadratic formula,

$$y_{AM} = \frac{-1 + \sqrt{1 + 4K_{a,A_2}P}}{2K_{a,A_2}P}$$

At $P = 0.016$ bar, for acetic acid, $y_{AM} = 0.333$, $y_{A_2} = 0.667$, and even at this low pressure, Eqn. 15.26 gives $Z = 0.6$. For the fugacity coefficient, starting with Eqn. 15.11,

$$y_A\hat{\phi}_A P = y_{AM}\hat{\phi}_{AM}P$$

---

1. McDougall, F. H., *J. Amer.Chem.Soc.*, 58:2585(1936), 63:3420(1941).

Example 15.2 Dimerization of carboxylic acids (Continued)

Since the system is pure, $y_A = 1$ and the fugacity coefficient on the left-hand side will be for a pure specie. Since the model uses ideal chemical theory, $\hat{\varphi}_{AM} = 1$. Therefore

$$\varphi_A = y_{AM} = 0.333$$

The same procedure can be repeated for propionic acid, however it will be even more non-ideal. The answers are: $y_{AM} = 0.275$, $Z = 0.580$, $\varphi_A = 0.275$.

Example 15.3 Activity coefficients in a solvated system

1,4-dioxane (component B) is a cyclic 6-member ring, $C_4H_8O_2$, with oxygens in the 1 and 4 positions. When mixed with chloroform (component A) the oxygens provide solvation sites for the hydrogen on chloroform. Since there are two sites on 1,4-dioxane, two complexes are possible, AB and A_2B. McGlashan and Rastogi[1] have studied this system and report the liquid phase can be modeled with ideal chemical theory using $K_{a,\,AB} = 1.11$, $K_{a,\,A_2B} = 1.24$ at 50°C. Calculate the true mole fractions and activity coefficients across the composition range.

Solution: We will use the program IChemT.exe to solve Eqns. 15.21 and 15.22. The monomers will be the first two species in solution. The program provides user information and prompts for the compound information. For the AB compound, $a_i = 1$, $b_i = 1$, $K_{a,i} = 1.11$; for the A_2B compound, $a_i = 2$, $b_i = 1$, $K_{a,i} = 1.24$. Then we enter 0 0 0 to end compound information. For the compositions, we enter increments of 0.05 for the superficial compositions. Near the endpoints, we enter $x_A = 0.001$ and $x_A = 0.999$ to avoid numerical underflows and overflows. The results are displayed on the screen at each composition and written to an output file. This file is easily imported into Excel for analysis and plotting. The activity coefficients are easily calculated using $\gamma_A = x_{AM}/x_A$, $\gamma_B = x_{BM}/x_B$ since neither component exhibits association. The results are shown in Fig. 15.5. Note that the solvation causes negative deviation from Raoult's law. Also note the relation between the complex stoichiometry and the maxima in the complex concentration. Can you rationalize why the infinite dilution activity coefficient of 1,4-dioxane is smaller than the infinite dilution activity coefficient of chloroform?

IChemT.exe

---

2. McGlashan, M. L., Rastogi, R. P., *Trans. Faraday Soc.*, 54:496 (1958).

Example 15.3   Activity coefficients in a solvated system (Continued)

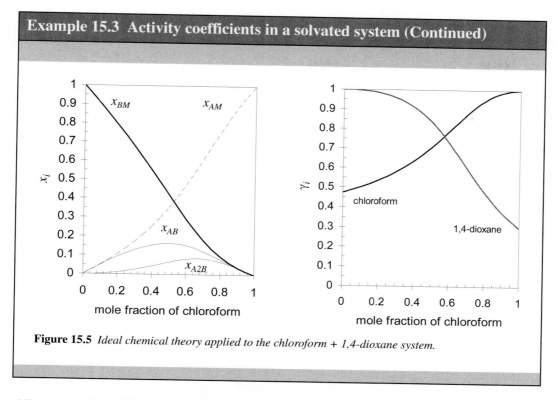

Figure 15.5 *Ideal chemical theory applied to the chloroform + 1,4-dioxane system.*

15.5   CHEMICAL-PHYSICAL THEORY

The assumptions of ideal chemical theory are known to be oversimplifications for many systems and physical interactions must be included. For a liquid phase, the activity coefficients of the true species can be reintroduced. Then $x_i = \dfrac{K_{a,\,i}(x_{AM}\alpha_{AM})^{a_i}(x_{BM}\alpha_{BM})^{b_i}}{\alpha_i}$. Utilizing this result with Eqns. 15.18 and 15.20, the following equations are obtained:

$$\sum_i \frac{K_{a,\,i}(x_{AM}\alpha_{AM})^{a_i}(x_{BM}\alpha_{BM})^{b_i}}{\alpha_i} - 1 = 0 \qquad\qquad 15.27$$

$$\sum_i (b_i x_A - a_i x_B)\frac{K_{a,\,i}(x_{AM}\alpha_{AM})^{a_i}(x_{BM}\alpha_{BM})^{b_i}}{\alpha_i} = 0 \qquad\qquad 15.28$$

where for the monomers, $K_{a,i} = 1$. Since most activity coefficient models require two parameters per pair of molecules, the number of parameters is now becoming large. In addition, any parameters for the complex must be estimated or fit to experiment since the complex cannot isolated. Solution of the equations is more challenging because the true activity coefficients must be updated with each iteration on x_{AM} and x_{BM}.

For chemical-physical theory applied to the vapor phase,

$$iy_i = \frac{K_{a,i} P^{(a_i+b_i-1)} (y_{AM}\hat{\phi}_{AM})^{a_i} (y_{BM}\hat{\phi}_{BM})^b}{\hat{\phi}_i}. \text{ Eqns. 15.18 and 15.20 then become}$$

$$\sum_i \frac{K_{a,i} P^{(a_i+b_i-1)} (y_{AM}\hat{\phi}_{AM})^{a_i} (y_{BM}\hat{\phi}_{BM})^{b_i}}{\hat{\phi}_i} - 1 = 0 \qquad 15.29$$

$$\sum_i (b_i x_A - a_i x_B) \frac{K_{a,i} P^{(a_i+b_i-1)} (y_{AM}\hat{\phi}_{AM})^{a_i} (y_{BM}\hat{\phi}_{BM})^{b_i}}{\hat{\phi}_i} = 0 \qquad 15.30$$

The physical properties of the complex must also be modeled with this approach, and the same challenges for solving the equations are present as discussed above for chemical-physical theory of liquid phases.

An interesting study has been performed by Harris[1] for acetylene in *n*-hexane, butyrolactone, and *n*-methyl pyrrolidone at 25°C. In this study, a simplified van Laar model was used to model the physical deviations, which resulted in one physical parameter. Naturally, the acetylene + *n*-hexane does not exhibit solvation, but the other binaries do, with the pyrrolidone showing the strongest complexation. Further, the *n*-hexane system has positive deviations from Raoult's law across the composition range, the pyrrolidone shows negative deviations, and the lactone shows both positive and negative deviations. All three systems are accurately modeled using two parameters each—one chemical parameter and one physical parameter.

15.6   PURE SPECIES WITH LINEAR ASSOCIATION

Some species like water and alcohols form extended networks in solutions. The hydrogen bonding reactions usually occur so quickly that they can be assumed to be at equilibrium all the time.[2] It is desirable to investigate the implications of a more rigorous approach for pure species. Although the approach at first appears overwhelming, we always have just enough reaction constraints that the Gibbs phase rule works out the same for an associating component as for a single component.

We begin by describing the premise on which the model is based. We assume that association will occur by formation of linear chains of species. Dimers, trimers, tetramers, etc. will all form in a pure fluid, and the chains that form can be of infinite length. This does not require that long chains exist; the assumption simply does not forbid such behavior. Whether such behavior is found will be determined by the resulting model. Further, it should be recognized that the association that we describe is not static. The associations are truly reversible, and undergo frequent formation/decompositions; the equilibrium distributions simply tell us how many of a certain species will exist at a given instant. As one associated complex decomposes, another forms.

1. Harris, H. G., Prausnitz, J. M., *Ind. Eng. Chem. Fundam.*, 8:180 (1969).
2. An exception is the reaction of formaldehyde with water, c. Hasse and Maurer, *Fluid Phase Equilibria* 64:185 (1991).

We need to model the chemical equilibria occurring in a given phase. The chemical formation of a dimer would be represented by

$$A + A = A_2 \qquad K_{a2}$$

where the equilibrium constant is K_{a2}. The use of subscript 2 reminds us that this equilibrium constant is for formation of dimer. We will define the true mole fraction as the mole fraction of each specie (monomer, dimer, etc.) that exists in solution. The superficial composition is the overall mole fraction in solution based on all species being monomer, which for a pure fluid is unity. The monomer and dimer appear very often in the derivations that follow so we also use subscripts M and D, respectively. However, the use of numerical subscripts is also preserved for use in generalized formulas, and occasionally both notation schemes are used within a given equation. From the reaction equilibrium constraint,

$$x_D = x_2 = x_M \left(\frac{x_M P \hat{\varphi}_M^2 K_{a2}}{\hat{\varphi}_D} \right) = x_M x_M \left(\frac{K_{a2} P \hat{\varphi}_M \hat{\varphi}_1}{\hat{\varphi}_2} \right) \qquad 15.31$$

where x_M represents the true mole fraction of monomer and x_D represents the true mole fraction of dimer. Other reactions that can occur include

$$A + A_2 = A_3 \qquad K_{a3} \qquad x_3 = \frac{x_M \hat{\varphi}_M x_2 \hat{\varphi}_2 P K_{a3}}{\hat{\varphi}_3} = x_M x_2 \left(\frac{K_{a3} P \hat{\varphi}_M \hat{\varphi}_2}{\hat{\varphi}_3} \right)$$

$$A + A_3 = A_4 \qquad K_{a4} \qquad x_4 = \frac{x_M \hat{\varphi}_M x_3 \hat{\varphi}_3 P K_{a4}}{\hat{\varphi}_4} = x_M x_3 \left(\frac{K_{a4} P \hat{\varphi}_M \hat{\varphi}_3}{\hat{\varphi}_4} \right)$$

$$A + A_4 = A_5 \qquad K_{a5} \qquad x_5 = \frac{x_M \hat{\varphi}_M x_4 \hat{\varphi}_4 P K_{a5}}{\hat{\varphi}_5} = x_M x_4 \left(\frac{K_{a5} P \hat{\varphi}_M \hat{\varphi}_4}{\hat{\varphi}_5} \right)$$

and linear association is assumed to continue to increasing chain sizes.

Let us *assume* that the ratio of $K_{ai} P \dfrac{\hat{\varphi}_M \hat{\varphi}_{i-1}}{\hat{\varphi}_i}$ is the same for all i. Note that the $\hat{\varphi}_i$ are based on the true mole fractions of oligomers in the mixture not the superficial mole fraction one would obtain by assuming presence of just monomer. Empirical calculations show that this is a reasonable assumption. It basically corresponds to the equilibrium being the same for each step in the oligomerization process. But *it does represent a significant assumption* that may not be accurate for all situations. This ratio appears often in the following derivation so it is helpful to abbreviate it as

$$K_{ai} P \frac{\hat{\varphi}_M \hat{\varphi}_{i-1}}{\hat{\varphi}_i} = \alpha q \qquad 15.32$$

where α and q are variables defined below.[1] The point of this assumption is that it facilitates several substantial simplifications through our material balance relations. Including this assumption, Eqn. 15.31 becomes $x_2 = x_M (x_M \alpha q)$. Substitution of the expression for x_2 into the formula for x_3, yields $x_3 = x_M (x_M \alpha q)^2$. Substitution continues to higher oligomers resulting in the general formula,

$$x_{i+1} = x_M (x_M \alpha q)^i \qquad 15.33$$

Mass Balances

We can use material balances to obtain two simple relations between the true number of moles in the solution, n_T, and the superficial number of moles that we would expect if there was no associa- tion,[2] n_o. Note that n_o is the number of moles one would compute based on dividing the mass of solution by the molecular weight of a monomer as taught in introductory chemistry. For example, in $100 cm^3$ of water one would estimate

$$n_o = 100 \ cm^3 \cdot 1.0 \ g/cm^3/(18 \ g/mole) = 5.556 \ moles$$

But how many moles of H_2O monomer do you think truly exist in that beaker of water? We will return to this question shortly. Note that each i-mer contains "i" monomers, such that the contribu- tion to the superficial number of moles is $i \cdot n_i$. Note also that the true mole fractions, x_i, are given by n_i/n_T, but it may not look so simple at first.

$$n_o = \sum_i i \cdot n_i = n_T \sum_i i \cdot x_i \qquad 15.34$$

Substituting Eqn. 15.33

$$n_o = x_m n_T \sum_i i(\alpha q x_m)^{(i-1)} = x_M n_T [1 + 2(\alpha q x_M) + 3(\alpha q x_M)^2 + 4(\alpha q x_M)^3 + ...]$$

This series may not appear familiar but it is a common converging series. Referring to series formu- las in a math handbook, we find that

$$n_T x_M [1 + 2(\alpha q x_M) + 3(\alpha q x_M)^2 + 4(\alpha q x_M)^3 + ...] = n_T x_M [1/(1 - \alpha q x_M)^2]$$

$$\boxed{\frac{n_o}{n_T} = \frac{x_M}{(1 - x_M \alpha q)^2} = \sum_i i \cdot x_i} \qquad 15.35$$

Since the mole fractions must sum to unity, we can write a second balance, and using Eqn. 15.33 for x_i,

$$1 = \sum_i x_i = x_m [1 + (\alpha q x_M) + (\alpha q x_M)^2 + (\alpha q x_M)^3 + ...] \qquad 15.36$$

---

1. The reason for defining this set of variables in terms of two will become clear later when we show that $q = x_M$. Note that the α term is unrelated to α_{AM} defined by Eqn. 15.14.
2. Here we choose to use subscript o to clearly distinguish the notation for superficial moles, even though it would be the quantity normally reported from a macroscopic experiment.

and again recognizing the series,

$$1 = x_M [1 + (\alpha q x_M) + (\alpha q x_M)^2 + (\alpha q x_M)^3 + \ldots] = x_M [1/(1 - \alpha q x_M)]$$

$$\boxed{x_M = (1 - x_M \alpha q)} \qquad \text{15.37}$$

Substituting x_M for $(1 - x_M \alpha q)$ in Eqn. 15.35 results in,

$$n_o / n_T = x_M / x_M{}^2 = 1/x_M$$

$$\boxed{\frac{n_T}{n_o} = x_M} \qquad \text{15.38}$$

This equation turns out to be extremely important. It makes clear that the properties of the mixture are closely related to the properties of the monomer. The next step in our derivation is to reconsider Eqn. 15.32, this time focusing on the left-hand side. The pressure term and its relation to the equation of state are particularly interesting. Before we take this step, let's work through some of the implications of what we have derived thus far. There are a number of insights that can be gained based simply on the material balances.

Interpretations of Molar Densities

At this stage we get to some of the really confusing aspects of associating fluids. For instance, what is density? Mass density is the number of grams divided by the volume. But what about the molar density? We should take the true number of moles and divide by the volume. The true number of moles in our system depends on the degree of association, $n_T = x_M n_o$. We do not know the number of moles until we know x_M. What we do know is that what we call the superficial molar density, i.e., the molar density that we would get if we divided the mass density by the molecular weight of the monomer. *The superficial molar density is the molar density that would be reported from an experiment* where all species are assumed to be monomer, which is the commonly applied measurement technique. To relate the true molar density and the superficial molar density we seek a relationship between x_M, T, and ρ, where ρ is the superficial density. We develop this relation by means of the equation of state. In order to develop the equation of state, we must make additional assumptions about the combining rules.

$$\boxed{\frac{n_o}{V} = \rho, \text{ superficial molar density, experimentally reported}}$$

$$\frac{n_T}{V} = \rho_T, \text{ true molar density}$$

$$\boxed{\rho_T = \rho \frac{n_T}{n_o} = \rho x_M} \qquad \text{15.39}$$

Combining/Mixing Rules

Before we get into the specifics of developing an equation of state, we can anticipate the need for determining equation of state parameters for the mixture. These relations can be developed before the actual equation of state, and are independent of the actual equation of state. Heidemann and Prausnitz[1] show that some very simple relations result when simplifying assumptions are made on the attractive and size parameters for conventional equation of state mixing rules. These assumptions and simplifications are presented here.

Size parameter, b. To develop an equation of state, we must make approximations about the properties of the associating species. For instance, we can *assume* that the volume of a dimer is twice that of the monomer. The volume of the trimer would be three times the volume of a monomer. In terms of the molecular size parameter

$$b_i = i \cdot b_M \qquad \text{15.40}$$

Inserting this into the normal mixing rule for a mixture of monomers, dimers, trimers, etc., and incorporating Eqn. 15.35,

$$b = \sum_i x_i b_i = b_M \sum_i i \cdot x_i$$

$$\boxed{b = b_M \frac{n_o}{n_T} = \frac{b_M}{x_M}} \qquad \text{15.41}$$

Packing fraction, $b_M\rho$. The representation of density in an equation of state can always be rearranged into the dimensionless packing fraction. Can you anticipate the special property exhibited by the packing fraction? Let's see what happens when we combine Eqn. 15.41 with Eqn. 15.39.

$$\boxed{b\rho_T = \left(\frac{b_M}{x_M}\right)\rho x_M = b_M\rho} \qquad \text{15.42}$$

Note that $b_M\rho$ is the packing fraction we would have computed if we neglected association entirely. Thus we are free to apply either form interchangeably. The packing fraction is entirely independent of the extent of association. This result might seem obvious if you remember the definition of packing fraction. It is the volume occupied by molecules divided by the total volume. By setting the volume of an i-mer to be i times the volume of the monomer, we are assuming that there is no overlap caused by association. So the volumes of the molecules are the same before or after association. When you look at it this way, it should be no surprise that the packing fraction is constant.

Attractive parameter, a. If the molecular interactions are pairwise additive, then it is reasonable to assume that the interaction energy for a pair of dimers should be four times that of a monomer-monomer interaction, and the trimer should be nine times, as shown in Fig. 15.6. Therefore, $a_D = 4a_M$, $a_3 = 9a_M$, $a_4 = 16a_M$ etc. This means $a_i = i^2 a_M$, or

$$\sqrt{a_i} = i \cdot \sqrt{a_M} \qquad \text{15.43}$$

---

1. Heidemann, R. A., Prausnitz, J. M., *Proc. Nat. Acad. Sci.,* 73:1773(1976).

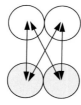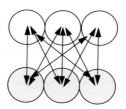

Figure 15.6 *Illustration that $a_{DD} = 4\,a_{MM}$ and $a_{33} = 9\,a_{MM}$ are reasonable by adding the number of pair interactions.*

The normal combining rule is $a_{ij} = \sqrt{a_i a_j}(1 - k_{ij})$, where it is reasonable to set $k_{ij} = 0$ for all ij. This means,

$$a_{ij} = \sqrt{a_i a_j} = \sqrt{i^2 a_M\, j^2 a_M} = i \cdot j \cdot a_M \qquad 15.44$$

In addition, when $k_{ij} = 0$ for all ij, the normal mixing rule, $a = \sum_i \sum_j x_i x_j a_{ij}$, becomes $a = \left(\sum_i x_i \sqrt{a_i}\right)^2$, which by substituting Eqn. 15.43 becomes $a = a_M\left(\sum_i i \cdot x_i\right)^2$, which by substituting Eqn. 15.35 becomes,

$$a = a_M\left(\frac{n_o}{n_T}\right)^2 = \frac{a_M}{x_M^2} \qquad 15.45$$

15.7   A VAN DER WAALS H-BONDING MODEL

For purposes of illustration, we return to the van der Waals equation, as usual when considering a new application. The principal modification is to write out explicitly what is meant by terms like density and n_T. The pressure will be determined by the true number of moles and the true density. We introduce the true compressibility factor, $Z_T = P\,V_T / RT$.

$$Z_T = \frac{P\underline{V}}{n_T RT} = 1 + \frac{b\rho_T}{(1 - b\rho_T)} - \frac{a\rho_T}{RT} \qquad 15.46$$

where a and b are the mixture parameters given by the mixing rules above. We can obtain the superficial compressibility factor if we multiply both sides of Eqn. 15.46 by n_T/n_o. However, since $n_T/n_o = x_M$, we have

$$Z = \frac{P\underline{V}}{n_o RT} = x_M + \frac{x_M b\rho_T}{(1 - b\rho_T)} - \frac{a\rho_T x_M}{RT} = \frac{x_M}{(1 - b\rho_T)} - \frac{a\rho_T x_M}{RT} \qquad 15.47$$

Thus the equation of state differs from the original van der Waals (Eqn. 6.12) only in the incorporation of the factor x_M. However, let us investigate how the right-hand side can be written in terms of a_M, b_M, and ρ_T. Recalling the simplification that occurred for the packing fraction in the repulsive term, a similar simplification occurs for the attractive term using Eqn. 15.39 and Eqn. 15.45 becomes $a\rho_T x_M = \left(\dfrac{a_M}{x_M^2}\right)(\rho x_M)x_M = a_M\rho$. The equation of state then becomes

$$Z = \frac{P\underline{V}}{n_o RT} = \frac{x_M}{(1-b_M\rho)} - \frac{a_M\rho}{RT} = 1 + \frac{b_M\rho}{(1-b_M\rho)} - \frac{a_M\rho}{RT} - \frac{(1-x_M)}{(1-b_M\rho)}$$

15.48

In this form we may define contributions to the compressibility factor such that $Z \equiv 1 + Z^{rep} + Z^{att} + Z^{assoc}$ where there is a one-to-one correspondence with terms of the right-most expression in Eqn. 15.48. Z^{rep} and Z^{att} are exactly as we have used them before. We may clearly identify Z^{assoc}.

$$Z^{assoc} = \frac{-(1 - x_M)}{(1 - b_M\rho)}$$

15.49

Note that in a fluid that does not associate, $x_M = 1$, so the standard van der Waals equation is recovered.

Solving the Equation of State for Superficial Density

If we know the size and attractive parameters for the model, and the degree of association, we can solve the equation of state. To put the equation of state in a non-dimensional form, we define

$B = \dfrac{b_M P}{RT}$, and $A = \dfrac{a_M P}{(RT)^2}$. Multiplying Eqn. 15.48 by $b_M\rho$,

$$B = \frac{x_M(b_M\rho)}{(1 - b_M\rho)} - \frac{A}{B}(b_M\rho)^2$$

15.50

⚠ Our objective is to solve this equation, but first we need to determine x_M.

Rearrangement yields[1]

$$\frac{A}{B}(b_M\rho)^3 - \frac{A}{B}(b_M\rho)^2 + (x_M + B)(b_M\rho) - B = 0$$

15.51

If we knew x_M (or αq in Eqn. 15.37) in terms of temperature and density, then solving this problem would simplify to solving for the pressure of a mixture of known composition with known mixing rules. In other words, solving for Z is entirely analogous to our previous applications of equations of state. We would simply have the intermediate step of computing x_M before computing Z. It sounds more complicated obviously, but with a computer, one would hardly notice the difference in computational speed. Let's derive an approximate relationship for x_M. We now explore and develop the procedure to determine x_M by looking at the functionality of αq.

Fugacity Coefficient for an Associating Species

For any practical applications, we need the superficial fugacity coefficient, which can be found quite directly using Eqn. 15.11; if we calculate the fugacity of the monomer, this will also be the superficial fugacity. We have used this procedure already in Example 15.2 on page 539, but now we no longer need to assume the true species form an ideal solution. For a pure, associating species we

---

1. An alternate form of the same equation appears later as Eqn. 15.65.

start by adapting of Eqn. 10.31 by writing the equation in terms of the true mole fractions, and true densities,

$$\ln\hat{\varphi}_i = -\ln(1 - b_M\rho) + \frac{b_i\rho_T}{1 - b_M\rho} - \frac{2\rho_T}{RT}\sum_j x_j a_{ij} - \ln Z_T \qquad 15.52$$

where we have already substituted $b\rho_T = b_M\rho$. Now, for the monomer,

$$\ln\hat{\varphi}_M = -\ln(1 - b_M\rho) + \frac{b_M\rho_T}{1 - b_M\rho} - \frac{2\rho_T a_M}{RT}\sum_j j x_j - \ln Z_T \qquad 15.53$$

Now, recognize $Z_T = P\underline{V}/n_T RT = P\underline{V}/n_o RT \cdot n_o/n_T = Z/x_M$. Also, using Eqn. 15.39

$$\ln\hat{\varphi}_M = -\ln(1 - b_M\rho) + \frac{b_M\rho x_M}{1 - b_M\rho} - \frac{2\rho a_M}{RT} - \ln Z + \ln x_M \qquad 15.54$$

Now, applying Eqn. 15.11, and rearranging the repulsive term involving x_M,

$$\ln\varphi = \ln x_M\hat{\varphi}_M = -\ln(1 - b_M\rho) + \frac{b_M\rho x_M}{1 - b_M\rho} - \frac{2\rho a_M}{RT} - \ln Z + 2\ln x_M \qquad 15.55$$

Determination of x_M for van der Waals' Equation

Returning to the left-hand side of Eqn. 15.32, we see that the function αq involves the ratio $\dfrac{\hat{\varphi}_M\hat{\varphi}_{i-1}}{\hat{\varphi}_i}$. To evaluate this expression, we must next solve for the fugacity coefficient of a component in a mixture in accordance with our new EOS. Then we can substitute the expression back into the reaction equilibrium equation and solve for x_M. As a specific example of the formula, consider the conversion of monomer to dimer, the first step in the oligomerization, represented by $\dfrac{\hat{\varphi}_M\hat{\varphi}_M}{\hat{\varphi}_D}$.

Writing the true fugacity coefficients in terms of logarithms,

$$2\ln\hat{\varphi}_M - \ln\hat{\varphi}_D = 2\left[-\ln(1 - b_M\rho) + \frac{b_M\rho_T}{1 - b_M\rho} - \frac{2\rho_T}{RT}\sum_j x_j a_{Mj} - \ln Z_T\right]$$
$$-\left[-\ln(1 - b_M\rho) + \frac{b_D\rho_T}{1 - b_M\rho} - \frac{2\rho_T}{RT}\sum_j x_j a_{Dj} - \ln Z_T\right] \qquad 15.56$$

Recall that $b_D = 2 b_M$ by Eqn. 15.40 and $a_{Dj} = 2ja_M$, and $a_{Mj} = ja_M$ by Eqn. 15.44. Therefore

$$2\ln\hat{\varphi}_M - \ln\hat{\varphi}_D = \ln\left[\frac{n_T RT}{P\underline{V}(1 - b_M\rho)}\right] \qquad 15.57$$

$$\frac{\hat{\varphi}_M\hat{\varphi}_M}{\hat{\varphi}_D} = \frac{n_T RT}{P\underline{V}(1 - b_M\rho)} \qquad 15.58$$

An analogous derivation for other oligomers gives

$$\frac{\hat{\varphi}_M \hat{\varphi}_{i-1}}{\hat{\varphi}_i} = \frac{n_T RT}{P\underline{V}(1 - b_M \rho)}$$

15.59

Since this expression of fugacity coefficients is independent of i, referring back to Eqn. 15.32, K_a is the same for all species.

$$\text{Finally } \alpha q = \frac{\hat{\varphi}_M \hat{\varphi}_{i-1}}{\hat{\varphi}_i} PK_a = \frac{n_T RT}{P\underline{V}(1 - b_M \rho)} PK_a = \frac{RTK_a}{(1 - b_M \rho)} \frac{n_o n_T}{\underline{V} n_o} = \frac{RTK_a \rho}{(1 - b_M \rho)} x_M$$

15.60

Now we split our definition of αq, by defining

$$\alpha = \frac{RTK_a \rho}{(1 - b_M \rho)} \qquad \text{and} \qquad q = x_M$$

15.61

α is a function of T, ρ. αq can now be substituted back into Eqn. 15.37, while recognizing that $q = x_M$.

$$x_M = 1 - \alpha x_M^2$$

Solving the quadratic for the only reasonable root gives

$$\boxed{x_M = \frac{-1 + \sqrt{1 + 4\alpha}}{2\alpha}}$$

15.62

   In principle, our problem is now solved. Note that the only difference between Eqn. 15.48 and the original van der Waals equation is the factor of x_M in the repulsive term, and we now have an equation to determine x_M at a given temperature and density. Two new parameters must be introduced, however, to characterize α and its dependency on temperature. Since α is closely related to K_a, we will need to consider the van't Hoff relation to characterize the dependence of K_a on T. The temperature dependence of the hydrogen bonding may be given by the form of the van't Hoff equation, which is simplified to be

$$\boxed{K_a = \frac{(K_a)_c}{T_r} \exp\left[H\left(1 - \frac{1}{T_r}\right)\right]}$$

15.63

where $(K_a)_c$ is the value of the equilibrium constant at the critical point, and the constant H is found by fitting the vapor pressure curve. The explanation for this form of the equation is given in Section 15.8 but it is easy to see that the value approaches zero at infinite temperature and is equal to the critical value at the critical temperature. Substitution into Eqn. 15.61 gives

$$\boxed{\alpha = \left(\frac{RT_c(K_a)_c}{b_M}\right)\left(\frac{b_M \rho}{(1 - b_M \rho)}\right) \exp\left[H\left(1 - \frac{1}{T_r}\right)\right] = K_a'\left(\frac{b_M \rho}{(1 - b_M \rho)}\right) \exp\left[H\left(1 - \frac{1}{T_r}\right)\right]}$$

15.64

$K_a' = RT_c K_{ac}/b_M$ is the dimensionless form of $(K_a)_c$.

> ### Example 15.4 Molecules of H_2O in a 100-ml beaker
>
> Assuming α is about 100 at room temperature and $\rho = 1$ g/cm³, estimate the moles of H_2O monomer in a 100 ml beaker of liquid water.
>
> **Solution:** Note that the problem statement requests moles of H_2O not $(H_2O)_2$ or $(H_2O)_3$ etc., so we are interested in the true number of H_2O monomer moles. We know $n_0 = 5.556$ from our previous calculation (see "Mass Balances" on page 544), but $n_M = x_M \cdot n_T$, so we must convert from our superficial basis. Proceeding, using Eqn. 15.62
>
> $$x_M = [-1 + \sqrt{1 + 4\alpha}\,]/2\alpha = 0.095;\ n_T/n_0 = x_M = 0.095 = n_T/n_0$$
>
> $$\Rightarrow n_M = 0.095 \cdot (n_T/n_0) \cdot n_0 = 0.095^2 \cdot n_0 = 0.05 \text{ moles}$$
>
> Therefore, the true number of moles is 100 times less than the superficial number of moles.

Solving the Equation of State for Density

We can see that Eqn. 15.51 is not truly a cubic in density because x_M depends on density. Also, its functional dependence is nonlinear, which makes implementation more difficult. However, α is a monotonic function of density, and successive substitution of density has been found to quickly yield converged values for density. The algorithm is given in Fig. 15.7.

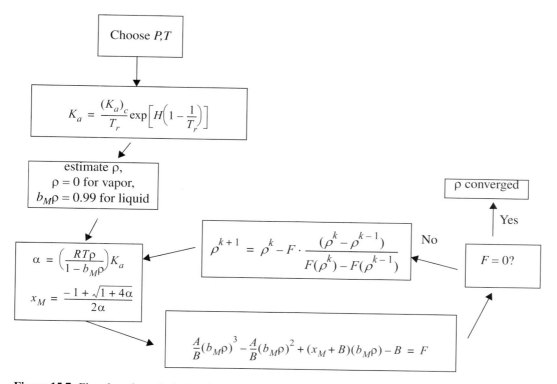

Figure 15.7 *Flowsheet for calculating density by the van der Waals associating fluid model.*

Fitting the Constants in the Equation of State

So far, we have assumed that the constants needed for a fluid are available. However, the associating fluid model must be used to obtain the parameter values. To obtain approximate hydrogen bonding parameters,[1] we can use the same approach applied for non-associating systems. Rearranging Eqn. 15.51 in terms of the compressibility factor.

$$Z^3 - (x_M + B)Z^2 + AZ - AB = 0 \qquad\qquad 15.65$$

Writing out $(Z - Z_c)^3$, which equals 0 at the calculated critical point:

$$Z^3 - (3Z_c)Z^2 + (3Z_c^2)Z - (Z_c^3) = 0 \qquad\qquad 15.66$$

Comparing term-by-term with the previous equation, we find at the critical point:

$$x_M + B_c = 3Z_c^{EOS} \qquad\qquad 15.67$$

$$A_c = 3(Z_c^{EOS})^2 \qquad\qquad 15.68$$

$$A_c B_c = (Z_c^{EOS})^3 \qquad\qquad 15.69$$

where the superscript *EOS* has been added to explicitly show that this is the value predicted by the equation of state. These three equations differ from the original van der Waals equation only by the x_M appearing in Eqn. 15.67. Plugging 15.68 into 15.69, we find

$$B_c = \frac{Z_c^{EOS}}{3} \qquad\qquad 15.70$$

Using this result in Eqn. 15.67,

$$Z_c^{EOS} = \frac{3}{8}(x_M)_c \qquad\qquad 15.71$$

where x_M would have a value of unity in a non-associating system. Plugging this into 15.70, we find

$$\boxed{b_M = \frac{RT_c}{8P_c}(x_M)_c} \qquad\qquad 15.72$$

and into 15.68,

$$\boxed{a_M = \frac{27}{64}\frac{(RT_c)^2}{P_c}(x_M)_c^2} \qquad\qquad 15.73$$

---

1. Note: Rigorously matching the critical point requires recognizing that x_M is a function of density, so the above method is only approximate in terms of matching the critical point. On the other hand, the characterization obtained in this way provides reasonable accuracy for an introduction. Problem 15.12 describes the route to rigorously matching the critical point.

Finding x_M from Z_c

Experimental values of Z_c cannot be used directly in Eqn. 15.71, because we know that the value of $Z_c = 0.375$ predicted by the van der Waals equation is significantly in error for non-associating species. For argon, krypton, and xenon, Z_c is approximately 0.29. Therefore, this equation is systematically in error even for spherical non-associating fluids. To account for the systematic error of the equation, one can assume the form of Eqn. 15.71 is correct, but the coefficient is incorrect; thus, we could preserve the proportionality $Z_c \propto (x_M)_c$. This implies that the ratio of the experimental Z_c to Z_c^{homo}, the critical compressibility factor of a non-associating homomorph[1] gives $(x_M)_c$,

$$\boxed{\frac{Z_c^{exp}}{Z_c^{homo}} = (x_M)_c}$$

15.74

where several common species and homomophs are listed in Table 15.3.

From $(x_M)_c$ we can use Eqns. 15.72 and 15.73 to determine a_M and b_M. We also can use Eqn. 15.37, and noting that $q = x_M$,

$$\alpha_c = \left(\frac{1-x_M}{qx_M}\right)_c = \left(\frac{1-x_M}{x_M^2}\right)_c$$

15.75

since

$$\alpha_c = \left(\frac{RT\rho}{1-b_M\rho}\right)_c (K_a)_c = \left(\frac{b_M\rho}{1-b_M\rho}\right)\left(\frac{RT_c K_{ac}}{b_M}\right) = \left(\frac{b_M\rho}{1-b_M\rho}\right)K_a'$$

15.76

where $K_a' = RT_c K_{ac}/b$. Noting from Eqn. 15.70 that $b_M\rho_c = B_c/Z_c = {}^1/_3$ for the van der Waals EOS gives:

$$2\alpha_c = K_a' = 2\left(\frac{1-x_{Mc}}{x_{Mc}^2}\right)$$

15.77

Table 15.3 *Homomorphs for several associating compounds, and $(x_M)_c$ for the associating van der Waals model.*

Specie	Z_c	Homomorph	Z_c^{homo}	$(x_M)_c$
ethanol	0.248	propane	0.281	0.883
methanol	0.224	ethane	0.284	0.789
water	0.233	methane	0.288	0.809

---

1. Homomorph means having the same shape. In this context it refers to a hydrocarbon having the same size and branching (e.g., isobutane is a homomorph of isopropanol).

This allows K_a' to be determined. The temperature dependence of the hydrogen bonding is given by the van't Hoff equation with $\Delta C_P = -R$, which is simplified to give Eqn. 15.63 above. The last constant, H in Eqn. 15.63, is found by fitting the vapor pressure curve. The justification for the form of the equation is given in Section 15.8. Non-polar fluids tend to have a slight downward curvature to the plot of $\ln P^{sat}$ versus $1/T$. Associating fluids actually have a little less curvature, and by fitting the vapor pressure curve, a parameter value for H can be obtained.

Helmholtz Energy and the Fugacity Coefficient for van der Waals

There is a connection between Helmholtz energy and the fugacity coefficient that we will stress before leaving the van der Waals equation. The relationship is important because when more complicated equations are developed in the next section, and extended to mixtures, we take advantage of a theoretical derivation for the Helmholtz energy to save a derivation of the fugacity coefficient. First, recall the way we have characterized the compressibility factor, $Z = 1 + Z^{rep} + Z^{att} + Z^{assoc}$. Therefore, we may directly associate the terms to their contributions to the fugacity coefficient. Recalling Eqn. 8.24

$$\frac{(G - G^{ig})}{RT} = \ln\left(\frac{f}{P}\right) = \int_0^\rho \frac{(Z-1)}{\rho}d\rho + (Z-1) - \ln Z = \frac{(A - A^{ig})_{TV}}{RT} + (Z-1) - \ln Z \qquad 8.24$$

where $(Z - 1)$ in the integral may be replaced with $Z^{rep} + Z^{att} + Z^{assoc}$ and each term integrated independently. A similar substitution can be made for the $Z - 1$ term following the integral.

$$\ln\varphi = \int_0^\rho \frac{(Z^{rep} + Z^{att} + Z^{assoc})}{\rho}d\rho + Z^{rep} + Z^{att} + Z^{assoc} - \ln Z$$

From the way we have defined contributions, we can define corresponding contributions to the fugacity coefficient resulting from the integral,

$$\ln\varphi = \ln\varphi^{rep} + \ln\varphi^{att} + \ln\varphi^{assoc} - \ln Z \qquad 15.78$$

where $\ln\varphi^{rep}$ is short hand notation for the two contributions of Z^{rep} to $\ln\varphi$

$$\frac{A^{rep}}{n_o RT} \equiv \frac{(A - A^{ig})_{TV}^{rep}}{RT} \equiv \int_0^\rho \frac{Z^{rep}}{\rho}d\rho, \qquad \ln\varphi^{rep} \equiv \frac{A^{rep}}{n_o RT} + Z^{rep}$$

and an analogous description applies to $\ln\varphi^{att}$ and $\ln\varphi^{assoc}$.

To see the desired relation for the association term, we need to rearrange Eqn. 15.55 to identify all the terms except $\ln\varphi^{assoc}$, and thus identify the relationship between the Helmholtz departure and the association. Adding and subtracting x_M in the form $x_M(1 - b_M\rho)/(1 - b_M\rho) - x_M$,

$$\ln\varphi = -\ln(1 - b_M\rho) + \frac{b_M\rho x_M}{1 - b_M\rho} + x_M\frac{(1 - b_M\rho)}{(1 - b_M\rho)} - \frac{2\rho a_M}{RT} - \ln Z + - x_M + 2\ln x_M \qquad 15.79$$

$$= -\ln(1 - b_M\rho) + \frac{x_M}{1 - b_M\rho} - \frac{2\rho a_M}{RT} - \ln Z - x_M + 2\ln x_M$$

Adding and subtracting one, in the form $1 - (1 - b_M\rho)/(1 - b_M\rho)$,

$$\ln\varphi = -\ln(1 - b_M\rho) + \frac{x_M}{1 - b_M\rho} - \frac{(1 - b_M\rho)}{(1 - b_M\rho)} - \frac{2\rho a_M}{RT} - \ln Z + 1 - x_M + 2\ln x_M \qquad 15.80$$

$$= -\ln(1 - b_M\rho) - \frac{\rho a_M}{RT} + \frac{b_M\rho}{(1 - b_M\rho)} - \frac{\rho a_M}{RT} \frac{1 - x_M}{1 - b_M\rho} + 1 - x_M + 2\ln x_M - \ln Z$$

Then matching the terms one-to-one with their origin:

$$\ln\varphi = \frac{A^{rep}}{n_oRT} + \frac{A^{att}}{n_oRT} + Z^{rep} + Z^{att} + Z^{assoc} + 1 - x_M + 2\ln x_M - \ln Z$$

Therefore, the remaining terms are the contribution to Helmholtz departure due to association,

$$\boxed{\frac{A^{assoc}}{n_oRT} = 1 - x_M + 2\ln x_M} \qquad 15.81$$

❶ The contribution of hydrogen bonding to the residual Helmholtz energy.

This derivation has been tedious, but the extension to mixtures and the adaptation to other equations of state should now be straightforward. We will take advantage of this fact when we turn to mixtures in Section 15.9.

Applications to Pure Fluids

Table 15.4 summarizes results of Eqns. 15.72–15.76 for three compounds. Note that $0.300/0.375 = 0.233/0.291$ maintaining the ratio of Eqn. 15.74. All that remains is to specify a value for the parameter H in Eqn. 15.63. One reasonable approach would be to set the energy of hydrogen bonding equal to a typical value like 20 kJ/mole. For water, this would result in $H = 3.716$. Other approaches to characterizing H are discussed in the homework problems. Given the values for these parameters, you should be able to implement the algorithm of Fig. 15.7 with no difficulty.

15.8   THE ESD EQUATION FOR ASSOCIATING FLUIDS

The van der Waals model has served well to introduce the methods used in developing a model for associating fluids, however, it has several shortcomings when considering quantitative modeling. One alternative would be to adapt the Peng-Robinson equation. This option is developed as a homework problem. That approach can provide improved accuracy for vapor-liquid equilibria, but a consistent

Table 15.4 *Role of association in depressing Z_c according to the vdW-HB EOS*

Component	$K_a' = K_{ac}RT_c/b$	x_{Mc}	a/bRT_c	Z_c (calculated)	Z_c (experiment)
Argon	0	1.000	3.375	0.375	0.291
Water	0.620	0.801	2.703	0.300	0.233
Acetonitrile	1.843	0.632	2.134	0.237	0.184

extension to liquid-liquid equilibria in systems like hydrocarbons with water has been difficult to develop.[1] Part of the problem with the Peng-Robinson equation is the inaccurate form of the van der Waals repulsive term. Heidemann and Prausnitz[2] corrected this shortcoming in their analysis of associating fluids, but their analysis applied only to pure fluids. A number of other authors have pursued adaptations which correct the repulsive term or address consistent models for vapor-liquid-liquid equilibria, but we will limit our discussion to two: the ESD equation,[3] and the SAFT equation.[4] In this section we introduce the ESD equation. The ESD equation has the following features:

1. The equation of state is cubic, which permits us to retain the principles we have developed for solving and applying the equation of state.

2. The equation of state explicitly represents the effect of shape for non-polar molecules. This means that extension of the equation to polymers is straightforward. This additional flexibility requires an additional characteristic parameter; however the shape parameter has been correlated in terms of the acentric factor for non-associating fluids, much like the parameter κ in the Peng-Robinson equation.

3. The equation of state has been developed by modeling computer simulations, and thus should capture the essential physics of size, shape, and hydrogen bonding.

Noting the detailed development for the van der Waals equation provided above, we will rapidly move through the adaptation to the ESD equation. All of the steps are essentially equivalent to those for the van der Waals equation. The one additional complication is that the shape parameter and H cannot be determined independently for an associating fluid. This is because the experimental acentric factor is a measure of both shape and hydrogen bonding for associating systems,[5] and thus the acentric factor is insufficient for determining the shape parameter. Other features include improvement of the assumption of a temperature-independent value for H in the van't Hoff Eqn. 15.63, which ignores the change in heat capacity due to temperature-dependent association, and the possibility of branching and its impact on the assumptions about $q\alpha$ and a_{ij}.

The equation proposed by Elliott et al. (1990) is:

$$\frac{PV}{RT} = 1 + \frac{4\langle c\eta \rangle}{1 - 1.9\eta} - \frac{9.5\langle qY\eta \rangle}{1 + 1.7745\langle Y\eta \rangle} \qquad \text{15.82}$$

where

$$b = \sum_i x_i b_i \text{ , therefore, } \eta = b\rho = \left(\sum_i x_i b_i\right)\rho$$

$$c = \sum_i x_i c_i \text{ , a "shape factor" which represents the effect of non-sphericity on the repulsive term.}$$

1. Raymond, M. B. Ph.D. Thesis, University of Akron, 1998.
2. Heidemann, R. A. and Prausnitz, J. M. *Proc Nat Acad Sci,* 73:17773 (1976).
3. Elliott, J. R., Suresh, S. J., Donohue, M. D. *Ind. Eng. Chem. Res.,* 29:1476 (1990).
4. Chapman, W. G., Jackson, G., Gubbins, K. E., Radosz, M., *Ind. Eng. Chem. Res.* 29:1709 (1990).
5. Recall that the acentric factor is a measure of the slope of the vapor pressure line. For non-polar substances this is determined by shape. For associating species, the association is also a function of temperature, so it also changes along the vapor pressure curve, also affecting the slope.

ESD.exe

$q = 1 + 1.90476(c - 1) = \sum_i x_i q_i$ is a shape factor which represents the effect of non-sphericity on the attractive term.

$$<c\eta> = c \cdot b \cdot \rho = \left(\sum_i x_i c_i \right) \cdot \left(\sum_j x_j b_j \right) \cdot \rho = \rho \sum_i \sum_j x_i x_j (b_i c_j + b_j c_i)/2$$

$$Y_{ij} = \exp(\varepsilon_{ij}/kT) - 1.0617$$

$$<qY\eta> = \rho<qYb> = \rho \sum_i \sum_j x_i x_j \; Y_{ij} \; (b_i q_j + b_j q_i)/2$$

$\varepsilon_{ij} = \sqrt{\varepsilon_i \varepsilon_j}(1 - k_{ij})$ the energy of disperse attraction (equivalent to well-depth of a square-well potential). As before k_{ij} is zero for monomer-oligomer interactions.

$$<Y\eta> = \rho<Yb> = \rho \sum_i x_i b_i Y_i$$

The most important modification over the van der Waals equation is the inclusion of the shape factor, c, in the repulsive term of the EOS. Elliott et al. illustrated that this form for the repulsive term significantly improves agreement with molecular simulation data for hard spheres[1] and chains[2] relative to the repulsive term assumed in the van der Waals or Peng-Robinson equations, as shown in Fig. 15.8. Note that the van der Waals and Peng-Robinson estimates are the same regardless of chain length whereas the ESD equation sets $c = 1 + (Nd - 1)/2$, where Nd is the chain length. In practice, the shape factor is actually correlated with the acentric factor obtained from experimental data. The attractive term of the ESD equation was derived from similar comparisons to square-well spheres[3]. Note that the effect of non-sphericity is stronger for the attractive term than the repulsive term by the factor of 1.90476.

Fugacity Coefficient

The fugacity coefficient is evaluated using Eqn. 10.16. The first step is to derive the Helmholtz departure,

$$\frac{A - A^{ig}}{RT} = n \int_0^\rho \frac{4cb}{1 - 1.9b\rho} d\rho - n \int_0^\rho \frac{9.5\langle qYb \rangle}{1 + 1.7745\langle Yb \rangle \rho} d\rho$$

$$= \frac{-4nc}{1.9} \ln(1 - 1.9\eta) - \frac{9.5n\langle qYb \rangle}{1.7745\langle Yb \rangle} \ln(1 + 1.7745\langle Y\eta \rangle)$$

15.83

1. Erpenpeck, J. J., Wood, W. W., *J.Stat. Phys.* 35:321 (1984).
2. Dickman, R., Hall, C. K., *J.Chem.Phys.* 89:3168 (1988).
3. Sandler, S. I., Lee, K. H., *Fluid Phase Equil.* 30:135 (1986).

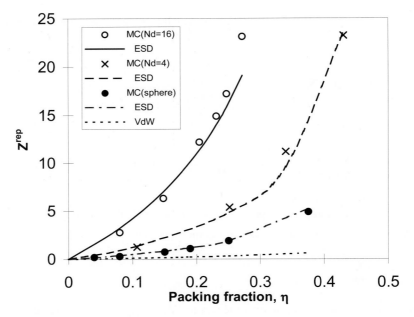

Figure 15.8 *Comparison of molecular simulations, the van der Waals equation, and the ESD equation of state for Z^{rep}. Nd is the number of spheres in a chain.*

Differentiating,

$$\left(\frac{\partial\left(\frac{A-A^{ig}}{RT}\right)}{\partial n_i}\right)_{T,\underline{V},n_{j\neq i}} = \frac{-4}{1.9}\ln(1-1.9\eta)\left(\frac{\partial(nc)}{\partial n_i}\right)_{T,\underline{V},n_{j\neq i}} + \frac{4nc}{(1-1.9\eta)}\left(\frac{\partial\eta}{\partial n_i}\right)_{T,\underline{V},n_{j\neq i}}$$

$$-\frac{9.5}{1.7745}\ln(1+1.7745\langle Y\eta\rangle)\left(\frac{\partial\left(\frac{n\langle qYb\rangle}{\langle Yb\rangle}\right)}{\partial n_i}\right)_{T,\underline{V},n_{j\neq i}} \qquad 15.84$$

$$-\frac{9.5n\langle qYb\rangle}{\langle Yb\rangle(1+1.7745\langle Y\eta\rangle)}\left(\frac{\partial\langle Y\eta\rangle}{\partial n_i}\right)_{T,\underline{V},n_{j\neq i}}$$

Determining the intermediate derivatives,

$$\left(\frac{\partial(nc)}{\partial n_i}\right)_{T,\underline{V},n_{j\neq i}} = c_i$$

$$\left(\frac{\partial\eta}{\partial n_i}\right)_{T,\underline{V},n_{j\neq i}} = \left(\frac{\partial(bn\rho)}{\partial n_i}\right)_{T,\underline{V},n_{j\neq i}} = b_i\rho \qquad 15.85$$

$$\left(\frac{\partial\langle Y\eta\rangle}{\partial n_i}\right)_{T,\underline{V},n_{j\neq i}} = b_iY_i\rho$$

$$\left(\frac{\partial\left(\frac{n\langle qYb\rangle}{\langle Yb\rangle}\right)}{\partial n_i}\right)_{T,\underline{V},n_{j\neq i}} = \left(\frac{\partial\left(\frac{n^2\langle qYb\rangle}{n\langle Yb\rangle}\right)}{\partial n_i}\right)_{T,\underline{V},n_{j\neq i}} =$$

$$\frac{1}{n\langle Yb\rangle}\left(\frac{\partial(n^2\langle qYb\rangle)}{\partial n_i}\right)_{T,\underline{V},n_{j\neq i}} - \frac{n^2\langle qYb\rangle}{n^2\langle Yb\rangle^2}\left(\frac{\partial(n\langle Yb\rangle)}{\partial n_i}\right)_{T,\underline{V},n_{j\neq i}}$$

15.86

The first of the derivatives in the right-most expression of Eqn. 15.86 is of the form of Eqn. 10.25, and the second is of the form of Eqn. 10.22, and they can be written by inspection. Combining,

$$\ln\hat{\varphi}_i = -\frac{4}{1.9}c_i\ln(1-1.9\eta) + \frac{4cb_i\rho}{1-1.9\eta}$$

$$-\frac{9.5}{1.7745}\frac{\ln(1+1.7745\langle Y\eta\rangle)}{\langle Yb\rangle}\left(\sum_j x_j Y_{ij}(b_iq_j + b_jq_i) - \frac{\langle qYb\rangle}{\langle Yb\rangle}Y_ib_i\right)$$

$$-\frac{9.5\langle qYb\rangle}{\langle Yb\rangle}\frac{Y_ib_i\rho}{1+1.7745\langle Y\eta\rangle} - \ln Z$$

15.87 ❗ Fugacity coefficient for a component in a non-associating mixture.

For a pure fluid, this simplifies to

$$\ln\varphi = -\frac{4}{1.9}c\ln(1-1.9\eta) + \frac{4cb\rho}{1-1.9\eta}$$

$$-\frac{9.5q}{1.7745}\ln(1+1.7745\langle Y\eta\rangle) - \frac{9.5qYb\rho}{1+1.7745\langle Y\eta\rangle} - \ln Z$$

15.88 ❗ Fugacity coefficient for a pure, non-associating fluid.

Extension to Associating Fluids

The extension to associating mixtures is analogous to the discussion above for the van der Waals equation; we start with the equation for non-associating fluids, and adapt it by multiplying by x_M, and writing the density in terms of true density.

$$Z_T = x_M + \frac{4x_M\langle c\eta_T\rangle}{1-1.9\eta} - \frac{9.5x_M\langle qY\eta_T\rangle}{1+1.7745\langle Y\eta_T\rangle}$$

15.89

where only the denominator of the second term is clearly unaffected; the other terms will be evaluated. We apply the same mixing rules for the size parameter, $b_i = i \cdot b_M$; so, like the van der Waals equation, $b = b_M/x_M$, and $b\rho_T = b_M\rho$.

For linear associating species, we assume[1]

$$c_i = c_M + (i-1)\cdot(c_M - 0.475) = i(c_M - 0.475) + 0.475$$

15.90

---

1. The motivation for this assumption is discussed in relation to Eqn. 15.97.

For the repulsive term in the EOS, the value of $<c\eta_T> = \left(\sum_i x_i c_i\right)\left(\sum_i x_i b_i\right)\rho_T$. Therefore,

$$4<c\eta_T> = 4\left(\sum_i x_i c_i\right)\left(\sum_i x_i b_i\right)\rho_T = 4\left(\sum_i x_i c_i\right)\eta$$

However,

$$\sum_i x_j c_j = \sum_i x_i \ [i\cdot(c_M - 0.475) + 0.475] = (c_M - 0.475)\ /x_M + 0.475$$

$$= c_M/x_M - 0.475(1/x_M - 1) \tag{15.91}$$

$$4c = 4c_M/x_M - 1.9(1/x_M - 1) \tag{15.92}$$

The value we need for the numerator of the second term of Eqn. 15.89 is $x_M<4c\eta>$

$$x_M<4c\eta> = 4\eta\ c_M - 1.9\eta(1 - x_M)$$

For the ESD equation, $Y_{ij} = Y$ for the monomer. Considering the attractive term, we will also need to evaluate q_i, which we will show equals $i\cdot q_M$. We keep the definition of $q_i = 1 + 1.90476\ (c_i - 1)$, which may be written $q_i = 1 + k_3\ (c_i - 1)$, where, $k_3 = 1.90476$. Note that $0.475 = (k_3 - 1)/\ k_3$, and inserting c from Eqn. 15.90

$$q_i = 1 + k_3\ \{c_i - 1\} = 1 + k_3 \cdot \{i(c_M - (k_3 - 1)/\ k_3) + (k_3 - 1)/\ k_3 - 1\} =$$

$$1 + \{ic_M k_3 - k_3 i + i + k_3 - 1 - k_3\} = i\cdot[1 + k_3(c_M - 1)] = i\cdot q_M$$

Using these results,

$$\langle qYb \rangle = \sum_i \sum_j x_i x_j Y_{ij} \frac{(ib_M j q_M + jb_M i q_M)}{2} = q_M b_M Y_M \sum_i \sum_j x_i x_j ij = \frac{q_M b_M Y_M}{x_M^2}$$

$$\langle qY\eta_T \rangle = \langle qYb \rangle \rho_T = \frac{q_M b_M Y_M}{x_M^2}\rho x_M = \frac{q_M b_M Y_M}{x_M}$$

and the numerator of the attractive term becomes

$$9.5x_M\langle qY\eta \rangle = 9.5q_M Y_M \eta$$

The term $\langle Y\eta \rangle$ in the denominator is independent of association,

$$\langle Yb \rangle = Y_M \sum_i x_i b_i = \frac{Y_M b_M}{x_M}, \ \langle Y\eta \rangle = Y_M\left(\sum_i x_i b_i\right)\rho_T = \frac{Y_M b_M}{x_M}\rho_T = Y_M b_M \rho = Y_M \eta$$

The final result is of the form $Z = 1 + Z^{rep} + Z^{att} + Z^{assoc}$.

$$Z = 1 + \frac{4c_M \eta}{1 - 1.9\eta} - \frac{9.5q_M Y_M \eta}{1 + 1.7745\langle Y\eta \rangle} - \left(\frac{1.9\eta(1 - x_M)}{1 - 1.9\eta} + 1 - x_M\right) \tag{15.93}$$

We can clearly identify

$$Z^{assoc} = \frac{-1.9\eta(1 - x_M)}{(1 - 1.9\eta)} - (1 - x_M) = \frac{-(1 - x_M)}{(1 - 1.9\eta)}$$

15.94

Pure Associating-Fluid Fugacity

For the fugacity of a true monomer, we adapt Eqn. 15.87,

$$
\begin{aligned}
\ln\hat{\varphi}_M = & -\frac{4}{1.9}c_M\ln(1 - 1.9\eta) + \frac{4cb_M\rho_T}{1 - 1.9\eta} \\
& -\frac{9.5}{1.7745}\frac{\ln(1 + 1.7745\langle Y\eta\rangle)}{\langle Yb\rangle}\left(Y_M\sum_j x_j(b_M q_j + b_j q_M) - \frac{\langle qYb\rangle}{\langle Yb\rangle}Y_M b_M\right) \\
& -\frac{9.5\langle qYb\rangle}{\langle Yb\rangle}\frac{Y_M b_M\rho_T}{1 + 1.7745\langle Y\eta\rangle} - \ln Z_T
\end{aligned}
$$

15.95

The summation term becomes $Y_M b_M\sum_i x_i q_M + Y_M q_M\sum_i ib_M = 2Y_M b_M q_M / x_M$. Substituting $4c$ from Eqn. 15.92,

$$
\begin{aligned}
\ln\hat{\varphi}_M = & -\frac{4}{1.9}c_M\ln(1 - 1.9\eta) + \frac{4c_M b_M\rho}{1 - 1.9\eta} \\
& -\frac{9.5}{1.7745}\ln(1 + 1.7745\langle Y\eta\rangle)(q_M) \\
& -\frac{9.5\langle qYb\rangle}{\langle Yb\rangle}\frac{Y_M b_M\rho}{1 + 1.7745\langle Y\eta\rangle} - \frac{1.9\eta(1 - x_M)}{1 - 1.9\eta} - \ln Z + \ln x_M
\end{aligned}
$$

15.96 ❶ True fugacity coefficient for monomer in a pure associating fluid.

Determination of x_M for the ESD Equation

Given the equation of state, we must next solve for the fugacity coefficient of a component in a mixture. Recalling the definitions of Eqn. 15.78, and noting that contributions to $\ln \varphi_D{}^{att}$ cancel with those of $\ln \varphi_M{}^{att,1}$

$$\ln\hat{\varphi}_i = -\frac{4c_i}{1.9}\ln(1 - 1.9\eta) + \frac{4cb_i}{1 - 1.9\eta} - \ln Z_T$$

15.97

Considering the monomer to dimer step,

$$
2\ln\hat{\varphi}_M - \ln\hat{\varphi}_D = 2\left[-\frac{4c_M}{1.9}\ln(1 - 1.9\eta) + \frac{4cb_M}{1 - 1.9\eta} - \ln Z_T\right] \\
-\left[-\frac{4c_D}{1.9}\ln(1 - 1.9\eta) + \frac{4cb_D}{1 - 1.9\eta} - \ln Z_T\right]
$$

15.98

---

1. In fact, the motivation for Eqn. 15.90 was to anticipate and ensure cancellation of $\ln \varphi_i{}^{att}$, and Eqn. 15.90 in turn motivated $k_3 = 1.90476$.

Recalling that $b_D = 2b_M$ and further assuming that $c_i = c_M + (i-1) \cdot (c_M - 0.475)$, we obtain

$$2 \ln \hat{\varphi}_M - \ln \hat{\varphi}_D = -\frac{4(2c_M - c_D)}{1.9} \ln(1 - 1.9\eta) - \ln Z_T$$

But, $c_D - 2 c_M = 2 c_M - 0.475 - 2 c_M = -0.475$ and $-4 \cdot (-0.475)/1.9 = 1$.

$$\boxed{\frac{\hat{\varphi}_M^2}{\hat{\varphi}_D} = \frac{1}{Z_T(1 - 1.9\eta)}}$$ 15.99

Noting that Eqn. 15.99 is analogous to Eqn. 15.58 then,

$$\boxed{x_M = \frac{-1 + \sqrt{1 + 4\alpha}}{2\alpha}}$$ 15.100

$$\alpha = \frac{\eta K_a'}{(1 - 1.9\eta)} \exp\left(H\left(1 - \frac{1}{T_r}\right)\right)$$ 15.101

⚠ Simplified form of α using the van't Hoff expression. The ESD program uses the form given by Eqn. 15.111.

Helmholtz Energy and the Fugacity Coefficient for the ESD Equation

As with van der Waals equation, we can seek the relationship between the hydrogen bonding and the Helmholtz departure. Writing Eqn. 15.96 for the superficial fugacity coefficient

$$\ln \varphi = \ln \hat{\varphi}_M x_M = -\frac{4}{1.9} c_M \ln(1 - 1.9\eta) + \frac{4c_M b_M \rho}{1 - 1.9\eta}$$
$$- \frac{9.5}{1.7745} \ln(1 + 1.7745\langle Y\eta \rangle)(q_M)$$ 15.102
$$- \frac{9.5\langle qYb \rangle}{\langle Yb \rangle} \frac{Y_M b_M \rho}{1 + 1.7745\langle Y\eta \rangle} - \frac{1.9\eta(1 - x_M)}{1 - 1.9\eta} - \ln Z + 2 \ln x_M$$

To obtain Z^{assoc} in the equation, we must subtract and add $(1 - x_M)$

⚠ Superficial fugacity coefficient of pure associating species.

$$\ln \varphi = \ln \hat{\varphi}_M x_M = -\frac{4}{1.9} c_M \ln(1 - 1.9\eta) + \frac{4c_M b_M \rho}{1 - 1.9\eta}$$
$$- \frac{9.5}{1.7745} \ln(1 + 1.7745\langle Y\eta \rangle)(q_M)$$
$$- \frac{9.5\langle qYb \rangle}{\langle Yb \rangle} \frac{Y_M b_M \rho}{1 + 1.7745\langle Y\eta \rangle} + \left(\frac{-1.9\eta(1 - x_M)}{1 - 1.9\eta} - (1 - x_M)\right)$$ 15.103
$$+ ((1 - x_M) + 2 \ln x_M) - \ln Z$$

From this equation, we find that the contribution to Helmholtz energy from association is the same as the van der Waals equation.

$$\boxed{\frac{A^{assoc}}{n_o RT} = 2 \ln x_M + (1 - x_M)}$$ 15.104

The Temperature Dependence of α

Hydrogen bonding is basically a simple exothermic reaction. As such, its behavior is described by the van't Hoff equation. As the temperature goes up, conversion decreases. Since it is a relatively weak reaction, hydrogen bonding can become very weak at elevated temperatures. To develop more quantitative expressions, we must analyze K_a in detail. We begin by assuming that the term ΔC_P is constant with respect to temperature. Then,

$$\ln K_a = \frac{-\Delta G_T^o}{RT} = \frac{-\Delta H_R^o - \Delta C_P(T - T_R)}{RT} + \frac{\Delta S_R^o + \Delta C_P \ln(T / T_R)}{R} \qquad 15.105$$

We are free to choose $T_R = T_c$. Then we have

$$\ln K_a = \frac{-\Delta G_{T_c}^o}{RT_c} + \frac{-\Delta H_{T_c}^o}{R}\left(\frac{1}{T} - \frac{1}{T_c}\right) - \frac{\Delta C_P(1 - 1 / T_r)}{R} + \frac{\Delta C_P \ln(T_r)}{R}$$

$$\ln K_a = \ln K_{ac} + \left(\frac{\Delta H_{T_c}^o}{RT_c} - \frac{\Delta C_P}{R}\right)\left(1 - \frac{1}{T_r}\right) + \frac{\Delta C_P \ln(T_r)}{R} \qquad 15.106$$

$$\ln\left(\frac{K_a}{K_{ac}}\right) = \ln(T_r)^{\Delta C_P / R} + \left(\frac{\Delta H_{T_c}^o}{RT_c} - \frac{\Delta C_P}{R}\right)\left(1 - \frac{1}{T_r}\right) \qquad 15.107$$

Setting $\Delta C_P / R = -1$ and defining a quantity $H = \left(\frac{\Delta H_{T_c}^o}{RT_c} - \frac{\Delta C_P}{R}\right)$, we have

$$\frac{T_r K_a}{K_{ac}} = \exp\left[H\left(1 - \frac{1}{T_r}\right)\right] \qquad 15.108$$

The ESD equation provides the following form of α, which after some minor rearrangements, we have

$$\alpha = \frac{\rho}{1 - 1.9\eta}K_a RT = \frac{\eta}{1 - 1.9\eta}\frac{K_a RT}{b_M} = \frac{\eta}{1 - 1.9\eta}K_{ac}'\frac{T_r K_a}{K_{ac}}$$

Where we have defined $K_a' = \frac{K_{ac}RT_c}{b_M}$. Substituting in the expression for $\frac{T_r K_a}{K_{ac}}$ from Eqn. 15.108, we obtain

$$\alpha = \frac{\eta}{1 - 1.9\eta}K_a'\exp\left[H\left(1 - \frac{1}{T_r}\right)\right] \qquad 15.109$$

α is thus characterized by the two parameters K_a' and H. These parameters can be determined by optimizing the fit to the critical properties and vapor pressure curve of the associating compound. For example K_a' can be determined from Z_c as given in Eqn. 15.74. Z_c, and H can be determined to fit the shape of the vapor pressure curve. The presence of association tends to raise the vapor pressure at low temperatures relative to a non-associating component of the same acentric factor. As an alternative, the values may be determined by optimizing the fit to other phase equilibrium data.

Suresh and Elliott[1] showed that the sensitivity of LLE calculations to the volumetric parameter, b_M, can be applied to optimize the fit of LLE for one binary pair then those parameters for c, ε/k, b_M, H, and K_a' can be applied to a large number of systems. Puhala and Elliott[2] illustrated the point by applying it to 320 binary systems, all correlated to less than 10% error.

There was a slight evolution in the assumed temperature dependence of α between 1990 and 1992. A closely related theory based on the statistical mechanics of intermolecular potentials designed to model hydrogen bonding[3] was investigated which suggested an alternative route to describing the thermodynamics of associating mixtures. Wertheim's theory offers especially powerful routes to relating the microscopic and macroscopic perspectives. We can build a bridge from the chemical theory to Wertheim's by examining the form assumed for $\Delta C_P /R$. Recall that we previously assumed $\Delta C_P /R = -1$ for the van der Waals associating fluid and ESD equations. For a molecule which only possesses 2 bonding sites, Wertheim's theory is exactly equivalent to the chemical theory if we assume a somewhat more complicated form for $\Delta C_P /R$. The Wertheim formula is:

$$\frac{\Delta C_P}{R} = \frac{\ln[\exp(\varepsilon_{HB} / RT) - 1] + (\varepsilon_{HB} / RT_c)(1 - 1 / T_r) - \ln(T_r)}{\ln(T_r) - (1 - 1 / T_r)} \qquad 15.110$$

whereby the formula for α takes on a remarkably subtle change.

$$\alpha = \frac{\eta}{(1 - 1.9\eta)} K^{AD}\left[\exp\left(\frac{\varepsilon_{HB}}{RT}\right) - 1\right] \qquad 15.111$$

The superscript "AD" indicates bond formation between a proton acceptor and a donor in the form of a linear chain. From an engineering perspective, there is little to distinguish the two forms of α other than analyzing which provides the basis for the most accurate correlations and predictions of engineering phase equilibrium data. Suresh and Elliott carried out extensive evaluations for hydrocarbon + water and hydrocarbon + alcohol systems including LLE as well as VLE. They found that the Wertheim form provided slightly greater accuracy.

The ESD program uses this formula for α.

Pure Component Parameters for the ESD Equation

The pure component parameters for a number of associating components are given in Table 15.5, and many more are given in the computer file esdparms.txt. For any component which does not self-associate (e.g. ethers, esters, most ketones, and halocarbons as well as hydrocarbons), the equation of state parameters may be estimated from T_c, P_c, and ω in much the same manner as applied in the Peng-Robinson equation:

$$c = 1.0 + 3.535\omega + 0.533\omega^2 \qquad 15.112$$

$$Z_c = (1 + 0.115/c^{1/2} - 0.186/c + 0.217/c^{3/2} - 0.173/c^2)/3 \qquad 15.113$$

$$b = \frac{RT_c}{P_c} \frac{Z_c^2\left[-(1.9k_1Z_c + 3a) + \sqrt{(1.9k_1Z_c + 3a)^2 + 4a(4c - 1.9)(9.5q - k_1) / Z_c}\right]}{2a(4c - 1.9)} \qquad 15.114$$

1. Suresh, S. J., Elliott, J.R. *Ind. Eng. Chem. Res.* 31:2783 (1992).
2. Puhala, A. S., Elliott, J. R. *Ind Eng. Chem Res.* 32:3174 (1993).
3. Wertheim, M. S. *J. Stat. Phys.* 42:477 (1986) and references therein.

Table 15.5 *Pure component parameters for a number of associating components according to the ESD equation. More parameters are available in the file* esdparms.txt *available on the Internet.*

Component	ε/k (K)	$b(cm^3/mole)$	c	ε_{HB}/RT_c	K_{AB}'
water	427.25	9.412	1.0053	4.00	0.1000
H2S	333.84	11.677	1.0416	2.00	0.0442
methanol	326.06	20.366	1.1202	5.17	0.0226
ethanol	269.72	23.540	1.5655	4.86	0.0283
1-propanol	242.51	25.124	2.7681	2.50	0.1000
2-propanol	236.54	27.701	2.3148	3.75	0.0500
phenol	354.33	29.996	2.0972	2.14	0.1220
acetone	247.70	30.273	2.1001	0.51	0.1000

$$Y_c = \frac{Z_c^3}{a}\left(\frac{RT_c}{bP_c}\right)^2 \tag{15.115}$$

$$\frac{\varepsilon}{k} = T_c \ln(Y_c + 1.0617) \tag{15.116}$$

where $a = 1.9(9.5q - k_1) + 4ck_1$ and $k_1 = 1.7745$

In cases of associating compounds, the ESD equation typically imposes the constraint that the critical temperature be matched and the error in vapor pressure be minimized. For some especially important compounds, like water, the optimization of pure component parameters is broadened to recognize that predictions for mixtures can be sensitive to the choice of pure component parameters. For example, predictions of water + hydrocarbon liquid-liquid equilibria are very sensitive to the volume parameter for water, b, but many values of b can give similar accuracy for the vapor pressure of water. Therefore, the value chosen for the b parameter is the one which optimizes both vapor pressure correlation and liquid-liquid equilibria correlation.

15.9   EXTENSION TO COMPLEX MIXTURES

The real motivation for physical chemical theory is to provide an alternative to local composition theory in the explanation of non-ideal mixture behavior. Our presentation would not be complete without an analysis of the mixture behavior. Fortunately, the groundwork has been laid such that there is no problem making this extension. Many of the derivations are entirely analogous to the derivations for pure fluids. The primary extension required is to generalize the A^{assoc} and mass balance relations to handle any number of associating species. At first this seems daunting. But we are lucky again in that the problem has been analyzed previously and we can take advantage of some of the results. Particularly valuable is the theory developed by Wertheim and its various adaptations by Chapman and coworkers. The formalism developed by Wertheim is very useful because it leads to very simple extensions to mixtures as soon as a small amount of groundwork is laid.

The primary adjustment required to apply Wertheim's theory is to rearrange the theory in terms of actual numbers of proton acceptor and proton donor sites in the fluid. The extent of association is then characterized in terms of the fraction of acceptor sites *not* bonded, X_j^A. This "fraction of acceptor sites not bonded" is closely related to the fraction monomer, x_M. To see the relationship, we must write the mass balances. We note that the linear association of chains that we developed in the previous section is consistent with the assumption of two bonding sites per molecule, one acceptor and one donor. We can easily count the number of acceptors and donors in such linear chains.

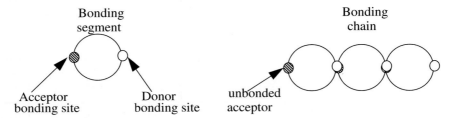

By noting that one unbonded acceptor is left in each bonded chain, we obtain the number of unbonded acceptors:

$$n^A = \sum_i n_i = n_T$$

But the total number of acceptors is given by noting that there are "i" total acceptors per i-mer,

$$n_o^A = \sum_i i \cdot n_i = n_o$$

Therefore,

$$X^A = n^A/n_o^A = n_T/n_o = x_M$$

There is a further simplification that results from treating the bonding sites instead of the bonding molecules. The fraction of sites bonded can be perceived as a simple product of the bonding probabilities. First, note that the fraction bonded and the fraction not bonded must sum to unity.

$$X^A + x^{AD} = 1 \qquad\qquad 15.117$$

The first term in Eqn. 15.117 is the fraction of acceptors not bonded. The second term is the fraction of acceptors that are bonded, whether they are bonded in monomers, dimers, trimers,... In principle, the second term is an infinite sum. From an acceptor site perspective, however, recall that we have assumed that the transition from the unbonded state to the bonded state is the same, regardless of the degree of polymerization for that i-mer. That transition can be represented by:

$$\frac{x^{AD}}{X^A X^D} = \alpha_{AD} \qquad\qquad 15.118$$

Combining these equations, we have

$$1 - X^A = X^A \cdot X^D \cdot \alpha_{AD}$$

The term on the left is the fraction of acceptors that are bonded, and the term on the right expresses the observation that acceptors and donors must be unbonded in order to be available for bonding.

By noting that one donor bonds for every acceptor, we see that $X^A = X^D$. Then we write Eqn. 15.118 in terms of X^A and α_{AD} and we can solve the quadratic equation to obtain

$$X^A = \frac{-1 + \sqrt{1 + 4\alpha_{AD}}}{2\alpha_{AD}}$$

15.119

The fact that this expression for X^A is identical to the one obtained previously confirms that the infinite sum can be replaced by 15.118. Anticipating the extension to mixtures, can you guess the simple form of this product formula for a mixture?

The extension of the mass balance relation becomes:

$$1 - X_i^A = X_i^A \sum_j x_j X_j^D Nd_j \alpha_{ij}$$

15.120

where x_i is the superficial mole fraction of component i. This equation simply states that the probability for an acceptor on the ith species to bond increases when there are donors on other molecules, and it decreases proportional to the mole fraction when the donor species are diluted by non-associating species. These suggestions are easy to accept. The next extension is to A^{assoc}. The result is:

$$\frac{A^{assoc}}{n_o RT} = \sum_i x_i Nd_i \left(2 \ln X_i^A + 1 - X_i^A\right)$$

15.121

where x_i is the superficial mole fraction of component i and Nd_i is the number of hydrogen bonding segments of component i. This expression suggests that the free energy for the mixture is simply the molar average of the pure component free energies. That seems possible, but free energies involve entropy and simple molar averages can be inappropriate when entropy is involved. On the other hand, there is a logarithmic term in the equation and that might take care of the entropic concerns. How can we confirm that a simple molar average makes sense? One way is to consider the impact that this form has on the pressure at low density. At low density, the primary impact of association is to reduce the number of molecules. This reduction is simply proportional to the extent of association for each species as evident in the expression for $Z_i^{assoc} \sim X_i^A$. In terms of n_T/n_0, the mole fraction weighted sum of X_i^A gives the total result for n_T/n_0. Then it becomes clear that $PV/n_0 RT = n_T/n_0$ at low density, and the molar average must be appropriate.

The molar average extensions to the mass balances and the free energy make it clear that the solution for mixtures can be solved in principle. We have enough equations to determine all our unknowns and we can take the necessary derivatives. One thing that remains is to define a combining rule for α_{ij}. We could imagine many possibilities, but Elliott[1] shows that many simplifications accrue if we choose a geometric mean. That is,

$$\alpha_{ij} = \sqrt{\alpha_{ii}\alpha_{jj}}$$

15.122

To develop an appreciation for these simplifications, consider the following argument. In principle, Eqn. 15.120 results in a non-linear system of equations since the subscript i cannot be factored out of the summation, but the geometric mean definition changes that. The fractions of donors can be

1. Elliott, J. R. *Ind. Eng. Chem. Res.* 35:1624 (1996)

instantly replaced by the fractions of acceptors because of symmetry. The key simplification is to rearrange Eqn. 15.120 such that,

$$\frac{1}{\sqrt{\alpha_{ii}}}\left(\frac{1}{X_i^A} - 1\right) = \sum_j x_j Nd_j X_j^A \sqrt{\alpha_{jj}}$$

15.123

Note that the summation on the right-hand side holds for all species, so

$$\frac{1}{\sqrt{\alpha_{ii}}}\left(\frac{1}{X_i^A} - 1\right) = \frac{1}{\sqrt{\alpha_{11}}}\left(\frac{1}{X_i^A} - 1\right) = \sum_j x_j Nd_j X_j^A \sqrt{\alpha_{jj}} \ \text{for all } i.$$

15.124

Rearranging:

$$X_j^A = \left[1 + \frac{\sqrt{\alpha_{jj}}}{\sqrt{\alpha_{ii}}}\left(\frac{1}{X_i^A} - 1\right)\right]^{-1}$$

15.125

Defining a quantity F and collecting a common denominator,

$$F \equiv \frac{1}{\sqrt{\alpha_{ii}}}\left(\frac{1}{X_i^A} - 1\right) = \sum_j \frac{x_j Nd_j \sqrt{\alpha_{jj}}}{1 + F\sqrt{\alpha_{jj}}}$$

15.126

This results in a simple equation for F which may be solved by iteration by taking $F = 0$ and $F = 1$ as initial guesses in a secant iteration. Note that for strongly associating species, $\alpha_{ii} \to \infty$. Thus,

$$\lim_{\alpha \to \infty} F \to \sum_j \frac{x_j Nd_j \sqrt{\alpha_{jj}}}{F\sqrt{\alpha_{jj}}} \Rightarrow F^2 \to \sum_j x_j Nd_j$$

15.127

In the limit of $\alpha \to 0$, $F \to 0$ by similar reasoning.

Given a value for F, it is straightforward to solve for X_i^A of all ith species. A similar set of relations exists for the derivatives of extents of bonding with respect to density and mole numbers. Furthermore, a set of simplifications which are specific to the ESD form of the equation of state make it possible to obtain an especially simple form for the final equations. These are detailed by Elliott,[1] and the interested reader can easily refer to that source for further information. The resulting equations are:

$$Z^{assoc} = \frac{-F^2}{1 - 1.9\eta}$$

15.128

$$\ln(\hat{\varphi}_k^{assoc}) = 2\ln(X_k^A) - F^2 \frac{1.9 b_k \rho}{1 - 1.9\eta}$$

15.129

1. Elliott, J. R. *Ind Eng. Chem. Res.*, 35:1624 (1996). Note that a contribution of "-1" was erroneously omitted from Eqns. 24, 26, and 27.

At first glance, this equation may not seem consistent with the definition of $\ln\varphi^{assoc}$ for a pure fluid implied by Eqn. 15.81 on page 555. To show that it is, note that $b_k\rho$ becomes η for the pure fluid, and

$$-F^2 \cdot 1.9\eta/(1 - 1.9\eta) = Z^{assoc} \cdot 1.9\eta = (1 - X^A)[1 - 1/(1 - 1.9\eta)] = (1 - X^A) + Z^{assoc}$$

Note the special property that X_i^A must go to unity when α_{ii} goes to zero, because of the cross-multiplication. This means that F need not be zero, even for non-associating components. The reason that the fugacity coefficient changes for those components is purely due to the change in the number of moles. The program for implementing the ESD equation is available on the Internet. It can be applied to pure fluids as well as mixtures.

15.10 STATISTICAL ASSOCIATING FLUID THEORY (SAFT)

The formulas for hydrogen bonding can be generalized to consider virtually any type of bond formation, not just linear acceptor-donor bond formation. The general expression becomes:

$$\boxed{\frac{A^{assoc}}{n_oRT} = \sum_{i,\,all\,B} x_i\left(\ln X_i^B + \frac{1 - X_i^B}{2}\right)} \qquad 15.130$$

where B indicates that the bonding site can be any type of bond that the user has in mind. Note that we obtain Eqn. 15.121 when bonding is restricted to sites of type A and D. Eqn. 15.130 indicates that the free energy due to bonding derives primarily from counting the number of bonds, and little else. To begin, imagine a type of bonding site C which can only bond with itself, with only one C site permitted per molecule. Then we could obtain dimers and nothing else. This is in fact the approach to modeling carboxylic acids typically applied to adaptations of Wertheim's theory. Now imagine types of bonding sites C_1 and C_2 which can only bond with themselves in an equimolar ternary mixture with C_1 on one component, C_2 on the second component, and one each on the third component. Then the largest species which could form would be a trimer. For our final step, imagine taking the limit as ε_{HB} approaches infinity. Then the extent of association would go to completion and, since we have a perfect balance of beginning, middle, and end segments, we would obtain a perfectly pure trimer chain fluid. Similarly, we could obtain pure 4-mers, 5-mers, etc. The elegance of this approach is that the thermodynamics for the transition from spheres to any length chain are completely described by Eqn. 15.130. We can now extend any equation of state for spheres to an equation of state for chains by simply adding the contribution for chain formation in the same way that we added the contribution for association. This is the essence of the SAFT equation of state. The SAFT equation was originally developed by Chapman et al. but immediately revised by Huang and Radosz.[1] Since then it has been adapted to a broad range of applications. The basic equations are:

$$Z = 1 + Z^{seg} + Z^{chain} + Z^{assoc} \qquad 15.131$$

$$Z^{seg} = m\left[\frac{4\eta - 2\eta^2}{(1 - \eta)^3} + \sum_i\sum_j jD_{ij}\left[\frac{u}{kT}\right]^i\left[\frac{\eta}{\tau}\right]^j\right] \qquad 15.132$$

1. Huang, S. and Radosz, M. *Ind. Eng. Chem. Res.* 29:2284 (1990); *Ind. Eng. Chem. Res.* 30:1994 (1991)

$$Z^{chain} = (1 - m)\left[\frac{2.5\eta - \eta^2}{(1 - \eta)(1 - 0.5\eta)}\right]$$

15.133

$$Z^{assoc} = \rho \frac{\partial A^{assoc}\big/RT}{\partial \rho}$$

15.134

$$\frac{A^{assoc}}{RT} = \sum_i x_i Nd_i\left[2 \ln(X_i^A) + 1 - X_i^A\right]$$

15.135

$$\eta \equiv \frac{\pi N_A}{6} \rho d^3 \sum_i x_i m_i$$

15.136

$$d^3 \equiv \frac{\displaystyle\sum_i \sum_j x_i x_j m_i m_j d_{ij}^3}{\left(\displaystyle\sum_i x_i m_i\right)^2}$$

15.137

$$d_{ij} = \left(d_{ii} + d_{jj}\right)/2$$

15.138

$$d_{ii} = \sigma_{ii}\left[1 - C \exp\left(\frac{-3u_{ii}^o}{kT}\right)\right]$$

15.139

$$\frac{u}{kT} \equiv \frac{\displaystyle\sum_i \sum_j x_i x_j m_i m_j \frac{u_{ij}}{kT} d_{ij}^3}{\displaystyle\sum_i \sum_j x_i x_j m_i m_j d_{ij}^3}$$

15.140

$$u_{ij} = \left(u_{ii} u_{jj}\right)^{1/2}(1 - k_{ij})$$

15.141

$$u_{ii} = u_{ii}^o\left[1 + \frac{e}{kT}\right]$$

15.142

$$m = \sum_i x_i m_i$$

15.143

where

m is a shape factor equal to the number of spherical segments per molecule.

Nd is the number of segments per molecule possessing hydrogen bonding sites.

σ is the segment diameter.

u^o is the characteristic dispersion energy per segment.

e, D_{ij}, C, τ are generalized constants given by Huang and Radosz (1990).

The SAFT equation is different from the ESD equation in that it applies the Carnahan-Starling equation for the hard sphere contribution and the attractive contribution of Chen and Kreglewski. But the Carnahan-Starling equation gives results very similar to those in Figure 15.8, and the attractive contribution of Chen and Kreglewski has its roots in molecular simulation data for the same square-well spheres treated by Sandler and Lee. The SAFT equation is especially appealing because of the self-consistent manner in which it derives the thermodynamics of polymer chain formation. Nevertheless, an implicit assumption of this approach is that all molecules are composed of chains of non-overlapping spherical segments. This is clearly not the case for aromatic species and it is even questionable for n-alkane chains since a typical bond length is 0.15 nm but typical values of σ are near 0.4 nm. This is why non-integer values for the SAFT shape parameter, m, are common.

One of the more important practical differences between the SAFT equation and the ESD equation is the manner in which the size, shape, and dispersion energy parameters are determined. The accepted procedure for SAFT is to simply optimize the representation of vapor pressure and liquid density data; the experimental value of the critical point is ignored. In practice, this provides more flexibility and simpler adaptation to polymers since the critical properties of polymers are not known and cannot be used in the generalized ESD characterizations. On the negative side, this also means that regressions must be performed for every component before the SAFT equation can be applied, and different researchers may obtain very different results depending on each exact characterization of the pure fluid parameters.

15.11   SUMMARY ANALYSIS OF ASSOCIATION MODELS

A simple way of remembering the qualitative conclusions of this analysis can be derived by considering the behavior of the fugacity coefficient. One can easily demonstrate that the fugacity coefficient of the monomeric species is insensitive to the extent of association if it is expressed on the basis of the true number of moles in the associated mixture. But all of our phase equilibrium algorithms are based on the fugacity divided by the superficial mole fraction; for example, the flash algorithm is the same for any equation of state. The relation between the two fugacity coefficients is given by

$$\hat{\varphi}_i = \frac{\hat{f}_i}{x_i P} = \frac{\hat{f}_i}{x_M P}\frac{x_M}{x_i} = \frac{\hat{\varphi}_M x_M}{x_i}$$

15.144

This means that we must simply multiply the fugacity coefficient from the usual equation of state expression by the ratio of true to superficial mole fraction. Since this ratio is always less than one, we see that the effect of association is to suppress the effective fugacity of the associating species. Eqn. 15.144 may be multiplied by P and rearranged to read

$$x_i \hat{\varphi}_i P = x_M \hat{\varphi}_M P$$
$$\hat{f}_i = \hat{f}_M$$

15.145

which tells us that the superficial fugacity is the same as the fugacity of the monomer in the true mixture.

For mixtures, elevation of the monomer mole fraction by breaking the association network accounts for VLE quite accurately. Fig. 15.9 that illustrates the benefit of a chemical physical model relative to a purely physical equation like the Peng-Robinson equation follows. The figure

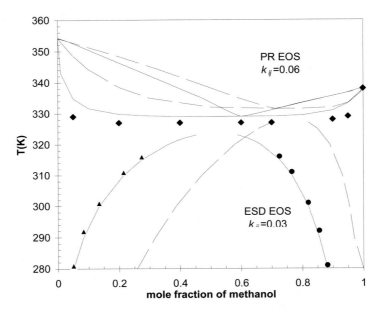

Figure 15.9 *T-x,y diagram for the system methanol + cyclohexane.*

depicting the methanol + cyclohexane system shows the improved accuracy in representing simultaneous LLE and VLE when hydrogen bonding is recognized. Notice the change in the skewness of the curves when hydrogen bonding is applied. The hydrogen bonding model is accomplishing this change in skewness as a clear and understandable explanation of the physics. By contrast, the van Laar model in Chapter 11 altered the skewness by simply ignoring the physics. We would expect that the stronger physical basis would provide greater capability for extrapolations to multicomponent mixtures. Unfortunately, remarkably few multicomponent studies have been performed to date. Hence, there is no single recommended method for treating non-ideal multicomponent solutions at this time.

From a theoretical perspective, however, we may still feel uncomfortable with having made several sweeping assumptions with little justification besides their making the equations easier to solve. This may not seem like much of an improvement over local composition theory. On the other hand, the assumptions could be reasonably accurate; they simply need to be tested. As in the case of local composition theory, molecular simulations provide an effective method of testing the assumptions implicit in the development of a theory. Fig. 15.10 shows a comparison of Eqn. 15.119 to molecular simulation results and to Wertheim's theory.[1] It can be seen that the above assumptions lead to reasonably accurate agreement with the molecular simulations and therefore they represent at least a self-consistent theory of molecular interactions.

This is not to say that the association model completely solves all problems. Local composition effects are real and should be incorporated into the mixing rules. Evidence supporting this step can be found in the anomalous behavior of the methane + hexane system. If such local composition effects are so prominent for non-associating solutions, they should be accounted for at all times. As an example of other problems, the association network of water seems to be different enough from

---

1. Liu, J-X., Elliott, J. R., *Ind. Eng. Chem. Res.* 35:1234 (1996).

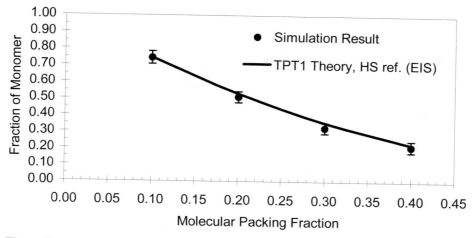

Figure 15.10 *DMD-B simulation of hard dumbbell methanol with reduced bond length l/σ = 0.4, at T = 300K and $N_A \varepsilon_{HB}/R = 2013K$. TPT1 theory is an adaptation of Wertheim's theory.*

that of alcohols that a more sophisticated model will be necessary to represent difficult solutions like hydrocarbons + water to the high degree of accuracy (ppm) required by organizations like the Environmental Protection Agency. Furthermore, the cross-associations between different species can be extremely complicated and require substantially more investigation to develop reliable engineering models. Finally, it is well-known that "non-additive" effects play a significant role in aqueous and alcoholic solutions.[1] That is, the energy of network formation changes in a way that cannot be understood based only on a simple potential model for a single water molecule. These peculiarities may seem esoteric, but they are key obstacles which prevent us from unshrowding many of the mysteries of biomolecular solutions. Other areas of application such as polymer solutions involving association, as in nylon, can also be imagined. These are the areas which remain to be explored. The methods for engaging in this exploration predominantly involve mathematically formalizing our treatment of the radial distribution function through applications of statistical mechanics. At this point, we leave this engagement to the "satisfaction and good fortune" of the reader.

15.12   HOMEWORK PROBLEMS

15.1 Consider a dilute isothermal mixing process of acetic acid(1) in benzene(2). For the dilute region (say up to 5 mol% acid), draw schematically curves for:

$\overline{S}^E_1$ versus x_1, $\overline{H}^E_1$ versus x_1, $\overline{G}^E_1$ versus x_1.

Briefly justify your schematic graphs with suitable explanations. Take standard states as the pure substances.

15.2 Acetic acid dimerizes in the vapor phase. Show that the fugacity of the dimer is proportional to the square of the fugacity of the monomer.

---

1. Hait et al., *Ind. Eng. Chem. Res.* 32:2905 (1993).

15.3 By assuming that the equilibrium constant for each successive hydrogen bond is equal in the generalized association approach developed in this chapter, what assumptions are being made about the Gibbs energy, enthalpy, and entropy for each successive hydrogen bond?

15.4 The value of the excess Gibbs energy at 298 K for an equimolar chloroform(1) + triethylamine(2) system is $G^E = -0.91$ kJ/mol. Assuming only a 1-1 compound is formed, model the excess Gibbs energy with ideal chemical theory, and plot the P-x-y diagram.

15.5 Suppose that, due to hydrogen bonding, the system $A + B$ forms a 1-1 complex in the vapor phase when mixed. Neither pure species self-associates in the vapor phase. The equilibrium constant for the solvation is $K_{AB} = 0.8$ bar^{-1} at 80°C. At 80°C, a mixture with a superficial (bulk) mole fraction of $x_A = 0.5$ is all vapor at 0.78 bar. Calculate the fugacity coefficient of A in the vapor phase using ideal chemical theory at this composition, temperature, and pressure. Use hand calculations.

15.6 At 143.5°C, the vapor pressure of acetic acid is 2.026 bar. The dimerization constant for acetic acid vapor at this temperature is 1.028 bar^{-1}. The molar liquid volume of acetic acid at this temperature is 57.2 cm^3/mol. Calculate the fugacity of pure acetic acid at 143.5°C and 10 bar. Use hand calculations.

15.7 An $A + B$ mixture exhibits solvation in the liquid phase, which is to be represented using ideal chemical theory. Because of a Lewis acid/base interaction, the system is expected to form a 1-1 compound.

(a) Which one of the following sets of true mole fractions are correct for the system using an equilibrium constant of 3.2 to represent the complex formation at a superficial composition $x_A = 0.4$?

	x_{AM}	x_{BM}	x_{AB}
Set I	0.2096	0.4731	0.3173
Set II	0.2646	0.3983	0.3372

(b) Based on your answer for part a, what are the superficial activity coefficients of A and B?

15.8 Water and acetic acid do not form an azeotrope at 760 mmHg. The normal boiling point of acetic acid is 118.5°C. Therefore, at 118.5°C and 760 mmHg, the mixture will exhibit only vapor behavior across the composition range. The following equilibrium constants have been fitted to represent the vapor-phase behavior:[1]

	dimer, $-\log_{10}K$ where K in (mmHg)$^{-1}$	trimer, $-\log_{10}K$ where K in (mmHg)$^{-2}$
acetic acid	10.108−3018/$T(K)$	18.63−4960/$T(K)$
water	6.881−808.2/$T(K)$	
acetic acid/water complex	same as water	

1. Tsonopoulos, C., Prausnitz, J. M., *Chem Eng. J.,* 1:273 (1970).

(a) Let compound A be acetic acid and B be water. Calculate the true mole fractions of all the species from $y_A = 0.05$ to $y_A = 0.95$. At what superficial mole fraction does each specie show a maximum true mole fraction? What is the relation of this superficial mole fraction with the compound's stoichiometry?

(b) Plot the fugacity coefficient of acetic acid and water as a function of acetic acid mole fraction. What is the physical interpretation of the rapid change of the acetic acid fugacity coefficient in the dilute region, if the water fugacity coefficient doesn't show such a dramatic trend in its dilute region?

15.9 Beginning with the Gibbs energy of mixing for a liquid binary system expressed in superficial mole fractions, $\Delta G_{mix}/RT = n_A \ln x_A \gamma_A + n_B \ln x_B \gamma_B$, introduce the concepts of chemical theory to show that the Gibbs energy of mixing is equivalently given by the ideal Gibbs energy of mixing for the true species plus the Gibbs energy of forming the complexes,

$$\Delta \underline{G}_{mix} = RT \sum_i n_i \ln x_i \alpha_i + \sum_i n_i \Delta G^o_{f,i},$$ where the summations are over true species. The

standard state is taken as the monomers in their respective pure components at the system T and P.

15.10 Furnish a proof that the concentration of true species i is maximum at composition $x_A^* = a_i/(a_i + b_i)$, $x_B^* = b_i/(a_i + b_i)$ where a_i and b_i are given in Eqn. 15.1. [Hint: The Gibbs-Duhem equation is useful for relating derivatives of activity.]

15.11 Use the ESD equation to model the monomer, dimer, and trimer in the vapor and liquid phases of saturated water at 373 K, 473 K, and 573 K. How does the monomer fraction of saturated vapor change with respect to temperature? How does monomer fractions of saturated liquid change?

15.12 Derive the equations for determining the critical point of the ESD equation based on ε_{HB} and K^{AD} being zero by noting that $dF/dZ = 0$ and $d^2F/dZ^2 = 0$, where $F = Z^3 + a_2Z^2 + a_1Z + a_0$ when hydrogen bonding is negligible.

15.13 Plot P against V at 647.3 K for water with the ESD equation using the characterization analogous to Eqns. 15.73–15.76. Apply the equal area rule and determine the vapor pressure at that temperature. Raise the temperature until the areas equal zero and compare this temperature to the true value of 647.3 K.

15.14 Apply the ESD equation to the methanol + benzene system and compare to the data in Perry's Handbook based on matching the bubble pressure at the azeotropic point. Prepare a T-x-y diagram and determine whether the ESD equation indicates a liquid-liquid phase split for any temperatures above 250 K. Perform the same analysis for the Peng-Robinson EOS. Do you see any differences? Compare to Fig. 10.3 on page 340.

15.15 Use the ESD equation to estimate the mutual LLE solubilities of methanol and n-hexane at 285.15 K, 295.15 K and 310.15 K. Use the value of $k_{ij} = 0.03$ as fitted to a similar system in Fig. 15.6 on page 547.

15.16 The hydrogen halides are unusual. For example, here are the critical properties of various hydrogen halides:

	MW	T_c(K)	P_c(bar)	Z_c	ω
HF	20.00	461.0	64.88	0.12	0.372
HCl	36.46	324.6	83.07	0.249	0.120
HBr	80.91	363.2	85.50	0.283	0.063

Experimental data for the vapor pressure and the apparent molecular weight of HF vapor are:

$T(K)$	P^{sat}	MW^{sat}	MW at 1 bar
227.3	0.0519	92.8	117.6
243.9	0.1265	85.0	112.5
277.8	0.5780	69.8	85.4
303	1.4353	58.4	43.0
322.6	2.6178	50.3	21.8

These apparent molecular weights have been found by measuring the mass density of the vapor and comparing with an ideal gas of molecular weight 20. Assuming that HF forms only monomers and hexamers, use the ESD EOS with $c = q = 1$ for both monomer and hexamer to match this value of Z_c, and fit the vapor density data as accurately as possible in the least squares sense.

15.17 (a) Compute the values of K_a', a/bRT_c, x_{Mc}, and $b\rho_c$ for methanol and ethanol according to the van der Waals hydrogen bonding equation of state.

(b) Assuming an enthalpy of hydrogen bonding of 24 kJ/mole and $\Delta C_P = -R$, calculate the acentric factors for methanol and ethanol according to the vdw-HB EOS.

15.18 Derive the association model for the Peng-Robinson model, using the van't Hoff formula with $\Delta C_P/R = -1$. Extend the homomorph concept by applying $\omega^{PR} = \omega^{homo}$, where ω^{homo} is the acentric factor for the nonassociating homomorph and ω^{PR} is the acentric factor substituting for the associating compound into the Peng-Robinson expression for a.

(a) For methanol, determine the values of K_a', b, a/bRT_c, x_{Mc} that match the critical point.

(b) Determine the vapor pressure at $T_r = 0.7$ for methanol assuming a hydrogen bonding energy of 15 kJ/mole and compare to the experimental value. Infer the acentric factor and compare to the experimental value.

(c) Plot log P_r^{sat} vs. T_r^{-1} for the Peng-Robinson EOS, the Peng-Robinson hydrogen bonding EOS, and experiment.

15.19 Acetic acid has a much stronger tendency to dimerize than any alcohol. Therefore, it is not reasonable to assume that $K_{a2} = K_{a3} = ...$ for acetic acid. The assumption is reasonable for $K_{a3} = K_{a4} = ...$, however. We can supplement the theory by adding a single additional equation for the dimerization reaction with an effective equilibrium constant equal to the ratio of the true K_{a2} divided by the linear association value. Assume that the linear association is negligible for the saturated vapor at ~300–350 K.

(a) Determine the value of K_{a2} that matches the saturated vapor compressibility factor in that range. Let $N_A \varepsilon_{HB}/R = 4000$ K for the dimerization.

(b) Determine the values of K_{AD}, b,α_c, ε_{HB} that match the critical point; and

(c) Determine the values of K_{AD}, b,α_c, ε_{HB} that match the vapor pressure at $T_r = 0.7$ for acetic acid.

15.20 Extend the ESD equation to compounds with more than one bonding segment.

(a) Consider ethylene glycol as a compound with both an associating head and tail. Extend the mixture analysis to treat this case with two bonding segments ($Nd = 2$).

(b) Treat water by the same model noting that water is merely a "very short glycol." Determine the acentric factor of the Peng-Robinson hydrogen bonding EOS with $\Delta H = 15$ kJ/mole.

15.21 Show that the result for Z^{assoc} is obtained by taking the appropriate derivative of A^{assoc}.

GLOSSARY

Adiabatic—condition of zero heat interaction at system boundaries.

Association—description of complex formation where all molecules in the complex are of the same type.

Azeotrope—mixture which does not change composition upon vapor-liquid phase change.

Barotropy—the state of a fluid in which surfaces of constant density (or temperature) are coincident with surfaces of constant pressure.

Binodal—condition of binary phase equilibrium.

Dead State—a description of the state of the system when it is in equilibrium with the surroundings, and no work can be obtained by interactions with the surroundings.

Diathermal—heat conducting, and without thermal resistance, but impermeable to mass.

Efficiency—see isentropic efficiency, thermodynamic efficiency, thermal efficiency.

EOS—Equation of State

Fugacity—characterizes the escaping tendency of a component, defined mathematically.

Heteroazeotrope—mixture that is not completely miscible in all proportions in the liquid phase and like an azeotrope cannot be separated by simple distillation. The heterazeotropic vapor condenses to two liquid phases, each with a different composition than the vapor. Upon partial or total vaporization, the original vapor composition is reproduced.

Infinite Dilution—description of a state where a component's composition approaches zero.

Irreversible—a process which generates entropy.

Isenthalpic—condition of constant enthalpy.

Isentropic—condition of constant entropy.

Isentropic efficiency—ratio characterizing actual work relative to ideal work for an isentropic process with the same inlet (or initial) state and the same outlet (or final) pressure. See also *thermodynamic efficiency, thermal efficiency*.

Isobaric—condition of constant pressure.

Isochore—condition of constant volume. See *isosteric*.

Isopiestic—constant or equal pressure.

Isopycnic—condition of equal or constant density.

Isolated—a system isolated from the surroundings, a closed system with no heat or work interactions across boundaries.

Isosteric—condition of constant density. See *isochore*.

Isothermal—condition of constant temperature.

LLE—liquid-liquid equilibria.

Master Equation—$U(V,T)$.

Measurable Properties—variables from the set $\{P,\ V,\ T,\ C_P,\ C_V\}$ and derivatives involving only $\{P,\ V,\ T\}$.

Metastable—signifies existence of a state which is non-equilibrium, but not unstable, e.g., superheated vapor, subcooled liquid, which may persist until a disturbance creates movement of the system towards equilibrium.

Nozzle—a specially designed device which nearly reversibly converts internal energy to kinetic energy. See *throttling*.

Polytropic exponent—The exponent n in the expression PV^n = constant.

Quality—the mass fraction of a vapor/liquid mixture that is vapor.

rdf—radical distribution function.

Reference State—A state for a pure substance at a specified (T,P) and type of phase (S,L,V). The reference state is invariant to the system (P,T) throughout an entire thermodynamic problem. A problem may have various standard states, but only one reference state. See also *Standard State*.

Sensible heat changes—heat effects accompanied by a temperature change.

Specific heat—another term for C_P or C_V.

Specific property—an intensive property per unit mass.

SLE—solid-liquid equilibria.

Solvation—description of complex formation where the molecules involved are of a different type.

Spinodal—condition of instability, beyond which metastability is impossible.

Standard Conditions—273.15 K and 1 atm, *standard temperature and pressure*.

Standard State—A state for a pure substance at a specified (T,P) and type of phase (S,L,V). The standard state T is always at the T of interest for a given calculation within a problem. As the T of the system changes, the standard state T changes. The standard state P may be a fixed P or may be the P of the system. Gibbs energies and chemical potentials are commonly calculated relative to the standard state. For reacting systems, enthalpies and Gibbs energies of formation are commonly tabulated at a fixed pressure of 1 bar and 298.15 K. A temperature correction must be applied to calculate the standard state value at the temperature of interest. A problem may have various standard states, but only one reference state. See also *Reference State*.

State of aggregation—solid, liquid, or gas.

Steady-state—open flow system with no accumulation of mass and where state variables do not change with time inside system boundaries.

STP—standard temperature and pressure, 273.15 K and 1 atm. Also referred to as *standard conditions*.

Subcooled—description of a state where the temperature is below the saturation temperature for the system pressure, e.g., subcooled vapor is metastable or unstable, subcooled liquid is stable relative to the bubble point temperature; superheated vapor is stable, superheated liquid is metastable or unstable relative to the dew point temperature; subcooled liquid is metastable or unstable relative to the fusion temperature.

Superheated—description of a state where the temperature is above the saturation temperature for the system pressure. See *Subcooled*.

Thermal efficiency—the ratio or work obtained to the heat input to a heat engine. No engine may have a higher thermal efficiency than a Carnot engine.

Thermodynamic efficiency—ratio characterizing actual work relative to reversible work obtainable for exactly the same change in state variables for a process. The heat transfer for the reversible process will differ from the actual heat transfer. See also *isentropic efficiency, thermal efficiency*.

Throttling—a pressure drop without significant change in kinetic energy across a valve, orifice, porous plug or restriction, which is generally irreversible. See *nozzle*.

Unstable—a state that violates thermodynamic stability, and cannot persist. See also *metastable, spinodal*.

VLE—vapor-liquid equilibrium.

Wet Steam—a mixture of water vapor and liquid.

APPENDIX

SUMMARY OF COMPUTER PROGRAMS

Several programs are furnished with the text to help you learn the material and to assist in repetitive and/or complex calculations. Programs are available for: TI and HP calculators, IBM PC-compatible computers, and Microsoft® Excel. There is some duplication of capabilities among the different operating platforms in the sense that programs for the personal computers may also be available for the calculators. To install the programs on your computer or calculator, see Sections A.6 and A.7. The software is periodically updated. Visit the website for the latest version of the software and appendix.

A.1 HP48 CALCULATOR PROGRAMS

Several programs are available in the file THRMO within the HP directory. The programs are organized in directories grouped as: EQ.S, for several commonly applied equations (e.g., C_P^{ig}, P^{sat}) that use the solver function; PRI, for Peng-Robinson calculations with one component; VLK for vapor-liquid K-value calculations, including bubble point pressure of multicomponent mixtures by the Peng-Robinson equation (k_{ij} specified by user for up to 5 components). Specifics for each program are presented below in accordance with the appropriate directory. See Section A.10 if you are a first-time HP48G user.

EQ.S (Push the key below this heading to enter this directory. These are referred to as "menu keys.")

1. INTRP—simple linear interpolation

    Push the menu key below **INTRP**, then **rightshift-SOLVE-ENTER** to start the equation solver. The menu provides input slots for y_1, y_2, x_1, x_2, and the desired value of x. Fill in as appropriate and move the cursor until the slot for y is highlighted. Press the menu key labeled **SOLVE**.

2. IG—this is a directory with ideal gas properties as a function of T and P.

    Push the menu key below **IG**. The following variables are in the directory.

    Heq, Seq—Equations for Solver to use to calculate ΔH, ΔS, T or P.

T2,T1,P2,P1—Temperature in K, and P for states 1 and 2. (Pressure units don't matter for ΔS as long as they are the same for both states, however see V1, V2).

V1, V2—molar volumes at state 1 and 2, using $R = 8.314$ J/mol-K. These menu keys give molar volume in cm^3/mol if T in K and P in MPa.

ΔH, ΔS—quick calculation of ΔH and ΔS from variables stored in T1, T2, P1, P2 and C_P.

CPA, CPB, CPC, CPD—heat capacity constants from the appendix.

CP2—calculated value of C_P at T2.

ΔHeq, ΔSeq—The last value of ΔH and ΔS found by Solver, *not necessarily* the values corresponding to the current state variables.

EQ—The last equation used by Solver.

To use Solver, press **SOLVE**, **solve equation**, and use the **CHOOSE** menu key, followed by the arrow keys to select Heq or Seq. Fill in as appropriate and move the cursor until the slot for desired unknown is highlighted. Press the menu key labeled **SOLVE**.

3. ANTOI—calculation of vapor pressure given Antoine coefficients

Push the menu key below **ANTOI**, then **rightshift-SOLVE-ENTER** to start the equation solver. Use the **CHOOSE** menu key to select the natural logarithm or common logarithm equation. The menu provides input slots for *A, B, C,* and the desired value of *T*. Fill in as appropriate and move the cursor until the slot for P^{sat} is highlighted. Press the menu key labeled **SOLVE**. Be a little careful about units on this one and especially careful about the limitations on the temperature range for the Antoine coefficients.

4. V.P—calculation of vapor pressure according to the shortcut vapor pressure equation.

Push the menu key below **V.P**, then **rightshift-SOLVE-ENTER** to start the equation solver. The menu provides input slots for T_c, P_c, ω, and the mole fraction value of *T*. Fill in as appropriate and move the cursor until the slot for P^{sat} is highlighted. Press the menu key labeled **SOLVE**. Enter P^{sat} and highlight *T* to solve for T^{sat} at a given pressure. This equation is accurate to within 10% when $0.5 < T_r < 1.0$.

5. WT%—calculation of weight fraction given mole fraction or vice versa.

Push the menu key below **WT%**, then **rightshift-SOLVE-ENTER** to start the equation solver. The menu provides input slots for M_1, M_2, and the mole fraction value of X_1. Fill in as appropriate and move the cursor until the slot for weight fraction W_1 is highlighted. Press the menu key labeled **SOLVE**. Enter W_1 and highlight X_1 to solve for mole fraction given weight fraction.

PRI—PVT and departure properties by the Peng-Robinson equation

1. INCRT—run this first to enter critical constants and the acentric factor. Example: For water, $T_c = 647.3$, $P_c = 22.12$, $\omega = 0.344$ used in examples.

2. PVTF—gives volume(cm^3/mole) and fugacity(MPa). Prompts for P and T, then offers the user the cubic equation real root(s) for the compound's constants previously entered by INCRT or REF(see below). Example: for superheated steam at 0.01 MPa and 323 K, run INCRT (or REF), then press PVTF, enter $T = 323$, $P = 0.01$, choose the vapor root ($Z = 0.9987$), gives $V = 268,208$ cm^3/mol, $f = 0.00999$MPa. Scroll up to review the Z value.

3. ZALT—allows selection of alternate roots and reruns PVTF without changing P,T. This program is not necessary if you know which root to use. For multiple roots, the center root is unstable, and among the others, the most stable root is the one with the lower fugacity.

4. DEPFUN—calculates U,H,S departures for the last root selected from PVTF. Example: following the PVTF example above, press DEPFU to find $U - U^{ig}/RT = -0.00196$, $H - H^{ig}/RT = -0.0032$, $S - S^{ig}/R = -0.00196$.

Property Programs

5. REF—enters the critical properties, acentric factor, and heat capacity constants for the compound and the reference state temperature, pressure and state of aggregation (vapor or liquid root). After running this program, run PVTF to set the P and T of interest and select the root (optionally followed by ZALT). Example: For water, $T_c = 647.3$, $P_c = 22.12$, $\omega = 0.344$, CPA = 32.24, CPB = 1.924E-3, CPC = 1.055E-5, CPD = −3.596E-9, Tref = 273, Pref = 0.0006, choose liquid reference state (Z = 5.5E-6), departures at reference state are displayed (*e.g.* $U - U^{ig}/RT = -19.675$). Then run PVTF.

6. UHSG—calculates U,H,S,G using the last root selected from PVTF. This program uses the heat capacities and reference state entered by the REF program. Example: After running REF and PVTF as above, press UHSG to find: $U = 45918$ J/mol, $H = 48600$ J/mol, $S = 152.261$ J/mol-K, $G = -580$ J/mol. Scroll up to review P, T, V.

Comparing the calculations with the steam tables, at the last state, $S = 152.26$ J/mole-K = 8.4589 J/g-K; at the same pressure and 100°C (373), we obtain $S = 8.7313$J/g-K. The value for ΔS is then 0.2724 J/g-K. This compares to a value of 0.2730 from a typical steam table. This may be a significant error for a design project situation, but accurate enough for many quick, order of magnitude calculations.

γ—activity coefficients by the van Laar equations

1. VLA—van Laar A_{ij} calculation given activity coefficients of a binary solution at a single composition.

   Hit the menu key labeled VLA to input the values of activity coefficients,γ_1 and γ_2, and desired composition of component 1, x_1, A_{12} and A_{21} are displayed on the screen.

2. VLAR—van Laar activity coefficient calculation given A_{12} and A_{21} of a binary solution. Hit the menu key labeled VLAR to input the values of A_{12} and A_{21} and desired composition of component 1, x_1, γ_1 and γ_2 are displayed on the screen.

PRMIX—vapor-liquid K-ratios, and bubble point pressure for mixtures by the Peng-Robinson equation. The number of components must be less than or equal to five.

1. INCRT—input critical properties of components for Peng-Robinson and bubble-point calculations.

   Hit the menu key labeled **INCRT** to input the number of components, critical temperatures (type all critical temperatures in a row with a blank space between each), critical pressures, acentric factors, and binary interaction coefficients (k_{ij}'s). For critical pressures, temperatures, and acentric factors, type all values sequentially with a blank space between. For the k_{ij}'s, the first row gives values for interactions with component 1, the second row for component 2, and so forth. The value of 0 should be entered for the k_{ij} of each component with itself. To start a new row, hit the rightshift-(down & left arrow). **You must run this program before you can run BPIS, or KVAL, or BUBP.** Example: nitrogen + methane. Number of components = 2, Tc = [126.2 190.2, (note the space between the two values), Pc = [33.94 46, w = [.04 .011, K = [[0 0 0 0. (The matrix is entered row by row. You can type the 0's for the k_{ij}'s straight in or include the line-feed to help you keep it straight for a bigger matrix. The result is the same.)

2. BPIS—bubble point pressure of an ideal solution from the shortcut vapor pressure equation. (Run INCRT first.)

Hit the menu key labeled **BPIS** to input the value of T and desired composition of all components. P is displayed on the screen. You must run this before running BUBP the first time. Example: nitrogen + methane at 100K, x = [.5 .5, P = 4.097 bars.

3. KVAL—vapor-liquid K-ratios from the Peng-Robinson equation given estimated phase compositions. (Run INCRT first.)

Hit the menu key labeled **KVAL** to input the values of T and desired compositions of all components in the liquid, x, and all components in the vapor, y. This calculation actually assumes that all binary interaction coefficients are zero. It takes a few seconds.

4. BUBP—bubble point pressure from the Peng-Robinson equation.

You must run INCRT and then run BPIS at least once to get an initial estimate of the pressure before running this program. Further calculation, as for phase envelopes, can proceed without calling BPIS if the previous result for the pressure provides a reasonable guess for the pressure at the new conditions. Hit the menu key labeled **BUBP** to input the values of T and desired compositions of all components in liquid, x. This calculation treats all binary interaction coefficients as given. It may take a minute. By changing values of k_{ij}, azeotropes can be characterized and studied much like the van Laar equation. Example: nitrogen + methane at 100K and x = [0.5 0.5, K1 = 1.89, K2 = 0.11, P = 4.250 bars.

UNIFAC—Activity coefficients for mixtures of user-specified chemical structures by the UNIFAC group contribution method. The number of components must be less than or equal to five but components may have any number of functional groups each.

1. INIT—initial specification of number of components and their chemical structure.

Hit the menu key labeled **INIT** to input the number of components and the chemical structure of each component. After you enter the number of components you will be prompted for the chemical structure of component 1 ("COMP1"). Type the number of occurrences of a given functional group, then press the menu key for that functional group (e.g., for pentane, press **3** [**CH2**] **2** [**CH3**][**INCPT**]). Some functional groups are listed on later windows that can be viewed by pressing the key [**NXT**]. When all functional groups have been specified for that component, press [**NXT**] to return to the first window and press [**INCPT**] to input that component. After a few seconds, you will be prompted for the next component. Continue until all components have been specified. (E.g., for methanol, press [**NXT**] 1 [**MEOH**] [**NXT**] [**NXT**] [**INCPT**]).

2. CONC—specification of concentration and output of activity coefficients. Run NEWT before this program to change the temperature.

Hit the menu key labeled **CONC** to input the mole fractions of components. At the x:[[prompt, type the mole fractions of the components, separated by spaces. (E.g., x:[0.5 0.5) After about 20–60 seconds the activity coefficients are displayed (e.g., 1.822, 1.789 for the pentane + methanol system at 80°C).

3. NEWT—specification of a new temperature.

Hit the menu key labeled **NEWT** to change the value of temperature. At the TDEGC: prompt, type the new temperature in degrees Celsius. After about 25 seconds the initial menu will reappear. The activity coefficients are not strong functions of temperature, so it may be sufficient to perform several calculations with the same value of temperature. Press **CONC** to recalculate the activity coefficients, or **INIT** to enter new components.

4. TDEGC—Use **TDEGC** to display the current value of temperature.

A.2 TI-85 PROGRAMS

Program PENG is documented in the Word6 file DEPFU4TI.DOC in the TI directory. This program provides Z and departure functions for the TI-85. The code is also provided.

A.3 PC PROGRAMS FOR PURE COMPONENT PROPERTIES

See Section A.9 for an introduction to Excel.

PREOS.XLS—(Same capabilities as PRPURE.EXE, but superior user interface). An Excel 5.0 workbook for calculating roots to the Peng-Robinson equation at sub and supercritical conditions. Also calculates departure functions and thermodynamic properties. You need to have the following information available:

Critical Temperature (K), Critical Pressure (MPa), Acentric factor

In addition, if you wish to calculate departure functions, you will need ideal gas heat capacity coefficients from Appendix B or Reid, Prausnitz and Poling.

PRPURE.EXE—FORTRAN code for Peng-Robinson equation calculation of vapor pressure, compressibility factor, fugacity, and fugacity coefficients. Simply type PRPURE at the DOS prompt to run the program or double-click the file's icon on file manager or explorer. PRPURE.FOR is the source code. The program has the same capabilities as PREOS.XLS, but advantages for generating an output file for repetitive calculations. Requires same input information as PREOS.XLS. See section A.9 for instructions on importing the output files into Excel for plotting, etc. PRPURE.EPS is a flowsheet of the program in postscript printer format.

STEAM.XLS—Steam property calculator. Same formulation implemented in Harvey, A. P., Peskin, A. P., Klein, S. A., NIST/ASME Stean Properties, Version 2-1, NIST Standard Reference Data Program, December 1997.

A.4 PC PROGRAMS FOR MIXTURE PHASE EQUILIBRIA

PRMIX.EXE—Compiled program which performs bubble, dew, vapor-liquid flash, and liquid-liquid flash calculations for a multicomponent mixture using the Peng-Robinson EOS. Double-click the file's icon on file manager or explorer. Components are specified using the index numbers in the endflap. A list is also available on screen during runtime. See the readme file to add more compounds to the database.

ESD.EXE—Compiled program analogous to PRMIX.EXE using the ESD (1990) EOS for associating mixtures. Double-click the file's icon on file manager or explorer.

ACTCOEFF.XLS—A workbook to calculate activity coefficients as a function of composition for the Margules, van Laar, Regular Solution, UNIQUAC, or UNIFAC models. These spreadsheets may be modified to calculate excess Gibbs energy, fugacities, and P-x-y diagrams.

1. MARGULES—A spreadsheet to use with the MARGULES activity coefficient model.

2. REGULAR—A spreadsheet to calculate VLE for methanol + benzene using van Laar and Scatchard-Hildebrand Theory.

3. UNIQUAC—A spreadsheet to use with the binary UNIQUAC activity coefficient model.

4. UNIQUAC5—A spreadsheet to use UNIQUAC with up to five components and can model LLE.

5. UNIFAC(VLE)—A spreadsheet to use with the UNIFAC activity coefficient model for up to five components to model VLE.

6. UNIFAC(LLE)—Two spreadsheets to use with the UNIFAC activity coefficient model to up to five components to model LLE.

7. ANTOINE—A table of Antoine parameters.

DEWCALC.XLS—Performs a dew temperature or dew pressure calculation for a binary system using activity coefficients and ideal gas phase. This spreadsheet demonstrates how to set up an iterative calculation on a spreadsheet.

GAMMAFIT.XLS—A spreadsheet for fitting activity coefficient parameters. Currently set up for the two-parameter Margules equation. A non-linear least squares technique is used via the Excel solver utility (make sure this has been installed in your PC's version of the Excel program). The spreadsheet can be converted to other activity coefficient models by editing the activity coefficient formulas.

FLSHR.XLS—A spreadsheet to calculate two-phase isothermal flash using Raoult's law. Currently set for a binary system.

IChemT.exe—A chemical theory program for calculating the true mole fractions at a given superficial mole fraction. The equilibrium constants need to known before using the program. The equilibrium constants for input are the concentration equilibrium constants, which for a vapor phase would be K_y of Eqn. 14.21, which is $K_y = (Ka \cdot P^{-\Sigma v})/K_\varphi$.

PRFUG.XLS—Worksheet to calculate component fugacities via the Peng-Robinson equation for up to three components. Useful for understanding mixing rules, and for manually following iterative steps for phase equilibria calculations.

RESIDUE.XLS—spreadsheet with macro for calculating residue curves for homogeneous systems using UNIQUAC for up to three components.

VIRIALMX.XLS—This spreadsheet calculates the second virial coefficient for a binary mixture using the critical temperature, pressure, and volume.

WAX.XLS—A spreadsheet to calculate wax solubilities.

A.5 REACTION EQUILIBRIA

RXNADIA.XLS—Adiabatic reaction temperature calculation for ammonia formation corresponding to the Example in the textbook. Instructions are given as a worksheet in the workbook.

RXNS.XLS—Workbook with spreadsheets used for multiple reaction equilibria in the text. Includes Gibbs energy minimization and simultaneous reaction and phase equilibria. Instructions are given as a worksheet in the workbook.

KCALC.XLS—Workbook to calculate equilibrium constants as a function of temperature.

CL2H2O.XLS—Worked electrolyte example.

A.6   HOW TO LOAD PROGRAMS

Download through the Internet from the website http://www.egr.msu.edu/~lira/thermtxt.htm, and follow the posted instructions.

A.7   DOWNLOADING HP PROGRAMS

To download to an HP48G you will need either: 1) a classmate who has the programs on their calculator, or; 2) a special cable sold by HP or another supplier such as Educalc [phone (800)677-7001, STOCK#2609] to hook to the serial port of a PC. If you are downloading from the PC you will need the communication program Kermit which has been furnished in the HP subdirectory. If you transfer from a classmate using the built in IR communications in the calculator, see the HP manual instructions. To download from the PC to your calculator:

1. Install the cable. It attaches to a COM port, possibly COM2, since a modem or mouse may use COM1.

2. Change to the PC directory HP containing Kermit and THRMO.

3. Execute Kermit (double click KERMIT.EXE in Windows).

   Type: set port com2 (or whatever COM port you're using).

   Then type: set baud 9600.

4. Turn on your calculator, press green **HOME** and **right-shift** I/O.

5. Select "Transfer." Set "PORT:" to "Wire," "XLAT" to "→255," "CHK" to "3."

6. Go back to the computer and give the command: SEND c:\thrmo.

7. Quickly press the menu key on the HP48 that corresponds to RECV.

8. You should get a computer screen that notes the number of packets sent, the % delivered, and other vital statistics. The THRMO file takes a couple minutes to transfer.

9. Press **HOME** on the HP. If you started Kermit from windows, type "exit" to close the DOS window.

A.8   USING FORTRAN PROGRAMS

The programs can be run from Windows by double clicking on them from the File Manager (or Explorer). However, some programs must be able to find input files in the same directory. It is recommended to run them from the DOS prompt unless you are an advanced Windows user.

Running from a DOS prompt

Some FORTRAN programs must be run with the drive containing data and output files as the default drive. Before you run the programs, type "dir" at the DOS prompt to be sure the input data files you need are on the default drive. If the directory does not contain the files you need, use "cd *directoryname*" to move to the desired directory. Some the output files from PRPURE.EXE are wide. It is recommended that you import wide files into Excel if you wish to print or manipulate them.

If you want to run with data and output files on drive A:\, insert the diskette with the input files in drive A:\, then enter A: at the DOS prompt. You should receive an A:> prompt. The program will look for files on the default drive (for this example on drive A:). To start a program from the C:\ drive, include the drive in the command, e.g., C:\prpure. An alternative is to copy the data files to the same drive as the program. Then run the program. When you are finished, remember to copy your data files to a floppy if you want to keep them.

A.9 NOTES ON EXCEL SPREADSHEETS

Using Excel spreadsheets

Start Excel. To open a data file, click on "File" on the menu bar, then choose "Open..." In the resulting dialog box, you can specify the drive for your data files. To print output, you can choose "Printer Setup..." to specify a printer, "Page Setup..." to specify margins, automatic scaling, etc., and "Print preview..." to preview output. To quit Excel and return to Windows, choose "File...", "Exit." To exit Windows, repeat the same commands from the File Manager screen.

*.XLS files

These workbooks are a starting point for homework problems. You may need to modify the existing spreadsheet to work a homework problem. The *.XLS files are provided in "document protected" format so that inadvertent modification will not occur. Only the unlocked cells which appear blue on the screen may be modified without turning the document protection off. (To change the document protection, select 'Protection....' from the 'Tools' menu. No passwords are used on the distributed spreadsheets. To change the protection status of individual cells, choose 'Cell Protection' from the 'Format' menu when document protection is turned off.) If you have access to other spreadsheet programs (like Lotus or Quattro) which can import Excel files, you may be able to use the Excel spreadsheets on your spreadsheet program. You may lose document or cell protection if you import the spreadsheet into a different spreadsheet program. Excel works like other spreadsheets. To enter a mathematical formula in a cell, start the formula with "=." Excel enables you to point, click and drag to modify your spreadsheet and produce plots.

HELP! What cells use a calculation? To find out the interdependencies of cells, first unprotect the sheet, then use Tools... Auditing.....

Importing to Excel

Data files from PRPURE, ZCHEM or other programs may be easily imported into Excel 5.0. Output from PRPURE has specifically been delimited with commas to facilitate importing. Excel will recognize it as a Windows ANSI file. Import it as *delimited*, using *commas* to identify fields, *text qualifier* set to " (double quote), and *treat consecutive delimiters as one* turned off. When prompted, accept the *General* format for importing. If desired, the cell widths can be adjusted after importing, although the default works well for most files. The data can be sorted easily in Excel and nice plots can be generated easily from sorted files. Output from ZCHEM is delimited by spaces, except text is identified with ".

Plotting

To create a graph, place the mouse in a corner of the cell range which you want to plot, hold the left mouse button down and drag the mouse over the cell ranges to be plotted. Release the mouse button to leave the area shaded. To use data in multiple areas that are not contiguous, press Ctrl while you drag the mouse over the second area. The wizards will help you create your plot. Generally the "x-y scatter plot" is the desired plot type. It is recommended that you add the plots on a second spreadsheet when prompted. Click on the portions of the graph to edit. If you get stuck, select 'Help' from any menu. Double click on the graph to edit and print.

You can turn on or off lines and/or data markers by *right* clicking on the data set on a plot. Select Format data series...., and follow the menus. To switch the x and y coordinates for a set of data, click the data set, then switch the cell ranges that appear in the formula bar. To add multiple data sets to an existing plot, highlight the data, then Copy...switch to the plot and choose Paste Special...Add cells as new series....and check the "Categories (X values in First Column") checkbox.

Using Solver with Excel

Excel includes an "add-in" feature called Solver which can be used to solve single or multiple equations. The installation of the feature is optional. If it is installed, it will be listed under the **Tools...** drop down menu. A related feature is called **Goal Seek...**, but it is less powerful since it can only solve a single objective function. If Solver is not installed, select **Add ins...** from the **Tools...** menu and follow the on-line instructions for installation. To solve an equation for a single variable, Goal Seek or Solver can be used. We will use Solver since it is the more general tool, and use of Goal Seek is simple if desired, once Solve has been used. **If you are using Solver with a spreadsheet that has been protected, it must first be unprotected using Tools...Protection....**

Suppose we wish to solve

$$x^2 + 2x = 1$$

Although the solution may be quickly found by the quadratic formula, a spreadsheet will be created to illustrate the technique that can be applied to more complex problems.

1. Create the following table, entering the labels in column A and the initial guess of 0 for x in cell $B1$.

	A	B
1	x	0
2	F(x)	

2. Enter the following formula in cell B2: $= B1\verb|^|2 + 2 \cdot B1 - 1$. (Note that you may click on B1 rather than typing the name as you enter the formula.)

3. Start Solver from the **Tools...** drop down menu. The Solver window will pop up as shown below. The objective function (Target cell) is entered in the top entry box. For this example, enter B2 (or click in the entry box, then click in B2). The radio buttons permit the objective cell to be maximized, minimized, or set to a specific value. In this example, select "Equal to: Value of:" and put the number 0 in the entry box. The **Options...** button controls the type of numerical technique that is applied, but we won't use that now. (For more information on the options, click the Help button from the "Options" subwindow). The next entry box specifies the cells to adjust in the search for the objective function. For this example, enter B1. (For multiple cells you can drag the mouse over the cell ranges.)

4. Click on **Solve**. Look closely at the information box that pops up when Solver has finished. Since numerical techniques are used, Solver may have difficulty finding a solution for poor initial guesses or poorly defined objective functions, and the box notifies of problems in this event. However, we are solving a simple example here, and the answer is quickly found. The answer should be 0.414.

5. If the other root is desired, a different initial guess should be entered in cell B1. Enter the value −2 and re-solve the equation. The answer −2.414 should result.

Excel may also be used to solve multiple equations. Consider the set of equations

$$x^2 + 2y = 10$$
$$x + y = 4$$

Create the spreadsheet

	A	B	C	D
1	x	1	F1	
2	y	3	F2	

The initial values in the second column satisfy the second equation. Let's call the first function F1. In cell D1 enter the formula $= B1^2 + 2*B2 - 10$ and in cell D2 enter $= B1 + B2 - 4$. At this point some trial and error adjustment of the variable in column B can help create a good initial guess, while judging the quality of the guess by watching column D. When you are ready, start Solver. Instruct Solver to Set D1 to a value of 0 by varying B1:B2 subject to the constraint D2 = 0. (Click the **Add...** button to specify the constraint.) One solution is x = −0.732, y = 4.732. Another solution is x = 2.732 and y = 1.268.

Solving by Successive Substitution

Occasionally, solutions are needed to complex equations. For example, the equation $2.5x - \exp(0.75x) = 0$ has two solutions. The solutions can be found by successive substitution. The technique works by rearranging the function in the form $x = f(x)$. For the example, there are two possibilities:

$$x = \frac{\exp(0.75x)}{2.5}$$ (A)

or

$$x = \frac{\ln(2.5x)}{0.75}$$ (B)

Successive substitution works by using $f(x)$ to generate a new value of x and then substituting back into $f(x)$. The technique will converge only if $|f'(x)| < 1$ at the value of x guessed. For the present example, there are two solutions: $x = \{0.652536, 2.37512\}$. To begin, enter the table into Excel.

	A	B
1	xnew = f(x)	
2	xold	1

Then for Eqn. A, enter into B1, $= \exp(0.75*B2)/2.5$. For the substitution, the following strategy is used:[1]

1. Copy cell B1.

2. Past the *value* into cell B2 by right-clicking on B2, selecting **Paste Special...** and then selecting the radio button for Paste **Values**. *A macro can easily be created to repeat these steps as described below.*

By repeating the substitution about 15 times, the iterations converge on $x = 0.6525$. Try to converge on 2.37 by starting from 2.2 or 2.6 in cell B2— you will find it impossible because the value of $f'(x) > 1$ at $x = 2.37$. Reprogram cell B1 to the right-hand side of Eqn. B. This time the successive substitution will converge to $x = 2.375$, but convergence to $x = 0.6524$ will be impossible.

Macros for Successive Substitution

Successive substitution can be tedious, but a macro can be quickly written to perform the task. Suppose the following spreadsheet is available for the above example, solving for Eqn. A, where cell B1 is calculated from cell B2.

	A	B
1	xnew = f(x)	0.8468
2	xold	1

---

1. Circular references can sometimes be created by entering =B1 into cell B2, after enabling iterations on the Tools... Options...Calculations.... window; however, the convergence or divergence can be unpredictable.

From the menu bar select **Tools... Record new Macro...** Click **Options>>** and give your macro a name and a shortcut keystroke. When you are ready to record, click **OK**. A Stop box may pop onto your screen. Every key stroke or mouse click that is now entered will be recorded.

1. Copy cell B1.
2. Paste the *value* into cell B2 by right-clicking on B2, selecting **Paste Special...** and then selecting Paste **Values**.
3. Click the **Stop** button, *or* select **Stop Recording** from the **Tools... Record Macro...** menu.

If all the steps went as planned, just press your shortcut key repeatedly. If you want to record over your macro, simply create a new macro with the same name.

Naming Variables

Variables may be named for use in formulas. Select the cell range to be named (for an array, highlight the entire array range), then select the name box in the left-most edge of the formula bar. Type in the desired name. Default cell names cannot be used, e.g., X1 is invalid, but _X1 is acceptable. See "named ranges" in online help for more information.

Array Operations

See Appendix B for an overview of matrices. To use a particular array element from an array, use the function INDEX(array,row_num,col_num). If the following named matrices exist in Excel:

$$B = \begin{bmatrix} 1 & 2 & 3 \\ 4 & 5 & 6 \end{bmatrix}, \quad \text{INDEX}(B,1,3) = 3, \quad \text{INDEX}(B,2,3) = 6 \quad X = [0.4\ 0.6], \quad \text{INDEX}(X,1) = 0.4,$$

INDEX(X,2) = 0.6

When entering a formula for a matrix operation that will involve matrix multiplication or results in an array, select the entire array output range (with the correct number of rows and columns) before entering the formula, then press CNTL+SHIFT+ENTER (or COMMAND+RETURN on a Macintosh) to tell Excel that this is a matrix formula. Excel inserts braces around the formula after it is entered. Failure to select the correct number of output rows or columns, or failure to press CNTL+SHIFT+ENTER, or failure to observe the requirements of matching the number of rows and columns in multiplicands can result in unexpected results or errors. Some common operations are shown below. See SUM and MMULT in online help for more details.

One-dimensional arrays:

(a) If both arrays appear as columns or both arrays appear as rows, use SUM(array1*array2);
(b) If array1 is a row and array2 is a column, use MMULT(array1,array2);
(c) If array1 is a column and array2 is a row, use MMULT(array2,array1).

Multidimensional arrays (select the entire output range before entering the formula):

(a) For $c_{ij} = \sum_k a_{ik} b_{kj}$, use MMULT(A,B);

(b) For $c_{ij} = \sum_k a_{ik} b_{jk}$, use MMULT(A,TRANSPOSE(B)).

A.10   NOTES ON HP CALCULATOR

Directory Structure

Moving about the HP is very similar to moving in the directories on a PC; however, the objects in a directory may be variables, equations, or other subdirectories. The objects in a directory can be quickly found by pressing the **VAR** key in the second row. The objects will be listed above the menu keys. (If there are more objects than menu keys, the next six objects can be found by pressing the **NXT** key in the second row.) The current directory will be displayed near the top of the display, and you can move down through the directory structure by using the menu keys. Directories have an overbar above the object name.

Sometimes you will need to move up through the directory structure. The top directory is called the home directory. You can always get to the home directory by pressing **HOME** in the third row. To move up one directory at a time, press the **UP** key in the third row.

System Halt

Occasionally the calculator processing flags are not reset properly. For example, if a program hangs or keys are pressed during execution when the hour-glass is present. If you notice that the calculator is slow when performing simple addition and subtraction, this is a symptom indicating the need to perform a system halt. Another symptom is an "out of memory" message when running a program. To perform a system halt, press **ON** and the menu **C** key simultaneously. The screen will momentarily blank. The program memory will be preserved. In the rare event that this fails to reset your calculator, see "system halt" in the index of the HP manual. After a system halt, press **VAR** to list the **HOME** directory objects.

Terminating a Program

If a program is executing, and you wish to terminate it, simply press **Cancel** at a prompt. You will want to perform a system halt as described above, which ends up moving you back to the **HOME** directory.

Equation Solver

To access the equation solver, press **right-shift SOLVE**, and select **Solve equation**. Select the **EDIT** menu key. Type in the equation to be solved using variable names. For practice, enter the equation $x^2 + 2x = y$ by entering the calculator keystrokes. (To enter variable names, press the α key. For a power, use the y^x key.) Press **ENTER** when finished. The HP will automatically identify the variables x and y. Use the cursor keys to move among the variables, setting the values of the known variables. (In this case, set $y = 1$. Now this equation will be the same one that was solved above in the Excel example.) Move the cursor to the variable which you want to find, and enter an initial guess. (Your answer will depend on your initial guess if multiple solutions exist. For this example, use 0 as an initial guess for x.) Press the menu **SOLVE** key when the cursor is in the correct field. (In this case, move to the field for x.) Solve again using $x = -2$ as the initial guess. Press **CANCEL** to exit solver.

Using Variables

There are many ways to create and use variables. This section summarizes only one method for each task. Consult the manual for more details.

(a) For a list of objects (including variables) in the current directory, press the **VAR** key.

(b) To create a new variable with a value: 1) enter the value in the stack; 2) press α and type the variable name; 3) press **STO**. The variable value is available by pressing the menu key. When a variable name is used in a program or equation, the current value will be automatically read from memory.

(c) To recall a variable value (or to enter the variable into a program or equation), just press the variable's menu key.

(d) To change the value of a variable in memory, enter the new value in the stack, then press **left-shift** followed by the variable's menu key.

(e) To delete a variable, press ' followed by the variable's menu key, then **PURG** (located right above the number pad).

(f) To create a new directory: 1) press **right-shift MEMORY**; 2) press menu key **NEW**; 3) enter the new directory name, and check the "directory" box, then press the **OK** menu key.

(g) To purge a non-empty directory, follow the procedure given for purging THRMO in the updating procedure below, but substitute the appropriate directory names in that procedure. Start the procedure in the parent directory where the directory variable that you wish to purge resides.

Updating THRMO

To update an existing THRMO directory, first make sure you move any of your own programs to a different directory. The calculator will not permit you to move directories; you will have to copy them, and then purge the old one. When you are ready to load the new THRMO, purge the old directory as outlined here. Follow the steps very carefully so that you do not purge something else accidentally. You will not be asked to confirm any step, and you cannot undo the purge. 1) move to {HOME}; 2) press ' followed by the THRMO menu key; 3) use the arrow keys, SPC and alpha keys to add the PGDIR command (PGDIR stands for *purge directory*)— your entry line should look like **'THRMO' PGDIR**; 4) Press enter. Now you can load the updated THRMO.

Programming the HP

First of all, do not use the THRMO directory for adding your programs. This is because the THRMO directory will need to be purged when it is updated, and *everything* in the directory will be deleted. Start a new directory with your programs.

Programming the HP can be quite powerful, and the manual is the best resource. However, simple programs are often the most useful, and don't require detailed reading of the manual. For example, a simple program to calculate area might be << 'H*W' →NUM >> where H is a variable for height and W is a variable for width. (You simply type the formula using the variables. The variable values will be read from memory when the program runs. The double angle brackets tell the HP that this is a program. The command →NUM is actually a key in the 3rd row, and to enter the command you just press the key—this command converts the result to a number when the program executes.)

To create this program, first create the variables AREA, W, and H with any values initially stored in them. (AREA will end up being the menu key that will be used to run the program.) Then begin typing the program << 'H*W' →NUM >> (The angle brackets can be found on the minus key. The single quotes are important. You can enter variable names in the program by just pressing the menu keys for the variable that you created above.) After the program is typed, press the enter key to put the program in the stack. Then store the program in AREA the same way you would store a number. Before you run the program, enter the desired variable values for H and W, then press the menu key AREA, and the resulting number for the area will appear on the stack. Change the variable values for H and W and run the program again to confirm your success. To edit a program after it has been written, place the variable name in the stack, then press right-shift EDIT.

You may browse through the THRMO directory for other programs to see how they are constructed. Recognize that the calculator *is not write protected*, so any change you make will be written to memory. Some of the features you will find are subroutines, prompts, and directory changes. Notice that variables or subroutines can be shared between directories if they are placed in the directory *above* the directories which will share them.

A.11   DISCLAIMER

The programs provided with this text are for educational use only. They are provided AS IS, without any warranty. They must not be sold under any circumstances.

APPENDIX B

MATHEMATICS

B.1 IMPORTANT RELATIONS

Algebra

Some functions like logarithms and exponentials appear so often in thermodynamics that it makes sense to summarize some of them here. Also, integrations and differentiations are frequently performed, so a few important formulae are presented.

$$\ln(ab) = \ln a + \ln b \qquad\qquad \text{B.1}$$

$$\ln\left(\frac{a}{b}\right) = \ln a - \ln b \qquad\qquad \text{B.2}$$

$$\ln a^y = y \ln a \qquad\qquad \text{B.3}$$

$$\exp(a + b) = e^a e^b \qquad\qquad \text{B.4}$$

$$\ln N! \approx N \ln N - N \text{ for large } N \text{ (Stirling's approximation)} \qquad \text{B.5}$$

The quadratic formula provides roots to the equation $ax^2 + bx + c = 0$:

$$x = \frac{-b \pm \sqrt{b^2 - 4ac}}{2a} \qquad\qquad \text{B.6}$$

Cubic equations are discussed in Section B.2.

Beginning in Chapter 10, summation notation is used extensively. Many of the formulas are easily programmed using matrices and linear algebra. A matrix is a rectangular representation of the elements of an array. The elements of an array are identified by subscripts. For example:

$$X = [0.4 \; 0.6], \; x_1 = 0.4, x_2 = 0.6$$

599

For a multidimensional array, the first element subscript represents the row and the second element subscript identifies the column, for example:

$$A = \begin{bmatrix} 1 & 3 \\ 2 & 4 \end{bmatrix} = \begin{bmatrix} a_{11} & a_{12} \\ a_{21} & a_{22} \end{bmatrix} \quad a_{11} = 1, \ a_{21} = 2, \ a_{12} = 3, \ a_{22} = 4.$$

The transpose of a matrix is obtained by exchanging a_{ij} with a_{ji}. In shorthand notation, the transpose is represented by a superscript T. The number of rows and columns interchange after the transpose operation:

$$A = \begin{bmatrix} 1 & 4 & 7 \\ 2 & 5 & 8 \\ 3 & 6 & 9 \end{bmatrix} \quad A^T = \begin{bmatrix} 1 & 2 & 3 \\ 4 & 5 & 6 \\ 7 & 8 & 9 \end{bmatrix},$$

$$X = \begin{bmatrix} 0.4 & 0.6 \end{bmatrix} \quad X^T = \begin{bmatrix} 0.4 \\ 0.6 \end{bmatrix}$$

$$B = \begin{bmatrix} 1 & 4 \\ 2 & 5 \\ 3 & 6 \end{bmatrix} \quad B^T = \begin{bmatrix} 1 & 2 & 3 \\ 4 & 5 & 6 \end{bmatrix}$$

Matrices can be multiplied. The product of array1 and array2 becomes array3 and:

(a) the number of columns in array1 must equal the number of rows in array2.

(b) array3 has the same number of rows as array1 and the same number of columns as array2.

(c) element *ij* of array3 is obtained by multiplying the elements in the *i*th row of array1 by the elements in the *j*th column of array2 and summing the products.

For one-dimensional arrays,

$$XB = \begin{bmatrix} x_1 & x_2 \end{bmatrix} \begin{bmatrix} b_1 \\ b_2 \end{bmatrix} = \sum_{i=1}^{2} x_i b_i$$

which is the linear mixing rule, Eqn. 10.9. An example of a one- and two-dimensional array is:

$$YA = \begin{bmatrix} y_1 & y_2 \end{bmatrix} \begin{bmatrix} a_{11} & a_{12} \\ a_{21} & a_{22} \end{bmatrix} = \begin{bmatrix} y_1 a_{11} + y_2 a_{21} & y_1 a_{12} + y_2 a_{22} \end{bmatrix} = \begin{bmatrix} \sum_{i=1}^{2} y_i a_{i1} & \sum_{i=1}^{2} y_i a_{i2} \end{bmatrix}$$

Multiplying the result by Y^T:

$$YAY^T = \begin{bmatrix} y_1 a_{11} + y_2 a_{21} & y_1 a_{12} + y_2 a_{22} \end{bmatrix} \begin{bmatrix} y_1 \\ y_2 \end{bmatrix} = y_1(y_1 a_{11} + y_2 a_{21}) + y_2(y_1 a_{12} + y_2 a_{22})$$

when $a_{ij} = a_{ji}$, we may write

$$YAY^T = y_1^2 a_{11} + 2y_1 y_2 a_{12} + y_2^2 a_{22} = \sum_{i=1}^{2} \sum_{j=1}^{2} y_j y_i a_{ij}$$

which is the quadratic mixing rule, Eqn. 10.9. For an example using two multidimensional arrays:

$$AB = \begin{bmatrix} a_{11} & a_{12} \\ a_{21} & a_{22} \end{bmatrix} \begin{bmatrix} b_{11} & b_{12} & b_{13} \\ b_{21} & b_{22} & b_{23} \end{bmatrix} = \begin{bmatrix} c_{11} & c_{12} & c_{13} \\ c_{21} & c_{22} & c_{23} \end{bmatrix} = C$$

$$= \begin{bmatrix} (a_{11}b_{11} + a_{12}b_{21}) & (a_{11}b_{12} + a_{12}b_{22}) & (a_{11}b_{13} + a_{12}b_{23}) \\ (a_{21}b_{11} + a_{22}b_{21}) & (a_{21}b_{12} + a_{22}b_{22}) & (a_{21}b_{13} + a_{22}b_{23}) \end{bmatrix}$$

For an overview of programming arrays in Excel, see Appendix P.

Calculus

Differentiation

$$\frac{d[\exp(f(x))]}{dx} = \exp(f(x))\frac{d[f(x)]}{dx} \qquad \text{B.7}$$

$$\frac{d(\ln[f(x)])}{dx} = \frac{1}{f(x)}\frac{d[f(x)]}{dx} \qquad \text{B.8}$$

General differentiation of composite functions:

(Product rule)

$$\frac{d[f(x) \cdot g(x) \cdot h(x)]}{dx} = f(x) \cdot h(x)\frac{d[g(x)]}{dx} + g(x) \cdot h(x)\frac{d[f(x)]}{dx} + f(x) \cdot g(x)\frac{d[h(x)]}{dx} \qquad \text{B.9}$$

$$d[f(x, y, z)] = \left(\frac{\partial f}{\partial x}\right)_{y, z} dx + \left(\frac{\partial f}{\partial y}\right)_{x, z} dy + \left(\frac{\partial f}{\partial z}\right)_{x, y} dz \qquad \text{B.10}$$

$$\frac{d[f(x)^x]}{dx} = f(x)^x \left\{ \ln[f(x)] + \frac{d\{\ln[f(x)]\}}{d\{\ln[x]\}} \right\} \qquad \text{B.11}$$

Integration

$$\int \frac{1}{x} dx = \ln x \qquad \text{B.12}$$

$$\int \frac{1}{ax + b} dx = \frac{1}{a}\ln(ax + b) \qquad \text{B.13}$$

$$\int \frac{x\,dx}{ax + b} = \frac{x}{a} - \frac{b}{a^2}\ln(ax + b) \qquad \text{B.14}$$

$$\int \frac{x^2 dx}{ax + b} = \frac{(ax + b)^2}{2a^3} - \frac{2b(ax + b)}{a^3} + \frac{b^2}{a^3} \ln(ax + b)$$ B.15

$$\int \frac{x^3 dx}{ax + b} = \frac{(ax + b)^3}{3a^4} - \frac{3b(ax + b)^2}{2a^4} + \frac{3b^2(ax + b)}{a^4} - \frac{b^3}{a^4} \ln(ax + b)$$ B.16

$$\int \frac{dx}{x(ax + b)} = \frac{1}{b} \ln\left(\frac{x}{ax + b}\right)$$ B.17

$$\int \frac{dx}{x^2(ax + b)} = -\frac{1}{bx} - \frac{a}{b^2} \ln\left(\frac{x}{ax + b}\right)$$ B.18

$$\int \frac{xdx}{(ax + b)^2} = \frac{b}{a^2(ax + b)} + \frac{1}{a^2} \ln(ax + b)$$ B.19

$$\int \frac{x^2 dx}{(ax + b)^2} = \frac{(ax + b)}{a^3} - \frac{b^2}{a^3(ax + b)} - \frac{2b}{a^3} \ln(ax + b)$$ B.20

$$\int \frac{x^3 dx}{(ax + b)^2} = \frac{(ax + b)^2}{2a^4} - \frac{3b(ax + b)}{a^4} + \frac{b^3}{a^4(ax + b)} + \frac{3b^2}{a^4} \ln(ax + b)$$ B.21

$$\int \frac{xdx}{(ax + b)^3} = \frac{-1}{a^2(ax + b)} + \frac{b}{2a^2(ax + b)^2}$$ B.22

$$\int \frac{x^2 dx}{(ax + b)^3} = \frac{2b}{a^3(ax + b)} - \frac{b^2}{2a^3(ax + b)^2} + \frac{1}{a^3} \ln(ax + b)$$ B.23

$$\int \frac{x^3 dx}{(ax + b)^3} = \frac{x}{a^3} - \frac{3b^2}{a^4(ax + b)} + \frac{b^3}{2a^4(ax + b)^2} - \frac{3b}{a^4} \ln(ax + b)$$ B.24

$$\int \frac{dx}{a + bx + cx^2} = \frac{1}{\sqrt{-q}} \ln\frac{(2cx + b - \sqrt{-q})}{(2cx + b + \sqrt{-q})} \quad \text{for } q < 0 \text{ where } q \equiv 4ac - b^2.$$

$$= \frac{2}{\sqrt{q}} \arctan\frac{2ax + b}{\sqrt{q}} \text{ for } q > 0$$ B.25

Integration by parts $\int udv = uv - \int vdu$

Numerical integration by trapezoidal rule:

$$\int\limits_{Xo}^{Xn} f(x)dx = \Delta x \sum_{i=0}^{n} f(x_i) - \Delta x\left[\frac{(f(x_0) + f(x_n))}{2}\right]$$ B.26

where Δx is a constant step size between discrete values of $f(x)$

See also Chapter 5 for additional mathematical relationships.

B.2   SOLUTIONS TO CUBIC EQUATIONS

A cubic equation of state may be solved by trial and error or analytically. Below are summaries of two methods for solving analytically. These techniques are implemented for the Peng-Robinson equation in the spreadsheet PREOS.XLS. A cubic equation of the form

$$Z^3 + a_2 Z^2 + a_1 Z + a_0 = 0$$

B.27

can be reduced to the form

$$x^3 + px + q = 0$$

B.28

by substituting for Z the value

$$Z = x - \frac{a_2}{3}$$

B.29

The values of p and q for Eqn. B.28 will then be

$$p = \frac{1}{3}(3a_1 - a_2^2) \text{ and } q = \frac{1}{27}(2a_2^3 - 9a_2 a_1 + 27 a_0)$$

B.30

If a_2, a_1, and a_0 are real (which they are for an EOS), then defining

$$R \equiv \frac{q^2}{4} + \frac{p^3}{27}$$

B.31

There will be one real root and two conjugate roots if $R > 0$;

There will be three real roots, of which two are equal if $R = 0$;

There will be three real and unequal roots if $R < 0$.

Solution Method I   Algebraic solution

Let

$$P = \sqrt[3]{-\frac{q}{2} + \sqrt{R}}, \qquad Q = \sqrt[3]{-\frac{q}{2} - \sqrt{R}}$$

B.32

then the values of x are given by using Eqns. B.30, B.31, B.32

$$x = P + Q, \quad -\frac{P+Q}{2} + \frac{P-Q}{2}\sqrt{-3}, \quad -\frac{P+Q}{2} - \frac{P-Q}{2}\sqrt{-3}$$

B.33

Values of Z are then found with Eqn. B.29.

Solution Method II   Trigonometric solution

Let $x = m \cos \theta$, then

$$x^3 + px + q = m^3 \cos^3 \theta + pm \cos \theta + q = 4 \cos^3 \theta - 3 \cos \theta - \cos(3\theta) = 0$$

therefore

$$\frac{4}{m^3} = -\frac{3}{pm} = -\frac{\cos(3\theta)}{q} \qquad \text{B.34}$$

which leads to

$$m = 2\sqrt{-\frac{p}{3}}, \quad \cos(3\theta) = \frac{3q}{pm} \qquad \text{B.35}$$

therefore

$$\theta_1 = \frac{1}{3}\cos^{-1}\left(\frac{3q}{pm}\right) \qquad \text{B.36}$$

where θ_1 is in radians. By the functionality of the cosine function, two other solutions will be

$$\theta_1 + \frac{2\pi}{3} \quad \text{and} \quad \theta_1 + \frac{4\pi}{3} \qquad \text{B.37}$$

The values of x are given by using Eqns. B.30 and B.37

$$x = m\cos\theta_1, \ m\cos\left(\theta_1 + \frac{2\pi}{3}\right), \ m\cos\left(\theta_1 + \frac{4\pi}{3}\right) \qquad \text{B.38}$$

Values of Z are found with Eqn. B.29.

Sorting Roots

Meaningful roots for the Peng-Robinson and many other common EOS are of the form

$$P = P^{rep} + P^{attr}$$

where $P^{rep} = \dfrac{RT}{V-b}$. To assure that $P^{rep} > 0$, we must have $Z > B$, where $B = \dfrac{bP}{RT}$.

When three real, positive roots exist, the meaningful roots must satisfy $\kappa_T = -\dfrac{1}{V}\left(\dfrac{\partial V}{\partial P}\right)_T \geq 0$, i.e., the isothermal compressibility must be positive. The value of κ_T may also be used to determine whether a root is vapor or liquid in cases where only one root is found by identifying the phase with the larger isothermal compressibility as the vapor phase. κ_T is always greater for a vapor root than for a liquid root except at the critical point.[1]

Determination of Equation of State Constants from the Critical Point

Determination of equation of state constants from the critical conditions has been the prevalent method of characterizing fluids. When a cubic equation of state is fit to the critical point, the parameters may be determined in a couple of ways. First, we can evaluate the derivatives

1. Poling, B. E., Grens II, E. A., Prausnitz, J. M., *Ind. Eng. Chem. Process Des. Dev.*, 1981, 20, 127–130.

$$\left(\frac{\partial P}{\partial \rho}\right)_T = 0 \qquad \text{and} \qquad \left(\frac{\partial^2 P}{\partial \rho^2}\right)_T = 0 \tag{B.39}$$

applied at T_c. By differentiating the equation of state, the resulting equations can be simultaneously solved to find two equation of state constants. A simpler approach using significantly less calculus and algebra is to write out $(Z - Z_c)^3$, which goes to zero at the critical point.

$$Z^3 - (3Z_c)Z^2 + (3Z_c^2)Z - (Z_c^3) = 0 \tag{B.40}$$

If we compare Eqn. B.40 with Eqn. B.27 at the critical point, we find

$$\begin{aligned} -a_2 &= 3Z_c \\ a_1 &= 3Z_c^2 \\ -a_0 &= Z_c^3 \end{aligned} \tag{B.41}$$

For the van der Waals equation, Comparing B.40 with Example 6.7

$$1 + B_c = 3Z_c^{EOS} \tag{B.42}$$

$$A_c = 3(Z_c^{EOS})^2 \tag{B.43}$$

$$A_c B_c = (Z_c^{EOS})^3 \tag{B.44}$$

where the superscript *EOS* has been added to explicitly show that this is the value predicted by the equation of state. Plugging B.43 into B.44, we find

$$B_c = \frac{Z_c^{EOS}}{3} \tag{B.45}$$

Plugging into B.42,

$$\boxed{Z_c^{EOS} = \frac{3}{8}} \tag{B.46}$$

Thus the van der Waals equation predicts a universal value of $Z_c = 0.375$. Plugging this into B.45, we find

$$\boxed{b = \frac{RT_c}{8P_c}} \tag{B.47}$$

and into B.43,

$$\boxed{a = \frac{27}{64}\frac{(RT_c)^2}{P_c}} \tag{B.48}$$

B.3   THE DIRAC DELTA FUNCTION

Understanding many of the terms appearing in thermodynamic functions like the repulsive contribution to the equation of state requires an understanding of a somewhat peculiar mathematical function known is the Dirac delta function. This function has the property of "sifting" the value of a function being integrated and focusing attention on the function value at a particular value of the independent variable. That is, over the interval $x_1 < x_0 < x_2$

$$F(x_0) = \int_{x_1}^{x_2} F(x)\delta(x - x_0)dx \qquad\qquad B.49$$

where δ is the Dirac delta function defined by

$$\delta(x - x_0) = \begin{cases} 0 \text{ for } x < x_0 \\ \infty \text{ for } x = x_0 \\ 0 \text{ for } x > x_0 \end{cases}$$

and shown in the figure below. The following discussion should clarify the function.

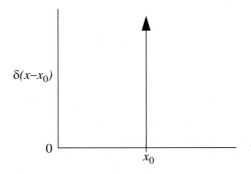

The Dirac delta function

Suppose for the moment that we approximate the delta function as a square pulse function, such that,

$$\delta(x) = \begin{cases} 0 & \text{for} & x < x_0 - 0.05 \\ 10 & \text{for } x_0 - 0.05 < & x < x_0 + 0.05 \\ 0 & \text{for} & x > x_0 + 0.05 \end{cases}$$

Then when we evaluate the integral,

$$I = \int_0^\infty \delta(x)\, F(x)\, dx$$

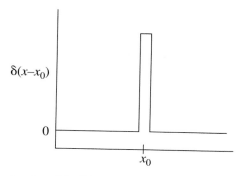

$\delta(x-x_0)$

Approximation of the Dirac delta function as a square pulse function.

a reasonable approximation would be that $F(x) = F(x_0)$ over the relatively short interval where δ holds a non-zero value. Since $F(x_0)$ is approximately constant over the short interval, it may be factored out of the integral, giving,

$$I = \int_0^\infty F(x)\delta(x)\,dx = F(x_0) \int_0^\infty \delta(x)\,dx = F(x_0)\,\{ \int_0^{x_0-0.05} 0\,dx + \int_{x_0-0.05}^{x_0+0.05} 10\,dx + \int_{x_0+0.05}^\infty 0\,dx\} =$$

$$= F(x_0) \cdot \{10\cdot[(x_0 + 0.05) - (x_0 + 0.05)]\} = F(x_0) \cdot \{10\cdot 0.10\} = F(x_0)$$

The next step is obvious. Let δ be zero except for an interval of ± 0.005 around x_0. Within the interval, set $\delta = 100$. Then our approximation of $F(x) = F(x_0)$ throughout the interval is even more reasonable. Proceeding in this fashion leads to an interval that is differentially deviating from x_0 with δ approaching infinity.

This seems like a crazy thing to do, since we have now defined the function value to be itself times an integral that must be unity. But note that our ultimate goal is to leave the function inside and we will show how this helps us momentarily. Looking at this final result, we see that the function δ has sifted out the value of $F(x)$ at a single value of x.

Despite its peculiarity, the δ function is a well-defined function and therefore integrable. To understand the importance of the δ function, consider taking the derivative of a step function, $H(x - x_0)$, shown below. The derivative of $H(x - x_0)$ is positively infinite at x_0 but zero on either side of x_0. It may help you to imagine slanting the corners of the step function and plotting the derivative first then making the corners systematically more square. If you plot this derivative function, you will see that it is the square pulse function and becomes the Dirac delta function in the limit of the Heaviside step function. *Therefore the Dirac delta function is the derivative of a Heaviside step function. And the integral of the Dirac delta function is equal to the Heaviside step function.* It is really this last observation that permits us to derive important results for the integrals of discontinuous functions.

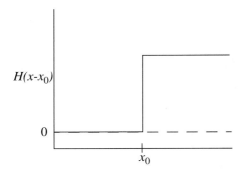

Heaviside step function.

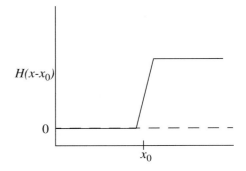

Approximation to the Heaviside step function whose derivative is the square pulse function explained above.

Example B.1 The hard sphere equation of state

A significant application of the Dirac delta function is suggested by the pressure equation developed for hard sphere fluids. The discussion in Unit II indicates that, at low density, the radial distribution function (rdf) is given by a simple Boltzmann weighting probability:

$$g(r) \sim exp(-u/kT)$$

Note that for hard sphere fluids, this function is discontinuous and behaves as the Heaviside step function at $r = \sigma$.

The pressure equation is:

$$\frac{P}{\rho RT} = 1 - \frac{N_A \rho}{6kT} \int_0^\infty r \frac{\partial u}{\partial r} g(r) 4\pi r^2 dr$$

If we were to substitute the hard sphere potential and a discontinuous rdf like the low-density one into the pressure equation, we would have a combination of discontinuities that we could not resolve. Note what happens if we postulate a function $y(r)$ such that

$$y(r) = exp(u/kT) \cdot g(r) \qquad \text{B.50}$$

Clearly, $y(r)$ would be continuous at low density and approaches a value of one. Furthermore, it turns out that $y(r)$ can be rigorously proven to be continuous for all densities and all potentials.[1] Apply these insights to develop the equation of state for a hard sphere fluid in terms of its rdf value at $r = \sigma$.

---

1. Hansen, J-P., and McDonald, I. R. *Theory of Simple Liquids,* 2ed. Academic Press, 1986.

Example B.1   The hard sphere equation of state (Continued)

Solution: Substituting $g(r) = y(r)\exp(-u/kT)$ and recognizing

$$\frac{\partial \exp\left(-\dfrac{u}{kT}\right)}{\partial r} = -\frac{1}{kT}\exp\left(-\frac{u}{kT}\right)\frac{\partial u}{\partial r} :$$

$$Z = \frac{P}{\rho RT} = 1 - \frac{N_A \rho}{6kT}\int_0^\infty \frac{\partial u}{\partial r}y(r)\exp\left(-\frac{u}{kT}\right)4\pi r^3 dr = 1 + \frac{N_A \rho}{6}\int_0^\infty \frac{\partial \exp\left(-\dfrac{u}{kT}\right)}{\partial r}y(r)4\pi r^3 dr$$

If you plot the function $\exp(-u/kT)$ versus r for the hard-sphere potential, you will see that it is a Heaviside step-function. This means that its derivative is a Dirac delta. So the pressure equation becomes,

$$Z = \frac{P^{HS}}{\rho RT} = 1 + \frac{N_A \rho}{6}\int_0^\infty \delta(r - \sigma)y^{HS}(r)4\pi r^3 dr$$

Applying B.49

$$Z = \frac{P^{HS}}{\rho RT} = 1 + \frac{4\pi\sigma^3 N_A \rho}{6}y^{HS}(\sigma)$$

Although it may seem that we don't know $y(\sigma)$, recall that it is a continuous function. Therefore the value of $y(\sigma)$ will be approximated by $y(\sigma^+)$. Further, recognizing that $u^{HS}(\sigma^+) = 0$, therefore $\exp(u^{HS}(\sigma+)/kT) = 1$ in Eqn. B.50, therefore $y^{HS}(\sigma+) = g^{HS}(\sigma^+)$.

$$Z = \frac{P^{HS}}{\rho RT} = 1 + \frac{4\pi\sigma^3 N_A \rho}{6}g^{HS}(\sigma^+) \qquad\qquad \text{B.51}$$

Defining $\eta = \frac{1}{6}\pi\sigma^3 N_A \rho$

$$\boxed{Z = \frac{P^{HS}}{\rho RT} = 1 + 4\eta g^{HS}(\sigma^+)}$$

B.52

The relationship between Z^{HS} and $g^{HS}(\sigma^+)$ is used often. It provides the functional form for building Z^{rep} in the equation of state. It also provides a basis for converting our approximate equation of state for Z^{rep} back into a quantitative estimate of $g^{HS}(\sigma)$. Recognize that Eqn. B.52 is a virial equation for hard spheres, $Z = 1 + B\rho$. At low density, $g^{HS}(\sigma^+) = \exp(-u^{HS}(\sigma^+)/kT) = 1$, resulting in Boltzmann's value for the second virial coefficient for hard spheres.

Example B.2 The square-well equation of state

The preceding example illustrates the use of all the tools necessary to develop any equation of state, but one more illustration may help to clarify the way that the tools can be used in combination to derive a wide variety of results. The problem statement is as follows: develop a formula for deriving the equation of state for any fluid described by the square-well potential ($R = 1.5$), given an estimate for the radial distribution function (rdf). Apply this formula to obtain the equation of state from the following approximation for the rdf.

$$g(r) \approx \begin{cases} 0 & r < \sigma \\ 1 + \dfrac{b\rho\varepsilon}{kT}\dfrac{1}{x^4} & r \geq \sigma \end{cases}$$

where $x = r/\sigma$, and $b = \pi N_A \sigma^3/6$. Before looking at the solution, think for yourself. Can you conceive of how to solve the problem without any help?

Solution: The trick is to realize that the exponential of the square-well potential is composed of two step functions, each of a different height. The step up is of height, $\exp(\varepsilon/kT)$ whereas the step down is of height, $\exp(\varepsilon/kT) - 1$.

$$\exp(-u/kT) = \exp(\varepsilon/kT)\, H(r - \sigma) - [\exp(\varepsilon/kT) - 1]\cdot H(r - R\sigma)$$

Taking the derivative of the Heaviside function gives the Dirac Delta in two places:

$$Z = 1 + \frac{N_A\rho}{6}\int \{\delta(r - \sigma)y(r)\exp(\varepsilon/kT) - \delta(r - R\sigma)y(r)[\exp(\varepsilon/kT) - 1]\}4\pi r^3 dr$$

$$Z = 1 + \frac{4\pi N_A\rho\sigma^3}{6}\{y(\sigma)\exp(\varepsilon/kT) - R^3 y(R\sigma)[\exp(\varepsilon/kT) - 1]\}$$

Noting that $y(r) = g(r)\exp(u/kT)$ and that $\exp(u/kT)$ is best evaluated inside the well:

$$Z = 1 + 4b\rho \{g(\sigma^+) - R^3[1 - \exp(-\varepsilon/kT)]\, g(R\sigma^-)\} \qquad \text{B.53}$$

This is valid for the square-well fluid with any g(r).

For the above expression: $g(\sigma^+) = 1 + b\rho\varepsilon/kT$ and $g(R\sigma^-) = g(1.5\sigma^-) = 1 + 0.198\, b\rho\varepsilon/kT$

$$Z = 1 + 4b\rho \{1 + b\rho\varepsilon/kT - 1.5^3[1 - \exp(-\varepsilon/kT)](1 + 0.198\, b\rho\varepsilon/kT)\} \qquad \text{B.54}$$

At the given conditions: $Z = 1 + 4(0.2)\{1 + 0.2 - 2.1333(1.0396)\} = 0.1858$

It would be straightforward at this point to develop the entire phase diagram for this new equation of state. The result would yield expressions for ε/k and b in terms of T_c and P_c. The value for the acentric factor would be a fixed value since there is no third parameter to affect it. Its numerical value could be determined in the same way that the value was determined for the van der Waals equation in Chapter 8.

Example B.2   The square-well equation of state (Continued)

At first glance, one may wonder why we have expended so much effort to represent this problem in terms of the rdf when we must approximate it anyway. It may seem fruitless to have translated our ignorance of the equation of state into ignorance of the radial distribution function. It turns out that the thermodynamic properties are fairly insensitive to details of the rdf. If you doubt this, reflect on what van der Waals achieved by effectively assuming that $g(r) = 1$ for all temperatures and densities. Furthermore, if you refer to the discussion of local composition theory in Chapter 11, you will see that the fundamental basis for virtually all of the activity coefficient models currently in use is: $g_{ij}(r)/g_{jj}(r) = \exp[(\varepsilon_{ij} - \varepsilon_{jj})/kT]$. As a slightly oversimplified summary of modern research in equilibrium thermodynamics, one could say that it is a search for better approximations of the radial distribution function. The thermodynamics of hydrogen bonding can be related to the rdf as discussed in Chapter 15. The thermodynamics of polymers and folding of proteins simply require generalization in terms of an intramolecular rdf known as the conformation. Electrolyte and solution thermodynamics are given very directly in terms of the rdf as discussed in Chapter 11. Adsorption and slit thermodynamics can be expressed in terms of the rdf between the fluid molecules and the adsorbent surface. Clearly, even rough approximations can be very useful when developed at the level of the rdf and carried to their logical conclusion. It may not be worth the effort for every chemical engineer, but it should not be beyond the grasp of many engineering students who sincerely want to understand the method behind the madness of fugacity estimation.

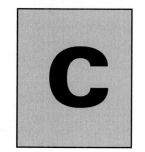

STRATEGY FOR SOLVING VLE PROBLEMS

In earlier chapters, we have discussed applications of VLE using simplified procedures such as bubble calculations using modified Raoult's law. This appendix summarizes one approach that can be used for solving the VLE problems when such simplifications are not possible. Compiled code has permitted solution of VLE using cubic equations without knowing the details of how the iterations are performed, and so a strategy is presented here also. There are many strategies throughout the literature, and the reader should be aware that other strategies are also successful.

The calculation procedures are summarized according to the VLE method. First the equation of state methods are presented, then the activity coefficient (sometimes called the gamma-phi) methods. For each approach, there are five flowcharts presented—bubble P, bubble T, dew P, dew T, isothermal flash. Specific routines may also be written for *vll* or other multiphase applications, which are not summarized here.

C.1 EOS METHODS

The equation that must be solved is: $y_i \hat{\varphi}_i^V P = x_i \hat{\varphi}_i^L P$

Bubble P

(The bubble pressure flowsheet is presented in Section 10.4)

Bubble T

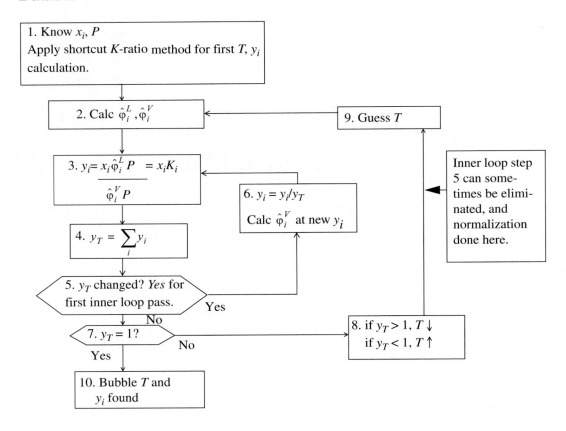

Dew P

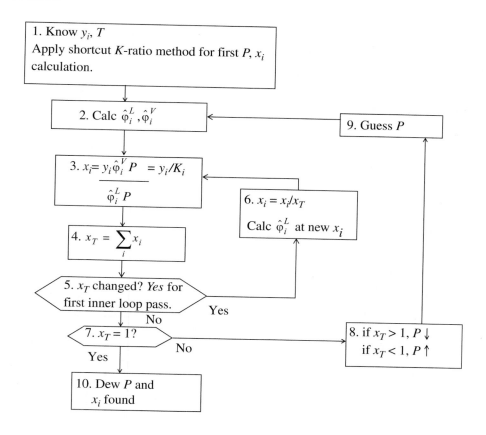

1. Know y_i, T
Apply shortcut K-ratio method for first P, x_i calculation.

2. Calc $\hat{\varphi}_i^L$, $\hat{\varphi}_i^V$

9. Guess P

3. $x_i = \dfrac{y_i \hat{\varphi}_i^V P}{\hat{\varphi}_i^L P} = y_i/K_i$

6. $x_i = x_i/x_T$
Calc $\hat{\varphi}_i^L$ at new x_i

4. $x_T = \sum_i x_i$

5. x_T changed? *Yes* for first inner loop pass.

Yes

No

7. $x_T = 1$?

No

Yes

8. if $x_T > 1$, $P \downarrow$
if $x_T < 1$, $P \uparrow$

10. Dew P and
x_i found

Dew T

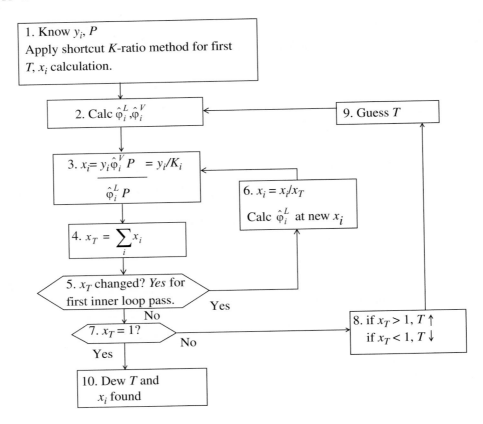

1. Know y_i, P
Apply shortcut K-ratio method for first T, x_i calculation.

2. Calc $\hat{\varphi}_i^L$, $\hat{\varphi}_i^V$

9. Guess T

3. $x_i = \dfrac{y_i \hat{\varphi}_i^V P}{\hat{\varphi}_i^L P} = y_i/K_i$

6. $x_i = x_i/x_T$
Calc $\hat{\varphi}_i^L$ at new x_i

4. $x_T = \sum_i x_i$

5. x_T changed? *Yes* for first inner loop pass.

Yes

No

7. $x_T = 1$?

No

Yes

8. if $x_T > 1$, $T \uparrow$
if $x_T < 1$, $T \downarrow$

10. Dew T and x_i found

Isothermal Flash

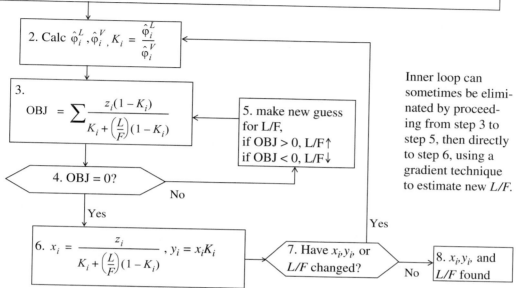

1. Know z_i, P, T, Perform bubble and dew to be sure system will flash at conditions given. Apply shortcut K-ratio method for first K_i calculation. Skip step 2 first time, set $x_i = y_i = z_i$ to force outer loop below to execute at least once.

2. Calc $\hat{\varphi}_i^L$, $\hat{\varphi}_i^V$, $K_i = \dfrac{\hat{\varphi}_i^L}{\hat{\varphi}_i^V}$

3.
$$OBJ = \sum \frac{z_i(1 - K_i)}{K_i + \left(\dfrac{L}{F}\right)(1 - K_i)}$$

4. OBJ = 0?

5. make new guess for L/F,
if OBJ > 0, L/F↑
if OBJ < 0, L/F↓

6. $x_i = \dfrac{z_i}{K_i + \left(\dfrac{L}{F}\right)(1 - K_i)}$, $y_i = x_i K_i$

7. Have x_i, y_i, or L/F changed?

8. x_i, y_i, and L/F found

No

Yes

Yes

No

Inner loop can sometimes be eliminated by proceeding from step 3 to step 5, then directly to step 6, using a gradient technique to estimate new L/F.

C.2   ACTIVITY COEFFICIENT (GAMMA-PHI) METHOD

The equation that must be solved is: $y_i \hat{\varphi}_i^V P = x_i \gamma_i \varphi_i^{sat} P_i^{sat} \exp(V_i^L (P - P_i^{sat})/RT)$

Bubble P

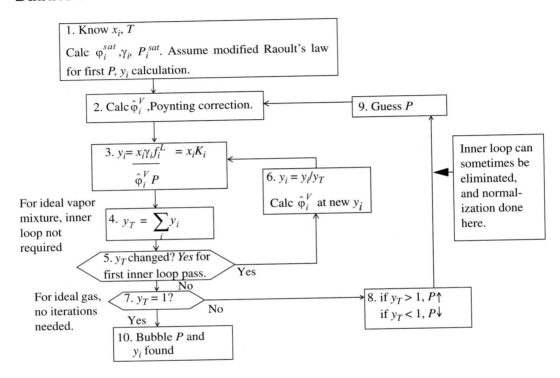

1. Know x_i, T

   Calc φ_i^{sat}, γ_i, P_i^{sat}. Assume modified Raoult's law for first P, y_i calculation.

2. Calc $\hat{\varphi}_i^V$, Poynting correction.

9. Guess P

Inner loop can sometimes be eliminated, and normalization done here.

3. $y_i = \dfrac{x_i \gamma_i f_i^L}{\hat{\varphi}_i^V P} = x_i K_i$

6. $y_i = y_i/y_T$

   Calc $\hat{\varphi}_i^V$ at new y_i

For ideal vapor mixture, inner loop not required

4. $y_T = \sum_i y_i$

5. y_T changed? *Yes* for first inner loop pass.   Yes

No

For ideal gas, no iterations needed.

7. $y_T = 1$?   No

Yes

8. if $y_T > 1$, $P\uparrow$
   if $y_T < 1$, $P\downarrow$

10. Bubble P and y_i found

Bubble T

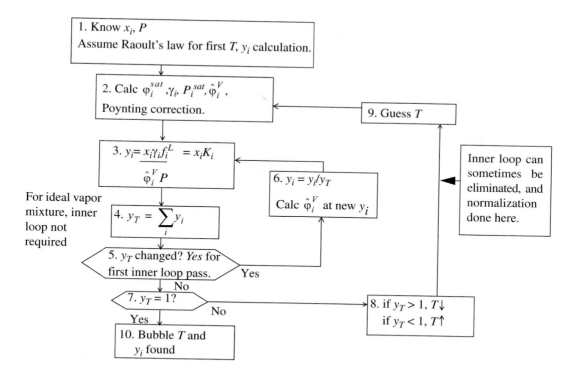

1. Know x_i, P
Assume Raoult's law for first T, y_i calculation.

2. Calc φ_i^{sat}, γ_i, P_i^{sat}, $\hat{\varphi}_i^V$, Poynting correction.

9. Guess T

3. $y_i = \dfrac{x_i \gamma_i f_i^L}{\hat{\varphi}_i^V P} = x_i K_i$

For ideal vapor mixture, inner loop not required

6. $y_i = y_i / y_T$
Calc $\hat{\varphi}_i^V$ at new y_i

Inner loop can sometimes be eliminated, and normalization done here.

4. $y_T = \sum_i y_i$

5. y_T changed? *Yes* for first inner loop pass.   Yes

No

7. $y_T = 1$?   No

Yes

8. if $y_T > 1$, $T\downarrow$
if $y_T < 1$, $T\uparrow$

10. Bubble T and y_i found

Dew P

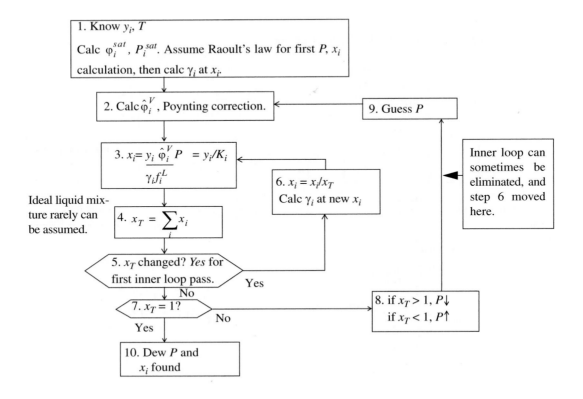

1. Know y_i, T

   Calc φ_i^{sat}, P_i^{sat}. Assume Raoult's law for first P, x_i calculation, then calc γ_i at x_i.

2. Calc $\hat{\varphi}_i^V$, Poynting correction.

9. Guess P

3. $x_i = \dfrac{y_i \, \hat{\varphi}_i^V \, P}{\gamma_i f_i^L} = y_i/K_i$

Ideal liquid mixture rarely can be assumed.

6. $x_i = x_i/x_T$
   Calc γ_i at new x_i

Inner loop can sometimes be eliminated, and step 6 moved here.

4. $x_T = \sum\limits_i x_i$

5. x_T changed? *Yes* for first inner loop pass.   Yes

No

7. $x_T = 1$?   No

Yes

8. if $x_T > 1$, $P\downarrow$
   if $x_T < 1$, $P\uparrow$

10. Dew P and x_i found

Dew T

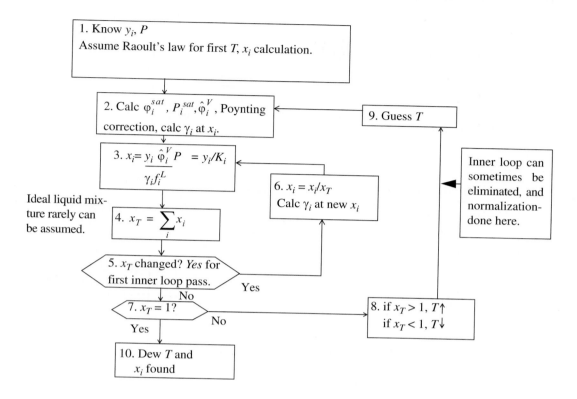

1. Know y_i, P
   Assume Raoult's law for first T, x_i calculation.

2. Calc φ_i^{sat}, P_i^{sat}, $\hat{\varphi}_i^V$, Poynting correction, calc γ_i at x_i.

9. Guess T

3. $x_i = \dfrac{y_i\,\hat{\varphi}_i^V\,P}{\gamma_i f_i^L} = y_i/K_i$

Inner loop can sometimes be eliminated, and normalization-done here.

Ideal liquid mixture rarely can be assumed.

6. $x_i = x_i/x_T$
   Calc γ_i at new x_i

4. $x_T = \displaystyle\sum_i x_i$

5. x_T changed? *Yes* for first inner loop pass.

Yes

No

7. $x_T = 1$?

No

8. if $x_T > 1$, $T\uparrow$
   if $x_T < 1$, $T\downarrow$

Yes

10. Dew T and x_i found

Isothermal Flash

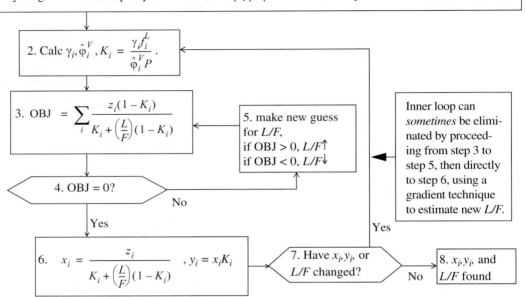

1. Know z_i, P, T

Apply shortcut K-ratio method for first K_i calculation. Calc φ_i^{sat}, P_i^{sat},

Poynting correction. Skip step 2 first time, set $x_i = y_i = z_i$ to force outer loop below to execute at least once.

2. Calc $\gamma_i, \hat{\varphi}_i^V$, $K_i = \dfrac{\gamma_i f_i^L}{\hat{\varphi}_i^V P}$.

3. OBJ $= \displaystyle\sum_i \dfrac{z_i(1 - K_i)}{K_i + \left(\dfrac{L}{F}\right)(1 - K_i)}$

4. OBJ = 0?

5. make new guess for L/F,
if OBJ > 0, $L/F\uparrow$
if OBJ < 0, $L/F\downarrow$

Inner loop can *sometimes* be eliminated by proceeding from step 3 to step 5, then directly to step 6, using a gradient technique to estimate new L/F.

No

Yes

Yes

6. $x_i = \dfrac{z_i}{K_i + \left(\dfrac{L}{F}\right)(1 - K_i)}$, $y_i = x_i K_i$

7. Have x_i, y_i, or L/F changed?

No

8. x_i, y_i, and L/F found

MODELS FOR PROCESS SIMULATORS

D.1 OVERVIEW

There are so many thermodynamic models commonly used in chemical process simulators that it would be overwhelming to cover all of them in great detail. This is why the discussion in the text focuses on a few representative models. Nevertheless, students interested in process engineering will often face the need to choose the most appropriate thermodynamic model, and the most appropriate model may not be one of those that we have covered in detail. Fortunately, the differences between many of the thermodynamic models and the ones that we have studied are generally quite small. In this appendix, we review some of most common thermodynamic models and put them into context with others that we have studied. This should help students to feel a bit more comfortable wading through the wealth of models from which to choose.

Students interested in becoming process simulation experts will be interested in reading the recent articles reviewing the selection of thermodynamic models. Schad (1998) and Carlson (1996) provide some significant examples and cite several relevant articles. A common thread throughout these articles is the emphasis on accurate application of thermodynamic principles. It is interesting to see the large number of examples in which practical engineering applications were so deeply affected by the fundamentals of thermodynamics.

D.2 EQUATIONS OF STATE

We have covered the Peng-Robinson and virial equation in fair detail, but there are many others. Some that we have mentioned but not treated in detail are the Redlich-Kwong (1949) equation (homework problem 6.9), the Lee-Kesler (1975) equation (Eqn. 6.11 on page 202), and a popular form of its extension to mixtures, the Lee-Kesler-Plocker equation (Plocker et al., 1978), the Soave (1972) equation (Eqn. 6.49 on page 225, also known as the Soave-Redlich-Kwong or SRK equation), the ESD equation (Elliott, et al.,1990) and the SAFT equation (Chapman et al., 1990). A slight variation on the Soave equation is the API equation (Graboski and Daubert, 1978); it changes only the value of κ as a function of acentric factor in order to obtain a slight improvement in the predicted vapor pressures of hydrocarbons. A specific implementation of the virial equation is the Hayden-O'Connell (1975) method. The Soave equation, Peng-Robinson equation, Lee-Kesler-Plocker, and the API equation are all very similar in their predictions of VLE behavior of hydrocarbon mixtures. They are accurate to within ~5% in correlations of bubble point pressures of hydrocarbons and gases (CO, CO2, N2, O2, H2S) and about ~15% for predictions based on estimated binary interaction parameters. The Lee-Kesler-Plocker equation can be slightly more accurate for enthalpy and liquid density for some hydrocarbon mixtures, but the advantage is generally slight with regard to enthalpy and there are better alternatives to equations of state for liquid densities if you want accurate values. The cubic equations have some convergence advantages for VLE near critical points and their relative simplicity makes them more popular choices for adaptations of semi-empirical mixing rules to tune in an accurate fit to the thermodynamics of a specific system of interest. The best choice among these is generally the one for which the binary interaction parameters have been determined with the greatest reliability. (Accurate reproduction of the most experimental data at the conditions of your specific interest wins.)

The primary role of equations of state is that they can predict thermodynamic properties at any conditions of temperature and pressure, including the critical region. The disadvantage is that they tend to be inaccurate for strongly hydrogen-bonding mixtures. This disadvantage is diminishing in importance with the development of hydrogen-bonding equations of state (like the SAFT and ESD equations), but it is not clear at this time whether these newer equations of state will displace any of the long-standing cubic equations of state with their semi-empirical modifications.

D.3   SOLUTION MODELS

We have covered many solution models in fair detail: the Margules equation, the Redlich-Kister expansion, the van Laar equation, Scatchard-Hildebrand theory (the most common implementation of regular solution theory), the Flory-Huggins equation, the Wilson equation, UNIQUAC, and UNIFAC. There is one more that deserves mention: the NRTL equation (Renon and Prausnitz, 1969). This a very slight modification of the Wilson equation which enables correlation of liquid-liquid equilibria as well as vapor-liquid equilibria. Once again, the best choice will most often depend on the availability of binary interaction parameters which are relevant to the specific conditions of interest.

The primary role of solution models is to provide semi-empirical models which have a greater degree of flexibility than equation of state models, owing to the greater number of adjustable parameters and their judicious choice such that both magnitude and skewness of the free energy curves can be accurately tuned.

D.4   HYBRID MODELS

Another set of models that have been developed relatively recently can be referred to as "hybrid" models in the sense that they combine equation of state models with solution models. The two most prevalent of these are the Modified Huron-Vidal (MHV) method and the Wong-Sandler mixing rules. The basic idea is to apply a solution model at high density or pressure to characterize the mixing rules of the equation of state and then interpolate from this result to the virial equation at low density. These methods tend to compete with the hydrogen bonding models in the sense that they enhance accuracy for non-ideal solutions at high temperatures and pressures. They are more empirical, but they tend to leverage the well-developed solution models (like UNIFAC) more directly. They also tend to be more efficient computationally than the hydrogen bonding equations of state.

D.5   RECOMMENDED DECISION TREE

When faced with choosing a thermodynamic model, it is helpful to at least have a logical procedure for deciding which model to try first. A decision tree is included in Fig. D.1. For non-polar fluids, an equation of state may suffice. For polar fluids, a fitted activity coefficient model is preferred, possibly in combination with the Hayden-O'Connell method or in combination with some other

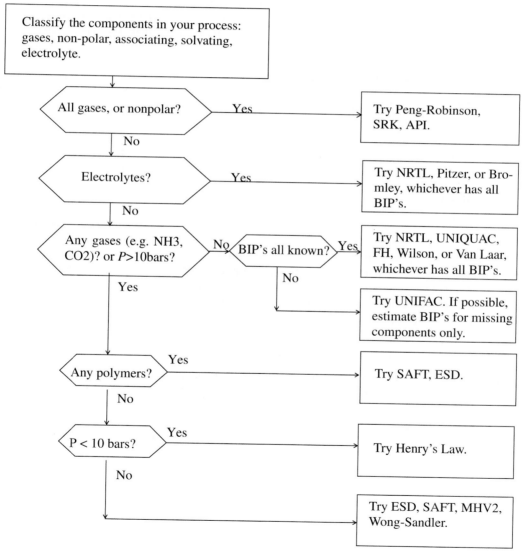

Figure D.1 *Flow chart to select the best thermodynamic model. The abbreviation BIP is used to mean binary interaction parameters.*

equation of state for the vapor phase (like the Peng-Robinson equation). This approach can often provide satisfactory predictions as long as the pressures are 10 bars or less. Predictions by this approach should be checked against literature data to the greatest extent possible. If there are no experimental data for one of the binary systems in this event, then UNIFAC can be used to generate "pseudo-data" that can be used to predict the Gibbs excess energy for that binary, and these pseudo-data can be used to regress UNIQUAC or NRTL parameters if desired (homework problem 11.40). Above 10 bars, the choices are not so obvious. The most obvious method to try if you are satisfied with the corelations below 10 bars is to apply the MHV or Wong-Sandler approach. If you need to predict phase behavior over a broad range of conditions based on few data in a narrow range of conditions, a hydrogen bonding equation might provide more reliable leverage in light of its clearer connection with the physical chemistry in the solution. If you are dealing with compounds which dissociate electrolytically or associate strongly and specifically in solution, then it will probably be necessary to apply a simultaneous reaction and phase equilibrium approach. These kinds of systems are common in gas strippers for compounds like CO_2, H_2S, and amines. For these systems, it is especially important to check your correlations against experimental data near the conditions of your specific interest.

D.6   THERMAL PROPERTIES OF MIXTURES

A problem not treated explicitly in the text is the performance of energy and entropy balances for mixtures. We mentioned that energy departures for mixtures are just like those for pure fluids, after the mixing rules are defined. But we never showed how that rigorous approach for treating the mixture properties can be used in solving practical engineering problems. Process simulators deal almost entirely with mixtures and energy balances play a major role in process simulation, so it is essential to have a framework for characterizing the thermal properties of mixtures as they flow through throttles, turbines, or reactors.

In discussing thermal properties of pure fluids, it was sufficient to define the reference state to be, say, the saturated liquid at its normal boiling point, such that every component had an independent reference state. In mixtures, especially mixtures that may change composition owing to reactions, reference states must account for energetic differences between components. This fact was recognized when dealing with heats of reaction, and especially in dealing with adiabatic reaction temperatures, but ideal gases were assumed in those examples. The extension to thermal properties of reacting mixtures requires a simple combination of the approaches used for pure fluids and reacting mixtures. Briefly, the standard states for all components can be set as the elements at 298 K and 1 bar in the ideal gas state. For example, the enthalpy of a flowing liquid benzene + styrene stream at 310 K includes (a) the heats of formation of the ideal gas components, (b) the integral of the ideal gas heat capacities, and (c) the enthalpy departure of the benzene + styrene mixture. The enthalpy of this stream will be large and negative, but not as large and negative as the same stream transformed into benzene + polystyrene at the same temperature, because the heat of polymerization must be removed to maintain constant temperature.

Example D.1 Contamination from a reactor leak

An equimolar mixture of benzene and styrene is contained in a reactor at 400 K and 10 bars when it begins to leak. The Environmental Protection Agency would like to know the concentration of benzene if the reactor contents mixed with air after release, forming a dispersion cloud. One way to model this process is as a series of isenthalpic flashes by slowly increasing the amount of air in the mixture to correspond with varying breadths of the dispersion cloud. Assuming 0.1 kmol of air is mixed with each one kmol of reactor leakage, use a process simulator to compute the temperature and vapor and liquid amounts and compositions, then check the simulator's calculations using the Peng-Robinson model. The binary interaction parameters can be taken as zero except for benzene + nitrogen ($k_{ij} = 0.1668$) and oxygen + nitrogen ($k_{ij} = -0.0119$).

Solution: This process can be modeled in a process simulator by feeding the reactor leakage and air stream separately into a "mixer," then sending the outlet from the mixer to an isothermal flash unit to obtain separate results for the vapor and liquid streams. If the mixer model is properly designed, it should check the energy balance to make sure the outlet enthalpy is the same as the inlet enthalpy, and in the process it must perform an isenthalpic flash. The isothermal flash then represents a simple splitter where the feed composition, temperature, and vapor/feed ratio are already known, and the vapor and liquid compositions are computed by the usual flash material balance. The heat requirement for the isothermal flash can be computed from an energy balance. The heats of formation can be taken as 82.88 kJ/mole for benzene and 147.36 kJ/mole for styrene. You can check whether these are being used as the reference conditions for your process simulator by typing 1 kmol for the stream flow with the mole fraction of the desired component set to unity, setting the temperature to 25°C and the pressure to some suitably low value, e.g., 1.E-6 bars. Similarly, the air inlet stream should be set to 25°C and 1 bar and the enthalpy of that stream should be near zero, since air is composed of ~79mole% nitrogen and 21% oxygen, both elements. If your simulator uses different values for the heats of formation at these conditions, write them down and substitute them appropriately in the calculations below.

The reactor contents at 400 K and 10 bars are actually liquid, so we need to integrate to obtain the ideal gas enthalpy at 400 K, then apply the enthalpy departure at 400 K and 10 bars. The ideal gas enthalpies can be computed for each component separately and then averaged together, since there is no heat of mixing ideal gases. The combined ideal gas enthalpy change from 298 to 400 K is 12.20. The FU option on the PRMIX.EXE program provides the necessary departure functions. The value of the departure function at 400 K and 1 MPa is $(H^L - H^{ig})/RT = -9.7906$, indicating $H^L - H^{ig} = -32.56$ kJ/mole. Combining the contributions from heats of formation, the ideal gas, and the departure function gives an enthalpy for the reactor stream of 94.76 kJ/mole compared to 94.769 from the process simulator.[1] Since the air stream is practically at the reference state, the total enthalpy is 94.76 for 1.1 total moles.

D.7 LITERATURE CITED

Abrams, D. S., Prausnitz, J. M., *AIChE J.* **1975**, 21, 116.

Carlson, E. C., *Chem. Eng. Prog.,* **1996,** (Oct.), 35.

---

1. In the interest of completeness, the simulator applied is ChemCAD Version 4.0.

Example D.2 Contamination from a Reactor Rupture (Continued)

In principle, the adiabatic flash can be computed using the PRMIX.EXE program by guessing an outlet temperature, performing the flash, computing the enthalpies of the vapor and liquid, and checking that the total enthalpy matches the value of 94.76. The guessing might become tedious if not for the fact that the process simulator can give us a very good hint. The simulator indicates a value of 362 K for the result from the isenthalpic flash. Let's assume 362 K as our guess. Applying PRMIX.EXE to the flash at 362 K and 0.1 MPa with feed composition of (0.4545, 0.4545, 0.0718, 0.0191) gives $V/F = 0.2686$, $x = (0.4201, 0.5796, 0.0002, 0.0001)$, $y = (0.5482, 0.1140, 0.2666, 0.0711)$ These compare to values from the simulator of $V/F = 0.2727$, $x = (0.4196, 0.5801, 0.000197, 0.000118)$, $y = (0.5480, 0.1187, 0.2635, 0.0700)$. The agreement on the flash calculation is less precise than that for the enthalpy calculation, but the values are close enough for this demonstration. Computing the liquid and vapor enthalpy departures: -34.74 and -0.12 for liquid and vapor. Using the results from PRMIX.EXE, the ideal gas enthalpy change and composition weighted heat of formation for the *liquid* are 7.03 and 120.23 for a total liquid enthalpy of 92.52 kJ/mole. The ideal gas enthalpy change and composition weighted heat of formation for the *vapor* are 6.03 and 62.23 for a total vapor enthalpy of 68.14 kJ/mole. Weighting the enthalpies by their molar amounts gives $H = (0.2686*68.14 + 0.7314*92.52)*1.1 = 94.57$ compared to 94.76. Considering the discrepancies between the flash results, the precision of the agreement on enthalpies is satisfactory.

We conclude by noting how nearly we have come to assembling all the building blocks of a steady-state process simulator through the subjects in this text. The tasks not covered in this text are: sizing heat exchangers (largely a matter of estimating heat transfer coefficients according to the heat and momentum transport), designing mass transfer equipment (a significant application of vapor-liquid and liquid-liquid equilibrium models), and sizing reactors (requiring the study of kinetics). Perusal of a typical chemical engineering curriculum should help to clarify how these additional subjects fit together. Although we have not covered everything, we have covered a substantial amount. A clever student should be able to see the light at the end of the tunnel even from here.

Chapman, W. G.; Gubbins, K. E.; Jackson, G.; Radosz, M. *Ind. Eng. Chem. Res.* **1990,** 29, 1709.

Elliott, J. R. Jr.; Suresh, S. J.; Donohue, M. D. *Ind. Eng. Chem. Res.* **1990,** 29, 1476.

Fredenslund, A., Jones, R. L., Prausnitz, J. M. *AIChE J.* **1975**, 21, 1086.

Graboski, M. S. and Daubert, T. E. *Ind. Eng.Chem.Proc.Des.Dev.* **1979**, 17, 448.

Hayden, J.G., O'Connell, J. P. *Ind. Eng.Chem.Proc.Des.Dev.* **1975**, 14, 3.

Lee, B. I., Kesler, M. G. *AIChE J.* **1975**, 21, 510.

Redlich, O, Kwong, J. N. S. *Chem. Rev.* **1949**, 44, 233.

Renon H., Prausnitz, J. M. *Ind. Eng.Chem.Proc.Des.Dev.* **1969**, 8, 413.

Peng, D.Y., Robinson, D. B. *Ind.Eng.Chem. Fundam.* **1976**, 15, 59.

Plocker, U., Knapp, H., Prausnitz, J. M. *Ind. Eng.Chem.Proc.Des.Dev.* **1978**, 17, 324.

Schad, *Chem. Eng. Prog.,* **1998,** 1, 94, 74.

Soave, G. *Chem. Eng. Sci.* **1972**, 27, 1197.

Wilson, G. M. *J. Am.Chem.Soc.* **1964**, 86, 127, 133.

Wong, D. S. H., Sandler, S. I., *AICHE J.* **1992,** 38, 671.

PURE COMPONENT PROPERTIES

E.1 IDEAL GAS HEAT CAPACITIES[1]

The heat capacity coefficients are used in the equation:

$$C_P = A + BT + CT^2 + DT^3$$

where T is in K and C_P is in J/(molK). The heat capacities tabulated on the endflap are only suitable for quick order of magnitude calculations. The full form of the heat capacity should be used when possible.

Caution: *Note that the value of the heat capacity at room temperature is not given by the first coefficient A. The polynomial should not be truncated.*

Compound		A	B	C	D
Paraffins					
1	Methane	19.25	5.213E-2	1.197E-5	−1.132E-8
2	Ethane	5.409	1.781E-1	−6.938E-5	8.713E-9
3	Propane	−4.224	3.063E-1	−1.586E-4	3.215E-8
4	n-Butane	9.487	3.313E-1	−1.108E-4	−2.822E-9
5	Isobutane	−1.390	3.847E-1	−1.846E-4	2.895E-8
7	n-Pentane	−3.626	4.873E-1	−2.580E-4	5.305E-8
8	Isopentane	−9.525	5.066E-1	−2.729E-4	5.723E-8
9	Neopentane (2,2-Dimethylpropane)	−16.59	5.552E-1	−3.306E-4	7.633E-8
11	n-Hexane	−4.413	5.280E-1	−3.119E-4	6.494E-8
17	n-Heptane	−5.146	6.762E-1	−3.651E-4	7.658E-8
27	n-Octane	−6.096	7.712E-1	−4.195E-4	8.855E-8
41	Isooctane (2,2,4-Trimethylpentane)	−7.461	7.779E-1	−4.287E-4	9.173E-8

1. Reid, R. C., Prausnitz, J. M., Poling, B. E. *The Properties of Gases and Liquids*, 4th ed. McGraw-Hill, 1987, p. 656–732.

Compound		A	B	C	D
46	n-Nonane	−8.374	8.729E-1	−4.823E-4	1.031E-7
56	n-Decane	−7.913	9.609E-1	−5.288E-4	1.131E-7
64	n-Dodecane	−9.328	1.149	−6.347E-4	1.359E-7
66	n-Tetradecane	−1.098E+1	1.338	−7.423E-4	1.598E-7
68	n-Hexadecane	−1.302E+1	1.529	−8.537E-4	1.850E-7
Naphthenes					
104	Cyclopentane	−5.362E+1	5.426E-1	−3.031E-4	6.485E-8
105	Methylcyclopentane	−5.011E+1	6.381E-1	−3.642E-4	8.014E-8
137	Cyclohexane	−5.454E+1	6.113E-1	−2.523E-4	1.321E-8
138	Methylcyclohexane	−6.192E+1	7.842E-1	−4.438E-4	9.366E-8
Olefins and Acetylene					
201	Ethylene	3.806	1.566E-1	−8.348E-5	1.755E-8
202	Propylene	3.710	2.345E-1	−1.160E-4	2.205E-8
204	1-Butene	−2.994	3.532E-1	−1.990E-4	4.463E-8
207	Isobutene (Isobutylene)	1.605E+1	2.804E-1	−1.091E-4	9.098E-9
209	1-Pentene	−1.340E-1	4.329E-1	−2.317E-4	4.681E-8
401	Acetylene	2.682E+1	7.578E-2	−5.007E-5	1.412E-8
303	1,3-Butadiene	−1.687	3.419E-1	−2.340E-4	6.335E-8
309	2-Methyl-1,3-Butadiene (Isoprene)	−3.412	4.585E-1	−3.337E-4	1.000E-7
	Styrene	−2.825E+1	6.159E-1	−4.023E-4	9.935E-8
Aromatics					
501	Benzene	−3.392E+1	4.739E-1	−3.017E-4	7.130E-8
502	Toluene	−2.435E+1	5.125E-1	−2.765E-4	4.911E-8
504	Ethylbenzene	−4.310E+1	7.072E-1	−4.811E-4	1.301E-7
505	o-Xylene	−1.585E+1	5.962E-1	−3.443E-4	7.528E-8
506	m-Xylene	−2.917E+1	6.297E-1	−3.747E-4	8.478E-8
507	p-Xylene	−2.509+1	6.042E-1	−3.374E-4	6.820E-8
510	Isopropylbenzene (Cumene)	−3.3936E+1	7.842E-1	−5.087E-4	1.291E-7
558	Diphenyl	−9.707E+1	1.106	−8.855E-4	2.790E-7
563	Diphenylmethane				
701	Naphthalene	−6.880E+1	8.499E-1	−6.506E-4	1.981E-7
702	1-Methylnaphthalene	−6.482E+1	9.387E-1	−6.942E-4	2.016E-7
706	1,2,3,4-Tetrahydronaphthalene (Tetralin)				
Oxygenated Hydrocarbons					
1101	Methanol	21.15	7.092E-2	2.587E-5	−2.852E-8
1102	Ethanol	9.014	2.141E-1	−8.39E-5	1.373E-9
1103	Propanol	2.470	3.325E-1	−1.855E-4	4.296E-8
1104	Isopropanol	3.243E+1	1.885E-1	6.406E-5	−9.261E-8
1105	1-Butanol	3.266	4.180E-1	−2.242E-4	4.685E-8
1107	Isobutanol	−7.708	4.689E-1	−2.884E-4	7.231E-8
1479	Tetrahydrofuran(THF)	1.910E+1	5.162E-1	−4.132E-4	1.454E-7
1402	Diethyl Ether	2.142E+1	3.359E-1	−1.035E-4	−9.357E-9
	Ethylene Oxide	−7.519	2.222E-1	−1.256E-4	2.592E-8
1052	Methyl Ethyl Ketone	1.094E+1	3.559E-1	−1.900E-4	3.920E-8

Compound		A	B	C	D
Halocarbons					
1601	Freon-12(CCl2F2)	3.160E+1	1.782E-1	−1.509E-4	4.342E-8
	Freon-22(CClF2)	1.730E+1	1.618E-1	−1.170E-4	3.058E-8
	Freon-11(CCl3F)	4.098E+1	1.668E-1	−1.416E-4	4.146E-8
	Freon-113(C2Cl3F3)	6.114E+1	2.874E-1	−2.420E-4	6.904E-8
1502	Methyl Chloride	1.388E+1	1.014E-1	−3.889E-5	2.567E-9
1501	Carbon Tetrachloride	4.072E+1	2.049E-1	−2.270E-4	8.843E-8
1521	Chloroform(CHCl3)	2.400E+1	1.893E-1	−1.841E-4	6.657E-8
	Chlorobenzene	−3.389E+1	5.631E-1	−4.522E-4	1.426E-7
Gases					
914	Argon	2.080E+1			
922	Bromine	3.386E+1	1.125E-2	−1.192E-5	4.534E-9
918	Chlorine	2.693E+1	3.384E-2	−3.869E-5	1.547E-8
913	Helium-4	2.080E+1			
920	Krypton	2.080E+1			
919	Neon	2.080E+1			
959	Xenon	2.080E+1			
912	Nitric Oxide	2.935E+1	−9.378E-4	9.747E-6	−4.187E-9
899	Nitrous Oxide	2.162E+1	7.281E-2	−5.778E-5	1.830E-8
910	Sulfur Dioxide	2.385E+1	6.699E-2	−4.961E-5	1.328E-8
911	Sulfur Trioxide	1.921E+1	1.374E-1	−1.176E-4	3.700E-8
901	Oxygen	2.811E+1	−3.680E-6	1.746E-5	−1.065E-8
902	Hydrogen (equilibrium)	2.714E+1	9.274E-3	−1.381E-5	7.645E-9
905	Nitrogen	3.115E+1	−1.357E-2	2.680E-5	-1.168E-8
908	Carbon Monoxide	3.087E+1	−1.285E-2	2.789E-5	−1.272E-8
909	Carbon Dioxide	1.980E+1	7.344E-2	−5.602E-5	1.715E-8
Nitrogen and Sulfur Gases					
1922	Hydrogen Sulfide	3.194E+1	1.436E-3	2.432E-5	−1.176E-8
1938	Carbon Disulfide	2.744E+1	8.127E-2	−7.666E-5	2.673E-8
	Hydrazine(N2H4)	9.768	1.895E-1	−1.657E-4	6.025E-8
1904	Hydrogen Chloride	3.067E+1	−7.201E-3	1.246E-5	−3.898E-9
	Hydrogen Cyanide	2.186E+1	6.062E-2	−4.961E-5	1.815E-8
Miscellaneous Compounds					
1051	Acetone	6.301	2.606E-1	−1.253E-4	2.038E-8
1772	Acetonitrile	2.048E+1	1.196E-1	−4.492E-5	3.203E-9
1251	Acetic Acid	4.840	2.549E-1	−1.753E-4	4.949E-8
1911	Ammonia	2.731E+1	2.383E-2	1.707E-5	−1.185E-8
1921	Water	32.24	1.924E-3	1.055E-5	−3.596E-9

E.2 LIQUID HEAT CAPACITIES[1]

The following constants are for the equation:

$$\frac{C_P}{R} = A + BT + CT^2$$

where T in K, over the temperature range from 273.15 to 373.15 K

Species	A	B	C
Ammonia	22.626	1.0075E-01	1.9271E-04
Aniline	15.819	2.9030E-02	-1.5800E-05
Benzene	-0.747	6.7960E-02	-3.7780E-05
1,3-Butadiene	22.711	-8.7960E-02	2.0579E-04
Carbon tetrachloride	21.155	-4.8280E-02	1.0114E-04
Chlorobenzene	11.278	3.2860E-02	-3.1900E-05
Chloroform	19.215	-4.2890E-02	8.3010E-05
Cyclohexane	-9.048	1.4138E-01	-1.6162E-04
Ethanol	33.866	-1.7260E-01	3.4917E-04
Ethylene Oxide	21.039	-8.6410E-02	1.7228E-04
Methanol	13.431	-5.1280E-02	1.3113E-04
n-Propanol	41.653	-2.1032E-01	4.2720E-04
Sulfur Trioxide	-2.93	1.3708E-01	-8.4730E-05
Toluene	15.133	6.7900E-03	1.6350E-05
Water	8.712	1.2500E-03	-1.8000E-07

E.3 SOLID HEAT CAPACITIES

The following constants are for the equation:

$$\frac{C_P}{R} = A + BT + CT^2 + DT^3$$

where T is in K.

Species	A	B	C	D
Carbon (298-1300 K)	-3.9578	5.586E-2	-4.5482E-5	1.5171E-8

---

1. Smith, J. M., Van Ness, H. C., Abbott, M. M., *Introduction to Chemical Engineering Thermodynamics,* 5th edition, McGraw-Hill, New York, 1996, pg 639.

E.4   ANTOINE CONSTANTS

The following constants are for the equation

$$\log_{10} P^{sat} = A - \frac{B}{T + C}$$

where P^{sat} is in mmHg, and T is in Celsius. Additional Antoine constants are tabulated in ACTCO-EFF.XLS.

	A	B	C	T range (°C)	Source
Acetic Acid	8.02100	1936.01	258.451	18–118	3
Acetic Acid	8.26735	2258.22	300.97	118–227	3
Acetone	7.63130	1566.69	273.419	57–205	3
Acetone	7.11714	1210.595	229.664	−13–55	3
Acrolein (2-propenal)	8.62876	2158.49	323.36	2.5–52	1
Benzene	6.87987	1196.76	219.161	8–80	3
Benzyl Chloride	7.59716	1961.47	236.511	22–180	1
Biphenyl (solid)	13.5354	4993.37	296.072	20–40	5
1-Butanol	7.81028	1522.56	191.95	30–70	2
1-Butanol	7.75328	1506.07	191.593	70–120	2
2-Butanone	7.28066	1434.201	246.499	−6.5–80	1
Chloroform	6.95465	1170.966	226.232	−10–60	3
Ethanol	8.11220	1592.864	226.184	20–93	3
Hexane	6.91058	1189.64	226.28	−30–170	3
1-Propanol	8.37895	1788.02	227.438	−15–98	3
2-Propanol	8.87829	2010.33	252.636	−26–83	3
Methanol	8.08097	1582.271	239.726	15–84	3
Naphthalene (solid)	8.62233	2165.72	198.284	20–40	4
Pentane	6.87632	1075.78	233.205	−50–58	3
3-Pentanone	7.23064	1477.021	237.517	36–102	3
Toluene	6.95087	1342.31	219.187	−27–111	3
Water	8.07131	1730.63	233.426	1–100	3

1. Fit to data from D.R. Stull, in *Perry's Chemical Engineering Handbook*, 5th ed., McGraw-Hill, pp. 3–46 to 3–62.
2. Fit to data from *Handbook of Chemistry and Physics*, 56th ed, R.C. Weast, ed., CRC Press, 1974–75, pp. D191–D210.
3. Gmehling, J., *Vapor-liquid Equilibrium Data Collection*, Frankfort, Germany, DECHEMA, 1977.
4. Fit to data of Ambrose, D., Lawerenson, I. J., Sprake, C. H. S, *J. Chem. Therm.*, 7(1975)1173.
5. Timmermans, J., *Physio-Chemical Constants of Pure Organic Compounds*, Elsevier, New York, 1950.

E.5 LATENT HEATS

	$T_m(°C)$	ΔH^{fus} kJ/mol	$T^{sat}(°C)$ at 1.01325 bar	ΔH^{vap} kJ/mol
Acetic Acid	16.6	12.09	118.2	24.39
Acetone	−95.	5.69	56.	30.2
Anthracene	216.5	28.86		
Benzene	5.53	9.837	80.1	30.765
Biphenyl	69.2	18.58		
n-Butane	−138.3	4.661	−0.6	22.305
Cyclohexane	6.7	2.677		
Ethanol	−114.6	5.021	78.5	38.58
n-Hexane	−95.32	13.03	68.74	28.85
Naphthalene	80.2	18.80		
Phenanthrene	99.2	16.46		
Phenol	40.9	11.43		
Water	0.00	6.0095	100.0	40.656

E.6   ENTHALPIES AND GIBBS ENERGIES OF FORMATION

Ideal Gases

Standard State: Ideal Gas 298.15 K and 1 bar

USE PRESSURE IN BAR WHEN USING THESE PROPERTIES

		$\Delta H_{f, 298}$ kJ/mol	$\Delta G_{f, 298}$ kJ/mol
Paraffins			
1	Methane	−74.8936	−50.45
2	Ethane	−83.82	−31.86
3	Propane	−104.68	−24.29
4	*n*-Butane	−125.79	−16.57
5	Isobutane	−134.99	−20.8781
7	*n*-Pentane	−146.76	−8.65
11	*n*-Hexane	−166.92	−0.16736
17	*n*-Heptane	−187.8	8.2
27	*n*-Octane	−208.75	16.4013
41	Isooctane(2,2,4-Trimethylpentane)	−224.01	13.6817
46	*n*-Nonane	−228.74	24.97
56	*n*-Decane	−249.46	33.18
68	*n*-Hexadecane	−374.17	82.15
Naphthenes			
104	Cyclopentane	−78.4	38.8693
105	Methylcyclopentane	−106.023	35.7732
137	Cyclohexane	−123.3	31.7566
138	Methylcyclohexane	−154.766	27.2797
Olefins and Acetylene			
153	cis-Decalin	−168.95	85.8138
154	trans-Decalin	−182.297	73.4292
201	Ethylene	52.51	68.43
202	Propylene	19.71	62.14
204	1-Butene	−0.54	70.24
205	cis-2-Butene	−7.4	65.5
206	trans-2-Butene	−11	63.4
207	Isobutene	−17.1	57
216	1-Hexene	−41.67	87.4456
260	1-Decene	−124.69	119.83
270	Cyclohexene	−4.6	106.859
303	1,3-Butadiene	109.24	149.73
401	Acetylene	226.731	209.2
Aromatics			
501	Benzene	82.88	129.75
502	Toluene	50.17	122.29
504	Ethylbenzene	29.92	130.73
505	1,2-Dimethyl Benzene	19	122.22
506	1,3-Dimethyl Benzene	17.24	119
507	1,4-Dimethyl Benzene	17.95	121.26
510	Isopropylbenzene	3.93	137.15

		$\Delta H_{f, 298}$ kJ/mol	$\Delta G_{f, 298}$ kJ/mol
558	Biphenyl	182.42	281.08
601	Phenylethene	147.36	213.802
701	Naphthalene	150.959	223.593
706	Tetralin	24.2	167.1
803	Indene	163.28	233.97
805	Phenanthrene	206.9	308.1
Gases			
899	Nitrous Oxide	82.05	103.638
901	Oxygen	0	0
902	Hydrogen	0	0
905	Nitrogen	0	0
908	Carbon Monoxide	−110.53	−137.16
909	Carbon Dioxide	−393.51	−394.38
910	Sulfur Dioxide	−296.81	−300.14
912	Nitric Oxide	90.25	86.58
913	Helium	0	0
914	Argon	0	0
917	Fluorine	0	0
918	Chlorine	0	0
Oxygenated Hydrocarbons			
1001	Formaldehyde	−117.152	−112.968
1002	Acetaldehyde	−166.021	−133.302
1051	Acetone	−215.7	−151.2
1052	2-Butanone	−239	−151.9
1101	Methanol	−200.94	−162.24
1102	Ethanol	−234.95	−167.73
1103	Propanol	−255.2	−161.795
1104	2-Propanol	−272.295	−173.385
1105	Butanol	−274.6	−150.666
1114	1-Hexanol	−316.5	−135.562
1181	Phenol	−96.3993	−32.8988
1201	Ethylene Glycol	−392.878	−304.47
1211	Propylene Glycol	−425.429	−304.483
1251	Acetic Acid	−434.425	−376.685
1256	Butyric Acid	−475	−355
1281	Benzoic Acid	−294.1	−210.413
1289	Terephthalic Acid	−663.331	−597.179
1312	Methyl Acetate	−408.8	−321.5
1313	Ethyl Acetate	−444.5	−327.4
1381	Dimethyl Terephthalate	−643.583	−473.646
1402	Diethyl Ether	−252	−122
1403	Isopropyl Ether	−318.821	−121.88
Halocarbons			
1501	Carbon Tetrachloride	−95.8136	−60.6261
1502	Methyl Chloride	−80.7512	−62.8855
1503	Ethyl Chloride	−112.26	−59.9986
1511	Dichloromethane	−95.3952	−68.8687
1521	Chloroform	−103.345	−68.5339
1591	Vinylidene Chloride(1,1-C2H2Cl2)	2.38	25.39
1601	Freon12(CCl2F2)	−493.294	−453.964

		$\Delta H_{f,298}$ kJ/mol	$\Delta G_{f,298}$ kJ/mol
Nitrogen and Sulfur Compounds			
1701	Methylamine	−23	32.0913
1704	Ethylamine	−48.6599	37.2794
1801	Methanethiol	−22.6	−9.5
1802	Ethanethiol	−46	−4.5
1820	Dimethyl Sulfide	−37.2	7.4
1877	Urea	245.81	−152.716
Misc.			
1901	Sulfuric Acid	−740.568	−653.469
1902	Phosphoric Acid	−1250.22	−1106.42
1903	Nitric Acid	−133.863	−73.4459
1904	Hydrogen Chloride	−92.3074	−95.2864
1911	Ammonia	−45.94	−16.4013
1912	Sodium Hydroxide	−197.757	−200.46
1921	Water	−241.835	−228.614
1922	Hydrogen Sulfide	−20.63	−33.284
1938	Carbon Disulfide	116.943	66.9063

Solids

Standard State: pure solid at 298.15 K, 1 bar.

		$\Delta H_{f,298}$ kJ/mol	$\Delta G_{f,298}$ kJ/mol
	$CaCO_3$	−1206.92	−1128.79
	CaO	−635.09	−604.03

E.7 PROPERTIES OF WATER[1]

I. Saturation Temperature

T (°C)	P (MPa)	V^L m³/kg	V^V m³/kg	U^L kJ/kg	ΔU^{vap} kJ/kg	U^V kJ/kg	H^L kJ/kg	ΔH^{vap} kJ/kg	H^V kJ/kg	S^L kJ/kg-K	ΔS^{vap} kJ/kg-K	S^V kJ/kg-K
0.01	0.000612	0.001000	205.9912	0.00	2374.92	2374.92	0.00	2500.92	2500.92	0.0000	9.1555	9.1555
5	0.000873	0.001000	147.0113	21.02	2360.76	2381.78	21.02	2489.04	2510.06	0.0763	8.9485	9.0248
10	0.001228	0.001000	106.3032	42.02	2346.63	2388.65	42.02	2477.19	2519.21	0.1511	8.7487	8.8998
15	0.001706	0.001001	77.8755	62.98	2332.51	2395.49	62.98	2465.35	2528.33	0.2245	8.5558	8.7803
20	0.002339	0.001002	57.7567	83.91	2318.41	2402.32	83.91	2453.52	2537.43	0.2965	8.3695	8.6660
25	0.003170	0.001003	43.3373	104.83	2304.30	2409.13	104.83	2441.68	2546.51	0.3672	8.1894	8.5566
30	0.004247	0.001004	32.8783	125.73	2290.18	2415.91	125.73	2429.82	2555.55	0.4368	8.0152	8.4520
35	0.005629	0.001006	25.2053	146.63	2276.04	2422.67	146.63	2417.92	2564.55	0.5051	7.8466	8.3517
40	0.007385	0.001008	19.5151	167.53	2261.86	2429.39	167.53	2405.98	2573.51	0.5724	7.6831	8.2555
45	0.009595	0.001010	15.2521	188.43	2247.65	2436.08	188.43	2394.00	2582.43	0.6386	7.5247	8.1633
50	0.012400	0.001012	12.0269	209.33	2233.40	2442.73	209.34	2381.95	2591.29	0.7038	7.3710	8.0748
55	0.015800	0.001015	9.5643	230.24	2219.10	2449.34	230.26	2369.83	2600.09	0.7680	7.2218	7.9898
60	0.019900	0.001017	7.6672	251.16	2204.74	2455.90	251.18	2357.65	2608.83	0.8313	7.0768	7.9081
65	0.025000	0.001020	6.1935	272.09	2190.32	2462.41	272.12	2345.38	2617.50	0.8937	6.9359	7.8296
70	0.031200	0.001023	5.0395	293.03	2175.83	2468.86	293.07	2333.03	2626.10	0.9551	6.7989	7.7540
75	0.038600	0.001026	4.1289	313.99	2161.25	2475.24	314.03	2320.57	2634.60	1.0158	6.6654	7.6812
80	0.047400	0.001029	3.4052	334.96	2146.60	2481.56	335.01	2308.01	2643.02	1.0756	6.5355	7.6111
85	0.057900	0.001032	2.8258	355.95	2131.86	2487.81	356.01	2295.32	2651.33	1.1346	6.4088	7.5434
90	0.070200	0.001036	2.3591	376.97	2117.00	2493.97	377.04	2282.49	2659.53	1.1929	6.2852	7.4781
95	0.084600	0.001040	1.9806	398.00	2102.04	2500.04	398.09	2269.52	2667.61	1.2504	6.1647	7.4151
100	0.101400	0.001043	1.6718	419.06	2086.96	2506.02	419.17	2256.40	2675.57	1.3072	6.0469	7.3541
105	0.120900	0.001047	1.4184	440.15	2071.75	2511.90	440.27	2243.12	2683.39	1.3633	5.9319	7.2952
110	0.143400	0.001052	1.2093	461.26	2056.41	2517.67	461.42	2229.64	2691.06	1.4188	5.8193	7.2381
115	0.169200	0.001056	1.0358	482.41	2040.92	2523.33	482.59	2215.99	2698.58	1.4737	5.7091	7.1828
120	0.198700	0.001060	0.8912	503.60	2025.26	2528.86	503.81	2202.12	2705.93	1.5279	5.6012	7.1291
125	0.232200	0.001065	0.7700	524.83	2009.44	2534.27	525.07	2188.03	2713.10	1.5816	5.4954	7.0770
130	0.270300	0.001070	0.6680	546.09	1993.44	2539.53	546.38	2173.70	2720.08	1.6346	5.3918	7.0264
135	0.313300	0.001075	0.5817	567.41	1977.24	2544.65	567.74	2159.13	2726.87	1.6872	5.2900	6.9772
140	0.361500	0.001080	0.5085	588.77	1960.85	2549.62	589.16	2144.28	2733.44	1.7392	5.1901	6.9293
145	0.415700	0.001085	0.4460	610.19	1944.23	2554.42	610.64	2129.16	2739.80	1.7907	5.0919	6.8826
150	0.476200	0.001091	0.3925	631.66	1927.39	2559.05	632.18	2113.75	2745.93	1.8418	4.9953	6.8371
155	0.543500	0.001096	0.3465	653.19	1910.32	2563.51	653.79	2098.02	2751.81	1.8924	4.9002	6.7926
160	0.618200	0.001102	0.3068	674.79	1892.99	2567.78	675.47	2081.97	2757.44	1.9426	4.8065	6.7491
165	0.700900	0.001108	0.2724	696.46	1875.39	2571.85	697.24	2065.57	2762.81	1.9923	4.7143	6.7066
170	0.792200	0.001114	0.2426	718.20	1857.53	2575.73	719.08	2048.82	2767.90	2.0417	4.6233	6.6650
175	0.892600	0.001121	0.2166	740.02	1839.37	2579.39	741.02	2031.69	2772.71	2.0906	4.5335	6.6241
180	1.002800	0.001127	0.1938	761.92	1820.91	2582.83	763.05	2014.16	2777.21	2.1392	4.4448	6.5840
185	1.123500	0.001134	0.1739	783.91	1802.13	2586.04	785.19	1996.22	2781.41	2.1875	4.3572	6.5447

1. Harvey, A. P., Peskin, A. P., Klein, S. A., NIST/ASME Steam Properties, Version 2.1, NIST Standard Reference Data Program, December 1997.

Saturation Temperature (cont.)

T (°C)	P (MPa)	V^L m³/kg	V^V m³/kg	U^L kJ/kg	ΔU^{vap} kJ/kg	U^V kJ/kg	H^L kJ/kg	ΔH^{vap} kJ/kg	H^V kJ/kg	S^L kJ/kg-K	ΔS^{vap} kJ/kg-K	S^V kJ/kg-K
190	1.25520	0.001141	0.1564	806.00	1783.01	2589.01	807.43	1977.85	2785.28	2.2355	4.2704	6.5059
195	1.39880	0.001149	0.1409	828.18	1763.56	2591.74	829.79	1959.03	2788.82	2.2832	4.1846	6.4678
200	1.55490	0.001157	0.1272	850.47	1743.73	2594.20	852.27	1939.74	2792.01	2.3305	4.0997	6.4302
205	1.72430	0.001164	0.1151	872.87	1723.53	2596.40	874.88	1919.95	2794.83	2.3777	4.0153	6.3930
210	1.90770	0.001173	0.1043	895.39	1702.92	2598.31	897.63	1899.64	2797.27	2.4245	3.9318	6.3563
215	2.10580	0.001181	0.0947	918.04	1681.90	2599.94	920.53	1878.79	2799.32	2.4712	3.8488	6.3200
220	2.31960	0.001190	0.0861	940.82	1660.43	2601.25	943.58	1857.37	2800.95	2.5177	3.7663	6.2840
225	2.54970	0.001199	0.0784	963.74	1638.50	2602.24	966.80	1835.35	2802.15	2.5640	3.6843	6.2483
230	2.79710	0.001209	0.0715	986.81	1616.09	2602.90	990.19	1812.71	2802.90	2.6101	3.6027	6.2128
235	3.06250	0.001219	0.0653	1010.04	1593.16	2603.20	1013.77	1789.40	2803.17	2.6561	3.5214	6.1775
240	3.34690	0.001229	0.0597	1033.44	1569.69	2603.13	1037.55	1765.41	2802.96	2.7020	3.4403	6.1423
245	3.65120	0.001240	0.0547	1057.02	1545.65	2602.67	1061.55	1740.67	2802.22	2.7478	3.3594	6.1072
250	3.97620	0.001252	0.0501	1080.79	1521.00	2601.79	1085.77	1715.16	2800.93	2.7935	3.2786	6.0721
255	4.32290	0.001264	0.0459	1104.77	1495.72	2600.49	1110.23	1688.84	2799.07	2.8392	3.1977	6.0369
260	4.69230	0.001276	0.0422	1128.97	1469.75	2598.72	1134.96	1661.64	2796.60	2.8849	3.1167	6.0016
265	5.08530	0.001289	0.0387	1153.41	1443.04	2596.45	1159.96	1633.53	2793.49	2.9307	3.0354	5.9661
270	5.50300	0.001303	0.0356	1178.10	1415.57	2593.67	1185.27	1604.42	2789.69	2.9765	2.9539	5.9304
275	5.94640	0.001318	0.0328	1203.07	1387.26	2590.33	1210.90	1574.27	2785.17	3.0224	2.8720	5.8944
280	6.41660	0.001333	0.0302	1228.33	1358.06	2586.39	1236.88	1542.99	2779.87	3.0685	2.7894	5.8579
285	6.91470	0.001349	0.0278	1253.92	1327.89	2581.81	1263.25	1510.48	2773.73	3.1147	2.7062	5.8209
290	7.44180	0.001366	0.0256	1279.86	1296.67	2576.53	1290.03	1476.67	2766.70	3.1612	2.6222	5.7834
295	7.99910	0.001385	0.0235	1306.19	1264.30	2570.49	1317.27	1441.43	2758.70	3.2080	2.5371	5.7451
300	8.58790	0.001404	0.0217	1332.95	1230.67	2563.01	1345.01	1404.63	2749.64	3.2552	2.4507	5.7059
305	9.20940	0.001425	0.0199	1360.18	1195.67	2555.85	1373.30	1366.13	2739.43	3.3028	2.3629	5.6657
310	9.86510	0.001448	0.0183	1387.93	1159.14	2547.07	1402.22	1325.73	2727.95	3.3510	2.2734	5.6244
315	10.55620	0.001472	0.0169	1416.28	1120.89	2537.17	1431.83	1283.22	2715.05	3.3998	2.1818	5.5816
320	11.28430	0.001499	0.0155	1445.31	1080.70	2526.01	1462.22	1238.37	2700.59	3.4494	2.0878	5.5372
325	12.05100	0.001528	0.0142	1475.11	1038.30	2513.41	1493.52	1190.81	2684.33	3.5000	1.9908	5.4908
330	12.85810	0.001561	0.0130	1505.80	993.35	2499.15	1525.87	1140.16	2666.03	3.5518	1.8904	5.4422
335	13.70730	0.001597	0.0118	1537.56	945.40	2482.96	1559.45	1085.90	2645.35	3.6050	1.7856	5.3906
340	14.60070	0.001638	0.0108	1570.62	893.82	2464.44	1594.53	1027.32	2621.85	3.6601	1.6755	5.3356
345	15.54060	0.001685	0.0098	1605.30	837.79	2443.09	1631.48	963.42	2594.90	3.7176	1.5586	5.2762
350	16.52940	0.001740	0.0088	1642.13	776.01	2418.14	1670.89	892.75	2563.64	3.7784	1.4326	5.2110
355	17.57010	0.001808	0.0079	1681.96	706.44	2388.40	1713.72	812.93	2526.65	3.8439	1.2941	5.1380
360	18.66600	0.001895	0.0069	1726.28	625.50	2351.78	1761.66	719.83	2481.49	3.9167	1.1369	5.0536
365	19.82140	0.002017	0.0060	1777.79	526.00	2303.79	1817.77	605.18	2422.95	4.0014	0.9483	4.9497
370	21.04360	0.002215	0.0050	1844.07	386.19	2230.26	1890.69	443.83	2334.52	4.1112	0.6900	4.8012
373.95	22.06400	0.003106	0.0031	2015.73	0.00	2015.73	2084.26	0.00	2084.26	4.4070	0.0000	4.4070

II. Saturation Pressure

T (°C)	P (MPa)	V^L m³/kg	V^V m³/kg	U^L kJ/kg	ΔU^{vap} kJ/kg	U^V kJ/kg	H^L kJ/kg	ΔH^{vap} kJ/kg	H^V kJ/kg	S^L kJ/kg-K	ΔS^{vap} kJ/kg-K	S^V kJ/kg-K
6.97	0.001	0.001000	129.1780	29.30	2355.19	2384.49	29.30	2484.37	2513.67	0.1059	8.8690	8.9749
17.50	0.002	0.001001	66.9869	73.43	2325.47	2398.90	73.43	2459.45	2532.88	0.2606	8.4620	8.7226
24.08	0.003	0.001003	45.6532	100.98	2306.90	2407.88	100.98	2443.86	2544.84	0.3543	8.2221	8.5764
28.96	0.004	0.001004	34.7911	121.38	2293.12	2414.50	121.39	2432.28	2553.67	0.4224	8.0510	8.4734
32.87	0.005	0.001005	28.1853	137.74	2282.06	2419.80	137.75	2422.98	2560.73	0.4762	7.9176	8.3938
36.16	0.006	0.001006	23.7334	151.47	2272.76	2424.23	151.48	2415.15	2566.63	0.5208	7.8082	8.3290
39.00	0.007	0.001008	20.5245	163.34	2264.71	2428.05	163.35	2408.37	2571.72	0.5590	7.7155	8.2745
41.51	0.008	0.001008	18.0989	173.83	2257.58	2431.41	173.84	2402.37	2576.21	0.5925	7.6348	8.2273
43.76	0.009	0.001009	16.1992	183.24	2251.19	2434.43	183.25	2396.97	2580.22	0.6223	7.5635	8.1858
45.81	0.01	0.001010	14.6701	191.40	2245.36	2437.16	191.81	2392.05	2583.86	0.6492	7.4996	8.1488
60.06	0.02	0.001017	7.6480	251.40	2204.58	2455.98	251.42	2357.52	2608.94	0.8320	7.0052	7.9072
69.10	0.03	0.001022	5.2284	289.24	2178.46	2467.70	289.27	2335.28	2624.55	0.9441	6.8234	7.7675
75.86	0.04	0.001026	3.9930	317.58	2158.75	2476.33	317.62	2318.43	2636.05	1.0261	6.6429	7.6690
81.32	0.05	0.001030	3.2400	340.49	2142.72	2483.21	340.54	2304.68	2645.22	1.0912	6.5018	7.5930
85.93	0.06	0.001033	2.7317	359.85	2129.10	2488.95	359.91	2292.95	2652.86	1.1455	6.3856	7.5311
89.93	0.07	0.001036	2.3648	376.68	2117.20	2493.88	376.75	2282.67	2659.42	1.1921	6.2869	7.4790
93.49	0.08	0.001039	2.0871	391.63	2106.58	2498.21	391.71	2273.47	2665.18	1.2330	6.2009	7.4339
96.69	0.09	0.001041	1.8694	405.10	2096.97	2502.07	405.20	2265.11	2670.31	1.2696	6.1247	7.3943
99.61	0.1	0.001043	1.6939	417.40	2088.15	2505.55	417.50	2257.45	2674.95	1.3028	6.0561	7.3589
120.21	0.2	0.001061	0.8857	504.49	2024.60	2529.09	504.70	2201.53	2706.23	1.5302	5.5967	7.1269
133.52	0.3	0.001073	0.6058	561.11	1982.04	2543.15	561.43	2163.45	2724.88	1.6717	5.3199	6.9916
143.61	0.4	0.001084	0.4624	604.22	1948.88	2553.10	604.66	2133.39	2738.05	1.7765	5.1190	6.8955
151.83	0.5	0.001093	0.3748	639.54	1921.17	2560.71	640.09	2108.02	2748.11	1.8604	4.9603	6.8207
158.83	0.6	0.001101	0.3156	669.72	1897.07	2566.79	670.38	2085.76	2756.14	1.9308	4.8285	6.7593
164.95	0.7	0.001108	0.2728	696.23	1875.58	2571.81	697.00	2065.75	2762.75	1.9918	4.7153	6.7071
170.41	0.8	0.001115	0.2403	719.97	1856.06	2576.03	720.86	2047.44	2768.30	2.0457	4.6159	6.6616
175.35	0.9	0.001121	0.2149	741.55	1838.09	2579.64	742.56	2030.47	2773.03	2.0941	4.5272	6.6213
179.88	1	0.001127	0.1944	761.39	1821.36	2582.75	762.52	2014.59	2777.11	2.1381	4.4469	6.5850
187.96	1.2	0.001139	0.1633	796.96	1790.87	2587.83	798.33	1985.41	2783.74	2.2159	4.3058	6.5217
195.04	1.4	0.001149	0.1408	828.36	1763.40	2591.76	829.97	1958.88	2788.85	2.2835	4.1840	6.4675
201.37	1.6	0.001159	0.1237	856.60	1738.23	2594.83	858.46	1934.36	2792.82	2.3435	4.0764	6.4199
207.11	1.8	0.001168	0.1104	882.37	1714.87	2597.24	884.47	1911.44	2795.91	2.3975	3.9800	6.3775
212.38	2	0.001177	0.0996	906.15	1692.97	2599.12	908.50	1889.79	2798.29	2.4468	3.8922	6.3390
223.95	2.5	0.001197	0.0799	958.91	1643.15	2602.06	961.91	1840.02	2801.93	2.5543	3.7015	6.2558
233.85	3	0.001217	0.0667	1004.69	1598.47	2603.16	1008.34	1794.81	2803.15	2.6456	3.5400	6.1856
242.56	3.5	0.001235	0.0571	1045.47	1557.47	2602.94	1049.80	1752.84	2802.64	2.7254	3.3989	6.1243
250.35	4	0.001253	0.0498	1082.48	1519.24	2601.72	1087.49	1713.33	2800.82	2.7968	3.2728	6.0696
257.44	4.5	0.001270	0.0441	1116.53	1483.15	2599.68	1122.25	1675.70	2797.95	2.8615	3.1582	6.0197
263.94	5	0.001286	0.0394	1148.21	1448.77	2596.98	1154.64	1639.57	2794.21	2.9210	3.0527	5.9737
275.59	6	0.001319	0.0324	1206.01	1383.89	2589.90	1213.92	1570.67	2784.59	3.0278	2.8623	5.8901
285.83	7	0.001352	0.0274	1258.20	1322.78	2580.98	1267.66	1504.97	2772.63	3.1224	2.6924	5.8148
295.01	8	0.001385	0.0235	1306.23	1264.25	2570.48	1317.31	1441.37	2758.68	3.2081	2.5369	5.7450

T (°C)	P (MPa)	V^L m³/kg	V^V m³/kg	U^L kJ/kg	ΔU^{vap} kJ/kg	U^V kJ/kg	H^L kJ/kg	ΔH^{vap} kJ/kg	H^V kJ/kg	S^L kJ/kg-K	ΔS^{vap} kJ/kg-K	S^V kJ/kg-K
303.35	9	0.001418	0.0205	1351.11	1207.42	2558.53	1363.87	1379.07	2742.94	3.2870	2.3921	5.6791
311.00	10	0.001453	0.0180	1393.54	1151.65	2545.19	1408.06	1317.43	2725.49	3.3607	2.2553	5.6160
327.81	12.5	0.001546	0.0135	1492.26	1013.35	2505.61	1511.58	1162.73	2674.31	3.5290	1.9348	5.4638
342.16	15	0.001657	0.0103	1585.35	870.27	2455.62	1610.20	1000.50	2610.70	3.6846	1.6260	5.3106
354.67	17.5	0.001803	0.0079	1679.22	711.32	2390.54	1710.77	818.53	2529.30	3.8394	1.3037	5.1431
365.75	20	0.002040	0.0059	1786.41	508.63	2295.04	1827.21	585.14	2412.35	4.0156	0.9159	4.9315
373.95	22.06400	0.003106	0.0031	2015.73	0.00	2015.73	2084.26	0.00	2084.26	4.4070	0.0000	4.4070

III. Superheated Steam

P = 0.01MPa (45.8)

T(°C)	V(m³/kg)	U(kJ/kg)	H(kJ/kg)	S(kJ/kg-K)
45.8	14.6701	2437.2	2583.9	8.1488
50	14.9139	2443.3	2592.4	8.1755
100	17.1964	2515.5	2687.5	8.4489
150	19.5132	2587.9	2783.0	8.6892
200	21.8256	2661.3	2879.6	8.9049
250	24.1361	2736.1	2977.4	9.1015
300	26.4456	2812.3	3076.7	9.2827
350	28.7545	2890.0	3177.5	9.4513
400	31.0631	2969.3	3279.9	9.6094
450	33.3714	3050.3	3384.0	9.7584
500	35.6796	3132.9	3489.7	9.8998
550	37.9876	3217.2	3597.1	10.0344
600	40.2956	3303.3	3706.3	10.1631
650	42.6035	3391.2	3817.2	10.2866
700	44.9113	3480.8	3929.9	10.4055
750	47.2191	3572.2	4044.4	10.5202
800	49.5269	3665.3	4160.6	10.6311
850	51.8347	3760.1	4278.6	10.7386
900	54.1424	3856.9	4398.3	10.8429
950	56.4501	3955.2	4519.7	10.9442
1000	58.7578	4055.2	4642.8	11.0428
1050	61.0655	4156.8	4767.5	11.1389

P = 0.05MPa (81.3)

T(°C)	V(m³/kg)	U(kJ/kg)	H(kJ/kg)	S(kJ/kg-K)
81.3	3.2400	2483.2	2645.2	7.5930
100	3.4187	2511.5	2682.4	7.6953
150	3.8897	2585.7	2780.2	7.9413
200	4.3562	2660.0	2877.8	8.1592
250	4.8206	2735.1	2976.1	8.3568
300	5.2840	2811.6	3075.8	8.5386
350	5.7469	2889.4	3176.8	8.7076
400	6.2094	2968.9	3279.3	8.8659
450	6.6717	3049.9	3383.5	9.0151
500	7.1338	3132.6	3489.3	9.1566
550	7.5957	3217.0	3596.8	9.2913
600	8.0576	3303.1	3706.0	9.4201
650	8.5195	3391.0	3816.9	9.5436
700	8.9812	3480.6	3929.7	9.6625
750	9.4430	3572.0	4044.2	9.7773
800	9.9047	3665.2	4160.4	9.8882
850	10.3663	3760.1	4278.5	9.9957
900	10.8280	3856.8	4398.2	10.1000
950	11.2896	3955.1	4519.6	10.2014
1000	11.7513	4055.1	4642.7	10.3000
1050	12.2129	4156.8	4767.4	10.3960

P = 0.10MPa (99.6)

T(°C)	V(m³/kg)	U(kJ/kg)	H(kJ/kg)	S(kJ/kg-K)
99.6	1.6939	2505.6	2675.0	7.3588
100	1.6959	2506.2	2675.8	7.3610
150	1.9367	2582.9	2776.6	7.6148
200	2.1724	2658.2	2875.5	7.8356
250	2.4062	2733.9	2974.5	8.0346
300	2.6388	2810.6	3074.5	8.2172
350	2.8710	2888.7	3175.8	8.3866
400	3.1027	2968.3	3278.6	8.5452
450	3.3342	3049.4	3382.8	8.6946
500	3.5655	3132.2	3488.7	8.8361
550	3.7968	3216.6	3596.3	8.9709
600	4.0279	3302.8	3705.6	9.0998
650	4.2590	3390.7	3816.6	9.2234
700	4.4900	3480.4	3929.4	9.3424
750	4.7209	3571.8	4043.9	9.4572
800	4.9519	3665.0	4160.2	9.5681
850	5.1828	3760.0	4278.2	9.6757
900	5.4137	3856.6	4398.0	9.7800
950	5.6446	3955.0	4519.5	9.8813
1000	5.8754	4055.0	4642.6	9.9800
1050	6.1063	4156.6	4767.3	10.0761

Continuation rows (1100–1300°C):

T(°C)	V(m³/kg)	U(kJ/kg)	H(kJ/kg)	S(kJ/kg-K)
1100	63.3732	4260.0	4893.7	11.2325
1150	65.6808	4364.7	5021.5	11.3239
1200	67.9885	4470.9	5150.7	11.4132
1250	70.2961	4578.4	5281.4	11.5004
1300	72.6038	4687.4	5413.4	11.5857

T(°C)	V(m³/kg)	U(kJ/kg)	H(kJ/kg)	S(kJ/kg-K)
1100	12.6745	4259.9	4893.7	10.4897
1150	13.1361	4364.6	5021.4	10.5811
1200	13.5977	4470.8	5150.7	10.6703
1250	14.0592	4578.4	5281.3	10.7576
1300	14.5208	4687.3	5413.3	10.8428

T(°C)	V(m³/kg)	U(kJ/kg)	H(kJ/kg)	S(kJ/kg-K)
1100	6.3371	4259.8	4893.5	10.1697
1150	6.5680	4364.5	5021.3	10.2611
1200	6.7988	4470.7	5150.6	10.3504
1250	7.0296	4578.3	5281.2	10.4376
1300	7.2604	4687.2	5413.2	10.5229

P = 0.20MPa (120.3)

T(°C)	V(m³/kg)	U(kJ/kg)	H(kJ/kg)	S(kJ/kg-K)
120.3	0.8857	2529.1	2706.2	7.1269
150	0.9599	2577.1	2769.1	7.2810
200	1.0805	2654.6	2870.7	7.5081
250	1.1989	2731.4	2971.2	7.7100
300	1.3162	2808.8	3072.1	7.8941
350	1.4330	2887.3	3173.9	8.0644
400	1.5493	2967.1	3277.0	8.2236
450	1.6655	3048.5	3381.6	8.3734
500	1.7814	3131.4	3487.7	8.5152
550	1.8973	3215.9	3595.4	8.6502
600	2.0130	3302.2	3704.8	8.7792
650	2.1287	3390.2	3815.9	8.9030
700	2.2443	3479.9	3928.8	9.0220
750	2.3599	3571.4	4043.4	9.1369
800	2.4755	3664.7	4159.8	9.2479
850	2.5910	3759.6	4277.8	9.3555
900	2.7066	3856.3	4397.6	9.4598
950	2.8221	3954.7	4519.1	9.5612
1000	2.9375	4054.8	4642.3	9.6599
1050	3.0530	4156.4	4767.0	9.7560
1100	3.1685	4259.6	4893.3	9.8497
1150	3.2839	4364.3	5021.1	9.9411
1200	3.3994	4470.5	5150.4	10.0304
1250	3.5148	4578.1	5281.1	10.1176
1300	3.6302	4687.0	5413.1	10.2029

P = 0.30MPa (133.5)

T(°C)	V(m³/kg)	U(kJ/kg)	H(kJ/kg)	S(kJ/kg-K)
133.5	0.6058	2543.2	2724.9	6.9916
150	0.6340	2571.0	2761.2	7.0791
200	0.7164	2651.0	2865.9	7.3131
250	0.7964	2728.9	2967.9	7.5180
300	0.8753	2807.0	3069.6	7.7037
350	0.9536	2885.9	3172.0	7.8750
400	1.0315	2966.0	3275.5	8.0347
450	1.1092	3047.5	3380.3	8.1849
500	1.1867	3130.6	3486.6	8.3271
550	1.2641	3215.3	3594.5	8.4623
600	1.3414	3301.6	3704.0	8.5914
650	1.4186	3389.7	3815.3	8.7153
700	1.4958	3479.5	3928.2	8.8344
750	1.5729	3571.0	4042.9	8.9494
800	1.6500	3664.3	4159.3	9.0604
850	1.7271	3759.3	4277.4	9.1680
900	1.8042	3856.0	4397.3	9.2724
950	1.8812	3954.4	4518.8	9.3739
1000	1.9582	4054.5	4642.0	9.4726
1050	2.0352	4156.2	4766.7	9.5687
1100	2.1122	4259.4	4893.1	9.6624
1150	2.1892	4364.1	5020.9	9.7538
1200	2.2662	4470.3	5150.2	9.8431
1250	2.3432	4577.9	5280.9	9.9303
1300	2.4202	4686.9	5412.9	10.0156

P = 0.40MPa (143.6)

T(°C)	V(m³/kg)	U(kJ/kg)	H(kJ/kg)	S(kJ/kg-K)
143.6	0.4624	2553.1	2738.1	6.8955
150	0.4709	2564.4	2752.8	6.9306
200	0.5343	2647.2	2860.9	7.1723
250	0.5952	2726.4	2964.5	7.3804
300	0.6549	2805.1	3067.1	7.5677
350	0.7140	2884.4	3170.0	7.7399
400	0.7726	2964.9	3273.9	7.9002
450	0.8311	3046.6	3379.0	8.0508
500	0.8894	3129.8	3485.5	8.1933
550	0.9475	3214.6	3593.6	8.3287
600	1.0056	3301.0	3703.2	8.4580
650	1.0636	3389.1	3814.6	8.5820
700	1.1215	3479.0	3927.6	8.7012
750	1.1794	3570.6	4042.4	8.8162
800	1.2373	3663.9	4158.8	8.9273
850	1.2951	3759.0	4277.0	9.0350
900	1.3530	3855.7	4396.9	9.1394
950	1.4108	3954.2	4518.5	9.2409
1000	1.4686	4054.3	4641.7	9.3396
1050	1.5264	4155.9	4766.5	9.4357
1100	1.5841	4259.2	4892.8	9.5295
1150	1.6419	4363.9	5020.7	9.6209
1200	1.6997	4470.1	5150.0	9.7102
1250	1.7574	4577.8	5280.7	9.7975
1300	1.8152	4686.7	5412.8	9.8828

P = 0.50MPa (151.8)

T(°C)	V(m³/kg)	U(kJ/kg)	H(kJ/kg)	S(kJ/kg-K)
151.8	0.3748	2560.7	2748.1	6.8207
200	0.4250	2643.3	2855.8	7.0610
250	0.4744	2723.8	2961.0	7.2614
300	0.5226	2803.2	3064.6	7.4614
350	0.5702	2883.0	3168.1	7.6346
400	0.6173	2963.7	3272.3	7.7955
450	0.6642	3045.6	3377.7	7.9465
500	0.7109	3129.0	3484.5	8.0892
550	0.7576	3213.9	3592.7	8.2249
600	0.8041	3300.4	3702.5	8.3543
650	0.8505	3388.6	3813.9	8.4784
700	0.8970	3478.5	3927.0	8.5977

P = 0.60MPa (158.8)

T(°C)	V(m³/kg)	U(kJ/kg)	H(kJ/kg)	S(kJ/kg-K)
158.8	0.3156	2566.8	2756.1	6.7593
200	0.3521	2639.3	2850.6	6.9683
250	0.3939	2721.2	2957.6	7.1832
300	0.4344	2801.4	3062.0	7.3740
350	0.4743	2881.6	3166.1	7.5481
400	0.5137	2962.5	3270.8	7.7097
450	0.5530	3044.7	3376.5	7.8611
500	0.5920	3128.2	3483.4	8.0041
550	0.6309	3213.2	3591.8	8.1399
600	0.6698	3299.8	3701.7	8.2695
650	0.7085	3388.1	3813.2	8.3937
700	0.7472	3478.1	3926.4	8.5131

P = 0.80MPa (170.4)

T(°C)	V(m³/kg)	U(kJ/kg)	H(kJ/kg)	S(kJ/kg-K)
170.4	0.2403	2576.0	2768.3	6.6616
200	0.2609	2631.0	2839.7	6.8176
250	0.2932	2715.9	2950.4	7.0401
300	0.3242	2797.5	3056.9	7.2345
350	0.3544	2878.6	3162.2	7.4106
400	0.3843	2960.2	3267.6	7.5734
450	0.4139	3042.8	3373.9	7.7257
500	0.4433	3126.6	3481.3	7.8692
550	0.4726	3211.9	3590.0	8.0054
600	0.5019	3298.7	3700.1	8.1354
650	0.5310	3387.1	3811.9	8.2598
700	0.5601	3477.2	3925.3	8.3794

(table continued from previous page)

T(°C)	V(m³/kg)	U(kJ/kg)	H(kJ/kg)	S(kJ/kg-K)
750	0.9433	3570.2	4041.8	8.7128
800	0.9897	3663.6	4158.4	8.8240
850	1.0360	3758.6	4276.6	8.9317
900	1.0823	3855.4	4396.6	9.0362
950	1.1285	3953.9	4518.2	9.1377
1000	1.1748	4054.0	4641.4	9.2364
1050	1.2210	4155.7	4766.2	9.3326
1100	1.2673	4259.0	4892.6	9.4263
1150	1.3135	4363.7	5020.5	9.5178
1200	1.3597	4470.0	5149.8	9.6071
1250	1.4059	4577.6	5280.5	9.6944
1300	1.4521	4686.6	5412.6	9.7797

P = 1.00MPa (179.9)

T(°C)	V(m³/kg)	U(kJ/kg)	H(kJ/kg)	S(kJ/kg-K)
179.9	0.1944	2582.8	2777.1	6.5850
200	0.2060	2622.2	2828.3	6.6955
250	0.2327	2710.4	2943.1	6.9265
300	0.2580	2793.6	3051.6	7.1246
350	0.2825	2875.7	3158.2	7.3029
400	0.3066	2957.9	3264.5	7.4669
450	0.3304	3040.9	3371.3	7.6200
500	0.3541	3125.0	3479.1	7.7641
550	0.3777	3210.5	3588.1	7.9008
600	0.4011	3297.5	3698.6	8.0310
650	0.4245	3386.0	3810.5	8.1557
700	0.4478	3476.2	3924.1	8.2755
750	0.4711	3568.1	4039.3	8.3909
800	0.4944	3661.7	4156.1	8.5024
850	0.5176	3757.0	4274.6	8.6103
900	0.5408	3853.9	4394.8	8.7150
950	0.5640	3952.5	4516.5	8.8166
1000	0.5872	4052.7	4639.9	8.9155
1050	0.6104	4154.5	4764.9	9.0118
1100	0.6335	4257.9	4891.4	9.1056
1150	0.6567	4362.7	5019.4	9.1972
1200	0.6798	4469.0	5148.9	9.2866
1250	0.7030	4576.7	5279.7	9.3739
1300	0.7261	4685.8	5411.9	9.4593

(table continued from previous page)

T(°C)	V(m³/kg)	U(kJ/kg)	H(kJ/kg)	S(kJ/kg-K)
750	0.7859	3569.8	4041.3	8.6283
800	0.8246	3663.2	4157.9	8.7395
850	0.8632	3758.3	4276.2	8.8472
900	0.9018	3855.1	4396.2	8.9518
950	0.9404	3953.6	4517.8	9.0533
1000	0.9789	4053.7	4641.1	9.1521
1050	1.0175	4155.5	4766.0	9.2482
1100	1.0560	4258.7	4892.4	9.3420
1150	1.0946	4363.5	5020.3	9.4335
1200	1.1331	4469.8	5149.6	9.5228
1250	1.1716	4577.4	5280.4	9.6101
1300	1.2101	4686.4	5412.5	9.6954

P = 1.20MPa (188.0)

T(°C)	V(m³/kg)	U(kJ/kg)	H(kJ/kg)	S(kJ/kg-K)
188.0	0.1633	2587.8	2783.7	6.5217
200	0.1693	2612.9	2816.1	6.5909
250	0.1924	2704.7	2935.6	6.8313
300	0.2139	2789.7	3046.3	7.0335
350	0.2346	2872.7	3154.2	7.2139
400	0.2548	2955.5	3261.3	7.3793
450	0.2748	3038.9	3368.7	7.5332
500	0.2946	3123.4	3476.9	7.6779
550	0.3143	3209.1	3586.3	7.8150
600	0.3339	3296.3	3697.0	7.9455
650	0.3535	3385.0	3809.2	8.0704
700	0.3730	3475.3	3922.9	8.1904
750	0.3924	3567.3	4038.2	8.3060
800	0.4118	3661.0	4155.2	8.4176
850	0.4312	3756.3	4273.8	8.5256
900	0.4506	3853.3	4394.0	8.6303
950	0.4699	3952.0	4515.9	8.7320
1000	0.4893	4052.2	4639.4	8.8310
1050	0.5086	4154.1	4764.4	8.9273
1100	0.5279	4257.5	4891.0	9.0212
1150	0.5472	4362.3	5019.0	9.1128
1200	0.5665	4468.7	5148.5	9.2022
1250	0.5858	4576.4	5279.3	9.2895
1300	0.6051	4685.4	5411.5	9.3749

(table continued from previous page)

T(°C)	V(m³/kg)	U(kJ/kg)	H(kJ/kg)	S(kJ/kg-K)
750	0.5892	3569.0	4040.3	8.4947
800	0.6182	3662.4	4157.0	8.6061
850	0.6472	3757.6	4275.4	8.7139
900	0.6762	3854.5	4395.5	8.8185
950	0.7052	3953.1	4517.2	8.9201
1000	0.7341	4053.2	4640.5	9.0189
1050	0.7630	4155.0	4765.4	9.1151
1100	0.7920	4258.3	4891.9	9.2089
1150	0.8209	4363.1	5019.8	9.3004
1200	0.8498	4469.4	5149.2	9.3898
1250	0.8787	4577.1	5280.0	9.4771
1300	0.9076	4686.1	5412.2	9.5625

P = 1.40MPa (195.0)

T(°C)	V(m³/kg)	U(kJ/kg)	H(kJ/kg)	S(kJ/kg-K)
195.0	0.1408	2591.8	2788.9	6.4675
200	0.1430	2602.7	2803.0	6.4975
250	0.1636	2698.9	2927.9	6.7488
300	0.1823	2785.7	3040.9	6.9552
350	0.2003	2869.7	3150.1	7.1379
400	0.2178	2953.1	3258.1	7.3046
450	0.2351	3037.0	3366.1	7.4594
500	0.2522	3121.8	3474.8	7.6047
550	0.2691	3207.7	3584.5	7.7422
600	0.2860	3295.1	3695.4	7.8730
650	0.3028	3384.0	3807.8	7.9982
700	0.3195	3474.4	3921.7	8.1183
750	0.3362	3566.5	4037.2	8.2340
800	0.3529	3660.2	4154.3	8.3457
850	0.3695	3755.6	4273.0	8.4538
900	0.3861	3852.7	4393.3	8.5587
950	0.4027	3951.4	4515.2	8.6604
1000	0.4193	4051.7	4638.8	8.7594
1050	0.4359	4153.6	4763.9	8.8558
1100	0.4525	4257.0	4890.5	8.9497
1150	0.4690	4361.9	5018.6	9.0413
1200	0.4856	4468.3	5148.1	9.1308
1250	0.5021	4576.0	5279.0	9.2182
1300	0.5187	4685.1	5411.2	9.3036

P = 2.00MPa (212.4)

T(°C)	V(m³/kg)	U(kJ/kg)	H(kJ/kg)	S(kJ/kg-K)
212.4	0.0996	2599.1	2798.3	6.3390
250	0.1115	2680.2	2903.2	6.5475
300	0.1255	2773.2	3024.2	6.7684
350	0.1386	2860.5	3137.7	6.9583
400	0.1512	2945.9	3248.3	7.1292
450	0.1635	3031.1	3358.2	7.2866

P = 1.60MPa (201.4)

T(°C)	V(m³/kg)	U(kJ/kg)	H(kJ/kg)	S(kJ/kg-K)
201.4	0.1237	2594.8	2792.8	6.4199
250	0.1419	2692.9	2919.9	6.6753
300	0.1587	2781.6	3035.4	6.8863
350	0.1746	2866.6	3146.0	7.0713
400	0.1901	2950.7	3254.9	7.2394
450	0.2053	3035.0	3363.5	7.3950

P = 1.80MPa (207.1)

T(°C)	V(m³/kg)	U(kJ/kg)	H(kJ/kg)	S(kJ/kg-K)
207.1	0.1104	2597.2	2795.9	6.3775
250	0.1250	2686.7	2911.7	6.6087
300	0.1402	2777.4	3029.9	6.8246
350	0.1546	2863.6	3141.8	7.0120
400	0.1685	2948.3	3251.6	7.1814
450	0.1821	3033.1	3360.9	7.3380

(continuation tables, high-temperature rows)

T(°C)	V(m³/kg)	U(kJ/kg)	H(kJ/kg)	S(kJ/kg-K)
500	0.2203	3120.1	3472.6	7.5409
550	0.2352	3206.3	3582.6	7.6788
600	0.2500	3293.9	3693.9	7.8100
650	0.2647	3382.9	3806.5	7.9354
700	0.2794	3473.5	3920.5	8.0557
750	0.2940	3565.7	4036.1	8.1716
800	0.3087	3659.5	4153.3	8.2834
850	0.3232	3755.0	4272.2	8.3916
900	0.3378	3852.1	4392.6	8.4965
950	0.3523	3950.9	4514.6	8.5984
1000	0.3669	4051.2	4638.2	8.6975
1050	0.3814	4153.1	4763.4	8.7938
1100	0.3959	4256.6	4890.0	8.8878
1150	0.4104	4361.5	5018.2	8.9794
1200	0.4249	4467.9	5147.7	9.0689
1250	0.4394	4575.1	5278.7	9.1563
1300	0.4538	4684.8	5410.9	9.2417

T(°C)	V(m³/kg)	U(kJ/kg)	H(kJ/kg)	S(kJ/kg-K)
500	0.1955	3118.5	3470.4	7.4845
550	0.2088	3205.0	3580.8	7.6228
600	0.2220	3292.7	3692.3	7.7543
650	0.2351	3381.9	3805.1	7.8799
700	0.2482	3472.6	3919.4	8.0004
750	0.2613	3564.9	4035.1	8.1164
800	0.2743	3658.8	4152.4	8.2284
850	0.2872	3754.3	4271.3	8.3367
900	0.3002	3851.5	4391.9	8.4416
950	0.3131	3950.3	4514.0	8.5435
1000	0.3261	4050.7	4637.6	8.6426
1050	0.3390	4152.7	4762.8	8.7391
1100	0.3519	4256.2	4889.5	8.8331
1150	0.3648	4361.1	5017.7	8.9248
1200	0.3777	4467.5	5147.3	9.0143
1250	0.3905	4575.3	5278.3	9.1017
1300	0.4034	4684.5	5410.6	9.1872

T(°C)	V(m³/kg)	U(kJ/kg)	H(kJ/kg)	S(kJ/kg-K)
500	0.1757	3116.9	3468.2	7.4337
550	0.1877	3203.6	3579.0	7.5725
600	0.1996	3291.5	3690.7	7.7043
650	0.2115	3380.8	3803.8	7.8302
700	0.2233	3471.6	3918.2	7.9509
750	0.2350	3564.0	4034.1	8.0670
800	0.2467	3658.0	4151.5	8.1790
850	0.2584	3753.6	4270.5	8.2874
900	0.2701	3850.9	4391.1	8.3925
950	0.2818	3949.8	4513.3	8.4945
1000	0.2934	4050.2	4637.0	8.5936
1050	0.3051	4152.2	4762.3	8.6901
1100	0.3167	4255.7	4889.1	8.7842
1150	0.3283	4360.7	5017.3	8.8759
1200	0.3399	4467.2	5147.0	8.9654
1250	0.3515	4575.0	5278.0	9.0529
1300	0.3631	4684.1	5410.3	9.1384

P = 2.50MPa

T(°C)	V(m³/kg)	U(kJ/kg)	H(kJ/kg)	S(kJ/kg-K)
(224.0)	0.0799	2602.1	2801.9	6.2558
250	0.0871	2663.3	2880.9	6.4107
300	0.0989	2762.2	3009.6	6.6459
350	0.1098	2852.5	3127.0	6.8424
400	0.1201	2939.8	3240.1	7.0170
450	0.1302	3026.2	3351.6	7.1767
500	0.1400	3112.8	3462.7	7.3254
550	0.1497	3200.1	3574.3	7.4653
600	0.1593	3288.5	3686.8	7.5979
650	0.1689	3378.2	3800.4	7.7243
700	0.1783	3469.3	3915.2	7.8455
750	0.1878	3562.0	4031.5	7.9620
800	0.1972	3656.2	4149.2	8.0743
850	0.2066	3752.0	4268.5	8.1830
900	0.2160	3849.4	4389.3	8.2882
950	0.2253	3948.4	4511.7	8.3904
1000	0.2347	4048.9	4635.6	8.4896
1050	0.2440	4151.0	4761.0	8.5863
1100	0.2533	4254.7	4887.9	8.6804
1150	0.2626	4359.7	5016.2	8.7722
1200	0.2719	4466.2	5146.0	8.8618
1250	0.2812	4574.1	5277.1	8.9493
1300	0.2905	4683.3	5409.5	9.0349

P = 3.00MPa

T(°C)	V(m³/kg)	U(kJ/kg)	H(kJ/kg)	S(kJ/kg-K)
(233.9)	0.0667	2603.2	2803.2	6.1856
250	0.0706	2644.7	2856.5	6.2893
300	0.0812	2750.8	2994.3	6.5412
350	0.0906	2844.4	3116.1	6.7449
400	0.0994	2933.5	3231.7	6.9234
450	0.1079	3021.2	3344.8	7.0856
500	0.1162	3108.6	3457.2	7.2359
550	0.1244	3196.6	3569.7	7.3768
600	0.1324	3285.5	3682.8	7.5103
650	0.1405	3375.6	3796.9	7.6373
700	0.1484	3467.0	3912.2	7.7590
750	0.1563	3559.9	4028.9	7.8758
800	0.1642	3654.3	4146.9	7.9885
850	0.1720	3750.3	4266.5	8.0973
900	0.1799	3847.9	4387.5	8.2028
950	0.1877	3947.0	4510.1	8.3051
1000	0.1955	4047.7	4634.1	8.4045
1050	0.2033	4149.9	4759.7	8.5012
1100	0.2111	4253.6	4886.7	8.5955
1150	0.2188	4358.7	5015.2	8.6874
1200	0.2266	4465.3	5145.0	8.7770
1250	0.2343	4573.3	5276.2	8.8646
1300	0.2421	4682.5	5408.8	8.9502

P = 3.50MPa

T(°C)	V(m³/kg)	U(kJ/kg)	H(kJ/kg)	S(kJ/kg-K)
(242.6)	0.0571	2602.9	2802.6	6.1243
250	0.0588	2624.0	2829.7	6.1764
300	0.0685	2738.8	2978.4	6.4484
350	0.0768	2836.0	3104.8	6.6601
400	0.0846	2927.2	3223.2	6.8427
450	0.0920	3016.1	3338.0	7.0074
500	0.0992	3104.5	3451.6	7.1593
550	0.1063	3193.1	3565.0	7.3014
600	0.1133	3282.5	3678.9	7.4356
650	0.1202	3372.9	3793.5	7.5633
700	0.1270	3464.7	3909.3	7.6854
750	0.1338	3557.8	4026.3	7.8027
800	0.1406	3652.5	4144.6	7.9156
850	0.1474	3748.6	4264.4	8.0247
900	0.1541	3846.4	4385.7	8.1303
950	0.1608	3945.6	4508.4	8.2328
1000	0.1675	4046.4	4632.7	8.3324
1050	0.1742	4148.7	4758.4	8.4292
1100	0.1809	4252.5	4885.6	8.5235
1150	0.1875	4357.7	5014.1	8.6155
1200	0.1942	4464.4	5144.1	8.7053
1250	0.2009	4572.4	5275.4	8.7929
1300	0.2075	4681.7	5408.0	8.8785

P = 4.00MPa

T(°C)	V(m³/kg)	U(kJ/kg)	H(kJ/kg)	S(kJ/kg-K)
(250.4)	0.0498	2601.7	2800.8	6.0696

P = 4.50MPa

T(°C)	V(m³/kg)	U(kJ/kg)	H(kJ/kg)	S(kJ/kg-K)
(257.4)	0.0441	2599.7	2798.0	6.0197

P = 5.00MPa

T(°C)	V(m³/kg)	U(kJ/kg)	H(kJ/kg)	S(kJ/kg-K)
(263.9)	0.0394	2597.0	2794.2	5.9737

The following three tables (top of page) share the column structure T(°C), V(m³/kg), U(kJ/kg), H(kJ/kg), S(kJ/kg-K).

T(°C)	V(m³/kg)	U(kJ/kg)	H(kJ/kg)	S(kJ/kg-K)
300	0.0589	2726.2	2961.7	6.3639
350	0.0665	2827.4	3093.3	6.5843
400	0.0734	2920.7	3214.5	6.7714
450	0.0800	3011.0	3331.2	6.9386
500	0.0864	3100.3	3446.0	7.0922
550	0.0927	3189.5	3560.3	7.2355
600	0.0989	3279.4	3674.9	7.3705
650	0.1049	3370.3	3790.1	7.4988
700	0.1110	3462.4	3906.3	7.6214
750	0.1170	3555.5	4023.6	7.7390
800	0.1229	3650.6	4142.3	7.8523
850	0.1289	3747.0	4262.4	7.9616
900	0.1348	3844.8	4383.9	8.0674
950	0.1406	3944.2	4506.8	8.1701
1000	0.1465	4045.1	4631.2	8.2697
1050	0.1524	4147.5	4757.1	8.3667
1100	0.1582	4251.4	4884.4	8.4611
1150	0.1641	4356.7	5013.1	8.5532
1200	0.1699	4463.5	5143.1	8.6430
1250	0.1757	4571.5	5274.5	8.7307
1300	0.1816	4680.9	5407.2	8.8164

T(°C)	V(m³/kg)	U(kJ/kg)	H(kJ/kg)	S(kJ/kg-K)
300	0.0514	2713.0	2944.2	6.2854
350	0.0584	2818.6	3081.5	6.5153
400	0.0648	2914.2	3205.6	6.7070
450	0.0708	3005.8	3324.2	6.8770
500	0.0765	3096.0	3440.4	7.0323
550	0.0821	3186.4	3555.6	7.1767
600	0.0877	3276.4	3670.9	7.3127
650	0.0931	3367.7	3786.6	7.4416
700	0.0985	3460.0	3903.3	7.5646
750	0.1038	3553.7	4021.0	7.6826
800	0.1092	3648.8	4140.0	7.7962
850	0.1145	3745.3	4260.3	7.9057
900	0.1197	3843.3	4382.1	8.0118
950	0.1250	3942.8	4505.2	8.1146
1000	0.1302	4043.9	4629.8	8.2144
1050	0.1354	4146.4	4755.8	8.3115
1100	0.1406	4250.4	4883.2	8.4060
1150	0.1458	4355.8	5012.0	8.4981
1200	0.1510	4462.5	5142.2	8.5880
1250	0.1562	4570.7	5273.7	8.6758
1300	0.1614	4680.1	5406.4	8.7615

T(°C)	V(m³/kg)	U(kJ/kg)	H(kJ/kg)	S(kJ/kg-K)
300	0.0453	2699.0	2925.7	6.2110
350	0.0520	2809.5	3069.3	6.4516
400	0.0578	2907.5	3196.7	6.6483
450	0.0633	3000.6	3317.2	6.8210
500	0.0686	3091.7	3434.7	6.9781
550	0.0737	3182.4	3550.9	7.1237
600	0.0787	3273.3	3666.8	7.2605
650	0.0836	3365.0	3783.2	7.3901
700	0.0885	3457.7	3900.3	7.5136
750	0.0934	3551.6	4018.4	7.6320
800	0.0982	3646.9	4137.7	7.7458
850	0.1029	3743.6	4258.3	7.8556
900	0.1077	3841.8	4380.2	7.9618
950	0.1124	3941.5	4503.6	8.0648
1000	0.1171	4042.6	4628.3	8.1648
1050	0.1219	4145.2	4754.5	8.2620
1100	0.1266	4249.3	4882.0	8.3566
1150	0.1312	4354.8	5011.0	8.4488
1200	0.1359	4461.6	5141.2	8.5388
1250	0.1406	4569.8	5272.8	8.6266
1300	0.1453	4679.3	5405.7	8.7124

P = 6.00 MPa   (275.6)

T(°C)	V(m³/kg)	U(kJ/kg)	H(kJ/kg)	S(kJ/kg-K)
275.6	0.0324	2589.9	2784.6	5.8901
300	0.0362	2668.4	2885.5	6.0703
350	0.0423	2790.4	3043.9	6.3357
400	0.0474	2893.7	3178.2	6.5432
450	0.0522	2989.9	3302.9	6.7219
500	0.0567	3083.1	3423.1	6.8826
550	0.0610	3175.2	3541.3	7.0307
600	0.0653	3267.2	3658.7	7.1693
650	0.0694	3359.6	3776.2	7.3001
700	0.0735	3453.0	3894.3	7.4246
750	0.0776	3547.5	4013.2	7.5438
800	0.0816	3643.2	4133.1	7.6582
850	0.0857	3740.3	4254.2	7.7685
900	0.0896	3838.8	4376.6	7.8751
950	0.0936	3938.7	4500.3	7.9784
1000	0.0976	4040.1	4625.4	8.0786
1050	0.1015	4142.9	4751.9	8.1760
1100	0.1054	4247.1	4879.7	8.2709
1150	0.1093	4352.8	5008.9	8.3632

P = 7.00 MPa   (285.8)

T(°C)	V(m³/kg)	U(kJ/kg)	H(kJ/kg)	S(kJ/kg-K)
285.8	0.0274	2581.0	2772.6	5.8148
300	0.0295	2633.5	2839.9	5.9337
350	0.0353	2770.1	3016.9	6.2304
400	0.0400	2879.5	3159.2	6.4502
450	0.0442	2979.0	3288.3	6.6353
500	0.0482	3074.3	3411.4	6.8000
550	0.0520	3167.9	3531.6	6.9506
600	0.0557	3260.9	3650.6	7.0910
650	0.0593	3354.3	3769.3	7.2231
700	0.0629	3448.3	3888.2	7.3486
750	0.0664	3543.3	4007.7	7.4685
800	0.0699	3639.5	4128.4	7.5836
850	0.0733	3736.9	4250.1	7.6944
900	0.0768	3835.7	4373.0	7.8014
950	0.0802	3935.9	4497.1	7.9050
1000	0.0836	4037.5	4622.5	8.0055
1050	0.0870	4140.5	4749.3	8.1031
1100	0.0903	4245.0	4877.3	8.1981
1150	0.0937	4350.8	5006.7	8.2907

P = 8.00 MPa   (295.0)

T(°C)	V(m³/kg)	U(kJ/kg)	H(kJ/kg)	S(kJ/kg-K)
295.0	0.0235	2570.5	2758.7	5.7450
300	0.0243	2592.3	2786.5	5.7937
350	0.0300	2748.3	2988.1	6.1321
400	0.0343	2864.6	3139.4	6.3658
450	0.0382	2967.8	3273.3	6.5579
500	0.0418	3065.4	3399.5	6.7266
550	0.0452	3160.5	3521.8	6.8799
600	0.0485	3254.7	3642.4	7.0221
650	0.0517	3348.9	3762.3	7.1556
700	0.0548	3443.6	3882.2	7.2821
750	0.0579	3539.1	4002.6	7.4028
800	0.0610	3635.7	4123.8	7.5184
850	0.0641	3733.5	4246.0	7.6297
900	0.0671	3832.6	4369.3	7.7371
950	0.0701	3933.1	4493.8	7.8411
1000	0.0731	4035.0	4619.6	7.9419
1050	0.0761	4138.2	4746.7	8.0397
1100	0.0790	4242.8	4875.0	8.1350
1150	0.0820	4348.8	5004.6	8.2277

(continued)

T(°C)	V(m³/kg)	U(kJ/kg)	H(kJ/kg)	S(kJ/kg-K)
1200	0.1133	4459.8	5139.3	8.4534
1250	0.1172	4568.1	5271.1	8.5413
1300	0.1211	4677.7	5404.1	8.6272

T(°C)	V(m³/kg)	U(kJ/kg)	H(kJ/kg)	S(kJ/kg-K)
1200	0.0971	4457.9	5137.4	8.3810
1250	0.1004	4566.4	5269.4	8.4690
1300	0.1038	4676.1	5402.6	8.5551

T(°C)	V(m³/kg)	U(kJ/kg)	H(kJ/kg)	S(kJ/kg-K)
1200	0.0849	4456.1	5135.5	8.3181
1250	0.0879	4564.6	5267.7	8.4063
1300	0.0908	4674.5	5401.0	8.4924

P = 9.00MPa

T(°C)	V(m³/kg)	U(kJ/kg)	H(kJ/kg)	S(kJ/kg-K)
303.4 (303.4)	0.0205	2558.5	2742.9	5.6791
350	0.0258	2724.9	2957.3	6.0380
400	0.0300	2849.2	3118.8	6.2876
450	0.0335	2956.3	3258.0	6.4872
500	0.0368	3056.3	3387.4	6.6603
550	0.0399	3153.0	3512.0	6.8164
600	0.0429	3248.4	3634.1	6.9605
650	0.0458	3343.4	3755.2	7.0953
700	0.0486	3438.8	3876.1	7.2229
750	0.0514	3534.9	3997.3	7.3443
800	0.0541	3632.0	4119.1	7.4606
850	0.0569	3730.2	4241.9	7.5724
900	0.0596	3829.6	4365.7	7.6802
950	0.0622	3930.3	4490.6	7.7844
1000	0.0649	4032.4	4616.7	7.8855
1050	0.0676	4135.9	4744.0	7.9836
1100	0.0702	4240.6	4872.7	8.0790
1150	0.0729	4346.8	5002.5	8.1719
1200	0.0755	4454.2	5133.6	8.2625
1250	0.0781	4562.9	5266.0	8.3508
1300	0.0807	4672.9	5399.5	8.4370

P = 10.00MPa

T(°C)	V(m³/kg)	U(kJ/kg)	H(kJ/kg)	S(kJ/kg-K)
311.0 (311.0)	0.0180	2545.2	2725.5	5.6160
350	0.0224	2699.6	2924.0	5.9459
400	0.0264	2833.1	3097.4	6.2141
450	0.0298	2944.5	3242.3	6.4219
500	0.0328	3047.0	3375.1	6.5995
550	0.0357	3145.4	3502.0	6.7585
600	0.0384	3242.0	3625.8	6.9045
650	0.0410	3337.9	3748.1	7.0408
700	0.0436	3434.0	3870.0	7.1693
750	0.0461	3530.7	3992.0	7.2916
800	0.0486	3628.2	4114.5	7.4085
850	0.0511	3726.8	4237.8	7.5207
900	0.0535	3826.5	4362.0	7.6290
950	0.0560	3927.5	4487.3	7.7335
1000	0.0584	4029.9	4613.8	7.8349
1050	0.0608	4133.5	4741.4	7.9332
1100	0.0632	4238.5	4870.3	8.0288
1150	0.0656	4344.8	5000.4	8.1219
1200	0.0679	4452.3	5131.7	8.2126
1250	0.0703	4561.2	5264.2	8.3010
1300	0.0727	4671.3	5397.9	8.3874

P = 12.50MPa

T(°C)	V(m³/kg)	U(kJ/kg)	H(kJ/kg)	S(kJ/kg-K)
327.8 (327.8)	0.0135	2505.61	2674.31	5.4638
350	0.0161	2624.8	2826.6	5.7130
400	0.0200	2789.6	3040.0	6.0433
450	0.0230	2913.7	3201.4	6.2749
500	0.0256	3023.2	3343.6	6.4650
550	0.0280	3126.1	3476.5	6.6317
600	0.0303	3225.8	3604.6	6.7828
650	0.0325	3324.1	3730.2	6.9227
700	0.0346	3422.0	3854.6	7.0539
750	0.0367	3520.1	3978.6	7.1782
800	0.0387	3618.7	4102.8	7.2967
850	0.0407	3718.3	4227.5	7.4102
900	0.0427	3818.9	4352.9	7.5194
950	0.0447	3920.6	4479.2	7.6249
1000	0.0466	4023.5	4606.5	7.7269
1050	0.0486	4127.7	4734.9	7.8258
1100	0.0505	4233.1	4864.5	7.9219
1150	0.0524	4339.8	4995.1	8.0154
1200	0.0543	4447.7	5127.0	8.1065
1250	0.0562	4556.9	5260.0	8.1952
1300	0.0581	4667.3	5394.1	8.2819

P = 15.00MPa

T(°C)	V(m³/kg)	U(kJ/kg)	H(kJ/kg)	S(kJ/kg-K)
342.2 (342.2)	0.0103	2455.6	2610.7	5.3106
350	0.0115	2520.9	2693.1	5.4437
400	0.0157	2740.6	2975.7	5.8819
450	0.0185	2880.7	3157.9	6.1434
500	0.0208	2998.4	3310.8	6.3480
550	0.0229	3106.2	3450.4	6.5230
600	0.0249	3209.3	3583.1	6.6796
650	0.0268	3310.1	3712.1	6.8233
700	0.0286	3409.8	3839.1	6.9572
750	0.0304	3509.4	3965.2	7.0836
800	0.0321	3609.2	4091.1	7.2037
850	0.0338	3709.8	4217.1	7.3185
900	0.0355	3811.2	4343.7	7.4288
950	0.0372	3913.6	4471.0	7.5350
1000	0.0388	4017.1	4599.2	7.6378
1050	0.0404	4121.8	4728.4	7.7373
1100	0.0421	4227.7	4858.6	7.8339
1150	0.0437	4334.8	4989.9	7.9278

P = 17.50MPa

T(°C)	V(m³/kg)	U(kJ/kg)	H(kJ/kg)	S(kJ/kg-K)
354.7 (354.7)	0.0079	2390.5	2529.3	5.1431
400	0.0125	2684.3	2902.4	5.7211
450	0.0152	2845.4	3111.4	6.0212
500	0.0174	2972.4	3276.7	6.2244
550	0.0193	3085.8	3423.6	6.4266
600	0.0211	3192.5	3561.3	6.5890
650	0.0227	3295.8	3693.8	6.7366
700	0.0243	3397.5	3823.5	6.8734
750	0.0259	3498.6	3951.7	7.0019
800	0.0274	3599.7	4079.3	7.1236
850	0.0289	3701.2	4206.8	7.2398
900	0.0303	3803.4	4334.5	7.3511
950	0.0318	3906.6	4462.9	7.4582
1000	0.0332	4010.7	4592.0	7.5616
1050	0.0346	4115.9	4721.9	7.6617
1100	0.0360	4222.3	4852.8	7.7588
1150	0.0374	4329.8	4984.6	7.8531

P = 20.00MPa

T(°C)	V(m³/kg)	U(kJ/kg)	H(kJ/kg)	S(kJ/kg-K)
365.8 (365.8)	0.0059	2295.0	2412.4	4.9315
400	0.0100	2617.9	2816.9	5.5525
450	0.0127	2807.2	3061.7	5.9043
500	0.0148	2945.3	3241.2	6.1446
550	0.0166	3064.7	3396.1	6.3389
600	0.0182	3175.3	3539.0	6.5075
650	0.0197	3281.4	3675.3	6.6593
700	0.0211	3385.1	3807.8	6.7990
750	0.0225	3487.7	3938.1	6.9297
800	0.0239	3590.1	4067.5	7.0531
850	0.0252	3692.6	4196.4	7.1705
900	0.0265	3795.7	4325.4	7.2829
950	0.0278	3899.5	4454.7	7.3909
1000	0.0290	4004.3	4584.7	7.4950
1050	0.0303	4110.0	4715.4	7.5957
1100	0.0315	4216.9	4846.9	7.6933
1150	0.0327	4324.8	4979.4	7.7880

1200	0.0453	4443.1	5122.3	8.0192
1250	0.0469	4552.6	5255.7	8.1083
1300	0.0485	4663.2	5390.3	8.1952

P = 25.00MPa

T(°C)	V(m³/kg)	U(kJ/kg)	H(kJ/kg)	S(kJ/kg-K)
400	0.0060	2428.5	2578.7	5.1400
450	0.0092	2721.2	2950.6	5.6759
500	0.0111	2887.3	3165.9	5.9642
550	0.0127	3020.8	3339.2	6.1816
600	0.0141	3140.0	3493.5	6.3637
650	0.0154	3251.9	3637.7	6.5242
700	0.0166	3359.9	3776.0	6.6702
750	0.0178	3465.8	3910.9	6.8054
800	0.0189	3570.7	4043.8	6.9322
850	0.0200	3675.4	4175.6	7.0523
900	0.0211	3780.2	4307.1	7.1668
950	0.0221	3885.5	4438.5	7.2765
1000	0.0232	3991.5	4570.2	7.3820
1050	0.0242	4098.3	4702.5	7.4839
1100	0.0252	4206.0	4835.4	7.5825
1150	0.0262	4314.8	4969.0	7.6781
1200	0.0272	4424.6	5103.5	7.7710
1250	0.0281	4535.4	5238.8	7.8613
1300	0.0291	4647.2	5375.1	7.9493

P = 40.00MPa

T(°C)	V(m³/kg)	U(kJ/kg)	H(kJ/kg)	S(kJ/kg-K)
400	0.0019	1854.9	1931.4	4.1145
450	0.0037	2364.2	2511.8	4.9449
500	0.0056	2681.6	2906.5	5.4744
550	0.0070	2875.0	3154.4	5.7857
600	0.0081	3026.8	3350.4	6.0170
650	0.0091	3159.5	3521.6	6.2078
700	0.0099	3282.0	3679.1	6.3740
750	0.0107	3398.6	3828.4	6.5236
800	0.0115	3511.8	3972.6	6.6612
850	0.0123	3623.1	4113.6	6.7896
900	0.0130	3733.3	4252.5	6.9106
950	0.0137	3843.1	4390.2	7.0256
1000	0.0144	3952.9	4527.3	7.1355
1050	0.0150	4063.0	4664.2	7.2409
1100	0.0157	4173.7	4801.1	7.3425
1150	0.0163	4284.9	4938.3	7.4406
1200	0.0170	4396.9	5075.9	7.5357
1250	0.0176	4509.7	5214.1	7.6279
1300	0.0182	4623.3	5352.8	7.7175

1200	0.0388	4438.4	5117.5	7.9449
1250	0.0402	4548.3	5251.5	8.0343
1300	0.0416	4659.2	5386.4	8.1215

P = 30.00MPa

T(°C)	V(m³/kg)	U(kJ/kg)	H(kJ/kg)	S(kJ/kg-K)
400	0.0028	2071.9	2156.2	4.4808
450	0.0067	2618.9	2821.0	5.4421
500	0.0087	2824.0	3084.7	5.7956
550	0.0102	2974.5	3279.7	6.0402
600	0.0114	3103.4	3446.7	6.2373
650	0.0126	3221.7	3599.4	6.4074
700	0.0137	3334.3	3743.9	6.5598
750	0.0147	3443.6	3883.4	6.6997
800	0.0156	3551.2	4020.0	6.8300
850	0.0166	3658.0	4154.9	6.9529
900	0.0175	3764.6	4288.8	7.0695
950	0.0184	3871.4	4422.3	7.1810
1000	0.0192	3978.6	4555.8	7.2880
1050	0.0201	4086.5	4689.6	7.3910
1100	0.0210	4195.2	4823.8	7.4906
1150	0.0218	4304.8	4958.7	7.5871
1200	0.0226	4415.3	5094.2	7.6807
1250	0.0235	4526.8	5230.5	7.7716
1300	0.0243	4639.2	5367.6	7.8602

P = 50.00MPa

T(°C)	V(m³/kg)	U(kJ/kg)	H(kJ/kg)	S(kJ/kg-K)
400	0.0017	1787.8	1874.4	4.0029
450	0.0025	2160.3	2284.7	4.5896
500	0.0039	2528.1	2722.6	5.1762
550	0.0051	2769.5	3025.3	5.5563
600	0.0061	2947.1	3252.5	5.8245
650	0.0070	3095.6	3443.4	6.0373
700	0.0077	3228.7	3614.6	6.2178
750	0.0084	3353.1	3773.9	6.3775
800	0.0091	3472.2	3925.8	6.5225
850	0.0097	3588.0	4072.9	6.6565
900	0.0103	3702.0	4216.8	6.7819
950	0.0109	3814.9	4358.7	6.9004
1000	0.0114	3927.3	4499.4	7.0131
1050	0.0120	4039.7	4639.3	7.1209
1100	0.0125	4152.2	4778.9	7.2244
1150	0.0131	4265.1	4918.4	7.3242
1200	0.0136	4378.6	5058.1	7.4207
1250	0.0141	4492.7	5198.1	7.5141
1300	0.0146	4607.4	5338.4	7.6048

1200	0.0340	4433.8	5112.8	7.8802
1250	0.0352	4544.0	5247.2	7.9699
1300	0.0364	4655.2	5382.6	8.0574

P = 35.00MPa

T(°C)	V(m³/kg)	U(kJ/kg)	H(kJ/kg)	S(kJ/kg-K)
400	0.0021	1914.8	1988.5	4.2142
450	0.0050	2497.5	2671.0	5.1945
500	0.0069	2755.3	2997.9	5.6331
550	0.0083	2925.8	3218.0	5.9092
600	0.0095	3065.0	3398.9	6.1228
650	0.0106	3190.9	3560.7	6.3030
700	0.0115	3308.3	3711.6	6.4622
750	0.0124	3421.2	3855.9	6.6069
800	0.0133	3531.5	3996.3	6.7409
850	0.0141	3640.5	4134.2	6.8665
900	0.0149	3748.9	4270.6	6.9853
950	0.0157	3857.2	4406.2	7.0985
1000	0.0165	3965.8	4541.5	7.2069
1050	0.0172	4074.8	4676.8	7.3112
1100	0.0179	4184.4	4812.4	7.4118
1150	0.0187	4294.8	4948.4	7.5091
1200	0.0194	4406.1	5085.0	7.6034
1250	0.0201	4518.2	5222.2	7.6950
1300	0.0208	4631.2	5360.1	7.7841

P = 60.00MPa

T(°C)	V(m³/kg)	U(kJ/kg)	H(kJ/kg)	S(kJ/kg-K)
400	0.0016	1745.2	1843.2	3.9317
450	0.0021	2055.1	2180.2	4.4140
500	0.0030	2393.2	2570.3	4.9356
550	0.0040	2664.5	2901.9	5.3517
600	0.0048	2866.8	3156.8	5.6527
650	0.0056	3031.3	3366.7	5.8867
700	0.0063	3175.4	3551.3	6.0814
750	0.0069	3307.6	3720.5	6.2510
800	0.0075	3432.6	3880.0	6.4033
850	0.0080	3553.2	4033.1	6.5428
900	0.0085	3670.9	4182.0	6.6725
950	0.0090	3786.9	4328.1	6.7944
1000	0.0095	3901.9	4472.2	6.9099
1050	0.0100	4016.5	4615.1	7.0200
1100	0.0104	4130.9	4757.3	7.1255
1150	0.0109	4245.5	4899.1	7.2269
1200	0.0113	4360.4	5040.8	7.3248
1250	0.0118	4475.8	5182.5	7.4194
1300	0.0122	4591.8	5324.5	7.5111

IV. Compressed Liquid

P = 5 MPa

T(°C)	V(m³/kg)	U(kJ/kg)	H(kJ/kg)	S(kJ/kg-K)
0	0.000998	0.0	5.0	0.0001
20	0.001000	83.6	88.6	0.2954
40	0.001006	166.9	172.0	0.5705
60	0.001015	250.2	255.4	0.8287
80	0.001027	333.8	339.0	1.0723
100	0.001041	417.6	422.9	1.3034
120	0.001058	501.9	507.2	1.5236
140	0.001077	586.8	592.2	1.7344
160	0.001099	672.5	678.0	1.9374
180	0.001124	759.5	765.1	2.1338
200	0.001153	847.9	853.7	2.3251
220	0.001187	938.4	944.3	2.5127
240	0.001227	1031.6	1037.7	2.6983
260	0.001275	1128.5	1134.9	2.8841

P = 10 MPa

T(°C)	V(m³/kg)	U(kJ/kg)	H(kJ/kg)	S(kJ/kg-K)
0	0.000995	0.1	10.1	0.0003
20	0.000997	83.3	93.3	0.2943
40	0.001003	166.3	176.4	0.5685
60	0.001013	249.4	259.6	0.8260
80	0.001024	332.7	342.9	1.0691
100	0.001038	416.2	426.6	1.2996
120	0.001055	500.2	510.7	1.5191
140	0.001074	584.7	595.5	1.7293
160	0.001095	670.1	681.0	1.9315
180	0.001120	756.5	767.7	2.1271
200	0.001148	844.3	855.8	2.3174
220	0.001181	934.0	945.8	2.5037
240	0.001219	1026.1	1038.3	2.6876
260	0.001265	1121.6	1134.3	2.8710
280	0.001323	1221.8	1235.0	3.0565
300	0.001398	1329.4	1343.3	3.2488

P = 15 MPa

T(°C)	V(m³/kg)	U(kJ/kg)	H(kJ/kg)	S(kJ/kg-K)
0	0.000993	0.2	15.1	0.0004
20	0.000995	83.0	97.9	0.2932
40	0.001001	165.7	180.8	0.5666
60	0.001011	248.6	263.7	0.8234
80	0.001022	331.6	346.9	1.0659
100	0.001036	414.8	430.4	1.2958
120	0.001052	498.5	514.3	1.5148
140	0.001071	582.7	598.7	1.7243
160	0.001092	667.6	684.0	1.9259
180	0.001116	753.6	770.3	2.1206
200	0.001144	840.8	858.0	2.3100
220	0.001175	929.8	947.4	2.4951
240	0.001212	1021.0	1039.2	2.6774
260	0.001256	1115.1	1134.0	2.8586
280	0.001310	1213.4	1233.0	3.0409
300	0.001378	1317.6	1338.3	3.2279
320	0.001473	1431.9	1454.0	3.4263
340	0.001631	1567.9	1592.4	3.6555

P = 20 MPa

T(°C)	V(m³/kg)	U(kJ/kg)	H(kJ/kg)	S(kJ/kg-K)
0	0.000990	0.2	20.0	0.0005
20	0.000993	82.7	102.6	0.2921
40	0.000999	165.2	185.2	0.5646
60	0.001008	247.8	267.9	0.8208
80	0.001020	330.5	350.9	1.0627
100	0.001034	413.5	434.2	1.2920
120	0.001050	496.8	517.8	1.5105
140	0.001068	580.7	602.1	1.7194
160	0.001089	665.3	687.0	1.9203
180	0.001112	750.8	773.0	2.1143
200	0.001139	837.5	860.3	2.3027
220	0.001170	925.8	949.2	2.4867
240	0.001205	1016.1	1040.2	2.6676
260	0.001247	1109.0	1134.0	2.8469
280	0.001298	1205.5	1231.5	3.0265
300	0.001361	1307.1	1334.4	3.2091
320	0.001445	1416.6	1445.5	3.3996
340	0.001569	1540.2	1571.6	3.6086
360	0.001825	1703.6	1740.1	3.8787

P = 50 MPa

T(°C)	V(m³/kg)	U(kJ/kg)	H(kJ/kg)	S(kJ/kg-K)
0	0.000977	0.3	49.1	-0.0010
20	0.000980	80.9	130.0	0.2845
40	0.000987	161.9	211.3	0.5528
60	0.000996	243.1	292.9	0.8055
80	0.001007	324.4	374.8	1.0442
100	0.001020	405.9	456.9	1.2705
120	0.001035	487.7	539.4	1.4859
140	0.001052	569.8	622.4	1.6916
160	0.001070	652.3	705.8	1.8889
180	0.001091	735.5	790.1	2.0790
200	0.001115	819.7	875.2	2.2628
220	0.001141	904.4	961.4	2.4414
240	0.001171	990.6	1049.1	2.6156
260	0.001204	1078.2	1138.4	2.7864
280	0.001243	1167.7	1229.9	2.9547
300	0.001288	1259.6	1324.0	3.1218
320	0.001341	1354.3	1421.4	3.2888
340	0.001405	1452.9	1523.1	3.4575
360	0.001485	1556.5	1630.7	3.6301
380	0.001588	1667.1	1746.5	3.8101

P = 100.0 MPa

T(°C)	V(m³/kg)	U(kJ/kg)	H(kJ/kg)	S(kJ/kg-K)
0	0.000957	-0.3	95.4	-0.0085
20	0.000962	78.0	174.2	0.2699
40	0.000969	157.0	253.9	0.5328
60	0.000978	236.2	334.0	0.7809
80	0.000988	315.6	414.5	1.0153
100	0.001000	395.1	495.1	1.2375
120	0.001014	474.6	576.0	1.4487
140	0.001028	554.4	657.2	1.6501
160	0.001045	634.3	738.8	1.8429
180	0.001063	714.5	820.8	2.0280
200	0.001083	795.1	903.4	2.2064
220	0.001104	876.3	986.7	2.3788
240	0.001128	958.0	1070.8	2.5459
260	0.001154	1040.3	1155.8	2.7084
280	0.001183	1123.5	1241.8	2.8869
300	0.001215	1207.6	1329.1	3.0219
320	0.001250	1292.8	1417.8	3.1740
340	0.001290	1379.1	1508.2	3.3238
360	0.001335	1466.8	1600.3	3.4717
380	0.001385	1556.0	1694.5	3.6182

E.8   PRESSURE-ENTHALPY DIAGRAM FOR METHANE

(Source: NIST, Thermophysics Division, Boulder, CO, USA, used with permission.
)

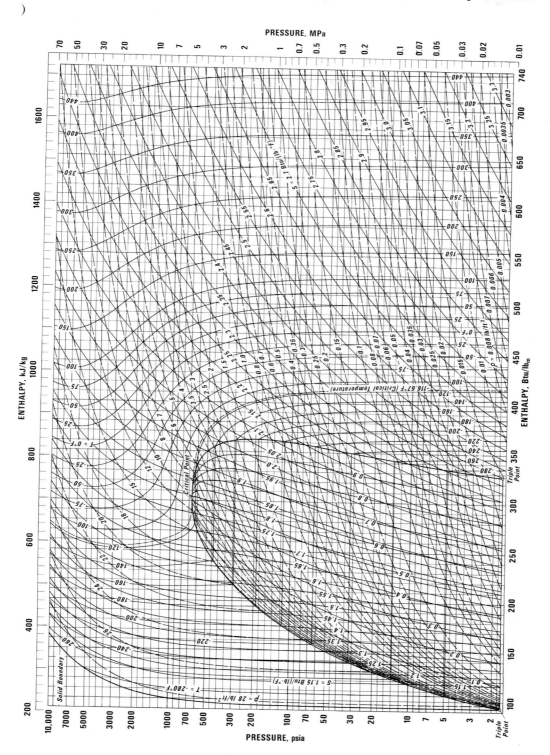

E.9   PRESSURE-ENTHALPY DIAGRAM FOR PROPANE

(Source: NIST, Thermophysics Division, Boulder, CO, USA, used with permission.)

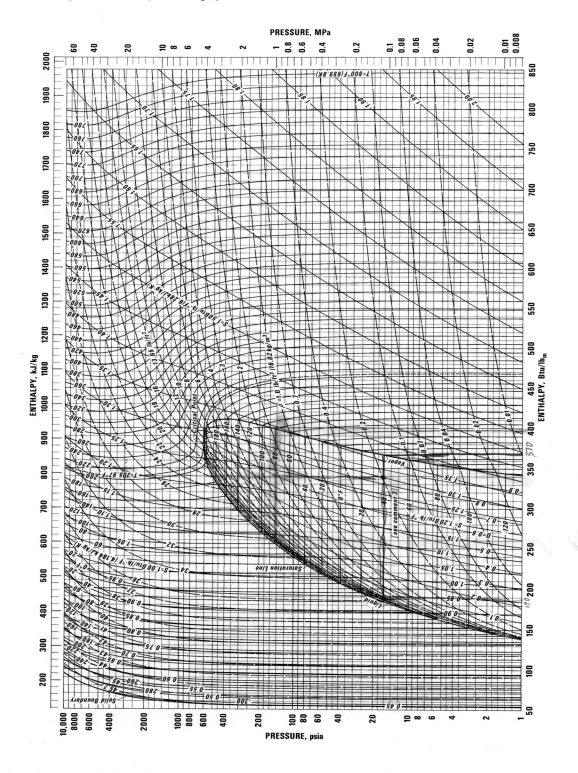

E.10 PRESSURE-ENTHALPY DIAGRAM FOR R134A (1,1,1,2-TETRAFLOUROETHANE)

(Source: NIST, Thermophysics Division, Boulder, CO, USA, used with permission.)

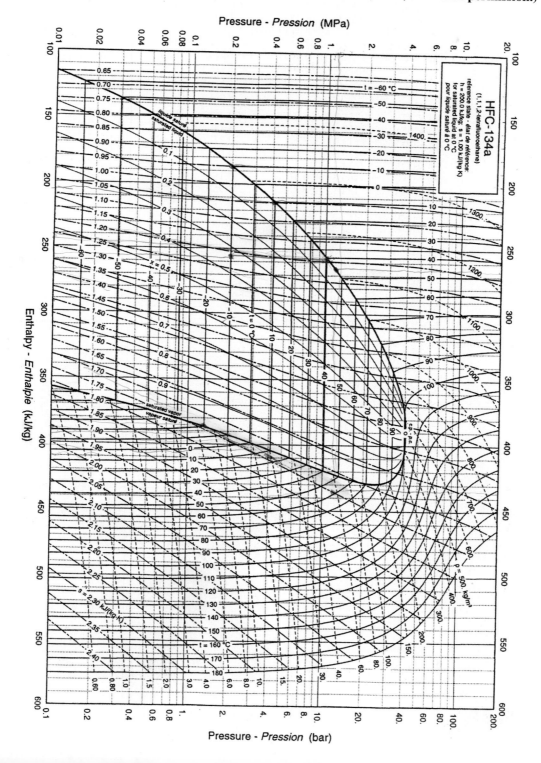

Properties of Saturated HFC-134a.

T	P	ρ^L	ρ^V	H^L	H^V	S^L	S^V
K	MPa	kg/m^3	kg/m^3	kJ/kg	kJ/kg	kJ/kg-K	kJ/kg-K
240	0.07248	1397.7	3.8367	156.78	378.33	0.8320	1.7552
244	0.08784	1385.8	4.5965	161.87	380.85	0.8530	1.7505
248	0.10568	1373.8	5.4707	166.99	383.35	0.8738	1.7462
252	0.12627	1361.7	6.4715	172.14	385.84	0.8943	1.7423
256	0.14989	1349.5	7.6117	177.33	388.31	0.9147	1.7388
260	0.17684	1337.0	8.9051	182.55	390.75	0.9348	1.7356
264	0.20742	1324.4	10.3660	187.81	393.17	0.9548	1.7327
268	0.24197	1311.6	12.0110	193.11	395.56	0.9747	1.7301
272	0.28080	1298.5	13.8570	198.45	397.93	0.9943	1.7277
276	0.32426	1285.3	15.9230	203.84	400.25	1.0139	1.7255
280	0.37271	1271.7	18.2270	209.26	402.54	1.0332	1.7235
284	0.42651	1258.0	20.7940	214.74	404.79	1.0525	1.7217
288	0.48603	1243.9	23.6450	220.27	406.99	1.0717	1.7200
292	0.55165	1229.5	26.8080	225.85	409.14	1.0907	1.7184
296	0.62378	1214.7	30.3130	231.49	411.23	1.1097	1.7169
300	0.70282	1199.6	34.1920	237.18	413.26	1.1286	1.7155
304	0.78918	1184.1	38.4830	242.95	415.22	1.1475	1.7142
308	0.88330	1168.1	43.2280	248.78	417.11	1.1663	1.7128
312	0.98560	1151.5	48.4750	254.69	418.92	1.1850	1.7114
316	1.09650	1134.5	54.2820	260.68	420.63	1.2038	1.7100
320	1.21660	1116.7	60.7140	266.76	422.25	1.2226	1.7085
324	1.34620	1098.3	67.8510	272.94	423.74	1.2414	1.7068
328	1.48600	1079.0	75.7890	279.23	425.10	1.2603	1.7050
332	1.63640	1058.8	84.6440	285.63	426.31	1.2793	1.7030
336	1.79810	1037.5	94.5630	292.18	427.34	1.2984	1.7007
340	1.97150	1015.0	105.7300	298.88	428.17	1.3177	1.6980

Abstracted from R. Tillner-Roth; H. D. Baehr, *J. Phys. Chem. Ref. Data,* 23:657 (1994)

INDEX

PROPERTIES OF SELECTED COMPOUNDS

*Heat capacities are values for **ideal gas** at 298 K and should be used for **order of magnitude calculations** only. See appendices for temperature dependent formulas and constants.*

Compound		T_c(K)	P_c(MPa)	ω	Z_c	C_P/R
Paraffins						
1	METHANE	190.6	4.604	0.011	0.288	4.298
2	ETHANE	305.4	4.880	0.099	0.284	6.312
3	PROPANE	369.8	4.249	0.152	0.281	8.851
4	*n*-BUTANE	425.2	3.797	0.193	0.274	11.890
5	ISOBUTANE	408.1	3.648	0.177	0.282	11.695
7	*n*-PENTANE	469.7	3.369	0.249	0.269	14.446
8	ISOPENTANE	460.4	3.381	0.228	0.270	14.279
9	NEOPENTANE	433.8	3.199	0.196	0.269	14.616
11	*n*-HEXANE	507.4	3.012	0.305	0.264	17.212
17	*n*-HEPTANE	540.3	2.736	0.349	0.263	19.954
27	*n*-OCTANE	568.8	2.486	0.396	0.259	22.697
46	*n*-NONANE	595.7	2.306	0.437	0.255	25.451
56	*n*-DECANE	618.5	2.123	0.484	0.249	28.217
64	*n*-DODECANE	658.2	1.824	0.575	0.238	33.714
66	*n*-TETRADECANE	696.9	1.438	0.570	0.203	39.216
68	*n*-HEXADECANE	720.6	1.419	0.747	0.220	44.539
Naphthenes						
104	CYCLOPENTANE	511.8	4.502	0.194	0.273	9.974
105	METHYLCYCLOPENTANE	532.8	3.785	0.230	0.272	13.209
137	CYCLOHEXANE	553.5	4.075	0.215	0.273	12.738
138	METHYLCYCLOHEXANE	572.2	3.471	0.235	0.269	16.250
Olefins and Acetylene						
201	ETHYLENE	282.4	5.032	0.085	0.277	5.260
202	PROPYLENE	364.8	4.613	0.142	0.275	7.688
204	1-BUTENE	419.6	4.020	0.187	0.276	10.306
207	ISOBUTENE	417.9	3.999	0.189	0.275	10.724
209	1-PENTENE	464.8	3.529	0.233	0.270	13.174
401	ACETYLENE	308.3	6.139	0.187	0.271	5.320
303	1,3-BUTADIENE	425.4	4.330	0.193	0.270	9.560
309	ISOPRENE	484	3.850	0.158	0.264	12.777
Aromatics						
501	BENZENE	562.2	4.898	0.211	0.271	9.822
502	TOLUENE	591.8	4.109	0.264	0.264	12.485
504	ETHYLBENZENE	617.2	3.609	0.304	0.263	15.444
505	*o*-XYLENE	630.4	3.734	0.313	0.263	16.033
506	*m*-XYLENE	617.1	3.541	0.326	0.259	15.348
507	*p*-XYLENE	616.3	3.511	0.326	0.260	15.263
510	CUMENE	631.2	3.209	0.338	0.262	18.246
558	BIPHENYL	789.3	3.847	0.366	0.294	19.521
563	DIPHENYLMETHANE	768	2.920	0.461	0.250	21.865
701	NAPHTHALENE	748.4	4.051	0.302	0.269	16.033
702	1-METHYLNAPHTHALENE	772	3.650	0.292	0.297	19.075
706	TETRALIN	720.2	3.300	0.286	0.243	18.631
Oxygenated Hydrocarbons						
1101	METHANOL	512.6	8.096	0.566	0.224	5.283
1102	ETHANOL	516.4	6.384	0.637	0.248	7.879
1103	PROPANOL	536.7	5.170	0.628	0.253	10.496

continued on next page